Circuits and Systems Tutorials

IEEE Press
445 Hoes Lane, P.O. Box 1331
Piscataway, NJ 08855-1331

Circuits and Systems Tutorials

Editor
Chris Toumazou
Imperial College
London, U. K.

Assistant Editors
Nick Battersby
Northern Telecom
Ontario, Canada

Sonia Porta
Oxford Brookes University
Oxford, U. K.

IEEE PRESS

The Institute of Electrical and Electronics Engineers, Inc., New York

This book may be purchased at a discount from the publisher when ordered in bulk quantities. For more information contact:

IEEE Press Marketing
Attn: Special Sales
445 Hoes Lane, P.O. Box 1331
Piscataway, NJ 08855-1331
Fax: (908) 981-9334

Originally produced by LTP Electronics Ltd.,
Oxford UK on behalf of IEEE ISCAS '94.

ISBN 0-7803-1170-1
IEEE Order Number: PP5596

Library of Congress Cataloging-in-Publication Data

Circuits and systems tutorials / editor, Chris Toumazou : assistant editors, Nick Battersby & Sonia Porta.
p. cm.
"IEEE order number PP5596"—T.p. verso.
Based on tutorials presented at the 1994 IEEE International Symposium on Circuits and Systems held in London.
Includes bibliographical references and index.
ISBN 0-7803-1170-1 (pbk.)
1. Signal processing—Congresses. 2. Electronic circuits—Congresses. 3. Telecommunication systems—Congresses. 4. Electric filters—Congresses. I. Toumazou, C. (Chris) II. Battersby, N. C. (Nicholas C.) III. Porta, Sonia.
TK5102.9.C49 1996
621.382—dc20 95-42332
CIP

To the late Nick Battersby

"Through your enthusiastic, unassuming and positive nature, you instilled in all of us an inspiration for excellence and for the love of life..."

**Your friends in Otawa,
Jean-Marc, Vivek, Samina, Cindy and Doug**

"A young engineer with a dedicated, brilliant and inquisitive mind. A great loss to Electronics..."

Chris Toumazou on behalf of Imperial College

Contents

Tutorial 3

List of Contributors

Bryan Ackland
AT&T Bell Labs
Crawfords Corner Rd.,
P.O. Box 3030
Holmdel, NJ 07733-3030, USA

Robert Adams
Analog Devices Semiconductor
181 Ballardvale St.
Willmington, MA 01887, USA

Kiyharu Aizawa
Dept. of Electrical Engineering
The University of Tokyo, 7-3-1 Hongo
Bunkyo-ku, Tokyo 113, Japan

Timo Ali-Vehmas
Nokia Research Centre
Heikiläntie, 7 PL 4S
FIN-00211, Helsinki, Finland

Nicholas Battersby
Northern Telecom
185 Corkstown Rd.
Nepean, Ontario, Canada K2H 8V4

Ian Bell
Dept. of Electronic Engineering
University of Hull
Hull, U.K.

James Bryant
European Applications Manager
Analog Devices, U.K.

Gisle Bjoentegaard
Norwegian Tel. Research
Forskningsinstitutt. Postboks 83
2007 Kjeller, Norway

Erik Bruun
Technical University of Denmark
Electronics Institute
Building 349, DK-2800 Lyngby,
Denmark

Peter Chadwick
GEC Plessey, Cheney Manor
Swindon Wilts, SN2 2QW, U.K.

Nicholas Courtis
Nokia Mobile Phones
Penbroke Broadway
Camberley, Surrey GU15 3SP

Colin Cowan
Loughborough University of
Technology
Loughborough, LE11 3TU, U.K.

Edward J. Coyle
School of Electrical Engineering
Purdue University
West Lafayette, IN 47907, USA

Antonio de Queiroz
Dept. of Electronic Engineering
CP 68504, 21945-970
Rio de Janeiro, Brazil

Kevin Eckersall
Dept. of Electronic Engineering
University of Hull, Hull, U.K.

Thierry Fautier
Laboratoires d'Electronique Philips
(LEP) France

Jose Franca
Instituto Superior Tecnico,
Av. Rovisco
Pais, 1-1096 Lisboa Codex, Portugal

Moncef Gabbouj
Tampere University of Technology
Signal Processing Lab, P.O. Box 553
FIN-33101, Tampere, Finland

Barrie Gilbert
Analog Devices Inc.
1100 NW Compton Drive, Suite 301
Beaverton, OR 97006, USA

Martin Hasler
Swiss Federal Institute of Technology
CH-1015 Lausanne, Switzerland

Arnaud Jacquin
AT&T Bell Labs, 600 Mountain Ave.
P.O. Box 636, Murray Hill, NJ 07974,
USA

Peter Kennedy
Electronic & Electrical Engineering
University College, Dublin 4, Ireland

Bosco Leung
University of Waterloo
Waterloo, Ontario, Canada N2L 3G1

John Lidgey
Oxford Brookes University
School of Engineering
Headington, Oxford OX3 0BP, U.K.

V. John Mathews
The University of Utah
Dept. of Electrical Engineering
Salt Lake City, Utah 84112, USA

Sanjit Mitra
University of California-Santa Barbara
Dept. of Electrical/Computer
Engineering
Santa Barbara, CA, USA

Peter Mole
Analog Design Group in VLSI
Subsystems
BNR Europe, Ltd., London Road
Harlow, Essex CM17 9NA, U.K.

John Morris
Philips Research Labs
Cross Oak Lane, Redhill, U.K.

Hans G. Musmann
University of Hannover
Appelstrasse 9A, 3000 Hanover 1,
Germany

Yrjö Neuvo
Nokia Mobile Phone,
P.O. Box 226
FIN-00101 Helsinki, Finland

Maciej Ogorzalek
Dept. of Electrical Engineering
Univ. of Mining and Metallurgy
Mickiewicza 30/B1
30-059 Krakow, Poland

Alison Payne
Dept. of Electrical & Electronic
Engineering
Imperial College,
Exhibition Road
London SW7 2BT, U.K.

Peter Pirsch
University of Hannover
Applestr. 9A, D 30167 Hannover
Germany

Ioannis Pitas
Dept. of Electrical Engineering
University of Thessaloniki
Thessaloniki 54006, P.O. Box 430
Greece

Sitthichai Pookaiyaudom
Mahanakorn University of Technology
51 Cheum-Sampan Rd., P.O. Box 10
Nong-Chok District
Bangkok 10530, Thailand

Sonia Porta
Oxford Brookes University
School of Engineering
Headington, Oxford OX3 0BP, U.K.

Jan Rabaey
ECCS, Cory Hall
University of Berkeley, CA 94720,
USA

David Rhodes
Filtronics (UK) Ltd.,
Airedale House
Acorn Park, Charleston, Shipley
Yorkshire BD17 7SW, U.K.

Gordon Roberts
McGill University,
Electrical Engineering
McConnell Engineering Building
3480 University Street
Montreal, Quebec, Canada H3A 2A7

Angel Rodriguez-Vazquez
Centro Nacional de Microelectronica
University of Seville,
Avda Reina Mercedes
(Edificio CICA) 41012 Sevilla, Spain

Geir Edvard Saether
Dept. of Electrical & Electronic Engineering
Imperial College, Exhibition Road
London SW7 2BT, U.K.

Peter Shah
Dept. of Electrical & Electronic Engineering
Imperial College, Exhibition Road
London SW7 2BT, U.K.

Peter Saul
Elettronica (UK) Ltd.
8 Vincent Avenue
Crownhill, Milton Keynes,
MK8 0AB, U.K.

Ken Searle
Analog Design Group—VLSI Subsystems
BNR Europe, London Road, Harlow
Essex CM17 9NA, U.K.

G. L. Sicuranza
Elettrotecnica Elettronica Informtica
University of Trieste, Via A. Valerio 10
34127 Trieste, Italy

Ming-Ting Sun
Bellcore Communications Research
331 Newman Springs Rd.
Red Bank NJ 07701-5699, USA

Gaynor Taylor
Leeds Metropolitan University
Dean of Faculty
Leeds, U.K.

Gabor Temes
Oregon State University
Electrical Engineering Building, # 246
Corvallis, OR 97331, USA

Chris Toumazou (editor)
Electrical & Electronic Engineering
Imperial College, Exhibition Road
London SW7 2BT, U.K.

Phil Tudor
BBC Research & Development Dept.
Kingswood Warren, Tadworth, U.K.

Eric Vittoz
Swiss Center for Electronics and Microtechnology, Case Postale 41
CH-2007 Neuchatel, Switzerland

Doug Wadsworth
Northern Telecom Microelectronics Centre
185 Corkstown Rd., Nepean
Ontario, Canada K2H 8V4

Nicholas Wells
BBC Research & Development Dept.
Kingswood Warren, Tadworth, U.K.

Masakazu Yamashina
NEC Corporation
1120 Shimokuzawa, Sagamihara 229
Japan

Foreword

On mowing one's lawn

Keeping up to date in one's professional field is a little like mowing one's lawn: the problem is always there, it can be hard work, and you neglect it at your peril. Fortunately for those who attended the ISCAS'94 Tutorials, Chris Toumazou's infectious enthusiasm has persuaded many international known experts not only to deliver the tutorials, but to provide the material for this excellent book.

Keep mowing !

Bob Spence
General Chairman, ISCAS'94

Preface

Acquiring a basic knowledge outside their immediate specialisation can be difficult and time-consuming for engineers and so a number of 1-day tutorial courses on Circuits and Systems was launched at the 1994 IEEE International Symposium on Circuits and Systems (ISCAS 94) in London, UK, and this text was used as course material. It was decided that the clearest way of disseminating course material during the tutorials would be to have a text written by the speakers, covering their individual tutorial topic. Their efforts are most appreciated.

There are 10 tutorials broken down into 44 chapters with some 54 authors associated with the book, from all over the world. This is truly an international effort and it is admirable to see so much collaboration between authors. We are pleased to note that 24 authors are from industry and the rest from research institutions and universities. The timescale from book conception to final camera-ready copy was six months, which made considerable demands on the authors in the final rush to get the material ready.

One of the book editors, Dr. Nick Battersby, died tragically during the production of the book. His professionalism and dedication in assisting to get the text together will long be remembered.

This book is a valuable reference text for engineers intending to pursue research in one or more of the tutorial areas.

Chris Toumazou
Tutorial Chair ISCAS'94

Acknowledgments

The typesetting was carried-out by ourselves using material supplied by the authors on computer disks. Most of the initial typesetting was carried out by Dr. Nick Batterbsy while working for Northern Telecom, Ottawa, Canada. When Nick died, at the end of March 1994, his parents gave permission to recover the book material and this difficult task was carried out meticulously and at great speed by Doug Wadsworth, a colleague of Nick's at the Northern Telecom Microelectronics Centre, who forwarded it to us in the UK. Without Doug's efforts, this final book would not have been possible. Thank you Doug!

We would like to thank Professor John Lidgey for his technical support in proof reading many chapters and helping to get the final book together, and also Ann Toumazou for proof-reading. Many thanks are also due to Dr Jens Tingleff for his vital contribution and efforts in electronically converting many files into a standard word-processing format and assistance in the editing of some of the chapters.

The absolutely first class secretarial and administrative support of Mrs. Wiesia Hsissen at Imperial College is also gratefully acknowledged.

We would like to thank the staff at LTP Electronics Ltd., UK, for producing the book, Short-Run Press, UK, for arranging a very quick printing schedule and Professor Robert Spence, General Chairman at ISCAS 94, for permission to try out the new venture.

Finally, last but not least, many thanks to the tolerant Toumazou family, for their understanding and support.

Chris Toumazou
Nick Battersby
Sonia Porta

Introduction

Chris Toumazou, Sonia Porta and Nick Battersby

Over the past decade the international circuits and systems community has witnessed the conception of many new, diverse and exciting topic areas and engineering disciplines from circuit theory through to future communications systems. This book is a collection of 44 chapters based upon 10 key topic areas of the upmost international notoriety and importance in the field of circuits and systems. Each chapter was written by experts in the specific fields and is designed to provide non-specialists with an essential understanding of the key fundamentals and status of the subject. There are some 54 authors associated with the book, from all over the world, Asia, North and South America and Europe. This is truly an International effort and it is admirable to see so much collaboration between authors (Figure 1)

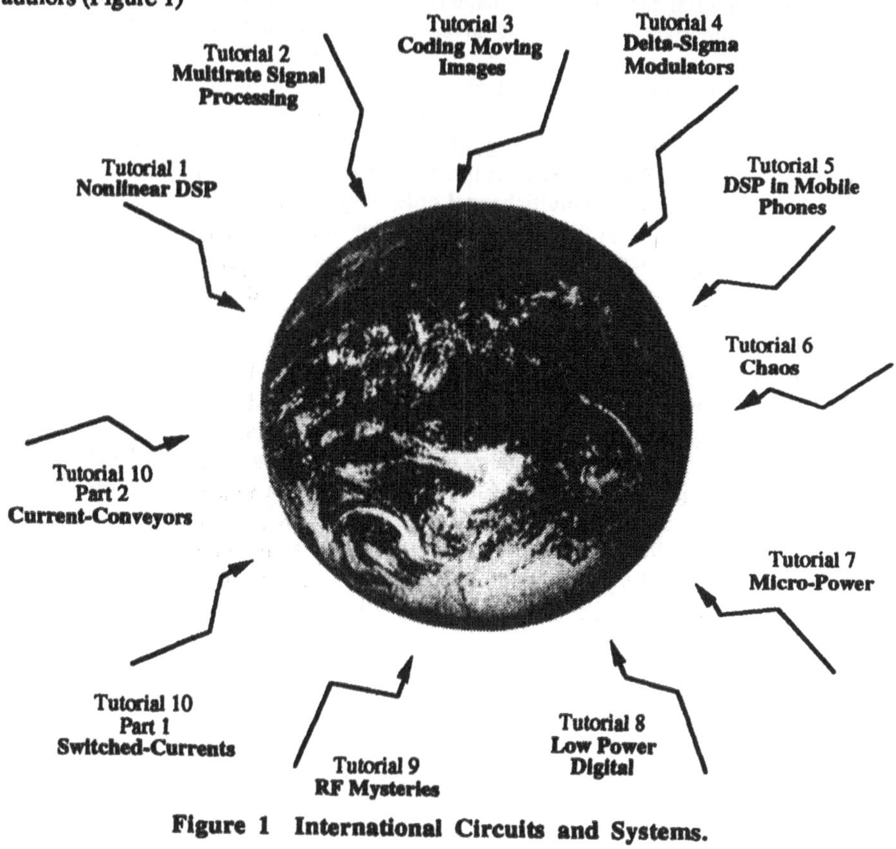

Figure 1 International Circuits and Systems.

Nonlinear DSP, multi-rate signal processing, moving images, delta-sigma modulators, mobile phones, low-power digital design, micro-power design, RF design and current-mode design are just some of the topics presented.

This book has been arranged into three major sections, Digital Signal Processing, Mixed-Signal Processing and Analog Signal Processing. Although each topic area is not intended to explicitly fit into any one section, the final category chosen was the most appropriate in the context of the rest of the book. The field of communications appears predominantly throughout the text and in fact it may have been more appropriate to have called the text circuits and systems for communications. The book covers a whole range of fields from visual communications; image coding and coding standards for TV broadcast; mobile phones and telephony and implementation issues; multirate signal processing and sigma delta converters; nonlinear signal processing and current-mode analog signal processing both switched-currents and current-conveyors. If one considers performance requirements of many of the aforementioned topics then high speed, low power, low cost relationship starts to dictate. The book, therefore, includes several chapters on low-power design techniques and RF design methods for both analog and digital systems.

In Tutorial 1, some fundamentals of nonlinear signal processing are presented. This is a very timely tutorial comprising six chapters. The field of Nonlinear DSP is becoming more and more important with the rapid advances in digital communications, computing and VLSI. Nonlinear analysis, although essential, can be very daunting to the practical engineer and so the tutorial has been written to provide the reader with an essential grasp of some fundamentals. The tutorial concentrates upon two classes of nonlinear techniques; the first is statistical based processing while the other considers polynomial signal processing techniques. The first part has three chapters based on weighted median filters (1.11), stack filters (1.12), and multichannel order statistics median filters (1.13).This is followed by a second part on the theory and approximation of polynomial filters (1.21), adaptive polynomial filters (1.22) and the application of adaptive volterra filters to equalisation (1.23).

Multi-rate processing is becoming very important for mixed-mode integrated systems as more of the integrated communication system becomes digital. Decimation and interpolation techniques applied to oversampling data converters are a very popular example. Other signal processors such as charge transport devices, switched capacitors etc. have led to the development of analog multi-rate processing techniques, filtering and video coding/decoding applications. Chapter 2 presents an excellent treatment of the subject of multirate signal processing from basic foundations to the design of advanced modern analog/digital multirate systems. The tutorial is an excellent introduction for the new-comer and includes both telephony and video systems and other functions such as signal detectors, multi-dimensional filters and bank filters.

Visual communications is a very fast growing area for telecommunications and the computer and media industries. In addition to the developments in the technology of digital storage media, image displays, workstations and communication networks are the backbone of video communications. Tutorial 3 comprises eight chapters. Four are concerned with video coding algorithms for MPEG2 (3.11- 3.14), two are concerned with VLSI architectures and implementation of MPEG (3.21, 3.22) and the final two chapters are concerned with advanced analysis and synthesis for the coding of very low bit rates

(3.31, 3.32). There is a strong industrial content in this tutorial and it provides an excellent introduction to the topic.

In mixed-mode VLSI, the importance of developing low cost, high-performance analog/digital interface circuits in a primarily digital environment becomes very challenging. The oversampling converter technique is becoming more and more widely accepted as the technique which ensures very high resolution with less stringent design facing the analog parts. What you always wanted to know about delta-sigma converters but were afraid to ask, is appropriately the title of Tutorial 4. This three chapter tutorial covers the basic operating principles of delta-sigma data converters as well as implementation issues, including circuit design, practical problems and their solutions, advantages and disadvantages of various oversampled architectures are considered. The first chapter (4.1) is a thorough review of the oversampling technique as a method of performing high resolution analog to digital data conversion. The chapter provides an excellent mathematical treatment of the subject with a thorough bibliography of the area. The next chapter (4.2) gives a brief summary of the reasons for introducing oversampling as a method of improving the accuracy of digital to analog converters and architectures available for the realization of digital noise shaping loops. Some practical examples of integrated converters are also discussed. In the final chapter (4.3) many practical aspects of high-order delta-sigma designs are considered together with common commercial design practices. The chapter is a mini practical designers 'cook book' to delta-sigma A/D converters and concludes with a description of a commercially available fifth order modulator.

During the past one and a half decade the world has witnessed a revolution in mobile telephony. The mobile phone has developed from a cumbersome manually operated vehicle mounted radio telephone, to an elegant low-power sophisticated piece of portable electronics. Tutorial 5 is devoted to this important emerging area with three chapters from the industry covering a general over-view of mobile telephony (5.1), DSP in mobile phones (5.2) and high speed DSP and low power circuits for future mobile phone systems (5.3). DSP is used widely in mobile telephony both in the terminals and the network. The aim of the tutorial is to give a basic understanding of how DSP enters mobile telephony by concentrating largely upon speech processing. Various techniques for low power consumption are also predominant throughout the chapter.

Tutorial 6 is devoted to the aspiring area of chaos from basic concepts to engineering applications. The tutorial gives an excellent treatment of the subject within four related yet stand alone chapters. The first chapter (6.1) takes us through some of the nonlinear dynamics of chaos and combines both a mathematical and intuitive treatment of the subject. This is followed by a chapter (6.2) based on synchronization principles and applications. The chapter considers various methods to synchronize chaotic systems and methods for transmitting information hidden in a chaotic signal, and discusses associated problems. The treatment is not too mathematical and is an excellent guide to the beginner in the area. The next chapter (6.3) is devoted to control concepts developed for systems operating in a chaotic regime. It starts by describing some of the most primitive concepts such as parameter variations, to quite sophisticated ones like stabilisation of unstable periodic orbits embedded within the strange attractor. Advantages and drawbacks of these methods are fully described. The final chapter (6.4) gives a detailed account of the area of fractals in signal processing, and their relationship with chaos. The chapter introduces various notions of fractal objects and points to where and why fractals and chaotic systems are commonly found in nature, and the possibility for modelling different types of real world signals. Particular emphasis is placed on the actual process of fractal modelling of natural scenes with very compact (i.e. very low bit rate) fractal signal processing. The chapter is aimed at engineers who are interested in an overview of fractals and chaos rather than mathematical detail.

Interest is growing in the field of micropower design. From some of the original developments more than 20 years ago for electronic watches, micropower circuits now have applications in areas such as implanted portable medical devices, hearing aids and paging receivers. The interest in micropower design has also increased in view of the need for portable/mobile communication products. Also there is much interest in micropower circuits for analog VLSI implementations of neural networks. Tutorial 7 is entirely devoted to the subject of micropower design. The related chapter 7 describes fundamental device and circuit aspects of low-voltage very low-current integrated circuits. It considers MOS transistor modes of operation and various analog building blocks for VLSI. Many of the 'pros' and 'cons' of micropower computational systems are discussed throughout the chapter with an excellent treatment of design techniques.

Most of the research and development in the area of digital electronics has been geared towards high speed and high complexity. As a consequence, however, digital chips are consuming more and more power. Power minimisation techniques are thus essential, and this is the main subject of Tutorial 8. This tutorial comprises one chapter devoted to low power digital design. The first part of the chapter is a detailed analysis of the sources of power dissipation in digital design, and the second part presents some recent power minimisation techniques both from a technological and circuit perspective. The chapter concludes with a glimpse of low power design tools.

As the world gets faster the need for RF designers is becoming crucial. Good RF designers are few and far between. Tutorial 9 is appropriately entitled 'taking the mystery out of RF design'. The entire tutorial comprises six chapters which have been written by industrialists, some with experience in RF design spanning many years. In chapter (9.1) a philosophical view is taken on the subject with some enlightening anecdotes and home-truths about RF design. In view of the physical reality of RF circuits the author coins the term 'Newtonian' electronics. This chapter is followed by an excellent treatment of direct digital synthesis (9.2). The chapter describes the mathematical basis of the technique and shows how it relates to practical devices. DDS techniques are already in use in radio systems, primarily in combination DDS/PLL synthesizers.

As in all good mysteries, in RF design things are never quite what they seem. For example, by including so-called parasitics, the design becomes closer to reality. A chapter (9.3) is thus devoted to RF design in BiCMOS and GaAs technology. The chapter is an excellent treatment of circuit techniques and methods aimed to compensate for parasitic effects introduced at RF.

Spread spectrum techniques have become very popular for military system applications, and types range from fast frequency hopping, slow frequency hopping, direct sequence and chirp. Chapter (9.4) gives an overview of these techniques and summarizes the advantages and disadvantages of each. Spread spectrum techniques are also becoming popular for particular application areas where communication without cables is required.

Chapter (9.5) is the engineer's cook-book to RF design, it is practical and gives simple tips. The author details the so-called basic laws of RF Ohm's law, Kirchoff's law, Faraday's law, Lenz's law and Murphy's law. 'If anything can go wrong, it will', which is an abbreviated definition of Murphy's law, is emphasised as the key law of RF design. Many examples with simple mathematics and simple circuits (single-components in many places) are used to demonstrate that by some appreciation of the aforementioned laws many of the RF design anomalies can be eliminated. Another important candidate for RF design is the receiver for cellular telephone base stations and this is the subject of the final chapter (9.6) in Tutorial 9. The chapter gives an overview of the UK 900 MHz cellular system and the USA 800 MHz system. There are no guard bands between the cellular operator both in the USA and UK and so the frequency oscillators are interleaved. This involves the design of receivers implemented as a cascode of filters and amplifiers in order to minimise the noise figure for sensitivity and to increase dynamic range by attenuating strong signals. Methods of receiver design are fully described throughout the chapter.

Analog IC design is receiving a tremendous boost from the design and application of current-mode analog signal processing. It is clear from Tutorial 10 that the switched-current technique is heralding a new era in analog sampled-data signal processing. Unlike switched-capacitor techniques, switched-current techniques do not require linear floating capacitors, so can fully exploit digital CMOS VLSI technology. The first chapter (10.1) in the tutorial reviews some basic research work aimed at enhanced memory cell performance in a realistic mixed-signal environment. Design issues and performance specifications for various SI cells are considered for use in sigma-delta modulators. The next chapter (10.2) considers various performance trade-offs of the switched-current cell, in particular speed versus dynamic range and analytical techniques and methods clearly indicating how each performance may be optimised. One important area of research and development in switched-currents is that of high resolution data converters for high speed low power applications. Another chapter (9.3) is devoted to this work and gives the novice a clear introduction to the underlying principles of SI data conversion, from algorithm to circuit realisation. The next chapter (10.4) considers a new application of switched-currents to realise cellular neural networks for image processing. The chapter summarises the use of current-mode circuits to realise nonlinear state equations. The chapter is essentially a bibliography of the research work on various implementations of cellular neural networks and nonlinear systems. In chapter (10.5) some analysis methods are presented for switched-current filters. The chapter describes a procedure for the frequency-domain analysis of switched current filters and is implemented in a computer programme named ASIZ.

The concept of current-monitoring has recently been extended from digital devices into the analog and mixed-analog and digital field. This important area of current-monitoring is reviewed in chapter (10.6), where current-monitoring self-test of switched-current systems appears to be natural for switched-current cell operation.

The current-conveyor is a versatile analog device which is intended to be used with other circuit components to implement many analog signal processing functions. Although the current-conveyor concept has been around for a very long time it is only in the past five years that a general purpose monolithic device has been made available. Tutorial 11 comprises five chapters comprehensively covering all aspects of the current-conveyor. The first chapter (11.1) considers basic principles and applications, followed by a chapter on bipolar integrated circuit realisations (11.2). Commercial industrial applications are presented in chapter (11.3) for both telephony and audio applications, followed by a chapter (11.4) specifically looking at current-conveyor/current-mode based oscillators. To conclude, in chapter (11.5), trends towards VLSI implementations of current-conveyors in CMOS technology are presented, showing promise for ASIC applications.

This book started as a collection of tutorial slides to be presented at the IEEE 1994 ISCAS. It has now turned into an authorative text-book covering numerous state-of-the-art topic areas in the field of circuits and systems. This tutorial text book has attempted to capture material both at an introductory and advanced level and should become a valuable reference text for engineers intending to pursue reseach in any one or more of these tutorial areas.

Tutorial 1

Fundamentals of nonlinear digital signal processing

Organised by: Ioannis Pitas, *University of Thessaloniki*
Moncef Gabbouj, *Tampere University of Technology*

Chapter 1.0

Introduction to Nonlinear Signal Processing

Colin F. N. Cowan, Edward J. Coyle, Moncef Gabbouj, V. John Mathews, Ioannis Pitas and Giovanni Sicuranza

1.0.1 Introduction

Many engineers have an unfounded fear towards systems involving *nonlinear* components. They try to avoid them if possible, and if they cannot, they will approximate them with *linear* components. Of course, they may be sacrificing performance, but they have a linear system! While it is generally true that nonlinear systems are more complex than linear ones, there are many real systems and tasks which are or require nonlinear processing. The purpose of this tutorial is to reduce the bias in electrical engineering education towards linear algorithms for signal processing. It serves as an eye opener to graduate students and practitioners to the wide, still virgin, and most importantly promising, field of nonlinear signal processing. The technological advances in today's computing and integrated circuits are readier than ever to challenge complex nonlinear problems. Along with the dedicated work of a growing number of researchers and scientists who are striving to build the necessary basic structures, develop the underlying theories and bridge the gap between the latter and the real-world applications, it is expected that the area of nonlinear signal processing will become not just popular and attractive, but also a fact of life.

The fact that anything that is not linear is nonlinear does not give nonlinear systems and algorithms a unifying basis. Far from it, there does not exist, and will probably never be, a unifying approach to all systems which are not linear. But let us not let this distract us from wandering into nonlinear systems and techniques. Let us consider individual sets of applications and look for a family of nonlinear techniques, hopefully a large one, which can effectively deal with each set. This, in itself, is a great accomplishment when successful.

1.0.2 Overview

In the area of signal and image processing, we encountered a number of system nonlinearities which call for nonlinear processing. A non-exhaustive list of the major nonlinearities is provided next.

- ***Signal and image features, non-stationary signals.*** Digital image statistics change from one image region to the other. Adaptive (often nonlinear) operators are used to keep track of the varying image statistics.

- ***Non-Gaussian, non-additive noise.*** A primary example is impulsive noise filtering that frequently occurs in telecommunication channels, in TV broadcasting and in information losses in decoding or due to packet losses.

- ***Signal and image acquisition and transmission.*** Several acquisition devices are highly nonlinear systems (e.g. photographic film, video cameras), whose nonlinearities are of polynomial form. Furthermore, nonlinear (usually analog) preprocessing is used sometimes in order to increase the dynamic range of the acquisition, or for compensation reasons (e.g. *gamma* correction in television).

- ***Characteristics of the human perception system.*** In many occasions, the final judge of a digital signal/image processing system is human perception. It is widely known that perception mechanisms are very difficult to model. However, it has been found that certain parts have nonlinear models. For example, it is widely accepted that the human vision has logarithmic nonlinearities. Furthermore, high-level knowledge mechanisms that are decision-driven interfere with the low-level perception mechanisms. Decisions appear as hard nonlinearities (e.g. decision on edge detection) that lead to decision-directed nonlinear digital signal/image processing methods. In other cases, the characteristics of human visual perception lead to highly nonlinear digital image processing operations (e.g. histogram equalization, dithering).

- ***Optimization of nonlinear objective measures.*** In several cases, nonlinear measures and/or features express fairly well certain signal/image characteristics. Image entropy is a primary example of a nonlinear measure of image information content (and tentatively its quality). Optimization of such measures leads to nonlinear digital signal processing (e.g. maximum entropy image restoration).

- ***Nonlinearities in digital image analysis.*** Many digital image analysis operations use hard nonlinearities (e.g. maximum/minimum selection, thresholding, comparisons, quadratic nonlinearities), because image analysis leads to symbolic descriptions that are based on decisions. For example, decisions are needed for edge detection and for region split/merge. Furthermore, shape features (e.g. skeletons, convex hulls, linear piecewise approximations) are usually nonlinear functions of the respective digital images.

The research in nonlinear signal processing has led to the identification of four major, partially overlapping, families of nonlinear techniques:

- Order statistics based techniques
- Polynomial based algorithms
- Morphological methods
- Homomorphic systems

In this tutorial, we shall concentrate on the first two classes of nonlinear techniques. The chapter is composed of two major sections. Section 1.1 deals with order statistics based signal processing; while, section 1.2 exposes the area of polynomial signal processing techniques. Each section contains three contributions from researchers working in the respective fields. Three major order statistics based subclasses are reviewed in section 1.1. These are weighted median filters (chapter 1.11), stack filters (chapter 1.12) and multichannel order statistics filters (chapter 1.13). Each subclass is formally defined and its major properties and potential applications are reviewed. In the area of polynomial filtering, we shall present also three contributions. First, in chapter 1.21, the theory and approximation of polynomial filters are reviewed. Second, the theory, with hints to possible applications, of adaptive polynomial filters is reviewed in chapter 1.22. Finally, an important application of adaptive Volterra filters to equalization is presented in chapter 1.23. An elaborate bibliography is included at the end of each chapter to guide the reader through the chapter and offer further readings.

1.0.3 Order statistics based signal processing

The area of order statistics has very deep roots in the statistics literature, e.g., the maximum likelihood estimation of a constant buried in Laplacian noise leads to the order statistic estimator known as the median filter; the use of least-median of squares in robust regression, the max, the min and threshold operations used in decision theory. The pragmatic example of order statistics based filters is the median filter. The step response of the median filter is a step (ideal for preserving stepwise abrupt changes in signals) and the impulse response is zero. Though one cannot use the impulse response to characterize the filter, it is an indication of how effective the median filter is for combating impulsive type noise. Weighted Median (WM) filters inherent the robustness and edge preserving capability of the classical median filter and resemble linear FIR filters in certain properties. Furthermore, WM filters belong to the broad class of nonlinear filters called stack filters. This enables the use of the tools developed for the latter class in characterizing and analyzing the behaviour and properties of WM filters, e.g. noise attenuation capability.

In 1.11, we briefly review WM filters, their deterministic and statistical properties. We point out the striking analogies existing between WM and FIR filters and summarize two important applications of WM filters in signal and image processing.

Two basic approaches to problems in signal and image processing can be identified. They can be called the *regularization approach* and the *filter-class approach*. We show how results in both of these areas point to the class of filters called stack filters for certain problems in signal and image processing. This presentation brings together ideas from the areas of image modelling, mathematical morphology, and order statistic filters.

Chapter 1.12 reviews the regularization and filter-class approaches, and specific results leading to stack filters are stated. Then the basics of a design theory for stack filters are reviewed. The chapter closes with a description of current research activities and applications of stack filters.

A research and development area that has been particularly active in the past five years is colour image processing. Its significance stems primarily from technological innovations, namely computer graphics, digital image printing, graphics arts, digital video, High Definition TV and multimedia systems. A very important research trend within colour image processing and analysis is based on nonlinear techniques in general and on multichannel order statistics in particular.

Chapter 1.13 presents multivariate data ordering and its use in colour image filtering. Various types of ordering (e.g. marginal ordering, reduced ordering) and the related multi-channel filters are presented in detail. Most of the filters that are presented are non-trivial extensions of the single-channel filters based on order statistics (e.g. median filters, trimmed mean filters). The use of polar coordinates in nonlinear multichannel signal processing is also covered. Angular statistics are studied and their application in colour image analysis and processing is exposed.

1.0.4 Polynomial signal processing techniques

The second major area of nonlinear signal processing covered in this tutorial, Section 1.2, is polynomial based techniques. Among the various mathematical representation of nonlinear systems, a popular description is given by the Volterra series expansions, studied by Vito Volterra in the 1880's as a generalization of the Taylor series of a function. This approach has been extensively applied to nonlinear system modelling and simulation in various fields of engineering and then extended to the discrete case. Following the seminal work of Wiener on characterizing nonlinear system behaviour using G-functionals, several researchers employed Volterra series expansion and related representations for estimation and nonlinear system identification. However, the interest in this approach dwindled, primarily because of the lack of adequate computational power to implement such techniques. With the availability of superior computational tools, there has been a resurgence in the study and application of Volterra system models since the early 1980's. This includes the development of the theory and approximations of polynomial filters, adaptive polynomial filtering and a set of elegant real-world applications.

In chapter 1.21, we briefly review the theory of the Volterra series for continuous systems and concentrate the rest of the discussion on the discrete case. We present the discrete Volterra series and the corresponding polynomial filters for one and multidimensional systems. We review the major attractive properties of the discrete Volterra series, and provide general and simplified polynomial models of practical interest to signal processing applications.

Chapter 1.22 discusses adaptive filters employing polynomial system models. We first review the basics of gradient-based and recursive least-squares adaptive Volterra filters with finite memory and finite order of nonlinearity. This chapter also discusses some of the fundamental issues involved in designing adaptive filters equipped with polynomial models containing nonlinear feedback.

Chapter 1.23 presents one of the most elegant applications of Volterra filters: adaptive equalisation for data communications channels. We discuss the optimum solution based on nonlinear clustering techniques. Although these techniques are seemingly computationally expensive, we show that an elegant solution emerges in the form of a simple second order Volterra filter which may be easily adapted using well known linear adaptation methods.

The chapters in this tutorial are self contained and each is followed by a list of references for those who are interested in the area. Some chapters point also to open problems in the field which could be of interest to graduate students searching for interesting research topics.

Chapter 1.11

Weighted Median Filtering — Striking Analogies to FIR Filters

Moncef Gabbouj

Abstract

Weighted Median (WM) filters have attracted a growing interest in the nonlinear signal processing community. They inherit the robustness and edge preserving capability of the classical median filter and resemble linear FIR filters in certain properties. Furthermore, WM filters belong to the broad class of nonlinear filters called stack filters. This enables the use of the tools developed for the latter class in characterizing and analyzing the behaviour and properties of WM filters, e.g. noise attenuation capability. The fact that WM filters are threshold functions allows the use of neural networks training methods to obtain adaptive WM filters.

In this chapter we briefly review WM filters, their deterministic and statistical properties. We point out the striking analogies existing between WM and FIR filters and summarize two important applications of WM filters in signal and image processing.

1.11.1 Introduction

Linear filters have been the dominating filter class through the history of signal processing, due mainly to the sound theoretical basis provided by the theory of linear systems and the computational efficiency of linear filtering algorithms. Despite the elegant linear system theory, not all signal processing problems can be satisfactorily addressed through the use of linear filters due to edge blurring and failure to remove heavy tailed distribution noise effectively [24, 25].

A nonlinear filter class that has been proven very useful is the class of median based filters whose theory and applications have been developing quite fast over the past decade. It is believed that median based filters provide a sound approach to nonlinear filtering. The success of median filters is based on two intrinsic properties: edge preservation and efficient noise attenuation with robustness against impulsive-type noise.

The median filter is not, however, a perfect filtering operation nor it is very flexible. It may cause edge jitter [4], streaking [5] and may remove important image details [1, 20]. The main reason is that the median filter uses only rank-order information of the input data within the filter window, and discards its original temporal-order information. For constant signals embedded in i.i.d. noise, temporal-order is irrelevant. This, however, is not true for non-constant signals corrupted by noise [23]. In order to utilize both rank- and temporal-order information of input data, several classes of rank order based filters have been developed in recent years, such as FIR-median hybrid (FMH) filters [14, 20], *Ll* filters [23], weighted median (WM) filters [6, 16, 46], and stack filters [8, 18, 31]. FMH filters and *Ll* filters make use of the desirable properties of both linear filters and median filters.

We shall concentrate on WM filters which are the simplest rank order based filters among those mentioned above. An N-length WM filter can be described by N parameters and implemented using a sorting operation with the same order of computations as the median filter. On the other hand, WM filters offer much greater flexibility in design specifications than the median filter. The weights control the filtering behaviour. The relation of WM filters to the median filter is comparable to that of linear FIR filters to the average filter.

The aim of this chapter is to summarize the recent developments in the theory and applications of WM filtering. In section 1.11.2, median and WM filters are introduced. Some very useful deterministic and statistical properties of median filters are summarized in the same section and the analogy between WM filters and linear FIR filters is highlighted. Section 1.11.3 explores the main deterministic and statistical properties of WM filters. The problem of optimal design of WM filters is addressed in section 1.11.4. Two applications of WM filters are outlined in section 1.11.5. Section 1.11.6 contains some conclusions. Interested readers are referred to the extensive bibliography for more detailed discussions about the theory, extensions, and applications of WM filters, in particular, [9, 11, 24, 39].

1.11.2 Median and weighted median filters

Following a brief discussion of the basic deterministic and statistical properties of median filters, WM filters are formally defined and an interesting analogy is drawn between the latter and linear FIR filters.

1.11.2.1 Median filtering

Since it was introduced by Tukey in 1974 [30], numerous papers, monographs, book chapters have explored the median operation, analyzing its behaviour and proposing new and viable extensions. To compute the output of a median filter, an odd number of sample values are sorted, and the middle or median value is used as the filter output. If the filter length is $N = 2K + 1$, the filtering procedure is denoted as

$$Y(n) = \text{MED}[X(n-K), ..., X(n), ..., X(n+K)] \tag{1}$$

where $X(n)$ and $Y(n)$ are the input and output sequences, respectively.

An immediate extension of the median filter is the class of rank order filters. These filters operate exactly like the median filter except that they put out the i-th largest sample in the window, the median being a special case:

$$Y(n) = i\text{-th largest value of the set } [X(n-K), ..., X(n), ..., X(n+K)] \tag{2}$$

Frequency analysis and impulse response have no meaning in median and rank order filtering; the impulse response of a median filter is zero! As a result, new tools are developed to analyze and characterize the behaviour of these nonlinear filters, deterministically and statistically. The basic descriptor of the deterministic properties of median filters is their root signal set, that is a set of signals which are invariant to further filtering. The basic statistical descriptor of median filters is the set of output distributions which are used to study the noise attenuation properties of median filters. These properties of median filters are briefly reviewed next.

(i) Deterministic properties of median filters: root signals

Since the output of the median filter is always one of the input samples, it is conceivable that certain signals could pass through the median filter unaltered. This has been shown to be true for median and many median-based filters. Since these signals define the signature of a filter, they are referred to as root signals.

The fact that some signals are invariant to median filtering offers interesting possibilities. In noise filtering, a problem is often how to preserve some desired signal features, while attenuating noise. An optimal situation would arise if the filter could be designed so that the desired features were invariant to the filtering operation and only noise would be affected. Since the median filter is a nonlinear filter, and the superposition principle does not apply, this of course can never be fully obtained. However, when a signal consists of constant areas and stepwise changes between these areas, a similar effect is achieved. Noise will be attenuated, but stepwise changes will remain. In image filtering, a common approach is to design a median filter such that certain image patterns, e.g. continuous lines, are root signals and thus not disturbed by the filtering operation.

The characterization of root signals is based on local signal structures, as defined in [12]. Based on these concepts, Gallagher and Wise proved that an arbitrary finite length signal is a median root if it only consists of constant neighbourhoods (region of at least $K+1$ consecutive identically valued samples) and edges (monotonically rising or falling set of samples surrounded on both sides by constant neighbourhoods of different values) [12]. They also proved that repeated median filtering of any finite length signal will result in a root signal after a finite number of passes. Root signals of median filters have been used for speech and image coding in [2] and [3].

(ii) Statistical properties of median filters

The recently developed tools make the analysis of the statistical behaviour of median filters possible and intuitive. Assuming that the samples inside the filter window are a combination of a constant (desired) signal corrupted by some additive white noise, as done often in the analysis of linear filters, one can evaluate the output distribution of the filter and hence estimate the output variance. Filter optimization is possible based on this approach, but yields even better results when combined with other types of constraints (e.g., structural constraints on the filter's behaviour [10]) and applied to extensions of median filters which allow much more design flexibility than just the filter length. We shall first review these statistical properties of median filters and then those of weighted median filters.

Let the input signal of a median filter be white noise, modelled here by i.i.d. random variables with mean μ and variance σ^2. The corresponding distribution and density functions are denoted by $\Phi(t)$ and $\phi(t)$, respectively. The distribution, $\Psi_{med}(t)$, and the density, $\psi_{med}(t)$, of the median filter output are

$$\Psi_{med}(t) = \sum_{i=K+1}^{N} \binom{N}{i} \Phi(t)^i (1 - \Phi(t))^{N-i} \tag{3}$$

and

$$\psi_{med}(t) = \frac{N!}{K!K!} \phi(t) \Phi(t)^K (1 - \Phi(t))^K \tag{4}$$

respectively. These two expressions provide the basis for quantitative analysis of the noise attenuation of median filters.

The good performance of the median filter with Laplacian distributed noise is due to an important optimality property of the median operation. The sample median is the maximum likelihood estimate for the mean value of the Laplacian density. This property shows that the median filter is optimal in the mean absolute error sense for attenuating double exponentially distributed noise. This analogy results from alternate definitions for

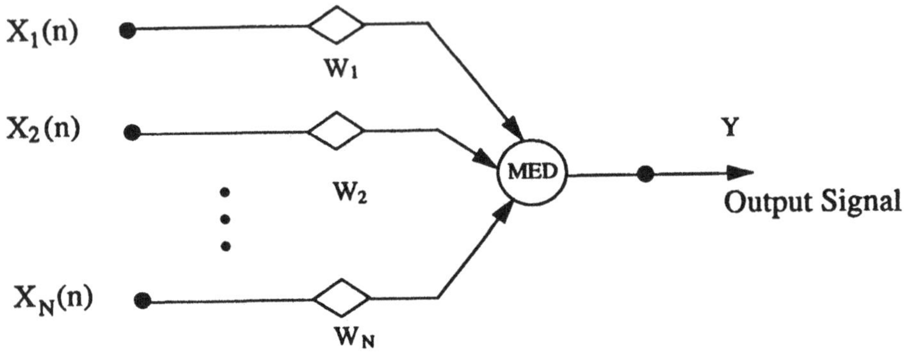

Figure 1 Weighted median filter.

the average and the median of N values $X_1, ..., X_N$. Consider the expression

$$\Phi(\beta) = \sum_{i=1}^{N} |X_i - \beta|^{\gamma} \tag{5}$$

The median of $X_1, ..., X_N$ can be defined as the value β minimizing $\Phi(\beta)$ in equation (5) when $\gamma = 1$. The sample average minimizes the same expression when $\gamma = 2$.

1.11.2.2 Weighted median filters

The weighted median (WM) filter was first introduced as a generalization of the standard median filter, where a non-negative integer weight is assigned to each position in the filter window [6, 16]. Here we define WM filters and discuss the analogy between WM filters and linear FIR filters.

As shown in Figure 1, the structure of a WM filter is quite similar to that of a linear FIR filter. For real-valued signals, WM filters can be defined in two different but equivalent ways.

Definition 1

For the discrete-time continuous-valued input vector $\underline{X} = [X_1, X_2, ..., X_N]$, the output Y of the WM filter of span N associated with the integer weights

$$\underline{W} = [W_1, W_2, ..., W_N] \tag{6}$$

is given by

$$Y = \text{MED}[W_1 \diamond X_1, W_2 \diamond X_2, ..., W_N \diamond X_N] \tag{7}$$

where MED[$\cdot$] denotes the median operation and $\diamond$ denotes duplication, i.e.,

$$K \diamond X = \overbrace{X, ..., X}^{K \text{ times}} \tag{8}$$

This filtering procedure can be stated as follows: sort the samples inside the filter window, duplicate each sample X_i to the number of the corresponding weight W_i and choose the median value from the new sequence. The second definition of the WM operation also allows positive non-integer weights to be used.

Definition 2

The weighted median of $\underline{X}$ is the value β minimizing the following expression

$$\Phi(\beta) = \sum_{i=1}^{N} W_i |X_i - \beta| \tag{9}$$

Here, β is guaranteed to be one of the samples X_i because $\Phi(\beta)$ is piecewise linear and convex, if $W_i \geq 0$ for all i.

The output of the WM filter for real positive weights can be calculated as follows: sort the samples inside the filter window; add up the corresponding weights from the upper end of the sorted set until the sum $\geq \frac{1}{2}\sum_{i=1}^{N} W_i$, the output of the WM filter is the sample corresponding to the last weight added.

(i) Analogy between WM filters and linear FIR filters

As the median filter is analogous to the average filter, there exists an interesting analogy between WM filters and linear FIR filters. Let $H_i, i = 1, 2, ..., N$, be the impulse response coefficients of a one-dimensional FIR filter of size N. If $H_i \geq 0$ for all i, $1 \leq i \leq N$ and $\sum_{i=1}^{N} H_i = 1$ then the output Y of the FIR filter can be calculated also as

$$Y = \sum_{i=1}^{N} H_i X_i = \text{average}\{W_1 \diamond X_1, W_2 \diamond X_2, ..., W_N \diamond X_N\} \tag{10}$$

where W_i, $1 \leq i \leq N$ are positive integers such that $H_i = W_i / \sum_{i=1}^{N} W_i$. If we replace the average in expression (10) with the median then the definition of the WM filter is obtained.

As mentioned earlier, the median is the maximum likelihood estimate of the signal level in the presence of uncorrelated additive biexponentially distributed noise; while, the arithmetic mean is that for Gaussian distributed noise. This means that the sample median of $X_i, i = 1, ..., N$ is the value β which minimizes expression (11) when $\gamma = 1$ and the sample mean is the value β which minimizes (11) when $\gamma = 2$:

$$\Phi(\beta) = \sum_{i=1}^{N} |X_i - \beta|^{\gamma} \tag{11}$$

If we introduce weights W_i in expression (11) as follows

$$\Phi(\beta) = \sum_{i=1}^{N} W_i |X_i - \beta|^{\gamma} \tag{12}$$

then, in the case $\gamma = 2$, the value β minimizing (12) can be expressed as

$$\beta = \sum_{i=1}^{N} W_i X_i / \sum_{i=1}^{N} W_i \tag{13}$$

which can be taken as a normalized FIR filter, i.e. a weighted average. Likewise, if $\gamma = 1$, the value β minimizing equation (12) is

$$\beta = \text{MED}[W_1 \diamond X_1, ..., W_N \diamond X_N] \tag{14}$$

i.e., a weighted median.

In the sequel, other interesting analogies will be drawn between WM filters and FIR filters. The similarity between them may suggest that WM filters would play as a fundamental role among median type filters as FIR filters among linear filters.

1.11.3 Properties of weighted median filters

Like median filters, WM filters are usually characterized by their deterministic and statistical properties. Based on the threshold function representation, Yli-Harja *et al* [46] analyzed the statistical properties of WM filters. The root signal properties of several subclasses of WM filters have been studied by several researchers, e.g. [13, 26, 27, 32, 33, 34, 38, 47].

1.11.3.1 Root properties of weighted median filters

Recall that a root of a filter is a signal that is invariant to this filter. The question of whether a WM filter will make a signal of a finite length converge to a root in a finite number of passes is crucial to the usefulness of that WM filter because a filter which does not possess this convergence property could have quite a chaotic filtering behaviour. Although the root properties of median filters are well known, root signals of WM filters are much more complicated than those of median filters. The convergence property of WM filters has only been proven for some special cases [7, 13, 27, 32, 38, 47]. The main reason may be that a WM filter with window width $2K+1$ can preserve details lasting less than $K+1$ samples. For example, consider the filter $[1,2,2,5,2,2,1]$. It is easy to check that the signal 111220200 is invariant to this filter.

Since symmetric WM filters are widely used in practice to avoid biased effects, we shall focus on the root properties of symmetric WM filters. A WM filter defined by

$$[W_{-K},\cdots,W_{-1},\mathbf{W}_0,W_1,\cdots,W_K] \tag{15}$$

is said to be symmetric if $W_{-i}=W_i \quad i=1,2,...,K$.

Based on the convergence properties of symmetric threshold automata, Wendt [32] proved that symmetric WM filters with $2K+1$ variables, which can preverve all the roots of median filter of window width $2K+1$, will make any appended input sequence converge to a root, or a cycle of period 2, in a finite number of passes.

In order to find conditions under which WM filters will converge, researchers have studied the root properties of several subclasses of WM filters, see [27, 38, 47]. In particular, Center WM filters, which give more weight only to the center weight in the window, are proven to possess the convergence property, [27].

1.11.3.2 Statistical properties of weighted median filters

As a class of robust smoothers, the performance of WM filters is usually described by their noise attenuation properties. These properties depend on the output distribution functions of WM filters expressed in terms of input distributions. An expression of the output distribution of WM filters was derived in [46] based on the positive Boolean function (PBF) representation of WM filters in the binary domain (for more details about the connection between PBFs and WM filters, see [21, 46]). Recently, by introducing a set of parameters called M_i, Yang *et al* [33] derived an intuitive expression of the output distribution and output moments of WM filters, which are stated in the following theorems. The optimal design of WM filters with structural constraints has thus emerged [33, 36].

Theorem [33]

Let the inputs of a WM filter, with window width $N = 2K+1$, be i.i.d. with a common distribution function $\Phi(t)$. The output distribution of the WM filter $\Psi_{wm}(t)$ has the following form

$$\Psi_{wm}(t) = \Psi_{med}(t) + \sum_{i=1}^{K} M_i \left\{ \Phi(t)^i \left(1 - \Phi(t)\right)^{N-i} - \Phi(t)^{N-i}\left(1 - \Phi(t)\right)^i \right\} \quad (16)$$

where $\Psi_{med}(t)$ is the output distribution of the median filter with the same window width and M_i's represent the number of different weight vector combinations, in which there are i weights and the sum of these i weights is not less than the threshold T_h, for $i = 1, 2, ..., K$.

Example 2

Given a WM filter with $\underline{W} = [1, 3, 8, 2, 3]$, then the threshold $T_h = 9$. It is obvious that

$$M_1 = 0,$$

$$M_2 = 4 : [8,3]; [8,3]; [8,2]; [8,1],$$
$$M_3 = 6 : [8,3,3]; [8,3,2]; [8,3,1]; [8,3,2]; [8,3,1]; [8,2,1],$$
$$M_4 = 5 : [8,3,3,2]; [8,3,3,1]; [8,3,2,1]; [8,3,2,1]; [3,3,2,1],$$
$$M_5 = 1 : [8,3,3,2,1]$$

The output moments of WM filters can be obtained using their output distribution function.

Theorem [34]

Given a WM filter with window width $N = 2K+1$, for i.i.d. inputs with common distribution function $\Phi(t)$ and density function $\phi(t)$. The output γ-norm standard deviation σ_{wm}^{γ} of the WM filter can be expressed as:

$$\sigma_{wm}^{\gamma} = \sigma_{med}^{\gamma} + \sum_{i=1}^{K} M_i L_i(N, \Phi, \gamma) \quad (17)$$

where σ_{med}^{γ} is the γ-norm standard deviation of the standard median filter with the same window size,

$$L_i(N, \Phi, \gamma) = \int_{-\infty}^{+\infty} \Gamma(\Phi(t))|t - m_{wm}|^{\gamma} \phi(t) dt \geq 0 \quad (18)$$

for symmetric $\Phi(t)$, $\gamma \geq 0$, $i = 1, ..., K$, and

$$\Gamma(\Phi(t)) = \left(i - N\Phi(t)\right)\Phi(i)^{i-1}\left(1 - \Phi(t)\right)^{N-i-1} + \left(i - N\left(1 - \Phi(t)\right)\right)\Phi(i)^{N-i-1}\left(1 - \Phi(t)\right)^{i-1} \quad (19)$$

This theorem demonstrates that the output standard deviation of the WM filter consists of two parts: the first corresponds to the standard median filter and the second quantifies the contribution of the weights. If all weights are equal, the second term vanishes. It is also obvious that

$$\sigma_{wm}^{\gamma} \geq \sigma_{med}^{\gamma} \quad (20)$$

since M_i and $L_i(N, \Phi, \gamma)$ are non-negative. This fact states that for i.i.d. inputs, the standard median filter achieves the best noise attenuation. Recall that the running average filter has the best noise attenuation among linear FIR filters, another interesting analogy.

The last theorem is easily applied to evaluate the noise attenuation capability of WM filters [33]. Let the output variance be a measure of noise reduction. We recast equation(17) when $\gamma = 2$,

$$\sigma_{wm}{}^2 = \sigma_{med}{}^2 + \sum_{i=1}^{K} L_i(N, \Phi, 2) M_i \tag{21}$$

where $\sigma_{med}{}^2$ denotes the variance of the standard median. In a vector form, equation (21) becomes $\sigma_{wm}{}^2 = \sigma_{med}{}^2 + \underline{M}\ \underline{L}^T$, where $\underline{L} = [L_1\ L_2\ \cdots\ L_K]$ and $\underline{M} = [M_1\ M_2\ \cdots\ M_K]$ denote the L-vector and M-vector, respectively.

For input signals having the same noise distribution function Φ, one can calculate the M-vectors and L-vectors and compare the variances of different filters. Since the L-vectors are independent of the weights, they can be tabulated for different noise distributions and different filter window sizes. L-tables containing the L-vectors for window sizes $N = 3$ to $N = 25$ for three commonly used noise distributions, uniform, Gaussian, and Laplacian can be found in [33].

1.11.4 Optimal weighted median filtering

In order to find optimal WM filters under certain error criterion, two different approaches have been developed recently, the estimation approach [28, 29, 40, 41, 43] and the structural approach [22, 36]. Here we shall review the latter approach and refer interested readers to the references above for the former.

1.11.4.1 Problem formulation

The optimal filtering problem over the class of WM filters can be stated as in Figure 2. The process $R(n)$ at the input of a WM filter is assumed to be a corrupted version of some desired process $D(n)$. The corruption is caused by a noise process $N(n)$ through a memoryless, possibly nonlinear, operation $g(.,.)$.

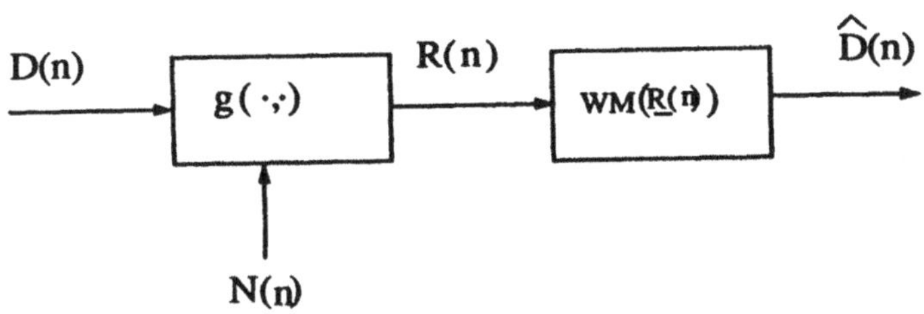

Figure 2 The optimal weighted median filtering problem.

A window of width N slides across the input process $R(n)$. Let $\underline{R}(n)$ be the vector containing the N samples in the window at time n,

$$\underline{R}(n) = [R(n-K), R(n-K+1), \ldots, R(n+K)] \tag{22}$$

where $2K + 1 = N$. The WM filter output is an estimate, called $\hat{D}(n)$, of the desired process $D(n)$; thus

$$\hat{D}(n) = \mathrm{MED}[W_1 \diamond R(n-K), W_2 \diamond R(n-K+1), ..., W_N \diamond R(n+K)] \tag{23}$$

1.11.4.2 Optimal WM filtering with structural constraints

It has often been observed that noise attenuation alone does not necessarily yield acceptable signal quality in images. It would be appropriate and very useful if this estimation approach would be somehow combined with the requirement to preserve certain image details. This approach was first developed in [10] for stack filters and called optimal filtering under structural constraints. This approach was later tuned and applied to WM filters in [22]. In case of WM filters, the signal structures can easily be expressed in terms of linear inequalities. In the special case of white noise corrupting a constant signal, optimal WM filters which preserve certain length signal details have been derived [36]. This approach is extended to 2-D filtering, where the image details to be preserved were lines in certain directions, areas and combinations, see [34, 35].

Using mathematical notations, we can express the problem of finding optimal WM filters under structural constraints as follows

$$J_\gamma(\underline{W}) = E[|D(n) - \hat{D}(n)|^\gamma], \quad \gamma > 0 \tag{24}$$

subject to

$$W_i \geq 0, \quad \text{for} \quad i = 1, 2, ..., N \tag{25}$$

$$\mathbf{A}\underline{W}^T \geq \underline{B}^T \tag{26}$$

where γ is a positive constant, $\gamma = 1, 2$ corresponding to the MAE and the MAE criteria, respectively. **A** is an $I \times N$ matrix, and $\underline{B}$ is an I dimensional vector of constants. Equation (26) denotes I inequalities which represent the structural constraints.

For example, given a WM filter with $\underline{W} = [W_1, W_2, \mathbf{W}_3, W_4, W_5]$. Suppose that constant pulses of length two or more shall be preserved, then the constraints on the weights should be

$$W_2 + W_3 \geq T_h,$$

$$W_3 + W_4 \geq T_h$$

Replacing T_h by $1/2 \sum_{i=1}^{5} W_i$, we can rewrite the above inequalities as follows

$$W_2 + W_3 \geq W_1 + W_4 + W_5,$$

$$W_3 + W_4 \geq W_1 + W_2 + W_5$$

In general, it is difficult to solve the optimization problem of equations (24)-(26). However, in certain cases optimal WM filters with structural constraints can be obtained even in closed form [34, 36].

When the desired signal is a constant signal, the MSE and the MAE of the WM filter are the variance and the absolute deviation of the output signal. Assume the constant signal is contaminated by i.i.d. noise with common distribution function $\Phi(t)$. Then the cost function of the optimization problem becomes

$$J_\gamma(\underline{W}) = \sigma_{med}^\gamma + \sum_{i=1}^{K} M_i L_i(N, \Phi, \gamma) \tag{27}$$

where $N = 2K + 1$.

It is straightforward to see that if there are no structural constraints on the weights, the optimal WM filter is the standard median filter.

Several algorithms are provided in [36, 37] for solving the optimization problem in equation (27). Using these algorithms, the optimal symmetric WM filters which preserve length

2 details and length 3 details have been found.

Given a WM filter with window size $N = 2K + 1$,

$$\underline{W} = [W_{-K}, ..., \mathbf{W_0}, ..., W_K]$$

The optimal WM filters preserving length 2 details are expressed by the following explicit form

$$\underline{W} = [1, 1, ..., 1, K, \mathbf{2K - 1}, K, 1, ..., 1, 1]$$

More importantly, this class of optimal filters is proven to have the following properties:

- These filters are optimal for different noise distributions.
- These filters are optimal for any positive γ. This means that they are optimal both in the sense of MSE criterion and in the sense of MAE criterion.

1.11.5 Applications of weighted median filters

With the development of WM filtering theory, the applications of WM filters have been explored in 1-D, 2-D, and 3-D signal processing. A WM filter is basically a smoother with unit gain and performs similar to an FIR filter with only positive coefficients summing up to unity. This is why they are mainly used as robust noise smoothers and predictors.

1.11.5.1 Speech signal processing

An early application of median filtering was introduced in [15], where Jayant studied the use of a short median filter for improving DPCM coded speech quality in the presence of transmission errors. He found that if the probability of a bit error is of the order of 1% or more, the result of smoothing with a 3-point (or a 5-point) median filter turns out to be perceptually desirable in spite of the speech smearing caused by the filter.

With small bit error rates, the smearing of the speech waveform caused by the 3-point median filter is unacceptable if compared to the advantage gained. However, even a bit error rate as low as 0.05% is very disturbing if the speech quality is otherwise good. Idempotent WM filters [13] have the property that apart from removing single impulses they have only a minimal effect on the signal. Therefore, they are good candidates for the current problem.

Haavisto *et al* [13] applied 1-D idempotent WM filters to PCM coded speech samples contaminated by random bit errors and compared their performance to that of the 3-point median filter. The 1-D idempotent WM filter they used can be represented in the following form:

$$\underline{W} = [\overbrace{1, 1, ..., 1}^{K\text{times}}, 2K - 1, \overbrace{1, ..., 1, 1}^{K\text{times}}] \tag{28}$$

where $2K + 1$ is the filter length.

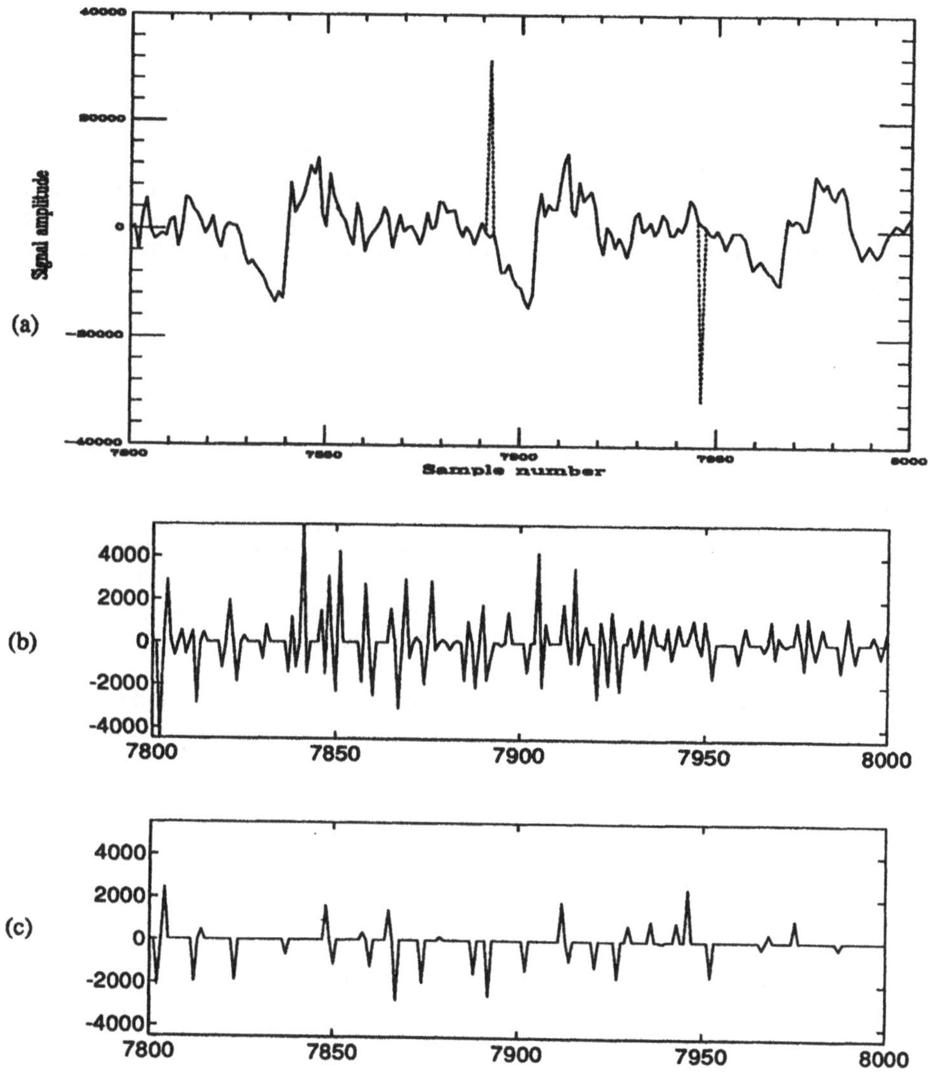

Figure 3 (a) 200 samples of a high quality speech sample. The speech sample (continuous line) is PCM coded and has a sampling frequency of 8 kHz. The length of the speech sample is 80000 samples, i.e. 10 seconds. Each sample is represented by 16 bits. Random bit errors were introduced to the speech sample with the probability 0.05% resulting in the signal shown with the dotted line. (b) The difference signal between the original and the filtered signal by a 3-point median filter. The smearing of the speech signal that is clearly visible and the speech quality has been severely decreased. (c) The difference signal between the original and the filtered signal by an idempotent filter given by (28). The smearing of the speech signal is minimal and the speech quality is very close to the original sample.

It was found that with $K = 5$, it was very difficult to distinguish between the original signal and the filtered signal. Therefore, only a minor additional advantage could be achieved by making the filter longer. With a bit error rate of 0.05%, this filter performed very well. The quality of the filtered speech signal was very close to the original; whereas, the smearing caused by the 3-point median filter was quit severe. This is illustrated in Figure 3. When the bit error rate is increased to 0.5%, the idempotent filter significantly improves the speech quality and is much better than the 3-point median filter. When the error rate is further increased to 2.5% the median filter performs better and a shorter idempotent filter should be used. However, in this example the 2.5% bit error rate means that there is on average at least one erroneous bit in every third data sample, and acceptable speech quality can no longer be achieved.

1.11.5.2 Optimal weighted median image filtering

An image consists of many signal structures, such as lines, corners, which are critical to perception. When filtering noisy images, care must be taken not to remove such important image details. The class of weighted median filters has proven to be a potential candidate for such tasks due to its capabilities in (impulsive) noise attenuation and detail preservation [6, 13, 17, 19, 27]. Several adaptive algorithms have been developed for weighted median and weighted order statistic filters, see [28, 29, 42, 45], all requiring training (ideal and noisy) signals. However, in many applications of interest, ideal signals may not be available. It is, therefore, necessary to develop an alternative scheme to design weighted median filters in these cases.

Important image structures, e.g. lines and corners, constitute (part) of the structural constraints which must be preserved by the WM filter. The task is to select a WM filter, among those which preserve the given set of structural constraints, which achieves maximum noise attenuation.

Consider WM filters with 3×3 square window

$$\underline{W} = \begin{pmatrix} W_1 & W_2 & W_3 \\ W_4 & W_5 & W_6 \\ W_7 & W_8 & W_9 \end{pmatrix} \tag{29}$$

In [33], we show that the weights of WM filters must satisfy the following set of linear inequalities in order for the filter to preserve vertical and horizontal lines in the image (other details are also considered in [33]).

Structural Constraints

$$\begin{cases} W_2 + W_5 + W_8 \geq T_h \\ W_4 + W_5 + W_6 \geq T_h \end{cases} \tag{30}$$

Solving this set of inequalities leads to the following optimal WM filter [34]

$$\mathrm{WM} = \begin{pmatrix} 1 & 3 & 1 \\ 3 & 5 & 3 \\ 1 & 3 & 1 \end{pmatrix}$$

Filter	Bridge		Lenna		Harbor	
	MSE	MAE	MSE	MAE	MSE	MAE
Wiener Filter	415	15.1	357	13.8	457	15.9
Standard Median	165	7.46	46.2	3.61	230	7.15
WM trained under MSE	141	6.39	46.4	3.88	178	5.84
WM trained under MAE	207	4.08	152	2.51	210	3.77
WM	124	5.58	42.2	2.83	164	5.33

Table 1 Filters' performance in impulsive noise.

The above optimal WM filter was used to filter image "Bridge-over-stream", which was corrupted by impulsive noise. The probability of impulses was 0.12 and the height of the impulses was set to ± 200. Using the original image and its corrupted version, the 3×3 FIR Wiener filter and the adaptive WM filters (which were trained under the MSE and MAE senses, respectively) [42] were applied to restore the corrupted image. The MSE and MAE values are listed in table 1. These filters were also applied to restore images "Lenna" and "Harbor" which were corrupted by the same noise in order to show the robustness of WM filters. Results are included in table 1. Note that the performance of the optimal WM filter (WM) is much better than that of the Wiener filter and the standard median filter, and is comparable to that of the adaptive WM filters. Remember that adaptive WM filters were trained using whole part of the ideal and noisy images.

1.11.6 Summary

The theory and major applications of weighted median filters are reviewed. The behaviour and performance of WM filters are assessed based on their deterministic and statistical properties. The former is related to the concept of root signals and deterministic convergence; while, the latter deals with e.g. output distributions and noise attenuation capability. Several interesting similarities between WM filters and linear FIR filters were pointed out.

Based on this threshold representation, optimal adaptive and non-adaptive weighted median filtering algorithms have been developed in the literature. The optimization goal is to minimize the mean absolute error or the mean square error (not reviewed here due to lack of space). Furthermore, under a given set of structural constraints, the resulting optimal WM filter will attenuate noise maximally while preserving certain desired signal structures.

To demonstrate the effectiveness of weighted median filtering, two important applications are reviewed, one in speech processing and the other in image restoration.

Even though a remarkable development has been achieved, the theory of WM filtering is far from mature. There are many open questions and research problems of considerable interest, e.g., the root structures of WM filters are not clear yet, except for several special cases. Even for symmetric WM filters, the limits on the number of passes to a root are not known. Full understanding of the root structures of WM filters will certainly make it possible to design optimal WM filters under root structural constraints.

Acknowledgement

The author wishes to thank Drs. L. Yin, J. Astola and Y. Neuvo, and particularly, Mr. R. Yang, for their contribution to this chapter.

References

[1] G. R. Arce and R. E. Foster, "Detailed preserving raked-order based filters for image processing," *IEEE Trans. Acoust., Speech, Signal Processing*, Vol. ASSP-37, pp. 83–98, Jan. 1989.

[2] G. R. Arce and N. C. Gallagher, Jr., "State description for root-signal set of median filters," *IEEE Trans. Acoust., Speech, Signal Processing*, Vol. ASSP-30, pp. 894–902, Dec. 1982.

[3] G. R. Arce and N. C. Gallagher, Jr., "BTC image coding using median filter root," *IEEE Trans. Communications*, Vol. COM-31, pp. 784–793, June 1983.

[4] A. C. Bovik, T. S. Huang and D. C. Munson, Jr., "The effect of median filtering on edge estimation and detection," *IEEE Trans. Pattern Analysis, Machine Intelligence*, Vol. PAMI-9, pp. 181–194, Mar. 1987.

[5] A. C. Bovik, "Streaking in median filtered images," *IEEE Trans. Acoust., Speech, Signal Processing*, Vol. ASSP-35, pp. 493–503, Apr. 1987.

[6] D. R. K. Brownrigg, "The weighted median filter," *Comm. of the ACM*, Vol. 27, p. 8, pp. 807-818, Aug. 1984.

[7] H. Chen, R. Yang and M. Gabbouj, "The convergence rate of symmetric weighted median filters," *European on Signal Processing Conference, EUSIPCO'94*, September 1994, accepted.

[8] E. J. Coyle and J. H. Lin, "Stack filters and the mean absolute error criterion," *IEEE Trans. Acoust., Speech and Signal Proc.*, Vol. ASSP-36, No. 8, pp. 1244–1254, Aug. 1988.

[9] E. R. Dougherty and J. Astola. *An introduction to nonlinear image processing*, SPIE Press, Vol. TT16, 1994.

[10] M. Gabbouj and E. J. Coyle, "Minimum mean absolute error stack filtering with structural constraints and goals," *IEEE Trans. on Acoust., Speech and Signal Processing*, Vol. ASSP-38, No. 6, pp. 955–968, June 1990.

[11] M. Gabbouj, E. J. Coyle and N. C. Gallagher, Jr., "An overview of median and stack filtering," *Circuits, Systems Signal Processing*, Vol. 11, No. 1, pp. 7–45, 1992.

[12] N. C. Gallagher, Jr. and G. W. Wise "A theoretical analysis of the properties of median filters," *IEEE Trans. on Acoust., Speech and Signal Proc.*, Vol. ASSP-29, pp. 1136–1141, Dec. 1981.

[13] P. Haavisto, M. Gabbouj and Y. Neuvo, "Median based idempotent filters," *Journal of Circuits, Systems and Computers*, Vol. 1, No. 2, pp. 125–148, June 1991.

[14] P. Heinonen and Y. Neuvo, "FIR-median hybrid filters," *IEEE Trans. on Acoustics, Speech and Signal Processing*, Vol. ASSP-35, No. 6, pp. 832–838, June 1987.

[15] N. S. Jayant, "Average and median-based smoothing technique for improving quality in the presence of transmission errors," *IEEE Trans. Comm.*, Vol. COM-24, pp. 1043–1045, Sept. 1976.

[16] B. I. Justusson, "Median filtering: statistical properties," *Two-dimensional digital signal processing II*, T.S. Huang, Ed., Springer-Verlag, 1981.

[17] S.-J. Ko and Y. H. Lee, "Center weighted median filters and their application to image enhencement," *IEEE Trans. on Circuits and Systems*, Vol. 38, No. 9, pp. 984–993, Sept. 1991.

[18] J. H. Lin and E. J. Coyle, "Minimum mean absolute error estimation over the class of generalized stack filters," *IEEE Trans. Acoustics, Speech and Signal Processing*, Vol. ASSP-38, No. 4, pp. 663–678, April 1990.

[19] T. Loupas, W. N. McDicken and P. L. Allan, "An adaptive weighted median filter for speckle suppression in medical ultrasonic images." *IEEE Trans. on Circuits and Systems*, Vol. CAS-36, pp. 129–135, Jan. 1989.

[20] A. Nieminen, P. Heinonen and Y. Neuvo, "A new class of detail-preserving filters for image processing," *IEEE Trans. Pattern Anal. Machine Intell.*, Vol. PAMI-9, Jan. 1987.

[21] J. Nieweglowski, M. Gabbouj and Y. Neuvo, "Weighted medians-positive Boolean functions conversion," *Signal Processing*, Vol. 34, No. 2, Nov. 1993.

[22] J. Nieweglowski, L. Yin, M. Gabbouj and Y. Neuvo, "Optimal WOS filters under structural constraints," *Proc. of IEEE Int. Symposium on Circuits and Systems*, San Diego, May 10–13, 1992.

[23] F. Palmieri and C. G. Boncelet, Jr., "Ll-filter - a new class od order statistic filters," *IEEE Trans. on Acoustics, Speech and Signal Processing*, Vol. ASSP-37, No. 5, pp. 691–701, May 1989.

[24] I. Pitas and A. N. Venetsanopoulos, *Nonlinear Digital filters: Principles and applications*, Kluwer Academic Publisher, 1990.

[25] W. K. Pratt, *Digital image processing*, New York, Wiley, 1991.

[26] M. K. Prasad and Y. H. Lee, "Weighted median filters: generation and properties," *Proc. IEEE Int. Symp. Circ. and Syst.*, pp. 425–428, 1989.

[27] T. Sun, M. Gabbouj and Y. Neuvo, "Center weighted median filters: some properties and application in image processing," *Signal Processing*, Vol. 35, No. 3.

[28] I. Tabus, M. Gabbouj and L. Yin, "Real domain adaptive WOS filtering using neural network approximations," *Proc. of IEEE Winter Workshop on Nonlinear Digital Signal Processing*, Tampere, Finland, pp. 7.2–1.1–7.2–1.6, Jan. 17–20, 1993.

[29] I. Tabus, M. Gabbouj and L. Yin, "Applications of neural networks training methods to optimal WOS image filtering," *Proc. of IEEE Workshop on Visual Signal Processing and Communications*, Melbourne, Australia, Sept. 20–23, 1993.

[30] J. W. Tukey, "Nonlinear (Nonsuperposable) methods for smoothing data," *Cong. Rec. EASCON' 74*, p. 673, 1974.

[31] P. D. Wendt, E. J. Coyle and N. C. Gallagher, Jr., "Stack filter," *IEEE Trans. Acoust., Speech, Signal Proc.*, Vol. ASSP-34, No. 4, pp. 898–911, Aug. 1986.

[32] P. D. Wendt, "Nonrecursive and recursive stack filters and their filtering behaviour," *IEEE Trans. on Acoustics, Speech and Signal Proc.*, Vol. ASSP-38, No. 12, pp. 2099–2107, Dec. 1990.

[33] R. Yang, L. Yin, M. Gabbouj, J. Astola and Y. Neuvo, "Output distribution of weighted median filters and optimality with structural constraints," *IEEE Trans. on Signal Processing*, to appear.

[34] R. Yang, L. Yin, M. Gabbouj and Y. Neuvo, "2-D optimal weighted median filters," *Proc. of 1992 IEEE Workshop on Visual Signal Processing and Communications*, North Carolina State University, Raleigh, North Carolina, pp. 36–39, Sept. 1992.

[35] R. Yang, L. Yin, M. Gabbouj, J. Astola and Y. Neuvo, "Performance of detail-preserving weighted median filters for image processing," *Proc. of IEEE Winter Workshop on Nonlinear Digital Signal Processing*, Tampere, Finland, pp.5.1–3.1–5.1–3.6, Jan. 1993.

[36] R. Yang, L. Yin, M. Gabbouj, J. Astola and Y. Neuvo, "Optimal weighted median filters under structural constraints," *Proc. of IEEE Int. Symposium on Circuits and Systems*, Chicago, IL, pp. 942–945, May 2–6, 1993.

[37] R. Yang, L. Yin, M. Gabbouj and Y. Neuvo, "Optimal weighted median filtering for image restoration," *Proc. of Image Processing: Theory and Applications*, San Remo, Italy, G. Vernazza and A. N. Venetsanopoulos Ed., pp.283–286, June 14–16, 1993.

[38] R. Yang, M. Gabbouj and Y. Neuvo, "On convergence behaviour of symmetric weighted median filters," *Proc. of European Conference on Circuit Theory and Design*, Davos, Switzerland, pp. 1053–1058, 30 Aug.-3 Sept., 1993.

[39] L. Yin, R. Yang, M. Gabbouj and Y. Neuvo, "Weighted median filters: A tutorial," submitted to *IEEE Trans. CAS*, March 1994.

[40] L. Yin, J. Astola and Y. Neuvo, "Adaptive weighted order statistic filters," *Int. Workshop on Algorithms and Parallel VLSI Architectures*, Part B, Pont-a-Mousson, France, pp. 327–332, June 10–16, 1990; also included in *Algorithms and parallel VLSI architectures*, E. F. Deprettere, Ed.

[41] L. Yin, J. Astola and Y. Neuvo, "Optimal weighted order statistic filters under the mean absolute error criterion," *Proc. of Int. Conference on Acoustics, Speech and Signal Processing*, Vol. 4, pp. 2529–2532, Canada, May 1991.

[42] L. Yin, J. Astola and Y. Neuvo, "Adaptive stack filtering with application to image processing," *IEEE Trans. on Signal Processing*, Vol. 41, No. 1, pp. 162–184, Jan. 1993.

[43] L. Yin, J. Astola and Y. Neuvo, "Adaptive multistage weighted order statistic filters based on the back-propagation algorithm," submitted to *IEEE Trans. on Signal Processing*, April 1991.

[44] L. Yin, J. Astola and Y. Neuvo, "Adaptive weighted median filtering under the mean absolute error criterion," *Proc. of IEEE Workshop on Visual Signal Processing and Communications*, Hsinchu, Taiwan, pp.184–187, June 6–7, 1991.

[45] L. Yin, J. Astola and Y. Neuvo, "A new class of nonlinear filters - neural filters," *IEEE Trans. on Signal Processing*, Vol. 41, No. 3, pp. 1201–1222, March 1993.

[46] O. Yli-Harja, J. Astola and Y. Neuvo, "Analysis of the properties of median and weighted median filters using threshold logic and stack filter representation," *IEEE Trans. on Signal Processing* , Vol. 39, No. 2, pp. 395–410, Feb. 1991.

[47] P.-T. Yu and W.-H. Liao, "The convergence behaviour of symmetric weighted median filters," *Proc. of IEEE Winter Workshop on Nonlinear Digital Signal Processing*, pp. 5.1–4.1–5.1–4.6, Tampere, Finland, Jan. 1993.

Chapter 1.12

Stack Filters in Signal and Image Processing

Edward J. Coyle

1.12.1 Two approaches to signal and image processing

There are two fundamentally different approaches to solving problems in signal and image processing. As a useful simplification, they might be called the regularization approach and the filter-class approach.

In the regularization approach, which is outlined in [1], you first define a goal that is to be met by a solution of the problem. Certain constraints that a solution must satisfy are then formulated. These may be motivated by the "physics" of the problem, by the analytical pleasure gained when solutions are provably unique, and/or by a desire for stability — i.e., solutions vary slowly as the data is perturbed. Finally, you determine a solution which minimizes the error criterion while satisfying these constraints.

One advantage of this approach is that the solution to the problem is not restricted to those that can be obtained by applying one member of some parameterized class of operators to the data. The solution is thus the best that can be obtained in the sense that the only restrictions imposed are supposedly motivated by the problem itself.

This freedom from parameterization of the operation can, unfortunately, be a serious drawback. It often leads to situations in which a great deal of computation is required to obtain the solution. Furthermore, the actual operation being performed on the data can not often be explicitly stated. A relevant exception to this can be found in [2], which relates the recursive median filter [3] to the solution of a particular regularization problem. Another relevant example can be found in [4], which relates the solution of a particular regularization problem — specifically, edge-preserving MAP estimation — to recursive weighted median filters.

The second approach, the filter-class approach, has appeared on a case-by-case basis in a very large number of papers. A review of some of these cases can be found in the introductory sections of [5]. In this approach, you choose both an error criterion and a parameterized class of operators. These two choices are usually coupled, the goal being "easy" determination of the parameters of an optimal operator given the characteristics of the data and the error criterion.

The choice of the set of operators is based on compromises among the generality of the set of operators, the behaviour of the operators, the speed with which an optimal set of parameters can be determined, and the ease with which they can be implemented. The classic example illustrating this approach is linear minimum mean square error estimation — Wiener filtering [6].

One difficulty with this approach is that it may not be simple to impose certain desirable constraints on the output of the operators while still maintaining performance and ease of identifying an optimal operator. For example, it may be difficult to obtain robustness in the face of incompletely specified or possibly nonGaussian noise when the class of filters is linear. Furthermore, it may turn out that none of the operators in the chosen class yield acceptable performance in the given situation. This is the case, for example, when an attempt is made to smooth impulsive noise with linear, translation-invariant filters.

Of course, the best of all possible worlds is realized when these two approaches yield the same results. The regularization approach to a given problem may lead to a specific class of operators which not only always contains the optimal operator but is well-suited for implementation. Examples of this are again provided by [2, 4]. Likewise, the filter-class approach may point to the class of problems for which the operator which is optimal among all possible operators is contained within the given parameterized class. An example in which a given implementable filter class always contains the overall optimal operator is minimum mean square estimation in the presence of additive Gaussian noise [7].

1.12.2 Problems and performance measures in image processing

The goal is to quickly determine, or calculate, solutions which are optimal in some sense for specified problems in image processing. This is obviously a vaguely stated goal. The "problems" are not specified and neither is the sense in which optimality is to be determined. In the following we will become more specific on both of these matters until the problem can finally be formulated analytically – and the formulation will be carried out, to the extent possible in a brief tutorial, under both the regularization and the filter-class approaches. Ideally, both approaches will yield the same solution to whatever problem is finally stated.

Note that the set of problems and issues that are considered in what follows are not necessarily those that others would choose. The intent is only to show how a certain set of problems leads to the class of filters known as stack filters [8]. The following is the list of problems and issues being considered:

1. ***Detection and/or preservation of edges*** in an image during the processing of the image. This is necessary because of the importance of edges in image segmentation, object identification, and other tasks in visual perception.

2. ***Scale invariance.*** Because the scale of an image is usually not known in advance, it is very desirable that the operations to be performed on the image be scale-independent.

3. ***Suppression of nonGaussian noise.*** The noise corrupting images consists of quantization noise, noise to due bit errors in transmission or processing, interference effects due to the use of coherent imaging systems, line drop-outs, etc. None of these noise sources can be adequately modelled as additive Gaussian noise; some of them are definitely impulsive in nature.

4. ***High processing speed and efficient VLSI implementation*** for whatever operation is performed.

5. ***Understanding*** the effects of the operation being performed by direct analysis of the operation itself.

Specifying an appropriate measure of performance can be just as difficult as identifying the problems to be solved. This is especially true in image processing, where human judgement of the quality of an image must somehow be translated into a quantitative measure of the performance of whatever operator is derived or synthesized.

For instance, in noise suppression tasks it is important to avoid blurring edges in the image but also important to remove essentially all impulses that occur. But how should an impulse be defined so that it is distinguishable from details in the image? And is preserving the sharpness of edges so important that it should take precedence over the problem of edge jitter, however that is defined?

It is thus clear that choosing a performance measure is often difficult. The following is surely only a partial list of the issues that must be considered.

1. ***Robustness.*** If the statistics of the image and/or noise should change, the error measure should still lead to an acceptable solution of the problem. An excellent discussion of robustness of error measures can be found in [9]. Of course, robustness might be added as a constraint instead of building it into the error measure.

2. ***Ease of solution.*** The easier it is to find an optimal solution under a given performance measure, the more likely it will be used. This is true for both the regularization and filter-class approaches, but is perhaps easiest to see its effect in the filter class approach. The reason the mean squared error criterion is used with linear filters is that a closed form solution for the best linear filter can be found under this error criterion [6].

3. ***Efficacy.*** Does the error measure lead to the desired solution in terms of the nonanalytical criteria it is meant to mimic?

Finally, any constraints which are to be imposed must be formulated. They are often needed under both the regularization and the filter-class approach. The constraints may be required to eliminate undesirable tendencies induced in the solution by the performance measure, or to ensure certain desirable behaviour.

With the above problems and issues in mind, we can state the following facts, all of which point to the use of stack filters in a wide variety of image processing tasks. These facts are classified according to whether they arise from the regularization approach or the filter-class approach.

1.12.3 The regularization approach and stack filters

In [4] a Markov random field model which allows realistic edge modelling and stable maximum *a posteriori* (MAP) solutions is developed. The model satisfies the desirable regularity properties of uniqueness and stability and also the property of scale-invariance mentioned previously. When the potential function under which the the MAP estimate is obtained is $|\Delta|$, with Δ the difference between pairs of pixel values in the neighbourhood, the operation performed on the data to obtain the solution is a recursive weighted median filtering operation. Weighted median filters are contained within the class of stack filters.

Also of interest in [4] is that potential functions of the form $|\Delta|^p$ were investigated for values of p between 1 and 2. When $p = 1$, sharp and smooth edges were found to have the same cost. Also, $p = 1$ yielded the best performance when the task was image reconstruction in low-dosage transmission tomography.

In [2], optimal smoothing of a noisy signal under a regularized absolute norm is investigated. When the class of signals is restricted to those which are locally monotone, then the best estimate of the original signal can be obtained by applying a recursive median filter to the data. Again, median filters are contained in the class of stack filters. Also of interest is the fact that signals which are locally monotone are fixed points of both the recursive median [3] and median operators [10, 11].

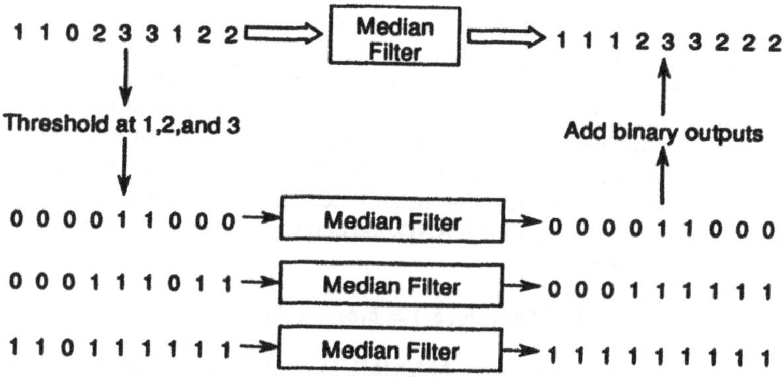

Figure 1 Median filtering by threshold decomposition.

1.12.4 The filter-class approach and stack filters

The median filter is in many ways analogous to the linear filter which computes the average of the data points inside a window. Since all other linear filters can be regarded as a generalization of the averaging filter, the question of how to generalize the median filter immediately arises. Many answers to this question have been offered, some better than others. A brief survey of some of the answers may be found in [5]; we will focus on only the generalization which leads to stack filters.

Any extension of the median filter should concentrate on specific properties of the median that are key to its operation. Two such properties are the threshold decomposition and stacking properties, which were first revealed in [12, 13].

The threshold decomposition property is a weak superposition property. It states that applying the median filter to a discrete-time signal which takes on the values 0 through M is equivalent to the following sequence of operations:

1. Decompose the signal into M binary signals by thresholding it at all values from 1 through M. Note that when these binary threshold signals are added together they sum to the original signal.

2. Apply the median filter to each of the binary signals obtained by thresholding the original signal. Since the output of the median filter is always equal to one of the sample values in the input window, the output signal will also be binary. The median filter with binary input is just a majority logic operator.

3. Add the filtered binary threshold signals together to obtain the output signal.

The importance of this property follows from the fact that an understanding of the effect of a median filter upon multi-valued signals can be obtained from a knowledge of its effect on binary signals. This dramatically simplifies the analysis of median filters and stack filters in many situations; particularly impressive examples of this fact can be found in the analysis of the convergence properties of median filters and stack filters [14, 15, 16].

This superposition property, which is illustrated with a particular signal in Figure 1, is a *weak* superposition property. It applies only when the input signal is decomposed via thresholding. If the filter were linear, the superposition would work no matter how the input signal was broken up into a sum of other signals. Note that all linear filters also possess the threshold decomposition property.

The stacking property is an ordering property. It states that if any point in the input signal is increased, then the output must not decrease. The importance of this property follows its implications for binary operators – recall that the median filter is a binary operator when its input is binary. A logical operator which possesses the stacking property is called a positive Boolean function. It always has a unique Minimum Sum of Products (MSP) expression which contains only the logical AND and OR operations; complementation is not allowed in the MSP expression. For instance, letting addition stand for the logical OR and multiplication stand for the logical AND, the output of the window width three median filter when the input is binary is

$$MED(b_1, b_2, b_3) = b_1b_2 + b_2b_3 + b_3b_1$$

Then, since the OR operation is just a MAXIMUM operation, and the AND is the MINIMUM operation [17], the median filter applied to any signal can be expressed as follows:

$$MED(x_1, x_2, x_3) = MAX(MIN(x_1, x_2), MIN(x_2, x_3), MIN(x_3, x_1))$$

It can be shown that all rank order operators, particularly the MIN and MAX in addition to the MED, possess the threshold decomposition and stacking properties [13]. It is now clear how the median filter can be generalized — consider all operators defined jointly by positive Boolean functions and the threshold decomposition. In other words, take any positive Boolean function, determine the multi–level representation of that function, and you have a filter possessing the threshold decomposition and stacking properties. For example, the positive Boolean function

$$AMED1(b_1, b_2, b_3) = b_2 + b_3b_1$$

defines the multi-level operator

$$AMED1(x_1, x_2, x_3) = MAX(x_2, MIN(x_3, x_1))$$

which is one of the window width three asymmetric median filters identified in [8]. It behaves like the median filter in most respects, except that it preserves positive-going impulses. Another difference with the median is that the asymmetric median is an idempotent operator; one pass of the filter over a signal yields a signal which is invariant to further passes of the filter.

As seen above, there is a one-to-one relationship between positive Boolean functions and stack filters and every stack filter possesses both the threshold decomposition and stacking properties. Many of the properties of stack filters are explained at a tutorial level in [5, 18], where summaries and relevant references are provided on the number of stack filters of a given window size, optimization over the class of stack filters, implementation of median and stack filters, and invariant signals of stack filters. There is also an interesting connection between stack filters and neural networks [19, 20, 21] since threshold logic operators with positive weights are a subset of stack filters. Also, stack filters have an interesting generalization to a more powerful class of filtering operations [22].

In summary, following two important properties of the median filter leads to the class of stack filters. Since this class of filters has very efficient, high-speed VLSI implementations — see [23] for the most recent improvements in implementation of stack filters — it is well-suited for the filter-class approach to signal and image processing.

The class of stack filters also results from a filter-class approach to image processing which starts at a much different point of view. This other point of view is known as the morphological approach to image processing [24, 25]. From very general principles, a class of filtering operations known as increasing morphological operators with flat structuring elements is arrived at. These operators are all stack filters as well [22, 26]. The connection between these filters and the shape filtering notions developed in morphology are thus important evidence that stack filters are appropriate for image processing.

Cellular automata are operators which have also been found to be useful in the field of image processing [27, 28, 29]. They are based on logical operators, as are stack filters. They also provide a connection between concepts in physics and stack filters [30].

In conclusion, the development and/or study of many different types of operations on images leads to the class of stack filters. They are thus very appropriate for the filter-class approach to image processing. The remaining piece that must be added is a tractable, fast method for choosing a stack filter which is a best solution to a given problem.

1.12.5 Notation and precise statement of stack filter properties

As stated above, stack filters are a class of nonlinear filters which satisfy two properties: the weak superposition property known as the threshold decomposition, and the ordering property called the stacking property.

To precisely define these properties, we must first define the threshold decomposition of an image. Images will be assumed to take values on a lattice S. Each point in the lattice will be denoted by $s \in S$, and, when necessary, the point will be denoted explicitly by its coordinates, $s = (s_1, s_2)$. A gray scale image, X, with pixel values ranging between 0 and M, may be represented as the sum of a series of binary-valued images,

$$X(s) = \sum_{l=1}^{M} x_l(s)$$

where, for each l, the binary image x_l is obtained by thresholding X at level l.

$$x_l(s) = \begin{cases} 1, & X(s) \geq l \\ 0, & X(s) < l \end{cases}$$

The threshold decomposition of the windowed field W_X is defined similarly:

$$W_X(s) = \sum_{l=1}^{M} w_{X,l}(s)$$

where $w_{X,l}(s)$ is the windowed field at position s of x_l.

Each stack filter $S_f(\cdot)$ is defined by a Boolean function $f(\cdot)$ which satisfies a *stacking* property: if the output of f applied to $w_{X,l}(s)$ is 1, then the outputs of f applied to threshold levels k where $k \leq l$ must also be 1's [8]. More formally, for all $k \leq l$

$$f(w_{X,k}(s)) \geq f(w_{X,l}(s)) \tag{1}$$

A Boolean function has this property if and only if it is positive [31].

A positive Boolean function is a logical operator whose minimum sum of products (MSP) expression does not have any complements; that is, it has only AND and OR operations. This implies that the multilevel stack filter may be formed by cascades and combinations of MAX and MIN operations [8, 17, 26].

For example, the window width three median filter is a stack filter defined by the

Boolean function

$$f(b_1, b_2, b_3) = b_1 b_2 + b_2 b_3 + b_3 b_1$$

in which the variables b_i are binary, "+" denotes the logical OR (binary MAX), and multiplication denotes the logical AND (binary MIN). The window width three multilevel median is thus defined as follows:

$$S_f(x_1, x_2, x_3) = MAX(MIN(x_1, x_2), MIN(x_2, x_3), MIN(x_3, x_1))$$

Since every stack filter has a MAX of MINS representation, every stack filter obeys the the superposition property called the *threshold decomposition* [8, 12, 13]:

$$Y(s) = S_f(W_X(s)) = S_f\left(\sum_{l=1}^{M} w_{X,l}(s)\right) = \sum_{l=1}^{M} S_f(w_{X,l}(s)) \quad (2)$$

1.12.6 Minimum mean absolute error stack filtering

Since each positive Boolean function defines a stack filter, a simple exhaustive search through the class is not feasible. Fortunately, the properties specified above may be used to develop computationally efficient algorithms for the design of optimal stack filters.

The goal of these algorithms is to determine a stack filter which minimizes the mean absolute error between the output of the filter and some desired image, given a corrupted version of the desired image as input to the filter [32, 33]. If X is the desired image, and $\tilde{X}$ is the corrupted version observed by the filter, then the error to be minimized by proper choice of f is

$$\text{MAE}_f = E\{\,|\,X(s) - S_f(W_{\tilde{X}}(s))\,|\,\}$$

The reason for the choice of the mean absolute error is primarily tractability; under this criterion the optimization problem reduces to a linear program (LP). The output of this LP is a positive Boolean function which defines an optimal stack filter. This breakdown of the filter design problem into the problem of choosing a positive Boolean function is a consequence of (1) and (2) [32]:

$$\begin{aligned}
\text{MAE}_f &= E\{|X(s) - S_f(W_{\tilde{X}}(s))|\} \\
&= E\left\{|\sum_{l=1}^{M} x_l(s) - S_f(w_{\tilde{X},l}(s))|\right\} \\
&= \sum_{l=1}^{M} E\left\{|x_l(s) - S_f(w_{\tilde{X},l}(s))|\right\} \\
&= \sum_{l=1}^{M} E\left\{|x_l(s) - f(w_{\tilde{X},l}(s))|\right\}
\end{aligned}$$

The primary problems with the linear program which finds an optimal stack filter are the number of constraints on f implied by (1) and the problem of precomputing the joint statistics of the image X and the process which corrupted it. These problems are both minimized by the adaptive stack filtering algorithms developed in [34, 35].

Figure 2(a) illustrates the procedure for adaptive stack filtering. The desired image, X, and its noise-corrupted version $\tilde{X}$ are used to train a positive Boolean function f to minimize the error between the filter output $\hat{X}$ and the desired image X when $\tilde{X}$ is the input to the filter S_f.

Since the composition of two stack filters is just a stack filter defined over a larger

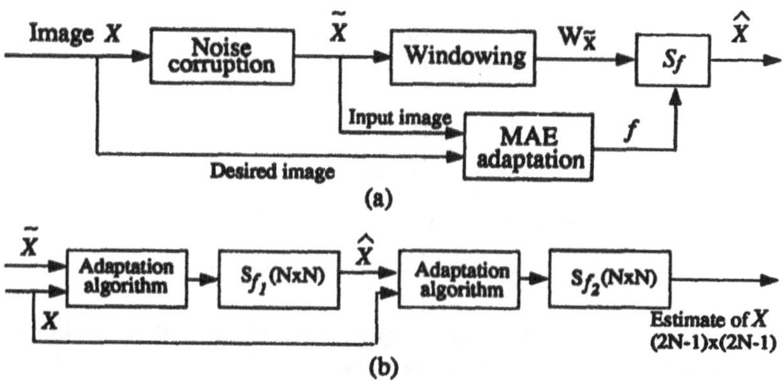

Figure 2 (a) Block diagram of optimal adaptive stack filtering algorithm. (b) Cascade of stack filters.

window, the computation required for the design of filters with large windows can be reduced by cascading filters with smaller windows. Figure 2(b) shows an example of a series connection of two stack filters with $N \times N$ input windows. The resulting filter is a stack filter with a $(2N-1) \times (2N-1)$ input window. The two filters are optimized sequentially as shown in the figure. The 5×5 filter designed for prefiltering is obtained as the cascade of two 3×3 filters.

1.12.7 Computational requirements of stack filters

The two key operations required in using stack filters are training and filtering. For this application, we believe that filtering is of more importance since training can be done in advance. However, both training and filtering algorithms are simple and fast since the updating of the Boolean function requires only compare, swap, and thresholding operations [34]. In general, the computation of filtering and training algorithms has empirically been found to scale linearly with the number of pixels in the window. The primary implementational limitation for stack filters is the size of the memory required to keep track of the Boolean function. For example, a 3×3 window requires 0.5 KBits of ram, while a 5×5 window requires approximately 32 MBits.

Table 1 lists the computational and memory requirements for implementing the filtering algorithm and one pass of the training algorithm, both using an $N \times N$ window. All experimental results used 10 passes of the training algorithm. This table assumes the use of a new fast, but simple method for computing the Boolean functions using a lookup table. The algorithm reduces computation by only evaluating the Boolean function at thresholds corresponding to values in the window.

The training algorithm potentially requires swapping operations for each table lookup. While the number of these swapping operations are limited by N^4 per pixel, in practice very few swapping operations actually occur; so the computation is dominated by table lookups.

	Table lookups per pixel	Swaps per pixel	Memory (Bits)
Filtering	N^2	N/A	2^{N^2+1}
Training (per pass)	N^2	$< N^4$	2^{N^2+1}

Table 1 The computation and memory requirements of stack filtering algorithms for an $N \times N$ window. All stack filters were trained using 10 passes of the training algorithm.

1.12.8 An application of stack filters: noise reduction

A new, very fast training algorithm that is well-suited for implementation on parallel computers has recently been developed [35]. It allows stack filters with small windows (say 9 points) to be trained within a matter of seconds. Filters with windows containing 25 points can be designed in approximately 30 minutes.

In Figs. 3 and 4, we show the results that can be obtained with these design algorithms.

All stack filters used in these experiments were trained to restore noise-free original images from noisy images. We used the aerial photo and the photo "Albert" shown in Figures 3a and 3b as noise-free original images. The aerial photo is a 256×256, 8 bits/pixel image; Albert is a 512×512, 8 bits/pixel image. We generate noise-corrupted aerial and Albert images by using impulsive noise with an occurrence probability of 0.1 and an absolute magnitude of 200. These noisy images are shown in Figures 3c and 3d and are denoted by $AerialI0.1$ and $AlbertI0.1$, respectively.

The results in Figure 4 show that a window no larger than 5×5 is required to remove impulsive noise from images. In fact, a 21 point window — a 5 with the corners chopped off — is sufficient. The training time for such filters is very reasonable, and robustness can be relied upon if the statistics of the image and/or noise should change. The issue of robustness is addressed in the next section and in [36].

1.12.9 An application of stack filters: edge detection

An intensity edge in an image is a region of rapid change of intensity with respect to distance. Such regions are characteristic of object boundaries and are therefore useful for segmentation and identification of objects in scenes. The detection of intensity edges is thus a fundamental problem in image analysis and has been addressed by many investigators.

Overviews of some of the schemes which have been proposed for intensity edge detection can be found in [37, 38, 39]. In general, they can be classified into one of the following approaches [40]: thresholding estimates of either the gradient or a set of directional derivatives [38, 41]; surface fitting followed by gradient estimation [42]; and sequential contour tracing [43, 44].

One goal which is common to all of these schemes is the reliable detection of edges even when the image has been corrupted by noise. In fact, the structure of the edge detector is often heavily influenced by the type of noise that is expected.

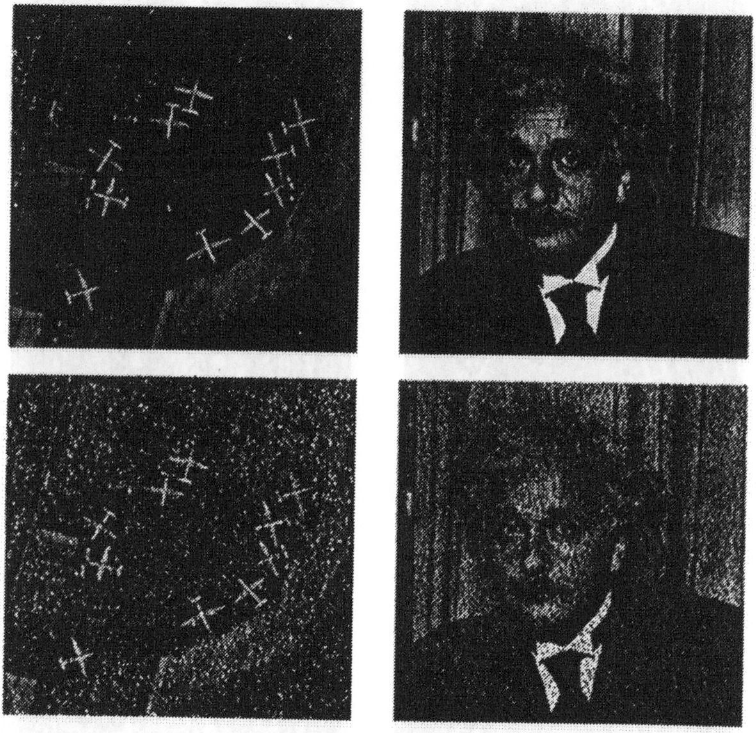

(a)	(b)
(c)	(d)

Figure 3 Images used in experiments; (a) Aerial photograph with 256 × 256 **resolution, (b) Albert with** 512 × 512 **resolution, (c)** *AerialI*0.1, **(d)** *AlbertI*0.1. **Noisy images in (c) and (d) are corrupted by impulse noise with an occurrence probability of 0.1.**

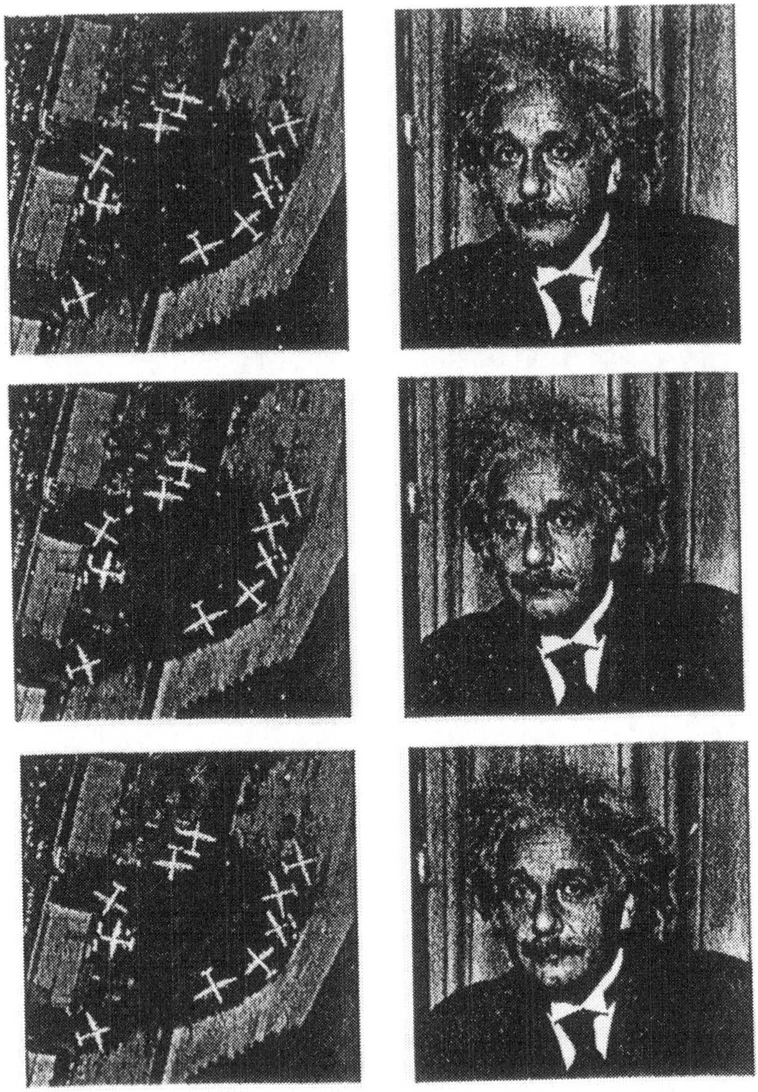

(a)	(b)
(c)	(d)
(e)	(f)

Figure 4 The filtered outputs by the new algorithm with the first method of enforcing the stacking property for (a) $Aerial I0.1$ with 3 × 3 window, (b) $Albert I0.1$ with 3 × 3 window, (c) $Aerial I0.1$ with 4 × 4 window, (d) $Albert I0.1$ with 4 × 4 window, (e) $Aerial I0.1$ with 5 × 5 window, and (f) $Albert I0.1$ with 5 × 5 window.

Suppose, for example, that the noise is additive and Gaussian, which occurs when thermal noise from the electronics in the imaging system is added to the acquired image. Then smoothing the image with a linear filter before applying a gradient estimator works very well [41]. In fact, such simple linear schemes as the Laplacian of a Gaussian ($\nabla^2 G$) are often near-optimal for images corrupted by Gaussian noise [45].

If the noise is impulsive, which can occur when bit errors corrupt either the acquisition or transmission of the image, then these linear schemes do not perform well. This is also the case if the noise is a mixture of Gaussian and impulsive or heavy-tailed noise. In either situation, an edge detector must incorporate some form of nonlinear filtering if it is to provide acceptable performance.

The solutions tried in the past include: the nonlinear dispersion estimators employed in [46]; a digital filter which is defined by Boolean functions involving median filters [47]; applying a linear edge detection scheme after the image has been smoothed with a median filter [48]; and prefiltering with other nonlinear filters, such as a hybrid of the median filter and linear filters proposed in [49]. A robust blurring technique for edge detection has also been proposed in [50]. Speckle is another type of noise that is often encountered in coherent imaging systems. An edge detection scheme tailored to the nonGaussian, nonadditive nature of speckle has also been proposed [51].

From the above discussion it is apparent that at least one edge detector has been developed for each commonly encountered noise environment. It also apparent, though, that there is no single detector which works well in all situations and, perhaps more importantly, that there is currently no methodology for edge detector design which is tractable for all types of noise that might be encountered. These observations were the motivation for this work.

The primary goal is the development of edge detection schemes which can produce accurate edge maps despite the presence of additive and/or impulsive noise whose distribution is poorly characterized. This distributional robustness and ability to preserve fine edge detail is obtained by applying the theory of minimum mean absolute error (MMAE) stack filtering [32] in two distinct approaches to edge detection.

1.12.10 Edge detection and stack filters

As stated before, stack filters are window-based nonlinear filters which are a generalization of median filters. The output of a stack filter when the window is at any given position is always one of the pixel values in the window. Which value is chosen is completely determined by the relative ranks and positions of the pixels in the window. Every stack filter is therefore a rank-order based filter which is also sensitive to time or spatial orderings. Thus, stack filters contain such other rank-order based nonlinear filters as center-weighted median filters, weighted median filters, and morphological filters with flat structuring elements [26].

The primary value of stack filters follows from an optimization theory which allows stack filters to be designed to preserve such image details as edges and fine lines while suppressing many different types of additive noise. Since rank-order operations are well-known to be robust with respect to outliers, these filters can be used when the true image and noise statistics differ from those used to design the filters. They are especially useful when the noise density is heavy-tailed or when the noise is a mixture of Gaussian and heavy-tailed noise.

Many of the stack filters designed in this section are obtained from a training algorithm [34]. It requires as input both an original, noise-free image and a noisy version of the same image. The filters trained with these images are then used to construct edge detectors for use on other images. The examples provided later in this paper illustrate how training filters with one type of image and noise produces filters which work well on images and noise with very different statistics.

1.12.11 The nonlinear prefiltering approach to edge detection

The most obvious application of stack filters in edge detection is to the nonlinear prefiltering approach. The goal is to train a stack filter to remove noise from an image while preserving as much detail as possible. If the resulting filter accomplishes this task, then a simple edge detection scheme can be used to detect edges in the filtered image.

As mentioned earlier, this approach has already been tried: a square-window median filter was used as a prefilter for the Robert's operator in [48]. The results obtained were quite good, in the sense that the number of spurious edges that the Robert's operator would have produced as a result of the noise was greatly reduced.

The difficulty with the median filter, though, is that it often eliminates or disrupts fine details in the image. Thin lines or small objects are often removed entirely. While some loss of detail is unavoidable when removing noise, the amount of detail removed by the median may often be excessive. A variety of median-type filters have been developed to solve this problem. They include center-weighted medians [52] and multistage medians [53, 54].

These generalizations of the median filter allow the filter designer some freedom to determine the tradeoff between noise reduction and detail preservation. The choice of a particular filter is usually made by trying several of them on the image of interest and then choosing the one which produces the best results.

Center-weighted median filters and multistage median filters are contained within the class of stack filters. The optimization schemes available for stack filters thus allow the designer to search a very large class of filters for the one offering the desired tradeoff between detail preservation and noise reduction. In the experiments discussed later, we compared the results of stack prefiltering with those of center-weighted and multistage median filters.

1.12.12 The difference of estimates approach to edge detection

The second application of stack filters to edge detection is based on a morphological filtering scheme which appears in [55, 56]. When the point of view of MMAE stack filtering is applied to this scheme, the result is a new approach to intensity edge detection which can be called the Difference of Estimates (DoE) approach. The structure of this new class of edge detectors follows from a formulation of the edge detection problem as an estimation problem. We illustrate this with the one-dimensional case since it is the simplest to describe yet has the same structure as the multidimensional case.

Consider a noise-free, one-dimensional signal $x(n)$. We would like to estimate the magnitude of the first difference, $|x(n+1)-x(n)|$, for this signal since the result may then be thresholded to determine if an edge is located at position n. We will estimate this first difference by applying a spatially invariant nonlinear filter to the noise-corrupted signal, $\tilde{x}(n)$. To do this we first define the windowing operation

$$W_{\tilde{X}}(n) = [\tilde{x}(n-k), \tilde{x}(n-k+1), \ldots, \tilde{x}(n), \ldots, \tilde{x}(n+k)]$$

The estimated first difference is then $F(W_{\tilde{X}}(n))$ where F is the nonlinear function to be determined. Our goal is then to determine the function F which minimizes

$$e(F) = E\Big[\big|F(W_{\tilde{X}}(n)) - |x(n+1) - x(n)|\big|\Big] \tag{3}$$

To determine the structure of F, we first rewrite the magnitude of the first difference by noting that

$$|x(n+1) - x(n)| = \max\{x(n+1), x(n)\} - \min\{x(n+1), x(n)\} \tag{4}$$

The magnitude of the first difference is then seen to be the difference between the windowed maximum and minimum, which are equivalent to the well-known dilation and erosion operations, respectively [25]. This suggests that we find two nonlinear filters which estimate the dilated and eroded signal from the noise-corrupted signal, and then form the difference between these estimates. This difference is then thresholded to form an edge map of the target image. This approach to edge detection is thus called the *Difference of Estimates* (DoE) approach.

We thus write

$$F(W_{\hat{X}}(n)) = S_d(W_{\hat{X}}(n)) - S_e(W_{\hat{X}}(n)) \tag{5}$$

where $S_d(\cdot)$ and $S_e(\cdot)$ are the nonlinear operations used to estimate the noise-free dilation and erosion, respectively. The error measure in (3) now becomes

$$\begin{aligned} e(d,e) &= E\left[|S_d(W_{\hat{X}}(n)) - \max\{x(n), x(n+1)\}\right. \\ &\quad \left. - [S_e(W_{\hat{X}}(n)) - \min\{x(n), x(n+1)\}]|\right] \\ &\leq E[|S_d(W_{\hat{X}}(n)) - \max\{x(n), x(n+1)\}|] \\ &\quad + E[|S_e(W_{\hat{X}}(n)) - \min\{x(n), x(n+1)\}|] \end{aligned}$$

Our strategy is then to minimize the overall error by choosing S_d and S_e to separately minimize each of the two expressions on the right hand side. In general, this is a very difficult problem in nonlinear estimation. However, if we restrict the solution to be from the class of stack filters, the problem can be solved using the techniques of MMAE filter design discussed above. The four reasons for the choice of stack filters are: they contain the minimum and maximum operators, which means that the optimal filters in the noise-free case are contained within the set of stack filters; stack filters preserve edges, lines and other image details; they are robust estimators; and, a tractable theory exists for the design of optimal stack filters.

1.12.13 Comparisons with other approaches

Very extensive experimental results are provided in [36] where the two stack filter based approaches reviewed above are compared with other approaches to edge detection. These experiments used both synthetic images and natural scenes and considered a wide variety of noise types. The conclusion reached in [36] is that the stack filter approach yields results comparable to those obtained with the Canny operator for images with additive Gaussian noise, but works much better when the noise is impulsive. Also, it was found that the robustness of stack filters leads to good performance for the target image, even when the statistics of the noise and/or image vary from those used in training.

1.12.14 Conclusions and further work

The purpose of this tutorial has been to show how the regularization and filter-class approaches to problems in image processing both lead to stack filters. Several examples showing how a design theory for stack filters can be used were also provided.

Other results of significance which were not mentioned above include: an approach which allows structural constraints and goals to be added to the basic goal of minimum mean absolute error stack filtering [57]; the design of stack filters with specific invariant signals [21]; and, the properties of stack filters which follow from the fact that they form a lattice [58].

References

[1] A. Tikhonov and V. Arsenin, *Solutions of Ill-posed Problems*, New York: Winston and Sons, 1977.

[2] S. Alliney, "Digital Filters as Absolute Norm Regularizers," *IEEE Trans. on Signal Processing*, Vol. 40, No. 6, pp. 1548-1562, June 1992.

[3] T. A. Nodes and N. C. Gallagher, "Median Filters: Some Modifications and Their Properties," *IEEE Trans. on Acoustics, Speech and Signal Processing*, Vol. 30, No. 5, pp. 739-746, October 1982.

[4] C. A. Bouman and K. Sauer, "A Generalized Gaussian Image Model for Edge-Preserving MAP Estimation," *IEEE Trans. on Image Processing*, Vol. 2, No. 3, pp. 296-310, July 1993.

[5] E. J. Coyle, J.-H. Lin, and M. Gabbouj, "Optimal Stack Filtering and the Estimation and Structural Approaches to Image Processing," Chapter appearing in *Digital Image Processing*, edited by Rama Chellappa, pp. 230-259, IEEE Computer Society Press, 1992.

[6] E. Wong and B. Hajek, *Stochastic Processes in Engineering Systems*, New York: Springer-Verlag, 1985.

[7] J. L. Doob, *Stochastic Processes*, New York: Wiley, 1953.

[8] P. D. Wendt, E. J. Coyle, and N. C. Gallagher, "Stack Filters," *IEEE Trans. on Acoustics, Speech, and Signal Processing*, Vol. ASSP-34, No. 4, pp. 898-911, August 1986.

[9] P. J. Huber, *Robust Statistics*, New York: Wiley, 1981.

[10] S. G. Tyan, "Median Filtering: Deterministic Properties," in *Two-Dimensional Digital Signal Processing, II: Transforms and Median Filters*, chap. 4, pp 197-217, Topics in Applied Physics, ed. T. S. Huang, New York: Springer-Verlag, 1981.

[11] N. C. Gallagher and G. L. Wise, "A Theoretical Analysis of the Properties of Median Filters," *IEEE Trans. on Acoustics, Speech, and Signal Processing*, Vol. ASSP-29, No. 6, pp. 1136-1141, December 1981.

[12] J. P. Fitch, E. J. Coyle, and N. C. Gallagher, "Median Filtering by Threshold Decomposition," *IEEE Trans. on Acoustics Speech and Signal Processing*, Vol. ASSP-32, No. 6, pp. 1183-1188, Dec. 1984.

[13] J. P. Fitch, E. J. Coyle, and N. C. Gallagher, "Threshold Decomposition of Multi-Dimensional Ranked Order Operations," *IEEE Trans. on Circuits and Systems*, Vol. CAS-32, No. 5, pp. 445-451, May 1985.

[14] J. P. Fitch, E. J. Coyle, and N. C. Gallagher, "Root Properties and Convergence Rates of Median Filters," *IEEE Trans. on Acoustics Speech and Signal Processing*, Vol. ASSP-33, No. 1, pp. 230-240, Feb. 1985.

[15] P. D. Wendt, E. J. Coyle, and N. C. Gallagher, "Some Convergence Properties of Median Filters," *IEEE Trans. on Circuits and Systems*, Vol. CAS-33, No. 3, pp. 276-286, March 1986.

[16] P.-T. Yu and E. J. Coyle, "Convergence Behaviour and N-Roots of Stack Filters," *IEEE Trans. on Acoustics, Speech and Signal Processing*, Vol. ASSP-38, No. 9, pp. 1529-1544, September 1990.

[17] Y. Nakagawa and A. Rosenfeld, "A Note on the Use of Local Min and Max Operations in Digital Picture Processing," *IEEE Trans. on Systems, Man, and Cybernetics*, Vol. 8, No. 8, pp. 632-635, August 1978.

[18] M. Gabbouj, E. J. Coyle, and N. C. Gallagher, "An Overview of Median and Stack Filtering," *Circuits, Systems, and Signal Processing*, Vol. 11, No. 1, pp. 7-45, January 1992.

[19] E. J. Coyle and N. C. Gallagher, "Stack Filters and Neural Networks," *Proceedings of the 1989 Int. Symposium on Circuits and Systems*, pp. 995-998, Portland, OR, May 9-11, 1989.

[20] P.-T. Yu and E. J. Coyle, "The Classification and Associative Memory Capability of Stack Filters," *IEEE Trans. on Signal Processing*, Vol. 40, pp. 2483-2497, Oct. 1992.

[21] P.-T. Yu and E. J. Coyle, "On the Existence and Design of the Best Stack Filter Based Associative Memory," *IEEE Trans. on Circuits and Systems II: Analog and Digital Signal Processing*, Vol. 39, No. 3, pp. 171-184, March 1992.

[22] J.-H. Lin and E. J. Coyle, "Minimum Mean Absolute Error Estimation over the Class of Generalized Stack Filters," *IEEE Trans. on Acoustics, Speech, and Signal Processing*, Vol. ASSP-38, No. 4, pp. 663-678, April 1990.

[23] G. B. Adams III, E. J. Coyle, L. C. Lin, L. Lucke, and K. K. Parhi, "Input Compression and Efficient VLSI Architectures for Rank Order and Stack Filters," *Signal Processing*, Special issue on Nonlinear Signal Processing; to appear in early 1994.

[24] J. Serra, *Image Analysis and Mathematical Morphology*, New York: Academic Press, 1982.

[25] J. Serra, *Image Analysis and Mathematical Morphology*, Vol. 2, New York: Academic Press, 1988.

[26] P. Maragos and R. W. Schafer, "Morphological Filters — Parts I and II," *IEEE Trans. on Acoustics, Speech, and Signal Processing*, Vol. ASSP-35, No. 8, pp. 1153-1184, August 1987.

[27] K. Preston, Jr., "Multidimensional Logical Transforms," *IEEE Trans. on Pattern Analysis and Machine Intelligence*, Vol. PAMI-5, No. 9, Sept. 1983.

[28] K. Preston, Jr., "Ξ Filters," *IEEE Trans. on Acoustics, Speech, and Signal Processing*, Vol. ASSP-31, No. 4, pp. 861-876, August 1983.

[29] K. Preston, Jr., *et al.*, "Basics of Cellular Logic with Some Applications in Medical Image Processing," *IEEE Proceedings*, Vol. 67, No. 5, pp. 826-856, May 1979.

[30] S. Wolfram, *Theory and Applications of Cellular Automata*, Singapore: World Scientific, 1986.

[31] E. N. Gilbert, "Lattice-theoretic properties of frontal switching functions," *J. Math Phys.*, Vol. 33, pp. 57-67, Apr. 1954.

[32] E. J. Coyle and J.-H. Lin, "Stack Filters and the Mean Absolute Error Criterion," *IEEE Trans. on Acoustics, Speech, and Signal Processing*, Vol. ASSP-36, No. 8, pp. 1244-1254, August 1988.

[33] E. J. Coyle, J.-H. Lin, and M. Gabbouj, "Optimal Stack Filtering and the Estimation and Structural Approaches to Image Processing," *IEEE Trans. on Acoustics, Speech and Signal Processing*, Vol. ASSP-37, No. 12, pp. 2037-2066, December 1989.

[34] J.-H. Lin, T. M. Sellke and E. J. Coyle, "Adaptive Stack Filtering under the Mean Absolute Error Criterion," *IEEE Transactions on Acoustics, Speech, and Signal Processing*, Vol. ASSP-38, No. 6, pp. 938-954, June 1990.

[35] J. Yoo, K. L. Fong, E. J. Coyle, and G. B. Adams, "Fast Algorithms for Designing Stack Filters," *Proceedings of the 31'st Annual Allerton Conference on Communications, Control, and Computing*, Monticello, IL, Sept. 29 - Oct. 1, 1993.

[36] J. Yoo, C. A. Bouman, E. J. Delp, and E. J. Coyle, "The Nonlinear Prefiltering and Difference of Estimates Approach to Edge Detection: Applications of Stack Filters," CVGIP: Graphical Models and Image Processing, Vol. 55, No. 2, pp. 140-159, March 1993.

[37] L. Davis, "A survey of edge detection techniques," *Computer Graphics and Image Processing* 7, pp. 375-390, 1978.

[38] D. Marr and E. C. Hildreth, "Theory of edge detection," *Proc. R. Soc. London Ser.* B. 207, pp.187-217, 1980.

[39] A. Rosenfeld and A. C. Kak, *Digital Picture Processing*, Vol. 2, New York : Academic Press, 1982.

[40] H. L. Tan, S. B. Gelfand, and E. J. Delp, "A comparative cost function approach to edge detection," *IEEE Trans. Syst. Man Cybern.*, Vol. SMC-19, No. 6, pp. 1337-1349, Nov./Dec., 1989.

[41] J. Canny, "A computational approach to edge detection," *IEEE Trans. Pattern Anal. Machine Intell.*, Vol. PAMI-8, pp. 679-698, Nov. 1986.

[42] R. M. Haralick, "Digital Step Edges from Zero Crossing of Second Directional Derivatives," *IEEE Trans. Pattern Anal. Machine Intell.*, Vol. PAMI-6, pp. 58-68, Jan. 1984.

[43] U. Montanari, "On the optimal detection of curves in noisy pictures," *Commu. ACM*, Vol. 14, No. 5, pp. 335-345, May 1971.

[44] P. H. Eichel and E. J. Delp, "A method for a fully automatic definition of coronary arterial edges from cineangiograms," *IEEE Trans. on Medical Imaging*, Vol. 7, No. 4, pp. 313-320, Dec. 1988.

[45] V. Torre and T. Poggio, "On Edge Detection," *IEEE Trans. Pat. An. Mach. Intell.*, Vol. PAMI-8, No. 2, pp. 147-163, March 1986.

[46] I. Pitas and A. N. Venetsanopoulos, "Edge detectors based on nonlinear filters," *IEEE Trans. Pattern Analysis and Machine Intelligence*, Vol. PAMI-8, No. 4, Jul. 1986.

[47] J. Astola, L. Koskinen, O. Yli-Harja and Y. Neuvo "Digital filters based on threshold decomposition and Boolean functions," *SPIE Visual Comm. and Image Proc. IV* Vol. 1199, 1989

[48] A. C. Bovik, T. S. Huang, and D. C. Munson, "The effect of median filtering on edge estimation and detection," *IEEE Trans. Pattern Analysis and Machine Intelligence*, Vol. PAMI-9, No. 2, Mar. 1987.

[49] Y. Neuvo, P. Heinonen and I. Defee, "Linear-Median Hybrid Edge Detectors," *IEEE Trans. Circuits Syst.*, Vol. CAS-34, No. 11, pp. 1337-1343, Nov. 1987.

[50] R. Feehs and G. R. Arce, "Multidimensional morphological edge detection," *Proc. of the SPIE Conference on Visual Communications and Image Processing*, Cambridge, MA, Oct. 1987.

[51] A. C. Bovik, "On detecting edges in speckle imagery," *IEEE Trans. Acoust., Speech, Signal Processing*, Vol. 36, No. 36, Oct. 1988.

[52] S.-J Ko, and Y. H. Lee, "Center-weighted median filters and their applications to Image Enhancement," *IEEE Trans. Circuits and Systems*, Vol. 38, No. 9, pp. 984-993, Sep. 1991.

[53] A. Niemenen, P. Heinonen, and Y. Neuvo, "A new class of detail preserving filters for image processing," *IEEE Trans. Pattern Anal. Machine Intell.*, Vol. 9, No. 1, pp. 74-90, Jan. 1987.

[54] G. R. Arce, and R. Foster, "Detail preserving ranked-order based filters for image processing," *IEEE Trans. Acoust., Speech, Signal Processing*, Vol. 37, No. 1, pp. 83-98, Jan. 1989.

[55] J. S. J. Lee, R. M. Haralick, and L. G. Shapiro, "Morphological edge detection," *IEEE Trans. Robotics Automat.*, Vol. RA-3, pp. 142-156, Apr. 1987.

[56] P. Maragos and R. D. Ziff, "Threshold superposition in morphological image analysis systems," *IEEE Trans. Pattern Anal. Machine Intell.*, Vol. 12, No. 5, May 1990.

[57] M. Gabbouj and E. J. Coyle, "Minimum Mean Absolute Error Stack Filtering with Structural Constraints and Goals," *IEEE Trans. on Acoustics, Speech and Signal Processing*, Vol. ASSP-38, No. 6, pp. 955-968, June 1990.

[58] L. C. Lin, G. B. Adams III, and E. J. Coyle, "Stack Filter Lattices," *Signal Processing*, Special issue on Nonlinear Signal Processing; to appear in early 1994.

Chapter 1.13

Multichannel Order Statistical Filtering

Ioannis Pitas

1.13.1 Introduction

Any colour can be reproduced by mixing an appropriate set of three primary colours [1] (usually red, green and blue). Therefore, any colour image can be considered as a two-dimensional three-channel sequence. Furthermore, in cases of constant brightness a two-variable scheme can be developed by factoring out luminance while keeping only the chromaticity attributes e.g., hue and saturation, thus projecting the three-dimensional colour space on a plane. In this case, we can treat colour image as a two-dimensional, two-channel sequence. Therefore, colour image processing is essentially multichannel two-dimensional signal processing. Until recently, little attention was paid to the multichannel nature of colour images. Thus, black and white (BW) image processing techniques were applied to each channel separately. Later, transformation techniques (e.g., the KarhuneN-Loeve transformation) have been used to decorrelate the three channels in order to apply BW image processing techniques afterwards. Although this is a valid approach, we think that a more natural way to deal with colour image processing is to apply multichannel signal processing techniques.

In the following, the use of ordering of multivariate data in multichannel signal processing is investigated, with particular application to colour image processing. We shall define a number of nonlinear multichannel digital filters that are based on multivariate data ordering and we shall investigate their properties in colour image processing/analysis and in motion field processing.

1.13.2 Multivariate data ordering

Let us denote by $\mathbf{X}$ a p-dimensional random variable, i.e., a p-dimensional vector of random variables $\mathbf{X} = [X_1, \ldots, X_p]^T$. We denote by $f(\mathbf{X})$ and $F(\mathbf{X})$ the probability density function (pdf) and the cumulative density function (cdf) of this p-dimensional random variable, respectively. Let $\mathbf{x}_1, ..., \mathbf{x}_n$ be n random samples from a p-dimensional distribution having mean $\mathbf{m}$ and dispersion matrix Σ [2]. The notion of data ordering cannot be extended in a straightforward manner from the one dimensional case to the case of multivariate data. An excellent treatment of multivariate data ordering was given in [7]. It is shown that there are several ways to order multivariate data but none of them is unambiguous nor universally agreeable. The following so-called subordering principles are discussed in the literature: (1) marginal ordering, (2) reduced (aggregate) ordering, (3) partial ordering, and (4) conditional ordering [3, 7].

In marginal ordering, the multivariate samples are ordered along each of the p-dimensions

$$
\begin{aligned}
x_{1(1)} \le x_{1(2)} \le \ldots \le x_{1(n)} \\
x_{2(1)} \le x_{2(2)} \le \ldots \le x_{2(n)} \\
\vdots \\
x_{p(1)} \le x_{p(2)} \le \ldots \le x_{p(n)}
\end{aligned} \tag{1}
$$

i.e., ordering is performed in each channel of the multichannel signal independently. $\mathbf{x}_{(1)} = [x_{1(1)}, x_{2(1)}, \ldots, x_{p(1)}]^T$ are the minimal elements in each dimension.

$\mathbf{x}_{(n)} = [x_{1(n)}, x_{2(n)}, \ldots, x_{p(n)}]^T$ are the maximal elements in each dimension. $\mathbf{x}_{(\nu+1)} = [x_{1(\nu+1)}, x_{2(\nu+1)}, \ldots, x_{p(\nu+1)}]^T$ is the (marginal) median of the multivariate data for $n = 2\nu + 1$. The i-th marginal order statistic is the vector $\mathbf{x}_{(i)} = [x_{1(i)}, x_{2(i)}, \ldots, x_{p(i)}]^T$. The vector $[x_{1(r_1)}, x_{2(r_2)}, \ldots, x_{p(r_p)}]^T$ for arbitrary $r_i, 1 \le r_i \le n, i = 1, \ldots, p$ is a marginal order statistic too. Needless to say that the median or any i-th marginal order statistic may not correspond to any of the samples $\mathbf{x}_1, \ldots, \mathbf{x}_n$. In contrast, in the one-dimensional case there is an one-to-one correspondence between the samples $x_1, \ldots, x_n$ and the order statistics $x_{(1)}, \ldots, x_{(n)}$ [4]. Early studies of the properties of marginal order statistics can be found in [8, 9, 10]. The investigation of the probability distributions and the variance of the marginal order statistics is presented in [11, 13].

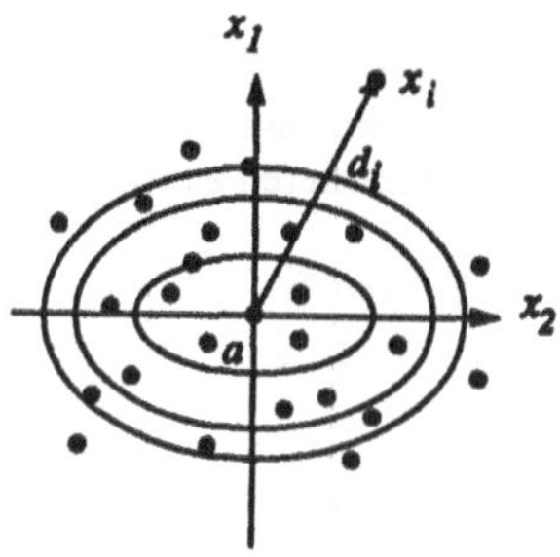

Figure 1 ***R*-ordering of multivariate data.**

The reduced ordering (*R*-ordering) is based on the generalized distance

$$d = (\mathbf{x} - \mathbf{a})^T \Gamma^{-1} (\mathbf{x} - \mathbf{a}) \tag{2}$$

of a sample $\mathbf{x}$ from a point $\mathbf{a}$ that may be either the origin or the sample arithmetic mean $\bar{\mathbf{x}}$

$$\bar{\mathbf{x}} = \frac{1}{n} \sum_{i=1}^{n} \mathbf{x}_i \tag{3}$$

or the marginal median $\mathbf{x}_{(\nu+1)}$. Γ may be the identity matrix, the dispersion matrix Σ or the sample dispersion matrix $\mathbf{S}$

$$\mathbf{S} = \frac{1}{n-1} \sum_{i=1}^{n} (\mathbf{x}_i - \bar{\mathbf{x}})(\mathbf{x}_i - \bar{\mathbf{x}})^T \tag{4}$$

The various data $\mathbf{x}_i$ are ordered according to their distances d_i from $\mathbf{a}$, as can be seen in

Figure 1. Thus, multivariate ordering is reduced to one-dimensional ordering.

Partial ordering (P-ordering) is based on the notion of the convex hull of the points $\mathbf{x}_1, \ldots, \mathbf{x}_n$, which is the minimum convex set which encloses all n samples. Some of the samples lie on the perimeter of the convex set. These samples are denoted c-order group 1 and are discarded. The process is repeated with the rest of points and produces the c-order group 2 and so on. The resulting P-ordering is shown in Figure 2.

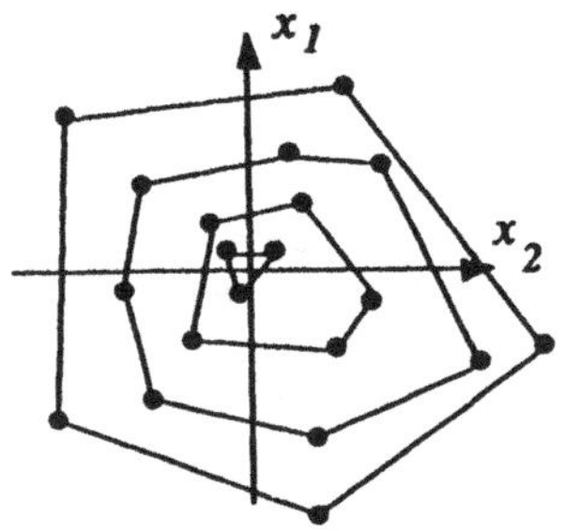

Figure 2 P-ordering of multivariate data.

The conditional ordering (C-ordering) is conducted on one of the marginal sets of observations conditional on ordering within the data in terms of other marginal sets of observations [7]. The multivariate samples are ordered as follows

$$\begin{aligned} x_{1(1)} &\leq x_{1(2)} \leq \cdots \leq x_{1(n)} \\ x_{2[1]} &\leq x_{2[2]} \leq \cdots \leq x_{2[n]} \\ &\vdots \\ x_{p[1]} &\leq x_{p[2]} \leq \cdots \leq x_{p[n]} \end{aligned} \tag{5}$$

$x_{1(i)},\ i = 1, \ldots, n$ are the marginal order statistics of the first dimension. $x_{j[i]},\ j = 2, \ldots, n,\ i = 1, \ldots, n$ are the quasi-ordered samples in dimensions $j = 2, \ldots, n$ conditional on the marginal ordering of the first dimension. In the two-dimensional case ($p = 2$) the statistics $x_{2(i)},\ i = 1, \ldots, n$ are called *concomitants* of the order statistics of X_1.

1.13.3 Multichannel filters based on vector ordering

One-dimensional order statistics (especially the median) and their linear combinations such as the α-trimmed mean and the L-Estimators, have been used extensively as estimators of one-dimensional location [3, 4]. The marginal median, has been studied in [11, 13]. Robustness measures (e.g. influence function, gross-error sensitivity) in the presence of outlying observations were presented in [11]. The influence function of the two-dimensional marginal median was also derived. It was found that the marginal median is a B-Robust estimator because its gross error sensitivity is finite [11]. The definitions of L-Estimators can be easily extended to the p-dimensional case by using marginal order statistics. Thus, the p-dimensional marginal L-Estimator is given by [11]

$$\mathbf{T}_n = \sum_{i_1=1}^{n} \cdots \sum_{i_p=1}^{n} \mathbf{A}_{i_1,\ldots,i_p} \mathbf{x}_{(i_1,\ldots,i_p)} \tag{6}$$

where $\mathbf{x}_{(i_1,\ldots,i_p)} = [x_{1(i_1)}, \ldots, x_{p(i_p)}]^T$ are the marginal order statistics and $\mathbf{A}_{i_1,\ldots,i_p}$ are $p \times p$ matrices. The p-dimensional marginal median is a special case of the L-Estimator. The output of a p-channel L-Filter of length N operating on the sequence of p-dimensional vectors $\mathbf{x}(k)$, for N odd, can be also be expressed as

$$\mathbf{y}(k) = \sum_{j=1}^{p} \mathbf{A}_j \tilde{\mathbf{x}}_j(k) \tag{7}$$

where $\mathbf{A}_j$ are appropriate $(p \times N)$ coefficient matrices and $\tilde{\mathbf{x}}_j(k) = (x_{j(1)}(k), \ldots, x_{j(N)}(k))^T$ are the $(N \times 1)$ vectors of the order statistics along each channel. L-estimators can be used in multichannel image filtering in the case of additive white noise

$$\mathbf{x} = \mathbf{s} + \mathbf{n} \tag{8}$$

where $\mathbf{s}$ is the desired signal, $\mathbf{n}$ is a white noise process and $\mathbf{x}$ is the corrupted signal. It has been proven that it behaves well in the case of additive white long-tailed noise and in the case of impulsive noise. The MSE design of p-channel marginal L-Filters that operate on the p-dimensional observed signal $\mathbf{x}(k)$ using $\mathbf{s}(k)$ as desired signal is described in [14, 15, 16]. Multichannel L-Filters have been developed for various multichannel distributions in [16] and have been compared to single channel L-Filters and to marginal filters (e.g. marginal and vector median filter and moving average filter). The designed filters outperform by far any other nonlinear single-channel or multichannel filter reported in the literature. The p-dimensional marginal α-trimmed mean is another special case of the p-dimensional L-Estimator and is defined as follows [13]

$$\mathbf{T}_n = \begin{bmatrix} \frac{1}{n(1-2\alpha_1)} \sum_{i=\alpha_1 n+1}^{n-\alpha_1 n} x_{1(i)} \\ \cdots \\ \frac{1}{n(1-2\alpha_p)} \sum_{i=\alpha_p n+1}^{n-\alpha_p n} x_{p(i)} \end{bmatrix} \tag{9}$$

is another special case of the p-dimensional L-estimator.

Another definition of the multichannel median called *vector median* has been proposed for colour image filtering [17, 18, 19, 20]. It is the vector $\mathbf{x}_{med}$ that minimizes the L_1 error norm

$$\sum_{i=1}^{n} |\mathbf{x}_i - \mathbf{x}_{med}| \longrightarrow \min \tag{10}$$

Other distance measures (e.g. L_2 norm, Mahalanobis distance) can be used in (10) instead of the L_1 norm. This definition of the multichannel median is a direct extension of the corresponding single channel median definition. Generally, it gives different results than the marginal median. The vector $\mathbf{x}_{med}$ may or may not be one of the data $\mathbf{x}_i$, $i = 1, \ldots, n$. The implicit definition (10) of the vector median makes difficult its application in colour image filtering because the use of an iterative optimization algorithm is required to produce each output image pixel. The computational load is reduced if the median $\mathbf{x}_{med}$ is forced to belong to the set $\mathbf{x}_i$, $i = 1, \ldots, n$. In this case, the L_1 norm (10) is computed for every $\mathbf{x}_{med} = \mathbf{x}_i$, $i = 1, \ldots, n$. The median is the one that minimizes [19]

$$\sum_{i=1}^{n} |\mathbf{x}_i - \mathbf{x}_{med}| \leq \sum_{i=1}^{n} |\mathbf{x}_i - \mathbf{x}_j| \quad j = 1, \ldots, n, \quad \mathbf{x}_{med} \in \{\mathbf{x}_1, \ldots, \mathbf{x}_n\} \tag{11}$$

Even in this case, the computational complexity of the vector median filter is relatively high, because we have to calculate $n(n-1)/2$ p-dimensional norms of the form $|\mathbf{x}_i - \mathbf{x}_j|$ and to perform $[n(n-1)/2]-1$ comparisons. The computation of the marginal median requires only $O(pnlogn)$ comparisons if the QUICKSORT algorithm is used for its computation on each channel [3]. The computations can be further reduced, both for the vector and the marginal median if running algorithms are employed [19, 21]. The computational complexity of the vector median filter is relatively small for $n=3$. Only the calculation of three p-dimensional norms as well as six comparisons are required.

The definition of the vector median (10) is equivalent to the definition of the marginal median only for the one dimensional case. Generally, marginal and vector medians are different for dimensions $p \geq 2$. However, the marginal and the vector median tend to have approximately the same values and similar behaviour for multivariate data that are strongly correlated among their different dimensions. For example, if all multivariate data samples lie on a line on the p-dimensional space, multivariate ordering is reduced to one-dimensional ordering. In this case both marginal median and vector median have the same value.

An important class of nonlinear filters for image processing can be introduced as a compromise between the median and the moving average filter. These filters are the modified trimmed mean filters [3]. The multichannel MTM filter is defined as follows [13]. First, the marginal median inside each window is determined along each channel and the vector of the medians $\mathbf{x}_{med}$ is formed. Then, the input vectors $\mathbf{x}_i$ are ordered by means of their generalized distance $d_i = (\mathbf{x}_i - \mathbf{x}_{med})^T \Gamma^{-1} (\mathbf{x}_i - \mathbf{x}_{med})$. Vectors lying in the filter window and having generalized distances d_i smaller than a threshold d are averaged and form the filter output

$$\mathbf{y}_{ij} = \frac{\sum_A \sum a_{rs} \mathbf{x}_{i+r,j+s}}{\sum_A \sum a_{rs}} \tag{12}$$

where

$$a_{rs} = \begin{cases} 1 & (\mathbf{x}_i - \mathbf{x}_{med})^T \Gamma^{-1} (\mathbf{x}_i - \mathbf{x}_{med}) < d \\ 0 & \text{otherwise} \end{cases} \tag{13}$$

The threshold value d may be selected as the average of the generalized distances d_i.

A similar generalization can be done in order to define the multichannel DW-MTM filter. As in the single channel case, two window sizes are used. The marginal median is computed first inside the small window. Then, the data vectors lying in the large window and having generalized distances d_i less than some value d are averaged to give the filter output.

The matrix Γ in the expression of the generalized distance d may be the identity matrix $\mathbf{I}$, or the sample dispersion matrix Σ.

A nonlinear filter that is based on multivariate distances is the R_E filter [22]. Let $\mathbf{x}(n)$ denote the central sample of the input (noisy) signal within the filter window W. The output $\mathbf{y}(n)$ of the R_E filter is given by

$$\mathbf{y}(n) = \begin{cases} \mathbf{x}(n) & \text{if } \|\bar{\mathbf{x}} - \mathbf{x}(n)\| \leq d \\ \mathbf{x}(k) \in W : \|\mathbf{x}(k) - \mathbf{x}(n)\| = \min & \text{otherwise} \end{cases} \tag{14}$$

d is a suitably chosen threshold and $\bar{\mathbf{x}}$ is the arithmetic mean within the filter window.

Another filter that is a compromise between the multivariate average and median filters is reported in [23]. Its output is given by

$$\mathbf{y}(n) = arg \min_{\mathbf{x}(k) \in W} \{\|\bar{\mathbf{x}} - \mathbf{x}(k)\| + \|\mathbf{x}_{med} - \mathbf{x}(k)\| + \|\mathbf{x}_{n^*} - \mathbf{x}(k)\|\} \tag{15}$$

where W is the filter window, $\bar{\mathbf{x}}$ is the local arithmetic mean, $\mathbf{x}_{med}$ is the local median and $\mathbf{x}_{n^*}$ is central sample in W.

1.13.4 Nonlinear filters of directional data

So far, multichannel data have been represented in cartesian coordinates. However, in many cases polar representation of multivariate data is very important, e.g. for motion field direction estimation or in hue estimation in the HSI colour space [1, 6]. Vector magnitude filtering by using L-Filters is reported in [24, 27]. In case of polar representation, vector directions should be treated using the theory of angular or directional statistics [5]. We may regard the angular observations as observations on a circle having unit radius. An observation θ, measured in degrees ($0^0 < \theta \leq 360^0$) from the positive x axis, represents a vector of unit length having direction θ. Such data are called *circular* or *directional*. Due to the periodic nature of the circular data, the definitions for the probability density function (pdf) and the cumulative density function (cdf) differ from those used on linear data. A simple way to define the cdf F for a circular random variable Θ would be the following

$$F(\theta) = Pr(0 < \Theta \leq \theta), \quad 0 < \theta \leq 2\pi \tag{16}$$

However, the domain of definition of this function cannot be limited to the interval $(0, 2\pi)$ because of the periodicity existing on the circle. Therefore, the domain of definition for the cdf should be extended by using the relation

$$F(\theta + 2\pi) - F(\theta) = 1, \quad -\infty < \theta < \infty \tag{17}$$

The following equations should also hold

$$\lim_{\theta \to \pm\infty} F(\theta) = \pm\infty, \quad F(0) = 0, \quad F(2\pi) = 1 \tag{18}$$

The probability density function f is periodic in θ with a period of 2π and the following relation holds

$$\int_0^{2\pi} f(\theta)d\theta = 1 \tag{19}$$

Among the many distributions that have been studied for the directional data, the von Mises distribution plays a key role. Its pdf has the following form

$$g(\theta; \mu_0, k) = \frac{1}{2\pi I_0(k)} \exp(k\cos(\theta - \mu_0)) \tag{20}$$

$$0 < \theta \leq 2\pi, \quad k > 0, \quad 0 \leq \mu_0 < 2\pi$$

where $I_0(k)$ is the modified Bessel function of the first kind and order zero. The parameter μ_0 is the mean direction (to be defined in the sequel), while the parameter k is the *concentration parameter*. It controls the dispersion of the distribution. The von Mises distribution is considered to be the "normal distribution on the circle ". Its behaviour is controlled by μ_0 and k in the same way that $N(\mu, \sigma)$ is governed by its parameters μ, σ.

The classical linear measures of location and dispersion depend on the choice of the zero direction on the circle. Therefore, new measures of location and dispersion must be defined. The *mean direction* μ_0 is defined as the phase angle of the first trigonometric moment $\alpha + j\beta = \rho e^{j\mu_0}$ where

$$\alpha = E[\cos\theta], \quad \beta = E[\sin\theta] \tag{21}$$

The mean direction is proven to be the maximum likelihood estimator of location for the von Mises distribution.

The *sample mean direction* $\overline{x_0}$ of n sample points P_i on the unit circle is defined to be the direction of the mean resultant $\overline{R}$ of the unit vectors $\overline{OP_1}\ldots\overline{OP_n}$, where O is the circle center. It is given by

$$x_0 = \arctan\left(\frac{\overline{S}}{\overline{C}}\right) \tag{22}$$

$$\overline{C} = \frac{1}{n}\sum_{i=1}^{n}\cos\theta_i \quad \overline{S} = \frac{1}{n}\sum_{i=1}^{n}\sin\theta_i \tag{23}$$

Another location estimator is the *median direction.* The median direction ξ_0 for a population distributed according to pdf f is defined to be the solution of the equation

$$\int_{\xi_0}^{\xi_0+\pi} f(\theta)d\theta = \int_{\xi_0+\pi}^{\xi_0+2\pi} f(\theta)d\theta = 1/2 \tag{24}$$

with the additional constraint that $f(\xi_0) > f(\xi_0+\pi)$. Let N denote the reciprocal point of M on the unit circle. Then MN is a diameter. The sample median is the point M that satisfies the following properties for a set of sample points around the unit circle

1. The diameter MN that passes through it, divides the sample points to two subsets, each one containing half of them.

2. The majority of the sample points are concentrated closer to M rather than to N.

The previous definition holds for an odd number of samples. Unlike the linear median, the circular median direction is not always unique. Uniqueness holds only in the case of unimodal distributions. The median direction defined as above is different from the median that can be evaluated if we consider the sample points to be on the line and use the corresponding definition.

In order to evaluate the median direction of a set of points θ_i the following property can be exploited. Let a be a fixed direction. Then, the median direction minimizes the following measure of dispersion around a

$$d_0 = \frac{1}{n}\sum_{i=1}^{n}\xi_i \tag{25}$$

or equivalently

$$d_0 = \pi - \frac{1}{n}\sum_{i=1}^{n}|\pi - |\theta_i - a|| \tag{26}$$

where ξ_i is the smallest of the two angles that are defined by the points a and θ_i on the unit circle. d_0 is called *circular mean deviation.* We can find the median of a set of points by evaluating d_0 for all the sample points and keeping the one that gives the smallest result. The method will give a unique median direction even if more than one median directions exist. This method of evaluation requires a large number of operations since the sum (26) needs to be evaluated n times.

Median filters are just a special case of the generalized order statistics filters. The definition of such filters in the case of directional data is not a simple extension of the corresponding definition in the linear case. Simple linear ordering cannot be used in this case, because it does not take under consideration the circular periodic nature of the data. Furthermore, as it was previously stated, the median evaluated using linear ordering does not coincide with the definition of the circular median direction. The ordering that will be used should be coherent with this definition, exploit the circular behaviour of the data and also allow for the calculation of the distributions of the circular order statistics. Having this in mind, we utilized the following ordering relation

Let us denote by $\theta_{[k]}$ the ordering on the circle. Let $\theta_1, \theta_2, \ldots, \theta_n$ be samples from a circular distribution, and $\theta_{(1)}, \theta_{(2)}, \ldots, \theta_{(n)}$ be the linearly ordered samples, i.e.

$$\theta_{(1)} \leq \theta_{(2)} \leq \ldots \leq \theta_{(n)} \tag{27}$$

We calculate the circular median $\theta_{med} \triangleq \theta_{[\frac{n+1}{2}]}$ by using its definition. Let us suppose that θ_{med} corresponds to $\theta_{(i)}$. The rest of the ordered samples can be defined by the following relation

$$\theta_{[k+\frac{n+1}{2}]} = \theta_{((i+k)mod(n))}, \quad -\frac{n-1}{2} \leq k \leq \frac{n-1}{2} \tag{28}$$

This definition of ordering on the circle is coherent to the median definition and also to the definition of the first and third quartile directions Q_1^0, Q_3^0 (the values of the $n/4$ observation from the median direction M_0 in the clockwise and the anticlockwise sense respectively [5]).

Having defined the way that the circular data are ordered, the derivation of the various order statistics filters is straightforward. For example, the α-trimmed mean filter [3] can be defined as

$$\overline{y_\alpha} = \frac{1}{N(1-2\alpha)} \sum_{j=\alpha N+1}^{N-\alpha N} x_{[j]} \quad 0 \leq \alpha < 0.5 \tag{29}$$

Parameter α controls the number of the smaller and the bigger samples in the filter window that will be rejected. Nonlinear dispersion measures on angular data have been defined as well [24, 25]. They have been used as edge detectors in the analysis of colour image hue [25].

Angle ranking has also been used in the vector ranking that is defined as follows [28, 29]. Let $\mathbf{x}_1, \ldots, \mathbf{x}_n$ be the input observations and $\alpha(\mathbf{x}_i, \mathbf{x}_j)$, $i, j = 1, \ldots, n$ the vector angles among them. The sums of all vector angles corresponding to vector $\mathbf{x}_i$

$$\alpha_i = \sum_{j=1}^{n} \alpha(\mathbf{x}_i, \mathbf{x}_j) \tag{30}$$

can be ordered according to the magnitude

$$\alpha_{(1)} \leq \alpha_{(2)} \leq \ldots \leq \alpha_{(n)} \tag{31}$$

This implies ordering of the corresponding vectors

$$\mathbf{x}_{(1)} \leq \mathbf{x}_{(2)} \leq \ldots \leq \mathbf{x}_{(n)} \tag{32}$$

The ordered vectors constitute the output of the generalized vector directional filters [28].

1.13.5 Discussion

Nonlinear multichannel filtering is a rather novel area. Several digital filter classes based on the notion of multivariate ordering have appeared so far, as has already been described in the previous sections. As expected, most of them (e.g. marginal and vector medians) perform very well in the case of impulsive noise removal. Some of them perform well in the case of white additive noise (Gaussian or Laplacian). A comparison of the performance of various nonlinear multivariate digital filters is presented in [13] but it is far from complete. Filters that have been designed under optimality criteria (e.g. optimal MSE L-filters) outperform by far filters that have a rather ad hoc structure. However, no complete comparative analysis of the properties of these filters has been presented so far. The main reason is the lack of realistic multivariate noise models that match noise characteristics that appear in practice (e.g. in TV channels). Thus, an effort has been made to design mathematical models of various types of multivariate noise [13, 15]. This effort should be continued. Regarding the performance of the multichannel filters versus the performance of their single-channel counterparts, the multichannel filters are superior to single channel ones in various cases (e.g. multichannel L-Filters) at the expense of higher computational complexity both in design and in implementation. The use of polar coordinates and directional statistics in multichannel signal processing is another important area whose potential is still unexplored. Their use is particularly important in colour image analysis.

References

[1] A. K. Jain, *Fundamentals of digital image processing*, Prentice Hall, 1989.

[2] G. A. F. Seber, *Multivariate observations*, J. Wiley, 1984. certain areas (e.g. motion field and colour image analysis).

[3] I. Pitas, A. N. Venetsanopoulos, *Nonlinear digital filters: principles and applications*, Kluwer Academic, 1990.

[4] H. A. David, *Order Statistics*, Wiley, 1981.

[5] K. V. Mardia, *Statistics of directional data*, Academic Press, 1972.

[6] I. Pitas, *Digital image processing algorithms*, Prentice Hall, 1993.

[7] V. Barnett, "The ordering of multivariate data," *Journal of the Royal Statistical Society A*, Vol. 139, Pt. 3, pp. 318–354, 1976.

[8] A. M. Mood, "On the joint distribution of the medians in samples from a multivariate population," *Annals of Mathematical Statistics*, Vol. 12, pp. 268–279, 1941.

[9] C. M. Mustafi, "A recurrence relation for distribution functions of order statistics from bivariate distributions," *Journal of the American Statistical Association*, Vol. 64, pp. 600–601, 1969.

[10] J. Galambos, "Order Statistics of samples from multivariate distributions," *Journal of the American Statistical Association*, Vol. 70, pp. 674–680, 1975.

[11] I. Pitas, "Marginal order statistics in colour image processing," *Optical Engineering*, Vol. 29, No.5, pp. 495–503, May 1990.

[12] I. Pitas, "Marginal order statistics in colour image processing," *Proc. European Conference on Signal Processing*, Barcelona, 1990.

[13] I. Pitas, P. Tsakalides, "Multivariate ordering in colour image restoration," *IEEE Transactions on Circuits and Systems for Video Technology*, in press, 1991.

[14] C. Kotropoulos, I. Pitas, "Multichannel L–Filters based on M-Ordering," *Proc. IEEE Winter Workshop on Nonlinear Digital Signal Processing*, Tampere, Finland, 1993.

[15] C. Kotropoulos and I. Pitas, "Multichannel L-Filter design based on marginal data ordering," *Proc. IEEE Int. Conf. on Acoustics, Speech and Signal Processing*, Adelaide, Australia, April 1994.

[16] C. Kotropoulos, I. Pitas, "Multichannel L-Filters based on marginal data ordering," *IEEE Transactions on Signal Processing*, in press, 1994.

[17] J. Astola, P. Haavisto, P. Heinonen, Y. Neuvo, "Median type filters for colour signals," *Proc. IEEE Int. Symp on Circuits and Systems*, pp. 1753–1756, 1988.

[18] K. Oistamo, P. Jarske, J. Astola, Y. Neuvo, "Vector median filters for complex signals," *Proc. IEEE Int. Conf. Acoust. Speech, Signal Processing*, pp. 813–816, Glasgow, 1989.

[19] J. Astola, P.H aavisto, Y. Neuvo, "Vector median filters," *Proceedings IEEE*, Vol. 78, No.4, pp. 678–690, April 1990.

[20] S. Sanwalka and A. N. Venetsanopoulos, "Vector Order Statistic Filtering of Colour Images," *Proceedings of the Thirteenth GRETSI Symposium on Signal and Image Processing*, Juan-les Pins, France, Sept.16–20, 1991.

[21] I. Pitas, "Fast algorithms for running ordering and max/min calculation," *IEEE Transactions on Circuits and Systems*, Vol. CAS-36, No. 6, pp. 795–804, June 1989.

[22] R. C. Hardie, G. R. Arce, "Ranking in R^p and its use in multivariate image estimation," *IEEE Transactions on Circuits and Systems for Video Technology*, Vol. CSVT-1, No. 2, pp. 197–209, June 1991.

[23] K. Tang, J. Astola, Y. Neuvo, "Multivariate Order Statistics in Colour Image Processing," *IEEE International Conf. on Systems Engineering*, Kobe, Japan, September 1992, pp. 584–587.

[24] N. Nikolaidis, I. Pitas, T. Kilindris, "Application of directional statistics to nonlinear vector field filtering and analysis," *Proc. IEEE Winter Workshop on Nonlinear digital signal processing*, pp. 1.2.4.1–1.2.4.6, Tampere, Finland, 1993.

[25] N. Nikolaidis, I. Pitas, "Directional statistics in nonlinear vector field filtering," *Signal Processing*, in press, 1994.

[26] N. Nikolaidis, I. Pitas, "Optimal L–Filters for vector magnitude filtering," *Proc. EUSIPCO 94*, Edinburgh, UK, 1994.

[27] N. Nikolaidis and I. Pitas, "Application of directional statistics in vector direction estimation," *Proc. IEEE Int. Conf. on Acoustics, Speech and Signal Processing*, Adelaide, Australia, April 1994.

[28] P. E. Trahanias, A. N. Venetsanopoulos, "Multichannel Image Processing Using Vector-Angle Ranking," *SPIE* Vol. 1902, Nonlinear Image Processing IV, 1993.

[29] P. E. Trahanias, I. Pitas, A. N. Venetsanopoulos, "Colour image processing," in *2D and 3D signal processing techniques and applications*, C. T. Leondes editor, Academic Press, in press, 1994.

Chapter 1.21

Theory and Approximation of Polynomial Filters

Giovanni L. Sicuranza

Abstract

After a brief review of the theory of the Volterra series for continuous systems, the discrete case is considered in detail and the general discrete Volterra series is introduced together with its main properties. General and simplified models of practical interest are presented, too.

1.21.1 The continuous-time case

While a nonlinear continuous time-invariant system without memory can be described, within some restrictions, by means of a Taylor series, a nonlinear continuous time-invariant system with memory can be represented by means of an extension of this series expansion known as the *Volterra series* [1, 2, 3, 4, 5]

$$y(t) = h_0 + \int_{-\infty}^{\infty} h_1(\tau_1)x(t-\tau_1)d\tau_1 +$$

$$\int_{-\infty}^{\infty}\int_{-\infty}^{\infty} h_2(\tau_1,\tau_2)x(t-\tau_1,t-\tau_2)d\tau_1 d\tau_2 + \cdots$$

$$\cdots + \int_{-\infty}^{\infty}\cdots\int_{-\infty}^{\infty} h_p(\tau_1,\ldots,\tau_p)x(t-\tau_1,\ldots,t-\tau_p)d\tau_1\cdots d\tau_p + \cdots \tag{1}$$

The nonlinear system is completely characterized by the multidimensional functions $h_p(t_1,\ldots,t_p)$, called the *Volterra kernels*. The zero-order kernel h_0 is a constant, while the higher-order kernels can be assumed, without loss in generality, as symmetric functions of their arguments so that any of the $p!$ possible permutations of $t_1,\ldots,t_p$ leaves $h_p(t_1,\ldots,t_p)$ unchanged.

For a causal system, the lower limits in the integral in equation (1) are set to zero. The upper limits of the integrals in equation (1) given as infinity indicate that the causal system has an infinite memory. If a (common) finite value is assigned to the upper limits, the system possesses a finite memory. If the p-dimensional integrals in equation (1) are used to define the p-th order Volterra operators $H_p[x(t)]$, the following compact relation is derived

$$y(t) = h_0 + \sum_{p=1}^{\infty} H_p[x(t)] \tag{2}$$

A truncated Volterra series is obtained by setting the upper limit of the summation in equation (2) equal to a finite integer value P, which is often called the *order*, or the *degree*, of the Volterra series. It is known [6] that any time-invariant, finite-memory system which is a continuous functional of its input can be uniformly approximated over a uniformly bounded, equicontinuous set of input signals by a Volterra series of appropriate finite order P. Moreover, equation (2) reveals the nature of power series intrinsic to the Volterra expansion, and as a consequence there are some limitations associated with the applications of the Volterra series to nonlinear problems. Because of convergence problems, the representation is not efficient for systems including strong nonlinearities, such as for example saturating elements, as in the case of the Taylor series representation of a strongly nonlinear function. Therefore, the Volterra series can be applied with good results only to systems with mild nonlinearities. Convergence of the Volterra series for both deterministic and stochastic inputs has been studied extensively in the literature.

In contrast to the linear case, the impulse response does not fully characterize a nonlinear system: in fact, by letting $x(t) = cu_0(t)$, the nonlinear system impulse response is

$$h(t) = h_0 + ch_1(t) + c^2h_2(t,t) + \cdots + c^p h_p(t,\ldots,t) + \cdots \tag{3}$$

The impulse response is determined by the diagonal values in the kernels h_p, i.e. for $t_1 = t_2 = \cdots = t_p$ and so it does not completely specify the p-th order Volterra kernel.

Moreover, the measurement of the Volterra kernels is possible only if the contribution of each Volterra operator can be separated from the total system response. If the order of the system is not finite, such contributions cannot be separated, in general, from the total system response. It is worth noting that Wiener overcomes these problems in his classical theory of nonlinear systems, by forming from the Volterra functionals $H_p[x(t)]$ the set of the so-called G-Functionals which are orthogonal when the input is a white Gaussian time function [4, 7].

Finally, multiple-valued nonlinearities cannot be dealt with the Volterra series since it is a single-value representation. Therefore, important phenomena such as hysteresis cannot be considered.

1.21.2 The discrete-time case

The first approaches to the nonlinear signal processing using the discrete Volterra model have been presented at the beginning of 80's [8, 9, 10, 11, 12], and then an increasing interest has been demonstrated for such a topic. In analogy to the continuous case, a nonlinear discrete time-invariant system with memory is described by means of the discrete Volterra series derived by sampling from equations (1) and (2) in the form

$$y(n) = h_0 + \sum_{p=1}^{\infty} \overline{h}_p[x(n)] \tag{4}$$

where $y(n)$ and $x(n)$ are the sampled versions of $y(t)$ and $x(t)$, respectively, (with normalized sampling interval $T = 1$), and

$$\overline{h}_p[x(n)] = \sum_{m_1=0}^{\infty} \cdots \sum_{m_p=0}^{\infty} h_p(m_1,\ldots,m_p)x(n-m_1)\cdots x(n-m_p) \tag{5}$$

since the discrete independent variables $m_1,\ldots,m_p$ are usually defined on a strictly causal support and thus

$$h_p(m_1,\ldots,m_p) = 0 \text{ for any } m_i < 0, \quad i = 1,\ldots,p \tag{6}$$

The constant h_0 is an offset term, $h_1(m_1)$ is the impulse response of a digital IIR filter, and $h_p(m_1,\ldots,m_p)$ can be considered as a generalized p-th order impulse response characterizing the nonlinear behaviour of the system. The upper limit in the summations in equation (5) given as infinity indicates that the discrete system has an infinite memory. As in the linear case, recursive models of finite order involving delayed output terms can be devised, and expansions given in terms of recursive polynomials can be provided. Very often, however, the required memory is finite and thus only the expansion of the input signal is considered. This simpler but still useful class of nonlinear system models is derived by limiting all the summations in equation (5) to a (common) finite value N, i.e. by defining the discrete independent variables $m_1, \ldots, m_p$ on a finite domain. In this case $h_1(m_1)$ represents an FIR filter, and the effect of the nonlinearity on the output depends only on the present and past input values defining the extension of the filter support. Moreover, it has been shown [13, 14] that a discrete time-invariant causal finite-memory system, which also has the property that *small* changes in the system input result in *small* changes in the system output, can be uniformly approximated, according to the Stone-Weierstrass theorem, over a uniformly bounded set of input signals by a truncated Volterra series, obtained by limiting the summation in equation (4) to a sufficiently large, but finite, value P. As a consequence, the expression of a finite-memory finite-order Volterra series is given by the equations

$$y(n) = h_0 + \sum_{p=1}^{P} \overline{h}_p[x(n)] \tag{7}$$

and

$$\overline{h}_p[x(n)] = \sum_{m_1=0}^{N-1} \cdots \sum_{m_p=0}^{N-1} h_p(m_1, \ldots, m_p) x(n-m_1) \cdots x(n-m_p) \tag{8}$$

The nonrecursive models have been extensively studied in the literature because of their relative simplicity at least in comparison with the recursive ones.

1.21.3 Properties of volterra filters

The Volterra series expansions have very interesting properties in both the continuous and the discrete cases: we directly refer in this section to the properties of the discrete Volterra expansions.

1.21.3.1 Structural properties of the kernels

As already mentioned in the continuous case, it is always possible to use in the Volterra expansions, without loss in generality, kernels which are symmetric functions of their arguments so that any of the $p!$ possible permutations of the indexes $m_1, \ldots, m_p$ leaves the kernel unchanged. This feature strongly alleviates the computation complexity of the Volterra expansions. The huge number of the elements involved in each kernel of the Volterra expansions is indeed one of the drawbacks of the representation. In fact, the complexity of the model immensely increases with the length of the memory and the order of the kernels. In principle, the p-th order kernel of the finite-memory Volterra expansion contains N^p coefficients, but they are immediately reduced, because of the symmetry property, to the number of combinations with repetition of N objects taken p at a time, expressed by the binomial factor

$$N_p = \binom{N+p-1}{p} \tag{9}$$

As it regards the identification of the Volterra kernels, it has been already mentioned that the impulse response alone is not sufficient to identify all the kernel elements. On the contrary, it is necessary in general to apply, as input excitation, a set of p products of unit impulses conveniently placed on the filter support. This problem has been considered in particular in [15] for 1-D and 2-D quadratic filters and the characteristics of the so-called *bi-impulse response* have been described.

Another property sometimes imposed on the Volterra filters in modelling and realization problems is the kernel separability. The complete kernel separability is expressed in the following form

$$h_p(m_1, \ldots, m_p) = \prod_{j=1}^{p} h_j(m_j) \tag{10}$$

This property is the basis of the multiplicative realizations.

1.21.3.2 Structural properties of the input-output relationship

These properties are very important in order to completely characterize the nonlinear behaviour of the Volterra filters.

The first relevant property is the linearity of the output with respect to the kernel coefficients, as pointed out by equations (4), (5), (7) and (8). In other words, the nonlinearity of the expansion is given by the multiple products of the delayed input values, while the filter coefficients appear linearly in the output expression. This peculiar property allows the extension of various concepts and procedures applied in the linear case to the nonlinear systems, as the orthogonal projection principle, the optimum filter theory, the derivation of adaptation algorithms, etc.

In addition, from the equations (4), (5), (7) and (8) it clearly appears that the nonlinearity of the system is mapped into multidimensional kernels whose dimensionality is equal to the number of the corresponding products of input values. Such a mapping can be interpreted as an appropriate convolution, and thus the output of a Volterra system can be expressed by sums of multidimensional convolutions. Let us refer to the p-th term of the expansion

$$y(n) = \bar{h}_p[x(n)] = \sum_{m_1=0}^{\infty} \cdots \sum_{m_p=0}^{\infty} h_p(m_1, \ldots, m_p) x(n-m_1) \cdots x(n-m_p) \tag{11}$$

and consider the P-D convolution

$$w(n_1 \ldots, n_p) = \sum_{m_1=0}^{\infty} \cdots \sum_{m_p=0}^{\infty} h_p(m_1, \ldots, m_p) v(n_1 - m_1, \ldots, n_p - m_p) \tag{12}$$

It is easy to see that for $v(n_1-m_1, \ldots, n_p-m_p) = x(n-m_1) \cdots x(n-m_p)$ and $n_1 = \cdots = n_p = n$ the outputs of the two previous equations coincide, i.e $y(n) = w(n, \ldots, n)$. This fact means that the output of a nonlinear filter characterized by the equation (11) is obtained as the trace of a multidimensional convolution on the principal diagonal of a corresponding multidimensional filter support. In other words, the nonlinearity of order p is associated with an array of identical dimensionality. Moreover, this interpretation suggests a possible realization for the Volterra filters: however, such a realization is not convenient in general since it requires an excessive amount of computations to calculate the multidimensional convolutions.

A frequency domain interpretation of the behaviour of the Volterra filters is now possible, based on the multidimensional convolution property. In fact, the frequency response of a P-D linear system can be exploited to derive the frequency behaviour of the output from a filter characterized by a generic p-th order kernel. Let us apply a P-D sampled complex sinusoid

$$v(n_1,\ldots,n_p) = e^{j\omega_1 n_1}\ldots e^{j\omega_p n_p} \tag{13}$$

as input sequence to a causal linear filter described by the P-D impulse response $h_p(m_1,\ldots,m_p)$. Thus, the output sequence is given by the equation

$$w(n_1,\ldots,n_p) = H_p(e^{j\omega_1},\ldots,e^{j\omega_p})e^{j\omega_1 n_1}\ldots e^{j\omega_p n_p} \tag{14}$$

where

$$H_p(e^{j\omega_1},\ldots,e^{j\omega_p}) = \sum_{m_1=0}^{\infty}\cdots\sum_{m_p=0}^{\infty} h_p(m_1,\ldots,m_p)e^{-j\omega_1 m_1}\ldots e^{-j\omega_p m_p} \tag{15}$$

is the continuous frequency response of the P-D linear filter. The output of a Volterra filter characterized by the p-th order kernel is then derived by equation (14), by assuming $v(n_1,\ldots,n_p) = x(n_1)\cdots x(n_p)$ and $n_1 = \cdots = n_p = n$

$$y(n) = w(n,\ldots,n) = H_p(e^{j\omega_1},\ldots,e^{j\omega_p})e^{j(\omega_1+\cdots+\omega_p)n} \tag{16}$$

so that the frequency contributions at the output are recovered at the angular frequency $\omega_1+\cdots+\omega_p$. This fact immediately shows that new frequencies appear at the output that are not present at the input. As an example, the response of a purely quadratic filter to a complex sinusoid $e^{j\omega_a n}$ is another complex sinusoid $y(n) = H_2(e^{j\omega_a},e^{j\omega_a})e^{j2\omega_a n}$ at the double angular frequency $2\omega_a$.
If the input signal has a continuous spectrum $X(e^{j\omega})$ it is then possible to derive the output signal by means of an integral armonic representation, still with $n_1 = \cdots = n_p = n$, in the form

$$y(n) = w(n,\ldots,n) =$$

$$= \frac{1}{(2\pi)^p}\int_{-\pi}^{\pi}\cdots\int_{-\pi}^{\pi} H_p(e^{j\omega_1},\ldots,e^{j\omega_p})X(e^{j\omega_1})\cdots X(e^{j\omega_p})e^{j(\omega_1+\cdots+\omega_p)n}d\omega_1\cdots d\omega_p \tag{17}$$

As it appears from equation (17), the angular frequencies in the argument of the exponential function are (algebraically) summed up: as a consequence, when the input signal is a sinusoid, multiples and sub-multiples components of that frequency may be present in the output signal. In presence of combinations of sinusoids at the input, intermodulations terms are present at the output, too. These general considerations confirm one of the peculiarities of the nonlinear behaviour of the Volterra filters: new frequencies appear in the output signal that are not present in the input signal.

1.21.3.3 Volterra series expansion for multidimensional systems

The discrete Volterra series expansion can be extended to the M-D signals, too. It is worth noting that in such a case the complexity of the expansion may be considered in general even bigger than in the 1-D case. However, in image processing applications in 2-D or 3-D, very often filters defined on very small supports are able to give remarkable effects. Moreover, a quadratic filter is often sufficient to obtain significant improvements with respect to the use of conventional linear filters. The arrangement of the independent variables in suitable vectors is at the basis of the representation of a M-D Volterra filter. Moreover, to derive a compact notation, it is necessary to conveniently order the input data and the filter coefficients [16].

1.21.4 Realizations and approximations of polynomial filters

In several applications it is known *a priori* that the nonlinear system to be modelled involves large time delays. However, it may frequently happen that only a few kernels in the expansion are present and, in each kernel, the nonzero coefficients are sparsely distributed. On the other hand, even when the Volterra series is not sparse, it may happen that a number of coefficients has not significant effects on the system output values. Since the number of the Volterra coefficients increases geometrically as the delays (or memory) increase, a great deal of computation effort will be wasted if general estimation methods of the complete set of the coefficients in the Volterra expansion are applied in these cases. Identification methods based on the optimum filter theory or the Least-Square criterion usually require the solution of a linear system of exceeding dimensionality. Suitable hypotheses on the statistics of the input signals, i.e. gaussianity, may only partially alleviate these difficulties. General methods for progressively building difference equation models of nonlinear systems and for modelling time-series data have been recently reviewed by Korenberg [18]. Since finite-order Volterra series are included in these models, kernel estimation can be carried out by using such methods. Among them, the Orthogonal Search Methods, that are able to expand the model by adding a further difference equation term one at a time, require in general an excessive amount of computer time since the pertinent computations must be done for all candidate terms. To increase the search speed, fast orthogonal methods have been devised and many variations have been considered, consisting for example in arranging the candidate difference equation terms into disjoint subsets, in using thresholds for discarding unuseful terms, etc.

To overcome the computational drawbacks of general approximation methods, two alternative approaches can be exploited:

i) the recursive polynomial system models;

ii) various forms of reduced-complexity approximations based on the truncated Volterra series expansion.

With reference to the first approach, it is worth noting that recently there has been a certain amount of efforts in such a field since, like linear IIR filters can approximate a large number of linear systems with great parsimony in the use of the coefficients, the recursive polynomial systems can also approximate a large number of nonlinear systems with good efficiency. The simplest of such models is the so-called bilinear system, in which the output linearly depends on the input values, on the previous output values and on the products of the input and output values. However, the main problem associated with recursive nonlinear systems is that their stability is not guaranteed: therefore, it is necessary to find out sufficient conditions for the input-dependent stability and then apply appropriate stability tests. A condition of this kind has been reported in [17], while further research activities are still ongoing.

As a consequence of these difficulties the second approach based on the truncated Volterra series, which indeed corresponds to the use of FIR filters in the linear case, is still valid and worth of further consideration. A first possibility consists in assuming a simplified model to describe the Volterra filters. Among the simplified models it is convenient to mention the so-called Hammerstein model, in which a memoryless nonlinearity is followed by an FIR filter, and the Wiener model in which the two operations are executed in a reverse order. To describe the corresponding sub-classes of Volterra filters, the nonlinearity is given as a polynomial function. Another sub-class of the Volterra filters is obtained by factorizing the kernel of order p in p separable linear FIR filters. The model consists of a parallel combination of p linear filters: the actual filter output is obtained by multiplication of the outputs from each linear filters. This model has been studied, for example, in [11] for a quadratic filter ($p = 2$). The drawbacks of this approach are that the decomposition is not unique and that in the identification procedures, based on the alternative adaptation of the coefficients of the linear filters, a problem of local minima may exist. As a consequence, the convergence through iterations is not easily established, in particular for higher-order kernels.

Coming back to the estimation of the kernels of a nonlinear system with lengthy memory, a very effective method relies on an approach based on a parallel-cascade structure. In fact, it has been shown in [14] that any causal, finite-memory, time-invariant, discrete-time system which is a mapping of its input so that small changes in the system input result in small changes in the system output can be uniformly approximated by a finite sum of a dynamic linear system, a static nonlinearity and a dynamic linear system. Moreover, it has also been shown that, in order to apply the parallel-cascade identification method it is sufficient to include in each parallel channel a dynamic linear system followed by a static nonlinearity [19], i.e. a Wiener structure. The suggested identification procedure consists in approximating the nonlinear system using a first channel, computing the residual between the system output and the channel output, and then approximating the residual using a second channel, and so on. It has been proved that, for a large class of systems, the output of the parallel-cascade structure can approximate the output of the system to be modelled with an arbitrary precision according to the Mean-Square Error criterion. The linear system in each channel is defined by using a first-order crosscorrelation function, or a slice of a second- or higher-order crosscorrelation, of the input with the residue remaining after the estimation of the previous channels; the static nonlinearities in each channel are then represented using polynomials. In addition to the rapid modelling capabilities of high-order nonlinear systems, the parallel-cascade structure enables accurate kernel estimation for systems with lengthy memory, too. It is worth noting that the parallel-cascade structure is able to represent a quadratic kernel exactly, but an additional multiplier on each channel. This result has been reported in [12], by using a general matrix decomposition technique for filters having fixed coefficients. In this case it has also been shown that the number of channels must equate the rank of the matrix formed with the coefficients of the quadratic kernel. A reduced-complexity representation can be achieved by first applying the Singular Value Decomposition to the coefficient matrix in order to obtain an approximation based on its most important eigenvectors. The approach has been then extended in [20] to adaptive structures, and thus it is possible to use such a technique for the identification of quadratic filters.

1.21.5 Applications of polynomial filters

In the recent years various applications of Volterra filters have been reported on. Very often simple quadratic filters are sufficient to achieve significant improvements with respect to linear techniques [21].

Applications can be found in tasks involving 1-D time-dependent signals, such as modelling of nonlinear systems, detection and estimation, echo cancellation, cancellation of nonlinear intersymbol interference, channel equalization in communications, nonlinear prediction, etc.

Applications involving spatial and spatio-temporal signals (images and sequences of images) can be found for image enhancement, edge preserving smoothing, processing of document images, texture discrimination, spatial and spatio-temporal prediction, nonlinear interpolation of image sequences, etc.

References

[1] V. Volterra, "Sopra le funzioni che dipendono da altre funzioni," *Rend. Regia Accademia dei Lincei, 2o Sem.*, pp. 97–105, 141–146, 153–158, 1887.

[2] V. Volterra, *Leçons sur les functions de lignes*, Paris, France: Gauthier-Villars, 1913.

[3] V. Volterra, *Theory of functionals and of integral and integro-differential equations*, New York: Dover, 1959.

[4] M. Schetzen, *The Volterra and Wiener theories of nonlinear systems*, New York: Wiley, 1980.

[5] W. J. Rugh, *Nonlinear system theory. The Volterra-Wiener approach*, Baltimore, MD: John Hopkins University Press, 1981.

[6] M. Frechet, "Sur les fonctionnels continues," *Ann. Sci. Ecole Normal Sup.*, Vol. 27, pp. 193–219, 1910

[7] N. Wiener, *Nonlinear problems in random theory*, New York: Wiley, 1958.

[8] E. Biglieri, "Theory of Volterra processors and some applications," in *Proc. ICASSP-82*, Paris, France, pp. 294–297, May 1982.

[9] B. Picinbono, "Quadratic filters," in *Proc. ICASSP-82*, Paris, France, pp. 298–301, May 1982.

[10] G. L. Sicuranza, "Nonlinear digital filter realization by distributed arithmetic," *IEEE Trans. on Acoust., Speech, Signal Processing*, Vol. ASSP-33, No. 4, Aug. 1985.

[11] T. Koh and E. J. Powers, "Second-order Volterra filtering and its application to nonlinear system identification," *IEEE Trans. Acoust., Speech, Signal Processing*, Vol. ASSP-33, pp. 1445–1455, Dec. 1985.

[12] H. H. Chang, C. L. Nikias, and A. N. Venetsanopoulos, "Efficient implementations of quadratic filters," *IEEE Trans. on Acoust., Speech, Signal Processing*, Vol. ASSP-34, pp. 1511–1528, Dec. 1986.

[13] J. Dieudonné, *Foundations of Modern Analysis*, Berlin, Heidelberg, New York: Springer, 1976.

[14] G. Palm, "On the Representation and Approximation of Nonlinear Systems, Part II: Discrete Time," *Biol. Cybern.*, Vol. 34, pp. 49–52, 1979.

[15] G. Ramponi, "Bi-impulse Response Design of Isotropic Quadratic Filters," *IEEE Proceedings*, Vol. 78, No. 4, pp. 665–677, April 1990.

[16] B. G. Mertzios, G. L. Sicuranza, A. N. Venetsanopoulos, "Efficient realizations of two-dimensional quadratic digital filters," *IEEE Trans. on Acoust., Speech, Signal Processing*, Vol. ASSP-37, No. 5, May 1989.

[17] J. Lee and V. J. Mathews, "A stability condition for certain bilinear systems," in *Proc. of IEEE Winter Workshop on Nonlinear Digital Signal Processing*, Tampere, Finland, Jan. 17–20, 1993.

[18] M. J. Korenberg and L. D. Paarmann, "Orthogonal Approaches to Time-Series Analysis and System Identification," *IEEE SP Magazine*, July 1991.

[19] M. J. Korenberg, "Parallel Cascade Identification and Kernel Estimation for Nonlinear Systems," *Ann. Biomed. Eng.*, Vol. 19, 1991.

[20] Y. Lou, C. L. Nikias and A. N. Venetsanopoulos, "Efficient VLSI array processing structures for adaptive quadratic digital filters," *J. Circuits Systems Signal Process.* Vol. 7, No. 2, pp. 253–273, 1988.

[21] G. L. Sicuranza, "Quadratic filters for signal processing," *Proc. IEEE* (invited paper), Vol. 80, No. 8, Aug. 1992.

Chapter 1.22

Adaptive Polynomial Filters

V. John Mathews

1.22.1 Introduction

There are a large number of situations involving nonlinear estimation problems in which the underlying nonlinear system is time-varying and/or the statistics of the signals involved are unknown *a priori*. Adaptive filters equipped with nonlinear system models are necessary in such situations. This section is devoted to a discussion of structures and capabilities of adaptive nonlinear filters that employ polynomial system models.

As mentioned in chapter 1.21, one of the biggest advantages of using polynomial models is the fact that the system output is linear in some polynomial functions of the input and output signals. Consequently, many linear filtering concepts can be extended to the polynomial filtering case. Much of the work in adaptive polynomial filtering makes use of this approach. This chapter is divided into two main parts. The first part discusses adaptive polynomial filters employing truncated Volterra system models. The system models have finite memory and finite order of nonlinearity. The second part involves the use of nonlinear system models with feedback in adaptive filtering applications.

1.22.2 Adaptive volterra filters

While the ideas discussed in this section are generally applicable, we will restrict ourselves to the quadratic model consisting of a finite memory, second order nonlinearity for pedagogic simplicity. Two of the most commonly employed classes of algorithms for adaptive filtering are gradient search algorithms and recursive least-squares algorithms.

1.22.2.1 Least-mean-square (LMS) quadratic filters

Consider the problem of adaptively estimating a desired response signal $d(n)$ as a quadratic expansion in the most recent N samples of the input signal $x(n)$, i.e., we want to estimate $d(n)$ as

$$\hat{d}(n) = \sum_{m_1=0}^{N-1} \sum_{m_2=m_1}^{N-1} h_2(m_1, m_2; n) x(n-m_1) x(n-m_2) \tag{1}$$

where $h_2(m_1, m_2; n)$ for different values of m_1 and m_2 are the coefficients of the adaptive filter at time n. The objective of the LMS quadratic filters [2, 7] is to iteratively adjust the coefficients of the adaptive filter so as to reduce

$$e^2(n) = \left(d(n) - \hat{d}(n)\right)^2 \tag{2}$$

at each iteration. The coefficient adjustment is done using a steepest descent approach as

$$\begin{aligned} h_2(m_1, m_2; n+1) &= h_2(m_1, m_2; n) - \frac{\mu}{2} \frac{\partial}{\partial h_2(m_1, m_2; n)} e^2(n) \\ &= h_2(m_1, m_2; n) + \mu e(n) x(n-m_1) x(n-m_2) \end{aligned} \tag{3}$$

where μ is a small positive constant called the step size of the adaptive filter. The step size controls the speed of convergence and the steady-state tracking capabilities.

The LMS quadratic filter can be analyzed in a manner similar to the analysis of LMS linear filters. Let

$$\begin{aligned} \mathbf{H}(n) = & [h_2(0,0;n), h_2(0,1;n), \ldots, h_2(0,N-1;n), \\ & h_2(1,1;n), \ldots \ldots h_2(N-1,N-1;n)]^T \end{aligned} \tag{4}$$

denote the coefficient vector for the adaptive filter at time n. Similarly, define the input vector at a time n as

$$\mathbf{X}(n) = [x^2(n), x(n)x(n-1), \ldots, x(n)x(n-N+1),\ x^2(n-1), \ldots, x^2(n-N+1)]^T \tag{5}$$

Then, one can show, using several simplifying assumptions commonly employed in the analysis of adaptive filters, that the adaptive filter will converge in the mean squared sense if

$$0 < \mu < \frac{2}{3tr(\mathbf{R_{xx}})} \tag{6}$$

where $tr(\cdot)$ denotes the trace of the matrix $(\cdot)$ and $\mathbf{R_{xx}}$ is the statistical autocorrelation matrix of the input vector $X(n)$. The above result assumes stationarity of the input signal. One can also show that the speed of convergence is controlled by the eigenvalue spread of the autocorrelation matrix.

The dependence of the adaptive filter on the statistics of the input signal is one major problem associated with LMS adaptive filters. This problem is even greater for the nonlinear case than for the linear case. This is because the eigenvalue spread of the autocorrelation matrix will in general be much larger for the nonlinear filtering problem than for the linear case. There are several approaches that have been employed in the past for partially or completely overcoming the dependence of the quadratic filters on the input signal statistics. We will briefly discuss some of these approaches now.

1.22.2.2 Variable step size algorithms

The most common approach for improving the performance of the LMS adaptive filters is that of employing a time-varying convergence factor. Several heuristic algorithms are available in the literature. We now discuss an algorithm that is derived based on more rigorous and systematic considerations [17]. In this approach, the step size is changed so that the change is guaranteed to bring about a reduction in the squared error. The algorithm can be implemented using the following equations:

$$\begin{aligned} e(n) &= d(n) - \mathbf{H}^T(n)\mathbf{X}(n) \\ \mu(n+1) &= \mu(n) + \rho e(n)\mathbf{X}^T(n)\mathbf{X}(n-1)e(n-1) \end{aligned} \tag{7}$$

and

$$\mathbf{H}(n+1) = \mathbf{H}(n) + \mu(n)\mathbf{X}(n)e(n) \tag{8}$$

In the above equations, ρ is a small positive constant that controls the rate of change of the step size. Care should be taken to ensure that the step size $\mu(n)$ is limited to the range of values that ensure stable operation of the adaptive filter. Experimental results [16] have indicated much faster convergence behaviour for the above adaptive step size algorithm than the basic LMS algorithm in Volterra filtering applications.

Initialization

$$\begin{aligned} \mathbf{H}(0) &= [0,0,\ldots,0]^T \\ \mathbf{R}^{-1}(0) &= \delta^{-1}I \\ \delta &= \text{a small positive constant} \end{aligned}$$

Algorithm

$$\begin{aligned} \mathbf{k}(n) &= \frac{\lambda^{-1}\mathbf{R}^{-1}(n-1)\mathbf{X}(n)}{1+\lambda^{-1}\mathbf{X}^T(n)\mathbf{R}^{-1}(n-1)\mathbf{X}(n)} \\ \epsilon(n) &= d(n) - \mathbf{H}^T(n-1)\mathbf{X}(n) \\ \mathbf{H}(n) &= \mathbf{H}(n-1) + \mathbf{k}(n)\epsilon(n) \\ \mathbf{R}^{-1}(n) &= \lambda^{-1}\mathbf{R}^{-1}(n-1) - \lambda^{-1}\mathbf{k}(n)\mathbf{X}^T(n)\mathbf{R}^{-1}(n-1) \\ e(n) &= d(n) - \mathbf{H}^T(n)\mathbf{X}(n) \end{aligned}$$

Table 1 The RLS Adaptive Quadratic Filter.

1.22.2.3 Recursive least-squares (RLS) quadratic filters

Most least-squares adaptive filters attempt to find the exact solution to the problem of minimizing a deterministic function of the squared error values. For example, the exponentially weighted RLS quadratic filter minimizes the cost function

$$J(n) = \sum_{k=0}^{n} \lambda^{n-k} \left(d(k) - \mathbf{H}^T(n)\mathbf{X}(k)\right)^2 \tag{9}$$

at each time. An iterative solution to the problem is given in table 1. The basic derivation is similar to the derivation of RLS adaptive linear filters given in [5].

The performance of the adaptive RLS Volterra filters has been analyzed in [10]. The analysis has shown that the convergence behaviour and the steady-state properties of the adaptive filter are independent of the signal statistics. Furthermore, experimental comparisons with gradient search and other competing algorithms have demonstrated that the RLS algorithm enjoys many superiorities over them in its capabilities [9, 10, 14].

One big drawback of the RLS adaptive quadratic filter in table 1 over its LMS counterparts is that the former requires an order of magnitude more computational complexity than the LMS algorithms. More efficient algorithms have been developed in the recent past. Derivations and performance evaluations can be found in [10].

The fast Volterra filter derivations transform the nonlinear filtering problem into an equivalent multi-channel, but linear filtering problem and then makes use of the theory of fast multichannel filters. The resultant algorithms require $O(N^3)$ arithmetic operations per sample to implement an adaptive RLS quadratic filter with N-sample memory in the system model. This compares with the $O(N^4)$ operations required by the algorithm in table 1. It should also be pointed out that LMS quadratic filters and its variations require only $O(N^2)$

operations per sample. The increased computational complexity even for the most efficient algorithm is the price we pay for the superior performance capabilities of the RLS Volterra filters.

The early versions of the fast RLS Volterra filters utilized a direct form structure for the system model and had relatively poor numerical properties. Very recently, fast algorithms involving lattice and other non-direct form structures have been developed [21, 22]. Particularly noteworthy is the fact that some of these algorithms are implemented solely using numerically robust Givens' rotations. Extensive experimental results involving finite precision (fixed-point) implementation of these filters have been presented in [21, 22].

1.22.2.4 *An overview of other approaches*

Adaptive Volterra filters based upon distributed arithmetic implementation are presented in [18, 20]. A gradient adaptive quadratic filter employing an LU decomposition of the coefficient matrix is discussed in [12]; this paper also discusses VLSI implementations of adaptive Volterra filters. An adaptive lattice structure for realizing second order Volterra filters specifically when the input signals are Gaussian has been presented in [6]. A computationally efficient, but approximate least-squares adaptive second order Volterra filter has been described in [3]. Another computationally efficient variable step size algorithm employing the sign algorithm [1, 13] was presented in [19].

1.22.3 Recursive nonlinear systems

Perhaps the biggest disadvantage associated with truncated Volterra series modeling of nonlinear systems is the extremely large number of coefficients such models require. Note that a very general Volterra system model involving P-th order nonlinearity and N-sample memory requires $O(N^P)$ coefficients. Even for moderately large values of N and P, the number of coefficients can become unreasonably large. One approach to dealing with this problem is to look for sparse Volterra filter approximations. We now briefly discuss an alternate approach to dealing with the large number of coefficients associated with truncated Volterra system models. This involves the use of recursive nonlinear system models.

The input-output relationship of a general polynomial system model with N-sample memory and up to P-th order nonlinear terms can be described using the following nonlinear difference equation

$$y(n) = \textstyle\sum_{i=0}^{P} P_i \left\{y(n-1), y(n-2), \ldots,\ y(n-N+1), x(n), \ldots (n-N+1)\right\} \tag{10}$$

Here, $P_i(\cdot)$ denotes polynomial involving only i-th order of terms in the signal samples within the brackets. The simplest of such recursive nonlinear system models is the bilinear system model [15] whose input-output relationship is given by

$$y(n) = \textstyle\sum_{i=1}^{N-1} b_i y(n-i) + \sum_{j=0}^{N-1} a_j x(n-j) + \sum_{i=1}^{N-1}\sum_{j=0}^{N-1} c_{ij} x(n-j) y(n-i) \tag{11}$$

It is known that bilinear system models can approximate a very large class of nonlinear systems with arbitrary precision. Much like linear IIR filters can approximate a large number of linear systems with great parsimony in the use of coefficients, bilinear and other recursive nonlinear models can also approximate a large number of nonlinear systems with good efficiency. Consequently, there has been a fair amount of recent effort to model nonlinear systems with polynomial models. We will not go into the details here, but will instead discuss the very important issue of stability of recursive nonlinear systems.

1.22.3.1 Stability of recursive systems

One of the biggest problems associated with recursive nonlinear systems is that almost all such systems are inherently unstable. To see this more clearly, let us rewrite the input-output relationship for a bilinear system as

$$y(n) = \sum_{i=1}^{N-1} \left\{ b_i + \sum_{j=0}^{N-1} c_{ij} x(n-j) \right\} y(n-i) + \sum_{j=0}^{N-1} a_j x(n-j) \tag{12}$$

It is relatively straightforward to infer from the above equation that one can almost always find bounded input signals that can cause the output of the system to diverge. However, there is also a class of signals for which a given recursive nonlinear system may produce very useful outputs. It is such class of signals and any associated conditions for input-dependent stability that we are interested in.

We will now describe a sufficient condition for guaranteeing that the output of a bilinear system described in (11) is bounded whenever the input signals are bounded by some fixed constant M_x.

Theorem [8]

For the bilinear system model represented by (11), let $p_1, p_2, \ldots,$ and p_{N-1} denote the zeros of the polynomial $q^{N-1}\left(1 - \sum_{i=1}^{N-1} b_i q^{-i}\right)$. Given a real, positive number M_x,

$$|p_i| < 1, for\ i = 1, 2, \ldots, N-1 \tag{13}$$

and

$$M_x \sum_{i=0}^{N-1} \sum_{j=1}^{N-1} |c_{i,j}| < \prod_{i=1}^{N-1} (1 - |p_i|) \tag{14}$$

constitute a sufficient condition for every $x(n)$ bounded by M_x, i.e., $|x(n)| \leq M_x$, to produce a bounded output $y(n)$.

It should be noted that when there is no nonlinearity in the system, the above theorem reduces to the well-known necessary and sufficient condition for the resultant linear time-invariant system to be stable in the bounded-input, bounded-output sense. A detailed derivation of the theorem is available in [8].

1.22.3.2 Adaptive bilinear filters

Adaptive systems to track bilinear or other recursive nonlinear system models can be devised in a relatively straightforward manner. For a tutorial overview of such filters, the reader is referred to [14]. The biggest issue associated with such systems is that of stability. When developing and using such filters, one must make sure that the system model produced by the adaptive filter is guaranteed to be stable, or if this cannot be achieved, one must equip the adaptive filters with continuous stability monitoring mechanisms. We will now briefly review some of the recent results in this area of nonlinear filtering research.

Extended least-squares adaptive polynomial filters are guaranteed to be stable in the sense that the time-averaged value of the squared estimation error is always bounded by a finite quantity [11]. However, this result does not imply that the adaptive filter coefficients will not fluctuate wildly for certain type of signals. Variations of exact least-squares adaptive filters and gradient search algorithms do not share the property of bounded average squared error value. In [4], this problem has been overcome by appending a Kalman filter to the basic adaptive filter to monitor the stability of the overall system. The Kalman filter senses the onset of instability and modifies the behaviour of the basic adaptive filter in such situations. Gradient search adaptive bilinear filters equipped with a projection facility can also be developed [9]. Such systems will continuously check the adaptive filter coefficients to ensure that the conditions of the theorem described earlier are satisfied. If the coefficients do not satisfy the above conditions, the projection mechanism projects the coefficients back into the space of parameters that satisfy the conditions of the theorem and thus avoids instability problems.

All the above-described results are of preliminary nature and much work in the way of performance evaluation and system development needs to be still done.

References

[1] T. A. C. M. Claasen and W. F. G. Mecklenbrauker, "Comparison of the convergence of two algorithms for for adaptive FIR digital filters," *IEEE Trans. Acoust., Speech, Signal Proc.*, Vol. ASSP-29, pp. 670-678, June 1981.

[2] M. J. Coker and D. N. Simkins, "A non-linear adaptive noise canceller," *Proc. IEEE Int. Conf. Acoust., Speech, Signal Proc.*, 1980.

[3] C. E. Davila, A. J. Welch and H. G. Rylander, III, "A second-order adaptive Volterra filter with rapid convergence," *IEEE Trans. Acoust., Speech, Signal Proc.*, Vol. ASSP-35, pp. 1259-1263, September 1987.

[4] F. Fnaiech and L. Ljung, "Recursive identification of bilinear systems," *International Journal of Control*, Vol. 45, No. 2, pp. 453-470, 1987.

[5] S. Haykin, *Adaptive Filter Theory* (second edition), Prentice Hall, Englewood Cliffs, New Jersey, 1991.

[6] T. Koh and E. J. Powers, "An adaptive non-linear digital filter with lattice orthogonalization," *Proc. IEEE Int. Conf. Acoust. Speech, Signal Proc.*, Boston, Massachusetts, April 1983.

[7] T. Koh and E. J. Powers, "Second-order Volterra filtering and its application to nonlinear system identification," *IEEE Trans. Acoust., Speech, Signal Proc.*, Vol. ASSP-33, pp. 1445-1455, December 1985.

[8] J. Lee and V. J. Mathews, "A stability condition for certain bilinear systems," *Proc. IEEE Winter Work Shop on Nonlinear DSP*, Tampere, Finland, January 1993.

[9] J. Lee, "Adaptive Polynomial Filtering Algorithms," PhD. Dissertation, University of Utah, August 1992.

[10] J. Lee and V. J. Mathews, "A fast, recursive least-squares second-order Volterra filter and its performance analysis," *IEEE Transactions on Signal Processing, IEEE Trans. Signal Proc.*, Vol. 41, pp. 1087-1102, March 1993.

[11] J. Lee and V. J. Mathews, "On the extended RLS adaptive bilinear filters," *Proc. ICASSP*, Minneapolis, Minnesota, April 1993.

[12] Y. Lou, C. L. Nikias, and A. N. Venetsanopoulos, "Efficient VLSI array processing structures for adaptive quadratic digital filters," *Journal Circuits Systems Signal Process.*, Vol. 7, No. 2, pp. 253-273, 1988.

[13] V. J. Mathews and S. H. Cho, "Improved convergence analysis of stochastic gradient adaptive filters employing the sign algorithm," *IEEE Trans. Acoust., Speech, Signal Proc.*, Vol. ASSP-35, pp. 450-454, April 1987.

[14] V. J. Mathews, "Adaptive polynomial filters," *IEEE Signal Processing Magazine*, Vol. 8, No. 3, pp. 10-26, July 1991.

[15] V. J. Mathews and Z. Xie, "Stochastic gradient adaptive filters with gradient adaptive step-sizes," *IEEE Int. Conf. Acoust., Speech, Signal Proc.*, Albuquerque, New Mexico, April 1990.

[16] V. J. Mathews and Z. Xie, "Stochastic gradient adaptive filters with gradient adaptive step-sizes," *IEEE Trans. Signal Proc.*, Vol. 41, pp. 2075-2087, June 1993.

[17] R. R. Mohler and W. J. Kolodziej, "An overview of bilinear system theory and applications," *IEEE Trans. Systems, Man and Cybernetics*, Vol. SMC-10, pp. 683-688, October 1980.

[18] G. L. Sicuranza and G. Ramponi, "Adaptive nonlinear digital filters using distributed arithmetic", *IEEE Trans. on Acoust., Speech, Signal Processing*, Vol. ASSP-34, No. 3, pp. 518-526, June 1986.

[19] G. L. Sicuranza and G. Ramponi, "A variable-step adaptation algorithm for memory oriented Volterra filters," *IEEE Trans. Acoust., Speech, Signal Proc.*, Vol. ASSP-35, October 1987.

[20] M. J. Smith, C. F. N. Cowan and P. F. Adams, "Nonlinear echo cancellers based on transpose distributed arithmetic," *IEEE Trans. Circuits and Systems*, Vol. CAS-35, pp. 6-18, January 1988.

[21] M. A. Syed and V. J. Mathews, "Lattice algorithms for recursive least squares adoptive Volterra filtering," to appear in *IEEE Trans. Circuits and Systems - II*, April 1994.

[22] M. A. Syed and V. J. Mathews, "QR-Decomposition based algorithms for adaptive Volterra filtering," *IEEE Trans. Circuits and Systems - I: Fundamental Theory and Applications*, Vol. 40, pp. 372-382, June 1993.

Chapter 1.23

Application of Adaptive Volterra Filters to Equalization

Colin F. N. Cowan

1.23.1 Introduction

One of the main motivations in the development of adaptive filter theory over the last few decades has been the application in telecommunications systems. In particular this has centered on equalization [1] and echo cancellation [2] for data communications. Linear adaptive systems provided adequate solutions for many data communications systems where the main problem to be faced was a lack of a-priori knowledge relating to the distortion introduced by the transmission channel and noise was not a major problem.

However, modern communications systems place ever greater demands on transmission channel performance and this has led to a situation where effects such as channel noise, time invariance and transmission non-linearity have great significance. This section examines the problems introduced in such data transmission systems and develops a non-linear adaptive solution for the equalization case. It is further shown how this solution may be efficiently treated as a closed form Volterra filter of order 2.

1.23.2 Adaptive equalization

The process of distortion in communications systems may, typically, be modelled as a linear process between the transmitter and the receiver. A common means of overcoming this problem has thus been to introduce a (linear) inverse filter into the receiver to "equalize" the channel (Figure 1). However, the designer does suffer from the disadvantage of having no a-priori knowledge of the channel transfer function and, is further assured that this function will be time varying. For these reasons such equalizers are commonly adaptive in nature; that is, they automatically adjust their parameters subject to some external (statistical) stimuli.

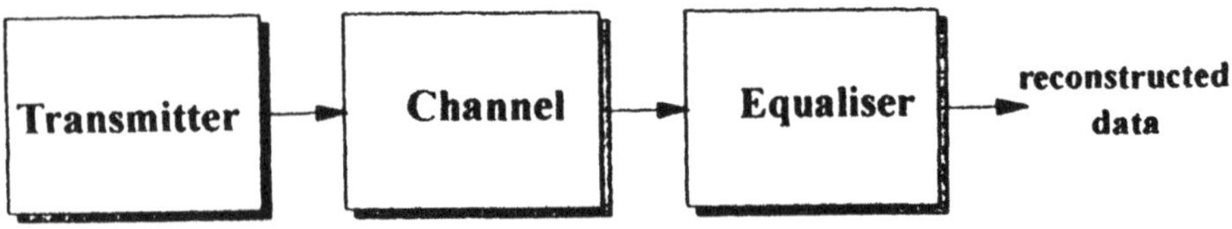

Figure 1 Basic communications equalization.

Common techniques which are applied to this problem deploy linear adaptive filters in an attempt to combat these problems [1]. Such techniques are popular due to the highly developed nature of linear adaptive filter theory [3], and the relative simplicity of such processes. However, it will be shown here that this problem is one which is inherently non-linear in nature and an alternative strategy for equalization will be evolved.

1.23.3 Equalization as classification

It has already been stated that equalization is frequently viewed as a linear filtering problem, and indeed it could be adequately stated in this light from the view-point of providing a complete inverse to the channel. However, data communications deal exclusively with information samples drawn from a finite alphabet, rather than continuous values. Therefore, we could state that the task of the receiver (by combining the task of equalizer and detector) is to determine the correct value of the currently transmitted symbol, given observations of the present and past channel outputs. This is, in fact, a classification process. Despite this alternative view, it may be shown, [4], that a linear equalizer is capable of performing this task, optimally, provided the channel is minimum phase and noise free. This may best be illustrated by using a simple example. In this case we propose a channel model

$$H(z) = 1 + 0.5z^{-1} \tag{1}$$

which is minimum phase. It is further assumed that additive noise is present and that the input set is {-1,1}. The equalizer to be used takes two successive observations of the channel output on which to base its classification.

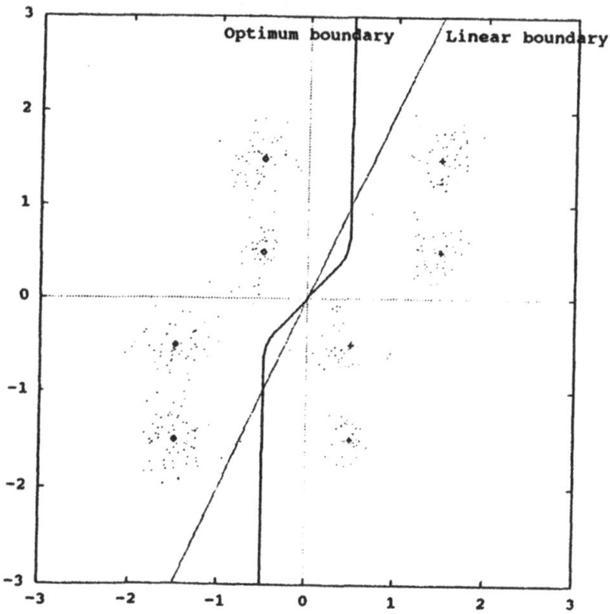

Figure 2 Received signal states given a 2-dimensional observation vector.

The diagram in Figure 2 shows the distribution of samples on the 2-dimensional observation plane. Samples denoted by a circle represent a transmitted symbol of -1, the crosses representing +1. The solid line represents the classification boundary achieved by a linear equalizer, which is obviously sufficient in the noise free case. The dashed line, however, is the optimum (non-linear) boundary for the case with additive noise. Clearly something other than a linear equalizer is needed.

An alternative way to view this problem is one of identifying the state of the transmission sequence at any given observation time instant. This would involve forming clusters centered on the possible noise free observation states. This, however, is an ill-conditioned formulation as it is not possible to say a-priori how many such states (centers) will exist (this is dependent on the channel time dispersion). A modification to the above strategy uses a single cluster to represent these distorted states. Since the distortion introduces spreading of the observed signal states this single cluster must be appropriately shaped to take account of the correlations introduced by the channel. Generally, the shape of such a cluster will be hyper-elliptic.

Figure 3 shows the result of applying such a technique to the example introduced here. The figure shows contours of equal distance around the 4 centers in our 2-dimensional example. Generally, the classification process is achieved by measuring the weighted distance of an observed symbol from each of these cluster centers. The signal is classified as belonging to the cluster it is nearest to. The distance measure applied is given by the Mahalanobis distance [5]

$$d_i = (\boldsymbol{x}(n) - \boldsymbol{m}_i)^t \, \Sigma_i^{-1} \, (\boldsymbol{x}(n) - \boldsymbol{m}_i) \tag{2}$$

where $\boldsymbol{m}_i$ is the mean of cluster i,

$\boldsymbol{x}(n)$ is the observation vector at time n,

and Σ_i is the covariance matrix associated with cluster i.

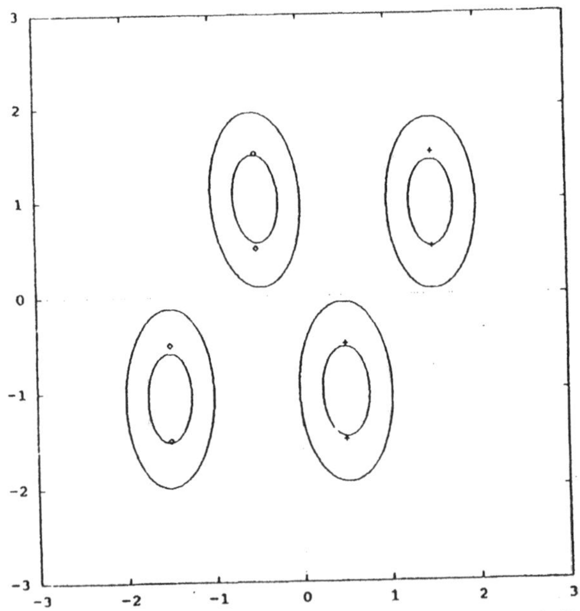

Figure 3 Elliptical clustering applied to the signal space defined in Figure 1.

This strategy does result in a realization of the optimum non-linear decision boundaries [6] which provides an improved bit error rate, relative to a linear equalizer, at low signal-to-noise ratios.

However, calculation of the distance, d_i, is a numerically sensitive operation when noise levels are low due to the effective collapse of the elliptical minor axis. Increasing the dimension of the observation space also leads to rapid growth of computational complexity, due to the matrix inverse operation. An alternative way of realizing this function is, therefore, desirable.

1.23.4 Adaptive Volterra equalization

A more general form of the distance classifier may be developed [7] by noting the similarity of a general Volterra series to the Mahalanobis distance of (2). A quadratic Volterra filter acting on an input vector $\underline{x}(n)$ to produce an output y_{volt} may be defined by

$$y_{volt} = h_o + H_1[x(n)] + H_2[x(n)] \tag{3}$$

where h_o is a bias term $= a_j$,

$H_1[x(n)] = x^t(n)\, b_j$,

$H_2[x(n)] = x^t(n)\, Q_j\, x(n)$,

b_j is a vector of linear coefficients and,

Q_j is a (diagonally symmetric) matrix of quadratic filter coefficients.

The equivalence of the last part of (3) to (2) is obvious. The effect of the linear term is merely to locate the centers of the clusters. We may, therefore, define a new distance measure

$$\begin{aligned} d_j &= f(y_{volt}) \\ &= f(a_j + x^t(n)b_j + x^t(n)Q_j x(n)) \end{aligned} \tag{4}$$

where

$$f(\theta) = \frac{1}{1 + e^{-\theta}}$$

The signoid function $f(\theta)$ is an additional feature of the classification process, which is not strictly necessary but does tend to yield better performance in the presence of noise [8]. The various parameters (a_j, b_j and Q_j) are simply updated using the stochastic gradient adaptation method since these parameters are linear with respect to the output.

The advantage of this formulation is that it leads to the same solution as the Mahalanobis classifier but there is no necessity to deal with a matrix inverse. Therefore the problem of numerical ill-conditioning does not exist.

The result shown in Figure 4 provides bit error rate curves for various equalization approaches. The channel used was the one previously used to generate Figure 3. The curves show the results obtained by a linear equalizer, the Mahalanobis classifier and a Euclidean classifier using the known cluster centers.

This last example, although not practical, provides the optimum achievable non-linear performance. The first point to note is that the elliptical technique obtains results which are essentially the same as the optimal classifier. The performance of the linear and non-linear equalizers converge at low levels of noise but the non-linear technique shows substantial advantage at high additive noise levels. The Volterra result is not shown explicitly here as it is the same as the Mahalanobis distance result.

The main points to note here are that the Volterra classifier does, in fact, exactly realize the optimum non-linear equalizer. The computational complexity is greatly reduced relative to the original formulation, particularly for higher input dimension. The final point is that the adaptation process is simple.

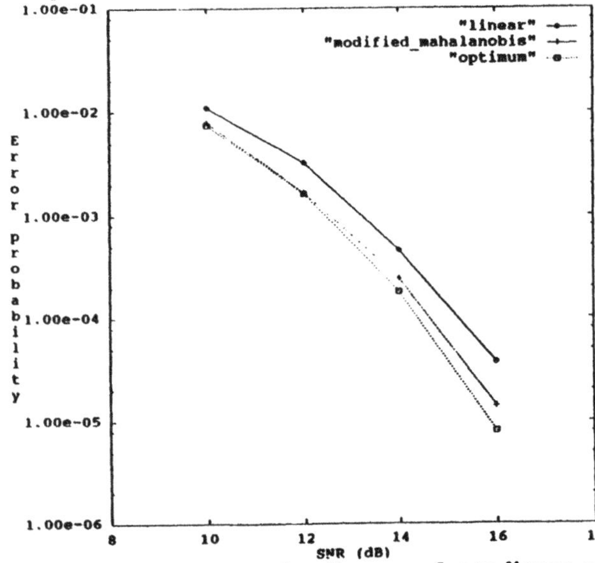

Figure 4 Bit error rate curves comparing linear and non-linear equalizers.

References

[1] S. U. H. Qureshi, "Adaptive Equalization," *Proc. IEEE*, Vol. 73, pp. 1349-1387, Sept. 1985.

[2] S. B. Weinstein, "Echo cancellation in the telephone network," *IEEE Comm. Soc. Mag.*, pp. 9-15, Jan. 1977.

[3] C. F. N. Cowan and P. M. Grant, *Adaptive Filters*, Prentice Hall: Englewood Cliffs, NJ, 1985.

[4] G. J. Gibson, S. Siu and C. F. N. Cowan, "The application of nonlinear structures to the reconstruction of binary signals," *IEEE Trans. Signal Processing*, Vol. 39, pp. 1877-1884, Aug. 1991.

[5] C. F. N. Cowan, "Communications equalization using non-linear adaptive networks," in *Proc. 25th IEEE Asilomar Conf. on Signals, Systems & Computers*, Pacific Grove, CA, pp. 199-202, Nov. 1991.

[6] S. Theodoridis, C. M. S. See and C. F. N. Cowan, "Nonlinear channel equalisation using clustering techniques," in *Proc. IEEE Int. Conf. on Communications*, Chicago, pp. 343.5.1-343.5.4, June 1992.

[7] C. P. Callender and C. F. N. Cowan, "Equalisers for digital communications using generalised distance measurement," in *Proc. 26th IEEE Asilomar Conf. on Signals, Systems & Computers*, Pacific Grove, CA, pp. 521-525, Oct. 1992.

[8] S. Chen, G. J. Gibson and C. F. N. Cowan, "Adaptive channel equalisation using a polynomial-perceptron structure," *Proc. IEE*, Pt. I, Vol. 137, pp. 257-264, Oct. 1990.

Tutorial 2

Design and applications of analog-digital multirate signal processing

Organised by: Sanjit Mitra, *University of California*
Jose Franca, *Instituto Superior Tecnico*

Chapter 2

Design and Applications of Analog-Digital Multirate Signal Processing

José E. Franca and Sanjit K. Mitra

2.1 A review of multirate concepts [1-3]

2.1.1 Basic components

In addition to the three basic components of a linear single-rate discrete-time system, namely, the unit delay, the adder and the multiplier, a linear multirate discrete-time system employs two other components: the down-sampler and the up-sampler.

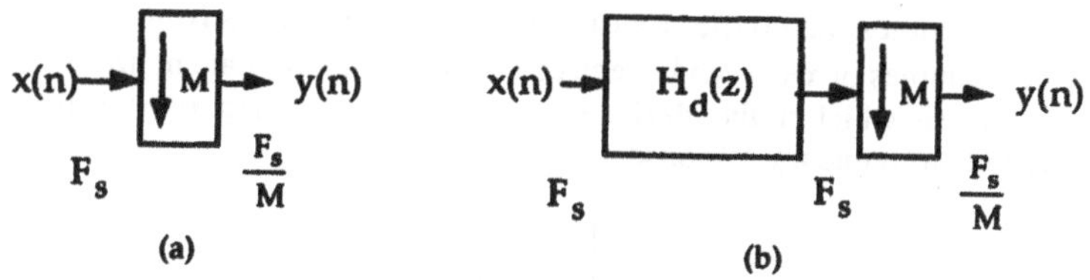

Figure 1 (a) Down-sampler and (b) decimator.

The *down-sampler*, shown symbolically in Figure 1a, is used to generate from a discrete-time sequence $x(n)$ another discrete-time sequence $y(n)$ with a lower sampling rate. The time-domain input-output relation of a down-sampler is given by

$$y(n) = x(Mn) \tag{1}$$

where the down-sampling factor M is a positive integer greater than 1. Thus, if the sampling rate of $x(n)$ is F_S, the sampling rate of $y(n)$ is F_S/M. In the frequency-domain, the input-output relation of a down-sampler is given by

$$Y(e^{j\omega}) = \frac{1}{M}\sum_{k=0}^{M-1} X(e^{j(\omega-2\pi k)/M}) \tag{2}$$

where $X(e^{j\omega})$ and $Y(e^{j\omega})$ denote the Fourier transforms of $x(n)$ and $y(n)$, respectively. As indicated by equation (2), $Y(e^{j\omega})$ is the sum of M frequency scaled and shifted images of $X(e^{j\omega})$ with adjacent images of $Y(e^{j\omega})$ separated by 2π. Thus, unless $X(e^{j\omega})$ is band-limited to the region $0 \le \omega \le \pi/M$, there will be overlap of adjacent terms in equation (2) causing aliasing. As a result, to prevent any aliasing that may be caused by the down-sampling process, $x(n)$ is usually passed through a lowpass filter approximating the ideal frequency response

$$H_d(e^{j\omega}) = \begin{cases} 1 & |\omega| \le \pi / M \\ 0 & \pi / M < \omega < \pi \end{cases} \tag{3}$$

before it is down-sampled as indicated in Figure 1b. The combination of an anti-aliasing filter and a down-sampler is usually called a *decimator* and the process of lowering the sampling rate by this arrangement is called *decimation*.

The device used to generate a discrete-time sequence $x^u(n)$ with a higher sampling rate from another discrete-time sequence $x(n)$, called the *up-sampler*, is shown in Figure 2a. The time-domain input-output relation of the up-sampler is given by

$$x^u(n) = \begin{cases} x\left(\frac{n}{L}\right) & n = 0, \pm L, \pm 2L, \ldots \\ 0 & \text{otherwise} \end{cases} \tag{4}$$

where the up-sampling factor L is a positive integer greater than 1. Thus, if F_S is the sampling rate of $x(n)$, the sampling rate of $x^u(n)$ is LF_S. In the frequency-domain the corresponding relation is given by

$$X^u(e^{j\omega}) = X(e^{j\omega L}) \tag{5}$$

where $X^u(e^{j\omega})$ denotes the Fourier transform of $x^u(n)$. It follows from equation (5) that the zero-valued samples of $x^u(n)$ can be replaced with their appropriate nonzero values by an interpolation process which in the ideal case is achieved by passing $x^u(n)$ through an ideal lowpass filter with a frequency response

$$H_u(e^{j\omega}) = \begin{cases} L & |\omega| \le \pi / L \\ 0 & \pi / L < \omega < \pi \end{cases} \tag{6}$$

The combination of an up-sampler and the lowpass filter, shown in Figure 2b, is usually called an *interpolator* and the process of increasing the sampling rate by this arrangement is called *interpolation*.

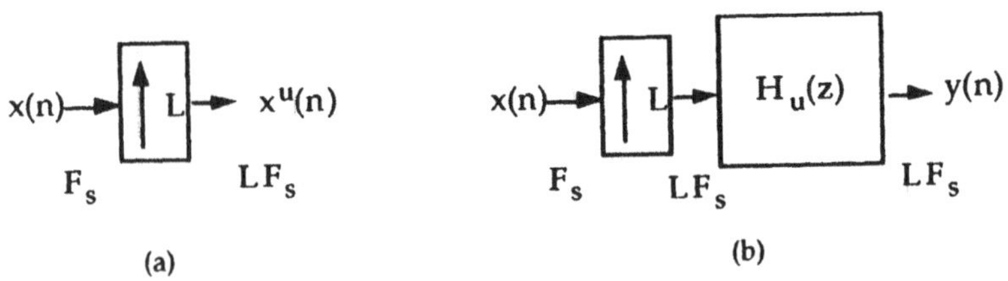

Figure 2 (a) Up-sampler and (b) interpolator.

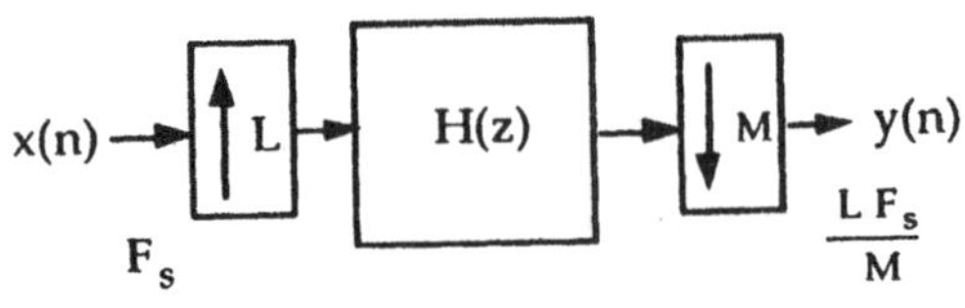

Figure 3 Sampling rate alteration by a rational factor.

A cascade of an up-sampler, a lowpass filter and a down-sampler, as indicated in Figure 3, can be employed to alter the sampling rate of a discrete-time signal $x(n)$ by any rational factor L/M.

2.1.2 Polyphase implementation [4]

In the implementation of the lowpass filter in the decimator of Figure 1b, the arithmetic operations are carried out at the sampling rate F_s of the input rather than at the lower sampling rate F_s/M of the output. A computationally efficient structure for the implementation of the decimator can be obtained by making use of the polyphase representation of the lowpass filter:

$$H_d(z) = \sum_{i=0}^{M-1} z^{-i} G_i(z^M) \tag{7}$$

resulting in the structure of Figure 4 where now the arithmetic operations are carried out at the lower sampling rate F_s/M. In the case of an FIR transfer function $H_d(z)$, the polyphase decomposition can be derived by inspection. On the other hand, in the case of an IIR transfer function $H_d(z) = N(z)/D(z)$, the polyphase decomposition is obtained by multiplying the numerator and the denominator with an appropriate factor $Q(z)$ such that the modified transfer function can be expressed as

$$H_d(z) = \frac{N(z)Q(z)}{D(z)Q(z)} = \frac{N'(z)}{D'(z^M)} = \frac{\sum_{i=0}^{M-1} z^{-i} N_i(z^M)}{D'(z^M)} \tag{8}$$

For certain IIR transfer functions, a more efficient polyphase type of implementation can be obtained with $G_i(z)$ as stable allpass transfer functions [5, 6]. A similar structure can be derived for the efficient implementation of an interpolator.

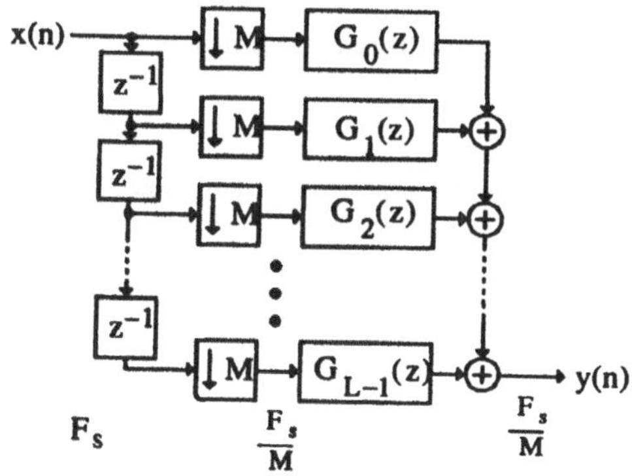

Figure 4 Computationally efficient implementation of a down-sampler.

2.1.3 Quadrature-mirror filter (QMF) bank

A novel multirate structure, developed primarily for efficient data compression, is the M-channel quadrature mirror filter (QMF) bank shown in Figure 5. In this structure, the input discrete-time signal $x(n)$ is passed through an analysis filter bank consisting of a set of M band pass filters $H_i(z)$ decomposing $x(n)$ into a set of M subband signals $x_i(n)$ occupying contiguous nonoverlapping frequency bands of width $2\pi/M$. The subband signals are then down-sampled by a factor of M. Next, these down-sampled signals are up-sampled by a factor of M and passed through a synthesis filter bank composed of a set of M bandpass filters $F_i(z)$ to remove the images resulting in an output signal $y(n)$ whose sampling rate is the same as that of the input $x(n)$. The input-output relation in the z-domain is given by

$$Y(z)=\frac{1}{N}\sum_{k=0}^{N-1} F_k(z)\sum_{i=0}^{N-1} H_k(zW_N^i)X(zW_N^i) \tag{9}$$

The analysis and synthesis filters can be designed to achieve a perfect reconstruction at the output implying that the output is a delayed and scaled replica of the input with complete aliasing cancellation [2]. In this case, equation (9) reduces to $Y(z)=z^{-n_0}X(z)$.

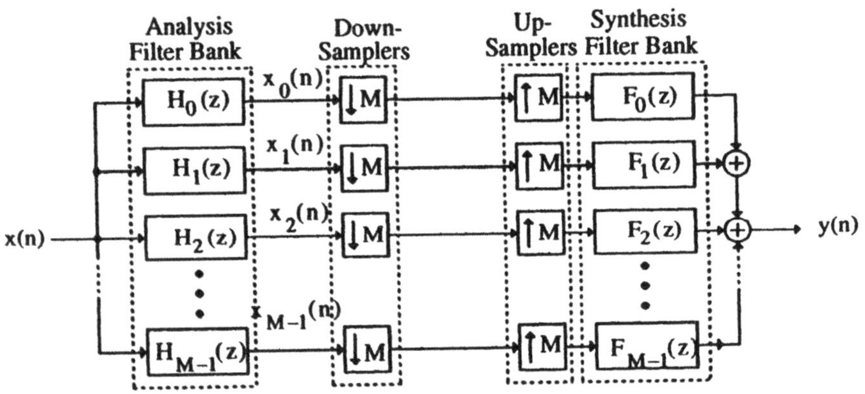

Figure 5 Quadrature-mirror filter (QMF) bank.

2.2 Basic aspects of analog multirate techniques

2.2.1 *Introduction*

Switched-capacitor decimation and interpolation circuits were introduced in the early 80's as a means of relaxing the requirements of continuous-time pre- and post-filters in the context of a traditional SC filter system [7 - 10]. Then, the systematic application of the concepts of multirate discrete-time signal processing led to the development of new classes of multirate SC circuits [11 - 19] and their utilisation for the realisation of a variety of signal processing functions [20 - 23].

Usually, such SC decimator and interpolator circuits employ more complex multiphase digital clock generation circuits than traditional single rate SC filters, but they may lead to considerable gains in the analog circuitry, namely with respect to silicon and power consumption and also to the performance behaviour under non-ideal characteristics of the amplifiers. There are even some applications whose practical implementation in IC form can only be achieved using such multirate SC circuits [24 - 27]. The need for analog and non-trivial digital circuitry for implementing multirate SC signal processing is compatible with the technologies which are currently available to enable the strategic objective of realising on a single chip all aspects of signal processing, both analog and digital.

In this section we provide an overview of the most popular circuit techniques and design methodologies for multirate SC building blocks. First we consider the design of finite impulse response (FIR) decimators and interpolators with single-amplifier architectures which are rather simple to design but which may have practical limitations considering the maximum speed of operation as well as the complexity of the IC realization. To overcome such limitations, we consider also SC decimator circuits based on multi-amplifier architectures which have much more favorable requirements with respect to component count, speed of the OA's, and complexity of the digital switching waveform generator.

The design of multirate SC filters with infinite impulse response (IIR) discrete-time transfer functions is also addressed, both for the case of decimating building blocks and for the case of the complementary interpolating building blocks. In later sections we also discuss the application of such building blocks for designing narrow and very narrow passband bandpass filters, video filters, and front-end filters for magnetic-disk read channels.

2.2.2 *Direct-form polyphase FIR SC decimators and interpolators [12]*

(i) Direct-form SC decimators

The simplest architecture for implementing a sampled-data decimator is the direct-form polyphase structure introduced in the previous section, and which is repeated in Figure 6 for convenience.

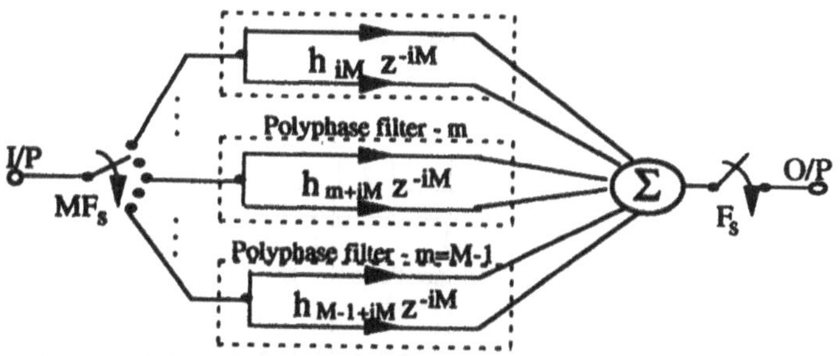

Figure 6 Direct-form polyphase structure for decimation.

The implementation in SC form of the polyphase structure in Figure 6 can be divided into two parts, namely the transmission factors and the accumulator. The implementation of the transmission factors is based on classical SC elements with only switches and capacitors. The simpler SC elements implement the terms with transmission factors h_m, and correspond to a situation where the input signal is sampled during the reference period. Such SC elements can be realised using conventional SC branches switched in the normal manner, which means that during the reference period the time slot for sampling precedes the time slot for charge transfer. The toggle-switch-inverter branch (TSI) in Figure 7a realises positive coefficients whereas the open-floating-resistor branch (OFR) in Figure 7b realises negative coefficients. An alternative realisation of the terms with negative coefficients uses the parasitic-compensated-toggle-switched-capacitor branch (PCTSC) in Figure 7c. The delays of these SC branches refer to the reference period in Figure 7d, and are determined with respect to the sampling slots. The SC elements in Figure 7 are canonic in the sense that they utilise only one SC branch with one capacitor in order to realize one term of the FIR impulse responses[1].

The SC elements that implement the terms with transmission factors $h_{m+M} \cdot z^{-M}$ correspond to a situation where the input signal is sampled in the period preceding the reference period. Such SC elements can be realised by inverting the sequence of sampling and charge transfer, during the reference period. This operation can only be achieved using TSI and PCTSC branches because in such branches sampling and charge transfer take place in different time slots.

The TSI in Figure 8a realises positive coefficients, while the PCTSC in Figure 8b realises negative coefficients. It is easy to see that these SC elements are also canonic, as before. Finally, in order to implement terms with transmission factors $h_{m+2M} \cdot z^{-2M}$, $h_{m+3M} \cdot z^{-3M}$, and so forth, it is necessary to introduce SC branches with reduced sampling rate that can sample the input signal at two, three or more periods before the reference period. But because of the reduced sampling rate it is also necessary to have multiple SC branches to ensure that such terms are realised every reference period. Such a scheme is illustrated in Figure 9 for a transmission factor $h_{m+2M} \cdot z^{-2M}$.

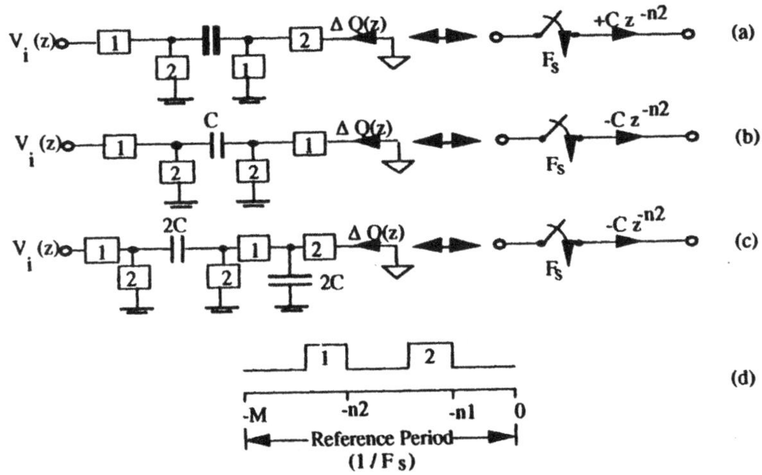

Figure 7 SC elements for the first term of the polyphase filters using a normal sequence of sampling followed by charge transfer.

The SC elements consist of two parallel SC branches, with reduced sampling rate $F_s/2$ and with an inverted sequence of charge transfer followed by sampling. We should note that although this SC element is noncanonic with respect to the number of SC branches and capacitors, it is still a direct-form implementation since it realizes only one term of the FIR impulse response.

Besides the transmission factors implemented by the SC branches discussed above, the polyphase structure of Figure 6 also includes an SC accumulator that realises the addition of the packets of charge $\Delta Q_n(z)$ produced by the SC elements, and can be implemented using the structure shown in Figure 10.

---

[1]We assume the PCTSC branch with two capacitors $2C$ as equivalent to an SC branch with one capacitor C.

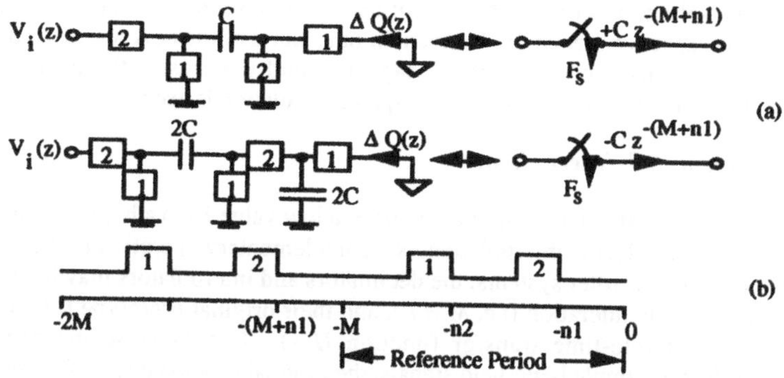

Figure 8 SC elements for the second term of the polyphase filters using an inverted sequence of charge transfer followed by sampling.

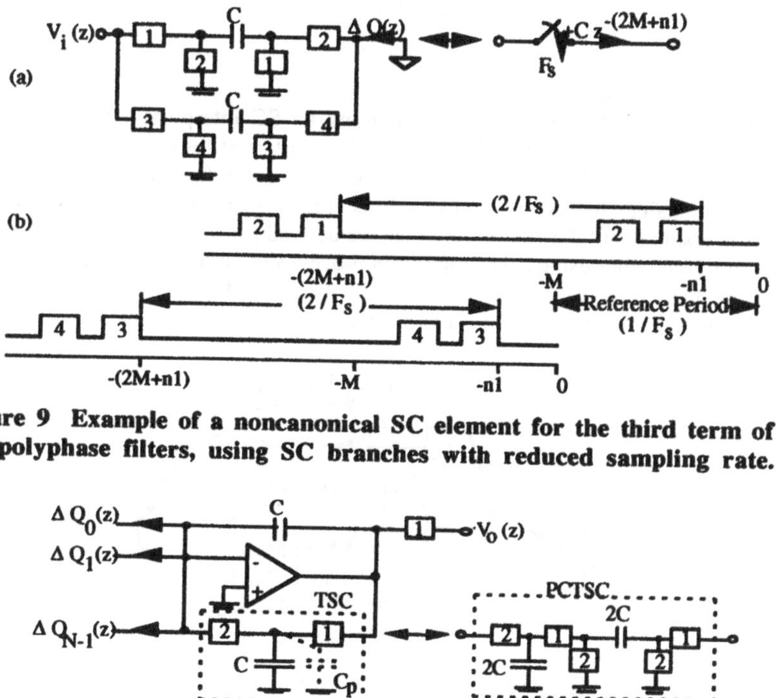

Figure 9 Example of a noncanonical SC element for the third term of the polyphase filters, using SC branches with reduced sampling rate.

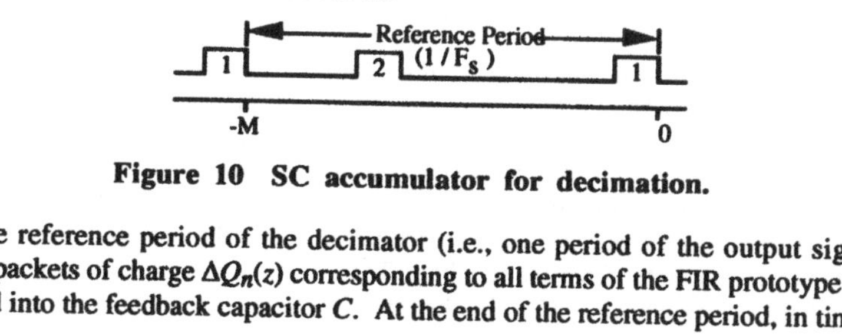

Figure 10 SC accumulator for decimation.

During the reference period of the decimator (i.e., one period of the output signal), the incoming packets of charge $\Delta Q_n(z)$ corresponding to all terms of the FIR prototype filter are transferred into the feedback capacitor C. At the end of the reference period, in time slot 1, the new output voltage is sampled by the following SC circuit and also by the feedback toggle-switched-capacitor branch (TSC); then, in time slot 2 of the next reference period, the TSC transfers its charge to the feedback capacitor and thus resets the previous output voltage. At the same time, a new cycle of charge accumulation is executed.

(ii) Direct-form SC interpolators

The operation of increasing the sampling rate from a low value F_S to a higher value LF_S (L is an integer greater than 1), i.e. interpolation, is a complementary operation to the decimation described above. In SC filter systems, the decimators and interpolators may have the same factor of sampling rate alteration (i.e. $M = L$), and their original prototype filters may also have the same discrete-time transfer function $H(z)$. But, because the operation of interpolation refers to sampled-and-held signals, the original prototype filter for interpolation should be modified according to

$$H'(z) = H(z) \sum_{l=0}^{L-1} z^{-1} \tag{10}$$

where the unit delay period is $1/LF_S$.

The simplest architecture for implementing an SC interpolator is the polyphase structure of Figure 11. Each one of the L polyphase filters at lower rate F_S produces one output sample determined by its transfer functions $H_l(z)$, and then the output signal of the interpolator is obtained by resampling the outputs of the polyphase filters at higher rate LF_S. The impulse responses of the polyphase filters are derived from the impulse response of the modified prototype filter using a decomposition process similar to that for the decimator. In this structure, the outputs of the polyphase filters only matter during the period of time when they are connected to the output of the interpolator.

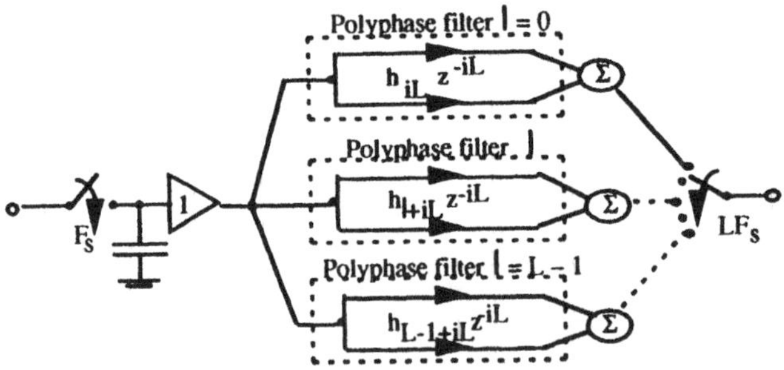

Figure 11 Polyphase interpolator structure with direct-form polyphase filters.

It is possible to implement the above polyphase interpolating structure using only one time-shared accumulator, as shown in Figure 12, for the simple case of $L = 2$. During one reference period of the interpolator (i.e., one period of the input signal), there are $L = 2$ cycles of the operation of the SC accumulator, each of which produces one sample of the output interpolated signal. Firstly, in time slot 0, the charge packets of the polyphase filter $l = 0$ are transferred to the output of the accumulator, while the previous output voltage is reset by the charge transfer from the feedback TSC. In time slot 2, the new output voltage is sampled by the feedback TSC. In the second cycle, the packets of charge of the polyphase filter $l = 1$ are transferred to the output of the accumulator in time slot 1, while the previous output voltage is reset by the charge transfer from the feedback TSC. This second cycle ends with time slot 3, when the new output voltage is sampled by the feedback TSC. With such an arrangement of the switch phasing the output of the interpolator is implicitly sampled and held, because the output voltage changes only during time slots 0 and 1. For implementing the terms of the polyphase filters in the structure of Figure 11 we can also employ the basic SC elements described in Figures 7, 8, and 9. However, we have now to consider that the delay terms are defined with respect to the charge transfer slots whereas in the previous decimating structures they were defined with respect to the sampling slots.

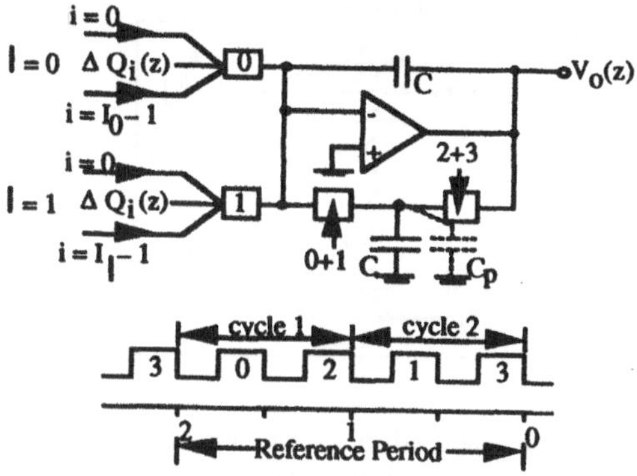

Figure 12 A simple time-shared SC accumulator for interpolation (L = 2).

2.2.3 *Active-delayed block polyphase FIR SC decimators [17]*

We observed above that an FIR polyphase decimator is characterised by the length N of its impulse response and the decimation factor M determining the number of polyphase filters. In the specific type of polyphase architectures to be considered now, the polyphase filters share a number of common blocks to realise the basic operations of coefficient multiplication, delay and accumulation using minimum hardware. They will, therefore, be referred to as active-delayed-block (ADB) polyphase architectures.

The ADB polyphase architecture schematically illustrated in Figure 13 consists of B common blocks [B is the smaller integer equal to or greater than (N / M)], each of which realises a *single term* of the associated transfer function of the polyphase filters. The first block b_0, realises the terms h_m of the polyphase filters ($m = 0, 2, ..., M - 1$) which have an overall zero delay. The second block b_1, realises the terms of the polyphase filters which have an overall delay of z^{-M} and whose coefficients are h_{m+M}. In general, the block b_j (j=0,1,..., [$B - 1$]) realises the terms of the polyphase filters which have an overall delay of z^{-jM} and whose coefficients are h_{m+jM}, subject to the condition $m + jM \leq N$ determined by the length N of the overall impulse response of the decimator. A total of [$B - 1$] common blocks, each with an overall delay of z^{-M}, and an additional common block directly connected to the output accumulator are required to implement this type of polyphase architecture. As in the previous decimating architectures, the input SC branches are usually implemented using either the TSI branch, for positive coefficients, or the PCTSC branch, for negative coefficients. Also as before, the typical SC implementation of an accumulator employs an OA with a feedback integrating capacitor (C_I) and reset network, such that the accumulation of the packets of charge produced by the input SC branches is scaled by the charge-to-voltage conversion factor ($- 1/C_I$).

By combining such an SC accumulator together with a simple output TSI branch, with equivalent capacitance value D, we obtain the typical implementation of an SC active-delayed block illustrated in Figure 14 whose charge-to-charge conversion factor becomes $(D/C_I)^2$.

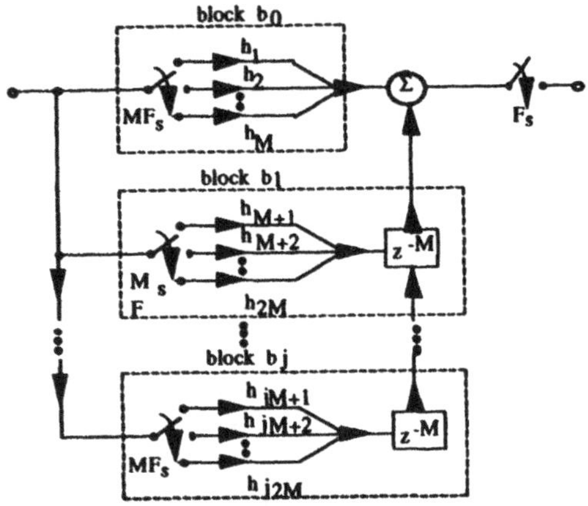

Figure 13 Schematic representation of the architecture of an ADB polyphase decimator.

Hence, based on such circuits we can see that the first block - b_0 of the ADB architecture is formed by M SC branches with capacitors C_m ($m = 0, ..., M - 1$), and are directly connected to the output accumulator with integrating capacitor C_{I0}. Each remaining block b_j is also formed by M input SC branches with capacitors C_{m+jM} and one SC active-delayed block with integrating capacitor C_{Ij} and output TSI branch with capacitor D_j. The design of such an architecture is illustrated below.

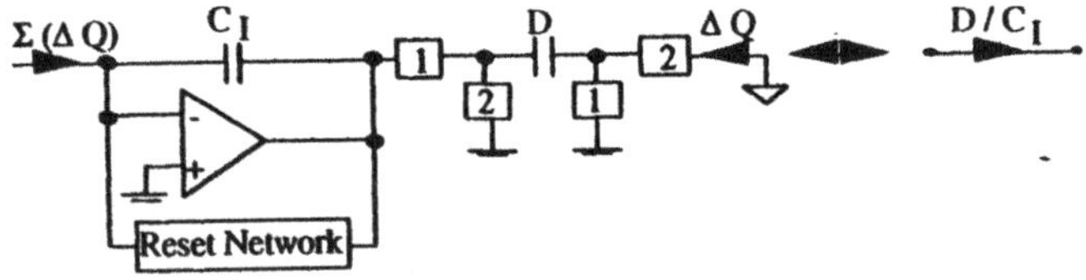

Figure 14 Charge-to-charge conversion factor of an SC active-delayed block.

In Figure 15 we illustrate the example of an ADB polyphase FIR SC decimator, with $N = 19$ and $M = 5$, and where five polyphase filters are implemented using four common blocks. The sign of the impulse response coefficients defines the type of SC branch to be employed at the input of the four common blocks, i.e., TSI branches for positive coefficients and PCTSC branches for negative coefficients. The capacitance values of the input SC branches, on the other hand, are obtained from the magnitude of the impulse response coefficients.

---

2 For simplicity, we disregard here the delay factors.

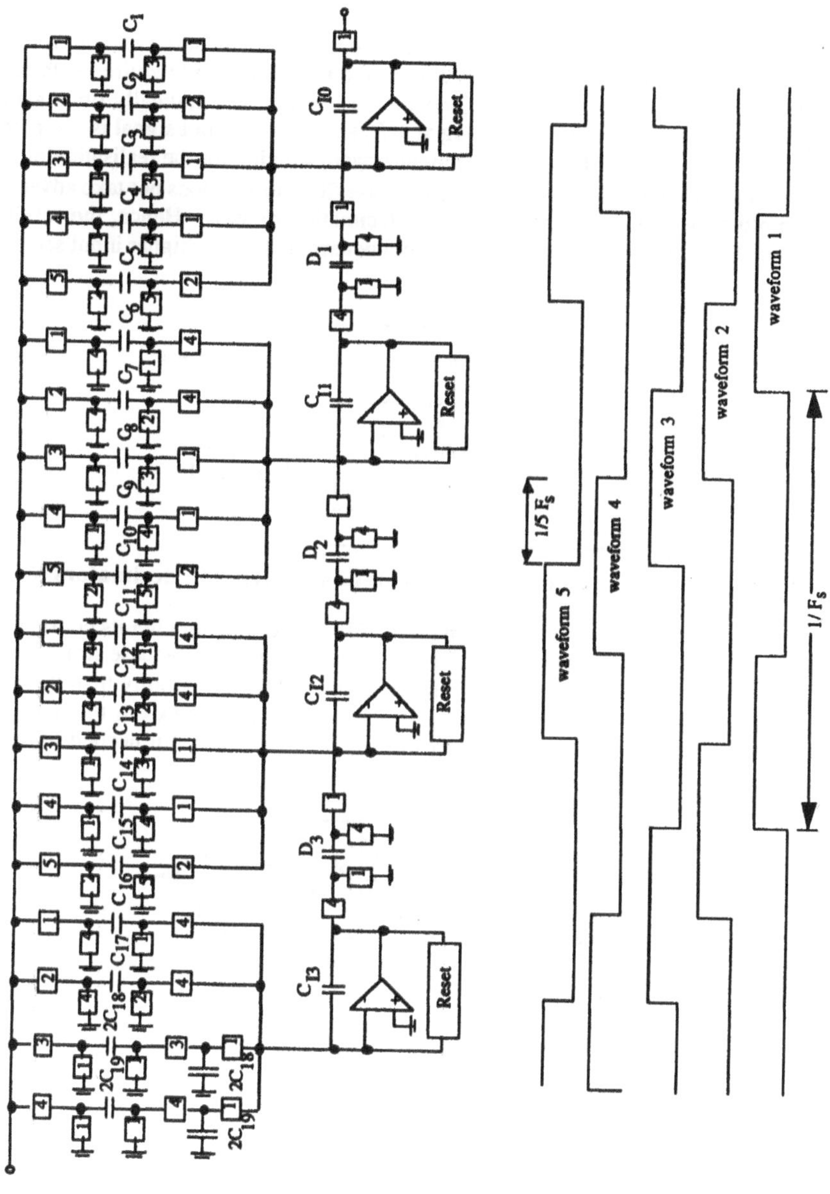

Figure 15 FIR SC decimator circuit with ADB polyphase architecture $N = 19$; $M = 5$.

2.2.4 IIR SC decimation and interpolation building blocks [13, 15]

(i) IIR SC decimators

Infinite impulse response (IIR) SC decimators can be implemented based on *non-optimum* and *sub-optimum* classes of circuits, according to the resulting speed requirements of the OA's. The non-optimum class of IIR SC decimator circuits, schematically illustrated in Figure 16, consists of an SC filter operating at MF_S and whose output signal is sampled at the lower rate F_S. Although this offers a straightforward solution for the realisation of high selectivity amplitude responses using simple bi-phase SC filters, it does not take advantage of the sampling rate reduction inherent to the decimation process. Hence, the resulting speed of the OA's and capacitance spread are both determined by the higher input sampling rate MF_S.

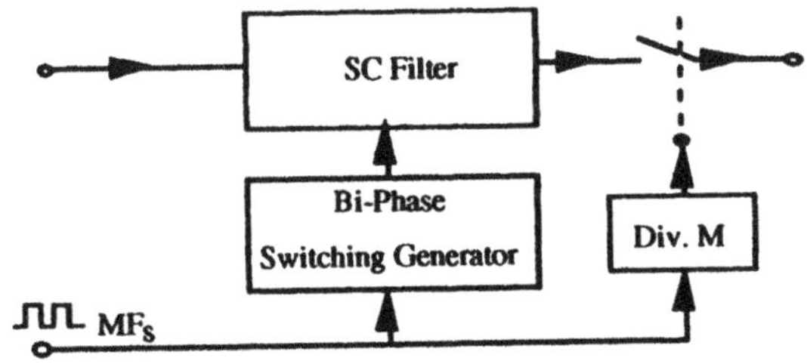

Figure 16 Non-optimum implementation of IIR SC decimators.

In Figure 17 we represent the sub-optimum class of IIR SC decimator circuits consisting of the cascading of an SC filter operating at MF_S and a non-recursive polyphase structure with a decimating factor of M. This architecture gives greater design flexibility than the above non-optimum class of IIR SC decimators, and also allows slower OA's in the polyphase structure with output sampling rate F_S. However, faster OA's are still needed in the SC filter operating at the higher sampling rate MF_S.

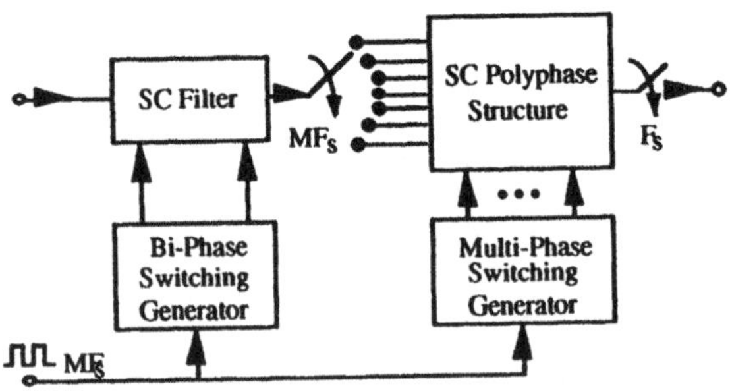

Figure 17 Sub-optimum implementation of IIR SC decimators.

In order to overcome the speed limitations of the above architectures we consider instead the optimum IIR SC decimator with a sampling rate reduction factor of M, schematically illustrated in Figure 18. This consists of a single network whose input sampling rate is MF_S and whose output sampling rate is F_S. The amplitude response of such network is designed to meet the desired passband specifications and, at the same time, achieve the required rejection of the unwanted alias signal components associated with the signal sampled at F_S.

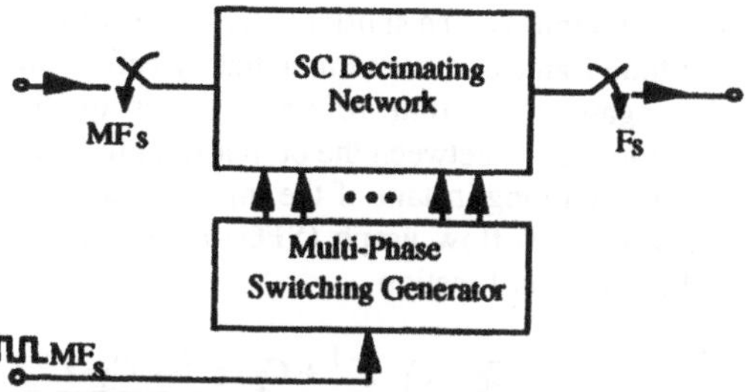

Figure 18 Optimum implementation of IIR SC decimators.

Such an optimum IIR SC decimator is illustrated in Figure 19a, together with the associated switch timing given in Figure 19b, for a simple example of a first-order SC lowpass decimator with a decimating factor of $M = 4$. The OA with integrating capacitor C_I and feedback SC branch C_F is viewed as a first-order charge-to-voltage converter, while the input SC branches C_i (i = 0 to 3) are viewed as voltage-to-charge converters and are organised as in a polyphase structure.

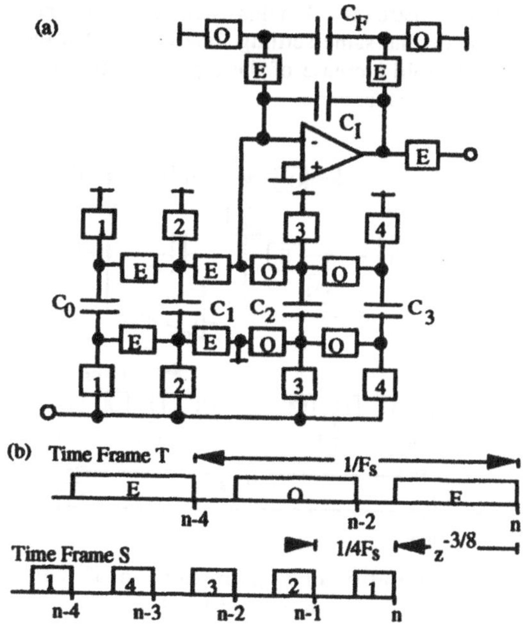

Figure 19 Example of an optimum first-order SC decimator with M = 4 (a) circuit (b) time frames.

We shall now discuss in more detail the adopted switch timing arrangement given in Figure 19b. Time frame T, for charge transfer, has two wide time slots like a conventional SC integrator operating at F_s, which determine the operation of the switches controlling the flow of charge through the virtual ground of the amplifier. On the other hand, the four time slots in time frame S, for signal sampling, control the switching operation of the switches which are responsible for the sampling of the input signal. The width of such time slots is simply limited by the equivalent RC time constants associated with the charging of each one of the input switched-capacitors, and thus can be substantially narrower than the time slots in time frame T. Time slot 1, in time frame S, occurs immediately before time slot E in time frame T. Hence, by referring again to the time reference given in Figure 19b, we have an unimportant flat delay factor of $z^{-3/8}$ between the output sampling instant (at the end of time slot E) and the most recent sampling instant of the input signal (at the end of time slot 1). By employing well-known signal flow graph (SFG) techniques for the analysis of SC networks, we arrive at the z-transfer function

$$T(z) = z^{-3/8} \cdot \frac{C_0 + C_1 \cdot z^{-1} + C_2 \cdot z^{-2} + C_3 \cdot z^{-3}}{(C_I + C_F) - C_I \cdot z^{-4}} \tag{11}$$

In the above SC decimator building block, the switching of the SC branches is devised such that the sampling of the input signal is performed during a short time interval, whereas the transfer of charge is performed during a much wider time interval to maximise the time allowed for amplifier settling. In comparison with the non-optimum implementation of this SC decimator, shown in Figure 20 for clarity, the SC decimator building block of Figure 19 employs a larger number of SC branches and switching waveforms. However, we now obtain a significant M-fold increase ($M = 4$ in this example) of the time allowed for amplifier settling. In other words, for the same settling time of the OA, the proposed circuit is capable of achieving an M-fold increase of the input sampling rate compared with the traditional non-optimum implementation.

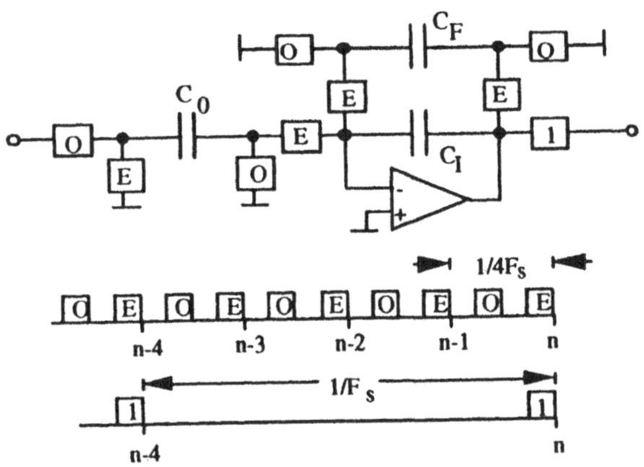

Figure 20 Non-optimum implementation of a first order SC decimator with M = 4.

The above technique can also be extended to the design of second-order SC decimator building block, as illustrated in Figure 21a. This employs a basic TIL structure which is responsible for the realisation of the denominator function, and where capacitor E (capacitive damping) and switched-capacitor F (resistive damping) represent two alternative forms of damping that can be utilised in the loop. In a similar fashion as before, this TIL structure is viewed as a second-order charge-to-voltage converter whose input packets of charge are produced by the two sets of SC branches, set-X and set-Y connected to the virtual ground of OA1 and OA2.

These SC branches, organised in the form of polyphase structures, are represented by polygonal symbols where we indicate the time slot for input signal sampling, the time slot for output signal sampling, and the capacitance value of the branch as well as the polarity of charge transfer. For greater design flexibility we shall always use different input and output time slots.

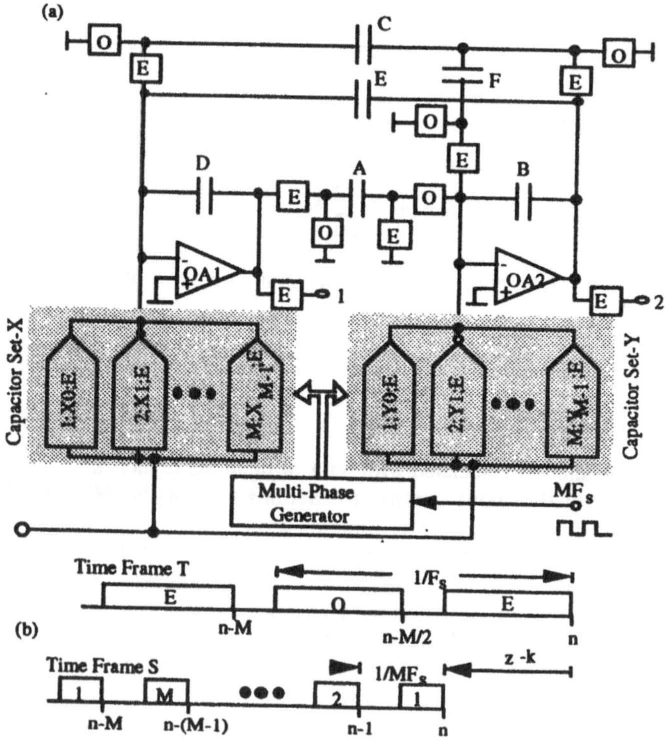

Figure 21 Optimum implementation of a second-order SC decimator (a) general structure (b) SC branches (c) time frames.

Hence, PCTSC branches are utilised for positive charge transfers whereas TSI branches are employed for negative charge transfers. The time frames controlling the operation of the SC network, in Figure 21b, are designed to maximise the time allowed for amplifier settling. Two wide time slots, in the charge transfer time frame *T*, control the flow of charge through the virtual ground of the amplifiers. The sampling of the input signal is controlled by the sampling time frame *S* with much narrower time slots that do not affect the speed requirements of the amplifiers. The delay terms associated with each input SC branch are determined by the sampling instants at the end of the respective input time slot *relative to the sampling instant at the output of the amplifier to which they are connected.* The term z^{-k} represents a flat delay between the first instant of input signal sampling at the end of time slot 1, in time frame *S*, relative to the output sampling instant at the end of time slot *E*, in time frame *T*. As before, by making use of SFG analysis techniques we arrive at the following overall *z*-transfer functions

$$T_1(z) = \frac{[(B+F)\cdot X(z)-(C+E)\cdot Y(z)]+[E\cdot Y(z)-B\cdot X(z)]\cdot z^{-M}}{(B\cdot D+D\cdot F)+(A\cdot C+A\cdot E-D\cdot F-2B\cdot D)\cdot z^{-M}+(B\cdot D-A\cdot E)\cdot z^{-2M}}\cdot z^{-k} \tag{12a}$$

for terminal 1, and

$$T_2(z) = \frac{[D \cdot Y(z)] + [A \cdot X(z) - D \cdot Y(z)] \cdot z^{-M}}{(B \cdot D + D \cdot F) + (A \cdot C + A \cdot E - D \cdot F - 2B \cdot D) \cdot z^{-M} + (B \cdot D - A \cdot E) \cdot z^{-2M}} \cdot z^{-k} \quad (12b)$$

for terminal 2. For conciseness, we have represented by

$$X(z) = \sum_{u=0}^{M-1} X_u \cdot z^{-u} \quad (13a)$$

the voltage-to-charge conversion function associated with set-X of input SC branches, and by

$$Y(z) = \sum_{v=0}^{M-1} Y_v \cdot z^{-v} \quad (13b)$$

the corresponding function associated with set-Y of input SC branches.

For each one of the above z-transfer functions (12) of the SC decimator building block, the coefficients of the z-transfer function are determined as functions of the input switched-capacitors X_u and Y_v, respectively, in capacitor set-X and capacitor set-Y, as well as the capacitors in the loop of the decimator. The adoption of a particular topology, including the selection of the output terminal and type of damping, as well as the design of capacitor sets X and Y, is carried out on a case-by-case basis to minimise the capacitance spread and total overall capacitor area and also to obtain reduced sensitivity of the amplitude response with respect to capacitance ratio errors. Designing for maximum signal handling capability is achieved by using the conventional technique of scaling the voltages at the output of the OA's.

(ii) IIR SC interpolators

IIR SC interpolators have been traditionally implemented based on the non-optimum and sub-optimum classes of circuits defined according to the resulting speed requirements of the operational amplifiers (OA's).

In the alternative type of SC interpolator circuits with optimum implementation, all OA's can operate instead at the lower sampling rate F_s and the realization of arbitrary interpolating functions can be achieved employing a reduced number of switching waveforms and requiring low capacitance spread and total capacitor area.

For conciseness, the type of SC interpolator circuit considered here will be described through the example illustrated in Figure 22a where the sampling rate increases by a factor of $L = 3$ and the interpolating function is of second order. Such SC interpolator consists of a TIL network whose amplifiers *operate at the low input sampling frequency* F_S, together with two output polyphase networks whose characteristics depend on the interpolation factor L and on the complexity of the numerator polynomial function.

The interpolated signal is obtained at the output of an SC accumulator which linearly combines the packets of charge transferred through the branches of the polyphase networks. Such accumulator employs a unity gain buffer for reduced power and silicon consumption and high frequency of operation.

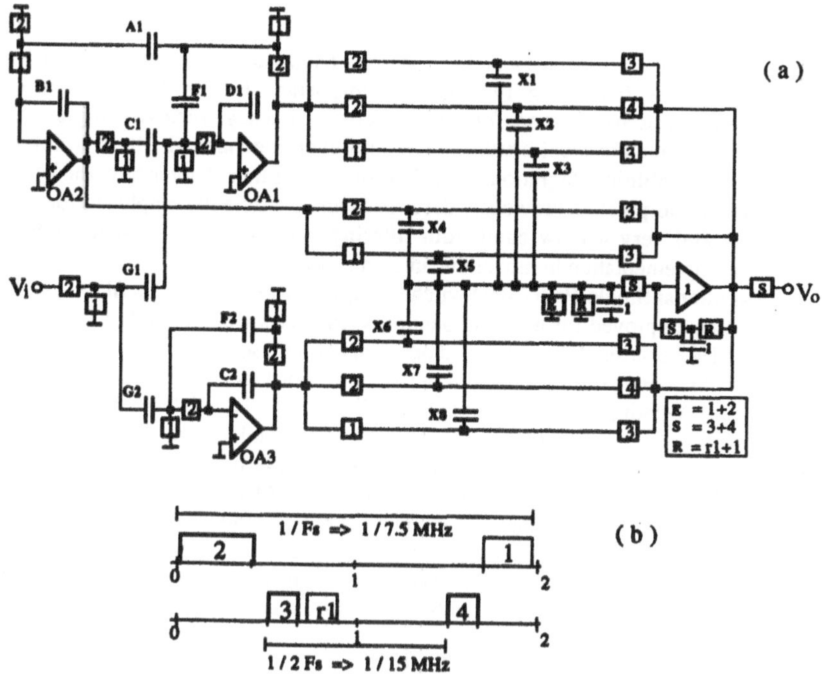

Figure 22 Second-order SC interpolator with $L = 3$ (a) circuit architecture (b) time frames.

The switch timing indicated in Figure 22b is divided into three sets of time slots corresponding to time frames S, T and R. Time slots E and O of time frame S define the input sampling instants and control the operation of the recursive part of the circuit in a similar way as in a conventional biquad. Such time slots are also utilised to define the input sampling of the polyphase networks and whose operation additionally requires the time slots of time frames T and R. While each type of the time slots 0, 1 and 2 of time frame T, used for charge transfer define an interpolated output sample, time slots R_0, R_1 and R_2 of time frame R are utilised to reset the previous output sample of the accumulator. By employing SFG analysis techniques, as before, it can be shown that the overall z-transfer, function is given by

$$T(z)=\frac{G\cdot z^{-1/16}}{D\cdot(B+F)}\cdot\frac{\left[\left(1-z^{-3}\right)\cdot\sum_{i=0}^{4}(X_i\cdot z^{-i})+A\cdot z^{-3}\cdot\sum_{i=0}^{4}(Y_i\cdot z^{-i})\right]}{1+\left(\frac{(A\cdot C-D\cdot F-2B\cdot D)}{D\cdot(B+F)}\right)\cdot z^{-L}+\left(\frac{B\cdot D}{D\cdot(B+F)}\right)\cdot z^{-2L}} \qquad (14)$$

where the unit delay period still refers to the output sampling rate $3F_S$. The factor $z^{-1/16}$ represents an unimportant flat delay between the end of time slot E and the end of time slot 0.

2.3 Multirate systems for high-selectivity filtering

2.3.1 Introduction

Designing high-selectivity, narrow and very narrow bandwidths filter systems in monolithic integrated form is becoming increasingly important for applications in modern communication systems because, for size and cost efficiency, such filtering systems are envisaged to replace classical crystal, SAW and even mechanical filters which are not aminable for monolithic integration. We shall discuss in this chapter how the type of SC decimators and interpolators previously described can be efficiently employed for designing such narrow and very narrow bandwidth filtering systems in a way that renders them attractive for implementation in monolithic integrated form.

Conventional Single-Path (SP) Switched-Capacitor (SC) bandpass filter systems (e.g. [11]) cannot be employed for realising bandpass responses when the relative bandwidth is below about 1% because of the combined problems of high sensitivity and large capacitance spread of the SC bandpass filter used in the system. For this range of relative bandwidths we may utilise *N*-path [28 - 33] or Single-Path Frequency-Translated (SPFT) [21, 22] SC filter systems, both of which make use of the alias and image frequency-translated responses of SC filters in order to achieve highly selective bandpass responses using low sensitivity SC filters with low capacitance spread. After reviewing this process of frequency translation in SC filters we examine typical architectures of the two types of systems and compare them with respect to their ability to achieve specified anti-aliasing and anti-imaging requirements. In order to overcome the disadvantages of the existing systems we then also discuss more sophisticated NP SC systems using bandpass path filters.

2.3.2 Frequency translation in SC filters

(i) Single-path filters

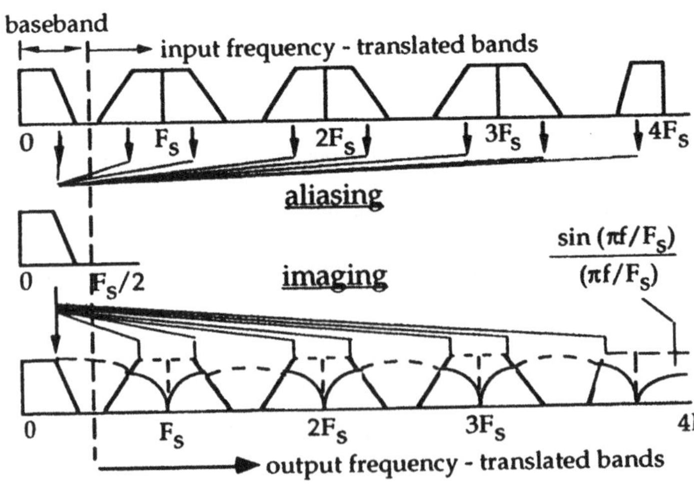

Figure 23 Alias and image frequency-translated responses of an SP SC filter.

A conventional SP SC filter with switching frequency F_S is designed to produce a specified frequency response below the Nyquist frequency $F_S/2$. Owing to its sampled-data nature, the frequency response of the SC filter shows two forms of ambiguity, *aliasing* and *imaging*, whereby at the input and at the output of the SC filter there are a multiplicity of frequency bands above $F_S/2$ that relate to the baseband below $F_S/2$, as illustrated in Figure 23. The output image response is multiplied by the function $[\sin(\pi f/F_s)/(\pi f/F_s)]$ because the output signal is sampled and held with period $1/F_S$. The SC filter is preceeded and followed, respectively, by an Anti-Aliasing Filter (AAF) and an Anti-Imaging Filter (AIF) to correspond only one output frequency band to only one input frequency band, each of which can be either the baseband below $F_S/2$ or any one of the frequency-translated bands above $F_S/2$. The AIF may also compensate for the sample and hold effect when this is necessary.

(ii) N-path filters

Conceptually, an NP SC filter, an example of which for $N = 3$ is shown in Figure 24a is formed by N identical SP SC filters switched in a way such that they are sequentially connected between the input and output terminals during a fraction $1/NF_S$ of the switching period $1/F_S$, and which possess alias and image frequency responses as above. It can be shown that below the Nyquist frequency $NF_S/2$, the multiband response of the NP SC filter is formed by the summation of the frequency-translated components with the same frequency at the input and at the output; the signals with different frequencies at the input and at the output correspond to unwanted frequency-translated components which are ideally eliminated by cancellation [34]. Above the Nyquist frequency $NF_S/2$, the multiband response of an NP SC filter is periodically repeated with NF_S yielding alias and image frequency-translated responses similar to an SP SC filter, as shown in Figure 24b.

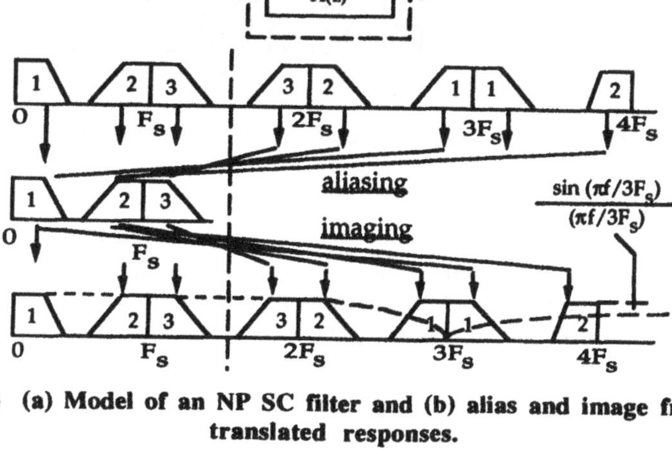

Figure 24 (a) Model of an NP SC filter and (b) alias and image frequency-translated responses.

The image response is multiplied by the function $[\sin(\pi f/NF_S)/(\pi f/NF_S)]$ because the output signal of an NP SC filter is sampled and held with period $1/NF_S$. An AIF and an AAF are also employed in order to select the desired frequency band, respectively at the output and at the input of an NP SC filter, and to reject the unwanted frequency bands that would give rise to ambiguous responses.

In an NP SC filter, the cancellation of the alias and image signals below the Nyquist frequency $NF_S/2$ is extremely sensitive to matching of the filter paths, which in practice is not perfect, giving rise to the generation of unwanted frequency-translated components[3]. The solution of pseudo NP filters [35] in theory eliminates such unwanted frequency-translated components, though in practice it only reduces them, but the pseudo NP SC filters that have been implemented [30, 32, 33, 36] have high noise figures and are not suitable for applications with high dynamic range.

2.3.3 *Review of SC bandpass filter systems*

(i) *General system architecture*

The general architecture of an SC bandpass filter system which produces a single bandpass response is shown in Figure 25[4]. The SC filter, which may be SP or NP, has to be designed bearing in mind the operation mode of the filter system in order to meet given specifications.

The AAF and the AIF comprising continuous-time filters and, respectively, an SC decimator and an SC interpolator, are designed in order to achieve a minimum level of rejection of the unwanted alias and image frequency-translated components, which is also specified.

The lowpass continuous-time filters (maximum 3rd order) attenuate the unwanted signals whose frequencies are much higher than the midband frequency of the system (typically 15 times or more). The SC decimator and the SC interpolator attenuate the unwanted frequency-translated components which are closer to the system midband frequency (e.g. [11, 21, 22]).

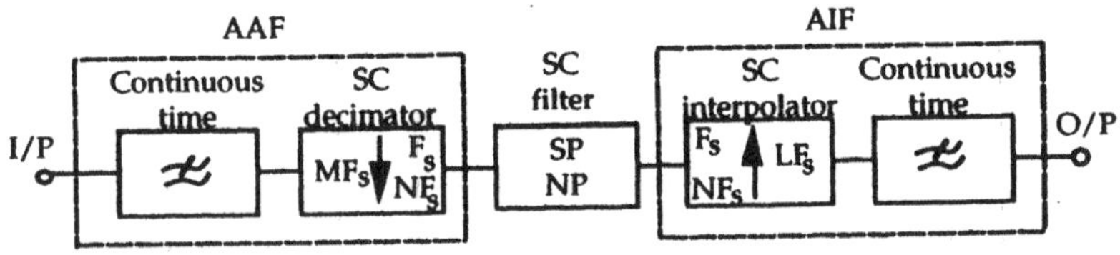

Figure 25 General architecture of an SC bandpass filter system.

(ii) *SP systems*

A conventional SP system using an SC bandpass filter is shown in Figure 26. The selectivity of the filter system corresponds to that of the SC bandpass filter and, as a result, such systems are utilised only for relative bandwidths greater than about 1% because of the problems of high sensitivity and large capacitance spread of the SC bandpass filter.

---

3 Often referred to as a mirror frequencies.

4 We do not consider the case of filter bank systems which produce multiple independent bandpass responses.

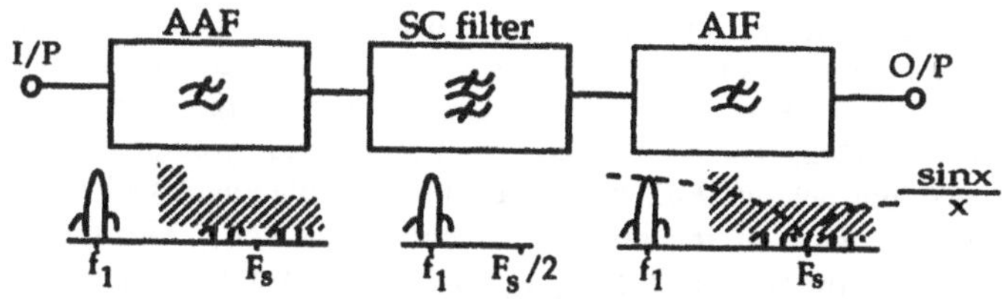

Figure 26 Conventional single-path system.

The SPFT system of Figure 27 implements bandpass responses with relative bandwidths below 1% [21, 22]. It utilises an AAF and an AIF with bandpass, rather than lowpass, responses in order to select an uppper (+) or a lower (–) frequency-translated band around a multiple n of the switching frequency F_s (n = 1, 2,...). For an SC bandpass filter with midband frequency $f_1 = F_s/m$ and Q-factor Q_1 the effective Q-factor of the frequency-translated responses is $Q_0 = K{\cdot}Q_1$, where $K = m{\cdot}n + 1$ is the Q-enhancement factor of the system. Due to the sample and hold effect, the level of the selected output image is multiplied by the factor $[\sin(K/M)/(K/m)]$. SPFT systems preserve the high dynamic range of the high quality low frequency SC bandpass filter that is employed to produce the frequency-translated response [22, 37].

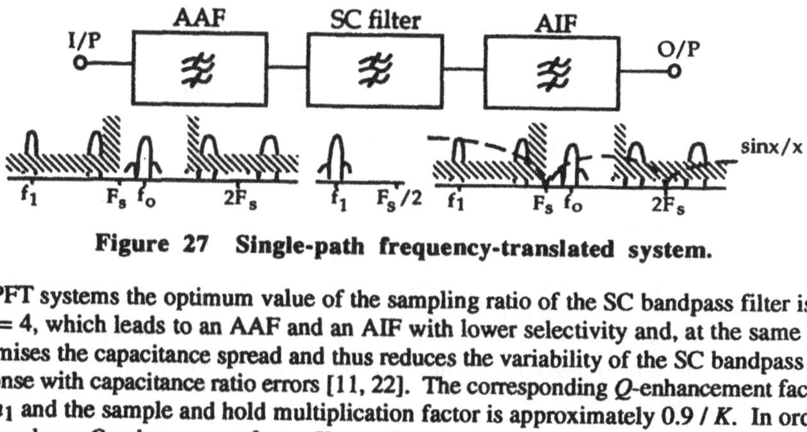

Figure 27 Single-path frequency-translated system.

In SPFT systems the optimum value of the sampling ratio of the SC bandpass filter is $m = F_s/f_1 = 4$, which leads to an AAF and an AIF with lower selectivity and, at the same time, minimises the capacitance spread and thus reduces the variability of the SC bandpass filter response with capacitance ratio errors [11, 22]. The corresponding Q-enhancement factor is K-$4{\cdot}n_1$ and the sample and hold multiplication factor is approximately 0.9 / K. In order to obtain a large Q-enhancement factor K we select the frequency-translated response around a high multiple of the switching frequency, i.e. we choose n high. This reduces the level of the selected output image, the reconstruction of which requires an AIF with higher selectivity. It is practical to design an AIF to compensate for a maximum sample and hold attenuation of about 30 dB ($\approx$ 0.32), and thus the maximum realistic Q-enhancement capability of SPFT systems is about $K_{max} = 28$. This allows the realisation of bandpass responses with relative bandwidths as low as 0.035% ($Q_0 = 2800$) by using a conventional SC bandpass filter with minimum relative bandwidth of only 1% ($Q_l = 100$).

(iii) NP systems using lowpass path filters

An NP SC filter using lowpass path filters [28 - 30] needs a minimum of three paths (N =3) in order to realise a bandpass response, as in the system shown in Figure 28. The multiband response of the NP SC filter consists of a lowpass response plus the bandpass response centred at $f_O = F_S$ and which is formed by two adjacent frequency-translated bands of the lowpass path filters. In general, the AIF is bandpass and selects the output bandpass response and attenuates the lowpass response as well as all frequency translation above $3F_S/2$, whereas the AAF is lowpass and attenuates the input frequency translated components above $3F_S/2$. The NP SC filter itself eliminates by outphasing the unwanted frequency translations bellow $3F_S/2$, but the level of cancellation of such components may not be adequate because of the problem of path mismatch. The mirror frequency translations of the form $(F_S - f_O)$ to $(F_S + f_O)$ and vice versa, are particularly serious because both of these components can lie in the passband (when f is small) and, consequently, they cannot be attenuated by the AAF and by the AIF. In addition, the passband response may also be corrupted by clock feedthrough that occurs at the midband frequency F_S depending on the SC implementation of the filter [28 - 30]. This system is suitable only for applications with low anti-aliasing and anti-imaging rejection requirements and low dynamic range.

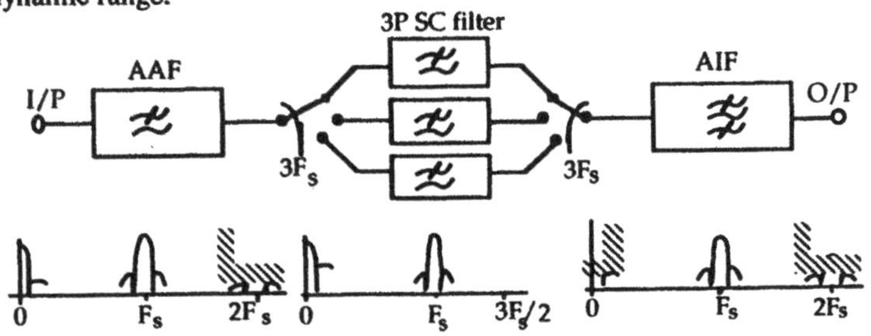

Figure 28 3-path system using lowpass path filters.

(iv) NP systems using highpass path filters

The realisation of NP SC filters with highpass path filters was proposed as a means of overcoming the problem of clock feedthrough in the passband [32, 33].

This may also yield more economical realisations of a system since only two paths (N= 2) are required to obtain a bandpass response centred at $f_O = F_S/2$, in which case the AIF may be lowpass, as shown in Figure 24.

However, due to path mismatch, this approach also suffers from the generation of unwanted frequency-translated components in the passband, and which cannot be attenuated further by the AAF and by the AIF.

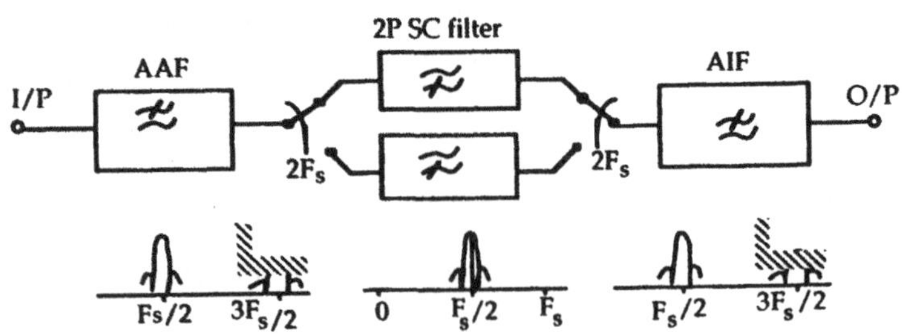

Figure 29 2-path system using highpass path filters.

The problem of inband mirror frequencies in the above NP systems arises because the passband is formed by the joining of two adjacent frequency translated bands produced by the SP SC lowpass or highpass filters. In SPFT systems, on the contrary, the passband is formed by only one frequency-translated band produced by an SP SC bandpass filter, and hence the unwanted frequency-translated bands can be sufficiently attenuated by the AAF and by the AIF.

On the other hand, the selected output image in SPFT systems is more attenuated because the signals are sampled and held with period $1/F_s$ rather than with period $1/NF_s$, as in NP systems, and thus the selectivity of the AIF must increase accordingly. Next, we shall also discuss alternative NP systems which are logical extensions of the above systems, and which consist of modifying the architecture of SPFT systems simply by replacing the SP SC bandpass filter by an NP SC filter using bandpass path filters.

2.3.4 NP systems using bandpass path filters

The system given in the example of Figure 30 comprises a 3-path SC filter using bandpass path filters, each with midband frequency f_1, Q-factor Q_1 and switching frequency $F_s = m \cdot f_1$. The bandpass AAF is similar in function and shape to that in an SPFT system (Figure 26), although it may have lower selectivity bearing in mind that the unwanted frequency-translated components from DC to $3F_s/2$, and their high frequency repetitions above $3F_s/2$, are partly attenuated by cancellation in the 3-path filter. The selectivity of the bandpass AIF is also lower than in SPFT systems because the frequency-translated images are multiplied by the function $[\sin(\pi f/3F_s)/(\pi f/3F_s)]$ as in NP systems.

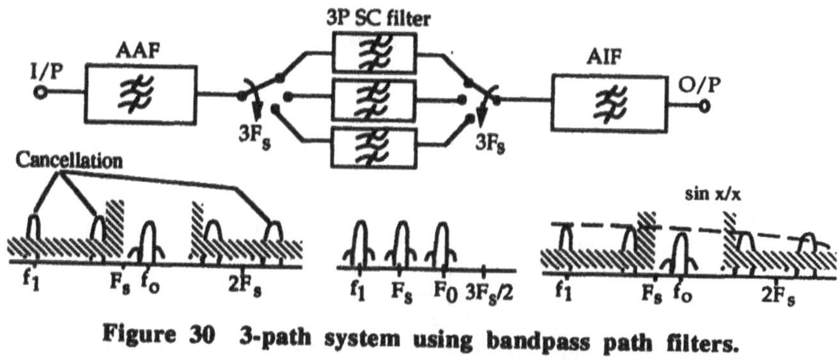

Figure 30 3-path system using bandpass path filters.

In the system in Figure 30, with N arbitrary, the Q-enhancement factor of the frequency-translated response is $K = m \cdot n + 1$ and the corresponding sample and hold attenuation factor is $[\sin(K/mN)/(K/mN)]$. In order to achieve a given value of K we can trade-off the number of paths N, the sampling ratio m and the multiple of the switching frequency n against the complexity and capacitance area of the NP SC filter and the selectivity of the AAF and AIF. The relation between n and N determines the operation mode of the filter system. If we choose $N > 2n$ or $N > 2n + 1$, respectively for a lower or an upper selected frequency-translated band, the NP system operates always below the Nyquist frequency $NF_s/2$ of the NP SC filter (see Figure 24b). Alternatively, if $N < 2n$ or $N < 2n+1$ then we have an N-path Frequency-Translated (NPFT) system which operates in a frequency-translated mode above $NF_s/2$, similarly to the case of SPFT systems (see again Figure 24b). Some of the trade-offs in such systems are illustrated in the simple examples shown in Figure 31 (only the AAF characteristic is shown). Figure 31a corresponds to an NP system operating below the Nyquist frequency of the NP SC filter, and which has the advantage of a low sample and hold attenuation (low selectivity AIF). However, the large number of paths ($N = 11$ in the NP SC filter implies considerable complexity and also the AAF has a more selective upper stopband. Figure 31b corresponds to an NPFT system using a much simpler NP SC filter with reduced number of paths ($N = 3$) which conveys a high frequency-translated band above $3F_s/2$. The design of the AAF may also be simplified owing to its approximately symmetrical characteristic but, on the other hand, the AIF has to be more selective because of the increased sample and hold attenuation.

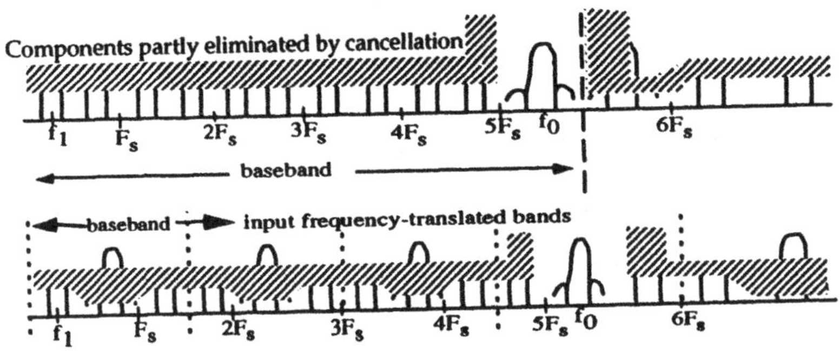

Figure 31 AAF characteristics of NP systems: $k = 21$; $n = 5$; $m = 4$
(a) NP system with $N = 11$ and $\alpha = 0\ 665$ (-3.54 dB)
(b) NPFT system with N=3 and $\alpha = 0.129$ (-17.81 dB)

2.3.5 *A 20 kHz 0.48% bandpass filter system [22]*

(i) Architecture

For practical illustration, we shall now consider the SPFT SC bandpass filter system (apart from the continuous-time filters) whose complete schematic is shown in Figure 32a. The conventional SC bandpass filter with coupled-biquad structure, 6th order elliptic bandpass response, centre frequency $f_O = 4$ kHz, 3 dB bandwidth of 80 Hz and switching frequency $F_s = 16$ kHz, is designed for minimum capacitance spread. The capacitance spread of this SC bandpass filter with 2.4% relative bandwidth is considered at the maximum acceptable limit for integration, and, for this overall capacitance spread, we are obtaining a bandpass response with a relative bandwidth of 0.48% which would otherwise require a grossly unacceptable spread. The SC bandpass decimator with $M = 24$ is formed by the cascade of an FIR SC lowpass decimator with $M = 2$, for reducing the sampling rate from $24F_s = 384$ kHz to $12F_s = 192$ kHz, and an IIR SC bandpass decimator with $M = 12$ which further reduces the sampling rate to $F_s = 16$ kHz. The complementary overall SC bandpass interpolator with $L = 24$ consists of an IIR SC bandpass interpolator with $L = 12$ followed by an FIR SC lowpass interpolator with $L = 2$. The anti-aliasing operation of the SC bandpass decimator, illustrated in Figure 32b, corresponds to the selection of the frequency-translated band centred at $f_O = 20$ kHz and the rejection of the unwanted alias signal components in the frequency range from DC to 380 kHz. The complementary operation of the SC bandpass interpolator corresponds to the selection of the frequency-translated band centred at $f_O = 20$ kHz and the rejection of the unwanted image signal components in the frequency range from DC to 380 kHz. Here, unlike in the case of the decimator, there is an additional attenuation provided by the sampled-and-held nature of the signals at the output of the SC circuits.

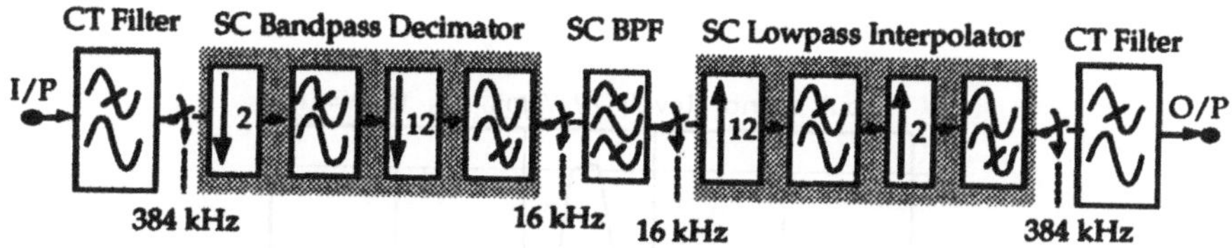

Figure 32 (a) Architecture of the 20 kHz 0.48% multirate bandpass filter system.

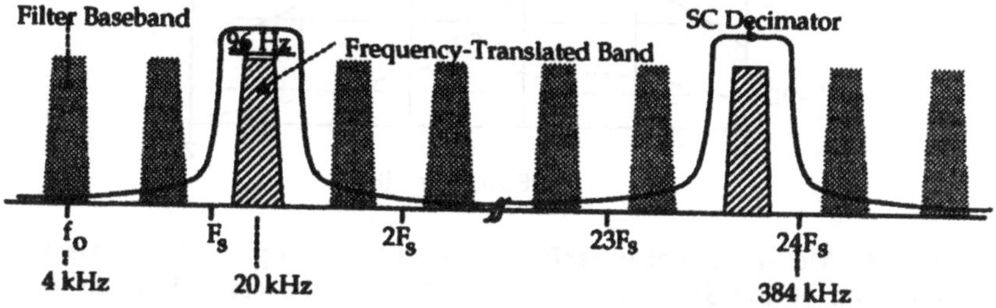

Figure 32 (b) Frequency-translated alias band selection in the multirate system for narrowband filtering using bandpass decimators.

(ii) Experimental characterization

A discrete component model of the system was constructed using CMOS 4016 analog switches, CMOS 7611 OA's for the SC bandpass filter and for the IIR SC decimator and interpolator circuits, BiMOS OA's for the FIR SC circuits, and standard CMOS-4000 logic circuits for the switching waveform generator.

The measured overall amplitude response is given in Figure 33. The measured noise performance of this SPFT SC bandpass filter system is given in Figure 34, showing that the noise floor in the passband of the system is very similar to that of the SC bandpass filter alone, despite the Q-enhancement factor $K = 5$ which reduces the relative bandwidth from 2.4% to 0.48%.

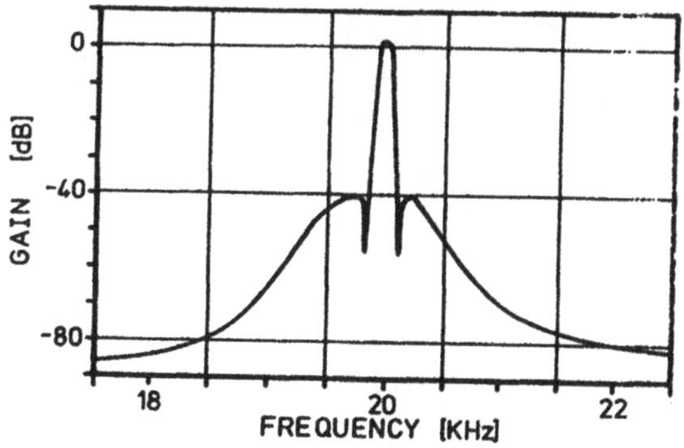

Figure 33 Measured overall amplitude response of the 20 kHz 0.48% bandpass filter system.

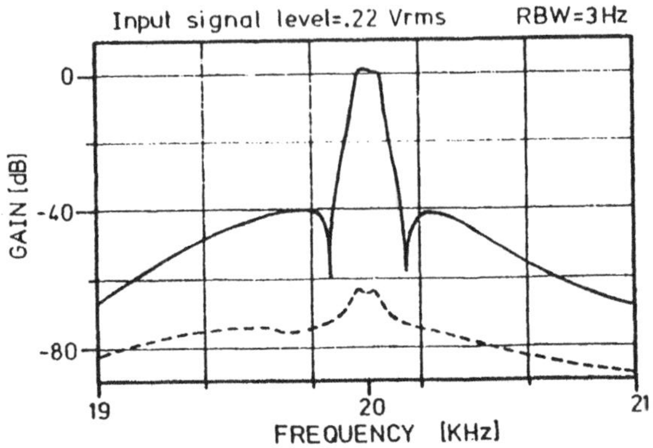

Figure 34 Measured noise performance of the complete bandpass filter system and of the SC bandpass filter alone at the core of the system.

2.4 Multirate filtering for high-frequency front-ends

2.4.1 Digitization of standard video (CCIR 601) [25, 38]

(i) Predominantly digital video interface system

The general architecture of a standard video coding system is presented in Figure 35. Due to the complex video signal characteristics the three equivalent coding paths need high amplitude selectivity and very stringent linear phase specifications.

In order to simplify the realisation of the AAF in typical coders, all the burden is placed on fast A/D converters followed by complex lowpass digital decimators. Recent implementations of the digital decimator alone at very high frequency required a large silicon area (22 mm^2) and power consumption (2 W) [39].

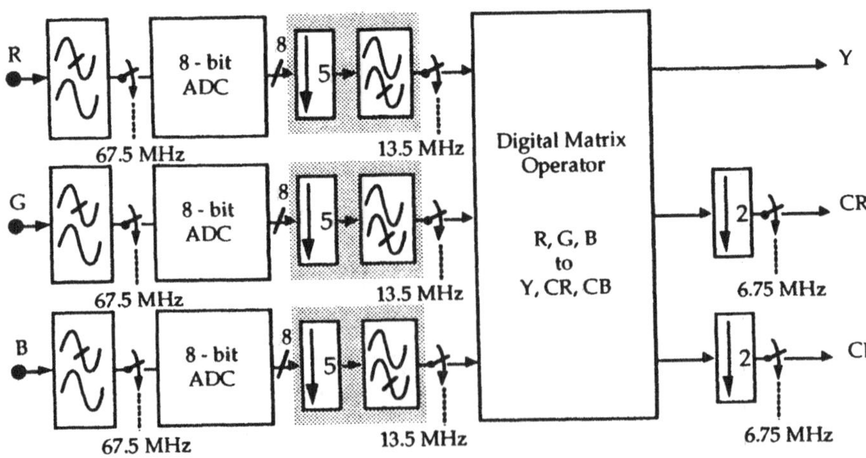

Figure 35 Predominantly digital video interface system.

(ii) Predominantly analog video interface system

The alternative system architecture illustrated in Figure 36 [25, 38] employs the type of multirate SC circuits described earlier in this chapter in order to relax the speed requirements of the A/D, simplify or even eliminate the digital filtering, and also relax the complexity of the front-end continuous-time filters.

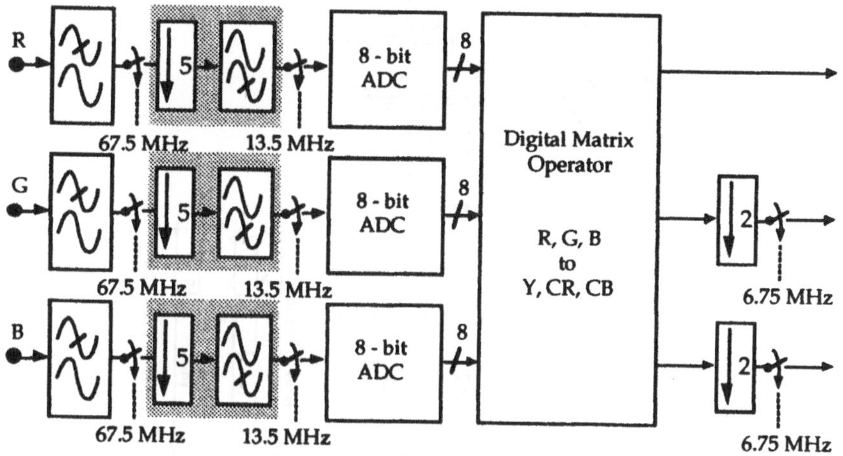

Figure 36 Predominantly analog video interface system.

In Figure 37 we can see a 5th order elliptic lowpass IIR SC decimator which may be adequate for the above predominatly analog video coder and whose input and output sampling frequencies are, respectively, $3F_S$ = 40.5 MHz and F_S = 13.5 MHz. The amplitude response is tailored to obtain a passband ripple of 0.2 dB, cut-off frequency of f_C= 3.6 MHz, and minimum 35 dB rejection above 4.44 MHz. The circuit is based on an N-th order architecture and is implemented using a low selectivity input polyphase network together with a high selectivity recursive network. The resulting computer simulated amplitude response is presented in Figure 38. The speed of the operational amplifiers is also limited by the lower output sampling frequency at 13.5 MHz.

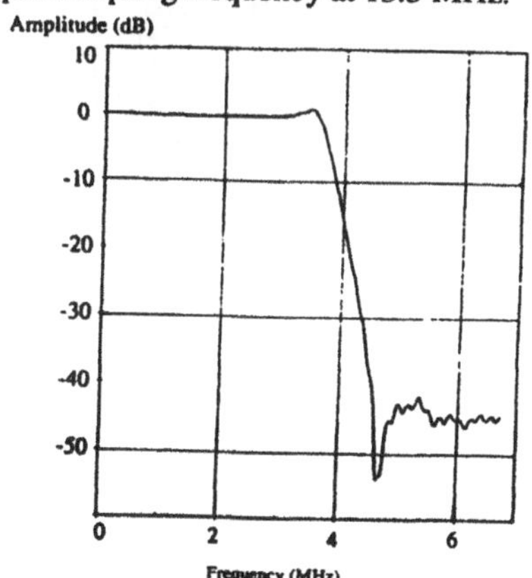

Figure 38 Computer simulated amplitude response of the SC video decimator of Figure 37.

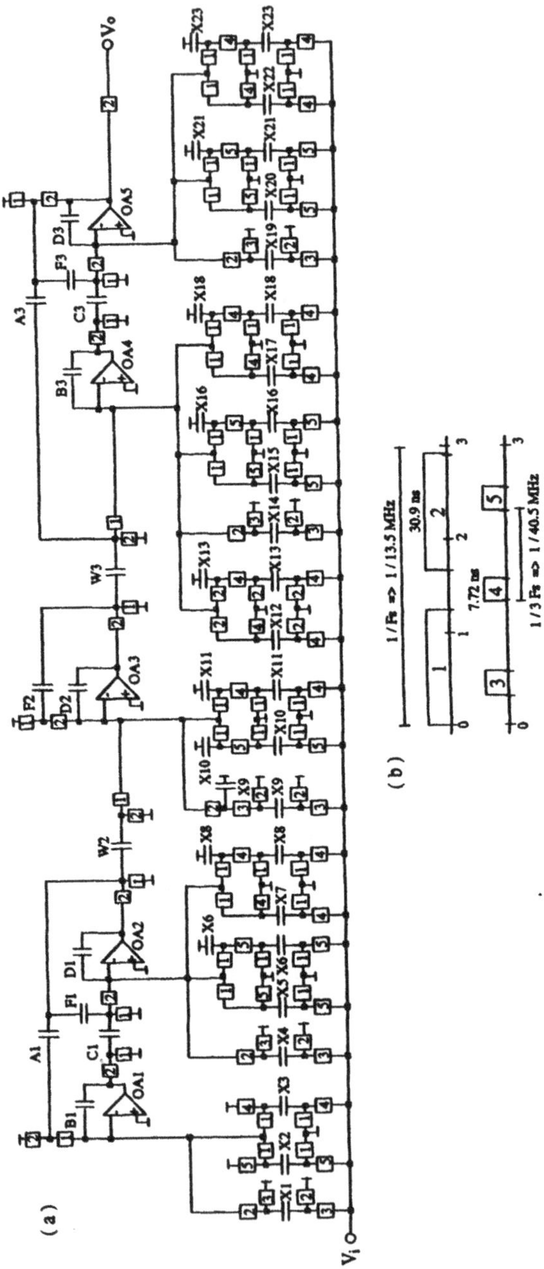

Figure 37 Lowpass 5th order elliptic video SC decimator with $M = 3$.

2.4.2 *Magnetic-disk read channels [27, 40]*

Another relevant application of predominantly analog multirate signal processing techniques is illustrated in the block diagram of Figure 39 corresponding to the A/D interface required for PRML magnetic-disk read channels [27].

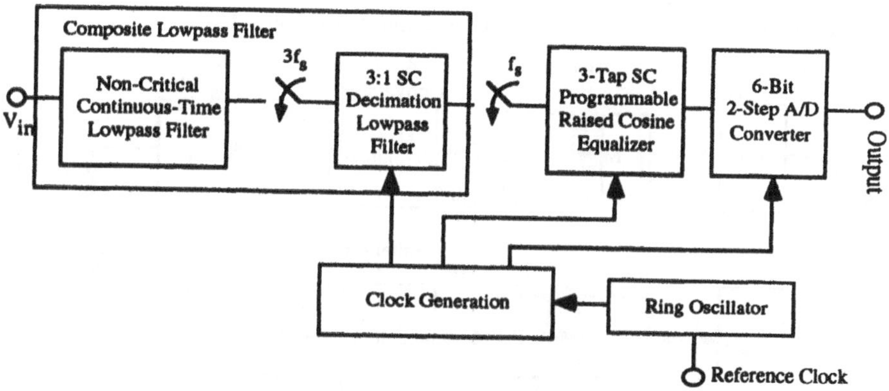

Figure 39 Block diagram of an A/D interface system for PRML magnetic-disk read channels.

The SC decimation filter shown in Figure 40, with $M = 3$, has 300 MHz input sampling and 100 MHz output rate. To achieve such high sampling rates the decimator employs 3 capacitor arrays operating in the staggered manner suggested in Figure 40. For maximum sampling rate a tapped ring oscillator must be employed allowing the phases to be spaced one gate delay appart [27, 40]. The sampling bandwidth can also be made as high as necessary by adopting an appropriate switching design strategy [40].

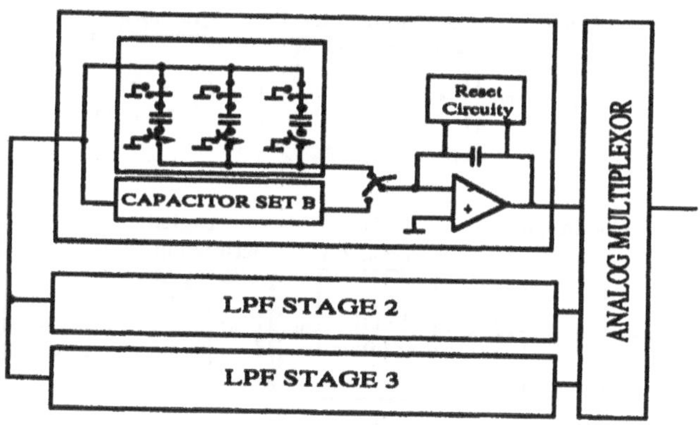

3:1 Switched - Capacitor
Decimation Lowpass Filter

Figure 40 Time-interleaved SC decimator for the magnetic-disk read channel interface.

2.5 Multirate techniques for data conversion

2.5.1 Generalized data conversion architecture

The generalized architecture of an integrated data converter viewed as a multirate system is schematically illustrated in Figure 41. In the most popular design solutions, only some of the building blocks are utilized to realize a complete system, as illustrated in the architecture of Figure 42a employing Nyquist-based quantization of the analog samples, or in the architecture of Figure 42b built around an oversampling A/D converter.

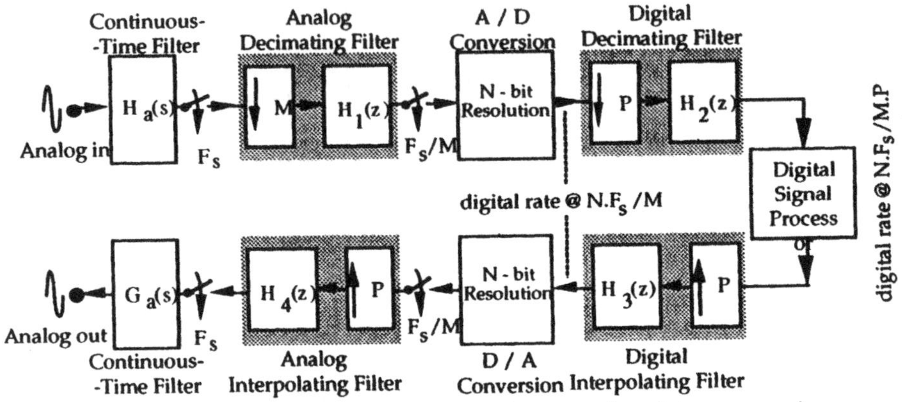

Figure 41 Generalized multirate architecture for data conversion.

Many other architectures can be developed by using the complete set of building blocks and trading-off their characteristics in order to achieve the highest performance specifications with the most economical design solution with respect to both silicon area and power consumption.

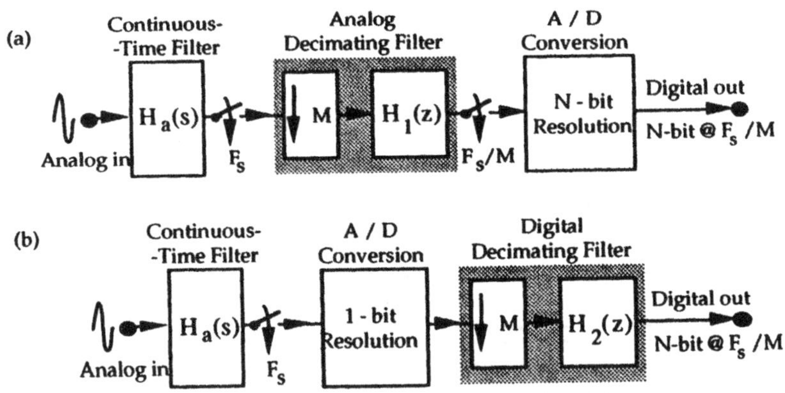

Figure 42 Multirate architecture for data conversion employing
(a) a Nyquist band converter using a fine *N*-bit quantizer
(b) an oversampling converter using a coarse 1-bit quantizer.

2.5.2 Oversampling sigma-delta converters

As mentioned above, the oversampling A/D converters are inherently multirate because they relate an input sequence of samples obtained at high frequency to an output sequence of digital words produced at much lower frequency. Such converters achieve high resolution by means of oversampling and noise shaping, and have been shown to be a good alternative for the implementation of high resolution, low-speed A/D converters mainly because they can accommodate the limitations associated with the implementation of its analog processing in digital MOS technology [41].

As illustrated in the schematic diagram of Figure 43, an analog input signal is sampled at a frequency significantly higher than f_{max}, the higher frequency present in the input signal.

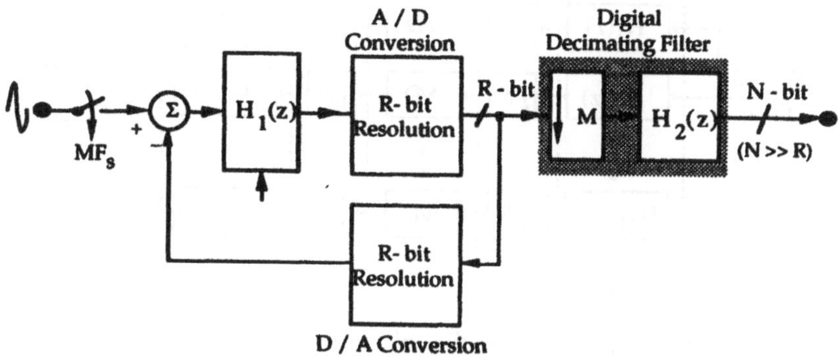

Figure 43 Typical architecture of a multirate analog-digital conversion system based on sigma-delta modulation.

The correspondingly high frequency bit stream produced by a coarse quantizer is then decimated to a much lower frequency digital signal until the required word length is formed. Such decimation is performed by means of a digital decimator consisting of a digital Nyquist lowpass filter together with a down-sampler.

A special case of Figure 43 is when $N = 1$. Here, the digital estimator is a simple voltage comparator and the D/A converter has one-bit resolution. The resulting A/D converter is called a delta-sigma A/D converter since it employs a delta-sigma modulator as the basic building block. The delta-sigma converter thus quantizes a very high sampling rate (e.g. 2 MHz) analog input sample sequence with very low resolution (e.g. 1 bit). With the aid of downsampling and digital filtering, the sampling rate is reduced (e.g. 8 kHz) and the resolution is increased (e.g. 10 bits).

The basic idea for SNR enhancement in a delta sigma converter is to reduce the quantization noise by a high-pass filter leaving only a small fraction of the noise power inside the frequency band of interest. Further SNR improvement can be obtained by increasing the order of the filter. Although its basic concept has been around for quite a few years, the integrated circuit technology needed to make the sigma-delta converters a viable solution has only recently become available. The delta-sigma A/D converter can be implemented on a single chip using MOS technology, with the analog part (delta sigma converter) usually implemented by SC circuits. The digital part consists of the decimator.

2.5.3 *QMF-based A/D converters*

In this recently introduced approach a quadrature-mirror filter (QMF) bank is used to implement a high speed A/D converter, suitable for applications in the video-rate range (up to 100 MHz) using a number of low-speed, low-cost A/D converters, as indicated in Figure 44 [42, 43]. Here, the analysis stage is an SC network composed of the analysis filter bank and the down-samplers, while the synthesis stage is a digital network composed of the synthesis bank filters and up-samplers. Basically, in this approach the input signal is first decomposed into a number of contiguous frequency bands (subbands) so that a specific A/D converter (subconverter) can be assigned to each subband signal.

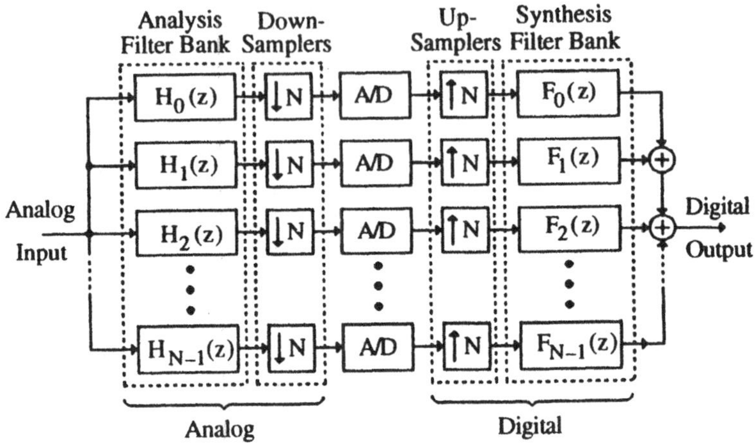

Figure 44 High-speed analog-digital conversion system based on the quadrature-mirror filter.

Although the structure of Figure 44 is a natural extension of the time-inter leaved A/D converter [44, 45] (in fact, the latter can be seen as a particular case, where the analysis filters $G_k(z) = z^{-k}$, and the synthesis filters $F_k(z) = z^{-(M-1-k)}$, the harmonic distortion due to mismatches among the subconverters is substantially reduced at a little additional cost [46].

The jitter problem due to uneven sample timing, which is another source of error in time-interleaved A/D converters, specially for high-frequency input signals, is reduced by the decimation stage in Figure 44. Similar idea has been used by Poulton et al [47] and has been called a two-rank architecture. The QMF-based converters also incorporate the advantages of subband coding: by appropriately specifying the resolution of the subconverters throughout the respective subbands, the quantization noise can be separately controlled in each band, and the shape of the reconstruction error spectrum can be controlled as a function of the frequency. This strategy has been used in many speech and image coding applications.

This new scheme has been simulated on a computer and tested in the laboratory using discrete components. Both simulation and experimental results have verified its good performance. Some of these results, as well as details about the circuit implementation of the analysis and filter banks can be found in [42, 43]. Efficient implementation of the SC filters that realize the analysis filter bank has been addressed in [48], where the design of the SC filters for high frequency operations has been discussed. The emphasis has been placed on the use of single-stage operational amplifiers with very high bandwidth and moderate gain. Theoretically, the overall resolution depends only on the resolution of the subconverters used. The noise generated by the SC circuits, however, places an upper bound of 11-12 bits [48], which is still well above the usual requirements for video applications. If successive approximation subconverters are used, then substantial savings in die area can be obtained when compared to flash converters.

A more efficient structure for implementation can be obtained by realizing the analysis and the synthesis filters in polyphase form, and then moving the down-samplers to the left of the polyphase sub-filters in the analysis bank and moving the up-samplers to the right of the polyphase sub-filters in the synthesis bank, as indicated in Figure 45 for a two-band decomposition.

Using a tree structure, QMF banks with more than two bands can be developed from the two-band structure of Figure 5 [49].

It should be noted that the approach described above can also be used for the design of a high-speed D/A converters using a bank of low-speed D/A converters, as indicated in the example of Figure 46.

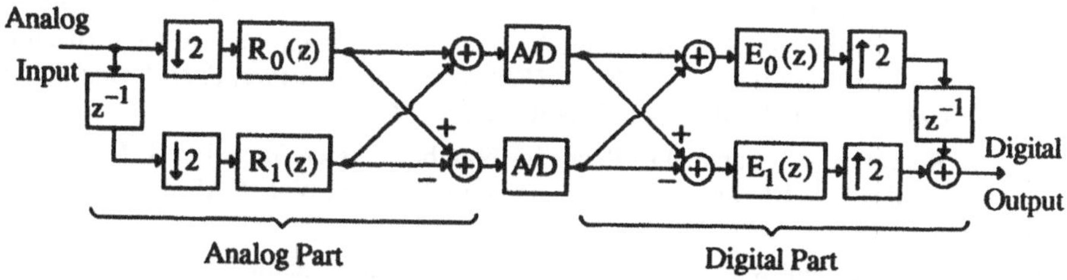

Figure 45 Example of a more efficient structure for implementation of a high-speed A/D conversion system based on the QMF bank.

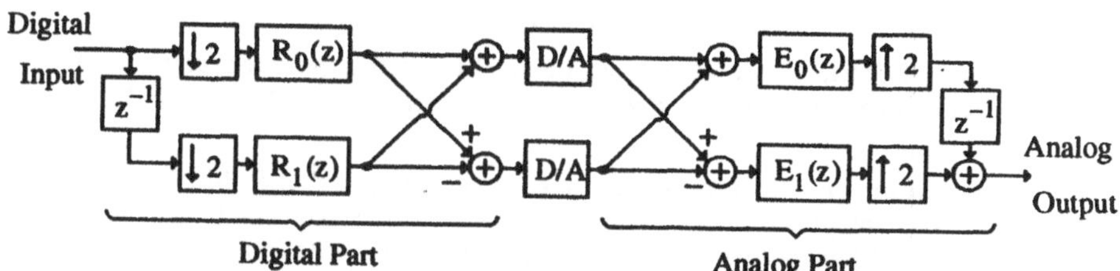

Figure 46 Example of a two-band structure for the implementation of a high-speed D/A conversion system based on the QMF bank.

2.6 Conclusions

This chapter presented an overview on the design and applications of analog-digital multirate signal processing. After reviewing some basic concepts of multirate systems, various design techniques for multirate SC building blocks were described, including both finite and infinite impulse response SC decimators and interpolators. The application of these design techniques and building blocks was then highlighted for realizing the type of high-selectivity filtering systems needed for example in many communications front-end applications, and multirate filtering for high-speed front-ends especially those required for video coders and magnetic-disk read channels. Bearing in mind the state-of-the-art technologies which allow the strategic objective of high quality analog-digital integration we then described various multirate systems for signal conversion, including the popular oversampling converters as well some new architectures based on the quadrature-mirror filter (QMF) banks.

References

[1] R. Crochiere and L. R. Rabiner, *Multirate Digital Signal Processing*, Prentice-Hall, Englewood Cliffs, NJ, 1983.

[2] P. P. Vaidyanathan, *Multirate Systems and Filter Banks*, Prentice-Hall, Englewood Cliffs, NJ, 1993.

[3] S. K. Mitra, "Some unconventional applications of multirate digital signal processing," *Proc. IEEE Int. Symp. on Circuits and Systems*, Singapore, pp. 13-16, June 1991.

[4] M. Bellanger, G. Bonnerot and M. Coudreuse, "Digital filtering by polyphase network: application to sample rate alteration and filter banks," *IEEE Trans. on Acoustics, Speech and Signal Processing*, Vol. ASSP-24, pp. 109-114, April 1976.

[5] P. P. Vaidyanathan, S. K. Mitra and Y. Neuvo, "A new approach to the realization of low sensitivity IIR digital filters," *IEEE Trans. on Acoustics, Speech, and Signal Processing*, Vol. ASSP-34, pp. 350-361, April 1986.

[6] M. Renfors and T. Saramaki, "Recursive N-th band digital filters Part 1: Design and properties," *IEEE Trans. on Circuits and Systems*, Vol. CAS-34, pp. 24-39, January 1987.

[7] R. Gregorian and W.E. Nicholson, "Switched-capacitor decimation and interpolation circuits," *IEEE Trans. on Circuits and Systems*, Vol. CAS-27, pp. 509-514, June 1980.

[8] D. C. Grunigen, *et al.*, "Simple switched-capacitor decimation circuits," *Electronics Letters*, Vol. 17, No.1, pp. 30-31, January 8, 1981.

[9] M. B. Ghaderi, G. C. Temes, S. Law, "Linear interpolation using CCD's or switched-capacitor filters," *Proc. Inst. Elect. Eng.*, Vol. 128-G, No. 4, pp. 213-215, August 1981.

[10] D. G. Grunigen, R. Sigg, M. Ludwig, U. W. Brugger, G.S. Moschytz, and H. Melchior, "Integrated switched-capacitor low-pass filter with combined anti-aliasing decimation filter for low frequencies," *IEEE J. Solid-State Circuits*, Vol. SC-17, pp. 1024-1029, December 1982.

[11] J. E. Franca, "Decimators and interpolators for narrowband switched-capacitor bandpass filter systems," *Proc. IEEE Int. Symp. on Circuits and Systems*, Montreal, Canada, pp. 789-793, May 1984.

[12] J. E. Franca, "Non-recursive polyphase switched-capacitor decimators and interpolators," *IEEE Trans. on Circuits and Systems*, Vol. CAS-32, pp. 877-887, September 1985.

[13] J. E. Franca, "Switched-capacitor decimators and interpolators with biquad-polyphase structures," *Proc. Midwest Symp. Circuits and Systems*, Syracuse, NY, pp. 1181-1184, August 1987.

[14] J.E. Franca and D. G. Haigh, "Optimum implementation of IIR SC decimators," *Proc. IEEE Int. Symp. on Circuits and Systems*, Philadelphia, PA, pp. 76-79, May 1987.

[15] J. E. Franca and R. P. Martins, "IIR Switched-capacitor decimator building blocks with optimum implementation," *IEEE Trans. on Circuits and Systems*, Vol. CAS-37, pp. 81-90, January 1990.

[16] R. P. Martins and J. E. Franca, "Optimum implementation of a multistage IIR SC bandpass decimator for a voiceband analogue interface system," *Proc. IEEE Int. Symp. on Circuits and Systems*, pp. 1661-1664, Helsinki, Finland, June 1988.

[17] J. E. Franca and S. Santos, "FIR switched-capacitor decimators with active-delayed block polyphase structures," *IEEE Trans. on Circuits and Systems*, Vol. CAS-35, No. 8, pp. 1033-1037, August 1988.

[18] R. P. Martins and J. E. Franca, "A novel N-th order IIR switched-capacitor decimator building block with optimum implementation," *Proc. IEEE Int. Symp. on Circuits and Systems,* Portland, Oregon, pp. 1471-1474, May 1989.

[19] R. P. Martins and J. E. Franca, "Infinite impulse response switched-capacitor interpolators with optimum implementation," *Proc. IEEE Int. Symp. on Circuits and Systems,* New Orleans, LA, pp. 2193-2197, May 1990.

[20] J. E. Franca, "A single-path frequency-translated switched-capacitor system for filtering and single sideband generation and detection," *Proc. IEEE Int. Symp. on Circuits and Systems,* Kyoto, Japan, pp. 1625-1628, June 1985.

[21] J. E. Franca, "A single-path frequency-translated switched-capacitor bandpass filter system," *IEEE Trans. on Circuits and Systems,* Vol. CAS-32, pp. 938-944, September 1985.

[22] J. E. Franca and D. G. Haigh, "Design and applications of single-path frequency-translated switched-capacitor Ssystems," *IEEE Trans. on Circuits and Systems,* Vol. CAS-35, pp. 394-408, April 1988.

[23] J. E. Franca, "Multirate switched-capacitor system approach to frequency division multiplexing", *Electronics Letters,* Vol. 24, No. 8, pp. 501-503, April 14, 1988.

[24] R. P. Martins and J. E. Franca, "An experimental 1.8μm CMOS anti-aliasing switched-capacitor decimator with high input sampling frequency," *Proc. European Solid State Circuits Conference - ESSCIRC'91*, Milan, Italy, pp. 13-16, September 1991.

[25] R. P. Martins and J. E. Franca, "A 2.4μm CMOS switched-capacitor video decimator with sampling rate reduction from 40.5MHz to 13.5MHz," *Proc. 1989 IEEE Custom Integrated Circuit Conference*, San Diego, California, May 1989.

[26] S. K. Berg, P. J. Hurst, S. H. Lewis and P. T. Wong, "A switched-capacitor filter in 2 μm CMOS using parallelism to sample at 80 MHz," *Digest of Technical Papers IEEE Int. Solid-State Circuits Conference 1994*, San Francisco, CA, pp. 62-63, February 1994.

[27] G. T. Uehara and P. R. Gray, "A 100 MHz output rate analog-to-digital interface for PRML magnetic-disk read channels in 1.2 μm CMOS," *Digest of Technical Papers IEEE Int. Solid-State Circuits Conference 1994*, San Francisco, CA, pp. 280-281, February 1994.

[28] D. J. Allstot and K. S. Tan, "A switched-capacitor N-path filter," *Proc. IEEE Int. Symp. Circuits and Systems 1980*, Houston, Texas, pp. 313-316.

[29] M. S. Lee and C. Chang, "Exact synthesis of N-path switched-capacitor filters," *Proc. IEEE Int. Symp. Circuits and Systems 1981*, Chicago, Illinois, pp. 166-169.

[30] M. B. Ghaderi *et al.*, "Narrow-band switched capacitor bandpass filters," *IEEE Trans. Circuits and Systems*, Vol. CAS-8, pp. 557-572, August 1982.

[31] D. C. Grunigen *et al.*, "An integrated CMOS switched-capacitor bandpass filter based on N-path and frequency-sampling principles," *IEEE J. Solid-State Circuits*, Vol. SC-18, pp. 753-761, December 1983.

[32] M. B. Ghaderi, *New Design Techniques for Switched-Capacitor Bandpass Filters, PhD Dissertation*, University of California, Los Angeles, CA, 1981.

[33] U. Kleine and J. Pandel, "Novel switched-capacitor pseudo N-path filters," *Electronics Letters*, Vol. 18, pp. 66-68, January 21, 1982.

[34] L. E. Franks and I. W. Sandberg, "An alternative approach to the realisation of network transfer functions: The N-path filter," *Bell System Technical Journal*, Vol. 39, pp. 1321-1350, September 1950.

[35] A. Fettweis and H. Wupper, "A solution to the balancing problem in N-path filters," *IEEE Trans. Circuit Theory*, Vol. CT-18, pp. 403-405, May 1971.

[36] J. A. Nossek and H.Weinrichter, "Noise figures of switched-capacitor N-path filters," *Proc. European Conference on Circuit Theory and Design 1983*, Stuttgart, Germany.

[37] K. Martin and A. Sedra, "Exact Design of Switched Capacitor Bandpass Filters Using Coupled-Biquad Structures," *IEEE Trans. Circuits and Systems*, Vol. CAS-27, pp. 469-475, June 1980.

[38] J. E. Franca and R. P. Martins, "Novel solutions for anti-aliasing and anti-imaging filtering in CMOS video interface systems," *Proc. IEEE Workshop on Visual Signal Processing and Communications*, Taiwan, Republic of China, June 1991.

[39] G. Chiappano and D. Raveglia, "Anti-aliasing digital filters for video signal coders," *Proc. IEEE Int. Symp. on Circuits and Systems*, Helsinki, Finland, pp. 709-713, June 1988.

[40] G. Uehara and P. Gray, "Practical aspects of high-speed decimation filter implementation," *Proc. IEEE Int. Symp. on Circuits and Systems*, San Diego, CA, pp. 2308-2311, May 1992.

[41] J. C. Candy and G. C. Temes, Editors, *Oversampling Sigma-Delta Converters: Theory, Design and Simulation*, IEEE Press, New York, NY, 1992.

[42] A. Petraglia and S. K. Mitra, "High speed A/D conversion incorporating a QMF bank," *IEEE Trans. on Instrumentation and Measurement*, Vol. IM-41, pp. 427-431, June 1992.

[43] A. Petraglia and S. K. Mitra, "Design of magnitude-preserving A/D converter," *IEICE Trans. on Fundamentals of Electronics, Communications and Computer Sciences*, Special Issue on High-Speed Analog Circuits and Signal Processing, Vol. E76A, pp. 149-155, February 1993.

[44] W. C. Black and D. A. Hodges, "Time-interleaved converter arrays," *IEEE Journal on Solid-State Circuits*, Vol. SC-15, pp. 1022-1029, December 1980.

[45] W. C. Black, *High Speed CMOS A/D Conversion Techniques*, PhD. Dissertation, University of California, Berkeley, CA, 1980.

[46] A. Petraglia and S. K. Mitra, "Effect of mismatches among A/D converters in a time-interleaved digitizer," *IEEE Trans. on Instrumentation and Measurement*, Vol. IM-40, pp. 831-836, October 1991.

[47] K. Poulton, J. J. Corcoran, and T. Hornak, "A 1-GHz 6-bit ADC system," *IEEE Journal on Solid-State Circuits*, Vol. SC-22, pp. 962-970, December 1987.

[48] A. Petraglia, F. Maloberti, and S. K. Mitra, "QMF-based A/D converters: Overview and new results," *Proc. Int. Conference on Analog to Digital and Digital to Analog Conversion*", Swansea, Wales, U.K., pp. 112-117, September 1991.

[49] P. A. Regalia, S. K. Mitra, P. P. Vaidyanathan, Y. Neuvo and M. Renfors, "Tree-structured complementary filter banks using allpass sections," *IEEE Trans. on Circuits and Systems*, Vol. CAS-34, pp. 1470-1484, December 1987.

Tutorial 3

Advanced algorithms and implementations for coding of moving images

Organised by: Peter Pirsch, *University of Hannover*

Tutorial 3. Introduction

Visual communications is a rapidly evolving field for telecommunications, computer and media industries. The progress in this field is supported by the availability of digital transmission channels and digital storage media. In addition to the available narrowband ISDN, new digital transmission channels such as broadband ISDN, digital satellite channels and digital terrestrial TV broadcasting will be introduced shortly. Moreover, personal computers and workstations have become important platforms for multimedia interactive applications. Communication based applications include ISDN videophone, videoconference systems, digital broadcast TV/HDTV and remote surveillance. Storage based audiovisual applications include training education, entertainment, advertising, video mail and document annotation.

Essential for the introduction of new image communications services is low cost. Bit rate compression schemes are employed for transmission and storage cost reduction. High reduction factors on bit rate require very sophisticated coding schemes which will result in hardware systems of high complexity. This calls for VLSI implementation. New architectural approaches adapted to the signal processing schemes and recent advances in VLSI technology relieve the hardware problems. In addition, the developments in the technology of digital storage media, image displays, workstations and communication networks make video communications economically viable.

To facilitate world wide interchange of digitally encoded audiovisual data there is a need for international standards for the coding methods and transmission formats. International standarization commitees have been working on the specification of several compression algorithms. The Joint Photographic Experts Group (JPEG) of the International Standards Organisation (ISO) have specified an algorithm for compression of still images. The CCITT proposed the H.261 standard for video telephony and the Motion Picture Experts Group (MPEG) completed its first standars MPEG-1, which will be used to store video comparable to VCR quality. MPEG made substantial progress for the second phase of standards MPEG-2, which will provide audiovisual quality of both broadcast TV and HDTV. Because of the wide field of applications MPEG-2 is a family of standards with different profiles and levels. Future work in audiovisual coding at very low bitrates (MPEG-4) employs model-based techniques by exploiting more directly the structural information of the image content.

Basic pricipes of video coding schemes applied by MPEG, the actual status of standarization and recent results of VLSI implementations for video codecs will be presented in the following sections. Four contributors are directed to video coding algorithms for MPEG-2. Then two contributors describe VLSI architectures and implementations of the MPEG coding schemes. The last two contributors are devoted to advanced analysis-synthesis coding schemes for very low bit ratre coding.

Peter Pirsch

Chapter 3.11

MPEG-2: The Main Profile

John Morris

3.11.1 MPEG-2 - What is it?

The rapid development of digital technology in the 1980's made compression of digital video signals a practical possibility. This opened up the possibility of a whole range of new services and products. To facilitate these new services, standardization has become a high priority to ensure that interworking between products is possible and to encourage investment in standard technologies so that the technology becomes cost-effective for consumer applications.

The Moving Pictures Expert Group (MPEG) is a working group within ISO/IEC that defines standards for digital compression of audio-visual signals. MPEG's standards are generic, meaning that they are capable of being used by a wide range of applications, covering storage-based multimedia systems, television broadcasting, and communications applications such as Video-on-Demand and face-to-face communication.

MPEG's first project was MPEG-1, published in 1993 as ISO/IEC 11172 [1,2]. It is a three part standard defining audio and video compression coding method and a multiplexing system for interleaving audio and video data so that it can be played back in close synchronization. It has been applied in the CD-i system and Video-CD for publishing full-screen motion video on CD-ROM. It also forms the basis of a number of field trials for Video-on-Demand services, and was used in one of the demonstrations of HDTV systems in the USA. Although MPEG-1 has found many applications, it principally supports video coding at bit-rates only up to about 1.5 Mbit/s using a non-interlaced format. It is not optimal for interlaced signals. During 1990, therefore, MPEG recognized the need for a second, related standard for coding video at higher data rates in an interlaced format (the essential technical difference is that MPEG-1 is designed for coding progressive pictures whereas MPEG-2 is optimized for interlaced pictures - MPEG-1 bit-streams are decodable on MPEG-2 decoders). The MPEG-2 standard [3] is capable of coding standard definition television at bit-rates from about 4-9 Mbit/s and high definition television at 15-25 Mbit/s.

Because MPEG has taken a generic approach to standardization, its standards provide techniques to support different applications with differing requirements. Analyzing these differing requirements led MPEG to realize that it is not possible to provide a single, unique method for all the different problems. Instead it has followed a "tool-kit" approach in which an extensive set of algorithmic "tools" are defined. For instance coding modes are provided both for scalable and non-scalable coding systems. The coding syntax that MPEG has defined provides tools to cover all the different applications it considered, and parameters can be chosen to allow working at different bit-rates and picture sizes and resolutions.

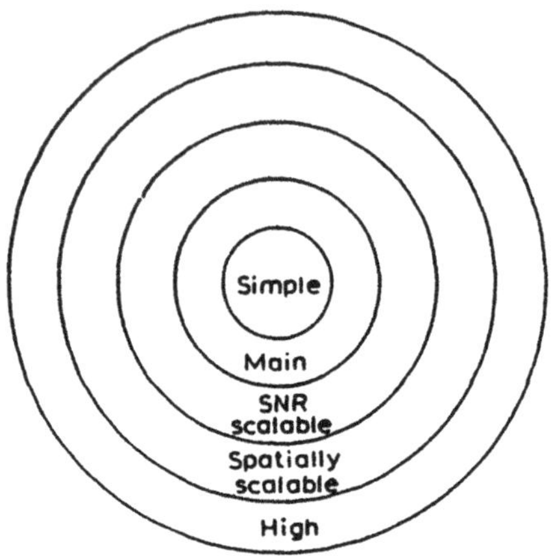

Figure 1 MPEG-2 profiles.

It is neither cost effective nor an efficient use of bandwidth to support all the features of the standard in all applications. In order to make the standard practically useful and enforce interoperability between different implementations of the standard, MPEG has defined *profiles* and *levels* of the full standard. Roughly speaking, a profile is a sub-set, suitable for a particular application, of the full possible range of algorithmic tools, and a level is a defined range of parameter values (such as picture size for instance) that are reasonable to implement and practically useful. Figure 1 illustrates the relationship between the different profiles that MPEG has adopted. The more complex profiles are super-sets of the less complex ones. Therefore a decoder capable of decoding the scalable profiles will also be able to decode data generated for the Main Profile (for non-hierarchical, non-scalable coding).

This chapter describes only the Main Profile of MPEG. Other papers in the series will consider the scalable options, and methods of providing low-delay performance typically needed for conversational or interactive telecommunication services [4, 5].

3.11.2 Schedule and method of working

MPEG began work on MPEG-2 in the summer of 1990. The Main Profile algorithm was frozen in March 1993 so that no further changes would take place, and the whole specification completed in November 1993. Currently it is going through the ISO/IEC approval process of balloting and possible revision. It is intended that this process will be completed by November 1994 and that the final text will be published during 1995. Early implementations of the standard are already available.

The MPEG group worked in three phases. Firstly, requirements of different applications were analyzed. A consequence of these requirements, and the technical implications of the conclusions reached, was MPEG's decision to adopt a tool-kit approach with profiles and levels. Once the requirements were clearer, the detailed definition of the algorithm began with a competitive phase during which many proposals from different laboratories were put through subjective tests aimed at identifying promising techniques. Finally, a convergence phase of collaborative work, doing well-defined experiments on a single *Test Model* (a full description of an encoder/decoder), progressively enhanced the performance of the system as the Test Model was updated based on the experimental results. At the end of this phase the Test Model was an embodiment of an MPEG-2 coding and decoding system on which consensus had been reached and from which the final specification is derived.

3.11.3 Requirements

MPEG-2 aims to be applicable to a wide range of applications and services covering many bit-rates, resolutions, and picture qualities. The applications range from interactive storage-based multimedia systems, to television broadcasting by cable, satellite or terrestrial networks, or personal communication systems. These applications require that a range of features should be supported:

Picture quality - The aim was to achieve normal TV picture quality at around 5 Mbit/s and studio quality at around 9 Mbit/s.

Delay - For some applications the encoder-decoder end-to-end delay is a very critical parameter. For others it is not at all important. A system with low-delay has a penalty in picture bandwidth efficiency, and so MPEG has introduced optional modes that an encoder can use to support low-delay.

Random access - Most applications require that it should be possible to start decoding either at defined points in the stream or a short time after acquiring the channel (channel hopping).

Error resilience - All applications need the system to recover if there are residual bit-errors after error correction. In different applications the residual error rate may differ by many orders of magnitude. MPEG has chosen to build in a certain amount of error robustness, and to allow a much higher degree to be used optionally by the encoder.

Data security - Many applications will use encryption to control access. MPEG has not defined encryption methods. It has provided methods for transporting information that allows unscrambling of scrambled data. This information is carried in the system multiplex, and is not discussed further in this paper.

Trick Modes - Applications that store video, either on tape or on disc, need to able to play fast forwards or backwards. MPEG has provided optional support for these modes.

3.11.4 What is standardized

The Main Profile of the MPEG-2 standard allows encoders and decoders from different manufacturers to operate together. The interface between the two is the compressed bit-stream that represents the coded video. To achieve interoperability it is only necessary to standardize the structure, content and meaning of the bit-stream and the way that it should be decoded to reconstruct the desired picture. There is no need to standardize the encoder. This is the approach that MPEG has taken as it has the advantage of leaving considerable freedom to encoder manufacturers to improve their encoding strategies as more is learnt about encoding, or to address different market segments with different tradeoffs of cost and complexity against picture quality.

3.11.5 Compression methods

The full MPEG-2 Video standard can be divided into two broad categories. The non-scalable syntax, which is structured as a super-set of the MPEG-1 standard, forms the Main

Profile. It provides a set of compression methods for coding interlaced video efficiently. The second major category is the scalable coding system, which has the key property of making it possible to recover useful pictures at a reduced quality or sampling rate from a part of the whole bit-stream. This is done by structuring the data in a number of hierarchical layers. This subject will be covered in more detail in another paper [4]. The Main Profile has the advantage that it is less complex to implement and is more efficient than scalable coding. Scalability is nevertheless important for some applications.

3.11.5.1 *Colour sampling*

A number of different colour sampling structures are supported by MPEG-2. They are all based on the luminance and two colour difference model of colour representation (Y, C_r, C_b) that is used in television systems. The 4:2:2 and 4:4:4 colour sub-sampling pattern defined in CCIR 601 are both supported [5]. However, the so-called 4:2:0 format (see Figure 2) is expected to be more widely used for consumer applications. 4:2:0 colour sub-sampling reduces the chrominance sampling rate by a factor of 2:1 with respect to the luminance in both the horizontal and vertical directions. This leads to a significant reduction in the size of the frame-stores needed in the decoder.

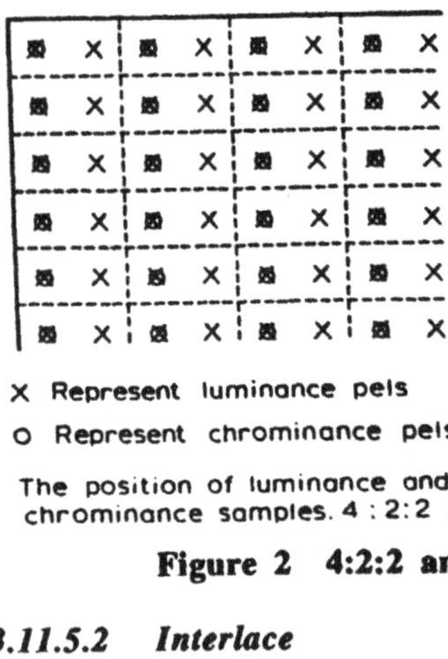

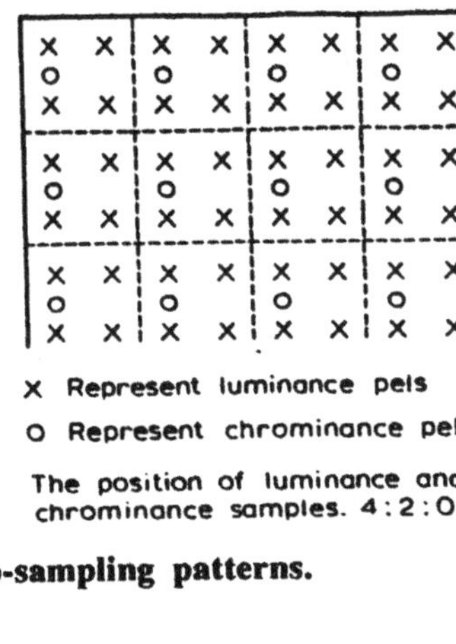

Figure 2 4:2:2 and 4:2:0 colour sub-sampling patterns.

3.11.5.2 *Interlace*

The most important difference between MPEG-1 and MPEG-2 is full support for coding interlaced video material. Each frame of interlaced video consists of two fields separated in time by a field period (20 ms in 50 Hz countries and 16.67 ms in 60 Hz countries). The vertical position of the lines in field one fall midway between the lines of field two. Together the two fields sample the image at the full vertical resolution of the frame, though at different time periods (see Figure 3).

On slowly moving video material the difference between the information on each field is small. In this case it is most efficient to combine the two fields into one frame and code the combined frame as one unit. However, when the motion is large, the information on each field is very different and it is more efficient to encode each field independently. For this reason it is possible in MPEG-2 to switch adaptively between field coding and frame coding modes on a frame-by-frame basis. Further, in the frame coding mode it is possible to switch into field coding modes locally within blocks. These coding modes form the basis of the system for highly efficient coding of interlaced video.

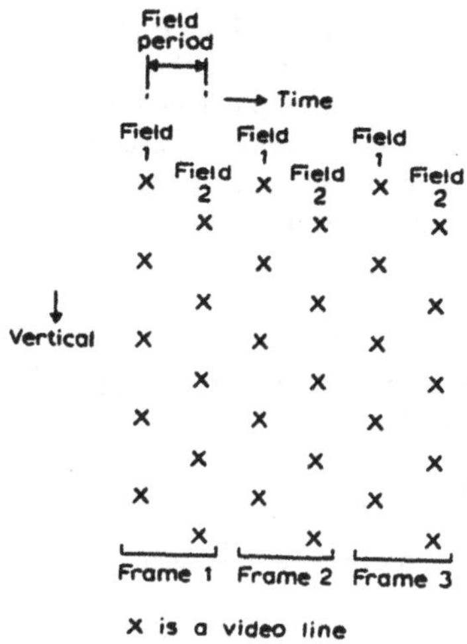

Figure 3 Vertical-temporal position of video data in an interlaced system.

3.11.5.3 Overview of the algorithm

MPEG achieves a high compression ratio, while preserving good picture quality and the necessary features of random accessibility and robustness to errors, by reducing the redundancy in the signal and also discarding information that is irrelevant to the observer. It is not a lossless coding scheme - some data is irreversibly discarded during encoding. It is possible to trade-off the subjective picture quality on one hand and the bit-rate on the other. The final bit-rate is a free parameter in MPEG that the encoder is able to select.

In the design of the algorithm it is necessary to balance the conflicting requirements of high compression and random access. High compression means that we need to remove the temporal redundancy in sequences of pictures by using motion compensated prediction from frame to frame. However, the need for random accessibility means that there must be breaks in the sequence of frame-to-frame prediction. Therefore there has to be a careful balance between coding intra-frame information without using any knowledge of previous pictures, and using recursive and non-recursive motion compensated coding.

A number of different compression methods are combined in sequence. The first set of methods uses block based motion compensation. Motion compensation is a process where pictures that have already been coded are combined with data representing motion vectors to form a prediction of the current picture. The encoder searches the previous picture to find a block of data that is the best match of the block that is currently being coded. The offset to this block is coded as a motion vector, and the best matching block itself is used as a prediction of the current block by subtracting the prediction from the sample values of the block. The residual prediction error is then quantized and encoded. At the decoder the process is inverted to recover the picture. In order to capture the different levels of redundancy that can be present in video signals MPEG has resorted to several different prediction modes. The basic block size for motion compensation is 16 pixels by 16 lines. These are known as macroblocks. Three main picture types have been defined. Intraframe coded pictures (I-pictures) are coded without reference to any previous pictures and are provided to allow access to the sequence of coded pictures at the cost of a reduction in the coding efficiency. Two types of motion compensated pictures are defined - predictive pictures (P-pictures) and bi-directional pictures (B-pictures). P-pictures use causal prediction from previously decoded pictures. Both P-pictures and previous I-pictures may be used as the basis for prediction of P-pictures. Thus P-pictures use recursive prediction. B-pictures on the other hand use non-causal prediction from pictures that will be displayed in either the past or the future and never form the basis of further predictions. To use bi-directional prediction the pictures must be re-ordered so that they are transmitted in a different order from the order they are displayed in (see Figure 4).

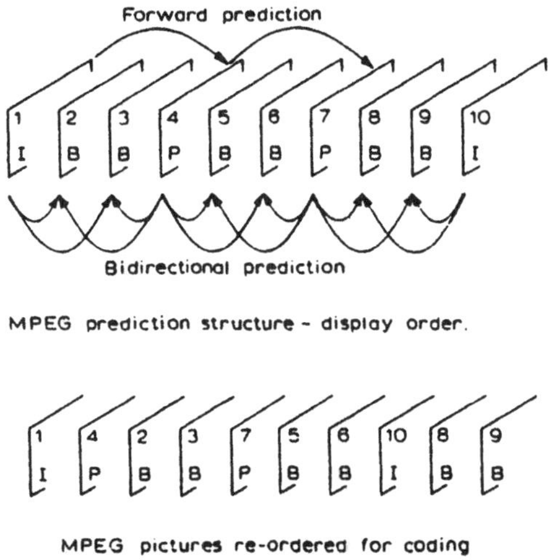

Figure 4 MPEG picture reordering.

As discussed previously, pictures may consist either of single fields or single frames (the two fields combined together). If the picture is a frame picture then the encoder may choose to form the prediction by using a vector for the entire block, or by using separate vectors for each field. Thus in B-pictures, where the motion compensation may be made from both past and future pictures in either field or frame mode, it may be necessary to transmit information for up to four motion vectors. It is important therefore to encode the motion vector information as efficiently as possible.

Motion compensation generates a prediction error signal that has eliminated most of the temporal redundancy. The error signal is then encoded by a set of techniques to reduce the spatial redundancy of the information. Whereas motion compensation is based on a 16 pixel by 16 line macroblock, spatial redundancy reduction is based on an 8 by 8 block structure. The data in the 8 by 8 blocks are transformed by the discrete cosine transform (DCT), quantized and then variable length coded. This coded data is then multiplexed with the side information - the prediction modes, motion vectors and other necessary information - to give the compressed bit-stream that is transmitted or stored.

A decoder reads the compressed bit-stream, separates out the different types of data, uses the mode and motion vector information to create a prediction of the current frame from previously decoded pictures, and combines this prediction with the information it recovers by decoding the compressed DCT data to form the final picture.

3.11.5.4 Motion compensation

Of the three picture types defined by MPEG-2, I-pictures do not use motion compensation. P-pictures use causal prediction from previous pictures in a recursive manner. B-pictures use non-recursive, non-causal prediction from pictures that have been decoded but will be displayed both before and after the current picture.

I-pictures may be used to break the recursive sequence of pictures, and so can be used to enable random access. Both P- and B-pictures are coded using motion compensation techniques. B-pictures are more efficiently coded than P-pictures for three main reasons. The bi-directional motion compensation improves prediction of areas of the picture uncovered by the motion; the averaging of data from two pictures has some noise reduction properties; and because the prediction of B-pictures is not recursive, it is possible to reduce the quality of the B-pictures relative to P-pictures, saving bits, and hence increasing the overall quality of other pictures, leading to a net increase in the quality of the sequence. Within these three picture types a range of coding modes are defined.

(i) Prediction modes

In the case of I-pictures and intra coded macroblocks, no prediction is required. However intra coded macroblocks may carry motion vector information. In normal circumstances it is not used. Its function is to provide a means of concealing decoding errors when data errors in the bit-stream make it impossible to decode the data correctly. This is one of the optional error concealment strategies that has been defined.

All pictures may either be structured as fields or frames. In field structure, predictions are made independently for each field of data from one or more of the previously decoded fields. In frame structure, the predictions are made either from one or more previously decoded frames, or fields. The fields and frames that are used to construct the predictions may themselves have been decoded either as field structured pictures or frame structured pictures (Figure 5).

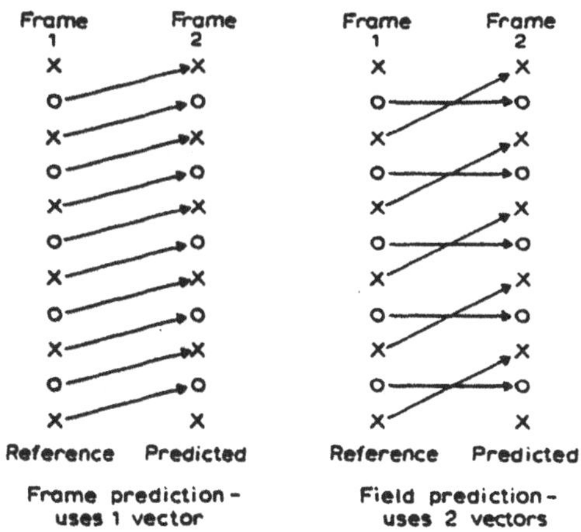

Figure 5 Frame prediction.

Generally for image sequences where the motion is slow, frame-prediction coding is more efficient. However, when motion speed increases, field structured coding becomes more attractive. Field structure also can be used to reduce the intrinsic delay of the system because in this case the basic period of operation of the system is a number of field periods rather than frame periods.

In addition to the two basic modes of frame prediction and field prediction, two further modes have been defined.

16 by 8 motion compensation uses two motion vectors for each macroblock. The first vector is used for the upper 16 by 8 region of the macroblock and the second vector for the lower half. In the case of B-pictures this may lead to the use of a total of four vectors for coding a macroblock (since there may be two references each to both past and future pictures). This mode is only permitted in field structured pictures. It is provided in field structured pictures so that the spatial area referred to by a vector is comparable to the area that a 16 by 16 block covers in a frame picture (fields have half the vertical sampling density of frames, therefore 8-field lines cover the same picture area as 16-frame lines). This mode is defined to improve the efficiency of field structured coding.

The so-called *dual-prime* mode is only permitted in P-pictures, though it may be used for both field and frame structured coding. For both field and frame structured pictures one vector is transmitted in the bit-stream, together with a small differential vector. In the case of field pictures, two motion vectors are derived from this data and are used to form two predictions from two reference fields (one of even and one of odd parity). These two predictions are averaged to form the final prediction. In the case of frame pictures this process is repeated for each of the two fields. Each field is predicted separately. This gives rise to a total of four field predictions that are combined to form the final predictions (see Figure 6). Dual prime prediction is relatively complex but has been found to be nearly as efficient as the bi-directional coding modes. Since it refers only to past pictures it is possible to use this mode to achieve nearly the same quality as can be achieved with B-pictures without the need for re-ordering the sequence and therefore saving a frame store in the decoder, possibly saving memory, and reducing the intrinsic system delay.

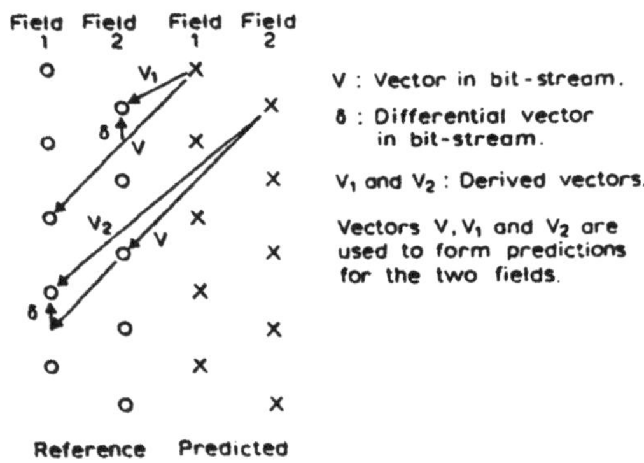

Figure 6 Dual prime predictions.

(ii) Motion estimation

Motion vectors are explicitly coded and transmitted to the decoder in the data. Therefore there is no need to standardize the methods used to find the vectors. MPEG simply defines how to code the vectors and how the decoder should interpret them. It is the responsibility of the encoder to choose the motion vectors within the constraints defined in the standard. Block matching techniques and various hierarchical schemes have been widely used. They are computationally expensive, and it is important to search a large enough range to capture the motion in the scene. The complexity of motion vector estimation is one of the major reasons why encoders are significantly more complex than decoders.

(iii) Motion representation

The basic coding unit for representing motion and other side information is the 16 by 16 macroblock. The choice of 16 by 16 is a trade-off between the need to limit the amount of overhead data in the stream and the desire to have a small basic unit for motion representation and to allow local control of the adaptation of the algorithm.

All motion vectors are coded differentially with respect to previously coded vectors. A decoder therefore has to maintain a set of predictors that record the last value of the appropriate motion vector that was decoded. Each vector is represented as the combination of the prediction and an update value. The update value is coded by variable length code for efficiency. One of the most important requirements for the choice of motion vectors is that their range is large enough to cover the actual motion present in the scene. Again, it is the task of the encoder to search a wide enough area to find good vectors.

3.11.5.5 Spatial coding

I-pictures and the residual pictures in P- and B-pictures after motion compensated prediction contain a lot of redundancy. The methods for removing this redundancy are block-based and use the DCT (discrete cosine transform).

(i) DCT

The DCT is an orthogonal transform which is exactly invertible and has a fast algorithm. Experiments have shown that for natural pictures it is a close approximation to the optimal transform that compacts the energy present in pictures in the most efficient way. It has been widely used for image compression. When the DCT is applied to pictures it is found that nearly all the energy is concentrated in a small number of low frequency coefficients.

The choice of a block size of 8 by 8 provides a trade-off between the need to use a large area for the transform so that the energy compaction is most efficient with the fact that the characteristics of a picture are spatially varying, which tends to suggest that a smaller block size should be used.

In frame structured coding MPEG-2 gives the encoder a choice of DCT modes. Data from the two fields may be combined into frames, so that the 8-lines of data in a DCT block are drawn from both fields before the DCT is performed, or the data from the two fields may be kept separate (see Figure 7). The choice depends on the speed of motion and how effective the prediction has been.

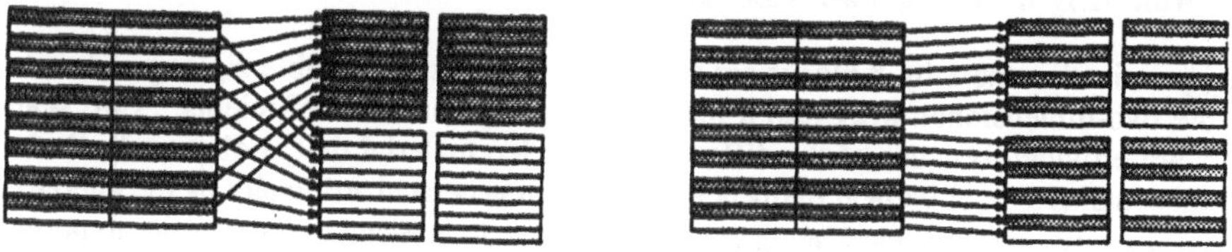

Luminance macroblock structure in field DCT coding modes.

Luminance macroblock structure in frame DCT coding modes.

Figure 7 Luminance macroblock structures.

(ii) Quantization

Quantization of the DCT coefficients is an essential operation where much of the bit-rate reduction is achieved. The DCT concentrates the energy but does not in itself reduce the amount of data needed to represent the picture. The amount of data needed is only reduced when the accuracy of the representation of the DCT coefficients is limited and the resulting values encoded. After quantization the number of non-zero DCT coefficients is usually greatly reduced.

Experiments have shown that the subjective perception of quantization errors introduced into the coefficients of block DCT data depends strongly on the frequency of the coefficient being quantized. It is necessary to quantize the low frequency coefficients accurately, but the high frequency ones can be quantized with greater errors because these errors are much less visible to the human visual system. MPEG therefore provides *weighting matrices* that allow the encoder to control the relative accuracy of the quantization of the different coefficients. Because I-pictures have signal energy at most frequencies and motion compensated pictures (P- and B-pictures) tend to have more high frequency energy, different weighting matrices can be used for different picture types. In all cases the encoder can choose appropriate matrices and download the values to the decoder as part of the compressed bit-stream.

The quantization levels are uniformly spaced except in the case of motion compensated pictures around zero, where there is a dead-zone (see Figure 8). The effect of the dead-zone is to suppress small deviations around zero. The deviations are generally caused by noise in the signal and so suppressing these values improves the overall picture quality.

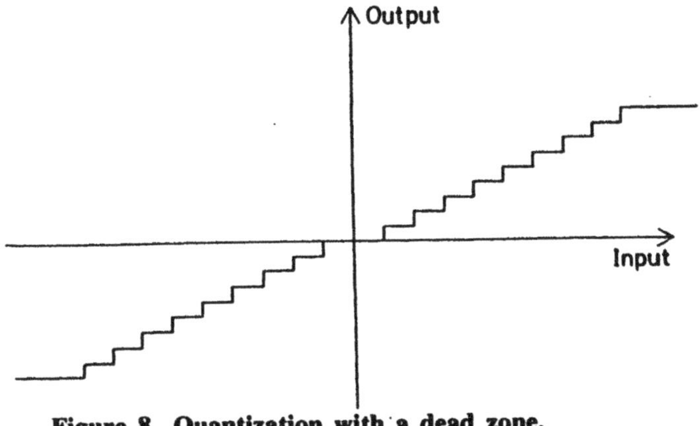

Figure 8 Quantization with a dead zone.

(iii) Run length coding and VLC coding

After quantization the 8 by 8 blocks are scanned in one of two possible zig-zag orders to turn the two-dimensional array into a linear string of quantized coefficients. Two scan orders are provided because of the differences in video material that is observed. Some material contains a lot of vertical frequencies, some of it generated artificially by the interlace picture structure. For this material there is an advantage in biassing the scan order to prefer the vertical direction. For other material it is better not to include this bias. The encoder chooses the scan order and signals its choice to the decoder in the bit-stream.

The string of coefficients is run-length coded by counting the number of zero coefficients preceding each non-zero coefficient. The combination of the run of zeros and the value of the non-zero coefficient is encoded using a table of variable length codes. Common events use a short code-word and less frequent events use longer code-words. Very unlikely events are coded by the combination of an escape length code and a fixed length code to indicate the event (see Figure 9).

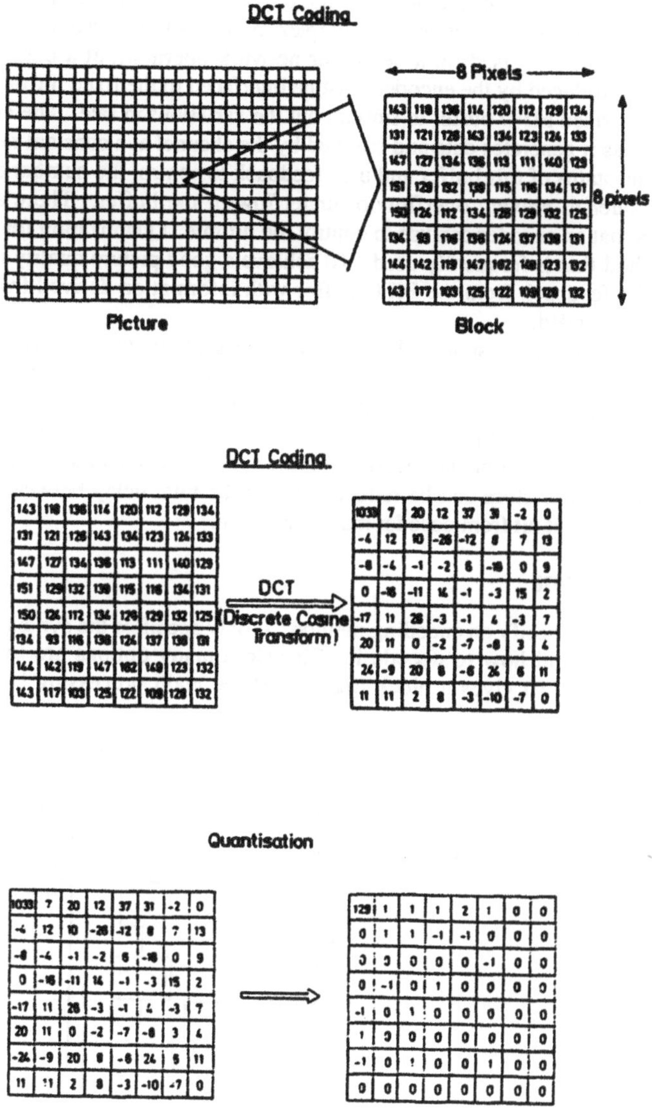

Figure 9 DCT coding and quantization.

(iv) Constant and variable bit-rate coding and buffering

The MPEG-2 system may be used for applications using either a constant bit-rate or variable bit-rate. The coding algorithm produces a variable number of bits for each picture. The number varies because the length of the run-length code depends on how many non-zero coefficients are produced, and the variable length code generates a different number of bits depending on the values of the non-zero coefficients and the lengths of the runs. In effect the number of bits generated per picture depends on the complexity and motion in the scene. Simple scenes with regular motions and little detail will produce very few bits. Scenes with many objects, at a high resolution with complex motion will generate many more bits.

If the system is used on a device or network operating at a constant bit-rate, the varying rate produced by the encoder must be interfaced to the fixed bit-rate of the network by a suitably sized FIFO buffer to smooth out the variations in rate. The buffer is filled by the encoder as it generates data and emptied by the channel at a constant rate. The encoder monitors the status of this buffer and uses feedback to control the number of bits generated by the encoder to prevent overflow or underflow of the buffer. There are a number of parameters that an encoder can use to control the bit-rate. Quantization for example gives a direct method that is frequently used. To achieve a good picture quality it is important to combine the feedback from the FIFO buffer with information derived from analyzing the pictures being coded.

Variable bit-rate systems allow a degree of variation about an average bit-rate up to some defined peak. This makes it possible to increase the bit-rate above its average value during sections of the video that are hard to encode for a period of time, and to compensate for this by reducing the bit-rate below its average value during easy sections. The aim is to achieve a more constant picture quality. Furthermore if the medium itself runs at a varying rate then the buffer store can be reduced and so the total delay between the input to the encoder and the output from decoder may be reduced.

3.11.6 Encoding

MPEG does not specify an encoding method. It only specifies the syntax and semantics of the bit-streams and how to decode it. Therefore many options are left to the encoder designer. Figure 10 gives a block diagram of the main processing blocks that will be present.

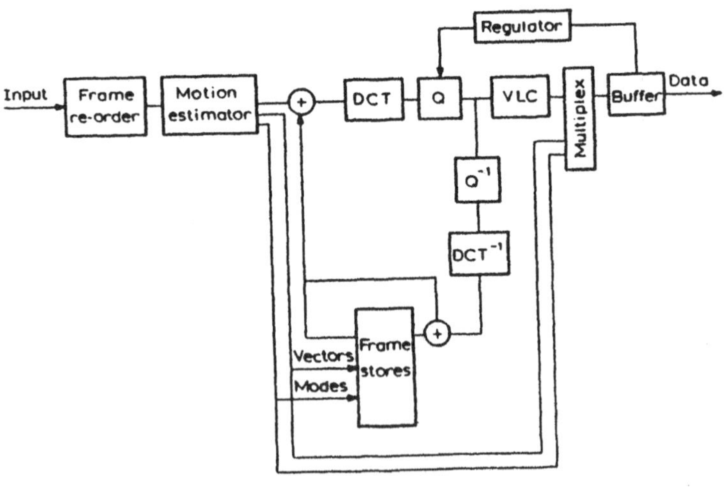

Figure 10 Simplified block diagram of an MPEG video encoder.

The input signal must first be digitized and converted to the desired colour space (Y, C_r, C_b) and colour sub-sampling ratio (4:2:0 for the Main Profile). It may improve the picture quality if some prefiltering is applied to the signal at this point.

The encoder chooses a picture type for each frame and whether it is to be coded as a field-structured picture or a frame-structured picture. Having defined this basic structure it has to estimate motion vectors for each 16 by 16 macroblock. The number of vectors depends on the picture type and the coding mode that will be chosen for each block. If B-pictures are being used the pictures must be re-ordered.

The basic coding unit in a picture is the macroblock. Each macroblock in the picture is coded in sequence from left to right and top to bottom. The macroblock consists of four 8 by 8 blocks of luminance data and blocks of chrominance data covering the same picture area. In the case of the 4:2:0 colour sub-sampling used for the Main Profile there are two chrominance blocks in a macroblock.

For each macroblock the coding mode is chosen. The choices available depend on the picture type, the picture structure and the effectiveness of the different prediction modes. Having chosen the coding mode, the motion compensated prediction of the contents of the block is made from past and/or future reference pictures. The prediction is subtracted from the actual data in the macroblock to form a prediction error signal. In a frame structured picture the encoder now has to choose whether to do a frame-based DCT or a field-based DCT. This mainly depends on how much motion there is between the two fields and how well the motion compensation has managed to predict it. Depending on this decision, the prediction error signal may have to be re-ordered. The error signal is then separated into 8 by 8 blocks and a DCT is applied to each block. Each resulting block of coefficients is quantized and scanned to form a one-dimensional string of coefficients. The side information needed by the decoder to reconstruct the block (coding mode, motion vectors etc.) is encoded and then the quantized coefficients are run-length using a variable length code table.

The rate control procedures monitor the state of the output FIFO that interfaces the encoder to the channel. This information is used to control the number of bits that the encoder will generate for subsequent blocks and pictures.

At this stage in the process a complete bit-stream has been generated that can be used by a decoder to recover the picture. The encoder has one further task to perform. It needs to store the I- and P-pictures that the decoder will recover so that it can use them as reference pictures for further motion compensated prediction. Therefore the quantized DCT coefficients are dequantized and passed through an inverse DCT transform. The prediction error signal is then added back to the prediction signal and limited to the required range to give the reconstructed reference picture.

3.11.7 Decoding

Decoding is the inverse operation. It is much simpler than encoding as there is no need to do any motion estimation and there are many fewer options. Figure 11 illustrates a block diagram of a basic system.

The input buffer is filled from the channel. The decoder reads the bit-stream until it finds the start of a coded picture and the type and structure of the picture. It decodes each macroblock in turn, using the macroblock type and motion vectors to construct a prediction of the block based on the reference pictures that are stored in the decoder. The coefficient data are read, decoded and dequantized. Each 8 by 8 block of data is transformed by the inverse DCT and the result is added to the prediction.

Once all the macroblocks have been processed, the picture has been reconstructed. If it is an I-picture or a P-picture it may be reference picture for future predictions and so must be stored, replacing the older reference picture stored in the decoder (only two reference picture need to be stored at any one time). The final stage is to reorder the pictures into the correct sequence before any post-processing and display.

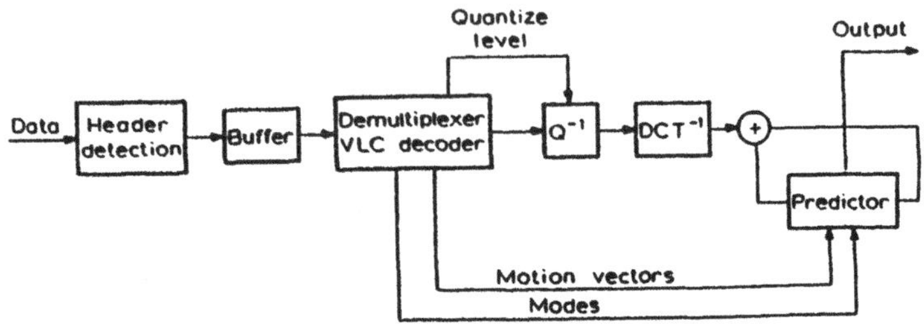

Figure 11 Block diagram of a decoder.

3.11.8 Conclusion

The MPEG-1 standard is becoming established as the method for publishing digital video on disc-based multimedia systems. Several products are now available using the principal capabilities of MPEG-1.

The MPEG-2 standard is now finalized in draft form and the path to publication as a full International Standard is well defined. It is expected that final approval will be obtained at the end of 1994.

Chip-sets and products based on MPEG-2 are now being developed, and early versions are already available. Trial services are being planned, and in some cases are already being started. It is expected that MPEG-2 will have a major impact on industry. The amount of compression that can be achieved means that bandwidth can be used very much more efficiently. This will lead to a huge increase in the number of TV channels available using conventional means of distribution, will allow new industries to enter the TV market offering new services, and open up the possibility of starting a digital HDTV service.

References

[1] ISO/IEC 11172, "Coding of moving pictures and associated audio for digital storage media at up to about 1,5 Mbit/s".

[2] D. Le Gall, "MPEG: A video compression standard for multimedia applications," *Communications of the ACM*, Vol. 34, No. 4, April 1991.

[3] ISO/IEC CD 13818, "Generic coding of moving pictures and associated audio," (MPEG-2).

[4] N. Wells and P. Tudor, "Standardization of scalable coding schemes", in *Proc. IEEE ISCAS 94 Tutorials*, May 1994.

[5] G. Bjoentegaard, "MPEG-2 coding schemes for low delay requirements", in *Proc. ISCAS 94 Tutorials*, May 1994.

[6] CCIR Recommendation 601-2, "Encoding parameters of digital television for studios".

Chapter 3.12

Standardization of Scalable Coding Schemes

Nick D. Wells and P. N. Tudor

Abstract

Scalable coding schemes are being considered in particular for HDTV in order to provide for compatibility with lower resolution scanning formats and to provide for graceful degradation of the picture quality in the presence of transmission errors. This chapter describes some of the scalable coding schemes considered within the Moving Pictures Expert Group (MPEG) and discusses the advantages and disadvantages of the different schemes which may be classified as frequency scalable, spatial scalable and SNR scalable. The reasons for selecting particular scalable schemes for inclusion in the MPEG-2 standard are described.

3.12.1 Introduction

A scalable video coding scheme is one which produces a bitstream that can be usefully decoded in parts. Decoding part of the bitstream will produce decoded pictures with less resolution or more noise than pictures decoded using the complete bitstream.

Scalable video coding schemes have a number of applications in digital communication systems. For telecommunications, scalable coding can provide compatibility with existing video transmission standards and assist concealment of transmission errors; for computer based video systems, scalable coding can generate a low resolution video preview without decoding a full resolution picture; for broadcasting applications, scalable coding can provide a range of picture qualities suited to different types of receiver and provide for graceful degradation of the picture quality in the presence of transmission errors.

Different applications have different quality requirements for the different layers within the bitstream. For example, in a broadcasting application in which a lower layer is being received continuously by portable receivers, optimization of the quality of the lower layer may be as important as optimization of the quality of the higher layer. However, in future computer applications involving the retrieval of stored video information, having a simple method for the rapid decoding and viewing of a low resolution preview picture may be more important than maintaining the quality of this picture.

3.12.2 Non-Scalable algorithm

The scalable coding schemes considered within MPEG [1] are all extensions of the non-scalable codec shown schematically in Figure 1. The coder generates a motion compensated prediction for a picture based on previous locally decoded pictures. The locally decoded picture is reconstructed exactly as it would be in the decoder. The difference between the prediction and the source picture is transformed using a Discrete Cosine Transform (DCT) which exploits the correlation between elements of each picture. The resulting DCT coefficients are quantized according to visibility weightings, variable length coded and transmitted to the decoder.

Within the decoder, a motion compensated prediction based on previously decoded pictures is formed, identical to the one in the coder loop. The prediction error sent from the coder is added to this prediction to produce the decoded output.

To allow the decoder to start decoding, the coder periodically transmits an 'intra' picture that does not require a motion compensated prediction in the decoder. Once this picture is decoded, the decoder is able to generate the motion compensated predictions necessary to decode future pictures.

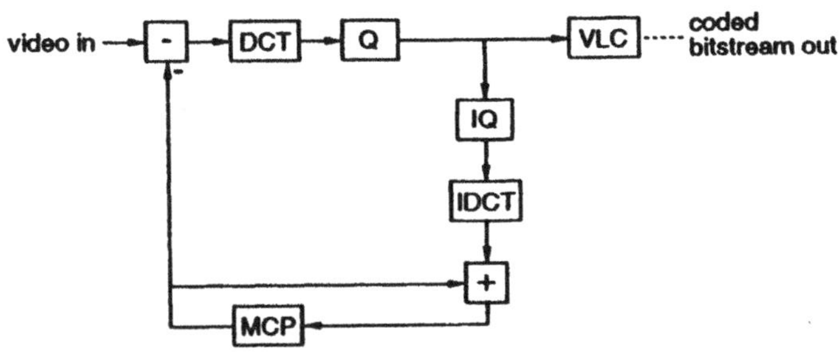

Figure 1a A non-scalable coder.

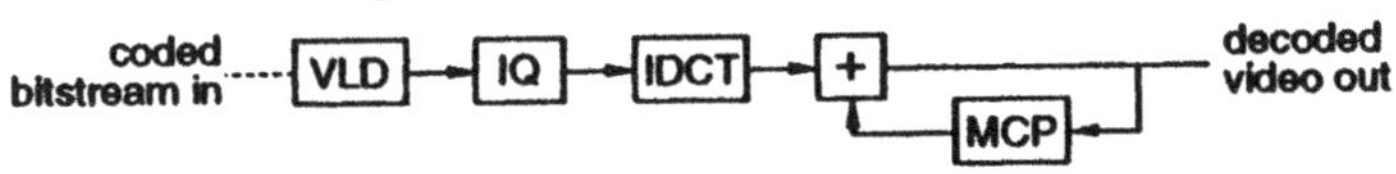

Figure 1b A non-scalable decoder*.

3.12.3 Scalable schemes

The scalable coding schemes considered within MPEG extend the basic codec architecture to provide *layers* of different resolution and *levels* of different picture quality (within a layer). The main schemes considered can be classified as frequency scalable, spatial scalable and noise (SNR) scalable.

3.12.3.1 Frequency scalable

Frequency scalable schemes are characterized by the use of the DCT to split the signal into different resolution layers and by the prediction of DCT coefficients between layers. The principle underlying this approach is outlined in Figure 2 which may be explained as follows.

Each block of the image, consisting of 8 pixels by 8 lines, is transformed using the DCT. This process gives 64 coefficients which may be transformed back to the original image using the inverse DCT process. The DCT is closely related to the Fourier Transform and we find, for example, that high horizontal frequencies present in the block tend to give high horizontal frequency DCT coefficients. Similarly, high vertical frequencies in the image block tend to give high vertical frequency coefficients.

---

* In the codec diagrams, the following abbreviations are used:

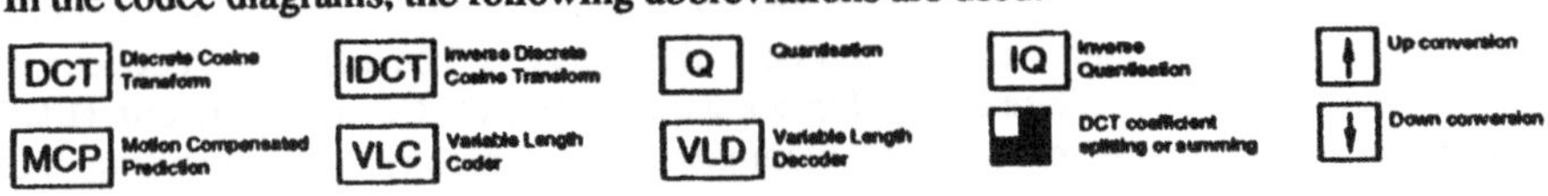

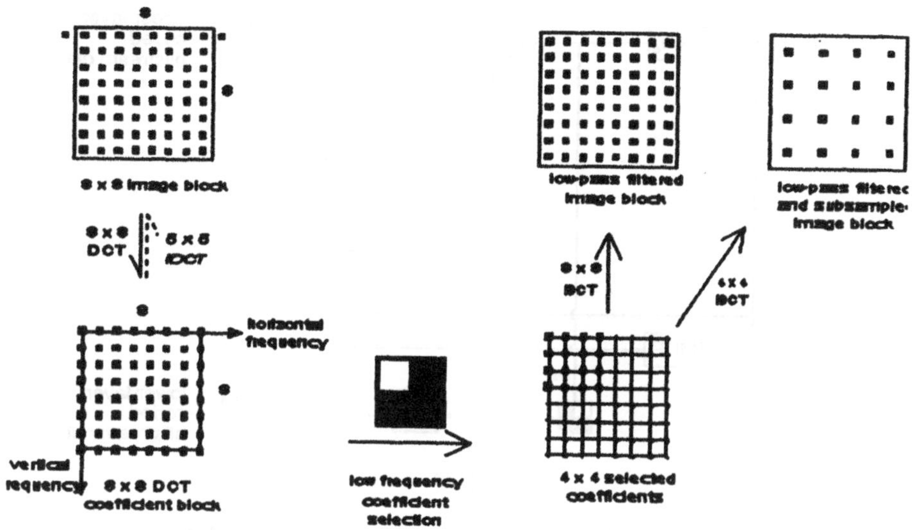

Figure 2 Principles of frequency scalable coding.

Having transformed a block, if all the coefficients are replaced by zero except the 4x4 lowest frequency coefficients, high frequency information is removed (in the main) from the block. Then, if a full 8x8 inverse transform is performed (including the 48 zero coefficients), an approximation to a low pass filtered version of the original image block is obtained. However, it is also possible to do a 4x4 inverse transform using only the 4x4 non-zero coefficients and generate a 4x4 pixel image block. The 4x4 image block corresponds (approximately) to a low-pass filtered and subsampled version of the original 8x8 image block. Figure 3 shows a two-layer frequency-scalable video coder and decoder which uses this principle.

The coder loop operates in a manner similar to the non-scalable coder, using an 8x8 DCT to split the prediction error signal into 64 coefficients. A block of 4x4 low frequency coefficients is removed from the DCT coefficient matrix, quantized, and transmitted in a lower layer. The upper layer transmits refinements to these coefficients and the remaining high frequency coefficients. By adjusting the quantization in each layer, the distortion and bit rate of the layers can be balanced by the coder. The best picture quality in the upper layer is achieved when minimal refinement of the low frequency coefficients is required. In this case, the upper layer picture quality is similar to the non-scalable case. For this to be achieved, a high proportion of the bit rate must be used for the lower layer. If the bit rate in the lower layer is reduced, additional refinement of the low frequency coefficients in the upper layer is required and the upper layer picture quality tends to be reduced.

In an upper layer decoder, the quantization refinement is added to the 16 coefficients of the lower layer and these are merged with the other 48 higher frequency coefficients. Then, the full 8x8 inverse DCT is performed to give a full resolution prediction error, which is added to the full resolution motion compensated prediction to give the upper layer output. In a lower layer decoder, only the lower layer coefficients are received and a 4x4 inverse DCT is used to reconstruct a low resolution picture. The decoded lower layer has half the resolution and sample rate of the upper layer in both horizontal and vertical directions.

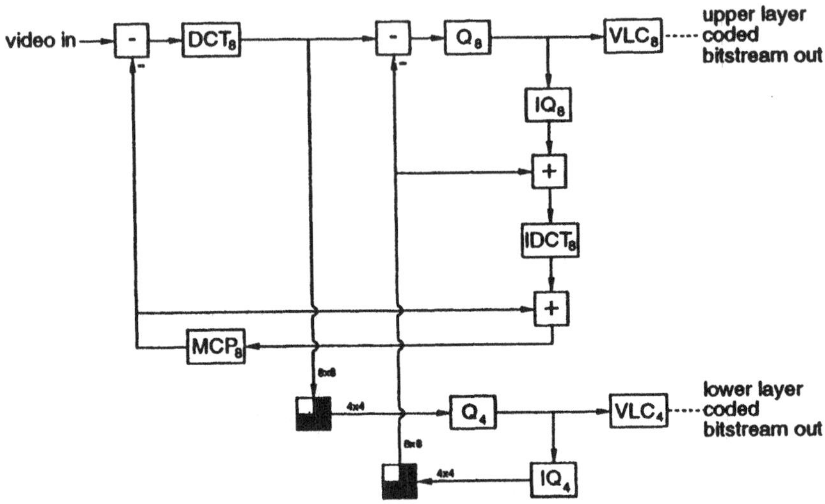

Figure 3a A two layer frequency-scalable coder.

An advantage of this frequency-scalable scheme is that only one motion compensated prediction loop is needed to decode each layer. However, there are a few limitations to such a scalable scheme. Firstly the down-conversion to the lower layer is by low frequency coefficient selection which is far from optimum when the upper and lower layers are interlaced, as is the case for HDTV in the upper layer and TV in the lower layer. This is because the high frequency coefficients are actually carrying information about the motion between adjacent interlaced fields. Methods to optimize the lower layer quality have been tried, such as selecting coefficients from field blocks rather than frame blocks and such as retaining some high vertical frequency coefficients in the 4x4 coefficient selection. However, such methods have not managed to remove visible down-conversion artifacts from the lower resolution picture.

Secondly, the quality of the decoded lower layer is degraded by what is termed "drift" in the lower layer. Drift is a progressive deterioration in the quality of the decoded picture which arises when the prediction in the decoder is not being constructed in exactly the same way as the prediction in the coder. In this case, the lower layer decoder loop is operating at a different resolution from the coder loop and these prediction loops are not exactly equivalent for the low frequency information. The predictions differ because the 8x8 and 4x4 DCTs do not split the signal perfectly into different frequency bands and because the motion vectors may not be applied to the same accuracy in the upper and lower resolution decoder loops. The seriousness and visibility of drift depends on many factors, including the scene content, the time interval between intra-coded pictures, the display size and the application for the decoded pictures.

The quality of the down-conversion can be improved and drift problems in the lower layer can be avoided by having a dual-loop coder as shown in Figure 4.

In this case the down-conversion is performed externally and can be optimized independently of the coding process. The lower layer loop then uses a 4 x 4 DCT process which matches that used in the lower layer decoder and so there will be no drift in the lower layer decoder. The 16 coefficients per block from the lower layer are used to predict the 16 low-frequency coefficients in the upper layer. The decoder is the same as the one shown in Figure 3b.

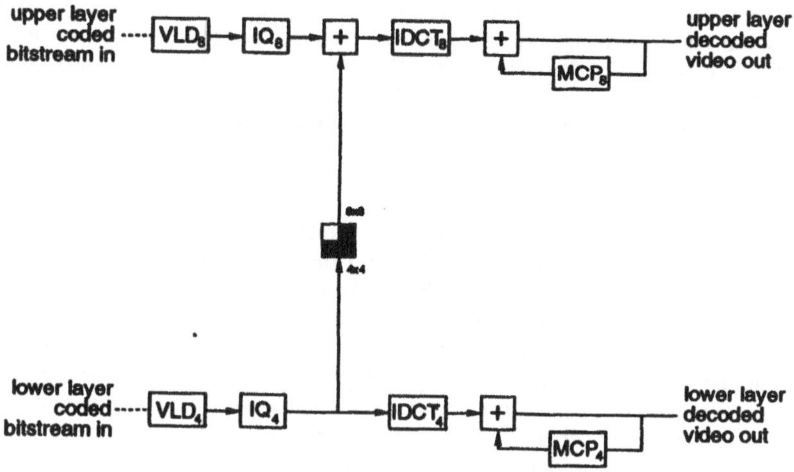

Figure 3b A two layer frequency-scalable decoder.

The disadvantages of using a dual-loop coder are that the complexity of the coder is increased and the upper layer decoded picture quality is reduced. This latter effect is because the lower layer loop coefficients make a less efficient prediction for the lower frequency coefficients of the upper layer loop, than in the single loop frequency-scalable case.

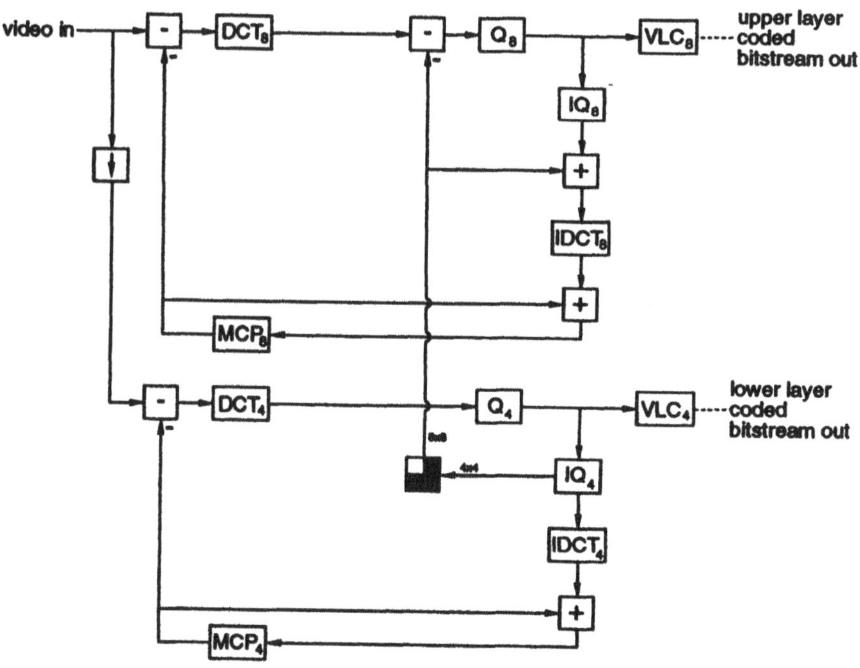

Figure 4 Two loop frequency-scalable coder with external down conversion.

In most proposed frequency-scalable schemes, the spatial resolution changes by a factor of two between layers. Changes in resolution by factors other than two are not normally considered. Note that in frequency-scalable schemes, lower layer decoders would be required to support 4x4 inverse DCT (and 2x2 inverse DCT for three resolution layers), rather than the 8x8 inverse DCT used in non-scalable MPEG decoders.

3.12.3.2 Spatial scalable

Spatial scalable schemes are characterized by the use of up converted pictures from a lower layer as a prediction in a higher layer. Figure 5 shows a block diagram of a two layer spatial scalable codec. There is no restriction on the ratio of spatial sampling frequencies in the different layers except for the complexity of the down and up conversion processes.

In the scheme shown in Figure 5, the prediction in the upper layer can be taken as the motion compensated prediction from the upper layer loop (termed the *temporal* prediction) or taken as the up-converted decoded picture from the lower layer loop (termed the *spatial* prediction) or a weighted combination of these two predictions. This selection is performed by the adaptive weighting function, *W*, in Figure 5.

In such a scheme, there is no drift in the decoded lower layer and the down-conversion to the lower layer can be optimized as required. Also, only 8x8 inverse DCT is required in the lower layer decoder.

The quality of the lower layer is not reduced in any way by the scalability, but the quality of the upper layer is reduced because the bit rate used on the lower layer is not completely compensated for by the improved efficiency of prediction available in the upper layer.

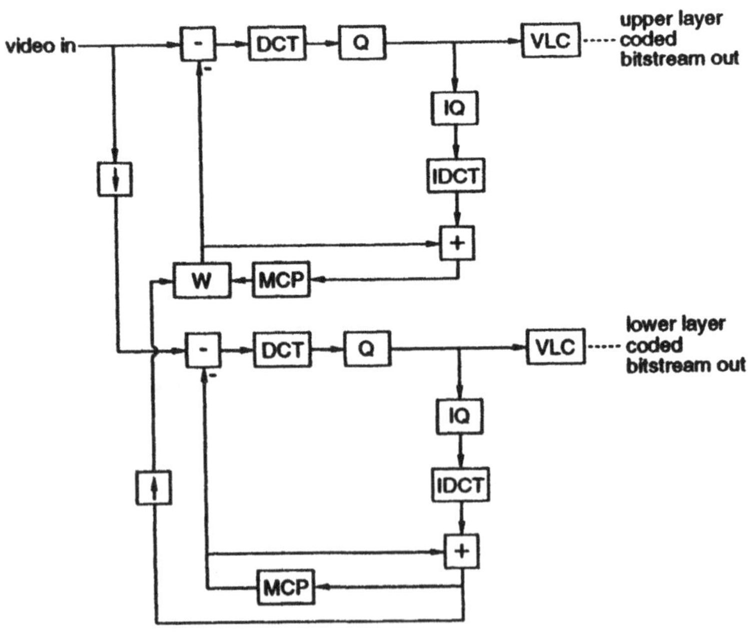

Figure 5a A two layer spatial scalable video coder.

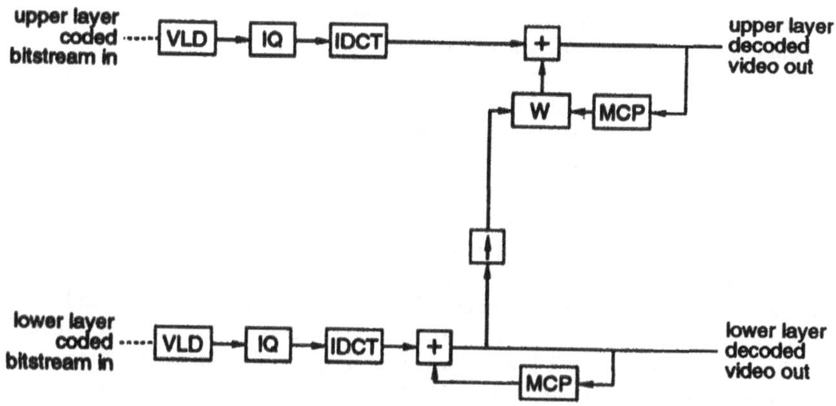

Figure 5b A two layer spatial scalable video decoder.

3.12.3.3 SNR scalable

SNR scalable schemes quantize the coefficients to different levels of accuracy in the different SNR levels. Figure 6 shows a block diagram of a two-level SNR-scalable video codec.

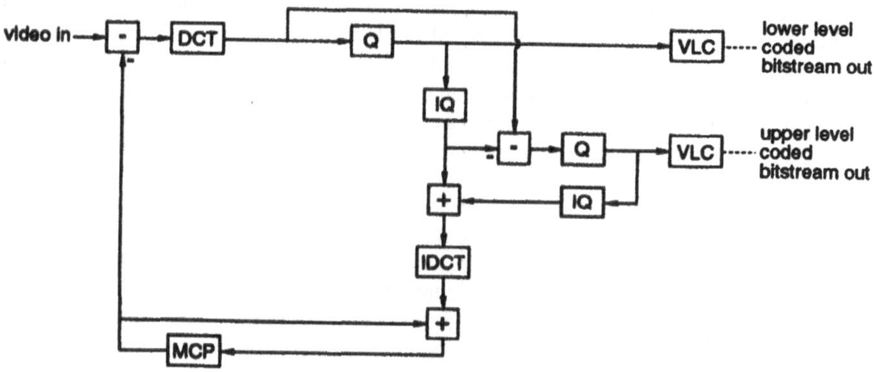

Figure 6a A two level SNR scalable video coder.

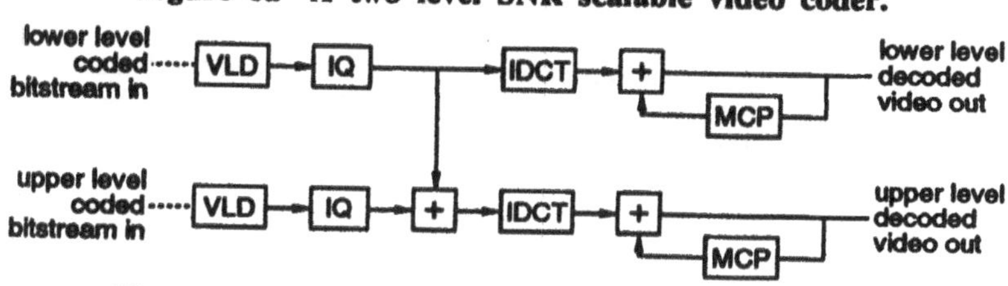

Figure 6b A two level SNR scalable video decoder.

The codec operates in a similar manner to the single-level codec, with the addition of an extra quantization stage. Two levels of decoded picture quality are produced, differing in SNR content but not resolution.

The lower level coder quantizes the DCT coefficients to a given accuracy and transmits these quantized coefficients in the lower level bitstream. The quantization error introduced by the lower level quantizer is requantized and sent in the upper level as a refinement to the lower level coefficients.

For an upper level decoder, both levels must be received and the upper level coefficient refinements are added to the lower level coefficient values following inverse quantization. The summed coefficients are then decoded using a single-loop decoder. The lower-level decoder receives just the coefficient values in the low level bitstream and decodes these directly using a single-loop decoder. In the upper level bitstream, as little data as possible is transmitted apart from the coefficient enhancement values. For example, motion vector information is only sent in the lower level bitstream.

Note that the lower level decoder output will have some drift because the prediction loop has at its input coarsely-quantized coefficients, whereas the coder loop (and the upper level decoder loop) use the more finely-quantized coefficients. However, experiments have shown that this drift in the SNR lower level is not very visible and the resulting quality is acceptable for many applications. Drift in the lower level can be corrected by using a dual-loop coder in a manner similar to that shown in Figure 4 for a frequency scalable coder except that lower and upper level loops work at the same picture resolution. However, a dual-loop coder is not as efficient, in terms of the upper level quality in a given bit rate, as a single-loop coder.

3.12.4 Chrominance scalability

MPEG is directed, in the main, towards systems in which the vertical sampling rate and resolution for the chrominance is half that of the luminance. However, for some studio-based applications it may be desirable to send a chrominance enhancement signal carrying the additional vertical chrominance resolution of the studio source pictures.

Spatial scalability was investigated as a means of sending an enhancement signal carrying the additional chrominance resolution. It was found that there was little to be gained by going to the complexity of spatial scalable coding compared with sending the whole of the full resolution chrominance component in addition to the lower resolution chrominance component (i.e. simulcasting the higher resolution component). However, it was found efficient to predict the DC coefficients of the high-resolution chrominance blocks using the appropriate DC coefficients of the lower resolution chrominance blocks. As little additional information as possible is sent in the enhancement layer and the syntax used to describe SNR scalability is used to transmit the chrominance "simulcast" enhancement.

3.12.5 Standards conversion

Satisfactory performance of resolution-scalable schemes depends on good quality standards conversion. Conversion between existing broadcast standards is problematical because of the interlaced scanning. Optimum performance for stationary pictures can be obtained if information from both input fields is used to derive each output field.

However, with fast movement, better results are obtained if each output field is derived using information from only one input field because of the changes in picture detail between fields.

Good quality conversion can be obtained using systems which adapt between inter-field and intra-field interpolation according to the amount of movement in the sequence. However, such methods can introduce visible switching effects.

More consistent results can be obtained using a non-adaptive system based on a combination of vertical and temporal interpolation [2]. The interpolation proeess can be described as a vertical-temporal filter removing unwanted spectral components from the input signal which would give rise to undesirable alias components at the output scanning standard. An important feature of the vertical-temporal interpolator is that the sum of the components from non-central fields is equal to zero which ensures good conversion of moving objects.

Three-field vertical temporal filters can give high-quality conversion between interlaced standards. The down-conversion process does not need to be defined for scalable schemes within the MPEG standard. However, in the case of spatial scalability where an up-converter is required in each coder and decoder, the up-conversion process must be precisely defined. The choice of up-converter was chosen taking the following factors into account:

i) The up-converter should be as simple as possible for mass-produced decoders.

ii) The up-converter should be intrafield or intraframe in order to avoid complications introduced by possible frame reordering for transmission. The frame order for transmission is changed when bi-directional prediction is used.

iii) The up-conversion process must be capable of simple extension to cases when the up-sampling ratio is other than 2:1.

iv) The upper layer picture quality is not critically dependent on the quality of the up-conversion.

v) The up-conversion should provide as good an up-converted picture quality as possible because, when graceful degradation is required in the presence of transmission errors, the up-converted lower layer is used as a replacement for the decoded upper layer picture.

The up-converter defined for spatial scalability within MPEG combines intraframe up-conversion with linear interpolation which can be applied for an arbitrary up-conversion ratio.

3.12.6 Discussion

Frequency and spatial scalable schemes provide layers of different resolution and SNR scalable schemes provide levels of different picture quality within a layer.

For broadcasting applications the picture quality of the lower layer is important and, for interlaced picture formats, external down-conversion is required which filters the picture in an optimum way according to the scan conversion. Frequency- and spatial-scalable schemes support external down-conversion, and algorithmically become similar with its inclusion. Also, the performance of the two schemes then becomes very similar in both layers.

It is expected that by the time an HDTV service starts there will be a large number of MPEG main profile TV decoders installed and compatibility with these will be important. MPEG main profile TV decoders support only 8x8 inverse DCT. Spatial scalability is then the only option for providing resolution scalability because a frequency-scalable scheme would require 4x4 or 2x2 inverse DCT processing which would not be compatible with main profile decoders. Therefore, the spatial scalable scheme was the one chosen within MPEG to provide for resolution scalability.

For computer-based video systems that require resolution scalability (in order to offer rapid decoding of picture previews), the quality of the lower layer may be less important. In such cases, frequency-scalable schemes can offer a low-complexity solution whilst maintaining high quality in the upper layer. If required, the SNR tools can be used to provide frequency scalable functionality.

3.12.7 Performance

One method of offering a range of different picture resolutions and qualities is to use the 'simulcast' method. With this method, independent coders operate in parallel to produce independent bitstreams intended for independent decoders. Each decoder uses only one bitstream and no attempt is made to use information intended for one decoder in another. Simulcast represents the lower coding performance threshold against which scalable schemes can be compared.

Using a two layer spatial scalable scheme with interlaced HDTV and TV, at bit rates of 14 Mbit/s and 6 Mbit/s respectively, the HD layer SNR is improved by between 0.5 dB and 1.5 dB relative to simulcast. The higher figure is approximately half way between the quality obtained by simulcasting and the non-scalable case where the entire bit rate is used for the HD layer. The improvement due to spatial scalability increases in scenes containing complex motion, for which the upper layer temporal prediction is poor. The main benefit of spatial scalability is in intra coded pictures where no temporal prediction is available. If a significant proportion of the available bit rate has been spent on coding the lower layer, then it is worthwhile to use spatial scalable coding to improve the efficiency of the upper layer coding.

SNR scalability works well. In a two level SNR scalable scheme, the loss of picture quality in the upper level is only slight compared to a non-scalable scheme using the entire bit rate. In the lower level, drift is barely perceptible and the picture quality is similar to that of a non-scalable scheme operating at the lower level bit rate.

3.12.8 Conclusions

The scalable coding schemes considered by MPEG have been discussed. Frequency- and spatial-scalable schemes offer scalable resolution and SNR scalability offers scalable picture coding quality.

For broadcasting applications, resolution scalability is best achieved using the spatial scalability method. This permits a high quality TV layer, supports a wide distribution of bit rate between the layers and is compatible with non-scalable main profile MPEG decoders in the lower layer.

Acknowledgments

The authors would like to thank the BBC for permission to present this material. This document is based on a paper first presented at the 6th Tirrenia International Workshop on Digital Communications on 6th September 1993 and published by Elsevier Science Publishers B. V.

References

[1] ISO/IEC JTC/SC29/WG11, MPEG-2 Third working draft, New York, July 1993.

[2] V. G. Devereux, "Standards conversion between 1250/50 and 625/50 TV systems," *IBC 1992. IEE Conference Publication*, No. 358, pp. 51-53, 1992.

Chapter 3.13

MPEG-2 Coding Schemes for Low Delay Requirements

Gisle Bjøntegaard

3.13.1 Introduction

Digital video compression techniques reduce the capacity needed to transmit or store the signal. The compression methods take considerable advantage of the correlation both within a picture and between neighbouring pictures. The use of this correlation in time typically introduces delay of a few picture periods. However, the process of generating efficient coding introduces different kinds of field/frame reordering, computational delay and buffering. The net result is that a delay of a few tens of sampling periods is introduced. This delay comes in addition to traditional delays such as that introduced by satellite transmission.

This chapter will focus on the contribution that different elements make to the total delay in MPEG-2 coding.

3.13.2 Applications requiring low delay

Applications requiring two way real time communication suffer considerably from delay introduced by the communication link. It is well known, from telephony, that the type of delay introduced by satellite transmission (typically 260 ms) is objectionable to the user. There is therefore good reason to believe that, for real time communication including video, a total delay of 200 - 300 ms will be annoying. Services like video telephony and video conferencing fall into the class of services that are particularly sensitive to additional delay. These services will also, to an increasing extent, be used over long distances where transmission may be via satellite. In this case any coding delay will come in addition to the 260 ms introduced by the satellite.

Concerning TV, added delay is not considered to be particularly critical. However, even here it is foreseen that low delay may be preferable in some cases when interaction is used between sources at different locations.

Another kind of delay that is more important for TV is the time it will take to receive a clean picture after entering the bitstream at an arbitrary point. This is referred to as the "channel hopping" delay because it indicates how long it will take from the time that you press a button for a new channel until you have a clear picture.

3.13.3 Two principal types of delay

3.13.3.1 Coder to decoder delay - D_{tot}

D_{tot} is the total delay from information retrieval at the source end to the presentation at the receiver end. D_{tot} is the important parameter in real time applications such as two way communication, where the total information delay is an important factor in characterizing the quality of service.

3.13.3.2 *Delay from entering a bitstream to getting a full quality signal - D_{ch}*

D_{ch} is of particular importance for "channel hopping" in the TV environment. It gives the time that it takes from entering an incoming bitstream until a full quality TV picture is presented. D_{ch} will typically vary between a minimum and a maximum value. Some of the same delay elements are included in D_{tot} and D_{ch}. Some delay elements have less influence on D_{ch} than on D_{tot} because D_{ch} is less influenced by actions on the encoder side (see note *,in table 1). It should be noted that D_{ch} may be smaller or larger than D_{tot}.

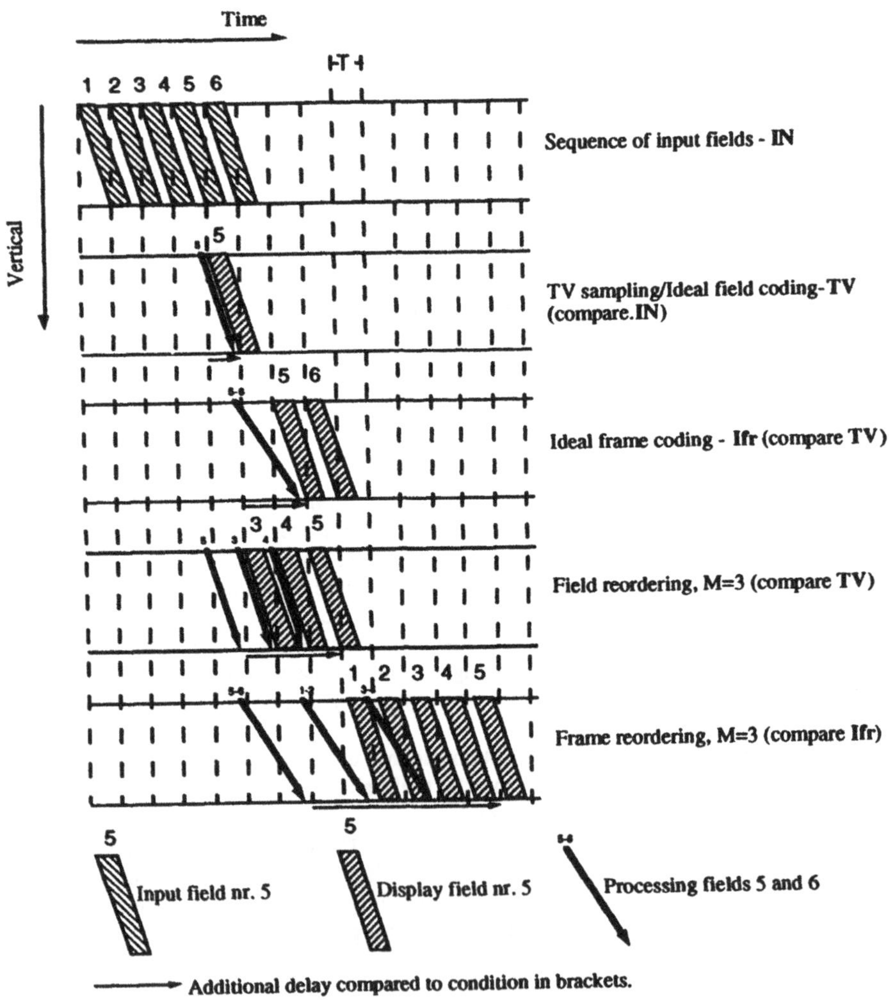

Figure 1 Illustration of coding delay elements.

3.13.4 Identification of different sources of delay in MPEG-2 coding

The total coding delay is composed of many separate elements. This section will try to identify the different parts and clarify their origin. The unit for delay is one field period - 16.7 ms for a 60 Hz system and 20 ms for a 625 line system.

Figure 1 illustrates the origin of the different delay elements. An explanation of how to read the illustrations is necessary.

The plots represent how the TV pictures are retrieved, coded, transmitted and displayed. The plots show the time and vertical dimensions. It is assumed that the retrieval and display processes are field based. The different shaded areas represent how a field is integrated in the camera sensor over a time period T and presented on a display unit for the same time period T. The magnitude of different delay elements can be read out from the plots by counting the number of time slots T between camera input and display. The total delay is obtained by the proper combination of the different delay elements. In the plots of Figure 1 neither transmission delay nor buffer delay are included.

3.13.4.1 Examples of total delay for different versions of MPEG-2 coding

Table 1 gives some examples of total delays taking into account all the elements discussed above. The following assumptions are made:

- Transmission delay is not included.
- Delay due to preprocessing of the signal is not included.
- The contribution from multiplexing is assumed to be well below the other elements and is therefore not included.

Comments on the parameters in the table:

- SAMP represents the sampling delay from Figure 1.
- The buffer delay is estimated. It reflects typical values from the simulation work. It may well be that hardware implementations will use somewhat larger buffer sizes.
- "Wait for intra" is the delay on channel hopping due to the separation of intra pictures. The size is set to $24T$ if intra pictures are used.

Coding mode	SAMP	Ideal frame	Reorder	Buffer	D_{tot}	Wait for intra	D_{ch}
1: Field M=1, intra slice	T			$3T$	$4T$		
2: Field M=3, intra slice	T		$3T$	$3T$	$7T$		
3: Frame M=1, intra slice	T	$2T$		$5T$	$8T$		
4: Frame M=3 intra slice	T	$2T$	$6T$	$5T$	$14T$		
5: Field M=1, intra field	T			$6T$	$7T$	$24T$	$(7\text{-}31)T$
6: Field M=3 intra field	T		$3T$	$6T$	$10T$	$24T$	$(10\text{-}34)T$
7: Frame M=1, intra frame	T	$2T(T)$*		$11T$	$14T$	$24T$	(13-37T
8: Frame M=3 intra frame	T	$2T(T)$*	$6T$	$11T$	$20T$	$24T$	$(19\text{-}43)T$

* The contribution to D_{tot} is $2T$ and to D_{ch} T.

Table 1 Examples of total delays - both end to end delay and "channel hopping delay" - for different coding modes.

3.13.4.2 Modes that fulfill the low delay requirement of 150 ms

The low delay limit is typically set to 150 ms. This represents $7.5T$ for a 50 Hz system and $9T$ for a 60 Hz system. Generally we may say that a "low delay" mode requires that D_{tot} stays below $8T$. This requirement is met by 4 of the coding modes in table 1.

- One of the possible low delay modes uses M=3.
- One of the possible low delay modes uses intra field instead of intra slices.

3.13.4.3 Channel hopping delay

Channel hopping delay is considered only in connection with intra field/frame update. (Intra slice updating is not expected to be used in applications where low channel hopping delay is important). It is furthermore given with minimum and maximum values. Aspects that influence channel hopping delay are:

- The main contribution comes from the separation between I-pictures.
- Using M>3, frame coding increases D_{ch} with MxT ms compared with field coding.
- Using frame coding, M>3 increases D_{ch} with $2M$xT ms compared with the use of M=1.

A target for maximum channel hopping delay has typically been 500 ms. This corresponds to $30T$ in a 60 Hz system and $25T$ in a 50 Hz system. From table 1 it is seen that this is possible only for some of the modes with update every 24 field period as assumed.

On the other hand, let us assume coding with mode8 (MP@ML, M=3, frame coding). N indicates how frequently I-pictures are used (N=10 means that every 10th frame is intra coded). From table 1 it is seen that the approximate relations between D_{tot} and D_{ch} (=500 ms) are:

50 Hz sequence:

$$\text{"Wait for intra delay"} \approx Dch - Dtot \approx 25T - 19T = 6T$$

60 Hz sequence:

$$\text{"Wait for intra delay"} \approx Dch - Dtot \approx 30T - 19T = 11T$$

The following requirements on N will be needed in order to obtain 500 ms channel hopping delay:

- 500 ms as maximum for 50 Hz system: intra frame every 120 ms (N=3).
- 500 ms as average for 50 Hz system: intra frame every 240 ms (N=6).
- 500 ms as maximum for 60 Hz system: intra frame every 167 ms (N=5).
- 500 ms as average for 60 Hz system: intra frame every 367 ms (N=11).

It is therefore seen that a low value of N has to be used unless the target channel hopping time is relaxed.

(i) Performance loss with frequent intra update

Intra pictures typically require five times as many bits as predicted pictures. This results in 25% increase in bit consumption if N is reduced from 12 to 5. This is a high penalty and other solutions, such as coding with $M=1$, would be preferable if low channel hopping delay gets very high priority.

3.13.5 Special tools for use in low delay

3.13.5.1 Intra slice update

Prediction between pictures largely improves coding performance. However, the sequence has to be regularly updated by intra coding, that is without prediction from previous pictures. Two of the main reasons for regular update are to prevent error propagation and to be able to enter the bitstream at a large number of points. The most common way of obtaining the desired function, is to code a whole picture intra. However, this picture requires several times as many bits as predicted pictures and as a result a large buffer (with corresponding coding delay) is needed to be able to feed a constant capacity channel and to present decoded pictures at a constant picture rate.

Alternatively, a small part of each picture may by intra coded so that the whole picture is refreshed after a certain time. One proposed method is to code two intra slices, a horizontal stripe of 16 lines per picture. With frame coding and CCIR 601 format this means that a total update takes 520 ms if the format is 525 lines and 720 ms if the format is 625 lines.

Figure 2 indicates how the slice updating is done. To guarantee that a complete update is obtained, some restrictions on the motion vectors must be made. Referring to Figure 2, this means that data in region 1 may never be predicted from data in region 2. The arrow in the figure represents such an "illegal" vector which may prevent a clean update.

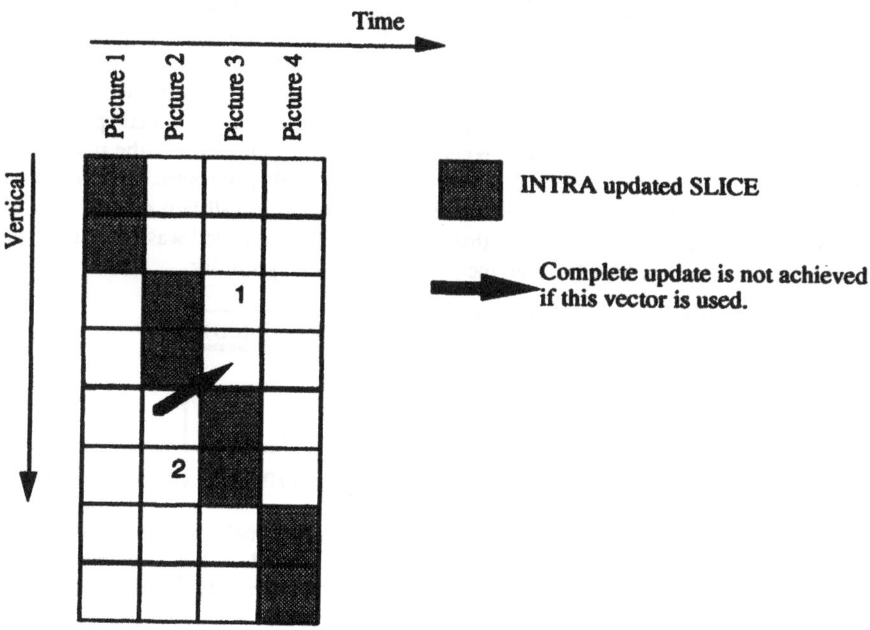

Figure 2 Intra slice updating scheme.

3.13.5.2 Dual′ prediction

One of the merits of B-pictures is that prediction is made from the average of two fields. In the case of B-pictures, this average is made up from one past and one future field. This averaging has a filtering effect in the time domain and this type of 2-tap filtering turns out to have a favorable effect on the prediction.

The dual′ prediction mode makes use of the last two decoded fields for prediction. The positive effect of averaging two fields is therefore obtained. The dual′ prediction mode could in principle be used for P- as well as for B-pictures. However, due to hardware considerations, it is only allowed to be used if B-pictures are not present. Dual′ prediction is therefore considered to be a coding tool to bring the performance of "low delay coding" closer to that of coding using B-pictures.

3.13.6 Coding performance at low delay coding

For real time applications, reduced delay increases the quality of the service. The optimum would be to have practically no delay at all (say less than 50 ms). This is possible within MPEG-2 using mode1 (field coding with M=1) in table 1. However, this increased quality of minimal delay would result in a noticeable reduction in picture quality.

The use of intra slices instead of intra pictures in low delay applications, has a slightly negative effect on coding performance. This is due to the limitations on motion vectors to obtain complete update (see Figure 2). The degradation in picture quality tends to stay below the "visibility level". Use of intra slices will normally increase the "channel hopping delay" by up to a factor of 2. This is not considered to cause a large problem in applications needing low delay.

If we stick to the "normal" requirement of a low delay of 150 ms, this may be obtained by different coding structures. Here I will focus on the mode3 in table 1. The coding performance of this mode is close to that of mode7. In the process of evaluating dual′ prediction, core experiments were set up to do comparisons between mode8 (considered to give the best picture quality) and mode7 with and without dual′ prediction. Several laboratories performed the core experiments. Table 2 summarizes the results that were presented. The SNR values for luma are used as the quality measure. Four different labs performed the experiment with eight different test sequences (all labs did not code all sequences). It should also be pointed out that the dual′ prediction tested was not exactly the same as the one defined in MPEG-2, but the performance is very similar.

Coding mode (see table 1)	Luminance SNR (dB)
7, no dual′	29.08
7, dual′	29.79
8	29.83

Table 2 Comparison of signal to noise ratios for different coding methods.

It is seen that, with SNR as a measure for performance, the dual′ prediction brings the performance of mode7 very close to that of mode8. SNR is just one measure of performance. The comparisons could therefore look somewhat different if subjective picture quality was assessed. Another aspect that should be kept in mind is that the difference in performance between M=1 and M=3 is larger for sequences where the camera introduces motion, especially panning. Many of the test sequences used in table 2 contains panning, whereas it would be expected that the camera would have fixed settings in a typical application requiring low delay.

For a typical real time application, where the camera is fixed and produces an interlace video, it may be concluded that close to optimum quality and end to end delay of less than 150 ms may be obtained within MPEG-2.

3.13.7 Conclusions

Two types of delay have been identified: "End to end delay" that is of importance for real time communication and "Channel hopping delay" that is of particular concern for TV applications. A number of sources for coder/decoder delay within the MPEG-2 standard have been discussed.

Some of the coding elements that increase compression performance have been shown to also increase delay. It has been shown that it is possible to obtain a good compromise, between coding performance and an end to end delay of less than 150 ms, within the MPEG-2 standard.

To obtain the target channel hopping delay in the MP@ML mode, frequent intra updating has to be used. This will reduce the coding performance. If intra updating is performed every 500 ms, the resulting channel hopping delay is typically in the range 700 - 900 ms.

Chapter 3.14

MPEG-2 Systems and the Transport over ATM

Ming-Ting Sun

Abstract

MPEG-2 (Moving Picture Experts Group - phase 2) and ATM (Asynchronous Transfer Mode) both are key enabling technologies for the emerging information superhighway. MPEG-2 is developed for high-quality general video applications including video on demand, two-way communications, and HDTV (High Definition TV). ATM is a cell relay technology which can provide the flexibility and high-speed needed for networked video services. For the transport of MPEG-2 video/audio over ATM, several issues including timing recovery, bit-error detection/correction, cell-loss detection/correction, and mapping of MPEG-2 bit-streams into ATM cells need to be carefully examined. Currently, these issues are being actively discussed in various forums and standards organizations including ATM Forum, ITU-T SG15, and ISO MPEG. This chapter provides a short tutorial on MPEG-2 systems and ATM, and discusses the issues related to the transport of MPEG-2 video/audio over ATM.

3.14.1 Introduction

MPEG-2 [1-3] and ATM [4] both are key enabling technologies for the emerging information superhighway. MPEG-2 is developed for high-quality general video applications including video on demand, two-way communications, and HDTV. ATM is a cell relay technology which can provide the flexibility and high-speed needed for networked video services.

The MPEG-2 standard includes four parts: Systems [1], Video [2], Audio [3], and Conformance Testing (the Conformance Testing specification is still under study). In this chapter we will focus on the MPEG-2 . For the transport of MPEG-2 video/audio over ATM, several issues related to MPEG-2 systems and ATM need to be carefully examined. For example, the MPEG-2 systems specification uses time-stamps and a constant delay timing model for decoder clock recovery. However, in ATM networks, this model is not valid due to the cell delay jitter. The cell delay jitter in ATM networks will introduce jitter in the recovered decoder system clock. It is important to make sure that the decoder system clock jitter will not cause visible degradation to the decoded video and audio. Other issues include the mapping of MPEG-2 bit-streams into ATM cells, bit-error detection/correction, and cell-loss detection/correction. Currently, these issues are being actively discussed in various forums and standards organizations including ATM Forum, ITU-T SG15, and ISO MPEG. A block diagram of some of the options under study by ITU-T SG15 for a broadband audiovisual terminal is shown in Figure 1 and will be explained in the following sections.

This chapter provides a short tutorial on MPEG-2 systems, ATM, and issues related to the transport of MPEG-2 video/audio over ATM. The purpose of this paper is to promote an end-to-end video delivery solution. Sections 3.14.2 and 3.14.3 provide a brief tutorial on MPEG-2 systems and ATM respectively. The issues related to the transport of MPEG-2 video/audio over ATM are discussed in section 3.14.4. Finally, a brief summary is presented in section 3.14.5.

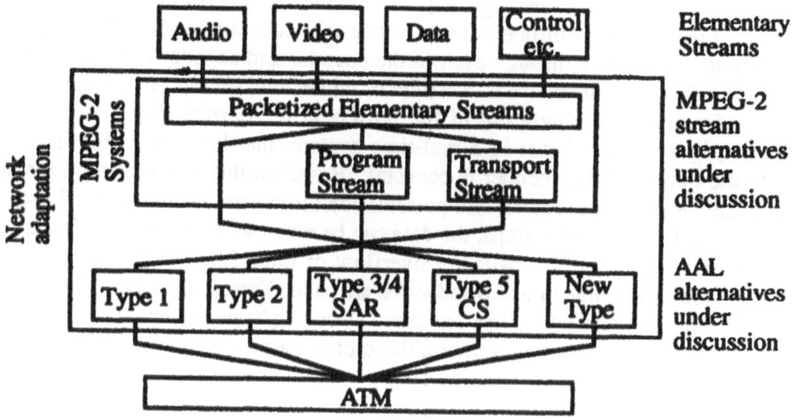

Figure 1 A block diagram of a broadband audiovisual terminal.

3.14.2 MPEG-2 systems

The MPEG-2 systems specification [1] defines the syntax for a valid MPEG-2 bit-stream. This syntax allows the support of functions which include combining multiple coded video, audio, data, and control streams into a single stream which is suitable for storage and transmission over different media, supplying timing information for decoder buffer management, clock recovery, and synchronized playback of video and audio, providing information for establishing resources needed for video/audio decoding, and providing hooks for conditional access and local insertion of advertisement.

In MPEG-2 systems, the coded video, audio, data, and control streams are called Elementary Streams. Elementary Streams are packetized into Packetized Elementary Streams (PES streams) which consist of PES packets. Each PES packet can contain variable number of coded bytes from one and only one elementary stream. Multiple PES streams can be multiplexed into a single MPEG-2 systems stream for storage or transmission. Two kinds of MPEG-2 systems streams are defined - Program Stream (PS) and Transport Stream (TS). Both stream definitions are packet-oriented multiplexes. The PS is intended for relatively error-free environment (or where the errors have been corrected as in CD-ROM applications). The PS packets can have variable and relatively long length (may be a few thousand bytes). The TS is suitable for a relatively error-prone environment. The TS packets have a fixed length of 188-bytes (4 byte of packet header and 184 byte of payload). This length was chosen based on the considerations from encryption (184-byte is a multiple of 8-byte which is the block-size of popular encryption algorithms), error correction needs (a length not greater than 255 byte is desirable for Reed-Solomon codes), and ATM adaptation (188 = 4 x 47 which can fit into 4 ATM cells as will be explained later). The Grand Alliance HDTV has also adopted the MPEG-2 TS structure. Having the same transport stream for all applications greatly promotes interoperability since the circuitry for packet processing may be standardized and no transcoding of formats is necessary between multiplexes in combined satellite/cable/telephony networks. It should be noted that the PS and TS are designed for different applications. It is possible and reasonable to convert from one to the other. However, one is not a subset or superset of the other. In this paper we will focus on the TS.

One of the important functions of the MPEG-2 systems is to supply the timing information for decoder buffer management. The MPEG-2 systems timing model is shown in Figure 2. For a constant bitrate channel, since the video and audio encoding produce variable bitrates, the delays through each of the encoder and decoder buffers are variable. To prevent the decoder buffers from overflow or underflow, timing information is needed for the decoder to read the data out of the buffers at the correct time. For this buffer management, it is observed that in a real-time system, the delay from the input of the encoder to the output of the decoder is constant. Based on this observation, time stamps which indicate the correct decoding and presentation timing are added into the bitstream in the encoder. With these time stamps, the data can be read out of the decoder buffers at the correct time so that the buffers will not overflow or underflow. The time stamps that indicate the presentation time of audio and video are called Presentation Time Stamps (PTS); those which indicate the decoding time are called Decoding Time Stamps (DTS). In addition to supporting decoder buffer management, these time stamps are also used to synchronize the playback of video and audio.

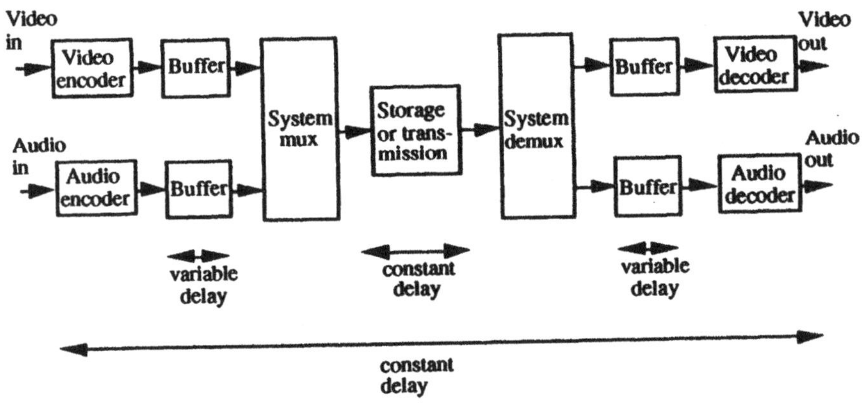

Figure 2 MPEG-2 systems timing model.

Another important function of the MPEG-2 systems is to supply the timing information for decoder timing recovery. In the MPEG-2 systems model, both the encoder and the decoder have a 27 MHz system clock. The system clocks are used to derive other clocks (such as the audio and the video sampling clocks) in the encoder and the decoder. For the decoder buffer not to overflow or underflow, the decoder system clock should be properly locked to the encoder system clock. To achieve this, the MPEG-2 timing model assumes the transmission channel has constant delay. Time stamps called Program Clock References (PCRs) which indicate the encoder system clock time are added to the bit-stream systematically. These time stamps arrive at the decoder with a fixed delay. They are compared with time stamps generated based on the decoder system clock. The differences between the time stamps from the encoder and those from the decoder are the error signals used to control the decoder phase lock loop to lock the decoder system clock to the encoder system clock. An example of using PCRs and a phase lock loop for the decoder clock recovery is shown in Figure 3 [5].

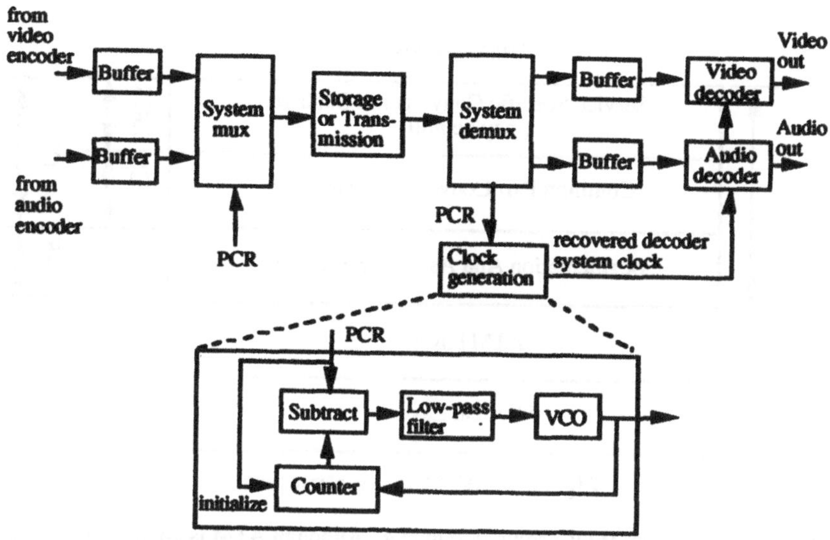

Figure 3 MPEG-2 timing recovery using PCR.

3.14.2.1 ATM

ATM [4] is a cell based technology supported by international standards. ATM is defined to operate over a number of physical layers ranging from mega bits per second to giga bits per second, and can support ATM connections offering service from a few bits per second to nearly the capacity of the underlying physical layers. An ATM cell consists of a 5 octet header and a 48 octet payload. All traffic is divided into cells and switched based on the information in the cell header. The switching speed is maximized with a fixed cell structure, pre-defined paths, and no link-to-link error recovery. The information field length is relatively small to limit the packetization delay and the queuing delay in a switching node as required by real-time interactive services. Cell sequence integrity is preserved in the cell delivery. ATM is an integrated solution to high-speed networking in that it supports the transmission and reception of voice, video, and data.

ATM is connection-oriented. Before information transfer, a logical/virtual connection set-up phase is required to reserve the necessary network resources. If sufficient resources are not available, the connection is refused. Each connection is associated with a VCI (Virtual Channel Identifier) and a VPI (Virtual Path Identifier) which are assigned at the call set-up. The VCI and VPI are used in an ATM switch to determine where to relay traffic to the next node. The VPI identifies a group of virtual channel links which follow the same virtual path. The VCI and VPI have only local significance on the link between ATM nodes and may be changed in the networks. When the connection is released, the VCI and VPI values on the involved links will be released, and can be reused by other connections.

The 5 octet header of an ATM cell for use at a User-Network Interface (UNI) contains a 4-bit GFC (General Flow Control), an 8-bit VPI, a 16-bit VCI, a 3-bit PTI (Payload Type Identifier, to identify the type of traffic, e.g. data, maintenance, etc.), an 1-bit CLP (Cell Loss Priority), and an 8-bit HEC (Header Error Control, allows the correction of a single bit error and the detection of multiple bit errors in the header). The 5 octet header for use at a Network-Network Interface (NNI) is similar to the UNI cell header except that it does not have the GFC, but has a 12-bit VPI.

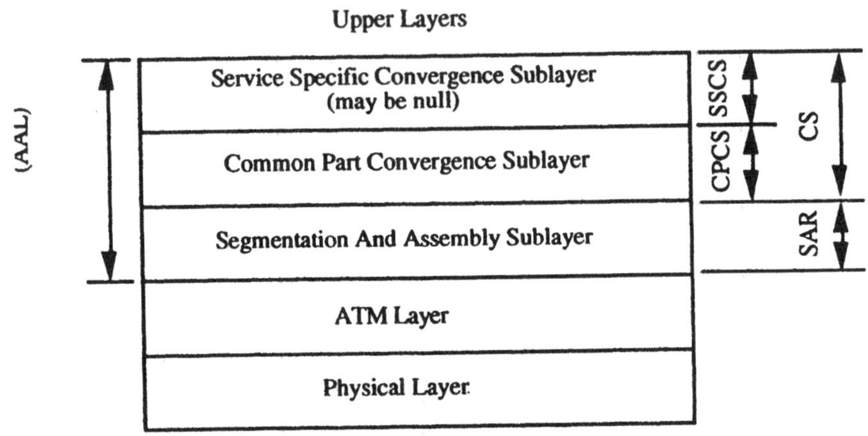

Figure 4 ATM layer structure.

The ATM layer structure [6] is shown in Figure 4. Although ATM is connection oriented, it supports both connection oriented and connectionless services. To accommodate various services, 4 types of ATM Adaptation Layer (AAL) [7, 8] have been defined to provide service specific functions (e.g. clock recovery, cell loss recovery, etc.). The AAL layer is responsible for acting as the interface between user applications and the ATM layer. The AAL is subdivided in 2 sublayers: the Segmentation And Reassembly sublayer (SAR) and the Convergence Sublayer (CS). The CS depends on the particular service and may support different functions such as clock recovery and data structure recovery. The SAR is responsible for segmentation and reassembly operations. Different combinations of SAR and CS sublayers provide different service access points to the layer above the AAL. In some applications the SAR and/or CS may be empty. The service classes supported by ATM and the AALs are summarized in Figure 5.

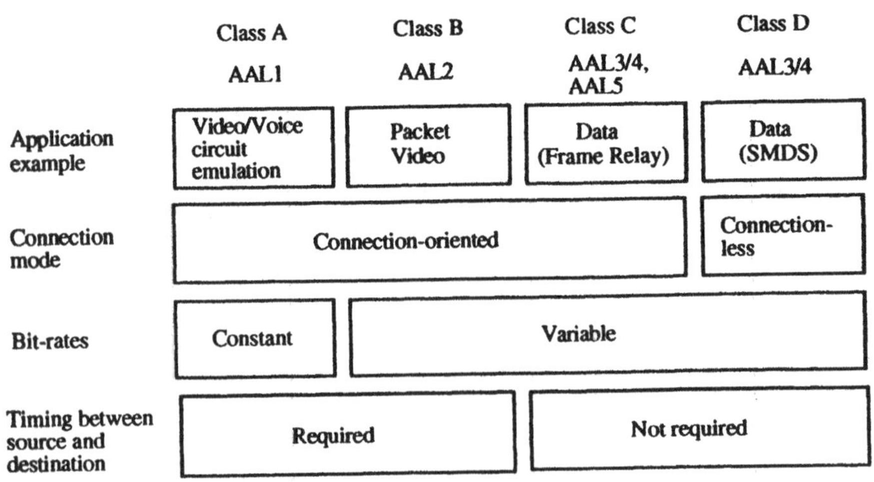

Figure 5 ATM service classes and AALs.

AAL1 is intended for constant bitrate circuit emulation services. It contains fields for SRTS (Synchronous Residual Time Stamp), cell sequence number, and error protection for the above fields. One byte per cell is used to support the above functions, which leaves 47 bytes for the payload. AAL2 is not decided yet. AAL3/4 SAR contains 2 bytes of header and 2 bytes of trailer per cell. The header contains segment type, sequence number, and multiplex identification. The trailer includes length indicator and CRC (Cyclic Redundancy Check). AAL5 requires no per cell overhead but the CS contains 8 bytes of trailer per user packet which include length indicator and CRC. Since different AALs support different functions and provide different sizes of payload to the user, their implications to the transport of MPEG-2 video/audio need to be carefully studied.

3.14.3 Technical issues

Several technical issues need to be examined in order to ensure the performance of transporting MPEG-2 video/audio over ATM is satisfactory. The issues include: timing recovery, bit-error detection/correction, cell-loss detection/correction, and MPEG-2 bit-streams to ATM cell mapping. It is desirable to have a standard solution to these issues in order to minimize the cost of the encoder and the decoder and to support interoperability. In the following, the issues are briefly explained.

3.14.3.1 Timing recovery

As discussed in section 3.14.2, the MPEG-2 systems model assumes constant channel delay for decoder timing recovery. When an MPEG bit-stream is transported over an ATM network, there are questions regarding the effect of the ATM cell delay jitter on the complexity and feasibility of clock recovery schemes using the MPEG-2 PCRs. The cell delay jitter may cause the PCRs to arrive at the decoder with a variable delay. Since the arrival time of the PCRs is used directly in deriving the error signals which control the phase lock loop, the PCR jitter may result in jitter in the recovered decoder system clock.

Since all the clocks in the decoder may be derived from the system clock, a jittered system clock may cause the derived clock to be jittery which may cause picture degradation. The SMPTE NTSC specification [9] specifies that the color subcarrier should have a frequency jitter less than 10 Hz and that the frequency drift should be less than 0.1 Hz per second. With typical cell delay jitter, it may be difficult to meet this specification using the PCRs and a low cost phase lock loop [10]. One proposal which is currently being discussed in the standards organizations is to use SRTS which gives a "jitter-free" clock at the receiver [11, 12]. SRTS encodes the frequency offset between the encoder system clock and a common network clock which is available to both the encoder and the decoder. From this frequency offset and the common network clock, the receiver can generate a jitter free clock locked to the encoder clock even in the presence of cell delay jitter. The main concern of using SRTS is that a common network clock may not be available in some networks.

It may be argued that the SMPTE NTSC specification is for studio applications and for the consumer grade video it may not be required. However, the questions of what kind of quality should the solution support and what is the suitable specification still need to be studied.

3.14.3.2 Bit-error detection/correction and cell-loss detection/correction

ATM can be carried over many different physical media including fiber, coaxial cable, and twisted pair copper. The Bit-Error-Rate (BER) depends on the physical media and the environment in which it operates. For a link with BER of 10E-9, a 4 Mb/s MPEG-2 bit-stream will encounter a bit-error about every 4 minutes. Due to the characteristics of the variable-length-codes used in the MPEG-2 video coding, a bit-error usually causes the decoding to lose synchronization and result in error propagation. The MPEG-2 bit-stream syntax reserves several start codes for the synchronization purpose. A start code cannot be emulated by any concatenations of other codewords and thus can be used to re-establish synchronization. The smallest decoding unit which contains a start-code is a slice which may contain up to a horizontal strip of macro blocks in a picture. Thus, a bit-error usually results in error propagation until the next video slice. Also, since MPEG-2 video uses inter-frame coding, the error in a slice may also propagate until the next intra-frame coded picture. An intra-frame coded picture typically appears once in every 15 frames. However, the actual frequency is totally up to the encoder. Error concealment techniques can be applied to make the degradation less objectionable. However, whether the result is acceptable or not depends on the applications and may need further study.

The Cell-Loss Rate (CLR) depends on the physical media and if the application is CBR or VBR. For VBR applications, it is more likely that the network will encounter congestion and drop cells. A cell loss may cause up to 48 bytes of data to be lost. However, the effect to video quality may be similar to a bit-error since it is likely to also cause error propagation until the next video slice. For a variable rate MPEG-2 bit-stream with an average bit rate of 4 Mb/s and a CLR of 10E-7, it will encounter a cell loss about every 16 minutes.

In considering the error correction codes, it should be noted that due to the use of self-synchronized scrambler and the line-codes in the physical layer, a single transmission bit-error may produce multiple correlated errors. Since the errors are correlated, efficient Reed-Solomon (RS) codes can be designed to correct these multiple correlated bit-errors [13]. If cell loss correction is needed, a block of cells may be octet interleaved and then protected by RS codes. This requires additional memory and hardware. It also results in extra overhead and delay. Several interleaving structures have been proposed and are being studied in ITU-T SG15 and SG13. The issues include: What BER and CLR should be considered for the transport of MPEG-2 audio/video over ATM? Can error detection and concealment in the video decoder provide satisfactory quality so that bit-error detection/correction is not needed? What is the suitable error detection/correction code if the bit-error detection/correction is needed? What is the suitable method for cell-loss detection/correction if it is needed? Currently there is no consensus on these issues.

3.14.3.3 MPEG-2 streams to ATM cells mapping

The issues include: Should the PES, PS, or TS be used in the transport of MPEG-2 over ATM? Which AAL should be used? The factors that should be considered include: delay, packing efficiency, error resilience, timing recovery, and implementation.

Instead of mapping MPEG-2 TS into ATM cells, there are proposals to use PES or PS in the mapping [14]. It is argued that this is more desirable in terms of packing efficiency and protocol processing since TS and ATM contain duplicate functions (such as packet delineation, multiplexing, and prioritization). Also, TS requires large PES packets for good packing efficiency since the MPEG-2 systems specifies that the PES packet headers need to be aligned with the TS packet header. For error resilience, a video slice is a natural unit for error control within MPEG-2 video. Alignment of the start of each video slice to the start of an ATM cell payload provides good spatial error localization. Thus, it is desirable to pack a video slice into a PES packet and map the PES packet directly into multiple ATM cells so that a transmission error will only affect a video slice. For a 720x576 picture at 30 frames/sec, with a slice being one horizontal strip of macro blocks, the average number of bytes per slice is 462 for a 4 Mb/s service. In this case, the efficiency of mapping such a short PES packet to TS packets and then to ATM cells is not very good.

Arguments supporting the use of TS in the mapping include the following. Most video services including video on demand and HDTV over satellite, cable, and broadband networks are expected to use MPEG-2 TS. Using TS promotes interoperability and eliminates the need for format transcoding. The overlapping of functionality between TS and ATM is mainly in the 4-byte TS header. Since a TS packet contains 188 bytes, the overhead of 4-byte header may not be very significant considering the header also supports many other MPEG-2 systems functions such as conditional access and local insertion of advertisement. In the development of MPEG-2 systems, PES is not intended for transport over networks. Initiating a new study to transport MPEG-2 PES over ATM may take longer time and delay the deployment of MPEG-2 over ATM. The probability of a transmission error affecting two video slices may not be high enough to require the alignment of a video slice with a PES packet. Also, if bit-error correction and cell-loss correction are used, the error rate should be low enough that it is not a concern.

The question of which AAL to use depends on the required functionality (timing recovery support, bit-error detection/correction, cell-loss detection/correction, etc.), mapping efficiency, and implementation. AAL1 can carry the TS packets efficiently (mapping a TS packet into 4 ATM cells) but has no room for carrying FEC (Forward Error Correction) codes. It supports SRTS, but the number of bits assigned for SRTS may be inadequate for variable bitrate applications. AAL2 is intended for general video applications but is currently undefined. AAL3/4 and AAL5 would require enhancement to provide timing recovery, and do not support the mapping of a TS packet into 4 ATM cells. There are also proposals to develop AAL2 for the transport of MPEG-2 video or to develop a new AAL type (AAL6). These proposals are currently actively discussed in the standards meetings. As to implementation, chips for AAL5 and AAL3/4 are now available. For a new AAL type, the availability will be much later compared to existing AALs.

3.14.4 Summary

Transport of MPEG-2 signal over ATM is important for the success of video services over broadband networks. In this paper, we provided a tutorial on MPEG-2 systems, ATM, and the issues related to the transport of MPEG-2 over ATM. Currently there is no agreement on how an MPEG-2 bit-stream should be carried over ATM. Technical issues and the arguments supporting different positions need to be carefully examined. It is hoped that agreements can be quickly reached to provide a timely solution for the deployment of MPEG-2 services over ATM.

Acknowledgment

The author would like to thank Scott Quinn for the comments and the careful review of this paper.

References

[1] ISO/IEC/JTC1/SC29/WG11/Draft CD 13818-1, "MPEG-2 Systems," November, 1993.

[2] ISO/IEC/JTC1/SC29/WG11/Draft CD 13818-2, "Recommendation H.262," Nov. 1993.

[3] ISO/IEC/JTC1/SC29/WG11/Draft CD 13818-3, "Draft CD on MPEG-2 Audio", Nov., 1993.

[4] Martin De Prycker, *Asynchronous Transfer Mode - Solution for Broadband ISDN*, Ellis Horwood, 1993.

[5] "MPEG Systems Timing Model and Application Implications," Informative Annex, MPEG-2 Systems, ISO/IEC/JTC1/SC29/WG11/Draft CD 13818-1, Nov. 1993.

[6] CCITT Recommendation I.361 - B-ISDN ATM Layer Specification, 1992.

[7] CCITT Recommendation I.362 - B-ISDN ATM Adaptation Layer (AAL) Functional Description, 1992.

[8] CCITT Recommendation I.363 - B-ISDN ATM Adaptation Layer (AAL) Specification, 1993.

[9] SMPTE 170M, "Proposed SMPTE Standard for Television: Composite Analog Video Signal - NTSC for Studio Applications," *SMPTE Journal*, Jan. 1993.

[10] M. Perkins and T. Lookabaugh, "Comments on Clock Recovery in the Presence of Jitter," MPEG contribution, Seoul, Nov. 1993.

[11] R. Lau, "SRTS for Video Clock Recovery," ATM Forum Contribution 93-879, Sept. 1993.

[12] R. Lau, "Use of SRTS in an Asynchronous Local ATM Network," ATM Forum Contribution 93-880, Sept. 1993.

[13] S. M. Lei, "Forward Error Correction Codes for MPEG-2 over ATM," ATM Forum Contribution 93-878, Sept. 1993.

[14] S. Dunstan, "ATM Network Adaptation Performance Parameters," MPEG contribution AVC-585, Oct. 1993.

Chapter 3.21

VLSI Architectures for Multimedia and Video Conferencing

Bryan Ackland

3.21.1 Introduction

The opportunity to develop applications based on compressed video is the result of progress in three areas:

The first is standards. The recent development of CCITT H.261 (also known as p*64) as a standard for videoconferencing, ISO JPEG as a standard for high quality still image storage and retrieval, ISO MPEG I as a standard for interactive video playback and MPEG II as a standard for entertainment quality video distribution provides a platform for interoperability which can support large, widespread markets.

The second is networks. The availability of ubiquitous, isochronous digital communication networks such as ISDN, 100VG ethernet and FDDI-II will allow transport of compressed video to a broad user community. Broadband networks based on ATM will provide even greater bandwidth and flexibility. Standards and available network bandwidth lead to a convenient partitioning of compressed video applications as shown in Figure 1.

Market	Bandwidth	Standards	Size	Frame Rate
AnalogVideoPhone	5 - 10 k	none	170×128	2 - 5
BasicRate Videotelephony	56 k - 128 k	H.261	174×144 to 352×288	5 - 10
Business Conferencing	384 k +	H.261	352×288	15 - 30
Interactive Multimedia	1 - 2 M	MPEG	up to 352×288	15 - 30
Digital NTSC	3 - 10 M	MPEG II	728×480	30
High Definition TV	15 M +	FCC	1200×800	60

Figure 1 Applications of compressed video.

The third important development is the availability of high performance VLSI technologies such as sub-micron CMOS.

As shown in Figure 2, real time video compression requires processing power in the range of 20 MOPS (analog video telephony) to 100,000 MOPS (HDTV). Many of these applications are well beyond the processing power of general purpose microprocessors. VLSI technologies have advanced, however, to the point where the processing power and memory required to perform these tasks can be incorporated into a few custom silicon chips.

In this chapter, we will explore the relationship between video compression and VLSI. From the video applications perspective, high performance, full custom VLSI is an enabling technology.

From the VLSI designers perspective, video is an application area that places extreme demands on the technology in terms of process, architecture and circuit design, and CAD tools.

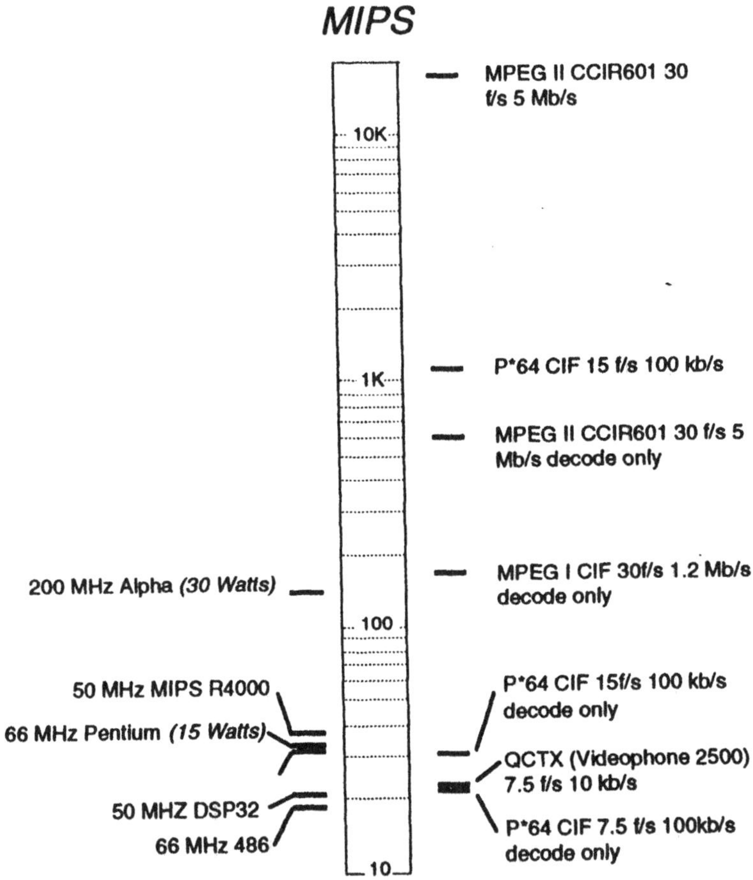

Figure 2 Computational power required to process compressed video.

3.21.2 Video compression

MPEG and H.261 both use a variety of techniques to isolate and remove redundancies in the image sequence and then code the remaining data efficiently according to information content. Some of the operations are relatively straightforward to implement in hardware; others place severe demands on the underlying VLSI technology. Before examining VLSI architectures, it is useful to review these video compression techniques and their ease of implementation. For a more detailed review of video coding techniques, the reader is referred to [1].

3.21.2.1 Chrominance sub-sampling

Because the eye is less sensitive to high frequency colour information, it is possible to code the chrominance channels at reduced resolution with little perceived loss in quality. Hence the input to a video encoder is usually assumed to be in a luminance/chrominance format known as $Y/C_r/C_b$ (Y=luminance, C_r=Y-red, C_b=Y-blue) in which the chrominance data is subsampled by a factor of two in each direction. This is known as 4:1:1 coding (each 2×2 group of pixels is described by 4 Y values, 1 C_r and 1 C_b value.

3.21.2.2 Transform coding

If we transform an image from the pixel domain into the frequency domain, we usually find large amounts of low spatial frequency information and relatively small amounts of high frequency information. This allows for more efficient coding of the image by allocating more transmitted bits to the low frequency portions of the image. Each frame is divided up into 8×8 pixel blocks. To each of these blocks, we apply a two dimensional Discrete Cosine Transform (2-D DCT). This produces an 8×8 block of 2-D spatial frequency components. Typically, many of these will be zero and need not be sent. Others will be near zero and can be efficiently coded using standard entropy coding techniques.

The DCT is the most demanding part of the coding/decoding task in terms of arithmetic operations. In a modest videotelephony application, a video codec may have to perform an 8×8 DCT over a QCIF (176×144) image at 10 frames per second. This requires over 6 million multiply adds per second. In a more demanding conferencing application based on a CIF image (352×288) at 30 frames per second, we require 73 million multiply adds per second. For an HDTV decoder, applying DCT over a 1280×720 image at 60 frames per second requires over 1.4 billion multiply adds/s.

3.21.2.3 Frame differencing

In a normal video sequence there is very little variation from one frame to the next. If, instead of coding each individual frame, we code the difference between the current frame and the previous frame, the amount of information needed to describe the current frame will be dramatically reduced.

3.21.2.4 Motion compensation

Much of the difference that does occur between successive frames can be characterized as a simple translational motion, caused either by moving objects in the scene or by a pan of the field of view. Coding efficiency can be improved by estimating the apparent offset of objects from one frame to another (motion estimation) and then frame differencing with respect to that offset (motion compensation). Motion estimation and compensation are performed on rectangular regions of the image called macroblocks. In MPEG and H.261, a macroblock consists of 16×16 luminance pixels. In motion estimation we search the previous frame in the vicinity of the current macroblock for an offset block of 16×16 pixels which most closely matches the macroblock in the current frame. Once we have identified the best match, we code the difference between the reference block in the current frame and the best match in the previous frame, along with the vector that describes the offset of that best match.

This search and compare process is very expensive. Figure 3 shows an exhaustive search operation for a 16×16 macroblock with a search range of ±15 pixels. For our modest videotelephony example, this type of motion estimation requires 240 million compare adds/s. A search range of ±32 pixels on a CIF image at 30 frames/s requires over 12.5 billion operations per second. This type of throughput can only be achieved with specialized systolic-like architectures.

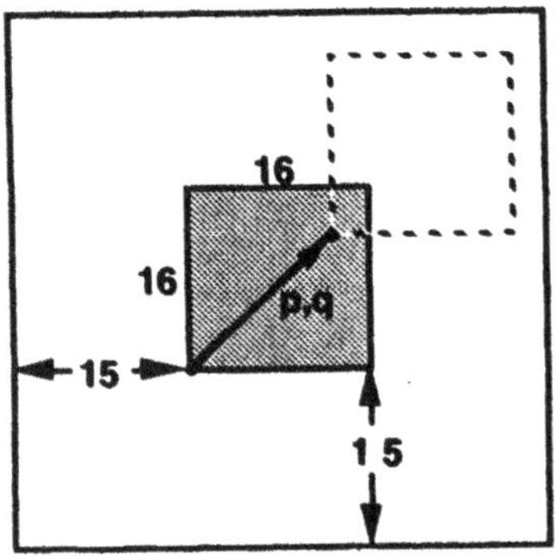

$$D_{p,q} = \sum_{i=1}^{16} \sum_{j=1}^{16} \left| R_{i,j} - S_{i+p,j+q} \right|$$

$$M = \{(x,y): \min D_{x,y}; \ -15 < x < 15, \ -15 < y < 15\}$$

Figure 3 Motion estimation.

An alternative solution is to reduce the operation count by developing a hierarchical search scheme. During the first pass, an approximate motion vector is obtained by performing a reduced resolution search. This can be either by a search in which not all vectors are evaluated (e.g. only checking even valued x and y displacements) or by a search in which not all pixels within the macroblock are used to determine the distance measure D (e.g. subsampling the reference and search spaces by a factor of two before calculating the distance). During the second pass, an exhaustive search is performed over a small region around the approximate vector to produce the final result. Hierarchical search reduces the throughput requirement at the cost of architectural complexity and picture quality. Studies have shown, however, that for most video material, a reduction in throughput of four times can be achieved with little or no visible degradation.

3.21.2.5 *Interpolation*

In MPEG, picture coding efficiency is further improved by predicting the value of an intermediate frame as the weighted average of a previous frame and a subsequent frame. Motion estimation and compensation are used to avoid artifacts caused by objects that have moved from one frame to another. Interpolated frames (also known as B frames) further increase computational demands since both forward and backward motion estimation is required. They also place heavy demands on bandwidth to the frame store memory. Encoding a CIF picture at 30 frames/s with a search range of ± 15 pixels, requires a peak bandwidth of over 60 Mbytes/s just to keep the motion estimator supplied with data.

3.21.2.6 *Entropy coding*

Once all the above steps have been performed, the resulting data will be very unevenly distributed across the range of possible numerical values. Small values of pixel DCT data will occur more frequently than large values. The same is true for the value of motion vectors. Entropy coders take advantage of this uneven distribution by using fewer bits to represent frequently occurring values. Hierarchical Huffman coding is the particular technique used in both H.261 and MPEG.

The Huffman decoder takes bits from the buffered communication channel and reconstructs them into fixed length symbols and data. Although the communication channel is constant rate, the bits are consumed at a very uneven rate according to picture content. An H.261 Huffman decoder operating in a 30 frames/s CIF application may have to consume up to 100 Mb/s (worst case) if it is to remain synchronized with the rest of the decoder on macroblock boundaries. Fortunately, the worst case is very unlikely to occur in normal operation. In addition, the peak processing power can be reduced by putting buffering on the output of the Huffman decoder. Nevertheless, bit processing rates far in excess of the communication channel bandwidth are required for reliable operation over a wide range of material.

A further property of these variable length codes is that the boundaries between the codewords cannot be determined until the codewords themselves are recognized. This requires a great deal of bit level processing (shift and mask). These are operations that are not well suited to conventional DSP and microprocessor instruction sets. Specialized hardware is usually required except for very low bandwidth applications. Figure 4 illustrates how the various techniques are combined to produce a H.261 encoder. Assume that the video data is presented to the encoder in 4:1:1 YC_rC_b format. The first step is to buffer the data to produce 16×16 macroblocks, each consisting of four 8×8 luminance blocks and two 8×8 chrominance blocks.

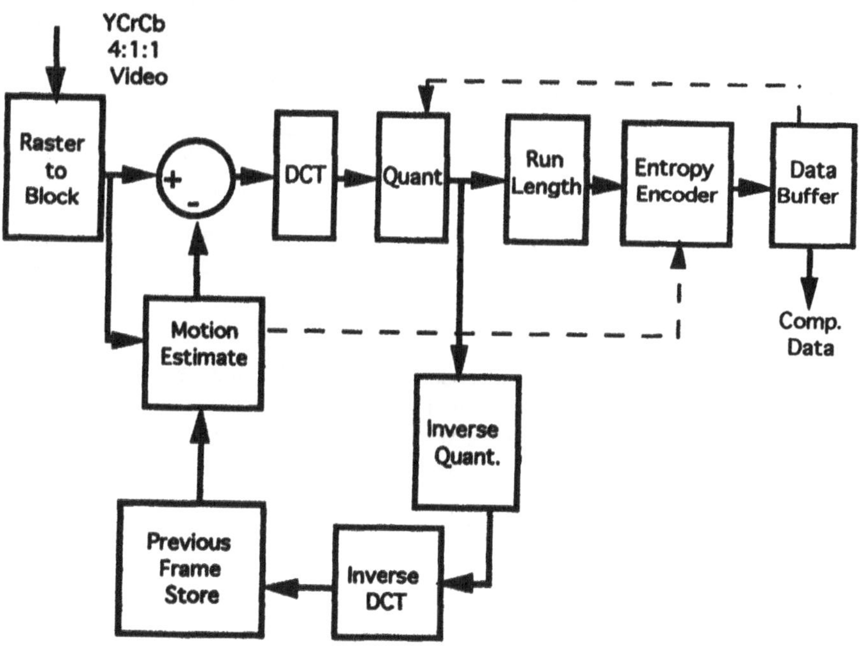

Figure 4 Video encoder architecture.

For each reference macroblock in the current frame, the motion estimator searches the previous frame buffer for an offset 16×16 block of pixels that most closely matches the reference. This best block is retrieved from the previous frame buffer and then subtracted from the reference to create a difference block. Next, the difference block is converted to the frequency domain using the DCT. The frequency components generated by the DCT are now quantized according to the number of bits available for coding. A run-length encoder removes zero valued coefficients. Finally, the entropy encoder converts the DCT values and motion vectors to variable length codes, and then outputs the coded sequence into a variable length buffer. The buffer is used to maintain a constant bit rate at the output. Buffer fullness is used to regulate the degree of quantization and, hence, the degree of compression.

Figure 5 shows the decoding operation. The compressed data stream is first applied to the entropy decoder which extracts DCT values, quantization levels and motion vectors. The DCT values are passed through an inverse quantizer that restores them to their correct level. The inverse DCT processor converts these back to real pixel difference values. The motion compensator takes the motion vector and uses it to calculate the address of the best match 16×16 pixel block in the previous frame. This block is then added to the different pixel values to produce the new decoded macroblock. YC_rC_b.

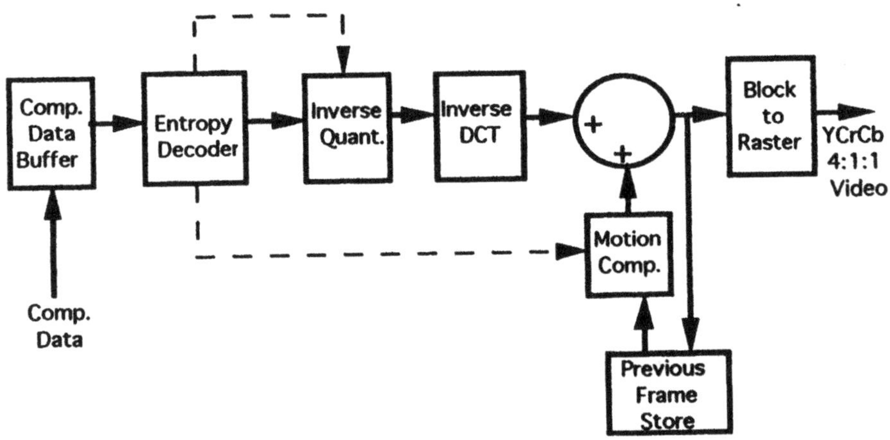

Figure 5 Video decoder architecture.

Note that a significant portion of the decoder is also included in the encoder architecture. The encoder transmits differential information that is referenced to the previous frame. It is therefore critical that both encoder and decoder have identical copies of the previous frame. The simplest way to achieve this is to include a decoding loop in the encoder to generate the previous frame as seen by the decoder.

3.21.3 VLSI architectures

In developing VLSI architectures for video coding, the designer is faced with a number of key choices. These include 1) full custom circuits and layout vs. cell based design, 2) programmable architectures vs. dedicated hardware structures, and 3) monolithic codecs vs. module based systems.

There is no one universally correct choice. The correct architecture will depend heavily on the application and its constraints on performance, power, flexibility, compatibility and of course cost.

Take, for example, the issue of programmability. Programmable designs are able to offer maximum flexibility at the cost of increased chip area and power for equivalent levels of performance. Dedicated architectures are most efficient in high performance applications but represent a very high risk in applications where the algorithms and standards are not yet well defined. As we move across the performance spectrum from analog videotelephony to HDTV, the performance constraint (measured in pixels per second) changes by a factor of over 200. At the same time, the cost constraint on these two systems varies no more than by a factor of 5. This naturally leads to a change in design style. Analog videophones, in which computational performance is low and standards are not yet defined, have to date been implemented using programmable processors. HDTV decoders, in which performance must be very high, cost constraints very low and for which standards will be set well before products become available, will likely be implemented using dedicated architectures.

Some of the tasks that must be implemented in silicon are not computationally demanding but are complex in terms of control and branching, for example DRAM interfaces that support a pixel, block and raster interface. Others in this class implement communication protocols defined in the standards (e.g. audio/video multiplexing and synchronization). They also are complex in terms of control and represent the part of the standard most likely to change as applications develop. These types of circuit are frequently best implemented using a flexible cell based design style.

Other tasks, particularly those involving large amounts of computation, belong to a class known as DSP algorithms. They are typically data flow algorithms in which a small number of basic arithmetic operations are applied to a large body of data in a data independent repetitious fashion (e.g. DCT and motion estimation). These algorithms can often be efficiently implemented using DSP architectures in which high throughput can be achieved using a combination of data path regularity, specialized arithmetic operators, parallelism and pipelining. VLSI architectures are, in turn, well suited to full custom VLSI implementation in which the layout is structured to directly capture the data and control flow employed in these special purpose architectures. If we survey currently available (or soon to be available) video codec chip sets, we find they fall into one of three architectural categories.

3.21.3.1 Video signal processors

This is an extension of the concept of general purpose digital signal processors (DSP) which were originally developed for audio applications. In this case, we develop a programmable processor whose architecture and instruction set are specially tuned to perform those operations required by video compression algorithms. An example is the two chip set developed by Integrated Information Technology [2]. The Vision Processor (VP) is responsible for compute intensive functions such as DCT and motion estimation. The Vision Controller (VC) performs I/O intensive tasks such as entropy encode/decode and memory interface. Both are 33 MHz microcodable processors. One VC plus one VP can encode H.261 at 15 frames/s or decode at 30 frames/s. Recently, IIT announced a new codec chip, called the VCP [3], which combines the functions of the VP and VC on to a single die and runs at 66 MHz. It is able to perform CIF H.261 coding and decoding at 30 frames/s. The VCP consists of a RISC processor for control and bitstream manipulation and a signal processor for arithmetic tasks. Another example of the VSP approach is the MVP video processor announced by Texas Instruments [4]. It consists of a RISC core supported by a number of general purpose DSP engines each with its own independent instruction cache.

The advantage of this approach is flexibility since codec function is determined completely by microcode. This allows product developers to rapidly respond to changes in standards and also improve and update compression algorithms after the silicon is built. The disadvantage is the cost in terms of silicon area. A programmable processor can take up to five times the area and dissipate considerably more power than its dedicated hardware counterpart. A user programmable part may also incur significant software development costs. In an application which requires standards compliance, a programmable part may also incur conformance and interoperability testing costs for the end user.

3.21.3.2 Building block

In the building block approach, each major video processing module is configured as a separate chip. There may be a DCT chip, a motion estimator chip, a variable length encoder chip etc. Each module is built using a dedicated hardware architecture. The interconnection of modules, however, is left to the designer. The advantage of this approach is that there is considerable flexibility in how the modules can be configured. The individual chips themselves are quite small and relatively cheap. The disadvantage of this modular approach is that it requires a large number of devices. LSI Logic's 64700 series [5] is a good example of this approach. Their devices have been used by a number of OEMs to assemble board level codec products. It takes nine VLSI devices plus a considerable amount of "glue logic" to assemble an encoder/decoder pair. This makes for a relatively expensive solution.

3.21.3.3 Monolithic codec

The third approach is to build a complete encoder or decoder (or both) on to a single die. This gives the best solution in terms of chip area and power dissipation. The disadvantage of this approach is lack of flexibility. If standards change or the application requires a new algorithm, a complete revision of the silicon may be required. This may be offset to some extent, however, by including a limited amount of programmability at key points in the architecture. An example of this approach is the VideoRISC architecture developed by C-Cube [6, 7]. It consists of a number of programmable processors, each having an architecture and instruction set optimized for a specific part of the coding algorithm, e.g., motion estimation or variable length coding. A single chip will perform MPEG and H.261 encoding or decoding at 30 frames/s or full duplex at 15 frames/s. A similar approach is seen in the AT&T AVP4000 chip set. This is described in some detail in the next section.

3.21.4 AVP 4000 multimedia codec

The AVP 4000 [8] is a multimedia chip set comprising three custom chips and the DSP3210. It is targeted at a broad range of intermediate bit rate applications from 40 kbit/s up to 4 Mbit/s. It therefore supports a wide range of video quality from "head & shoulders" videotelephony up to entertainment quality digital TV. A key emphasis has been to design a multimedia codec rather than just a video codec. This means a system that also supports audio and data together with all the framing, synchronization, error correction and communication interfaces that are required to turn a workstation or PC into an interactive multimedia terminal. Specifications include:

- Video picture size up to CIF (352×288). Stills up to 720×576.
- Frame rate variable up to 30 frames/s.
- Fully compliant with H.261, MPEG and JPEG.

- Audio coding to G.711 (3 kHz., 64 kb/s), G.722 (7 kHz., 64 kb/s) and G.728 (3 kHz., 16 kb/s). Also MPEG high fidelity audio.
- Audio / video / data multiplexing, synchronization and framing according to H.221 and MPEG Systems CD.
- Communication interfaces to basic and primary rate ISDN and its precursors (56 kb/s, fractional T1)

A block diagram of the overall codec architecture is shown in Figure 6. It consists of three custom chips and a DSP. The video encoder and decoder provide real time coding and decoding of motion video according to the H.261 and MPEG standards. The DSP3210 provides real time audio encoding and decoding, still picture coding and decoding according to JPEG, and echo suppression and cancellation for speakerphone applications (some combinations require two DSPs). The systems controller provides all the higher level protocol functions including synchronization, multiplexing, buffering and framing.

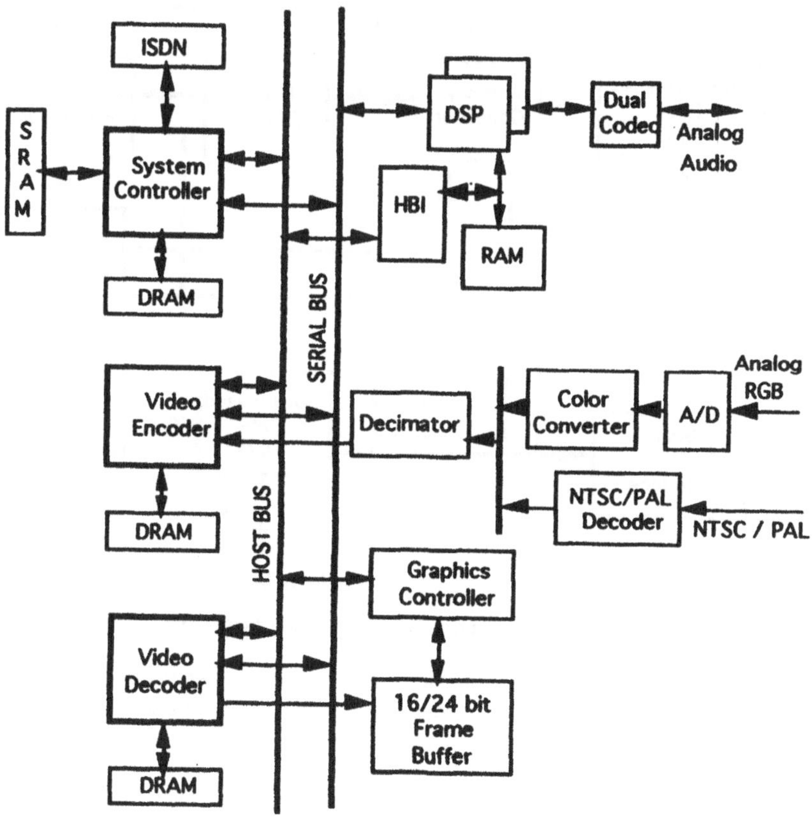

Figure 6 AVP multimedia codec.

3.21.4.1 Video encoder

The video encoder [9], shown in Figure 7, is the most demanding in terms of computational throughput. Accordingly it has been designed using full custom techniques throughout. It consists of a number of programmable engines that communicate through an internal data bus. Each of these engines performs a specific coding function (e.g., motion estimation, quantization, variable length encoding). Each has an instruction set and architecture tailored to its function. The programmability of these engines varies according to the flexibility and performance required. The motion estimator, for example, implements a well defined exhaustive search algorithm at rates up to 7 billion operations per second. These rates are achieved with a dedicated systolic-like architecture consisting of 256 simple processing elements (for computing and accumulating absolute value differences), 15 delay lines and a 1280×16 cache (for holding the pixels associated with the search area). This module has very limited programmability via a few constant registers that define the size and boundary conditions of the search area.

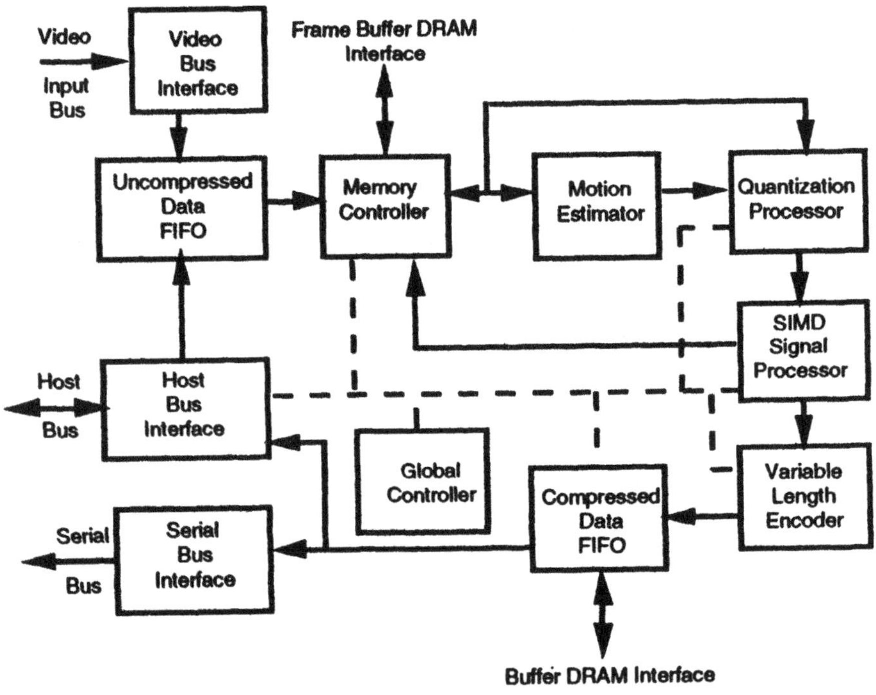

Figure 7a Video encoder chip - architecture.

The variable length encoder (VLE) implements a bit-stream syntax set by the standards. Although this is not expected to change, some flexibility may be required in future versions and so the VLE is implemented using a specialized RISC engine programmed via a 1024×25 mask programmed ROM. The quantization processor (QP) implements an algorithm that has great bearing on picture quality. It is expected to improve over time. The QP is therefore implemented using a more general processor with a 1024×28 downloadable instruction RAM.

The video encoder contains 1.6 million devices and performs approximately 7 billion motion estimation and 500 million arithmetic operations per second. It has been implemented in 0.9 μ CMOS in an area of 132 mm^2. Power dissipation at 31 MHz is approximately 3 Watts.

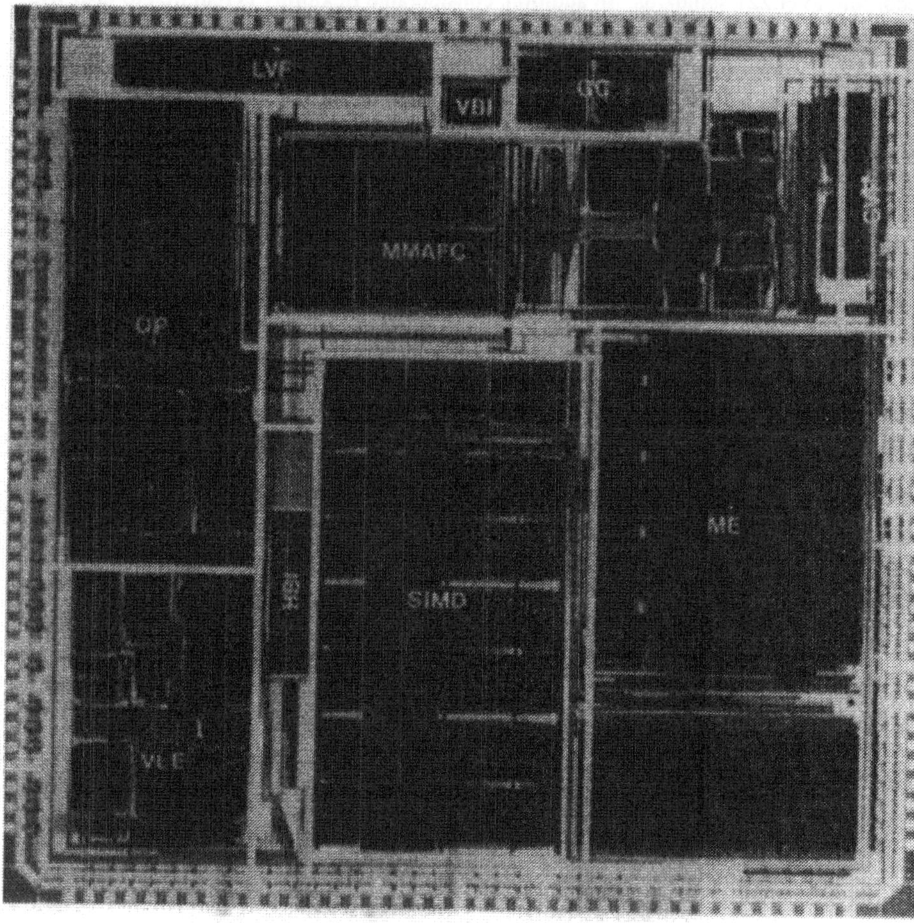

Figure 7b Video encoder chip - die photograph.

3.21.4.2 System controller

At the other end of the design spectrum, the system controller is the least demanding in terms of computational throughput but is the most demanding in terms of the complexity of the task that must be performed. The system controller implements the many layers of protocol that combine compressed video and audio data, provide error detection and correction, buffering and synchronization and enable a multimedia digital call to be set up according to H.221 standards. In addition, considerable flexibility is required to deal with anticipated changes in these, as yet, untested standards.

The architecture of the system controller is shown in Figure 8. This device is implemented as a RISC core processor whose instruction set is optimized for bit-level intensive operations. The processor is supported by a number of hardware engines designed to offload the most demanding tasks such as forward error correction. This chip also implements the many buffers that are employed to match and synchronize the bit rates of the various media (video, audio, data). This is achieved using a combination of on-chip FIFOs and external DRAM. The entire chip is implemented using 0.6 μ CMOS standard cells together with some custom memory modules. It occupies an area of 106 mm^2 and contains 450 K transistors.

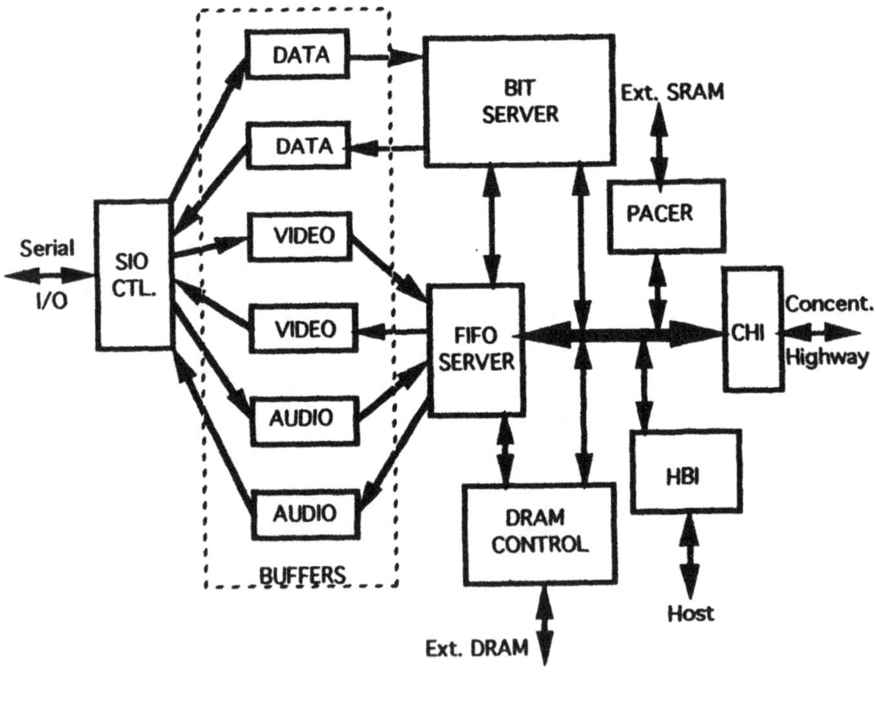

Figure 8a Systems controller chip - architecture.

A considerable portion of the "value added" for this device lies in the 45,000 lines of microcode that have been written for the RISC core. In addition to controlling the hardware engines that provide buffering, multiplexing and error correction, this microcode implements channel management, call setup and tear down according to the H.221, H.242

and H.230 specifications. It also provides combination and synchronization of up to three switched 56 kb/s or ISDN B channels that may experience different delays as they pass through the network.

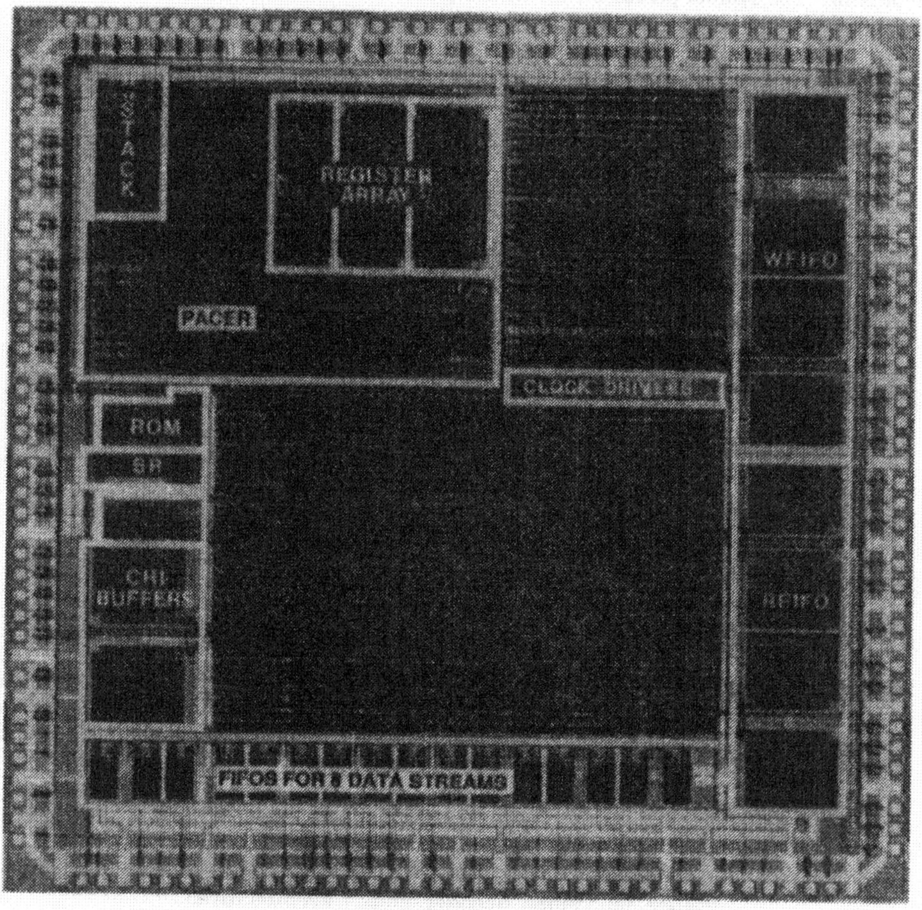

Figure 8(b) Systems controller chip - die photograph.

3.21.4.3 Video decoder

The video decoder [10], shown in Figure 9 is a mix of standard cell and full custom design styles. Compute intensive tasks such as the inverse DCT are performed on a full custom signal processor borrowed from the encoder design. Less demanding operations such as colour space conversion are implemented using standard cells. This device also uses a mix of hardwired and programmed operation. The memory controller, for example, which provides bit, block and raster access to the external DRAM has a well defined function and is implemented using standard cells. The inverse DCT is ROM coded in the SIMD signal processor. The symbol processor, which interprets the commands and data coming from the variable length decoder is responsible for initiating and synchronizing the operation of all other modules. Flexibility is achieved here by using a RAM programmed RISC processor approach. The decoder is implemented in 0.9 µ CMOS, contains 1.2 million transistors and occupies and area of 112 mm^2. It dissipates approximately 2.2 Watts at 40 MHz.

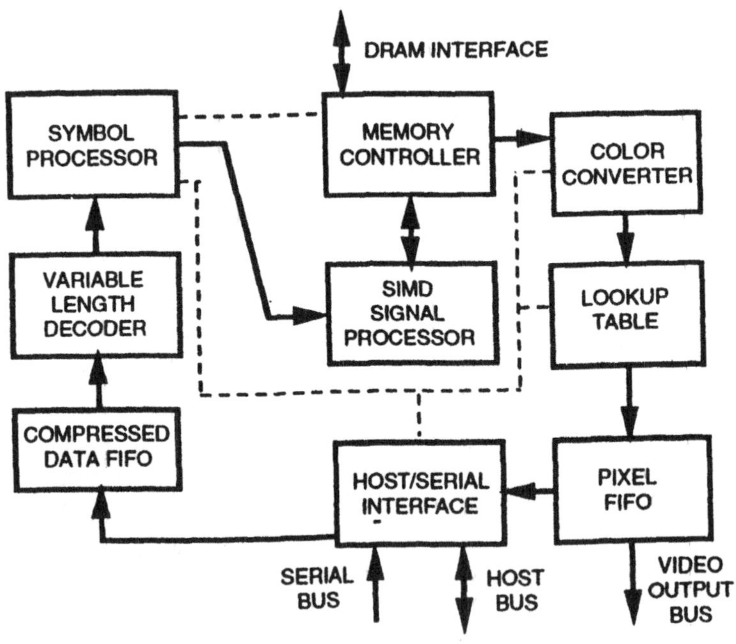

Figure 9a Video decoder chip - architecture.

3.21.5 Design challenges

The design of high performance full custom signal processing chips such as those described in the previous section is limited by more than just processing technology. Some of the more serious constraints that we encountered are described in the following subsections.

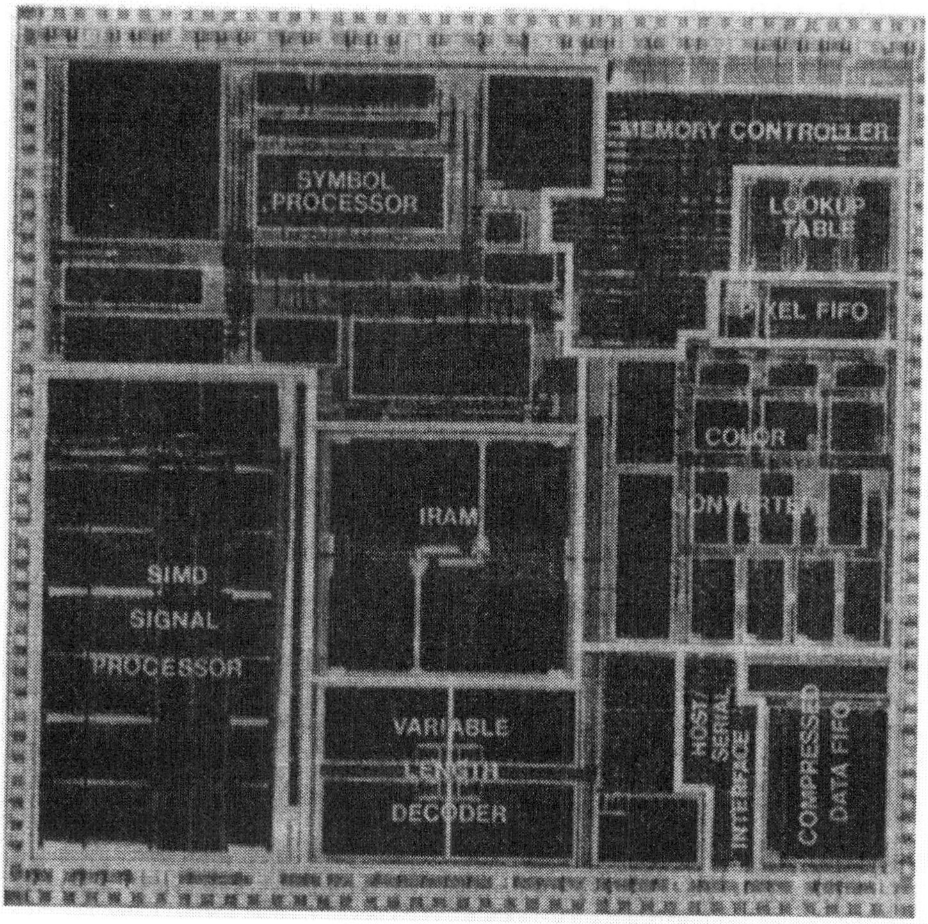

Figure 9b Video decoder chip - die photograph.

3.21.5.1 Algorithmic modelling

Prior to fabrication, all algorithmic studies are done via simulation. It takes about 1 minute to simulate a single frame of video compression on a high end workstation. This limits algorithmic evaluation to test sequences that are at most tens of seconds in length and are not necessarily representative of final application video material.

3.21.5.2 CAD tools and processors

Full custom design places particular demands on the CAD environment, especially in the areas of simulation and verification. Verification of video chips is further compounded by the large number of vectors that are required to simulate real operation. Processing a single video frame at 50 MHz, for example, takes 1.5 million vectors. Figure 10 shows some of the simulation times that were encountered in the design of the video encoder. Note that these times were obtained using a high end (70 MIP) workstation with 162 MB of memory and 2 GB of disk.

		Time to simulate			Typical Simulation Run	
Type of Simulation	Circuit Size	1 Clock	1 M-block	1 Frame	Vectors	Time
Algorithm	2600 lines	---	150 ms.	1 min.	250 frames	4 hours
C-model	11k lines	75 ms.	5 mins.	33 hours	3.2 M	67 hours
Gate	400 k gates	1.5 s.	100 min.	29 days	150 k	62 hours
Timing	1.6 M Tx.	15 mins.	40 days	44 years	200	50 hours

Figure 10 Encoder simulations.

3.21.5.3 Power and packaging

The extremely high throughput of these devices leads to significant power dissipation. This leads, in turn, to expensive packaging. Given that many of the most exciting applications of compressed video are in extremely cost sensitive markets, there is a continuing need to develop lower power circuit implementations and lower cost, high performance packaging.

3.21.5.4 Software development

Any real multimedia communications applications requires a considerable amount of software to provide simple, reliable communications and user interfaces. Some of this will be implemented as microcode on the various VLSI devices, some will be implemented as code on the host platform. The writing, debugging and testing (for reliability, standards conformance and interoperability) of this code represented a considerable portion of the overall design effort. Problems of software support are further magnified when the VLSI manufacturer must maintain a user programmable interface along with all the required compilers, debuggers and documentation.

3.21.6 Summary

The adoption of video coding standards, the deployment of digital communication services and the availability of high performance full custom VLSI will open up a wide range of exciting new multimedia products and services. These devices stretch the limits of our silicon technology but will continue to benefit from advances in processing, circuit design, packaging and CAD tools.

References

[1] R. Aravind *et. al.*, "Image and Video Coding Standards", *AT&T Technical Journal*, Vol. 72, No. 1, pp. 67-89, Jan. 1993.

[2] B. Fuller, "Video Conferencing Chips Cut Cost, Parts", *Electronic Engineering Times*, p. 10, Oct. 19 1992.

[3] "IIT VCP Preliminary Data Sheet", Integrated Information Technology, Santa Clara CA. 1993

[4] K. Guttag. "The Multiprocessor Video Processor, MVP", *Proceedings IEEE Hot Chips V*, Stanford CA, Aug. 1993.

[5] D. Pryce, "Monolithic Circuits Expedite Desktop Video", *EDN*, p. 67, Oct 24 1991.

[6] S. Bose. "A Single Chip Multistandard Video Codec," *Proceedings IEEE Hot Chips V*, Stanford CA, Aug. 1993.

[7] J. Yoshida. "C-Cube Rolls Real Time Video Compression Architecture", *Electronic Engineering Times*, p. 16, Oct. 4, 1993.

[8] B. Ackland *et. al.* "A Video Codec Chip Set for Multimedia Applications", *AT&T Technical Journal*, Vol. 72, No. 1, pp. 50-66, Jan 1993.

[9] S. Rao *et. al.* "A Real Time P*64/MPEG Video Encoder Chip", *Proceedings 1993 Intl. Solid State Circuits Conf.*, San Francisco CA, pp. 32-33, Feb. 1993.

[10] D. Brinthaupt *et. al.* "A Video Decoder for H.261 Videoteleconferencing and MPEG Stored Interactive Video Applications", *Proceedings 1993 Intl. Solid State Circuits Conf.*, San Francisco CA, pp. 34-35, Feb. 1993.

Chapter 3.22

VLSI Implementation of MPEG Decoders

Thierry Fautier

3.22.1 Introduction

ISO standards 11172 [1] and 13818 [2] known as MPEG1 and MPEG2 are becoming more and more popular in the industry. Their success will largely depend on the availability of low cost decoding systems that will be integrated in a wide range of products such as CD based entertainment machines (CD-I, 3DO, CD32,...), personal computers (PC, Mac,...), set top decoders (satellite, cable, telephone lines,...).

This chapter presents the different elements that drive the VLSI implementation of MPEG decoders by listing, in a qualitative and quantitative way, the requirements for decoding of MPEG video bitstreams.

It then introduces the architectures that are used in the VLSI design of MPEG video decoders, and gives a method for dimensioning of the different key architecture parameters of MPEG video decoders.

3.22.2 Overview of MPEG standards

3.22.2.1 MPEG1

MPEG1 was mainly designed to match the need of CDROM based applications. While its design was optimized for a bitrate of about 1.5 Mbits/s and for a SIF (352×288×25 or 352×240×30) format, it can be used for other bitrates and formats.

We will restrict the study of MPEG1 decoders to the constrained parameter set defined as follows

- maximum bitrate : 1.8 Mbits/s
- maximum pixel rate : 352×288×25 = 2453440 pixels/s

3.22.2.2 MPEG2

MPEG2 is a set of several profiles that can handle different resolutions at different layers. Each layer can be encoded from the other ones.

However, the first application of MPEG2 is a one layer coding scheme at the CCIR 601 resolution, called Main Profile at Main Level, with the following limitations

- maximum bitrate : 15 Mbits/s
- maximum pixel rate : 720×288×50 = 10368000 pixels/s

3.22.3 Requirements for MPEG video decoding

3.22.3.1 Global architecture

In all MPEG decoding systems, the common requirement is to have a single chip and single memory bank implementation for the MPEG decoding function. Therefore the system architecture of a MPEG VLSI decoder represented in Figure 1 comes naturally.

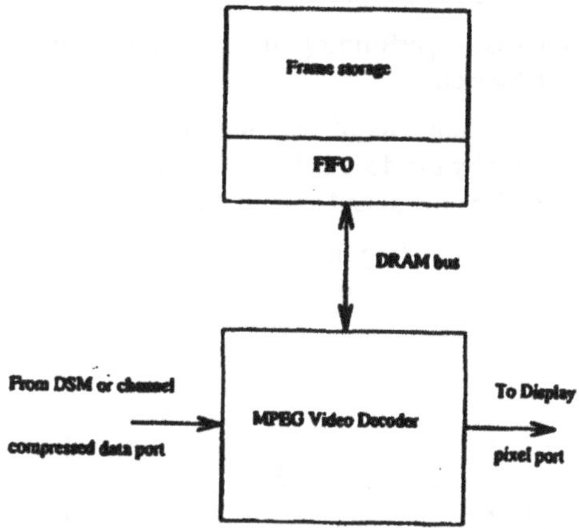

Figure 1 System architecture.

Compressed data are received from the Digital Storage Media (DSM) or from the channel and are stored in one part of the DRAM that is managed as a FIFO. Those data are decompressed by the MPEG video decoder and decoded pictures are written to the DRAM that serves as a frame storage for the predictions (I and P pictures) and for the display (I, P and B pictures). Pictures are decoded and presented on the pixel port in synchronisation with the video synchronisation signals.

3.22.3.2 Processing power

A MPEG decoder must have sufficient processing power to decode a picture in a picture period. Therefore from the pixel rate and from the maximum number of operations per pixel, one can compute the required processing power.

(i) Qualitative approach

The operations that a MPEG video decoder must perform to decode a picture are fully specified by ISO in an International Standard document for MPEG1 (IS 11172-2) [1] and in a Committee Draft document for MPEG2 (CD 13818-2) [2]. The operations can be splitted in different tasks that are depicted in the block diagram of Figure 2.

The different tasks are detailed below

- Start code decoding : consists of searching for MPEG video start codes (including error codes) and in synchronizing the decoding process.
- Fixed length decoding : consists of decoding all the fixed length headers. The decoded information is mainly of control type and is dispatched to all other blocks (dashed lines).

- Variable length decoding : consists of decoding the variable length Huffman coded data. This corresponds to macroblock headers (control information, such as motion vectors or decision) and to quantized pixels (data). The decoded control information is sent to all subsequent blocks (dashed lines), while the quantized pixels are sent to the inverse quantification function.
- Inverse quantizer : consists of multiplying the quantized pixels by the transmitted quantizer step and by the coefficients of a downloadable matrix.
- Inverse DCT : consists of performing the two dimensions inverse discrete cosine transform on 8 by 8 blocks.
- Motion compensation : consists of fetching blocks, which address in memory is given in half pixel unit by the decoded motion vectors. This operation can be done in forward direction (P pictures) and in both directions (B pictures).
- Interpolation : consists of adding the results of the inverse DCT with the results of the motion compensation and to write it back to memory.

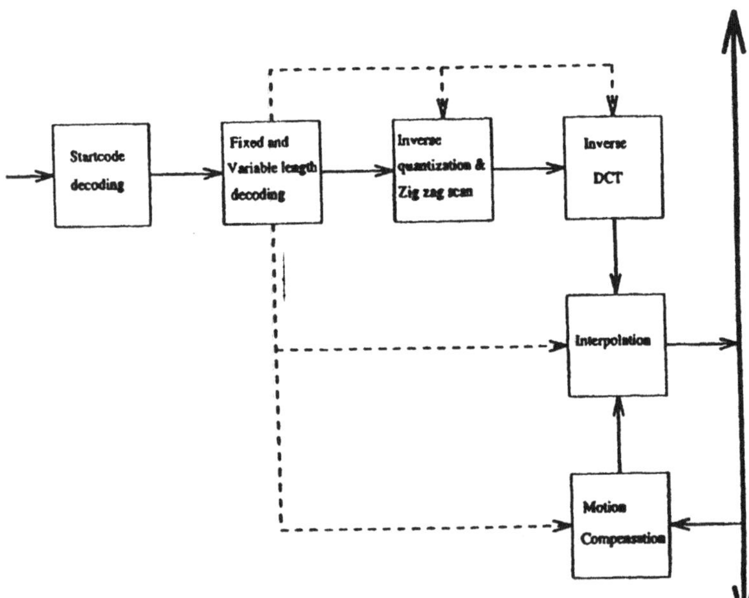

Figure 2 Block diagram of MPEG decoding.

(ii) Quantitative approach

MPEG video algorithms are designed such that all the above described functional blocks can be activated for the processing of a block. Therefore the architecture of a MPEG decoder must be dimensioned to be able to process the most critical case in terms of processing power. We give for the MPEG1 standard the computation load that a MPEG video decoder must support. This takes only into account the basic arithmetic operations and does not take into account the control of those operations.

Table 1 gives for each picture type the required processing per component (*Y*, *U* or *V*) required for each task. It has to be noted that in the worst case each component is coded with one *vlc* code, thus VLD has to process one code per component.

Task	Intra	Predictive	Bidirectional	Unit
VLD	1	1	1	Huffman decode
IQ	2	2	2	Multiply
IDCT	8	8	8	MAC
MC	0	4	8	Addition
Interp	0	1	2	Addition

Table 1

For a SIF picture, the total number of components is 1.5 times the pixel rate due to the 4:2:0 sampling format of MPEG (*U* and *V* sampled by a factor of 4). The worst case computation power is reached in B pictures.

Task	Operations	Unit
VLD	3.80 M	Huffman decode/s
IQ	7.60 M	Multiply/s
IDCT	30.41 M	MAC/s
MC	30.41 M	Addition/s
Interp	7.60 M	Addition/s

Table 2

3.22.3.3 Memory bandwidth

It is necessary that a MPEG decoder has enough memory bandwidth to perform all the DRAM accesses required by MPEG algorithm. Therefore from all the accesses that the video decoder must perform in a picture period, one can compute the necessary DRAM bandwidth.

(i) Qualitative approach

The different processes that need to access to the DRAM to perform the decoding and the display of MPEG coded pictures are represented in Figure 3.

The different processes are detailed below :

- Write Compressed Data (CD): transfers compressed data to the FIFO in DRAM.
- Read Compressed Data (CD): transfers compressed data from the FIFO in DRAM to the spatial decoding unit (VLD, IQ, IDCT).
- Read Motion Compensation (MC): reads the predicted blocks in DRAM.
- Write: writes the decoded blocks in DRAM.
- Display: reads decoded picture in line format from the DRAM.

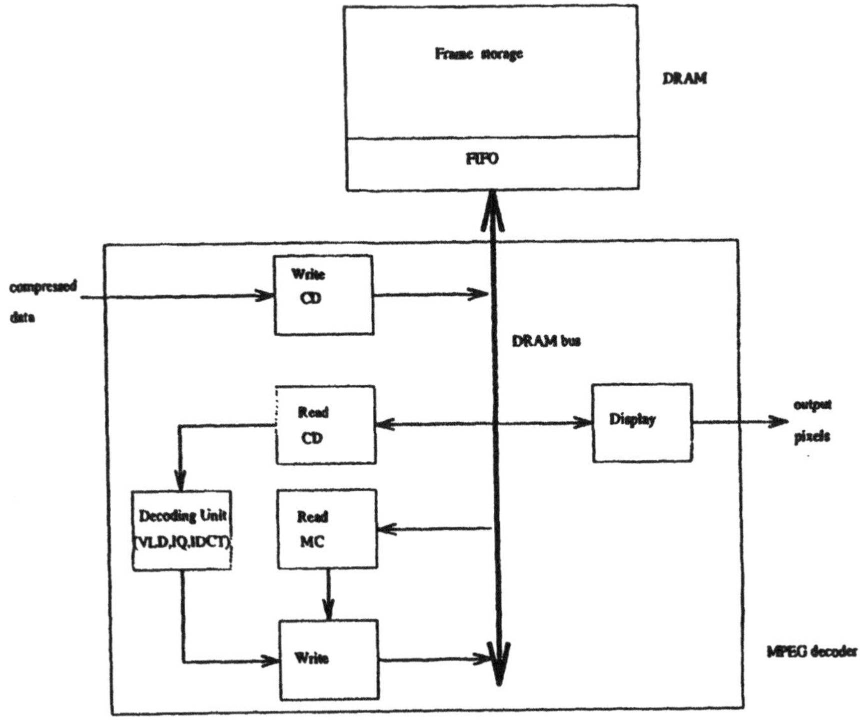

Figure 3 Process organisation.

(ii) Quantitative approach

We give for the MPEG1 standard the memory bandwidth for all the processes that access the DRAM. Table 3 gives for each picture the required bandwidth in bytes per second corresponding to the coding of SIF pictures at 1.2 Mbits/s. It has to be noted that to read a 8×8 block, 9×9 pixels must be read to perform the half pixel filtering.

Process	Intra	Predictive	Bidirectional
Write CD	150 k	150 k	150 k
Read CD	1 M	1 M	1 M
Read MC	0	4.8 M	9.6 M
Write	3.8 M	3.8 M	3.8 M
Display	3.8 M	3.8 M	3.8 M
TOTAL	8.75 M	13.55 M	18.35 M

Table 3

The maximum required bandwidth is 18.35 M bytes/s for B pictures. This computation gives a rough estimation of the DRAM bus load and does not take into account the overheads due to the DRAM internal organisation and to the arbitration between the different processes.

3.22.4 Dimensioning of the architecture

It is known that in VLSI implementations of video processing algorithms, processing power can be achieved provided that I/O bandwidth is sufficient. We will therefore address first the DRAM communication problem.

3.22.4.1 Access to the DRAM

(i) Strategy of access

In the case of MPEG1 video decoding, the maximum byte exchange rate is 18.35 MHz, which corresponds to a period time of 55 ns. Current commercial 60 ns DRAMs offer a minimum random access cycle time of 100 ns and a minimum page mode cycle time of 50 ns. All the processes that access the DRAM can access it by burst and therefore take advantage of page mode feature of the DRAM. This mode of communication is therefore preferred since it is twice as faster as the normal mode of access. Meanwhile the computed byte exchange rate does not take into account the overheads due to the page fault and to the bus arbitration. Thus, to take into account those overheads and to enable the choice of a DRAM clock speed lower than 20 MHz, a 16 bit DRAM organisation is often choosen for the implementation of MPEG1 decoders [3, 4].

This architecture choice requires a high speed DRAM protocol. [5] presents an implementation with a 32 bits DRAM bus with a low speed DRAM communication and more features with a high speed display interface (e.g. VGA). Access to the DRAM is organized in small bursts of 64 bytes that correspond to the size of a 8×8 block. The DRAM management is made by a central arbiter that grants accesses to DRAM to the requesting process with a fixed priority scheme. Deadlock cannot occur since each process releases the bus after accessing 64 bytes. Starvation is prevented because the overall DRAM bus load is far under 100 %.

(ii) Cache strategy

In order not to refuse incoming data that may come in burst on the compressed data port, a double buffer scheme is implemented in the Write CD process as shown in Figure 4. If VLD has only one buffer to interface with the FIFO in the Read CD process, it may be halted during the decoding of a block when this buffer is empty and waits for the DRAM bus to be filled. Also to prevent from a possible source of delay a double buffer scheme is implemented in the Read CD process.

Concerning the Read MC process, as the reading of the prediction is faster than the DCT decoding time, a single buffer scheme of 64 bytes is implemented for the prediction reading.

3.22.4.2 Processing units

(i) Data parsing

In the MPEG algorithm, the information to be processed in the bitstream can be splitt into three different classes:

- The MPEG headers (from sequence to slice) that are coded with a fixed length and a low rate. Those information do not require a specific processing and their decoding can be executed by a on chip CPU core.

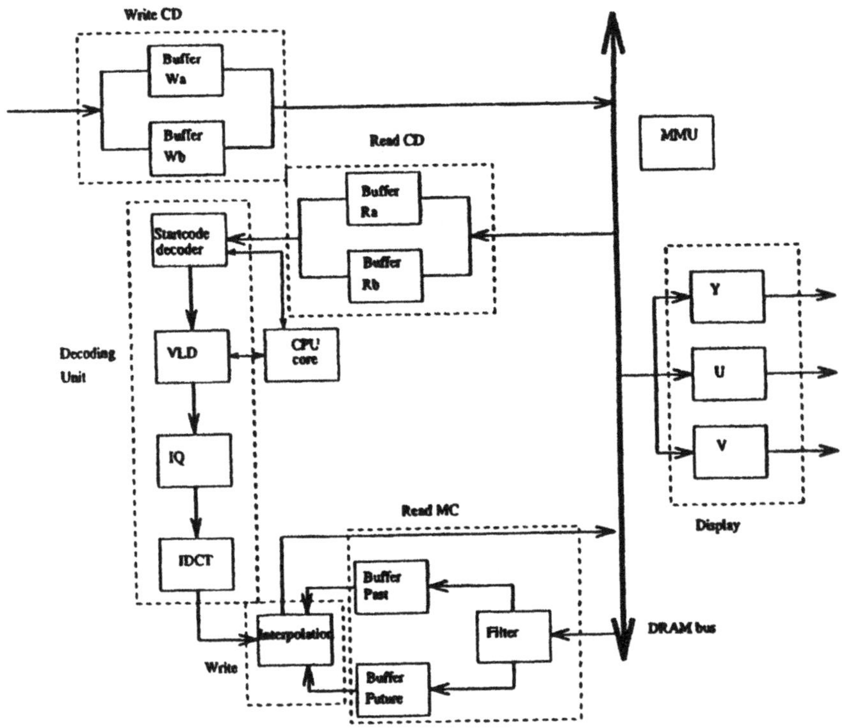

Figure 4 Architecture of a MPEG decoder.

- The macroblock headers (motion vectors, decisions,...) that are variable length coded at the macroblock rate. Those headers need specific hardware to be decoded fast enough. Once decoded they must be processed by the CPU core that will send the information to the destination blocks.
- The quantized pixels that are variable length coded at the pixel rate in the worst case. A specific hardware is needed for that task. Moreover a direct data path is needed to communicate with the inverse quantizer.

(ii) Start code detection

This function decodes the startcodes required for the control of the data parsing. This block works in parallel on the coming data flow and informs CPU core on the decoded start codes.

(iii) VLD

Many implementations of VLD have been proposed for Huffman decoding [6, 7]. A parallel implementation as described in Figure 5 is commonly used in the implementation of the VLD function. If decoding of one VLD word is done in p cycles, then the F_{vld} clock frequency, for that block should be at least $F_{vld} > p \times 3.8$ MHz (see Table 2).

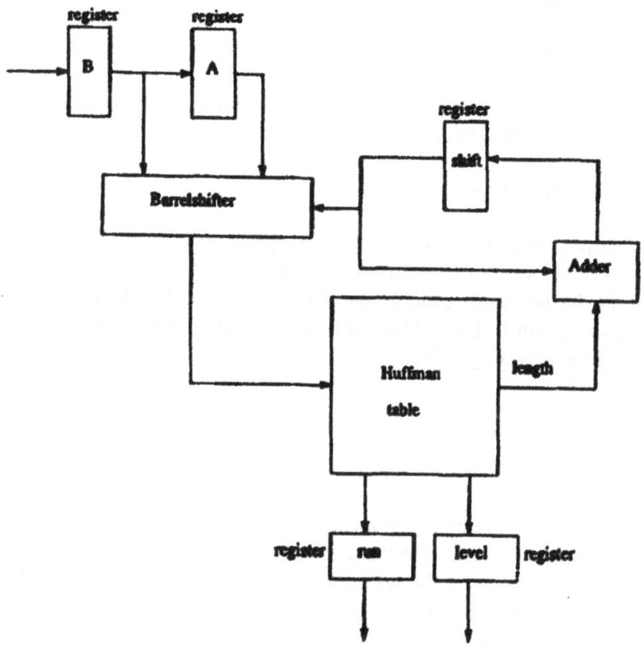

Figure 5 Variable length decoding.

(iv) Inverse quantization

This function requires two multiplications per pixels. If the same multiplier is used sequentially then the maximum frequency F_q for the IQ block should be at least $F_q > 2 \times 7.6$ MHz (see Table 2). This function is implemented with classical data path structures.

(v) Inverse DCT

DCT can be implemented either as a matrix product which is the most expensive solution, using a distributed arithmetic architecture [8], or by hardwiring a butterfly in a bit serial computation [9]. In the case of MPEG1, this block must process 59400 8×8 blocks per second.

(vi) Prediction

This block receives motion compensated blocks from the DRAM and performs on the fly a half pixel filtering as specified by MPEG. A typical signal processing structure is recommended for this block. This block and the Write block must work at the DRAM clock speed in order to have no wait state between the video decoder and the DRAM.

(vii) Display

This block reads back decoded pictures from the DRAM, resamples them if needed to the output format, presents them in a raster scan format on the pixel port, and works at the pixel frequency (13.5 MHz) and its dimensioning depends on the post processing features (windowing, resampling, resizing) to be included.

(viii) Memory management unit

This block is in charge of performing the address generation and the bus arbitration for the DRAM accesses. This block works at the DRAM access frequency.

3.22.4.3 Clock frequency

The most constraining requirement is the clock speed to access the DRAM. Once this frequency is fixed (after system simulation on worst cases), a multiple of this frequency is chosen to drive all the processing elements of the MPEG video decoder. It is at this step that the level of parallelism is defined for each processing element of the decoder in order to reach the specifications described above.

3.22.5 Conclusion

An overview of the different architecture features of MPEG video decoders has been given. LEP-Philips has designed a MPEG1 video decoder [5], DIVA (Decoder for Interactive Video Applications), following this approach. This circuit will be developed to an industrial standard by Philips Semiconductors in the course of 1994.

The method of synthesis of the architecture presented here for the MPEG1 standard can be reused for the MPEG2 standard at the CCIR resolution and also at the HDTV resolution. The processing power, (level of parallelism, clock speed) and the memory strategy access, (burst length, word depth, clock speed) are the parameters to be reconsidered in conjunction with the new possibilities of the microelectronics technology.

References

[1] "ISO IS 11172-2: Coding of Moving Pictures and Associated Audio for Digital Storage Media at Up to About 1.5 Mbit/s", *International Standards Organisation* March 93.

[2] "ISO CD 13818-2: Generic Coding of Moving Pictures and Associated Audio", *International Standards Organisation*, November 93.

[3] C-Cube, "CL450, MPEG video decoder user's guide".

[4] C. Schepens, "A MPEG decoder ASIC for CD-I", *Euro ASIC '92*.

[5] L. Bergonzi *et al.*, "DIVA: a MPEG1 video decoder for interactive applications", *Proc SPIE, EUROPTO, Video communications and PACS for Medical Applications*, 1993.

[6] J. Peake, "Decompaction", *IBM Technical Disclosure Bulletin*, Vol. 26 No. 9, Feb 1984.

[7] M. Sung *et al.*, " High speed programmable ICs for decoding Variable Length Codes, *Proc SPIE Application of Digital Image Processing XII*, Vol. 1153, 1989.

[8] M. Sun, "A concurrent architecture for VLSI implementation of Discrete Cosine Transform", *IEEE Trans Circuits and Systems*, Vol. CAS 34, No. 8, 1987

[9] F. Jutand *et al.*, "A 13.5 MHz single chip multiformat discrete cosine transform", *Proc SPIE Visual Communications and Image Processing II*, Vol 845, 1987.

Chapter 3.31

Object-Based Analysis-Synthesis Coding

Hans G. Musmann

Abstract

For encoding moving images at very low bit rates object-based analysis-synthesis coding has been proposed. According to this coding concept, each moving object of an image is described and encoded by three parameter sets defining the motion, the shape and the surface color of the moving object. The parameters to be coded depend on the source model describing the objects and their motion. Thus, the coding efficiency is also dependent on the source model applied. In the meantime object-based analysis-synthesis coding systems, which are based on source models of moving 2D- and 3D-objects, have been implemented and investigated.

In this tutorial the basic concepts of object-based analysis-synthesis coding and the optimization of such a coder will be explained. The results obtained with source models of moving 2D- and 3D-objects will be compared to those of a standard H.261 coder at a bit rate of 64 kbit/s. Also, first results obtained at 11 kbit/s will be presented.

3.31.1 Introduction

In order to encode moving video signals at the 64 kbit/s transmission rate, the standardized hybrid coding technique H.261 subdivides each image of a sequence into blocks of 16×6 picture elements (pels) and encodes each block by motion compensated predictive and transform coding [1].

A synthesized prediction image is composed of translatorically moved square blocks. Thus, this coding technique is based on the source model of moving square blocks. Two parameter sets have to be coded and transmitted, the motion parameter of a block represented by its displacement vector and the color information of the picture elements of the block encoded by transform coding of the prediction errors.

At very low bit rates block-based coding can generate visible distortions known as blocking and mosquito artifacts. To avoid these image distortions object-oriented analysis-synthesis coding [2] has been proposed. This coding technique is based on the source model of moving objects instead of moving square blocks. Each object is described and encoded by three parameter sets defining the motion, shape and surface color of the object.

The parameter sets to be coded vary with the kind of object model. Three kinds of source models have been investigated and will be described, rigid 2D-objects [3], flexible 2D-objects [4] and rigid 3D-objects [5].

In order to evaluate the coding efficiency of object-oriented analysis-synthesis coding, the encoded parameter bit rates of the cited source models are compared at a fixed picture quality. After an explanation of the basic components of an object-based analysis-synthesis coder in section 3.31.2 the coding of the parameter sets is described in section 3.31.3. Finally the influence of the various source models on the coding efficiency is discussed in section 3.31.4.

3.31.2 General structure of an object-oriented analysis-synthesis coder

The block diagram of an object-oriented analysis-synthesis coder is shown in Figure 1. Instead of a frame memory as in block-oriented coding techniques, object-oriented coding requires a memory to store the parameter sets $A = \{A_i\}$, $M = \{M_i\}$, $S = \{S_i\}$ defining the motion A_i, shape M_i and color S_i of each moving object i. The memory for object parameters of the coder and decoder contains the same parameter information and allows the coder and decoder to reconstruct a transmitted image by image synthesis. The reconstructed image I'_k is displayed at the decoder and used for image analysis of the next input image I_{k+1} at the coder.

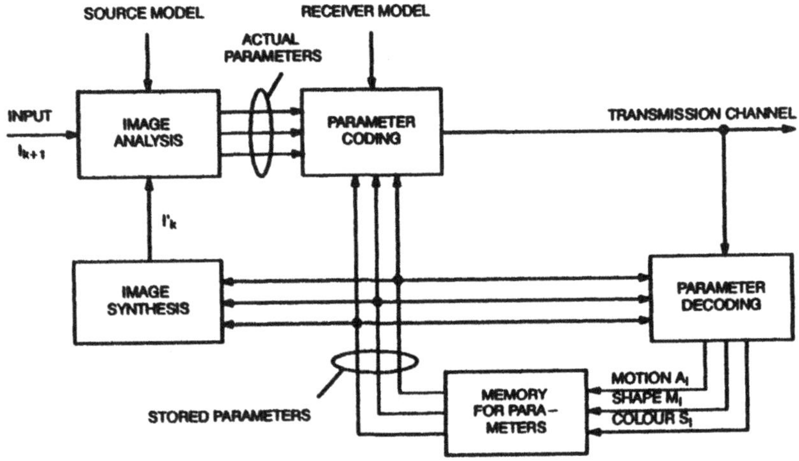

Figure 1 Block diagram of an object-oriented analysis-synthesis coder.

The task of the image analysis block in Figure 1 is to estimate the parameter sets A_i, M_i, S_i for each object i of the next input image I_{k+1} by use of the reconstructed image I'_k. At the output of the image analysis block, the parameter sets A_i, M_i, S_i are available in PCM representation.

Image analysis distinguishes between two types of objects. Model Compliance objects (MC-objects) can be reconstructed by transmitting only the motion A_i and shape M_i parameter sets and using the color S_i parameter set stored in the memory. Model Failure objects (MF-objects) are image areas where the motion description by the source model fails. MF-objects are encoded by the shape M_i and color S_i parameter sets, see Table 1.

	MC-object	MF-object
Motion A_i	A_i	
Shape M_i	M_i	M_i
Color S_i		S_i

Table 1 Parameter sets transmitted for MC- and MF-objects.

The actual parameter sets A_i, M_i, S_i of image I_{k+1} are encoded in the block parameter coding. For efficient coding, known predictive and transform coding techniques, which make use of the stored parameter sets [4, 5], are applied (see Figure 2). Coder mode control decides, with help of the MC/MF information, which parameter sets of each object will be transmitted.

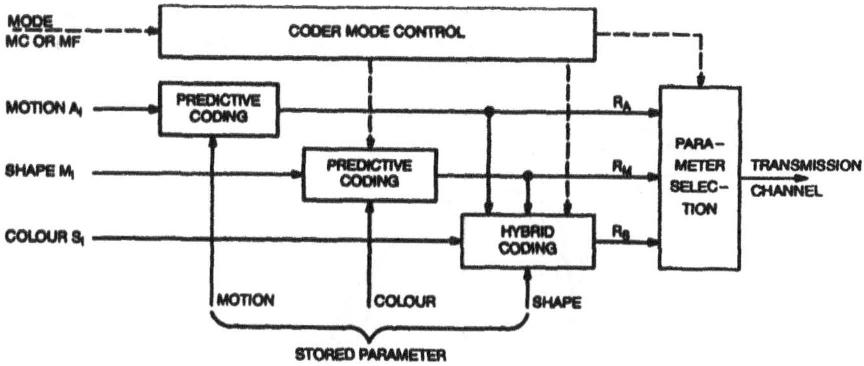

Figure 2 Block diagram of the parameter coder.

3.31.3 Parameter sets and parameter coding

The parameter sets to be coded depend on the source model as indicated in Table 2. In order to achieve efficient coding of the parameter sets, individual coding techniques are applied for each parameter set.

Motion parameter sets of MC-objects are predictively encoded using known DPCM techniques [3, 5]. In the case of rigid 2D-objects, three-dimensional motion is assumed. Therefore 8 mapping parameters $a_1 ... a_8$ describing the motion have to be encoded for each MC-object [2]. In the case of flexible 2D-objects, two-dimensional translatory motion of each block of an MC-object is assumed [4]. The motion of 3D-objects is described by 3 translatory motion a_1, a_2, a_3 and 3 rotary motion parameters a_4, a_5, a_6.

The shape parameters of 2D-objects describe the silhouette of a MC-object by polygon and spline approximation, see Figure 3. The vertices of this representation are predictively encoded using the actual motion and the stored shape information [3, 6]. The shape of 3D-objects is obtained only from silhouette and motion information [8]. Therefore only silhouette information has to be encoded.

	MC-object		MF-object	
	Motion A_i	Shape M_i	Color S_i	Shape M_i
Rigid 2D-objects 3D-motion	8 mapping parameters a_1, a_2, a_3, a_4 a_5, a_6, a_7, a_8	vertices of the object silhouette $\{m_1, m_2\}$	each pel of a MF-object $\{s_1, ..., s_N\}$	vertices of the boundary of the MF-area $\{m_1, m_2\}$
Flexible 2D-objects 2D-motion	2 translatory parameters for each block of an object $\{a_1, a_2\}$	vertices of the object silhouette $\{m_1, m_2\}$	each pel of a MF-object $\{s_1, ..., s_N\}$	vertices of the boundary of the MF-area $\{m_1, m_2\}$
Rigid 3D-objects 3D-motion	3 translatory and 3 rotatory parameters $a_1, a_2, a_3, a_4, a_5, a_6$	vertices of the object silhouette $\{m_1, m_2\}$	each pel of a MF-object $\{s_1, ..., s_N\}$	vertices of the boundary of the MF-area $\{m_1, m_2\}$

Table 2 Parameter sets to be coded per object for various source models.

The color information of MF-objects is encoded by a motion compensated block-based hybrid scheme where, for each block of 8×8 pel, a DPCM technique is used instead of a

DCT whenever it allows a more efficient coding [7]. This block-based hybrid scheme is also used as a fallback mode when object-based coding is less efficient.

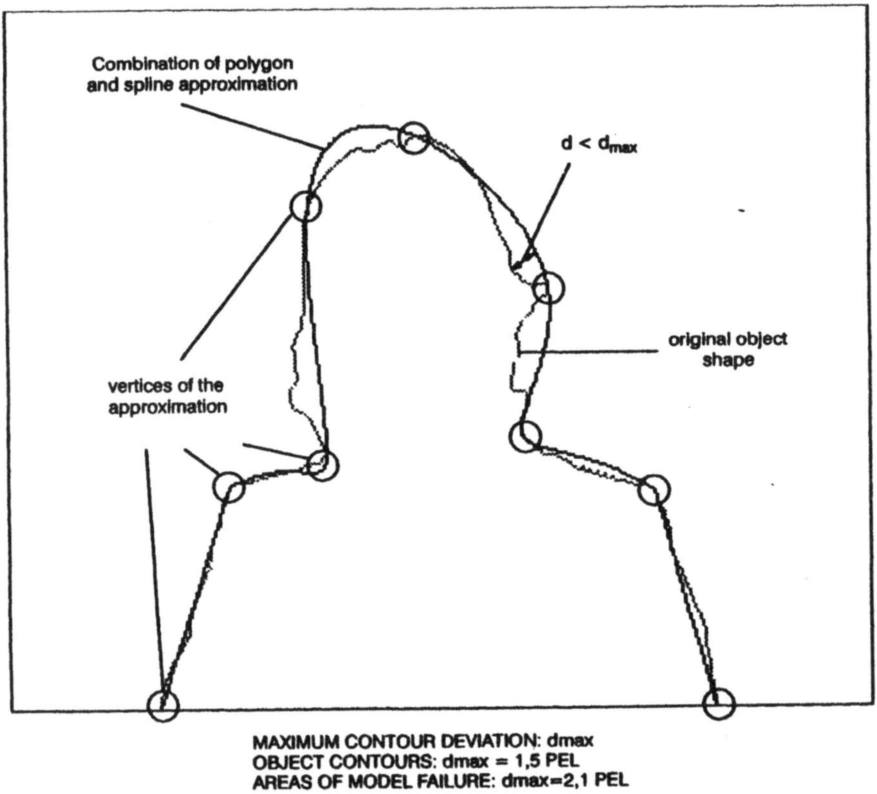

Figure 3 Representation of an object silhouette by polygons and splines.

The bit allocation to the three parameter sets for motion, shape and color has to be chosen in such a way that, for a given image quality, the overall data rate is minimum. The optimization procedure will be explained using the source model of flexible 2D-objects as an example.

The minimization problem is solved in two steps as proposed in [4]

1 optimization of the shape approximation,

2 optimization of the resolution for the motion description.

The accuracy of the shape approximation has to be adapted to the spatial image resolution and the camera aperture. The minimum allowable deviation of the shape approximation from the true shape contour is about 1.5 pel for an MC-object, in order to avoid visible distortion at object boundaries.

The resolution of the shape approximation of the MF-objects strongly influences the data amount for the coding of the color parameters. The more accurate the shape approximation, the less pels have to be updated by color coding. On the other hand, with increasing shape approximation accuracy, the data amount for the shape parameters of the MF-objects increases. In order to minimize the data rate for coding parameter sets of the

MF-objects, i.e. the sum of their color and shape information, the approximation error d_{max} is selected in such a way that the sum of the color and shape information is minimum for the MF-objects.

The experimental results of this optimization show that the data rate is minimum for an approximation error of $d_{max} = 2.1$ pel. The spatial resolution of the motion description is defined by the height and width of the measurement grid of the displacement vectors. The spatial resolution influences the efficiency of the motion compensated prediction used for color coding. Hence, the amount of data for the color information depends directly on the coding effort of the motion parameters. The finer the spatial and amplitude resolution, the less pels have to be coded by color and shape parameters. On the other hand with increasing spatial and amplitude resolution of the motion description, the amount of data for the motion information increases and more displacement vectors have to be coded and transmitted to the receiver. The results of the experimental investigations show that, for displacement estimation with 1/2 pel amplitude resolution, the optimum spatial resolution is 16×16 pels.

3.31.4 Coding efficiency

In order to compare the coding efficiency of object-oriented analysis-synthesis coders with different source models the bit rates R_A, R_M and R_S required for encoding the motion, shape and color parameter sets have to be analyzed. Table 3 shows the average bit rates R_A, R_M and R_S per image measured for the CCITT test sequences "Miss America" and "Claire" using a spatial resolution of 288 lines with 352 pel per line and 10 Hz frame frequency.

The bit rate R_M required for encoding the motion parameters of rigid objects is less than that for flexible objects, since many displacement vectors $\{a_1, a_2\}$ have to be encoded for only one MC-object.

	Motion information R_A	Shape information R_M	Color information $R_S = N\ r_s$
Rigid 2D-objects 3D-motion	600	1 300	15 000 r_s
Flexible 2D-objects 2D-motion	1 100	900	4 000 r_s
Rigid 3D-objects 3D-motion	200	1 640	4 000 r_s

Table 3 Average bit rates R_A, R_M, R_S in bit per image for various source models. N is the total number of pels of the MF-objects, r_s is the bit rate allocated per pel.

The bit rate R_M consists of two parts, one part for MC- the other for MF-objects. R_M for rigid 2D-objects is greater than that for flexible 2D-objects since the source model of rigid 2D-objects fails more often and therefore generates more MF-objects. This can also be recognized from the number of pels N of the MF-objects to be coded. The bit rate R_S for encoding the color parameters $\{S_1,...S_N\}$ results from the number of pels N to be coded multiplied by the bit rate r_s allocated for encoding each pel.

$$R_S = N \ @ \ r_s \tag{1}$$

With the choice of r_S the picture quality can be fixed. Thus N is a measure of the transmission rate required to code the color parameters and can be used for comparing the coding efficiency of different source models.

Table 3 shows that the two source models of 2D-objects generate almost the same bit rate $R_A + R_M$. Since N is much greater for the source model of rigid 2D-objects, the source model of flexible 2D-objects provides a higher coding efficiency.

Comparing the model of flexible 2D-objects with that of rigid 3D-objects, the bit rate R_A is smaller for rigid 3D-objects since the number of motion parameters is less. The shape information of MC-objects is the same for flexible 2D- and rigid 3D-objects since both are encoded by silhouettes. However, the shape information required for encoding the MF-objects is greater in the case of rigid 3D-objects since the total area N of MF-objects is split into more MF-objects of smaller size. Thus R_M is higher for rigid 3D-objects although N is the same.

To achieve an acceptable picture quality r_S should be greater than 1 bit per pel. Thus the source models of flexible 2D-and rigid 3D-objects allow coding with a total bit rate of 64 kbit/s. In a block oriented H.261 coder N is much greater in the range of 40000 pels.

Figure 4 represents the results of Table 3 in form of a graph. In addition first results obtained with the source model of flexible 3D-objects are included [9].

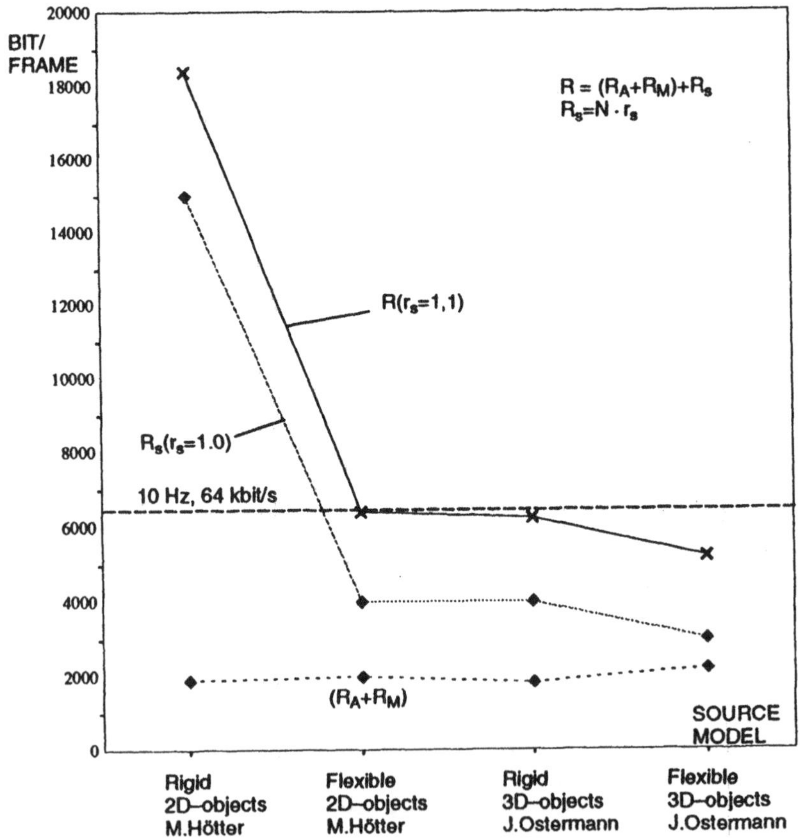

Figure 4 Experimental results of average bit rates $R_A + R_M$ and R in bits per CIF frame. r_s is the bit rate allocated per pel.

For encoding moving images at 8-16 kbit/s and QCIF image resolution the coding of the parameter sets has to be adapted to the changed image resolution [10].

3.31.5 Conclusion

It has been shown that there is only a relatively low degree of freedom for encoding the motion and shape parameters so that $R_A + R_M$ is almost fixed. R_S is linearly dependent on the number N of pels in the areas of model failure at a given picture quality.

Since coding of the areas of model failure (MF-objects) takes most of the bit rate for coding the color parameter set, it is important to reduce the areas of model failures. Consequently object-based analysis-synthesis coding using source models with flexible objects is more efficient than those with rigid objects.

In the case that object-based analysis-synthesis coding is inefficient, block-based hybrid coding is used as a fall back mode.

To improve the coding efficiency, predefined 3D object models can be introduced in future investigations.

References

[1] CCITT, "Draft revision of recommendation H.261: video codec for audio visual services at p×64 kbit/s", Study Group XV, WP/1/Q4, Specialist Group on Coding for Visual Telephony, Cambridge, USA. November 1988.

[2] H. G. Musmann, M. Hötter and J. Ostermann, "Object-oriented analysis-synthesis coding of moving images", *Signal Processing: Image Communications*, Vol. 1, No. 2, pp. 117-138, October 1989.

[3] M. Hötter, "Object-oriented analysis-synthesis coding based on moving two dimensional objects", *Signal Processing: Image Communications*, Vol. 2, pp. 409-428, Dec. 1990.

[4] M. Hötter, "Optimization and efficiency of an object-oriented analysis-synthesis coder", accepted for publication in *IEEE Trans. on Circuits and Systems for Video Technology*.

[5] J. Ostermann, "Object-oriented analysis-synthesis coding based on the source model of moving rigid 3D-objects", submitted for publication in *Signal Processing: Image Communications*.

[6] M. Hötter, "Predictive contour coding for an object-oriented analysis-synthesis coder", *IEEE International Symposium in Information Theory*, San Diego, California, USA, p. 75, January 1990.

[7] H. Schiller, M. Hötter, "Colour coding in an object-oriented analysis-synthesis coder", submitted for publication in *Signal Processing: Image Communication*.

[8] J. Ostermann, "Modelling of 3D-moving objects for an analysis-synthesis coder", in B. Girod, Ed., *SPIE/SPSE Symposium on Sensing and Reconstruction of 3D Objects and Scenes '90*, Proc. SPIE 1260, pp. 240-250, Santa Clara, Feb. 1990.

[9] J. Ostermann, "Analysis-synthesis coder based on moving flexible 3D-objects," accepted at *Picture Coding Symposium PCS'93*, Lausanne, March 1993.

[10] P. Gerken, "Object-based analysis-synthesis coding at very low bit rate", accepted for publication in *IEEE Trans. on Circuits and Systems for Video Technology*, Special Issue on Very Low Bit Rate Video Coding.

Chapter 3.32

Model-Based Image Coding

Kiyoharu Aizawa

3.32.1 Introduction

Historically, progress in image coding techniques has been through incorporating results from other fields such as information theory. Most of the existing coding methods such as predictive coding, transform coding, and vector quantization belong to information-theory-based methods in which image signals are considered as random signals and compressed exploiting their stochastic properties.

Apart from these information-theoretic coding methods, research on new approaches such as model-based coding (which is related to both image analysis and computer graphics) has recently intensified. An essential difference between conventional coding methods and these new approaches is the image model they assume. Contrary to conventional coding methods which efficiently represent 2-D waveforms of image signals, model-based coding represent image signals using structural image models which take into account the 3-D properties of the scene in some sense.

There have been two major approaches to model-based image coding. One approach uses general models such as planar patches and smooth surfaces to represent the 3-D properties of the objects [11, 19, 27, 30, 32, 35, 38, 39, 42]. The other approach utilizes detailed parameterized 3-D object models such as a parameterized facial model [2, 6, 13, 15, 17, 21, 23, 25, 31, 34, 40, 46-48, 50, 53].

The former approach has been discussed in the context of advanced motion compensation. The latter approach is clearly different from the conventional approaches, since it uses an explicit 3-D model and encodes images with computer-vision-oriented techniques and reproduces images with computer-graphics-oriented techniques. Both approaches will potentially achieve very low bit rate video compression.

In the following sections, image coding schemes are described from the point of view of their associated image models, and among our works done in Univ. of Tokyo [2-4, 8-10, 35, 39-41], 3-D model-based coding and 2-D deformable triangle based motion compensation which are related to those two approaches are described later.

3.32.2 Image coding schemes and their associated image models

In order to efficiently encode image signals it is necessary to assume an image model and apply it to representation of image signals. There are two aspects in image source model used for video coding: they are the segmentation model and the motion model.

Table 1 shows what kind of models are used in various coding schemes. Conventionally, from an information theory point of view, image signals have been modelled by using stochastic properties, that is, by exploiting statistical dependencies either in the spatial domain or the transform domain. Conventional coding schemes such as MC-

DCT usually divide a image into blocks in spatial domain and make use of statistical dependency. The motion is assumed to be limited in 2-D translation.

Image Source Models		Coding Schemes
Segmentation model	Motion model	
Pel		PCM
Statistically dependent pels block	2-D translation	MC-DCT etc.
2-D (model-based) approach 2-D features 2-D regions deformable triangle block, etc.	translation affine transform bilinear transform perspective transform, etc.	object oriented coding, 2-D deformable triangle based coding, etc.
3-D (model-based) approach 3-D model	3-D global motion 3-D local motion	3-D model-based coding

Table 1 Video coding schemes and their associated image source models.

On the other hand, from an image analysis point of view, images have been considered as having structural features such as contours and regions. Advanced image coding methods make use of variations of image source models, while the those of conventional coding methods are very limited.

There are 2-D approaches and 3-D approaches in advanced image coding schemes. 2-D approaches exploit 2-D structural information such as image features and regions to encode images at very low bit rates while keeping enough visible structures in the reconstruction. Their motion models are not limited to the simple translation model, but they exploit affine transform or perspective transform motion models etc. which better approximates motion field of a 3-D moving rigid object and linear deformation such as rotation and zooming.

Besides, in the last few years, this structural model based approach has adopted 3-D structural models of the scene. In this case, the coder and decoder have a 3-D model of an object; the coder analyzes input images and the decoder generates output images using the model. The motion models are naturally 3-dimensional. Similarity between the 2-D and 3-D approaches is that they tend to make use of structural features and properties of the image content in some sense. At present, 3-D models required for 3-D model-based coding are made beforehand because of difficulties of automatic modelling so that 3-D model-based coding currently deal with specific images such as facial images. 2-D approaches are more flexible in terms of scenes that they can handle.

In the following sections, our works related to 3-D model-based coding for facial images [2] and 2-D deformable triangle model based motion compensation [39] will be described.

3.32.3 3-D model-based coding for facial images

In this section, our contribution to model-based coding which images a person's face by utilizing a 3-D facial model is described as an example of model-based coding. Related works are also mentioned.

Our model-based coding system consists of three main components: a 3-D facial model, an encoder and a decoder. The description of this model-based coding system is illustrated in Figure 1.

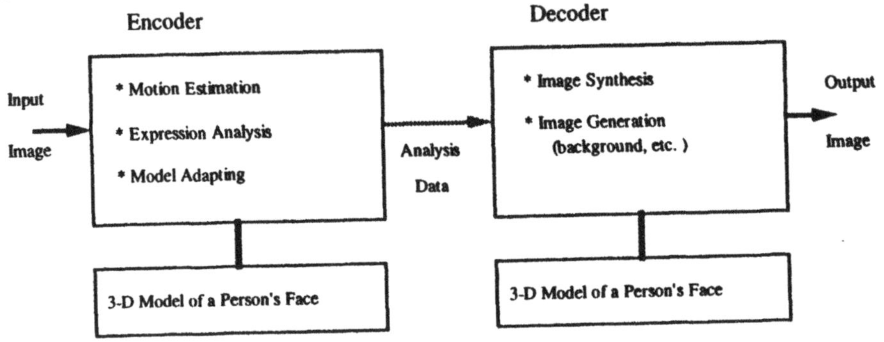

Figure 1 General description of a mode-based coding system.

The encoder separates the object from the object background, estimates the motion of the person's face, analyzes the facial expressions, and then transmits the necessary analysis parameters. The encoder will adapt new depth information and initially unseen portions of the object into the model by updating and correcting it if required. The decoder synthesizes and generates the output images by using a 3-D facial model and the received analysis parameters.

3.32.3.1 Modelling a person's face

Modelling an object is the most important part in model based coding because both analysis and synthesis methods strongly depend on the model they use.

For communication purposes, a person's face must be modelled in sufficient detail. Previous studies [44] which have attempted to image a human face with graphics have produced results which lack details and reality because the imaging methods that relied on computer graphics techniques used only wire frame models and shading techniques to reconstruct a human face.

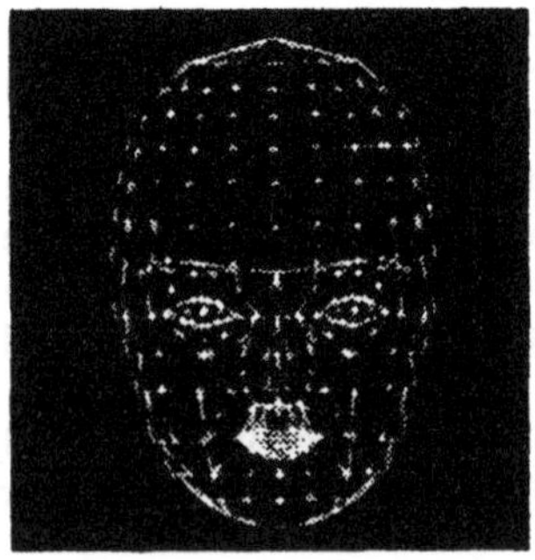

Figure 2 A generic wire frame model.

In order to develop an accurate model, original facial image and a texture mapping technique were utilized by us. That is, a 3-D wire frame generic face model (Figure 2) which approximately represents a human face and composed of several hundreds of triangles are deformed to fit its feature points positions and outmost contour lines to those of a frontal face image of the object person. Then, this original face image is mapped on the adjusted wire frame model.

Once a 3-D facial model has been obtained it can be easily rotated in any direction. The rotated images still appear natural, although the depth information on the created 3-D facial model is only a rough estimate.

Additional information such as side view of a face [48], continuous aspect view of a face [58] and range data [15], are used for increasing the accuracy of the initial 3-D facial model. Recently, a finer detail facial model which has three skin layers and muscles has been investigated [50].

As for the human body, a stick figure model has been proposed for describing and coding the motion [25].

3.32.3.2 Synthesis of facial movements

Texture mapping of original facial images onto a 3-D wire frame model gives rise to natural looking images. In addition, in order to synthesize naturally animating images, synthesis of facial movements (expressions) plays a very important role in the model-based coding system. In our system, two different methods for synthesizing facial expressions have been presented. One technique used a "clip-and-paste method" and the other a "facial structure deformation method".

The clip-and-paste method is performed by extracting specific expressive regions from an input image, transmitting them, and then placing them in the corresponding regions of the synthesized image. This method is suitable for the representation of fine facial movements such as eye expressions.

The facial structure deformation method deforms the structure of the 3-D wire frame general face model to simulate facial expressions. This method controls the locations of the vertices of the wire frame model according to "deformation rules" that are based on psychological and anatomical knowledge. In our contributions the "Facial Action Coding System"[12] has been adopted, in which facial actions are decomposed into combinations of "Action Units". Because there are many different ways in deforming 3-D model, other contributions have used different deformation methods [23, 34].

Figure 3 Clip and paste method. Eye and mouth area as clipped regions.

In order to improve the reconstruction quality, texture updating is required. Some approaches were proposed which additionally encodes differential signal between input images and synthesized images by using DCT [40], VQ [16] and contour coding [34].

The estimated bit transmission rates of both clip-and-paste method and structure deformation method is shown in Table 2.

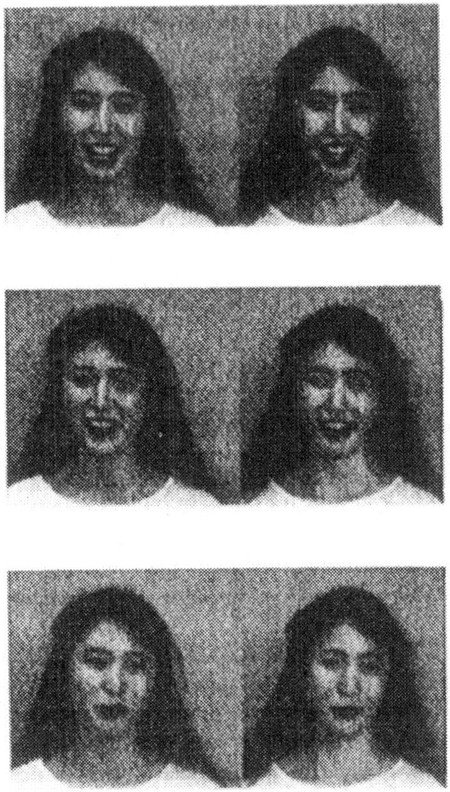

Figure 4 Three frames showing images synthesized by clip and paste method. The input images are on the left side and the reconstructed images are on the right side. "White marks" are placed on the input image for motion estimation.

Synthesis method	Information to be transmitted	Expected bit-rates
Clip-and-paste method	3-D motion parameters and clipped image signals	1 k - 10 k bit/s
Facial structure deformation method	3-D motion parameters and facial action unit parameters	< 1 k bit/s

Table 2 Expected bit-rates estimation of a 3-D model based coding scheme.

3.32.3.3 Analysis of facial images

Analysis problems are more difficult than synthesis problems. There seems to be no system, at present, which can work automatically both in modelling objects and in extracting model parameters even for restricted head-and-shoulder images.

The analysis methods strongly depend on what is assumed as the model and what is synthesized as output images. Besides, analysis problems include several different aspects such as segmentation of objects, estimation of global motion and estimation of local motion.

In case of facial images, global motion and local motion corresponds to head motion and facial expressions, respectively. Among the techniques which have been conducted for analysis of facial images are

- feature extraction with threshoulding [23, 40]
- motion estimation and tracking of marks plotted on the moving object [2]
- object segmentation by using stereo cameras [24]
- feature points extraction by using active contour model [21]
- facial muscle movement estimation based on optical flow [33]
- estimation of the action units by using feature points displacements [10]
- direct estimation of head motion and the action units by using spatio-temporal gradient and 3-D facial model [9, 31]
- global motion estimation of the head using 3-D model and displacement vector field obtained by the block matching technique [29]
- facial muscle movement estimation based on physical and anatomical models [47,50].

3.32.3.4 Implementations and applications

Several implementations and applications have been carried out. Among them are

(i) real-time synthesis of facial expressions which is aimed to be used by psychologists doing research on human facial expression [56]

(ii) speech driven facial animation synthesis system [36]

(iii) real-time model-based coding system [22]

(iv) application of 3-D model for converting images to stereo display [49].

From a practical point of view, a more general coding method can be included together with 3-D model based coding scheme in order to deal with unmodelled objects. In other words, the generality can be guaranteed by combining 3-D model and a more general image source model.

Nakaya of Univ. of Tokyo has worked on 2-D deformable triangular block based motion compensation [39]. In this section, his work is briefly reviewed and the combination of the general 2-D model and 3-D model based approaches is described.

3.32.4 2-D deformable triangular model for advanced motion compensation

3-D model-based coding is expected to have far broader applications than conventional waveform coding techniques. However, 3-D model based coding at present has problems as a complete coding scheme. The first important problem is its generality, that is, the object that can be handled by 3-D model based coding is restricted and it is not able to deal with unmodelled objects.

3.32.4.1 2-D deformable triangle based motion compensation

As mentioned in section 3.32.2, the conventional coding method segments images into square blocks and estimates 2-D translational displacement for each block.

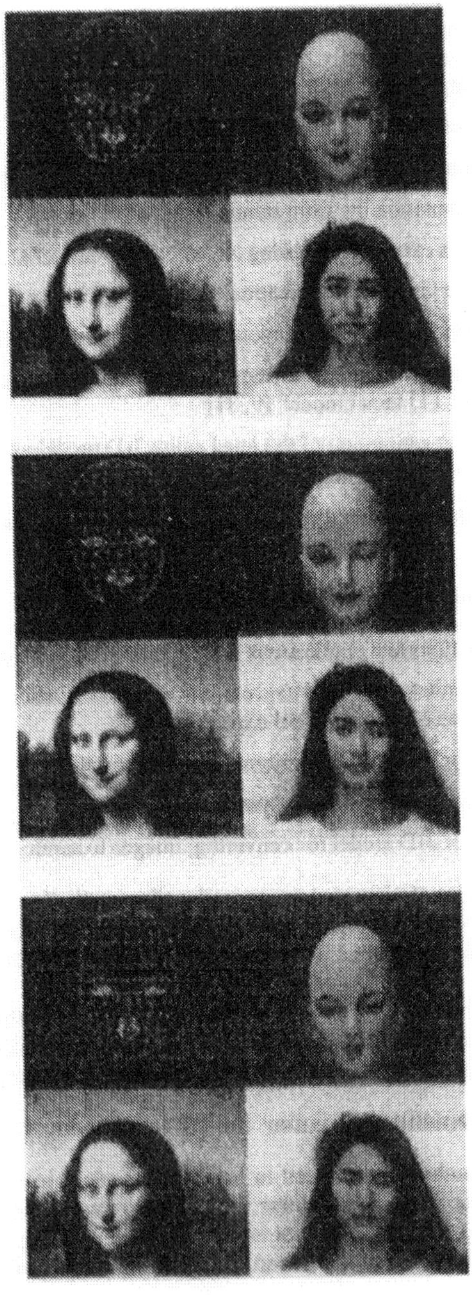

Figure 5 Images synthesized by structure deformation method. All our facial models are driven by the same combination of action units.

The proposed scheme is to divide images into triangular blocks and makes use of affine motion model; each triangle is deformable as follows.

$$\begin{pmatrix} x' \\ y' \end{pmatrix} = \begin{pmatrix} a & b \\ c & d \end{pmatrix} \begin{pmatrix} x \\ y \end{pmatrix}$$

The affine motion parameters of each triangles are uniquely determined by displacements of its grid points.

Motion compensation is performed by covering the current frame by triangular patches, estimating the motion of the grid points and synthesizing the prediction image by mapping (warping) the texture of the previous frame onto the corresponding patches of the current frame.

Estimation of displacements of grid points is done as follows: firstly initial displacements are computed by applying usual block matching at grid points, and secondly the initial displacements are adjusted successively so that the prediction error after warping is minimized.

The advantages of the triangle based motion compensation are listed below

- it can deal with linear deformation such as zooming and rotation
- it well approximates motion field of 3-D moving object
- it much alleviates blocking artifacts.

How the triangle motion prediction works is shown in Figure 6. The triangular segmentation of the previous frame and the segmentation of the current frame are shown for "a zooming circle" and "a rotating square" sequences. The prediction images of these examples are compared to that of the conventional block matching as shown in Figure 7. It shows that the triangle based method much alleviates blocking artifacts compared to the block-matching method.

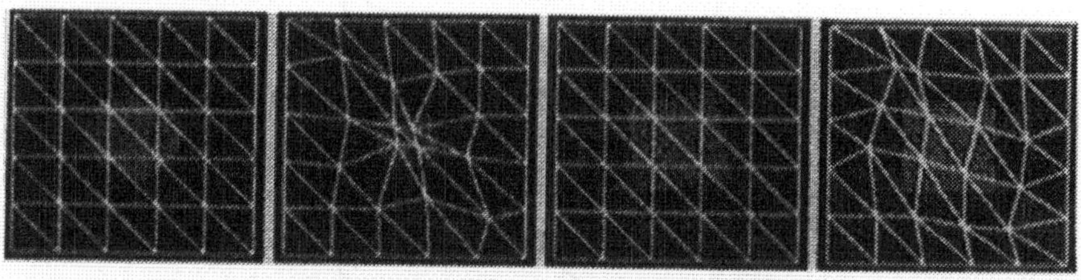

Figure 6 Motion estimation of the grid points. From left to right, current frame and previous frame of zooming circle sequence, and current frame and previous frame of rotating square sequence. Motion is estimated backward so that the current frame is divided into regular triangles and the previous frame into deformed triangles caused by motion.

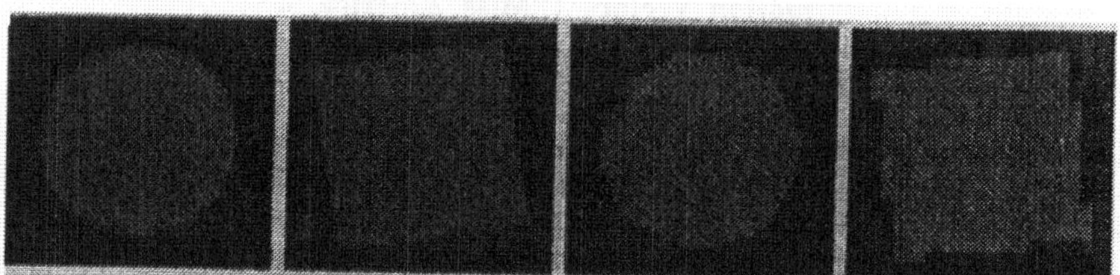

Figure 7 Prediction images. Left two images are prediction by the triangle based method, and right two by the conventional block matching method.

3.32.4.2 Coding at very low bit rates

(i) The number of motion vectors

Prediction performance of the triangle based method is compared with conventional block matching method in terms of the number of motion vectors in the image. (One of the segmentations used in this experiment is shown in Figure 8. When the number of motion vector is smaller, each segmentation is larger). The simulation is performed using the monochrome version of the "Miss America" of 352×288 pels (CIF) and the frame rate was 10 Hz. Original image was used for prediction. Averaged SNR is shown in Figure 9. The results indicate that the triangle based method needs less motion vectors compared to the conventional block matching.

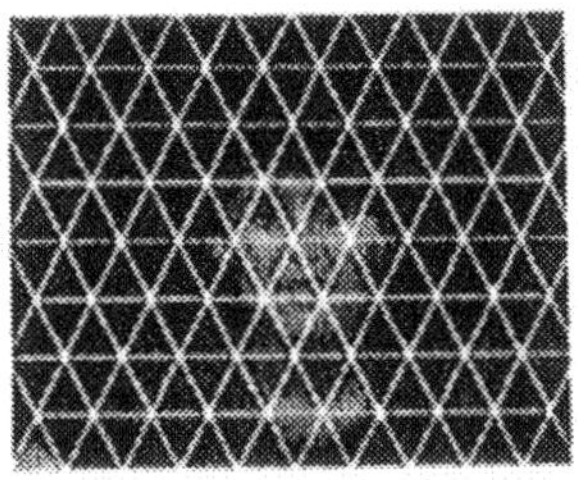

Figure 8 Segmentation used for 16 kbps coding by the triangle based MC. The number of motion vectors is 95 and 160 triangular patches.

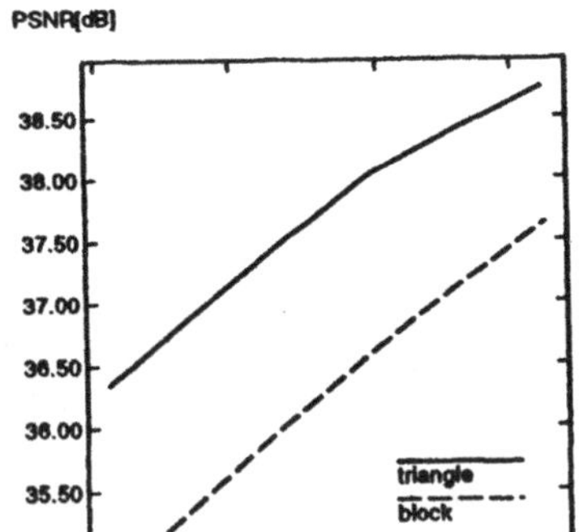

Figure 9 Prediction performance. Averaged SNR versus the number of motion vectors ("Miss America").

(ii) Coding at 16 kb/s

Color Sequence of "Miss America" (CIF format) at the frame rate 10 Hz was used for the experiments. Coding schemes in comparison are Triangle based MC-DCT, block-matching MC-DCT and RM8. The triangle based MC used the segmentation shown in Figure 8. The block matching MC-DCT used 99 blocks of 32×32 pixels. In both methods, the motion information was transmitted initially and error information was subsequently transmitted using the remaining channel capacity. The number of triangles and blocks for the triangle based method and block matching methods are optimized so that the best coding performance is obtained. RM8 was down-scaled and also tested at 16 kbps.

The coding performance of these methods is shown in Figure 10. It shows that the triangle based method performs best and the quality is kept good enough while block-matching MC-DCT an RM8 is degraded. Images coded by those methods are compared in Figure 11.

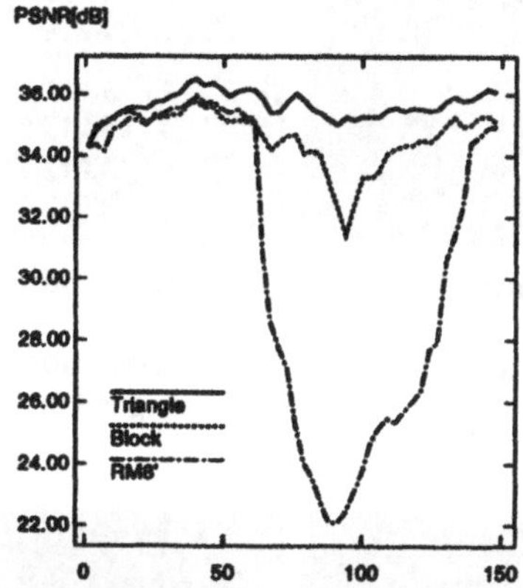

Figure 10 PSNR of the decoded images of Miss America.

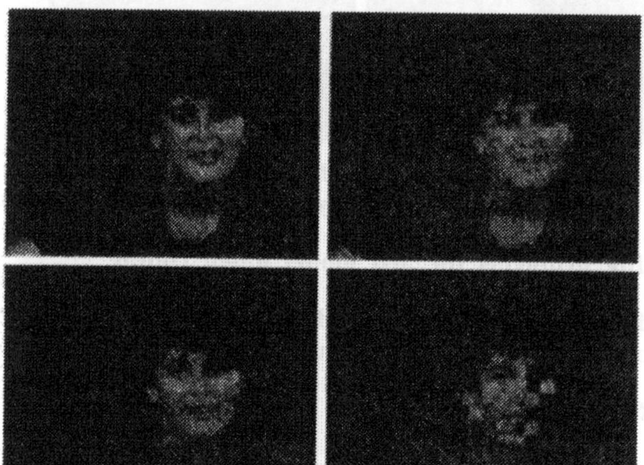

Figure 11 Decoded images. From upper left to lower right, original, the conventional block-matching MC-DCT, the triangle based MC-DCT, scale-down RM8.

3.32.4.3 Hybrid of 3-D and 2-D model-based coding schemes

Hybrid coding of 3-D model based coding and 2-D triangle motion compensation is investigated. As shown in Figure 12, in the hybrid scheme, face area is coded by 3-D model based coding and the other area is coded by 2-D triangle based motion compensation. Error signal is not coded to reduce bit rates. 3-D model is assumed to be made beforehand and positioned at the right place at the beginning of the sequence. Feature points such as corners of eyes and mouth are detected and they are automatically tracked in these experiments. The advantages of the hybrid scheme is that the face area is coded by very low bit rates by using 3-D model and more information can be assigned to other area.

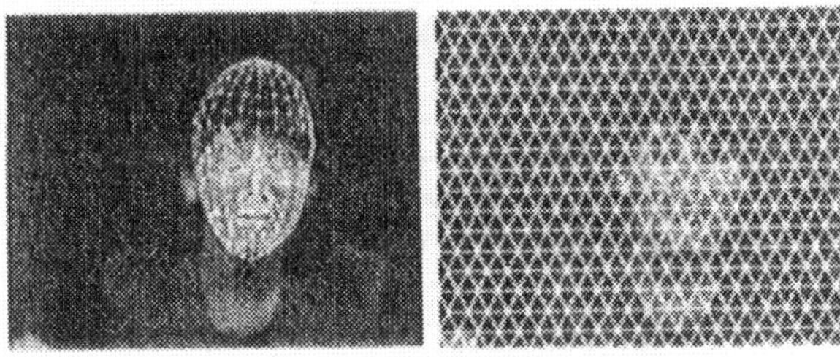

Figure 12 3-D/2-D hybrid coding.

The coded pictures are shown in Figure 13. The transmission bit-rate is 5 kbps: 1 kbps is assigned to 3-D model based coding and 4 kbps is assigned motion information for triangle based motion compensation.

Figure 13 Decoded image by the hybrid method. Left is original frame and right is the decoded images by using the hybrid method.

3.32.5 Remaining problems for 3-D model based coding

3-D Model-based coding is in its infancy and many problems remain to be solved.The following are, in our opinion, some of the difficult issues in 3-D model-based coding.

3.32.5.1 Modelling objects

Modelling objects is the most important issue in model based coding. Complexity of analysis and synthesis depends on the model adopted.

Currently well-approximated 3-D models which are obtained from a priori knowledge are used by most researchers. As mentioned at the beginning of section 3.32.4, a difficult problem is: how to deal with unknown objects? It seems necessary that a hierarchy of models in terms of approximation degree should be used. In this hierarchical modelling, every object is roughly modelled, and a priori 3-D models are further applied to known objects. Combination of model-based coding and waveform coding [16, 34, 41] or the hybrid scheme discussed in section 3.32.4 are considered practical approaches to this generality problem.

Another problem is caused by the complexity of the well-approximated model. If a model has finer details, it is more realistic for synthesis but more complex for analysis. It doesn't seem wise to use complex models for every object. Some objects such as faces had better be modelled by using well-approximated complex model, while the other objects had better be coded based on imprecise models. Model hierarchy will be helpful for this problem also.

3.32.5.2 Evaluation problem

For conventional waveform coding techniques a common evaluation measure is the mean squared error. Though it has often been claimed that it is not always a good criterion, it has been a force which guides the progress of waveform coding in the right way. What is a good quality criterion for model-based coding? There has been no discussion on this point. Though Pearson [45] deals with differences between an original image and an image synthesized by texture-mapped-model, squared error is still used for the quality measure. One of the attractive points of model-based coding is that it is free from squared error measure. There is no guidance at the present time on how to quantitatively evaluate the goodness of model-based coding systems.

References

[1] K. Aizawa and T. S. Huang, "Model-Based Image Coding", *Paper Summaries of Dynamic Scene Understanding*, 1991 IJCAI Workshop, Sydney, 1991.

[2] K. Aizawa, H. Harashima, and T. Saito, "Model-Based Analysis Synthesis Image Coding for a Person's Face", *Image Communication*, Vol. 1, No. 2, pp. 139-152, 1989.

[3] K. Aizawa, H. Harashima, and T. Saito, "Model-Based Synthetic Image Coding System", *Picture Coding Symposium '87*, Stockholm, 1987.

[4] K. Aizawa,T. Saito and H. Harashima, "Construction of a 3-dimensional personal face model for knowledge-based image data compression", *National Conference Record of IEICEJ*, Musashino, Japan, Sep. 3-6 1986, 542, p.I-221 (in Japanese).

[5] T. Akimoto and Y. Suenaga, "3-D Facial Model Creation Using Generic Model and Front and Side Views of Face", *IEICE Trans. Inf. & Sys.* Vol.75-D, No.2, pp.191-197, 1992.

[6] C. Braccini, S. Curinga and F. Lavagetto "Model-Based Coding of Facial Images: from Analysis to Synthesis", *Picture Coding Symposium '93*, Lausanne, Switzerland, 1993.

[7] Buck, "Segmentation of moving head-and-shoulder shapes", *Picture Coding Symposium 90*, 9.2, Boston, Mar.1990.

[8] C. S. Choi, T. Okazaki, H. Harahsima and T. Takebe, "A System of Analyzing and Synthesizing Facial Images", *IEEE International Symposium of Circuit and Systems*, pp.2665-2668, Singapore, 1991.

[9] C. S. Choi and T. Takebe, "Analysis and Synthesis of Facial Expressions in Knowledge-Based Coding System for Facial Image Sequence", *ICASSP91*, Toronto, 1991.

[10] C.S. Choi, K. Aizawa, H. Harashima, and T. Takebe, "Analysis and Synthesis of Facial Expressions in Model-Based Image Coding", *Picture Coding Symposium '90*, Boston, 1990.

[11] N. Diehl, "Object-Oriented Motion Estimation and Segmentation in Image Sequences", *Signal Processing: Image Communication*, Vol. 3, pp. 23-56, 1991.

[12] P. Ekman and W. V. Friesen, " Facial action coding system", *Consulting psychologists press*, 1977.

[13] R. Forchheimer and T. Kronander, "Image Coding-From Waveforms to Animation", *IEEE Trans. on ASSP*, Vol. 37, No. 12, pp. 2008-2023, December, 1989.

[14] R. Forchheimer, O. Fahlander, and T. Kronander, "Low Bit-Rate Coding Through Animation", *Picture Coding Symposium'83*, Davis, CA, pp. 113-114, March, 1983.

[15] T. Fukuhara, K. Asai, and T. Murakami, "Model-Based Image Coding Using Stereoscopic Images and Hierarchical Structuring of New 3-D Wire-Frame Model", *Picture Coding Symposium '91*, Tokyo, 1991.

[16] T. Fukuhara, K. Asai, and T. Murakami, "Hierarchical Division of 3-D Wire-Frame Model and Vector Quantization in a Model-Based Coding of Facial Images", *Picture Coding Symposium '90,* Boston, 1990.

[17] H. Harashima, K. Aizawa and T. Saito, "Model-based analysis synthesis coding of videophone images - conception and basic study of intelligent image coding", *IEICE Trans.* Vol. E72-5, pp.452-459 ,1989.

[18] M. Hötter and J. Ostermann, "Analysis Synthesis Coding Based on Planar Rigid Moving Objects", *Int. Workshop on 64 kbps Coding of Moving Video,* Hannover, 1988.

[19] M. Hötter, "Object-Oriented Analysis-Synthesis Coding Based on Moving Two-Dimensional Objects", *Image Communication,* Vol. 2, No. 4, pp. 409-428, Dec. 1990.

[20] H. Holtzman, "3-D video modeling", *CHI Conference on Human Factors in Computing Systems*, CA, May, 1992.

[21] T. S. Huang, S. Reddy, and K. Aizawa, "Human Facial Motion Modeling, Analysis and Synthesis for Video Compression", *Proc. Visual Communication and Image Processing SPIE*, pp. 234-241, Boston, 1991.

[22] M. Kaneko, A. Koike, and Y. Hatori, "Real-Time Analysis and Synthesis of Moving Facial Images Applied to Model-Based Image Coding", *Picture Coding Symposium '91*, Tokyo, 1991.

[23] M. Kaneko, A. Koike, and Y. Hatori, "Coding of Facial Image Sequence Based on a 3-D Model of the Head and Motion Detection.", *Journal of Visual Communication and Image Representation,* Vol. 2, No. 1, pp. 39-54, March, 1991.

[24] F. Kappei and G. Heipel, "3-D model based image coding", *Picture Coding Symposium 88*, Torino,1988.

[25] T. Kimoto and Y. Yasuda, "Hierarchical Representation of the Motion of a Walker and Motion Reconstruction for Model-Based Image Coding", *Optical Engineering,* Vol. 20, No. 7, pp. 888-903, 1991.

[26] F.Kishino and K.Yamashita, "Communication with realistic sensation applied to a teleconferencing system", *IEICE Tech. Rep.* IE89-35, 1989 (In Japanese).

[27] R. Koch, "Adaptation of a 3-D Facial Mask to Human Faces in Videophone Sequences Using Model Base Image Analysis", *Picture Coding Symposium '91*, Tokyo, 1991.

[28] R. Koch, "Dynamic 3-D Scene Analysis through Synthesis Feedback Control", *IEEE Trans. PAMI*, Vol.15, No.6, pp.556-568.

[29] A. Koike, M. Kaneko and Y. Hatori, "Model-based image coding with 3-D motion estimation and shape change detection", *Picture Coding Symposium 90*, Boston, 1990.

[30] M. Kunt, A.Ikonomopoulos and M.Kocher, "Second Generation Image Coding Techniques", *Proc. IEEE'73*, 4, pp.795-812, 1985.

[31] H. Li, P. Rovivainen, R. Forchheimer, "3-D Motion Estimation in Model-Based Faical Image Coding", *IEEE Trans. PAMI*, Vol.15, No.6, pp.545-555.

[32] H. D. Lin and D. G. Messerschmitt, "Video composition methods and their semantics", *ICASSP'91*, pp.2833-2836, Toronto, 1991.

[33] K. Mase, "An Application of Optical Flow - Extraction of Facial Expressions", *Proc. of MVA'90*, Tokyo, 1990.

[34] T. Minami, I. So, T. Mizuno, and O. Nakamura, "Knowledge-Based Coding of Facial Images", *Picture Coding Symposium '90*, Boston, 1990.

[35] H. Morikawa and H. Harashima, "3-D Structure Extraction Coding of Image Sequences", *Journal of Visual Communication and Image Representation*, Vol.2, No.4, pp.332-344 , 1991.

[36] S. Morishima, K. Aizawa and H. Harashima, "An Intelligent Facial Image Coding Driven by Speech and Phoneme", *ICASSP'89*, Glasgow, 1989.

[37] S.Morishima, "A human-machine interface using media conversion and model-based coding schemes", *CG International'92*, Springer-Verlarg.

[38] H. G. Musmann, M. Hötter and J. Ostermann, "Object-Oriented Analysis-Synthesis Coding of Moving Images", *Image Communication*, Vol. 1, No. 2, pp. 117-138, Oct. 1989.

[39] Y. Nakaya, H. Harashima, "Motion compensation based on spatial transformations", *IEEE CSVT* (to be published).

[40] Y. Nakaya and H. Harashima, "Model-based Waveform Hybrid Coding for Low-Rate Transmission of Facial Images", *IEICE Trans. Commun.*, Vol. E75-B, No.5, pp.377-384 ,1992.

[41] Y. Nakaya, K. Aizawa, and H. Harashima, "Texture Updating Methods in Model-Based Coding of Facial Images", *Picture Coding Symposium '90*, Boston, 1990.

[42] J. Ostermann, "Modelling of 3-D moving objects for an analysis-synthesis coder", *Proç. SPIE, Sensing and Reconstruction of Three-Dimensional Objects and Scenes*, Vol.1260, pp.240-249, 1990

[43] J. Ostermann, "An Analysis-Synthesis Coder Based on Moving Flexible 3-D-Objects", *Picture Coding Symposium '93*, Lausanne, Switzerland, 1993

[44] F. I. Parke, "Parameterized Models for Facial Animation", *IEEE Computer Graphics and Applications*, Vol. 12, pp. 61-68, Nov. 1982.

[45] D. E. Pearson, "Texture mapping in model-based image coding", *Image Communication*, 2, 4, pp.377-396, 1990.

[46] D. Pearson, "Model-Based Image Coding", *Proc. GLOBECOM '89*, Dallas, 16.1, pp. 554-558, Nov. 1989.

[47] J. Fischl, B. Miller and J. Robinson, "Parameter Tracking in a Muscle-Based Analysis/Synthesis Coding System", *Picture Coding Symposium '93*, Lausanne, Switzerland, 1993.

[48] L. Tang, M. Pouyat, K. Aizawa and T. S. Huang, "Accuracy of Modelling Human Face Using a Generic Model", *Picture Coding Symposium '93*, Lausanne, Switzerland, 1993.

[49] M. Tanimoto and S. Nakashima, "Basic experiment of 2D-3D conversion for a new 3D visual communication", *Picture Coding Symposium'90*, Boston, 1990.

[50] D. Terzopoulous, K. Waters, "Analysis and Synthesis of Facial Image Sequence Using Physical and Anatomical Models", *IEEE Trans. PAMI*, Vol.15 No.6, pp.569-579.

[51] H. Ueno, K. Dachiku, K. Ozeki, F. Sugiyama, "A study on facial region detection in the standard video coding method", *Int. Workshop. on 64 kbit/s coding of moving video*, 5-2, Sept. 1990.

[52] K. Waters and D. Terzopoulos, "Modeling and Animating Faces Using Scanned Data", *The Journal of Visualization and Computer Animation*, Vol. 2, pp. 129-131, 1991.

[53] W. J. Welsh, "Model-Based Coding of Videophone Images", *Electronics & Communication Engineering Journal*, Vol. 3, No. 1, pp. 29-36, Feb. 1991.

[54] W. J. Welsh, A. D. Simons, A. D. Hutchinson, R. A. Searby, "Synthetic face generation for enhancing a user interface", *Proc. Image Com'90*, pp.177-182, France, 1990.

[55] G. Xu, H. Agawa, Y. Nagashima, F. Kishino and Y. Kobayashi, "Three-Dimensional Face Modeling for Virtual Space Teleconferencing Systems", *IEICE Trans.*,Vol.E 73, No.10, 1990.

[56] H. Yamada, H. Chiba, K. Tsuda, and K. Maiya, "New approach to the research on the facial expressions: Model-based facial Action Synthesizing Computer, (MASC) system", *International Journal of Psychology*, Vol. 27, p.47, 1992.

[57] A. L. Yullie, P. W. Hallinan and D. S. Cohen, "Feature extraction from faces using deformable templates", *International Journal of Computer Vision*, Vol.8, No.2, pp.99-111,1992.

[58] J. Y. Zheng, Y. Nagashima, and F. Kishino, "3-D Modeling From Continuous Aspect Views , *Picture Coding Symposium'91*, Tokyo, 1991.

[59] In addition, many papers related to model-based coding have appeared in *Picture Coding Symposium'87,88,90,91,93*, *Picture Coding Symposium of Japan '86,87,88,89,90,91,92*.

Tutorial 4

What you always wanted to know about delta-sigma converters but were afraid to ask... !

Organised by: Gabor Temes, *Oregon State University*

Chapter 4.1

Theory of Σ–Δ Analog to Digital Converter

Bosco Leung

Terminology

- Oversampling converter: analog to digital interface that exploits oversampling technique, typically in conjunction with a loop filter, to improve resolution at the expense of speed reduction (e.g. straight oversampling, prediction, noise-shaping).
- Predictive converter: special form of oversampling converter that has the loop filter (low pass, bandpass) in the feedback path.
- Noise-shaping converter: special form of oversampling converter that has the loop filter (low pass, bandpass) in the forward path.
- Σ–Δ modulator: special form of noise-shaping converter that uses a coarse quantizer (1-bit or multi-bit).
- Σ–Δ or Δ–Σ are used interchangeably.
- Interpolative modulator: more advanced form of Σ–Δ converter that incorporates a more sophisticated loop filter but retains only 1 internal ADC and DAC (interpolative by Sodini).
- Cascaded Σ–Δ modulator or MASH (multi-stage noise-shaping) converter: Σ–Δ converter that achieves resolution improvement via conversion of quantization error and subsequent cancellation in the digital domain.
- Σ–Δ modulator (Σ–Δ M) or analog front end (AFE): the input stage which generates the fast-data-rate short-word-length digital input $y(kT)$ for the decimation filter.
- Decimation filter: the digital filter which low-pass filters and decimates the output signal $y(kT)$ of the Σ–Δ modulator, and lengthens its wordlength to provide the required resolution.
- Baseband frequency f_b : the passband width of the decimation filter; usually equal to the bandwidth (Nyquist frequency) of the input signal $x(t)$.
- Sampling frequency f_s : the fast sampling rate with which $x(t)$ is converted into $x(kT)$; the same as clock rate of $y(kT)$.
- Oversampling ratio: $\mathrm{OSR} = \frac{f_s}{2f_b}$. Usually, $\mathrm{OSR} = 2^M >> 1$, M integer.
- Instability in a Σ–Δ modulator: a condition in which the signal levels in the modulator grow indefinitely until one or more of the components saturate.

4.1.1 Introduction

As the cost of VLSI circuits decreases there is a greater demand on the analog-digital interfaces between these VLSI circuits and the analog world. The importance of developing low cost analog-digital interfaces in applications ranging from instrumentation, voice, audio and video coding cannot be overemphasized. Due to the large number of analog-digital interfaces in the VLSI systems, the development of low-cost monolithic MOS analog-digital conversion techniques has been an important element in reducing the cost of the total systems. As MOS technology feature sizes continue to fall, the high speed, programmability, and low production cost offered by VLSI technology are making digital signal processors a more attractive alternative to conventional analog signal processing. To exploit the significant computational power offered by these processors, high resolution analog to digital interfaces must be developed.

This tutorial reviews the oversampling technique as a method to perform high resolution data conversion. The technique is especially attractive for VLSI implementations because they can take advantage of the inherent speed advantage of the digital VLSI technology [1, 11]. In section 4.1.2, conventional ADCs (Analog to Digital Converters) operating at the Nyquist frequency will be reviewed, together with a summary of their limitations. Two broad classes of oversampled ADCs, based on prediction and noise shaping, will be compared in section 4.1.3. Section 4.1.4 concentrates the discussions on a particular class of oversampled ADCs, the sigma-delta modulator. Various techniques of realizing the accompanying decimation filters are presented in section 4.1.5.

4.1.2 Nyquist rate A/D converters

Shown in Figure 1 are the primary functions of an A/D (Analog to Digital) converter that converts a continuous time analog input signal $x(t)$ into a sequence of digital codes $y(kT)$ at a rate f_N.To prevent aliasing, the analog filter in front of the sampling block in Figure 1 limits the bandwidth of $x(t)$ to half the sampling rate. The sampled signal is then converted to digital codes $y(kT)$ by the quantizer.

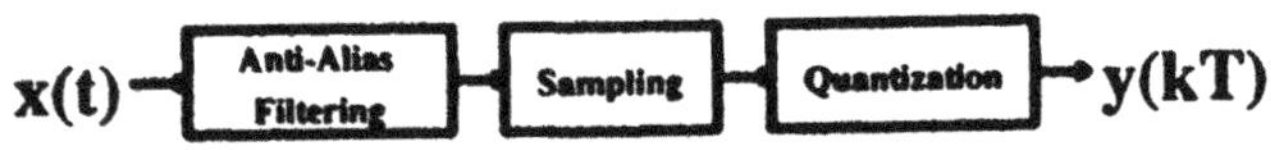

Figure 1 A/D converter block diagram.

A/D converter architectures can be subdivided into two groups based on the sampling rate of the converter. The Nyquist rate converters are defined to be converters that sample the analog signal at its Nyquist frequency $f_N = 2W$, where W is the signal bandwidth of the converter input. Oversampled converters are used to denote converters that sample at a much higher rate $f_S >> f_N$. The ratio of the sampling frequency to the Nyquist rate is defined as the oversampling ratio, OSR. The key idea is that for many applications, the bandwidth of the signal is small compared to the typical speed of operation available in VLSI circuits. In these cases, oversampled A/D converters offer the potential to trade this excess speed for improved resolution.

4.1.2.1 Classifications

The Nyquist rate architectures can be classified according to the number of clock cycles required to perform the quantization, which may vary from 1 to 2^N, where N is the resolution of the converter in bits. In a parallel A/D converter, commonly referred to as flash A/D converter, all of the possible quantization levels are simultaneously compared to the sampled analog input. To achieve the parallel operations 2^N-1 comparators are required for an N-bit converter. In practical implementations, the limitations in the resolution result from the exponential increase in the number of comparators as a function of the number of bits. For an additional bit of resolution the matching and accuracy requirements in the reference circuits will double. To date most flash A/D converters has been limited to about 10 bits of resolution.

To avoid the resolution limitations of fully parallel converters, the quantization operation may be distributed over extra clock cycles. For example, in a two-stage pipelined A/D converter, two steps are included in a conversion cycle. In the first pass, the conversion consists of a coarse quantization of the MSBs. The MSBs from the flash A/D subconverter are converted back to analog form by a D/A subconverter and subtracted from the input signal. The difference, or residue, is then subject to a fine quantization during the second pass, thus producing the LSBs. Another method, called subranging, includes performing coarse quantization during the first clock cycle to zoom in on the range of the input signal. In the second clock cycle this information is used to provide a suitable set of reference levels for the fine quantization process.

As a trade-off the number of comparators can be further reduced by performing the data conversion with a cascade of low-resolution stages. This reduction is accomplished at the expense of increased conversion time. In the extreme case, the stages can be pipelined to allow a one-bit conversion per stage, but with latency equal to the total number of bits.

In architectures in which one bit is resolved per clock cycle, only one comparator is required. Two distinct types of converters fall into this category. In successive approximation converters, starting from the MSB, a feedback voltage V_f is used to approximate the input analog voltage V_{in} in a sequence of successive steps. During each step, V_f is changed in accordance with the result of the previous comparison between V_{in} and V_f. At the same time the value of each bit is determined based on whether or not it causes the approximation to exceed the input. In algorithmic converters, the successive approximation algorithm is carried out serially in time. The difference between the converter input and a weighted value of the reference voltage, as determined by the previous bits, is multiplied by two. The difference is then compared with the reference voltage to determine the next bit.

At the lowest end of the speed ladder are Nyquist-rate A/D converter architectures that perform the comparison between the input and each of 2^N reference levels sequentially. They are sometimes referred to as indirect converters, as a quantity like frequency or time is used as an intermediate variable to obtain the final digital output code. The number of clock cycles it takes to resolve N bits is then 2^N.

4.1.2.2 Limitations of Nyquist rate A/D converters

(i) Anti-aliasing

Nyquist rate converters and oversampled converters differ in two fundamental aspects: anti-alias filtering and quantization. In the Nyquist rate converters, the filtering has to be done before the A/D converter, demanding the use of complex analog filters. In the case of oversampled converters the filtering can be done in multiple stages [3, 12]. The analog anti-aliasing filter preceding the A/D converter has a relaxed requirement due to the large "don't care" regions. The latter stages of filtering for subsequent decimation can be implemented digitally, thereby taking advantage of the enormous amount of digital signal processing capability provided by the VLSI technology. Shown in Figure 2 is a comparison of the requirements on the anti-aliasing filter of Nyquist rate and oversampled A/D converters. The Nyquist rate converter requires the use of an anti-aliasing filter with very sharp transition in order to provide adequate aliasing protection. The transition band of anti-aliasing filter of an oversampled A/D converter, on the other hand, is much wider than its passband, since anti-aliasing suppression is required only for a small band of frequencies around half the sampling rate, $\frac{f_s}{2}$. This simplification results in a considerably simpler anti-aliasing filter for oversampled A/D converters.

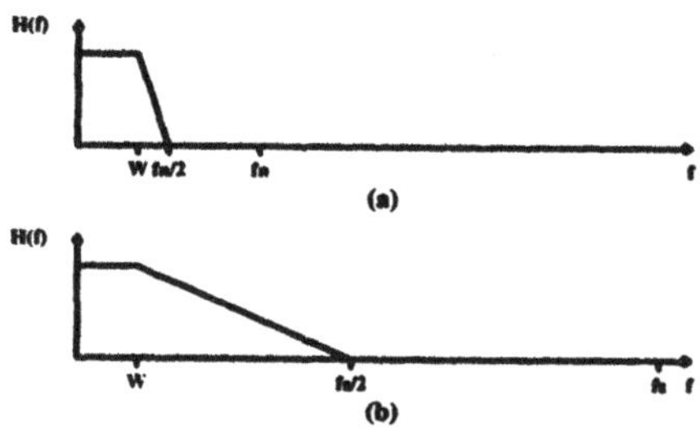

Figure 2 Requirements on the anti-alias filter (a) Nyquist rate ADC, (b) Oversampled ADC.

(ii) Quantization

Another difference between the two converters takes place in quantization. If the analog samples can be represented exactly, there is no apparent advantage to sampling at a rate higher than the Nyquist rate. However, during the analog to digital conversion quantization has to be performed, resulting in quantization errors. If more samples are taken, these errors can be reduced. For example, if the number of samples taken per unit time is doubled and each pair of samples produced is averaged to produce new samples at the original sample rate, the signal component of the samples will add linearly while the quantization noise component in the samples will add as random variables. If the noise is uncorrelated then the signal to noise ratio will improve by 3 dB. Consequently, in the case of oversampled A/D converters, a series of rough amplitude quantizations is made for every analog samples. This redundancy is combined with the averaging provided by a digital decimation filter to obtain a more precise estimate for the analog input [13]. On the other hand, for a Nyquist rate A/D converter quantization is performed for every sampling instant at the full precision of the converter. To appreciate more fully the advantage of the oversampled approach the quantization process in actual implementations will now be discussed.

In VLSI implementation, quantization is achieved by comparing the discrete time representation of the analog signal to a set of reference voltages generated by passive elements. The result of the comparisons is to generate a digital code with a nearest lower level to the input level. The resolution of a converter is defined by the spacing of the reference levels that can be generated and resolved. For a converter with 16 bit or higher resolution the spacing of these levels will have to be in the microvolt range. From the implementation viewpoint the spacing is determined by the matching properties of either the resistors or capacitors employed to generate the reference voltages. It has been shown in the past that the matching tolerance of these passive components is substantially lower than the above requirement [14]. In the past, techniques such as laser trimming or self-calibration have been utilized to improve the resolution of A/D converters. Oversampling can be viewed as another technique that will push the resolution of the converter beyond the component tolerances imposed by the technology.

4.1.3 Modulators for oversampled A/D conversion

The quantizer in an A/D converter performs quantization by approximating the analog input with the nearest lower level from a set of discrete reference levels. The quantization error e is defined as the difference between the quantizer input and the value of the output. If the input to the quantizer is sufficiently busy, this error can be modelled as uncorrelated noise. The variance of the noise is a measure of the quantizer's resolution and in a Nyquist rate converter is reduced by increasing the number of reference levels. In comparison, oversampled A/D converters reduce the quantization noise within the signal baseband, for a quantizer with a given resolution, through the use of feedback. Two approaches towards achieving this have been proposed: prediction and noise shaping [15, 16].

4.1.3.1 Predictive converter

The block diagram in Figure 3 illustrates the basic architecture of a predictive or differential modulator [17]. For a first order predictive converter, the transfer function $H(z)$ will be an integrator. It works on the principle that if one knows the course of a signal $x(t)$ up to a certain point in time, it is possible to make some inference $\hat{x}(t)$ about its continuation. This extrapolation process is called prediction. To achieve predictive encoding, the value of the present sample is used to calculate the next sample. The error $e(t)$ between the actual value and the estimation is quantized by the 1-bit ADC. The quantized value is then stored in the integrator. Since the error is encoded in digital form, to calculate the analog predicted value, a DAC (Digital to Analog Converter) is required in conjunction with either an analog or digital integrator. Typical waveforms corresponding to $x(t)$, $\hat{x}(t)$, and $e(t)$ are shown in Figure 3. As an analog to digital interface the advantage offered by prediction is that it reduces the dynamic range requirement on the ADC by subtracting $\hat{X}(t)$ from $x(t)$. The step size of the quantizer is adjusted to this reduced range, which results in a decrease of the quantization noise.

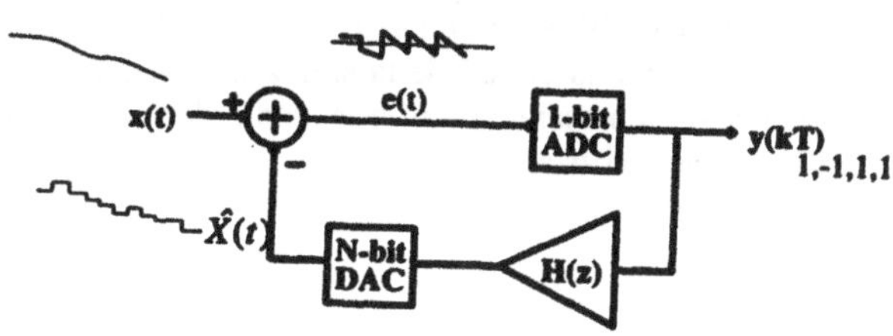

Figure 3 Block diagram of predictive converter.

4.1.3.2 Noise shaping convertor

Noise shaping converters do not reduce the magnitude of the quantization noise, but instead shape the power density spectrum of this error, moving the energy towards high frequencies. Provided that the analog input to the quantizer is oversampled, the high frequency quantization noise can be eliminated with a digital lowpass filter without affecting the signal band. Figure 4 shows the general power spectral density of a noise shaping converter of a first order modulator.

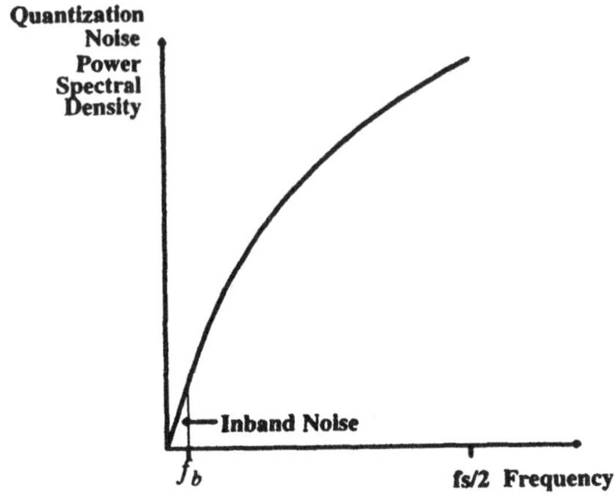

Figure 4 Power spectral density of a first-order noise-shaping converter.

4.1.3.3 Comparisons

A wide selection of implementations of both predictive and noise shaping converters have been reported. The former, commonly known as delta modulators, perform an implicit differentiation of the input signal. If there is a mismatch between the integrators in the modulator and demodulator, the signal can be distorted. In addition delta modulators exhibit slope overload for rapidly rising input signals and their performance is therefore dependent on the frequency of the input signal. Another potential shortcoming of the delta modulator is that with the integrator in the feedback loop, any non-ideality of the integrator will appear in the output [18]. Modulators that combine both the prediction and noise shaping techniques have been proposed. A multi-bit DAC is needed in the feedback loop that must be accurate because it determines the overall precision of the converter. As in the case of predictive converters, the performance of these converters is susceptible to the slope of the input signal.

Noise shaping converters encode the signal itself and their performance is insensitive to the rate of change of the signal. When the internal ADC and DAC are implemented with one bit resolution, the modulator is called delta-sigma modulator [19] or sigma-delta modulator [20]. The Z domain representation of a second order oversampled converter as shown in Figure 5 has been selected as an example.

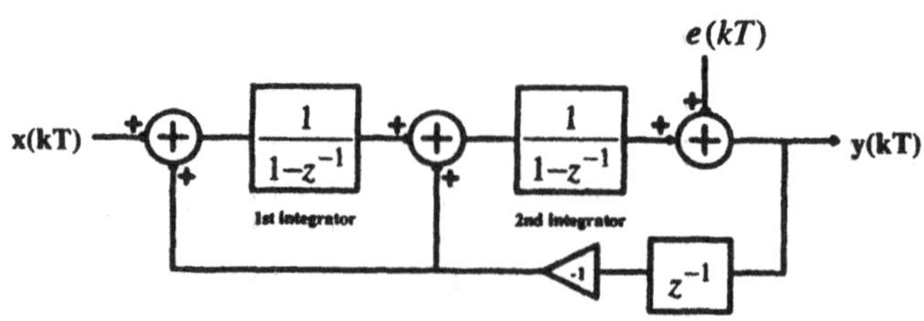

Figure 5a Second order noise shaping modulator.

Three configurations are shown: a 2nd order noise shaping, a 1st order predictive with 1st order noise shaping and a 2nd order predictive converter where the quantizer has been modelled as a unity gain element with additive white noise. The DAC has been represented by a unity delay. They will henceforth be denoted as structures *A*, *B* and *C* respectively. Signal transfer function (STF) and noise transfer function (NTF) for each structure are shown in Table 1. These correspond to transfer functions from the input $x(kT)$ to output $y(kT)$ and from the noise $e(kT)$ to output $y(kT)$ respectively.

Theoretically each structure has no major difference in terms of performance. However significant differences can arise in actual VLSI implementation.

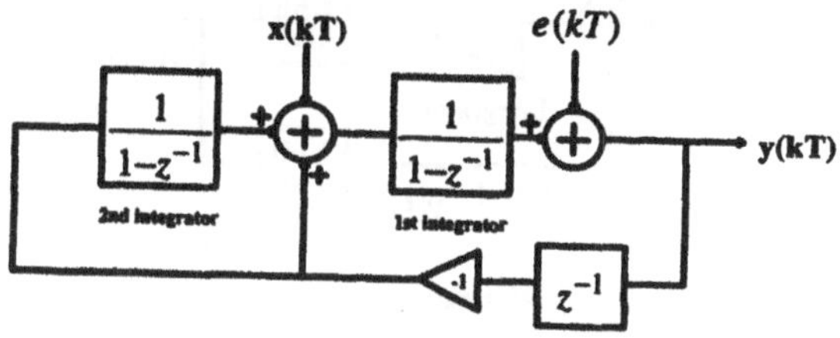

Figure 5b Oversampled modulator with first order noise shaping, first order prediction.

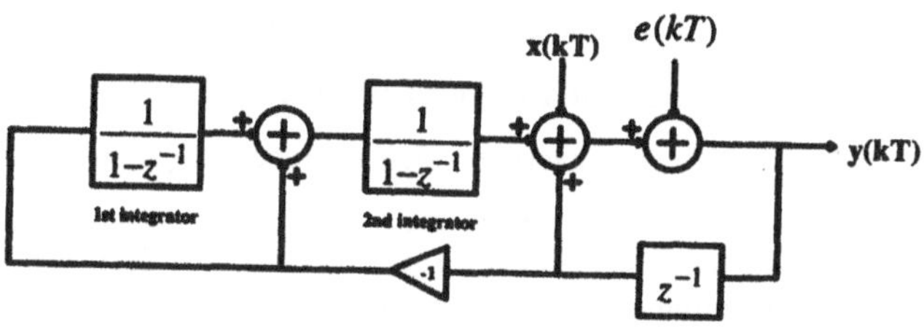

Figure 5c Second order predictive modulator.

Structure	STF	NTF
A	1	$(1-z^{-1})^2$
B	$(1-z^{-1})^1$	$(1-z^{-1})^2$
C	$(1-z^{-1})^2$	$(1-z^{-1})^2$

Table 1 Transfer functions of the three structures.

4.1.4 Sigma-Delta modulator architectures

4.1.4.1 First and second order sigma-delta modulators

(i) Principle of operation

The most widely used oversampled converter consists of a noise shaping converter with a one-bit internal quantizer, called the sigma-delta modulator. A block diagram of a first-order sigma-delta modulator is shown in Figure 6. It consists of a summing node, an integrator, and one-bit A/D and D/A converters in a feedback loop. Since the integrator has infinite gain at DC, the loop gain is infinite at DC and therefore the DC component or the average of the error signal is zero. Consequently the DC component or the average of the output from the feedback DAC will be identical to the DC component of the input signal. This means that even though the quantization error at every sample is large because a quantizer with only two levels is being used, the average of the quantized signal, and therefore the modulator output $y(kT)$, tracks the analog input signal $x(t)$. This average is computed by the decimation filter that follows the modulator.

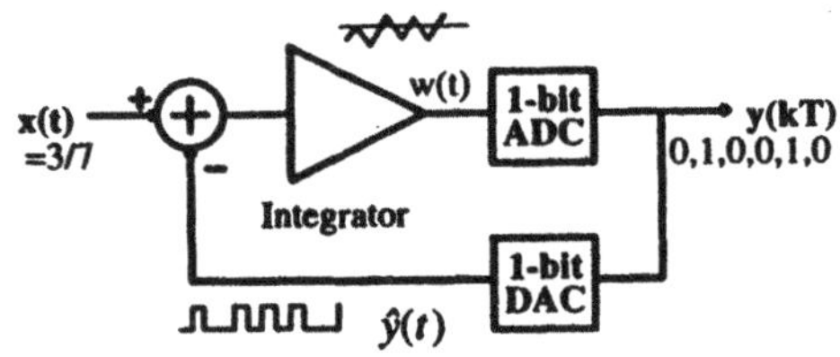

Figure 6 First order Σ−Δ modulator.

In general the integrator output $w(t)$ ramps up and down according to the value of the feedback DAC. Correspondingly the 1-bit ADC output $y(kT)$ is a bit stream of ones and zeroes which is a pulse density modulated representation of the input DC value. As an example when the input is $\frac{3}{7}$ and the initial condition of the integrator is at negative reference voltage (V_r), $y(kT)$ for the first 14 cycles is: "0", "1", "0", "0", "1", "0", "1", "0", "1","0", "0", "1", "0", "1". The average value approaches $\frac{3}{7}$. In general the amplitude resolution of the modulator increases when more samples are included in the averaging process, or as the OSR increases. However, the bandwidth is decreased at the same time. Consequently, the resolution of the modulator is a function of ratio of the sampling rate to the bandwidth of the modulator. The principle of operation of sigma-delta modulators relies on this fundamental trade-off between resolution and time.

To analyze mathematically the operation of a first-order sigma-delta converter a frequency domain analysis is simpler. The basic model of a typical first order modulator is shown in Figure 7a. The integrator usually has the delay in the forward path for ease of implementation using switched capacitor technique. The input of the modulator is defined to be $x(t)$, the output be $y(kT)$. The 1-bit ADC in the forward loop can be modeled as a unity gain block in conjunction with an additive noise source having error $e(kT)$, and the 1-bit DAC in the feedback loop can be represented by a unity gain element. The difference equation describing the entire system is

$$y(kT) = x(kT-T) + e(kT) - e(kT-T) \tag{1}$$

Note that when compared to Nyquist rate converter whose input-output relationship is described by

$$y(kT) = x(kT) + e(kT) \tag{2}$$

the error becomes a differential error. This is a first order difference relationship. In essence the modulator is trying to cancel the error by subtracting quantization errors from two adjacent samples. This principle of reducing the error source by exploiting the statistics between them can be extended to higher order modulators, where more past error samples are involved in the cancellation process to reduce the overall error. Viewed from a frequency domain this difference operation acts to attenuate the quantization noise at low frequencies, thus shaping the noise. The first order modulator can be proved to be inherently stable, provided that the input is within $[V_r, -V_r]$, where V_r is the output of the feedback DAC, or the reference voltage.

(ii) Second-order sigma-delta modulator

The procedure for increasing the resolution with feedback can be reiterated by replacing the internal quantizer of a first order modulator as shown in Figure 7a by an identical first order modulator. The resulting circuit is shown in Figure 7b [1]. As before the integrators may or may not contain delays. Usually one does and the other does not. The gains k_1, k_2 are used to optimize the output signal range of the integrators. Modulators with a second-order transfer function involve the cancellation of the two past samples and thus exhibits stronger attenuation at low frequencies. The noise shaping function of a first and second-order modulator are compared in Figure 8 [22].

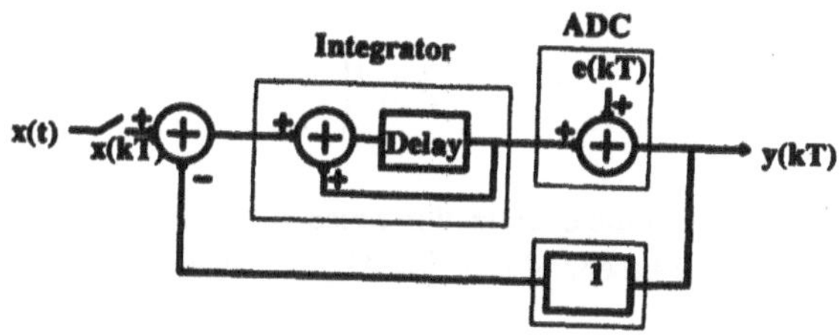

Figure 7a Model for first order Σ–Δ modulator.

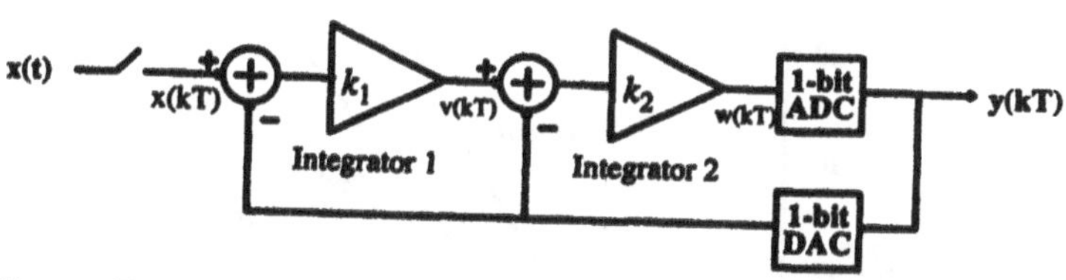

Figure 7b Modified architecture of second-order Σ–Δ modulator.

Second order modulator is shown to be conditional stable. The stability depends on total delay in feedback loop, signal amplitude and coefficient k_1, k_2. To avoid saturation, the maximum signal levels at the amplifier outputs should be adjusted by scaling.

The above analysis can be formalized to yield quantitative results for the resolution of sigma-delta modulators, provided that the spectral distribution of the quantization error $e(kT)$ can be assumed to be uncorrelated [23]. Then the modulator can be regarded as a linear system for which the power spectral density of the noise can be calculated [24].

For a first order system the noise transfer function NTF in discrete time domain is calculated as

$$\mathrm{NTF} = (1 - z^{-1}) \tag{3}$$

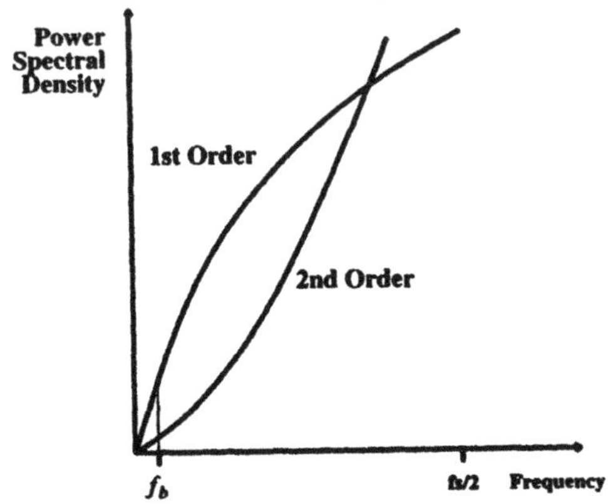

Figure 8 Power spectral density for first and second order sigma-delta modulators.

The power spectral density of the noise S_{nn} is given by

$$S_{nn} = 4\sigma_e^2 \sin^2\left(\pi \frac{f}{f_s}\right) \tag{4}$$

where σ_e^2 is the variance of the quantization noise due to the 1-bit quantizer and f_s is the sampling frequency. If the baseband is denoted by f_b where $f_s >> f_b$ the variance of the output noise is given by

$$\sigma_n^2 = \frac{8}{3}\pi^2 \sigma_e^2 \left(\frac{f_b}{f_s}\right)^3 \tag{5}$$

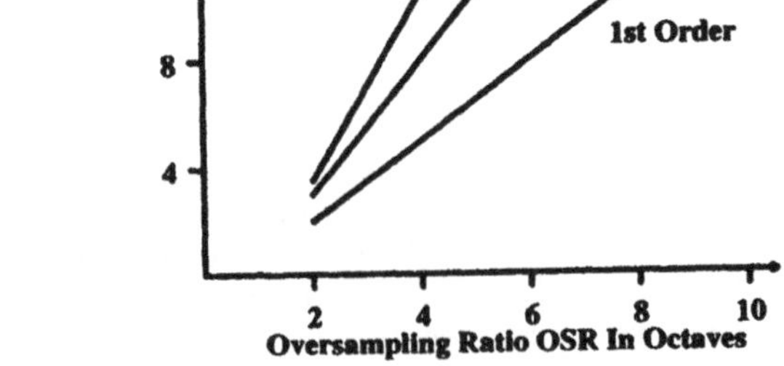

Figure 9 Dynamic range of Σ−Δ modulators.

In general the dynamic range (DR) of an A/D converter is defined as the ratio of the output power for a full scale sinusoidal input to the output signal power for a small input whose

SNR is unity (0 dB). Extending the above analysis to higher order sigma-delta modulator yields the following formula for DR

$$DR = \frac{3}{2}\frac{2L+1}{\pi^{2L}}(OSR)^{2L+1} \tag{6}$$

where L is the order of the sigma-delta modulator. Figure 9 shows the dynamic range of first, second, and third order sigma-delta modulators. The resolution is expressed as the equivalent number of bits of a Nyquist rate converter having the same dynamic range. It has been shown that the white noise model is not sufficient, in particular for 1st order modulator. Quantization noise spectrum is discrete, not white, and dithering is needed. Second order modulator is more randomized and dither is not always required.

(iii) Pattern noise

When the input x to the first order modulator is a DC signal, the quantized signal bounces between two levels, keeping its mean equal to the input. The frequency of repetition depends on the input level. When the repetition frequency lies in the signalband, the modulation is noisy, but when it does not, the modulation is quiet. In general the resampled modulator output $y(t)$ can be expressed as [27]

$$y(t) = Y' + \sum_{l}\sum_{k}\frac{\sin(\pi l x')}{\pi l}\exp\left(\pi j \frac{l x' + k}{\tau} t\right) \tag{7}$$

where Y' is the adjacent quantization level, $x' = \frac{x - Y'}{V_r}$, $j = \sqrt{-1}$. It can been seen the quantization error consists of discrete spectral lines and is not white.

It has been shown by [50] that if the input is a sinusoid, the output frequencies depend on the input frequency ω_0 and amplitude A and comprise all harmonics of the input frequency ω_0. In addition all shifts of these harmonics by $\frac{f_s}{2}$ are present. The output can be expressed as

$$S_m = \begin{cases} \frac{1}{2}; & m = 0 \\ \left[\frac{1}{\pi}\sum_{l=1}^{\infty}\frac{J_m(2\pi\alpha(2l-1))}{2l-1}(-1)^l\right]^2; & m = \text{even} \\ \left[\frac{1}{\pi}\sum_{l=1}^{\infty}\frac{J_m(4\pi\alpha l)}{2l}(-1)^l\right]^2; & m = \text{odd} \end{cases}$$

where J_m are Bessel functions, $\alpha = \frac{A}{2V_r \sin\left(\frac{\omega_0}{2}\right)}$. These results can be extended to a 2nd order modulator, provided that modulator does not overload. An alternative derivation based on developing an open loop model for the sigma-delta modulator of arbitrary order has been developed [56]. The specific application to a second order modulator is outlined below.

Consider the second order sigma-delta modulator as shown in Figure 7b with input $x(kT)$, output $y(kT)$ and initial conditions v_0 and w_0. Integrator 1 has delay in the feedback path and integrator 2 has delay in the forward path. If the quantizer never overloads, then for $k \geq 3$,

$$\begin{aligned} y(kT) &= x((k-1)T) + D^2 e_{int}(kT) \\ &= x((k-1)T) + D^2 e\left(\sum_{i=1}^{k-1}\sum_{j=1}^{i} x(jT) + k v_0 + w_0 + \gamma(kT)\right) \\ &= x((k-1)T) + D^2 e\left(v_0 + w_0 - 1 + \sum_{i=1}^{k-1}\left(v_0 - 1 + \sum_{j=1}^{i}(x(jT)-1)\right)\right) \end{aligned} \tag{9}$$

where

$$\gamma(kT) = \begin{cases} 0 \text{ if } k = 0 \pmod 4 \text{ or } k = 3 \pmod 4 \\ 1 \text{ if } k = 1 \pmod 4 \text{ or } k = 2 \pmod 4 \end{cases} \tag{10}$$

and $e_{int}(kT)$ is the internal quantization error, $e_{int}(kT) = y(kT) - w(kT)$.

With the double-loop modulator represented as an open-loop structure, a procedure [56] similar to that used in [25] has been adopted to determine the spectrum of the output quantization noise for a sinusoidal input, $x(kT) = A_i \sin(\omega k T + \phi_i)$. In this procedure, the quantization error function, $e(\cdot)$, is replaced by its Fourier series, which is possible since $e(\cdot)$ is periodic. The output quantization noise will then be a sum of exponentials, each with a sinusoidal term in the argument. Each of these terms can themselves be written as a sum of exponentials using Bessel's functions, and therefore the output spectrum can be expressed as a set of discrete spectral lines. Then the quantization error is

$$\begin{aligned} e(kT) = &\sum_{\substack{l \text{ even} \\ l \neq 0}} \sum_m \frac{4i\sin^2(\omega_{lm}/2)}{\pi l} J_m(\pi l A_0) \exp[i\omega_{lm}(k-1) + i\phi_{lm}] \\ &+ \sum_{l \text{ odd}} \sum_m \Bigg\{ \frac{4i\sin^2(\omega_{lm}^{+}/2)}{\sqrt{2}\,\pi l} J_m(\pi l A_0) \exp\left[i\omega_{lm}^{+}(n-1) + i\phi_{lm}^{+}\right] \\ &\qquad + \frac{4i\sin^2(\omega_{lm}^{-}/2)}{\sqrt{2}\,\pi l} J_m(\pi l A_0) \exp\left[i\omega_{lm}^{-}(n-1) + i\phi_{lm}^{-}\right] \Bigg\} \end{aligned} \tag{11}$$

for $n \geq 2$, where

$$\omega_{lm} = \omega m + \pi l B, \quad \omega_{lm}^{\pm} = \omega m + \pi l B \pm \pi/2 \tag{12}$$

$$\phi_{lm} = m\phi_0 + \pi l C, \quad \phi_{lm}^{\pm} = m\phi_0 + \pi l C \pm \pi/4 \tag{13}$$

$$A_0 = \frac{A_i}{4\sin^2(\omega/2)} \tag{14}$$

$$\phi_0 = \phi_i - \pi \tag{15}$$

$$B = A_i \frac{\cos(\omega/2+\phi_i)}{2\sin(\omega/2)} + v_0 \tag{16}$$

$$C = A_0 \sin\phi_i + w_0 \tag{17}$$

$$i = \sqrt{-1} \tag{18}$$

and $J_m(\cdot)$ is the m-th order Bessel function of the first kind. In equation (11), unless lower and upper bounds on summations are given, they are assumed to be $-\infty$ and $+\infty$ respectively. All the sums above converge absolutely.

If the frequency, $f = \omega/2\pi$, and amplitude term, B, are both irrational then all the spectral lines will be distinct. However, if they are not, then the lines will overlap. In the case when both f and B are rational, the overlapping will result in a finite number of spectral lines, whose magnitudes can be calculated without any infinite summations as shown below.

Using the assumptions and notation of equations (11-17), if $\omega = 2\pi r/K$ and $B = s/L$, where r, s, K and L are integers, then the output quantization error will be periodic with a period of LCM(4, L, K) (although this may not be the smallest period), where LCM is the least common multiple. Furthermore, the output will have finite spectral lines given by

$$e(nT) = \sum_{l=0}^{L-1}\sum_{m=0}^{K-1}\left\{-4\sin^2\left(\frac{\omega_{2l,m}}{2}\right)F_{l,m}\exp(i\omega_{2l,m}(n-1)) - \frac{4}{\sqrt{2}}\sin^2\left(\frac{\omega^{+}_{2l+1,m}}{2}\right)F_{l+1/2,m}\exp(i\omega^{+}_{2l+1,m}(n-1)-i\frac{\pi}{4}) - \frac{4}{\sqrt{2}}\sin^2\left(\frac{\omega^{-}_{2l+1,m}}{2}\right)F_{l+1/2,m}\exp(i\omega^{-}_{2l+1,m}(n-1)-i\frac{\pi}{4})\right\} \tag{19}$$

where ω_{lm} and $\omega^{\pm}_{lm}$ are defined in equations (11-17), and

$$F_{0,m} = \sum_{k=0}^{K-1}\frac{1}{2LK}e\left(2L\left(A_0\sin\left(\phi_0+\frac{2\pi k}{K}\right)+C\right)\right)\exp\left(\frac{2\pi ikm}{K}\right) \tag{20}$$

and

$$F_{z,m} = \sum_{k=0}^{K-1}\frac{1}{2iLK}\csc\left(\frac{\pi z}{L}\right)\exp\left(\frac{\pi iz}{L}\left(1-2\left(A_0\sin\left(\phi_0+\frac{2\pi km}{K}\right)\right)\right)+\frac{2\pi ikm}{K}\right) \tag{21}$$

for $z \neq 0$. In the previous equations, $e(\cdot) = 2[x/2] + 1 - x = 1 - 2 < x/2 >$ and $[a]$ denotes the greatest integer less than or equal to a. The frequency spectrum corresponding to equation (19) is shown in Figure 10. Extension to a second order 1-bit sigma-delta modulator with leaky integrator has also been proposed [51].

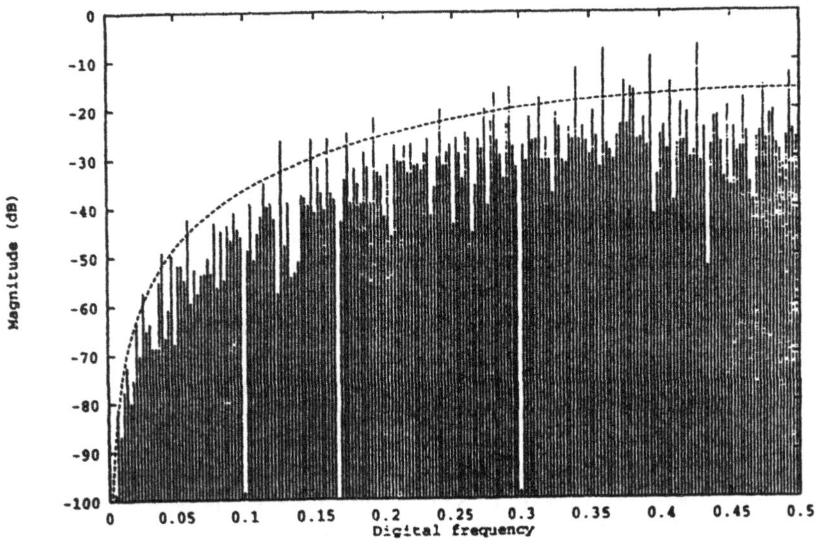

Figure 10 Calculated spectral lines with sinusoidal input with rational frequency and amplitude term. Solid lines are spectral lines, and dotted line is envelope predicted by white noise model.

4.1.4.2 Interpolative modulators

As one moves to higher order modulators [4, 9, 26] the spectral density of the quantization noise can be modified to improve the stability and enhance quantization noise reduction. The spectral density may no longer have a $\sin^L$ form, where L is the order of the modulator. An interesting example of these modulators is an interpolative modulator which incorporates both feedforward and feedback in the transfer function. A 4th-order example is shown in Figure 11 [9]. The system function where the ADC is modeled as an additive noise source with a delay z^{-1} is shown to be

$$Y(z) = H_X(z)X(z) + H_E(z)E(z) \tag{22}$$

The NFT is shown to be

$$|H_E(f)| = \left| \left(\frac{\pi f_b}{f_s}\right)^n \left(\frac{2}{A_N}\right) T_N\left(\frac{f}{f_b}\right) \right| \quad \text{for } f \leq f_b << f_s \tag{23}$$

Here, A_N is the feedforward coefficient as shown in Figure 11 where N is 4. $T_N(x)$ is the Chebyshev polynomials defined by

$$T_0(x) = 1$$
$$T_1(x) = x$$
$$T_N(x) = 2xT_{N-1}(x) - T_{N-2}(x) \quad (24)$$

In this design the poles are distributed throughout the signal band in order to lower the in-band noise. The zeros are chosen to flatten the filter response at high frequency, in order to reduce the high-frequency noise and prevent it from reducing the dynamic range. The noise spectrum obtained from such a modulator is given in Figure 11. Note the suppression provided by the poles throughout the passband.

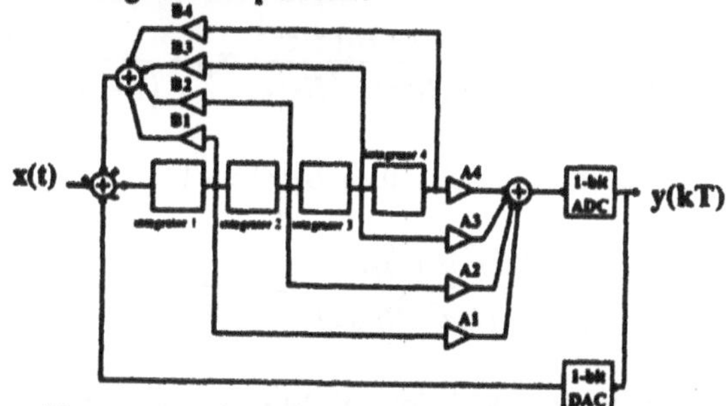

Figure 11 A fourth order interpolative modulator.

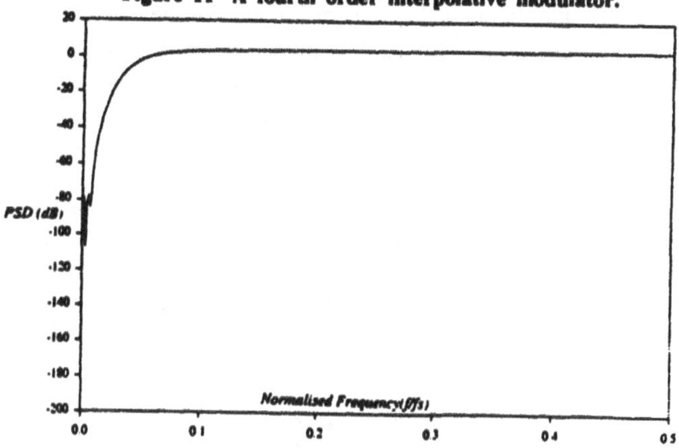

Figure 12 PSD of quantization noise of a 4th-order interpolative modulator.

4.1.4.3 Cascaded (MASH) architecture [54]

While high order sigma-delta modulators offer the potential to achieve higher resolution for a given oversampling ratio, modulators with more than two integrators exhibit the potential of being unstable. An architecture where several first-order or second order modulators are cascaded in order to achieve a performance level that is comparable to that of higher order modulators has been suggested as a means to overcome the stability problem. Another advantage of cascading is that it has been shown theoretically that if the number of stages is large, then the output is free from pattern noise as discussed above [27, 28]. Furthermore, stability can be as robust as the stage that involves the highest order. The Z domain representation of a two stage modulator's configuration is shown in Figure 13, [29, 30] where the quantizer can be exactly represented by an additive white noise source (N.B. [27,

28] points out that is true only when # of stages ≥ 3. For illustration purpose, this is assumed true here for a 2-stage structure).

Each stage consists of a first-order sigma-delta modulator. The first stage quantizes the input signal in the same way as a conventional sigma-delta modulator. The first stage quantization noise is fed into the second stage to be re-quantized and shaped. In conjunction with the differentiator the quantization noise coming out from the second stage would be identical to that of the first stage. Cancellation is achieved by subtracting this from the first stage output. If $H_1(z)$ is assumed equal to $H_2(z)$, the converter output $Y(z)$ is given by the equation

$$Y(z) = \frac{H_1(z)z^{-1}}{1+H_1(z)z^{-1}}X(z) + \frac{H_3(z)-H_1(z)z^{-1}}{1+H_1(z)z^{-1}H_3(z)}\sigma_{n1} + \frac{\sigma_{n2}}{\left(1+H_1(z)z^{-1}\right)H_3(z)} \tag{25}$$

If we concentrate on baseband frequencies, then $H_1(z) >> 1$

$$Y(Z) = X(Z) + \frac{\left(H_3(z)-H_1(z)\right)z^{-1}}{\left(1+H_1(z)z^{-1}\right)H_3(z)}\sigma_{n1} + \frac{\sigma_{n2}}{\left(1+H_1(z)z^{-1}\right)H_3(z)} \tag{26}$$

where σ_{n1} and σ_{n2} are standard deviations of the quantization noise of Quantizer 1 and Quantizer 2, respectively, and $H_3(z)$ is the transfer function of the differentiator. If $H_1(z)$ is equal to $H_3(z)$, the second term is cancelled and the third term is suppressed by $H_1(z)$ and $H_3(z)$. Consequently the resulting spectral density function of the noise is equivalent to embedding a second order transfer function in the feedback loop. In general this can be extended to an arbitrary number of stages. However, as shown above, if $H_1(z)$ and $H_3(z)$ do not match precisely, then there will be residual first order noise. In addition any tones generated from the first stage which consists of a first-order modulator will leak to the output. This requirement on precise gain matching between the individual first-order sections runs contrary to the original intention of designing A/D converters that are insensitive to component mismatches. In particular to match the integrator transfer function between the analog and digital path, the required opamp gain in a switched capacitor implementation has to be very high.

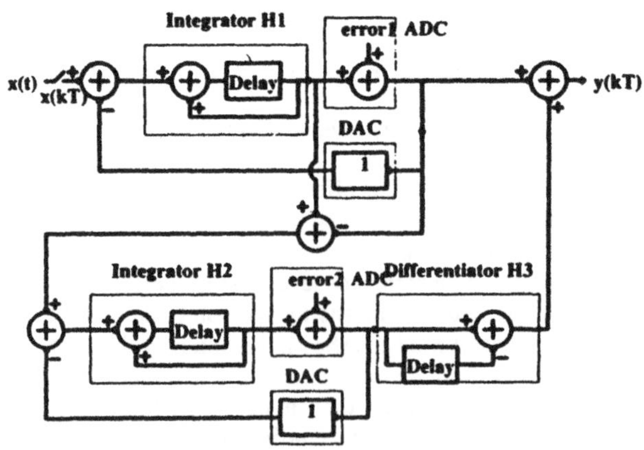

Figure 13 Two stage noise shaping modulator.

4.1.4.4 Single-path modulator with local feedback loop

To reduce the component mismatch and hence the high opamp gain requirement, a new architecture with compensation scheme is proposed which combines the advantages from the single path and cascaded architecture.

An example of the new scheme implemented with a third order structure is shown in Figure 14. Apart from the local feedback loops (including the overload detectors, 3-level DAC and l scale factors, and the cancellation circuits H_1, H_2 and H_3), the circuit resembles an ordinary triple loop modulator. The function of the overload detectors (OLDs) is to monitor the output voltage of the respective integrators and create local feedback signals through the 3-level DACs. In reality OLD is a tri-level A/D converter with non-uniformly spaced thresholds, or a window comparator. The DACs in the local feedback loops are tri-level DACs. The digital outputs of the OLDs are sent through digital blocks H_1, H_2 and H_3 to the output for cancellation. For example if the local feedback loop involving l_3, l_3 is activated, the output from the OLD is sent through l_3, l_3 to the output. To cancel this extra signal, the output from the OLD is sent through H_3 and subtracted from the modulator output digitally.

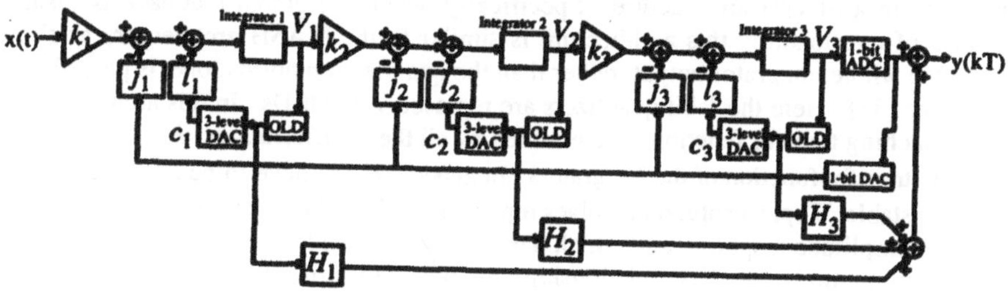

Figure 14 A third-order single-stage single-bit oversampled ADC with local feedback loops.

The function of the overload detector is explained in Figure 15. Output of the detector remains 0 until the input voltage goes beyond the normal operating range $(-V_{o1}, V_{o1})$. Then the detector outputs -1 or 1 depending on the polarity of the input voltage. When the integrator operates within the normal operating range, the OLD outputs are 0 and therefore the local feedback loops are not activated. The signal through the blocks H_1, H_2, and H_3 is also zero. If any of the integrators is overloaded, the overload detector initiates a proper local feedback signal which forces the integrator to return to the normal operating range through the tri-level DAC and the associated scale factor.

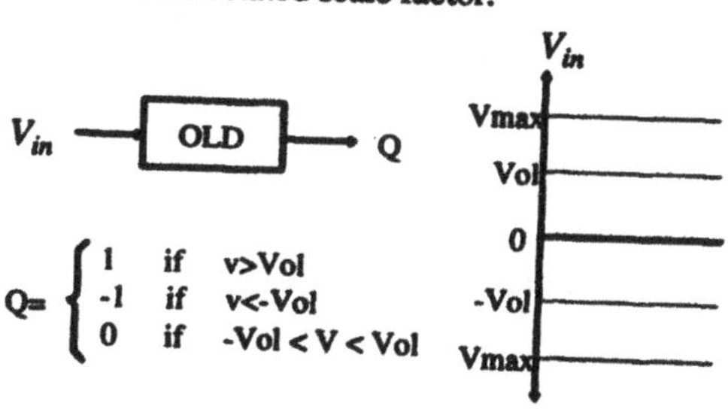

$$Q = \begin{cases} 1 & \text{if} \quad v > Vol \\ -1 & \text{if} \quad v < -Vol \\ 0 & \text{if} \quad -Vol < V < Vol \end{cases}$$

Figure 15 Overload detector.

In a differential architecture, a tri-level DAC can be implemented with a high degree of linearity. Consequently unlike other modulators using multi-bit internal quantizers, to a first order, the linearity of this modulator is not limited by the matching accuracy of the DAC elements, as will be shown in the next section. In addition, any nonlinearity remained due to switch mismatches and charge injection in this tri-level DAC will be in the signal path only when the local feedback loops are activated. Simulation results indicate that since the overloading frequency is small, this modulator has practically the same linearity performance as a single-bit modulator.

(i) Cancellation circuitry

In Figure 14 when the local feedback loops are activated, the effect of signals c_i have to be cancelled to restore the overall SNR. This can be achieved by sending the OLDs outputs through a signal path with an identical transfer function and then performing the cancellation at the output of the modulator. To achieve cancellation the digital transfer functions $H_i(z)$ have to match the transfer function of the modulator loop corresponding to the individual local feedback signals. Upon further investigation this cancellation mechanism, as shown in Figure 16, is found to be similar in principle to the way in which quantization noise is cancelled in a MASH architecture. Specifically, when the global feedback is disabled because of overloading, this architecture is similar to the MASH architecture as shown where the whole integrator output, rather than the quantization noise alone, is passed to the next stage [52]. Here the 1-bit quantizers are replaced with OLDs. In this architecture the exact matching transfer function is determined by all the coefficients j_i, k_i and l_i and also the exact transfer function of the integrators. In general $H_i(z)$ have to be IIR filters, which can be unstable. Typical solution involves replacing the IIR filter with a stable structure with the same amplitude response, and which has all its poles inside the unit circle. Because of the need to maintain the phase relationship, this scheme is not feasible in the present case. Again, due to a low overloading frequency, the cancellation circuitry need not be very accurate and can be greatly simplified.

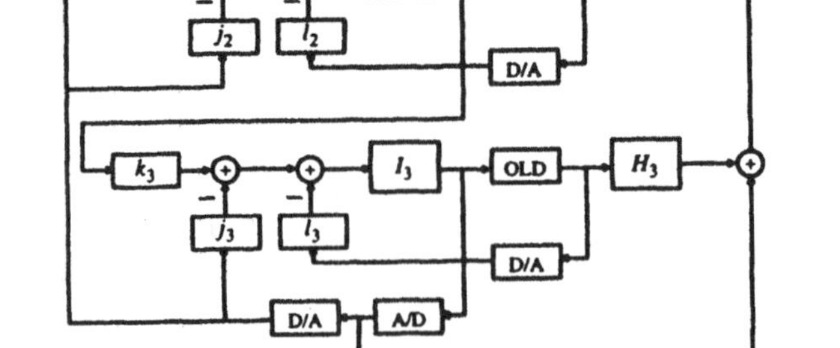

Figure 16 Architecture from Figure 14 redrawn to highlight its similarity to the MASH architecture (when global feedback is disabled due to overloading).

(ii) Selection of loop filter coefficients

The first consideration in the design of the modulator is to obtain the proper noise shaping function. To ensure stability, the parameters also have to satisfy the stability conditions. Under these assumptions the l coefficients of Figure 14 are selected to be

$$l_1 = l_2 = l_3 = 0.5 \tag{27}$$

Other considerations include the desire to confine the integrator's output voltage within $\pm V_{max}$, and to reduce the overload frequency. These together with the need to keep the implementation simple lead to values of k_2, k_3 being set to 0.25. Finally, to keep the overloading frequency low while making it insensitive to component mismatch, k_1 and the j coefficients are empirically chosen as

$$j_1 = 0.125,\ j_2 = 0.1,\ j_3 = 0.05,\ k_1 = 0.075, \tag{28}$$

4.1.4.5 Multi-bit sigma-delta modulators

Another approach to increasing the resolution of a sigma-delta modulator of a given order is to incorporate a multi-bit internal quantizer within the feedback loop. This approach can be applied to modulators of higher order (> 2), or cascaded architectures [54]. It has been shown in the past that multi-bit sigma-delta converters can improve the stability characteristics for higher order modulators, thus significantly lowering the quantization noise for a given oversampling ratio [10, 11, 31, 32]. The performance of the feedback loops using multi-bit quantizers can be predicted more accurately by theory that involves linearizing the quantizer than its single-bit counterpart. In addition for a fixed input signal range the quantization noise decreases exponentially with M, the number of bits used in the internal quantizer. This is due to the fact that the rms noise is proportional to the quantization step size δ, which in turn is proportional to 2^{-M}.

In general the multi-bit internal quantizer adds complexity. The principal drawback of multi-bit sigma-delta converters, though, has been the stringent absolute accuracy requirements placed on the low resolution feedback DAC. Past attempts to solve this problem resulted in trimming techniques which are expensive. The use of digital error correction on the DAC has also been reported [11]. The general principle is illustrated in Figure 17.

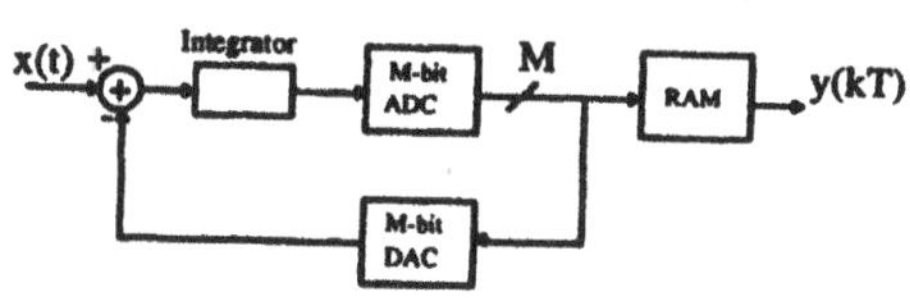

Figure 17 Multi-bit Σ–Δ modulators using digital error correction.

The feedback DAC and the RAM have the same inputs. Hence, if the RAM is programmed to generate the exact digital equivalent of the D/A analog output level for any digital input word, the D/A and RAM output signals will be identical to each other for any D/A output. In the baseband the D/A output signal spectrum is approximately the same as the input signal spectrum, due to the high-gain feedback loop. Hence, the baseband spectrum of the digital RAM output data will also be very close to the analog input spectrum. The programming of the RAM can be performed on-chip by reconfiguring of the modulator stages and feeding the system with a multi-bit digital ramp. This calibration may be performed at power-up time or continuously.

Another approach is the use of the dynamic element approach. This technique is applied to the DAC in the feedback loop by having different permutations on the arrangement of passive elements in generating a given voltage level. The use of random permutation involves randomly permuting the passive elements that generate the desired reference level, thereby translating the harmonic distortion due to mismatch into random noise [10, 31]. A variation of this technique consists of converting what would have been the distortion components caused by element mismatch in the feedback DAC to high frequency components lying outside the passband. A simple realization of this technique called the "clocked averaging approach", whereby the averaging is done by permuting the DAC elements in a periodic fashion, has been reported [33, 49]. This technique, however, can generate undesirable tones inside the passband for certain signal frequencies. For a given OSR and a L-th modulator then in order to ensure the tones do not fall inside the passband the following inequality has to be observed

$$\mathrm{OSR} = \frac{2^L + 1}{2} 2^L \tag{29}$$

As an example, L is 3 for a third order modulator. Substituting this into inequality (13) yields a minimum OSR of 36 [18]. In addition even if the above inequality is observed the tones can still become a problem from an anti-aliasing standpoint. To suppress the amplitude of the tone anti-aliasing has to be provided throughout the "don't care" regions in general. Finally, the nonlinearity introduced by the feedback DAC will modulate the large out-of-band quantization noise back to the baseband. An improved technique, called the individual level averaging approach, achieves the averaging by permuting the DAC elements on a per level basis.

As an example, this approach is applied to a first-order modulator as shown in Figure 6 that incorporates a feedback DAC with 2 unit elements. If the unit elements are identical then the DAC will have 3 levels as shown in Figure 18a(i). The input to the feedback DAC is the output of the Σ–Δ modulator $y(kT)$ and the DAC output is the feedback signal $\hat{y}(kT)$. If there is a mismatch between the two elements such that their values are "1.1" and "1.2", then the DAC will have 3 levels corresponding to: "0", "$\frac{1.1}{2.3}$" and "1" as shown in Figure 18a(ii). If the elements are swapped periodically between two phases ϕ_1, ϕ_2, then the DAC transfer characteristics will alternate between Figures 18a(ii) and (iii) at the sample rate f_s. The DAC levels corresponding to Figure 18a(iii) will be "0", "$\frac{1.2}{2.3}$" and "1". Figure 18b shows the four different cases when a DC input of "0.25" is applied to $x(t)$. Here $y(kT)$ for the first four cycles are assumed to be "00", "01", "00", "01". Note that in the individual level averaging case, both the $\frac{1.1}{2.3}$ and $\frac{1.2}{2.3}$ levels that correspond to the code "01" are exercised. Consequently the error is averaged out. The output sequences $\hat{y}(kT)$ for the different cases are now summarized in Table 2.

A 2nd order sigma-delta modulator with a 3-bit internal quantizer using individual level average approach has been implemented in a 1.2 μm CMOS technology. Testing results indicate a 16 bit dynamic range over a 20 kHz signal bandwidth and no tones are observed.

The clock frequency is 2.56 MHz which translates into an OSR of only 64. The active chip area is 3.1 mm^2 and the chip consumes a power of 67.5 mW from a 5 V power supply. A micro-photograph of the chip is shown in Figure 19.

$x(t) = 0.25$, First order Σ–Δ modulator, 2 unit elements						
Case	T	2T	3T	4T	Avg Value	Error
Ideal	0	0.5	0	0.5	0.25	0
No Averaging	0	$\frac{1.1}{2.3}$	0	$\frac{1.1}{2.3}$	0.239	-0.011
Clocked Averaging	0	$\frac{1.2}{2.3}$	0	$\frac{1.2}{2.3}$	0.261	0.011
Individual Level Averaging	0	$\frac{1.1}{2.3}$	0	$\frac{1.2}{2.3}$	0.25	0

Table 2 DAC output $\hat{y}(T)$ for different averaging algorithms.

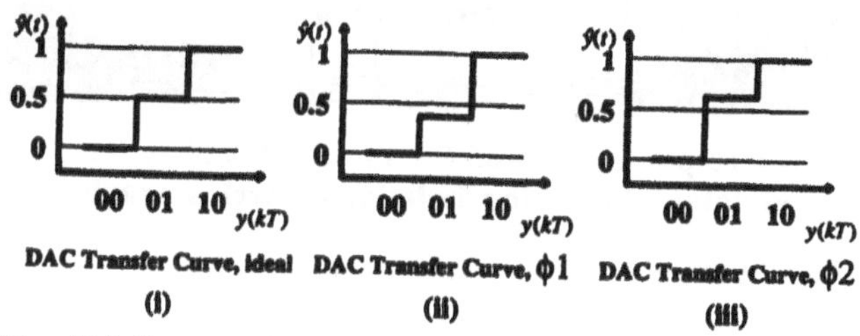

Figure 18a DAC transfer curves with ideal and non-ideal matching.

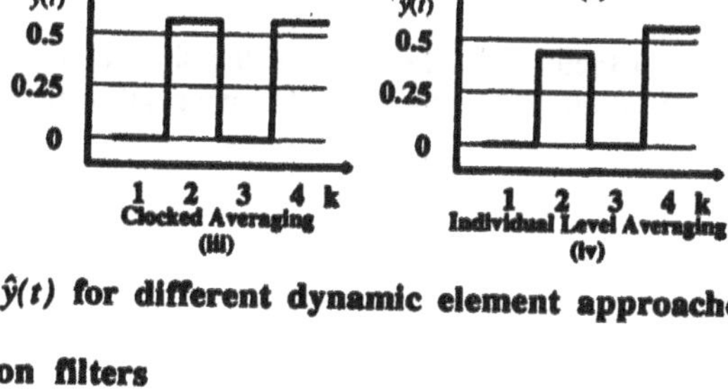

Figure 18b $\hat{y}(t)$ for different dynamic element approaches, $x(t) = 0.25$.

4.1.5 Decimation filters

The decimation filter in an oversampled A/D converter consists of a low pass filter and a resampler. The purpose of the filter is to remove out-of-band quantization noise and to suppress spurious out-of-band signals. Upon filtering, the signal is resampled at the Nyquist frequency. The rate reduction, or decimation, is usually performed in two or more steps to increase the ratio of the width of the transition band of the filter to the sampling rate, thereby decreasing the order of the individual filters. This decimation filter design differs from traditional decimation filter design in that the main desired objective is to suppress out of band quantization noise as opposed to just meeting a certain frequency response specification [34, 36]. Furthermore, if the output code is 1-bit only, then no multiplier is needed.

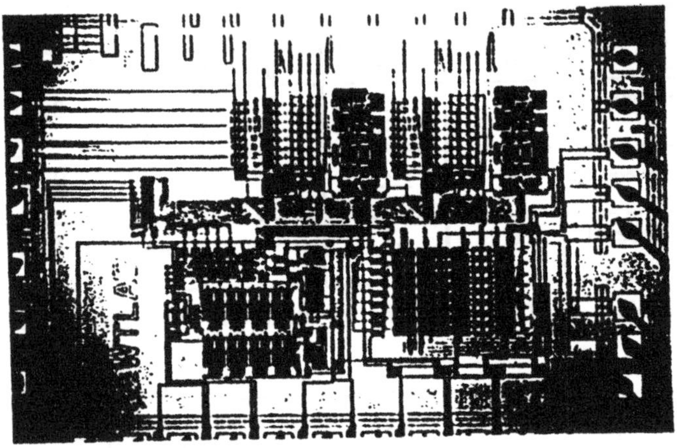

Figure 19 Micro-photograph of the multi-bit 2nd order sigma-delta modulator chip using individual level averaging.

4.1.5.1 Sinc decimators

For the sigma-delta front end whose power spectral density of the quantization noise has a sinusoidal response, it has been shown that the decimation filter can be implemented efficiently using a cascade of comb filters. The decimation filter will exhibit a sinc-type frequency response [12]. This type of filter can be implemented very efficiently because neither ROM nor multiplier is required. The filter transfer function is

$$\left(\sum_{i=0}^{D-1} z^{-i}\right)^k = \left(\frac{1}{D}\frac{1-z^{-D}}{1-z^{-1}}\right)^k \tag{30}$$

where D is the decimation factor, k is the order of the sinc filter. If the sigma-delta modulator is of order L, then setting k to $L+1$ would provide sufficient suppression of the quantization noise [12].

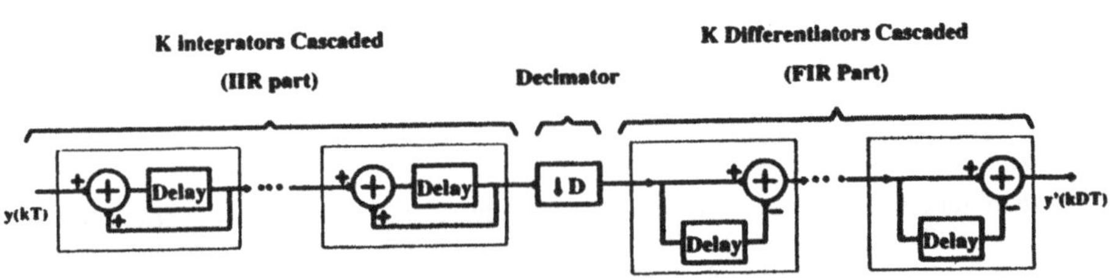

Figure 20 Decomposition of the comb filter with sinc response.

Notice if one factors out the denominator, then the numerator can be implemented at the decimated frequency [37-39] and the computation speed is greatly reduced. A block diagram of this implementation is shown in Figure 20. The attenuation of the filter in the passband can be found and compensated for in the second stage decimation filter. While this filter is simple to implement, it does not yield optimum results for modulators that have sharply rising noise spectral densities. Examples include the interpolative modulators described in section 4.1.4.2 above and the ones proposed in [26].

4.1.5.2 FIR decimation filters based on quadratic programming

To solve this problem another approach based on quadratic optimization of the out of band quantization noise has been reported [40]. The methodology tackles the FIR filter design problem by formulating a quadratic programming problem that minimizes the integral of the aliased noise subject to the passband and stopband constraints. This approach offers a design whose response is optimized to meet arbitrary quantization noise PSD (Power Spectral Density) and anti-alias requirements.

Conceptually the optimization process tries to minimize the noise energy that is the product of the FIR filter transfer function squared and the noise PSD, integrated over the desired bandwidth. The mathematical treatment will now be presented. There are four cases of FIR filters with exactly linear phases [41], of which two cases apply to low pass filters: Case 1 for filters with odd length and Case 2 for filters with even length. Without loss of generality, the Case 2 filters are assumed here. Given a Case 2 FIR filter with an impulse response $\{h(n)\}$, the frequency response is

$$H(e^{jw}) = \sum_{n=1}^{L/2} b(n)\cos\left(n-\frac{1}{2}\right)w \tag{31}$$

where

$$b(n) = 2h\left(\frac{L}{2}-n\right), \quad n = 1, \ldots, \frac{L}{2} \tag{32}$$

Defining the filter coefficient vector

$$X = \left[b(1)\ b(2)\ \ldots\ b\left(\frac{L}{2}\right)\right]^T \tag{33}$$

and the transform kernel vector

$$S(w) = \left[\cos\frac{1}{2}w\ \ldots\ \cos\left(\frac{L}{2}-\frac{1}{2}\right)w\right]^T \tag{34}$$

the zero phase response can be written as an inner vector product

$$H(e^{jw}) = X^T S(w) \tag{35}$$

Furthermore given the spectral density of the quantization noise from the oversampled ADC is $S_{nn}(e^{jw})$, then the quantization noise energy over a frequency interval $[a, b]$ is given by

$$E_s = \frac{1}{2\pi} \int_{w=a}^{w=b} \left|H(e^{jw})\right|^2 \left|S_{nn}(e^{jw})\right|^2 dw = \frac{1}{2\pi} X^T Q X \tag{36}$$

where the energy matrix

$$Q = \int_{w=a}^{w=b} S(w)\, S^T(w) \left|S_{nn}(e^{jw})\right|^2 dw \tag{37}$$

is symmetric.

The desired filter passband and stopband performance is attained by imposing magnitude constraints at a discrete set of points. The constraints are divided into equality and inequality constraints. The equality constraints would be $\left|H(e^{jw})\right|_{w=w_i} = k_i$ and the inequality constraints would be $\left|H(e^{jw})\right|_{w=w_j} < k_j$. Using vector notations we have $ST|_{w=w_i} X = K$ and $ST|_{w=w_j} X < K_1$ where $K = [k_0, \ldots, k_n]^T$ and $K_1 = [k_{10}, \ldots, k_{1n}]^T$. The design of the *quadratic* filter can be formulated as a quadratic programming problem:

Minimize E_s subject to $ST|_{w=w_i} X = K$ and $ST|_{w=w_j} X < K_1$

The optimization adopts the method of active set constraints [42].

From a physical point of view the quantity E_s represents energy and therefore is non-negative. Now E_s is defined by

$$E_s = \frac{1}{2\pi} X^T Q X \tag{38}$$

Therefore Q is positive definite. This implies that the projected Hessian matrix $Z^T{}_k Q Z_k$ is also positive definite at every iteration k [43], where Z_k is the matrix whose columns form the basis for every feasible direction at iteration k. Therefore the search direction is well defined at every iteration and the quadratic function has a unique minimum in the subspace defined by Z_k. Furthermore it has been shown that the function has a global minimum [44].

This approach offers design flexibility over the comb filter with a sinc type response but it requires ROM to store the impulse response coefficients. Specifically it can be used to design filters with more complicated noise-shaping functions [4, 9, 45]. It also provides a better trade-off between filter length and the amount of noise suppression provided. Reduction in filter length up to a factor of two can be achieved when compared to filters with sinc-type frequency response. Experimental results confirm that the optimization can be carried out successfully with quite arbitrary initial conditions (e.g. setting all the coefficients to zero).

4.1.5.3 *Polyphase IIR decimation filter based on optimization*

One advantage of designing decimating filter using FIR filters involves exploiting the 1-bit output nature of the oversampled modulator to simplify the computation. Furthermore, in the FIR decimation filter, the computation is done at the decimated frequency. However, in cascaded modulators (section 4.1.4.3), modulators with compensation circuitry (section 4.1.4.4), and modulators involving multi-bit internal quantizers (section 4.1.4.5), the modulator output will be multi-bit and the 1-bit advantage can no longer be exploited. In addition a method of transforming conventional IIR transfer function to a form whose denominator is a polynomial in Z^D only (i.e., polyphase structure), has been proposed [46]. Since D is the decimation ratio, the computation can now be done at the decimated frequency, making the IIR filter equally efficient. The filter transfer function is of the form

$$H(z) = \frac{\sum_{i=0}^{i=L} b_i z^{-i}}{1 - \sum_{j=0}^{j=K} a_j z^{-D}} \tag{39}$$

Here a_j and b_i are the coefficients of the filter transfer function. The multiplication rate has been shown to be [47]

$$\frac{K + 1 + \frac{L}{2}}{D} \tag{40}$$

Based on this polyphase architecture, a design methodology has been proposed whose objective is to minimize the out-of-band quantization noise, as well as the in-band phase distortion [48]. In the optimization process the degree of phase nonlinearity can be selected as one of the design specifications. To achieve minimal peak group delay deviation, the IIR filter should be realized by locating all the zeros in the Z plane on the unit circle and all the poles in the Z^D on the imaginary axis with the minimum possible radii. The restriction, in particular that of having a minimum radius, will reduce the achievable SNR. So upon optimization, the zeroes will not have minimum radii. The resulting polyphase IIR filters can achieve up to a factor of two reduction in multiplication rate over the quadratic programmed based FIR filters (section 4.1.5.2), with similar performance in quantization noise reduction. Because a rational function rather than a polynomial is being optimized, no global optimum exists. However designs that do not involve large number of coefficients (< 100) demonstrated that the algorithm is quite robust in obtaining a solution for arbitrary noise-shape function and initial conditions. Furthermore, the group delay is approximately constant.

References

[1] B. Boser, B. Wooley, "The design of Sigma-Delta Modulation Analog-to-Digital Converters", *IEEE Journal of Solid-State Circuits*, pp. 1298-1308, December 1988.

[2] M. Hauser, P. Hurst, R. Brodersen, "MOS ADC-Filter Combination that does not Require Precision Analog Components", *Digest of Technical Papers, International Solid-State Circuits Conference 1981*, pp. 80-81, February 1985.

[3] B. Leung, N. Robert, P. Gray, R. Brodersen, "Area-Efficient Multichannel Oversampled PCM Voice-Band Coder", *IEEE Journal of Solid-State Circuits*, pp. 1351-1357, December 1988.

[4] B. Signore, D. Kerth, N. Sooch, E. Swanson, "A Monolithic 20b Delta-Sigma A/D Converter", *Digest of the 1990 IEEE International Solid-State Circuits Conference*, pp. 170-171, February 1990.

[5] R. Koch *et al.*, "A 12-bit Sigma-Delta Analog-to-Digital Converter with 15-MHz Clock Rate", *IEEE Journal of Solid-State Circuits*, pp. 1003-1010, December 1986.

[6] P. Defraeye *et al.*, "A 3-µm CMOS Digital Codec with Programmable Echo Cancellation", *IEEE Journal of Solid-State Circuits*, pp. 679-688, June 1985.

[7] V. Friedman *et al.*, "A Dual-Channel Sigma-Delta Voiceband PCM Codec", *IEEE Journal of Solid-State Circuits*, pp. 274-280, April 1989.

[8] S. Norsworthy, I. Post, H. Fetterman, "A 14-bit 80kHz Sigma-Delta A/D Converter", *IEEE Journal of Solid-State Circuits*, pp. 256-266, April 1989.

[9] K. Chao, S. Nadeem, W. Lee, C. Sodini, "A Higher Order Topology for Interpolative Modulators for Oversampling A/D Converters", *IEEE Transactions on Circuits and Systems*, pp. 309-318, March 1990.

[10] R. Carley, "A Noise-Shaping Coder Topology for 15+ Bit Converters", *IEEE Journal of Solid-State Circuits*, pp. 267-273, April 1989.

[11] R. Walden, G. C. Temes *et al.*, "Architectures For High-Order Multibit Sigma-Delta Modulators", *Proceedings of the 1990 IEEE International Symposium on Circuits and Systems*, pp. 895-898, May 1990.

[12] J. C. Candy, "Decimation for Sigma Delta Modulation", *IEEE Transactions on Communications*, pp. 72-76, January 1986.

[13] J. C. Candy, "A Use Of Limit Cycle Oscillations to Obtain Robust Analog-to-Digital Converters", *IEEE Transactions on Communications*, pp. 298-305, March 1974.

[14] J. McCreary, "Matching Properties and Voltage and Temperature Dependence of MOS Capacitors", *IEEE Journal of Solid-State Circuits*, pp. 608-616, December 1981.

[15] S. Tewksbury, R. Hallock, "Oversampled, Linear Predictive and Noise-Shaping Coders of Order N > 1", *IEEE Transactions on Circuits and Systems*, pp. 436-447, July 1978.

[16] R. Steele, "SNR Formula For Linear Delta Modulation with Band-Limited Flat and RC-Shaped Gaussian Signals", *IEEE Transactions on Communications*, pp. 1977-1984, December 1980.

[17] K. Cattermole, *Principles of Pulse Code Modulation*, Iliffe Books Ltd., p. 201, 1969.

[18] B. Leung, S. Sutarja, "Multibit Sigma-Delta A/D Converter Incorporating a Novel Class of Dynamic Element Matching Techniques", *IEEE Transactions on Circuits and Systems II*, Vol. 39, No. 1, pp. 35-51, January 1992.

[19] H. Inose, Y. Yasuda, J. Murakami, "A Telemetering System by Code Modulation - Delta-Sigma Modulation", *Transactions on Space Electronics and Telemetry*, Vol. SET-8, pp. 204-209, September 1962.

[20] J. C. Candy, Y. C. Ching, D. S. Alexander, "Using Triangularly Weighted Interpolation to Get 13-Bit PCM from a Sigma-Delta Modulator", *IEEE Transactions on Communications*, pp. 1267-1275, November 1976.

[21] A. Yukawa, "An Oversampling A-to-D Converter Structure for VLSI Digital Codecs", *International Conference on Acoustics, Speech and Signal Processing Proceedings*, pp. 1400-1403, May 1985.

[22] J. C. Candy, "A Use of Double Integration in Sigma Delta Modulation", *IEEE Transactions on Communications*, pp. 249-258, March 1985.

[23] W. Bennett, "Spectra of quantized signals", *Bell System Technical Journal*, pp. 446-472, July 1948.

[24] B. Agrawal, K. Shenoi, "Design Methodology for Sigma-Delta Modulators", *IEEE Transactions on Communications*, pp. 360-369, March 1983.

[25] J. E. Iwerson, "Calculated Quantizing Noise of Single-Integration Delta-Modulation Coders", *Bell System Technical Journal*, pp. 2359-2389, September 1969.

[26] R. Adams, "Design and implementation of an audio 18-bit A/D converter using oversampling techniques", *J. Audio Engineering Society*, pp. 153-166, March 1986.

[27] J. C. Candy, "The Structure of Quantization Noise from Sigma-Delta Modulation", *IEEE Transactions on Communications*, pp. 1316-1323, September 1981.

[28] W. Chou, P. Wong, R. Gray, "Multi-Stage Sigma-Delta Modulation", *IEEE Transactions on Information Theory*, pp. 150-159, April 1990.

[29] K. Uchimura *et al.*, "Oversampling A-to-D and D-to-A Converters with Multistage Noise Shaping Modulators", *IEEE Transactions on Acoustics, Speech and Signal Processing*, pp. 1899-1905, December 1988.

[30] Y. Matsuya *et al.*, "A 16b Oversampling A/D Conversion Technology using Triple Integration Noise Shaping", *IEEE Journal of Solid-State Circuits*, pp. 921-929, December 1987.

[31] R. Carley, "An oversampling analog-to-digital converter topology for high resolution signal acquisition systems", *IEEE Transactions on Circuits and Systems*, pp. 83-91, January 1987.

[32] T. Cataltepe, G. C. Temes *et al.*, "Digitally corrected multi-bit Σ–Δ data converters", *Proceedings of the 1989 IEEE International Symposium on Circuits and Systems*, pp. 647-650, May 1990.

[33] Yashui Sakina, P. Gray, "Multi-bit Σ–Δ Analog to Digital Converters with Nonlinearity Correction Using Dynamic Barrel Shifting", Master Thesis, UC Berkeley, pp. 24-32, June 1990.

[34] E. B. Hogenauer, "An Economical Class of Digital Filers for Decimation and Interpolation", *IEEE Transactions on Acoustics, Speech, and Signal Processing*, pp. 155-162, April, 1981.

[35] A. N. Netravali, "Optimum Digital Filters for Interpolative A/D Converters", *Bell System Technical Journal*, pp.1629-1641, November 1977.

[36] A. Huber *et al.*, "FIR Lowpass Filter For Signal Decimation with 15 MHz Clock Frequency", *International Conference on Acoustics, Speech, and Signal Processing*, pp. 1533-1536, April 1986.

[37] T. Saramaki *et al.*, "Multiplier-Free Decimator Algorithms for Superresolution Oversampled Converters", *International Symposium on Circuits and Systems Conference Proceedings*, pp. 3275-3278, May 1990.

[38] T. Saramaki *et al.*, "Design of Optimal Multistage IIR and FIR Filters for Sampling Rate Alteration", *International Symposium on Circuits and Systems Conference Proceedings*, pp. 227-230, May 1986.

[39] E. Dijkstra *et al.*, "On the Use of Modulo Arithmetic Comb Filters in Sigma Delta Modulators", *International Symposium on Circuits and Systems Conference Proceedings*, pp. 2001-2004, May 1988.

[40] B. Leung, "Design Methodology of Decimation Filters for Oversampled ADC Based on Quadratic Programming", *IEEE Transactions on Circuits and Systems*, pp. 1121-1132, October 1991.

[41] L. Rabiner, B. Gold, *Theory and Application of Digital Signal Processing*, Prentice-Hall, 1975.

[42] J. Bunch, "A Computational Method for the Indefinite Quadratic Programming Problem", *Linear Algebra and its Applications*, 34, pp. 341-371, 1980.

[43] P. Gill, W. Murray, M. Wright, *Practical Optimization*, Academic Press, pp. 177-178, 1981.

[44] M. Avriel, *Nonlinear Programming, Analysis and Methods*, Prentice-Hall, pp. 434, 1976.

[45] R. Adams, "Companded Predictive Delta Modulation: a Low-Cost Conversion Technique for Digital Recording", *Journal of Audio Engineering Society*, pp. 659-672, September 1984.

[46] M. Bellanger, "Digital Filtering by Polyphase Network: Application to Sample-Rate Alteration and Filter Banks", *IEEE Transactions on Acoustics, Speech and Signal Processing*, pp. 109-114, April 1976.

[47] H. Martínez, "A Class of Infinite-Duration Impulse Response Digital Filter for Sampling Rate Reduction", *IEEE Transactions on Acoustics, Speech and Signal Processing*, pp. 154-162, April 1979.

[48] Z. Ma, B. Leung, "Polyphase IIR Decimation Filter Design For Oversampled A/D Converters with Approximately Linear Phase", *IEEE Transactions on Circuits and Systems II*, Vol. 39, No. 8, pp. 497-505, August 1992.

[49] B. Leung, "Architecture of Multi-bit Oversampled A/D Converters Employing Dynamic Element Matching Techniques", *IEEE International Symposium on Circuits and Systems Proceedings*, pp. 2004-2007, June 1991.

[50] R. Gray, "Quantization Noise Spectra", *IEEE Transactions on Information Theory*, Vol. IT-36, pp. 1220-1244, November 1990.

[51] A. Namdar, B. Leung, "Quantization noise of 1-bit Double Loop Sigma-Delta Modulator", accepted and to be presented at *IEEE International Symposium on Circuits and Systems*, London, June 1994.

[52] D. B. Ribner, "A Comparison of Modulator Networks for High-Order Oversampled Sigma-Delta Analog-to-Digital Converters", *IEEE Transactions on Circuits and Systems,* Vol CAS-38, pp. 145-159, February 1991.

[53] F. Chen, B. Leung, "A High Resolution Multi-Bit Sigma-Delta Modulator with Individual Level Averaging", accepted and to be presented at *IEEE 1994 Symposium on VLSI circuits*, Hawaii, June 1994.

[54] B. Brandt, B. Wooley, "A 50-MHz Multi-Bit Sigma-Delta Modulator for 12-b 2-MHz A/D Conversion", *IEEE Journal of Solid-State Circuits*, Vol. 26, No. 12, pp. 1147-1756, December 1991.

[55] S. Moussavi, B. Leung, "High-Order Single-Stage Single-Bit Oversampling A/D Converter Stabilized with Local Feedback Loops", *IEEE Transactions on Circuits and Systems II*, Vol. 41, No. 1, pp. 19-25, January 1994.

[56] S. Rangan, B. Leung, "Quantization Noise Spectrum of Double-Loop Sigma-Delta Converter with Sinusoidal Input", accepted and to appear in *IEEE Transactions on Circuits and Systems II*, February 1994.

Chapter 4.2

Delta-Sigma D/A Converters

Gabor C. Temes

This chapter will give a brief summary of the reasons for introducing oversampling as a method of improving the accuracy of DACs; also, it will describe the basic structures of Δ–Σ DACs and the architectures available for the realization of digital noise-shaping loops. The design issues arising in the practical design of Δ–Σ DACs and some examples of integrated Δ–Σ D/A converters will be discussed in chapter 4.3.

The motivation for introducing the Δ–Σ system for D/A conversion is readily understood if we analyze the conditions required for a DAC with, say, 16 bits of accuracy using a 3 Volt reference voltage (an increasingly common design task these days). The voltage corresponding to the permissible 1/2 least-significant-bit (LSB) error is then $2^{-17}\cdot 3$ V, or about 21 μV. This is a very small value; it is the voltage generated by about a dozen electrons stored in a 0.1 pF capacitor. It is comparable to the thermal noise present at the input of a typical MOS op-amp. The direct design and fabrication of a DAC with such accuracy normally requires expensive trimming and/or calibration procedures. As will be shown in this chapter, the use of oversampling techniques overcomes the analog accuracy problem by trading digital complexity and speed for insensitivity to analog nonidealities. Since state-of-the-art technologies offer fast and dense digital circuit realization, this is a desirable trade-off.

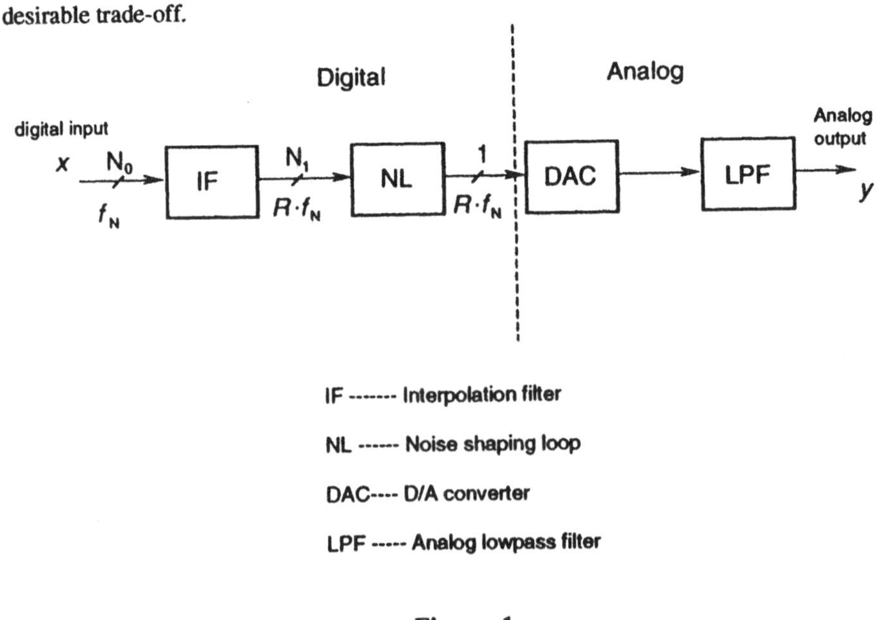

Figure 1

The basic system diagram of an oversampled noise-shaping DAC is shown in Figure 1. The input x to the converter is now a multibit digital signal: a stream of digital words with a wordlength N_0 and a data rate f_N. Usually, f_N is close to the Nyquist rate of the signal, i.e. close to $2 \cdot B$, where B is the signal bandwidth. This signal is processed by the interpolation filter IF, which changes the data rate to the oversampled value $R \cdot f_N$ (where R is the oversampling ratio), and which also suppresses the spectral replicas centered at f_N, $2 \cdot f_N, \ldots,$ $(R-1) \cdot f_N$. The first two curves of Figure 2 illustrate the spectra at the input and output of IF.

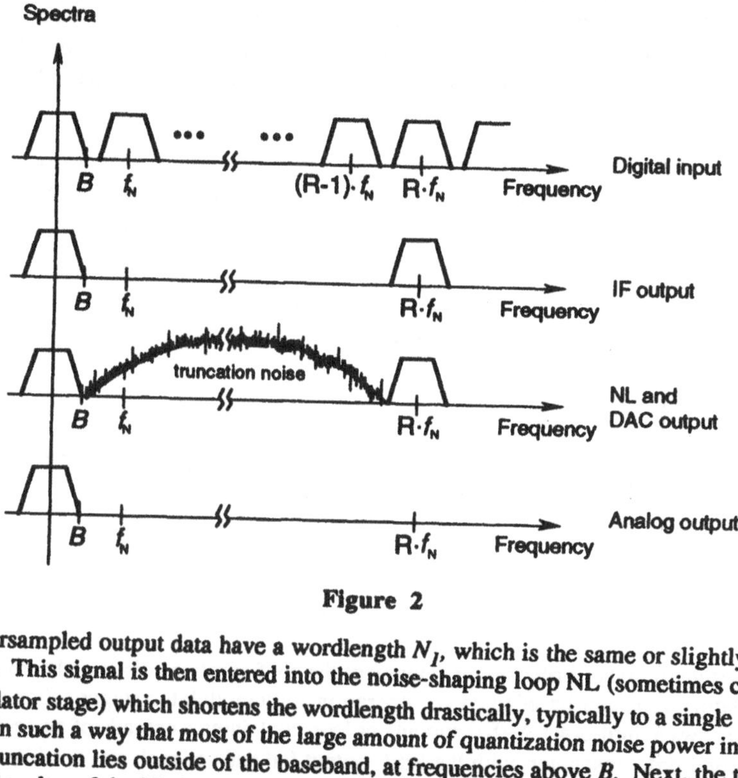

Figure 2

The oversampled output data have a wordlength N_1, which is the same or slightly smaller than N_0. This signal is then entered into the noise-shaping loop NL (sometimes called the demodulator stage) which shortens the wordlength drastically, typically to a single bit. This is done in such a way that most of the large amount of quantization noise power introduced by the truncation lies outside of the baseband, at frequencies above B. Next, the truncated output signal y of the NL is converted by the internal DAC, which is the first analog stage in the system. Since it is typically a 1-bit DAC, its realization is conceptually simple, and (as discussed earlier in connection with the 1-bit DAC used in $\Delta-\Sigma$ ADCs) it is inherently linear.

The analog output of the internal DAC contains a linear replica of the digital input signal x, plus a large amount of noise due to the quantization error e introduced in the NL. Since, as discussed above, most of this noise lies outside of the signal band, it can be nearly completely suppressed by the analog low-pass filter LPF following the DAC. The last two lines illustrate the spectra of the input and output signals of the LPF.

The overall performance of the overall $\Delta-\Sigma$ converter, for ideal operation, is determined by the noise-shaping loop. Its in-band signal gain characteristics and out-of-band quantization noise spectrum should be known before the digital and analog filters can be designed. Specifically, the in-band gain response of the loop must be known in advance, since often it has to be equalized by the digital filter; the out-of-band noise of the loop must be known before the digital filter can be designed, since it limits the amount of minimum stopband loss which one should specify for the digital IF. Also, the required stopband loss of the analog filter depends on this noise. Hence, the design of the NL block is the crucial part in the design of the DAC system.

Next, we shall describe and discuss some configurations for the noise-shaping loop (demodulator). The purpose of this stage is very similar to that of the analog noise-shaping loop of a Δ–Σ A/D converter: a finely-quantized input signal (in the case of the analog loop, an analog, i.e., infinitely finely quantized signal!) must be truncated to a very small wordlength, without introducing too much in-band truncation noise. Thus, the idea of using a digital structure similar to that of the Δ–Σ modulator is immediately obvious. The general form of this loop is shown in Figure 3. In the simplest case, the loop filter LPF may be simply an accumulator, resulting in a first-order noise shaping (Figure 4).

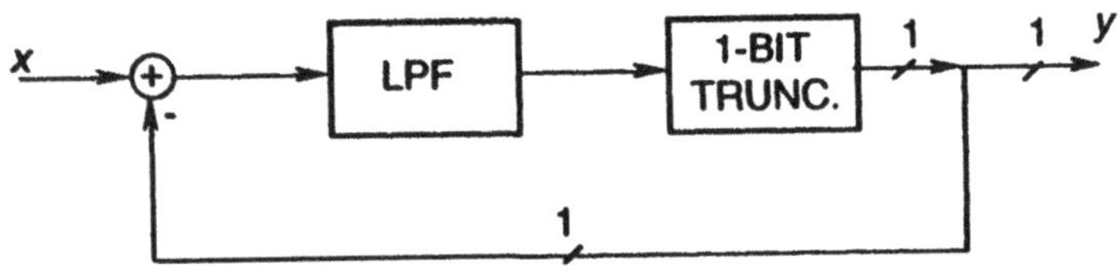

Figure 3

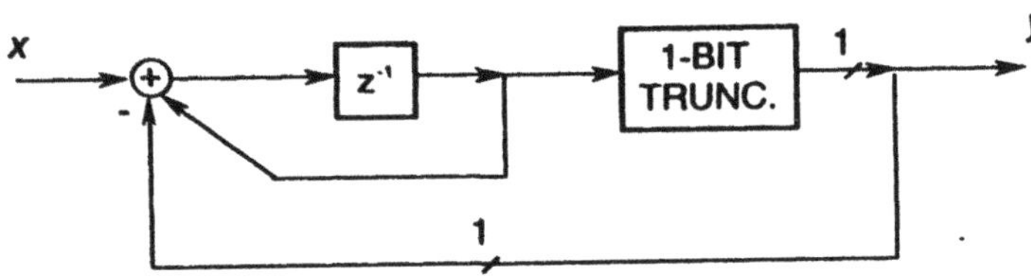

Figure 4

For the design of higher-order systems, many of the configurations discussed earlier for Δ–Σ modulator loops can be adapted for operation in demodulator stages. For example, Figure 5 shows a third-order Δ–Σ demodulator loop containing a cascade of an accumulator and a second-order resonator, with the MSB output fed back to each stage, while Figure 6 illustrates a fifth-order loop, containing two resonators and one accumulator, also with MSB feedback.

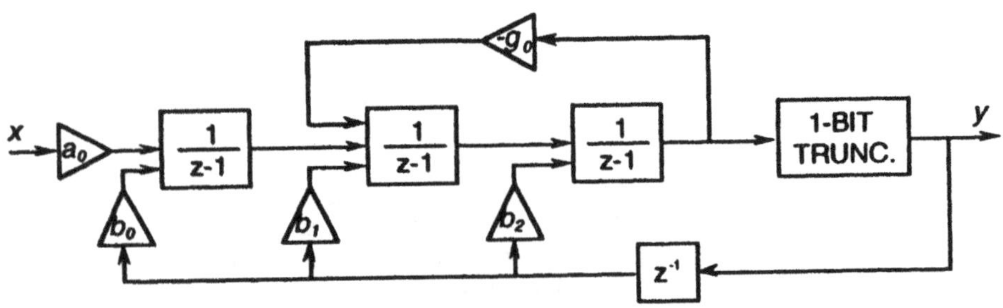

Figure 5

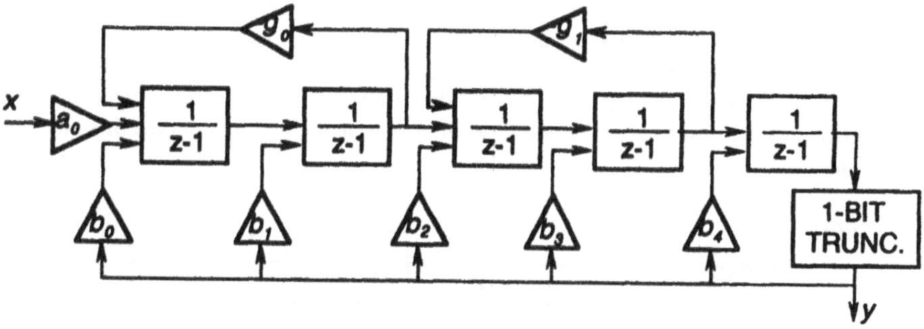

Figure 6

A similar modulator [1], in which the outputs of all stages are fed forward to the input of the quantizer, while the MSB output is fed back only to the input node of the loop, is shown in Figure 7. In these structures, hardware can be saved if the constant factors (e.g., a_0, b_0, b_1, etc. in Figure 5) are chosen as integer powers of 2. Such quantization of the coefficients may introduce a slope or even a peak into the nominally flat signal frequency response, but this amplitude distortion can be compensated in the digital filter characteristics. The number of bits carried in the loop can be reduced as the signal propagates from left to right. The analysis and design of these structures is similar to their analog counterparts, and therefore will not be discussed here.

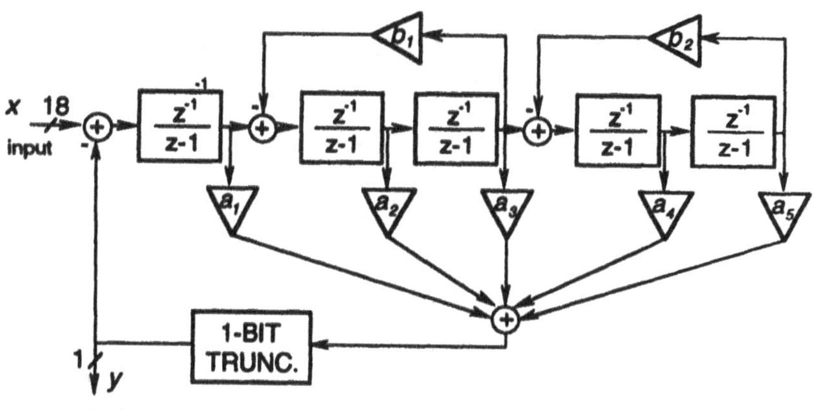

Figure 7

As was the case for the modulator loops, higher-order noise-shaping demodulators will be only conditionally stable. Thus, there is a limited range of permissible input amplitudes for stable operation which varies from structure to structure.

A useful alternative configuration for $\Delta-\Sigma$ demodulators is schematically illustrated in Figure 8. In this structure, instead of the MSB output, the negative of the truncation error (consisting of the neglected LSBs) is fed back to the input through a loop filter with a transfer function $H_1(z)$. Simple analysis shows that in the z-domain the output of the loop is given by

$$Y(z)=X(z)+[1-H_1(z)]E(z) \tag{1}$$

where Y, X and E are the z-transforms of the output y, the input x and the truncation error e of the loop, respectively. This so-called "error-feedback" structure usually leads to a simpler realization than the one illustrated in Figure 3, discussed above, and is hence often used for digital loops. For analog loops, the high sensitivity of the circuit to any imperfections of the components of the loop filter and of both adders makes this structure impractical.

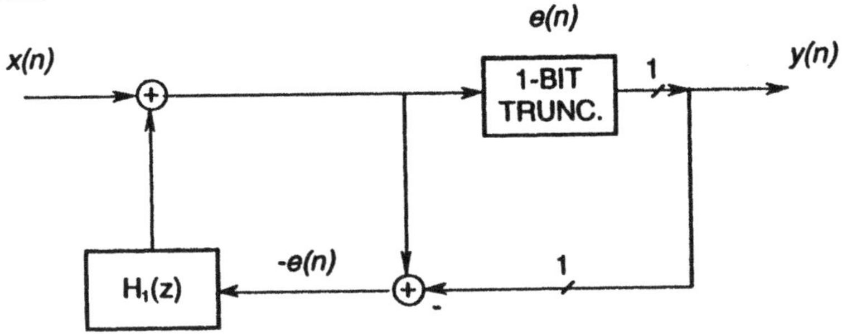

Figure 8 Error feedback scheme.

In the simplest case, $H_1(z)=z^{-1}$ can be chosen as the transfer function of the loop filter (Figure 9). Then, the noise transfer function (NTF) becomes

$$H(z)=1-H_1(z)=1-z^{-1} \tag{2}$$

and a first-order noise shaping is achieved. To obtain second-order noise shaping, we can select the loop filter transfer function as

$$H_1(z)=1-(1-z^{-1})^2=z^{-1}(2-z^{-1}) \tag{3}$$

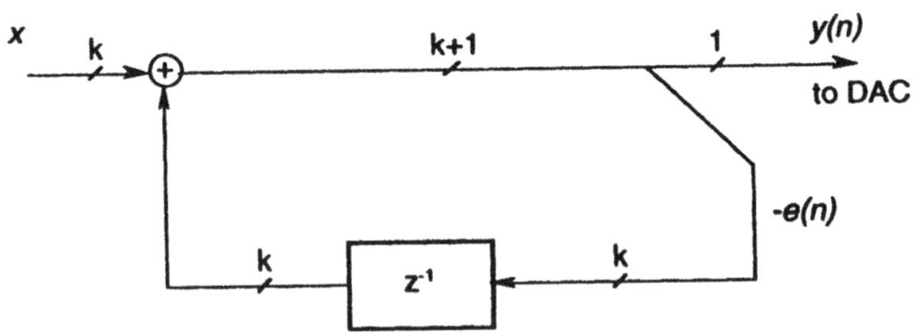

Figure 9 First-order noise-shaping loop.

Figure 10 shows the corresponding noise-shaping loop. Note that neither of these structures requires multiplication of multibit words; only delays, additions and shifts of the binary point are needed. To prevent overflow, a limiter may be added to the circuit [2], as shown in Figure 11. Alternatively, as shown in Figure 12, the circuit can be operated with a two-bit (rather than single-bit) output, and a second first-order loop can be added to generate the final 1-bit output data [3].

The overall performance of the system is slightly worse than if the input x would be directly fed to the input of the second loop; however, the complexity of the fast ($f_{clock}=4 \cdot f_s$) digital hardware is greatly reduced, at the cost of added slow ($f_{clock}=f_s$) components. In fact, the fast stage can be realized by only a small RAM and a shift register [3].

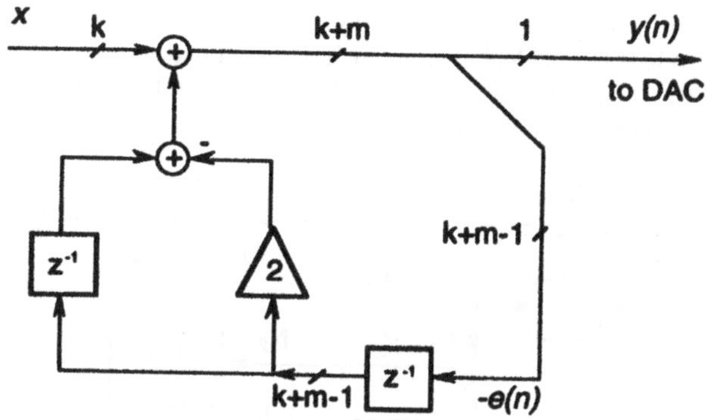

Figure 10 Second-order noise-shaping loop.

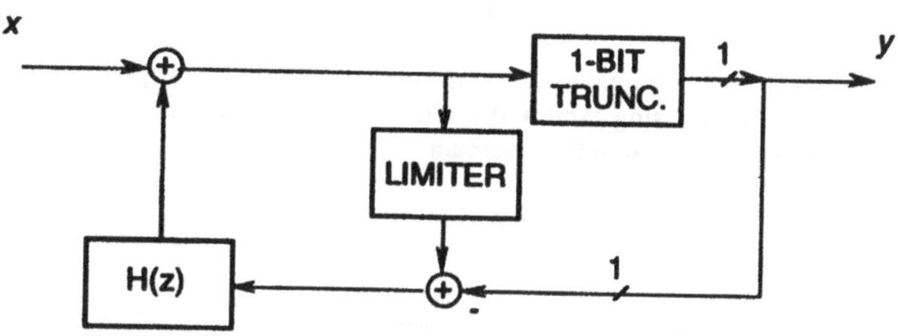

Figure 11

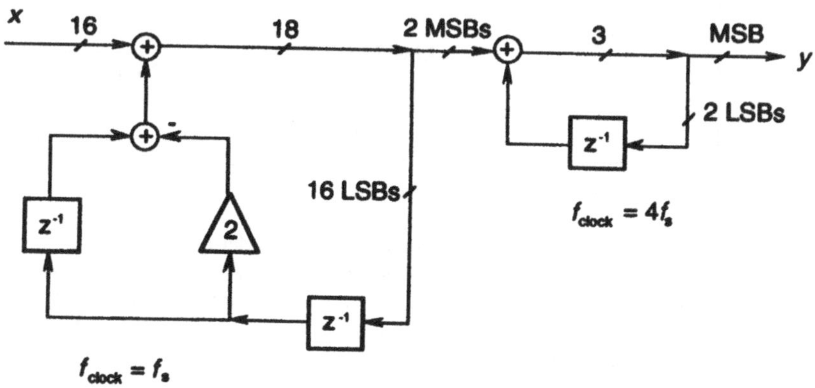

Figure 12

As was the case for modulator loops, the cascade (MASH) realization of digital noise-shaping loops is another option. Figure 13 shows a second-order demodulator realized as a cascade of two first-order loops [4]. Unfortunately, the output of the digital circuit is then the sum of two single-bit words, and hence a two-bit D/A converter is needed with the same linearity requirement as that of the overall system. To avoid this problem, it is possible to convert the outputs of the two loops individually into analog form using two 1-bit DACs, and combining their analog outputs in a summing stage. Imperfect gain matching between the two paths introduces noise leakage, but at least the signal will not be distorted since it passes through only a single-bit conversion. Alternatively, another simple Δ–Σ stage with a higher sampling frequency can be added to reduce the wordlength to 1 bit, as in Figure 12.

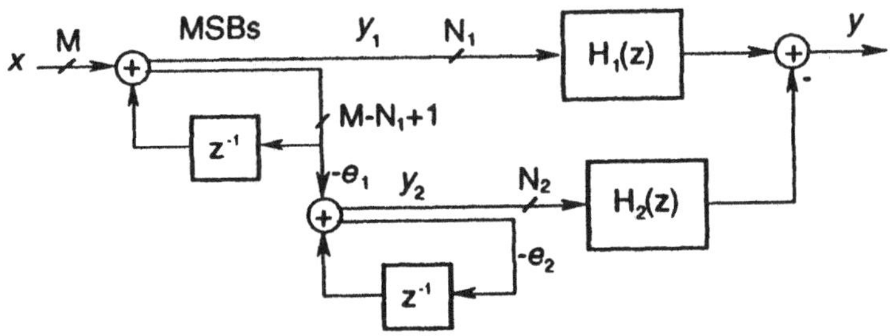

Figure 13

A structure in which both the quantization error and the single-bit output data are fed back (and which thus combines single-bit and multi-bit feedback) is shown in Figure 14 [5]. Analysis gives the output signal in the *z*-domain

$$Y(z) = z^{-2} \cdot X(z) + (1-z^{-1})^2 \cdot E(z) \tag{4}$$

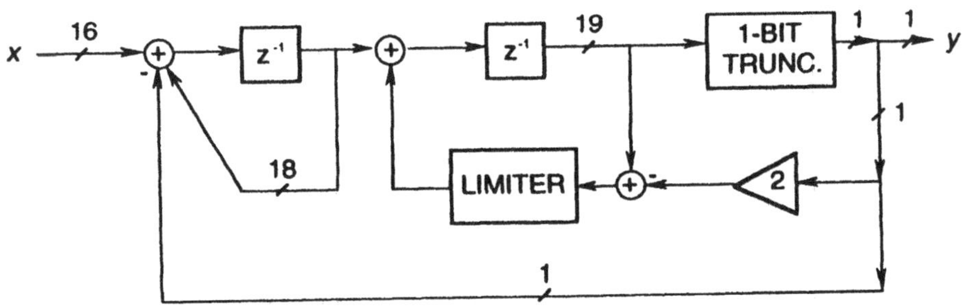

Figure 14

Thus, the circuit functions as a truncation loop with second-order noise shaping.

In chapter 4.1, the advantages of incorporating multi-bit, rather than single-bit, quantizers in Δ–Σ ADCs were discussed. These included improved stability properties, lower speed requirements, and reduced demands for the postfilter. As explained there, however, in order to obtain these improvements the detrimental effects of the inherent nonlinearities of the internal DAC had to be reduced, at the cost of some additional hardware. The situation is very similar for Δ–Σ DACs. As shown in Figure 1, the internal DAC is directly in the signal path, and hence the errors caused by its nonlinearities affect the signal without any noise filtering or attenuation. Hence, steps must be taken to reduce or cancel the distortion caused by these DAC nonlinearities.

One option available is to use one of the "element-swapping" methods discussed in chapter 4.1. As explained there, some of these techniques, such as the randomizing approach, convert the errors caused by the nonlinearities into random (or pseudo-random) noise. Others, such as the barrel-shifting or individual-level-averaging technique, achieve a reduction of in-band noise by modulating the nonlinearity-caused noise to a higher out-of-band frequency. The reader is referred to chapter 4.1 for details. These techniques remain applicable in the present application as well.

An alternative way of reducing the in-band noise and harmonic distortion caused by the nonlinearity of the internal DAC is to use continuously-calibrated current copiers [6] or capacitor averaging [7]. An in-band noise reduction corresponding to a linearity of 15-16 bits has been achieved this way.

The digital correction technique described in chapter 4.1 and [8] can also be applied in Δ–Σ DAC design. Figure 15 shows the block diagram of a digitally corrected D/A converter with a multibit internal DAC [8]. In this system, the RAM contains multibit words representing very accurately the actual output levels of the N-bit internal DAC. Since the internal DAC and the RAM have the same inputs, their outputs (digital and analog, respectively) are also corresponding values. Also, since the low-frequency (baseband) components of the RAM output track the digital input x very accurately due to the feedback loop, the same conclusion holds for the DAC output. In conclusion, the output y of the truncator is predistorted by the inclusion of the RAM in the feedback loop in such way that the low-frequency spectrum of the DAC output is an accurate analog replica of the digital input x.

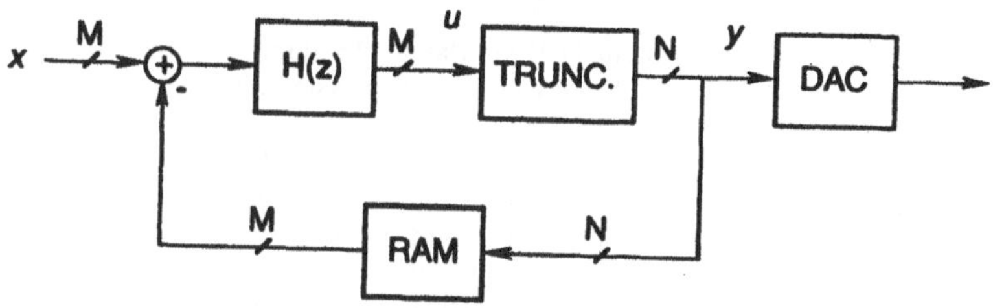

Figure 15

The data stored in the RAM in the system of Figure 15 can be acquired using the same calibration process as was used in the case of the digitally calibrated multibit ADC, described in [8]. To save some digital hardware, it is also possible to store the digital equivalents of the errors in the DAC levels, rather than the DAC levels themselves, and to combine the RAM's input and output signals [9] via an additional noise shaping loop.

The dual-quantization techniques used for Δ–Σ ADCs also have equivalents for Δ–Σ DACs. Figure 16 shows a dual-truncation DAC structure [10] somewhat similar to the Leslie-Singh ADC architecture [11]. In this system, the negative of the large quantization error e_1 due to the single-bit noise-shaping loop is fed into a correction path, where it is truncated to M bits (M>1). The truncated error is then converted into analog form, high-pass filtered in the analog filter $H_2(z)$, and added to the analog output of the 1-bit DAC to cancel (in practice, to reduce) e_1 in the output signal y.

One important advantage which this scheme (as well as the dual-truncation schemes described in the next paragraphs) has over the corresponding ADC architectures is that the DAC schemes require the accurate realization of analog *differentiators* in the correction filter $H_2(z)$, while the ADC systems require accurate analog *integrators*. An analog differentiator must realize a transmission zero at DC, which a capacitor can do very precisely, while the integrator needs a pole at DC, which cannot be realized accurately with practical finite-gain op-amps.

Under ideal conditions, the corrected output y no longer contains the noise-shaped e_1, but rather the M-bit truncation error e_M shaped the same way as e_1 was. This results in an improvement of nearly M bits in the *SNR*. In practice, there will always occur a leakage of filtered e_1 noise into y. However, due to the zeros at dc in the noise filter function, this effect need not be significant if the system is carefully designed. Similarly, the nonlinearity error of d_M of the M-bit DAC, (which is usually fairly small to start with) is high-pass filtered and thus further reduced by the analog filter $H_2(z)$.

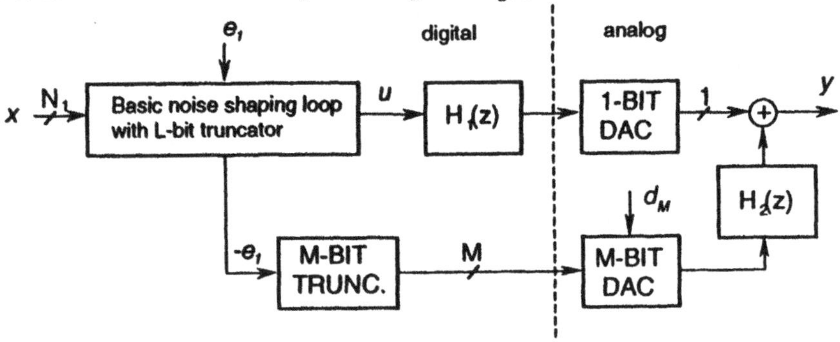

Figure 16

Figure 17 illustrates a DAC system [10] similar to the dual-quantizer MASH structure proposed for ADCs by Brandt and Wooley [12]. In this system, instead of simply truncating e_1 to M bits, a second noise-shaping loop is incorporated in the correction path which shapes the M-bit truncation error e_M. As a result, e_M is even more reduced; the order of the overall noise shaping is now the sum of the orders of the two noise-shaping loops.

Figure 18 shows the block diagram of a third-order DAC based on this scheme. In the diagram, C_1 and the associated switches realize the 1-bit DAC, while C_2, C_3 and C_4 and their switches perform the function of the $H_3(z)$ filter. Note that the transfer function of this filter will have a double zero at z=0 regardless of capacitor matching or finite op-amp gain.

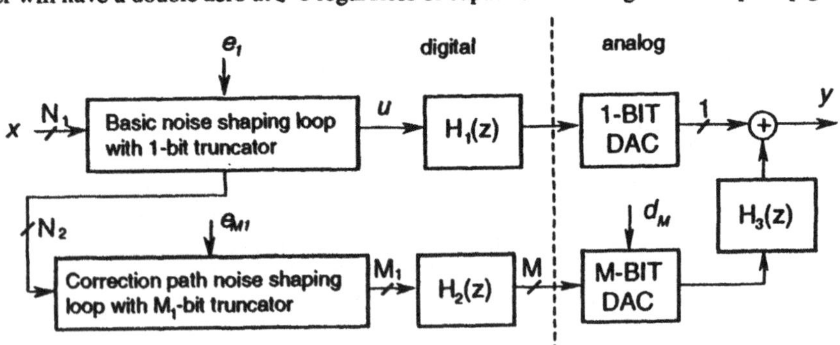

Figure 17

A different dual-truncation architecture, where the 1-bit and M-bit outputs are both fed back within a single loop, is shown in Figure 19. It is equivalent to the dual-feedback single-path ADC system of Hairapetian [13]. Here the $H_5(z)$ block is an analog high-pass filter, and hence the nonlinearity error of the *(M-1)*-bit DAC is suppressed in the baseband...

A problem common to all Δ–Σ DACs is the generation of tones [5]. This is particularly likely to occur in DACs, where all signals are rational multiples of the truncated output. In a high-order single-path system, such as that shown in Figure 6 or Figure 7, the internal states of the loop filter act as a dither signal and thus reduce the probability of tone generation.

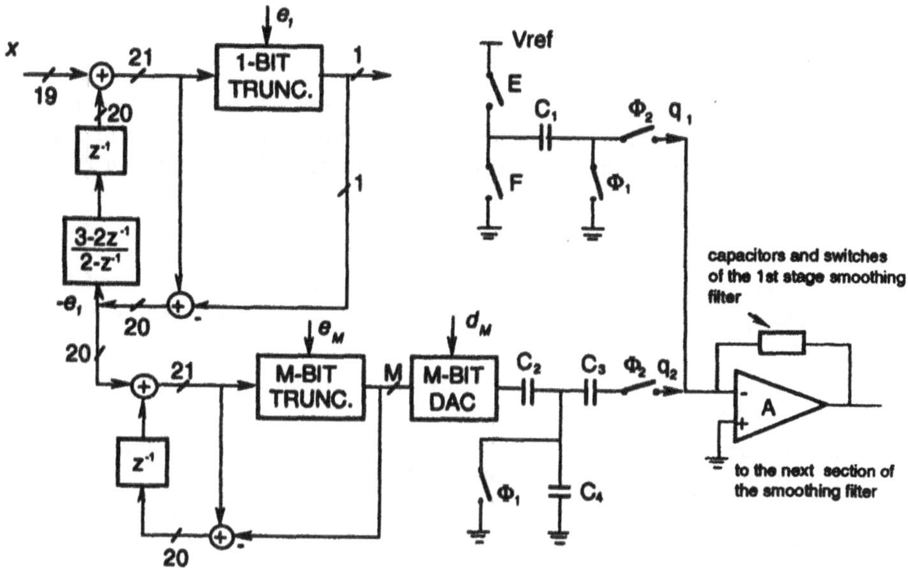

Figure 18

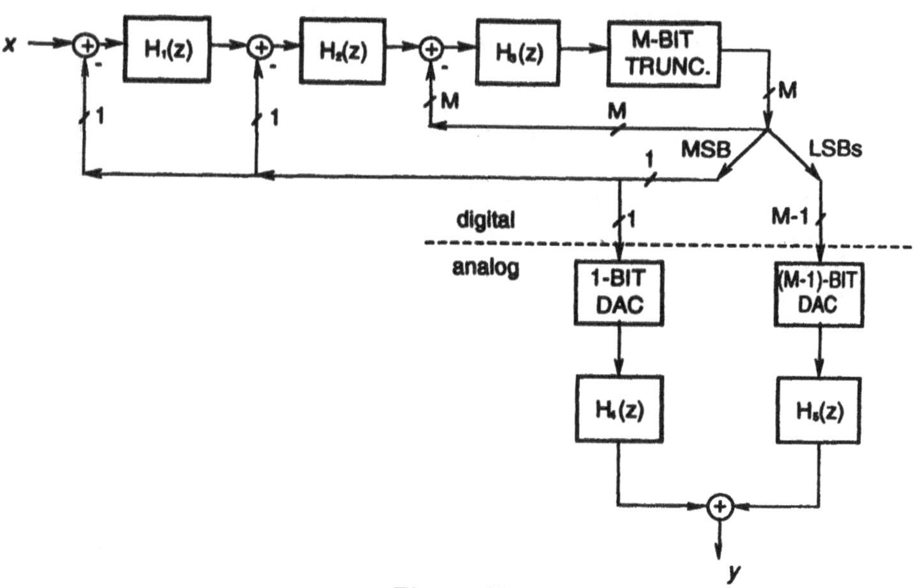

Figure 19

In a two-path structure, such as that shown in Figure 16 or Figure 17, the correction path processes the tone generated in the signal loop along with e_1, and subtracts it from the 1-bit output. If $M >> 1$ and conditions are ideal, the tone will be nearly cancelled; otherwise, it will only be attenuated in the output signal y. It is, of course, also possible to use digital dithering to prevent tones, e.g., by replacing the LSBs of the input signal with a random bit stream [4].

Acknowledgement

The authors of this chapter and chapter 4.3 would like to acknowledge IEEE Press for re-use of some of the material to be included in a new IEEE Press book, *Delta-Sigma Converters*, eds. S. Norsworthy, R. Schreier and G. Temes, to be published in 1994 or 1995.

References

[1] N. S. Sooch, J. W. Scott *et al.*, "18-bit stereo D/A converter with integrated digital and analog filters", *91st Convention of the Audio Engineering Society*, Preprint #3113, Oct. 1991.

[2] P. J. Naus, E. C. Dijkmans *et al.*, "A CMOS stereo 16-bit D/A converter for digital audio", *IEEE J. Solid-State Circuits*, Vol. SC-22, pp. 390-395, June 1987.

[3] H. G. Musmann and W. Korte, "Generalized interpolative method for digital/analog conversion of PCM signals", US Patent 4, 467, 316, Aug. 21, 1984.

[4] J. C. Candy and An-Ni Huynh, "Double interpolation for digital-to-analog conversion", *IEEE Trans. Commun.*, Vol. COM-34, pp. 77-81, Jan. 1986.

[5] J. C. Candy and G. C. Temes, "Oversampling methods for data conversion" in *Oversampling Delta-Sigma Data Converters*. New York: IEEE Press, 1992.

[6] H. Schouwenaars *et al.*, "An oversampling multi-bit CMOS DAC for digital audio with 115 dB dynamic range", *1991 IEEE Internat. Solid-State Circuits Conf. Digest*, pp. 72-73, Febr. 1991.

[7] B. Leung, "Pipelined multibit oversampled D/A converters with capacitor averaging", *Analog Integrated Circuits and Signal Processing*, Vol. 2, pp. 139-156, Apr. 1992.

[8] T. Cataltepe *et al.*, "Digitally corrected multi-bit sigma-delta data converters", *Proc. IEEE Internat. Symp. on Circuits and Systems Conf.*,, pp. 647-650, Portland, OR May 1989.

[9] M. Sarhang-Nejad and G. C. Temes, "A high-resolution multibit sigma-delta ADC with digital correction and relaxed amplifier requirements", *IEEE J. Solid-State Circuits*, Vol. 28, pp. 648-660, June 1993.

[10] X. F. Xu *et al.*, "The implementation of dual-truncation sigma-delta D/A converters", *Proc. IEEE Internat. Symp. on Circuits and Systems Conf.*, pp.597-600, San Diego, CA, May 1992.

[11] T. C. Leslie and B. Singh, "An improved sigma-delta modulator architecture", *Proc. IEEE Internat. Symp. on Circuits and Systems Conf.*, pp. 372-375, New Orleans, LA, May 1990.

[12] B. P. Brandt and B. A. Wooley, "A 50-MHz multibit sigma-delta modulator for 12-b 2-MHz A/D conversion", *IEEE J. Solid-State Circuits*, Vol. 21, pp.1746-1756, Dec. 1991.

[13] A. Hairapetian *et al.*, "Multibit sigma-delta modulator with reduced sensitivity to DAC nonlinearity", *El. Letters*, Vol. 27, pp. 990-991, May 1991.

Chapter 4.3

Design Aspects of High-Order Delta-Sigma A/D Converters

Robert Adams

4.3.1 Introduction

Until 1987 [1], it was widely believed that delta-sigma modulators with loop filters of order greater than two were impossible to stabilize. Indeed, any casual attempt to build such a modulator will result in impressive oscillations, which tend to discourage the fainthearted from further experimentation. This belief that high-order modulators could not be stabilized continued even after several commercial examples of such modulators were in production and appearing in everyday equipment built for the consumer audio market.

In this chapter we will examine many practical aspects of high-order delta-sigma designs, including loop filter topologies, maximum safe input range, prevention of idling tones, and common commercial design practices. Since most commercial delta-sigma ADCs use switched-capacitor circuitry, we will focus on those topologies that can be implemented using switched-capacitor circuits. A design example of a fourth-order modulator and a description of a commercially-available fifth-order modulator will be given.

4.3.2 Motivation for using high-order single-bit loops

The world of noise-shaping converters can be roughly divided into the following camps: single-bit single-loop low-order designs, single-bit single-loop high-order designs, multi-loop cascaded designs with feed-forward error cancellation and multi-bit noise-shapers (both low and high-order). The advantages and disadvantages of these topologies are summarized below.

4.3.2.1 Low-order single-loop single-bit

- Guaranteed stability.
- Simple loop-filter design.
- Simple circuit design.
- Input range may use almost the full range of 1s densities.
- Low SNR (except for high oversampling ratios).
- More prone to idling tones (dither may help).

4.3.2.2 High-order single-loop single-bit

- High SNR for modest oversampling ratios.
- Less prone to idling tones.

- Simple circuit design.
- Difficult loop filter design.
- Stability is signal-dependant.
- Max input range must be restricted to ensure stability.

4.3.2.3 Multi-loop cascade

- High SNR for modest oversampling ratios.
- Stability guaranteed.
- Max input range almost equal to the full range of 1's densities.
- Requires near-perfect matching between analog integrator and digital differentiator. Complex switched-capacitor circuits are required to insure matching.
- Imperfect matching may result in leakage of tones into baseband.
- Decimation filter must allow for multi-bit inputs.

4.3.2.4 Multi-bit

- High SNR for fairly low oversampling ratios.
- Stability much easier to achieve for high-order loops.
- Extra quantization levels allow for large dither signals at the quantizer input, eliminating idling tones.
- Imperfect matching of levels (in D/A half of quantizer) results in imperfect DC transfer function (INL errors). Auto-calibration or dynamic element matching may help.
- Decimation filter must allow for multi-bit inputs.
- More complex circuit design.

Single-bit high-order designs are most attractive whenever high SNR, simple circuit design and good idle-tone performance are important. The major obstacle to overcome is the issue of stability, but once a methodology has been established, designing a new loop filter is a simple matter of running a few programs and checking the result with a discretetime simulator.

For a given oversampling ratio, a high-order modulator is capable of much greater signalto-noise ratios than a simple second-order loop. This improvement is a function of the oversampling ratio; high oversampling ratios benefit the most from high-order loop filters, whereas low oversampling ratios (<16, for example) do not show as much improvement when high-order loop filters are used.

In the literature it is common to find graphs of SNR versus the oversampling ratio for various loop filter orders. These graphs should be used with great caution. In many cases, these results are derived from the linear model with a noise-shaping transfer function of

$$H(z) = \left(1 - z^{-1}\right)^n \tag{1}$$

where n is the order of the differentiation.

It is important to realize that the use of this noise-shaping transfer function does not result in stable systems for loop filters of order greater than two, and therefore the results are meaningless unless multi-bit quantizers are used. If we allow any arbitrary n^{th}-order noise-shaping transfer function, then the number of different stable noise-shaping transfer functions that can be designed is infinite. Therefore it is unwise to trust the results of another designer unless the coefficients of the loop filter are exactly given, and the results come from a discrete-time simulation and not a linear model. The only modulator for which there are no degrees of freedom is the lowly first-order modulator.

4.3.3 Design choices: SC or active-RC?

Perhaps the first choice that must be made in designing any delta-sigma modulator is a choice between a switched-capacitor (SC) implementation and a conventional active-*RC* (continuous-time) design. In general, most IC implementations of delta-sigma ADCs use switched-capacitor circuits, whereas most system-level or hybrid implementations use active-*RC* circuits. The reasons for this are summarized in below.

4.3.3.1 Advantages and disadvantages of SC and active-RC modulators

(i) Advantages

- Easily simulated.
- Compatible with VLSI CMOS process (extra poly layer desirable to make small-area linear caps).
- Insensitive to clock jitter, as long as full settling occurs.
- Insensitive to exact shape of op-amp settling waveform, as long as full settling occurs.
- Pole-zero locations are set by capacitor ratios, which are highly accurate.
- Easy to breadboard.
- Less prone to pick up digital noise (no true input samplers are used).
- Easy to drive from external sources - no switched-cap current pulses.
- SNR is not limited by cap size.

(ii) Disadvantages

- Large capacitors required for high SNR (kT/C noise limit).
- Switched-cap circuits are true samplers, potentially causing aliasing of out-of-band noise. They are thus more prone to picking up digital noise.
- Large spike currents drawn by capacitors are hard to drive from external sources (*RC* isolation circuits required).
- Very difficult to prototype (typical capacitor values are less than 1 pF, and are easily swamped by parasitics on a breadboard).
- Not as compatible with simple CMOS process. Needs large capacitors, linear high-value resistors, low-noise op-amps.
- Accurate *RC* time constants not possible for monolithic designs without laser trimming.

- SNR degraded by non-ideal comparator feedback signal. Sensitive to jitter, noise, and switching characteristics of 1-bit feedback waveform.
- Loop filter does not scale with clock frequency.
- Op-amps must remain linear at all times. It is NOT just the settled value that counts.
- Discrete-time simulation more difficult.

Figure 1 shows a comparison between the two design styles. In Figure 1a, a simple switched-cap circuit is shown with a two-phase clock. On phase 1, the input is sampled, and the comparator decision is made. On phase 2, the input is delivered to the integrator summing junction, along with the result of the comparator decision. A "1" comparator decision results in a charge of $C_f V_{ref}$ delivered to the integrator, whereas a "0" decision results in a charge of $-C_f V_{ref}$ being delivered. In many cases, the required inversion of the reference signal can be accomplished by appropriate phasing of the switches using only a single reference.

Figure 1b shows a simple continuous-time design. The output of the comparator is latched on every clock pulse, and the latched comparator decision is used to apply a $+V_{ref}$ or $-V_{ref}$ voltage to a resistor connected to the integrator summing junction. This voltage is continuously applied to the integrator during the entire clock period. Note that the amount of charge delivered to the integrator is a function of the width of the clock pulse, the rise and fall time of the voltage feedback waveform, and, in the case of unequal rise and fall times, the previous bit decision. The use of a "return-to-zero" scheme can in theory eliminate the effects of previous bit decisions and rise-fall mismatches, but not the sensitivity to jitter.

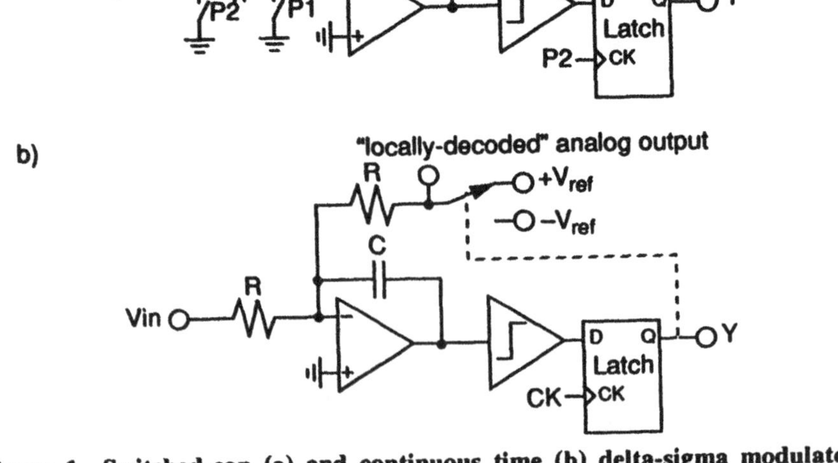

Figure 1 Switched-cap (a) and continuous time (b) delta-sigma modulators.

The sensitivity to these effects is quite severe; a typical audio converter with a 20 kHz bandwidth and an oversampling ratio of 128:1 might require jitter less than 20 ps peak-to-peak to achieve >90 dB of SNR! In one commercial example of a continuous-time loop, a multi-bit approach was used to add dynamic range "on top" of the jitter-limited noise floor [8]. This allowed an SNR of greater than 105 dB to be achieved, even though the theoretical SNR of the loop was much higher than this.

One common mistake made in continuous-time design is to assume that the V_{ref} feedback signal at point at the "locally decoded" output can be measured to determine the SNR of the loop. In the presence of non-ideal effects such as jitter, however, the effective value of a 1 or 0 comparator decision is changed by the presence of jitter. Since the loop is continuous-time, the feedback will insure that the analog version of the 1-bit feedback signal is a faithful representation of the input, regardless of non-ideal clocking effects. But the 1-bit digital signal that is passed to the decimator assumes the value of a 1 or 0 is invariant. It is very common to achieve high SNR when a measurement at the locally-decoded output point is made (with an analog spectrum analyzer, for instance), while the same measurement made at the decimated digital output exhibits excess noise.

4.3.4 Stability

There is as of yet no closed-form solution for the design of stable high-order loops. Instead, a "cookbook" design approach has been developed by the author, based on the idea that the "gain" of the comparator at low frequencies as seen by the delta-sigma loop is determined by the RMS magnitude of the "busy" time-domain pattern that appears at its input. The comparator gain as seen by the loop is also a function of the input signal, and falls at high input amplitudes that approach the V_{ref} feedback voltage. In high-order modulators, instability occurs for low comparator gains, and hence all high-order modulators become unstable for inputs that exceed a certain percentage of V_{ref}. Methods for insuring global stability will be discussed later.

Since the magnitude of the comparator input signal is determined by the loop filter design, and the loop filter is designed with an assumption of a particular comparator gain, we have a classic "chicken-and-egg" problem. The solution to this problem is to use rules of thumb that have proven accurate over many practical designs.

1) Design a high-pass discrete-time filter with the following characteristics:

   a) The high-frequency gain of the filter should be 3 dB.

   b) The first value of the impulse response of this filter must be 1.

Requirement "a" is derived from the fact that the high-frequency gain of the noiseshaping transfer function determines the amplitude of the comparator input signal, which in turn determines the low-frequency comparator gain as seen by the loop. With a value of 3 dB for the high-frequency gain, the comparator in-circuit gain is typically between values of 1 and 2. Requirement "b" results from the fact that delay-free loops cannot be physically implemented, and falls out very easily from ordinary loop algebra.

2) Simulate the loop using a discrete-time simulator (do NOT attempt to use SPICE!). Measure the "in-circuit" comparator gain by taking long-term averages of the comparator input and output for various DC levels.

3) If the comparator gain is incorrect by a large amount, choose a new noise-shaping transfer function with a different amount of high-frequency gain. Note that the causality

requirement does not allow independant choices of the high-frequency NTF gain and the NTF cutoff frequency; once the cutoff frequency is set, the high-frequency NTF gain is also set.

One of the most challenging aspects of stabilization analysis lies in establishing a "safe" input range. Dynamic signals often behave very differently from DC signals when it comes to stability.

4.3.4.1 *Stability and the uncontrolled input signal; a practical guide to safe operation*

In the previous section is was shown that large input signal approaching V_{ref} cause the comparator gain to drop, causing instability. The stable DC input range is related to the choice of NTFs; NTFs that are not very aggressive in the context of noise-shaping have larger usable input ranges. For systems that use Butterworth pole alignments (a good compromise between aggressive noise-shaping and stable operation), the stable input range is typically between 60 and 80 percent of the comparator feedback voltage. It is prudent to define the full-scale input to be less than the value at which instability actually occurs, as it is impossible to exhaustively simulate all the possible inputs that may be applied. Typically, a converter that is DC stable over an 80% bit density range will have a defined maximum input range of 65 to 70% of the comparator feedback voltage. This "safety factor" depends to a large extent on the Q of the poles in the NTF, as well as how well-behaved the input is.

4.3.5 Transient input signals and stability; the case for mild pre-filtering

The stability analysis based on the comparator gain argument is approximately valid for DC input signals, and for sinusoids that are low in frequency compared with the poles of the NTF. For example, a typical audio converter with a fourth-order loop filter might use a Butterworth NTF with a high-pass frequency of about 150 kHz and a clock frequency of 3 MHz, compared with a signal bandwidth of 20 kHz.

In real life, it is impossible to insure that the input signal is completely bandlimited, so one must consider what happens when transient or other wide-band signals are applied to the input. From our previous analysis of stability, we know that large comparator input signals cause reduced comparator gain and hence result in possible instability. We can then ask the question, "what input signal will cause the largest input to the comparator?". It is wellknown that the peak value at the output of a filter with a given impulse response is equal to the $H1$ norm (sum of absolute values) of the impulse response of the filter, assuming that the input is limited in amplitude to ± 1. The input signal that produces this peak output is simply a "sliced" version of the impulse response;

$$Input(n)_{worst-case} = \mathrm{sgn}(FILT(n)) \tag{2}$$

where

$$\mathrm{sgn}(x) = \begin{cases} 1, & x > 0 \\ -1, & x < 0 \end{cases}$$

and $FILT(n)$ is the impulse response of filter.

If we use the impulse response of the signal-transfer function (using the linear model), then we can obtain the input signal that causes the largest peak value at the comparator input.

This input signal is simply a signal that alternates between the positive and negative full-scale input of the converter depending on the sign of the impulse response. Note that the term "full-scale" here means the full-scale range defined at the input terminals, which results in less than 100% 1's density at the modulator output.

The comparator gain argument is a statistical argument that applies over the long term, and therefore it does not adequately address the situation where a single large comparator input occurs. It is intuitively apparent that repeated large values at the comparator input are more dangerous than single-cycle overloads. In the experience of the author, a good worst-case steady-state input signal is a square wave with amplitudes $\pm V_{max}$ and a frequency equal to that of the highest-Q complex pole-pair in the NTF. This tends to be roughly the same as the worst-case *H1* norm input signal derived above, only it continues forever. Note that this signal is always well outside of the normal signal bandwidth.

With no input bandlimiting, this worst-case input signal will result in instability occurring at much lower amplitude levels than for the case of normal in-band signals. Fortunately there is a simple and effective solution to this problem: pre-filter the input using a simple first or second-order lowpass filter. Since the frequency of the highest-Q pole is normally 5 to 10 times that of the useful signal bandwidth, a simple filter can effectively reduce the amplitude of out-of-band signals at the pole frequency without significantly affecting the in-band frequency response.

Exhaustive simulations with various proposed worst-case input signals have shown that the stable range of a pre-filtered delta-sigma loop is very close to the stable range of the same loop for DC signals.

If the NTF uses very high-Q poles resulting in a large response peak at the pole frequency, then more aggressive pre-filtering may be required to prevent out-of-band signals from causing instability.

If the input is highly bandlimited, simulations show that only a small "safety factor" needs to be observed. For example, if a particular modulator becomes unstable for DC values that exceed 1 Volt, it is unlikely that a bandlimited real-life input signal will cause instability at values less than 1 Volt. If, however, the input signal is not strongly bandlimited, the system may become unstable for signals considerably lower than 1 Volt. Bear in mind that these simulations need to be run for millions of modulator clock cycles; it is not at all uncommon to find a modulator that becomes unstable with a large DC input signal after many thousands of clock cycles.

4.3.6 Choices for the NTF

In this section, we will elaborate on how common choices for the noise-transfer function family impact the operation of a single-loop delta-sigma modulator.

4.3.6.1 n^{th}-order pure differentiation

Let us start with the simplest of noise-shaping functions, pure n^{th} -order differentiators

$$H(z) = (1 - z^{-1})^n \tag{3}$$

These functions meet the causality requirement that the first value of the impulse response must be 1. The magnitude response of *H(z)* for various values of *n* is shown in Figure 2. Note that at high values of *n*, we obtain more effective suppression of quantization noise at low frequencies, but have more gain at high frequencies.

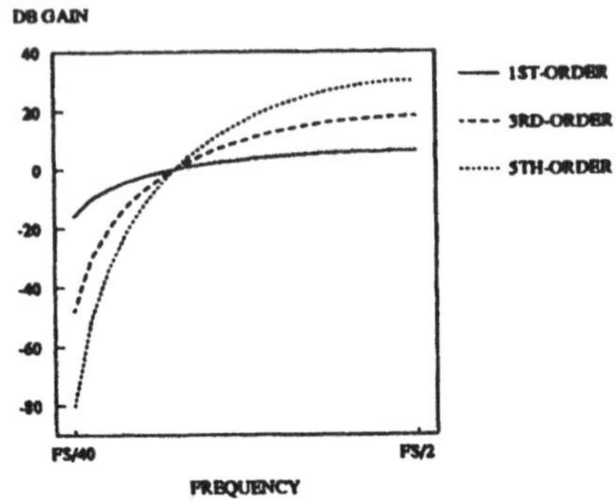

$NS(Z) = (1 - Z^{-1})^N$, N= 1,3, and 5

Figure 2 Noise-shaping function for various orders of pure differentiation.

Figure 3 shows a histogram of quantization levels used for idling only for various loop orders, assuming an infinite quantizer [6]. These plots are obtained by running a loop for thousands of cycles and counting the number of occurrences of each quantization level in the loop quantizer. Note that the number of levels used increases at a rate proportional to 2^n, which makes sense since the high-frequency noise-shaping gain also increases at this rate. From our previous discussion of comparator gain vs. assumed linear-model gain, it appears that this noise-shaping function is not a good candidate for loops with 1-bit quantizers. As a rule of thumb, any high-order loop that uses many levels for idling with an infinite quantizer will oscillate when the extra levels it needs are taken away and a 1-bit quantizer is used. The reason for this can be explained using the comparator gain argument outlined earlier. The fact that a loop uses many levels for idling implies that the random idling-waveform input to the quantizer is large. If we use only a 1-bit quantizer, the input to the comparator will be large, causing its in-circuit gain to be very small. This small gain causes oscillations to occur.

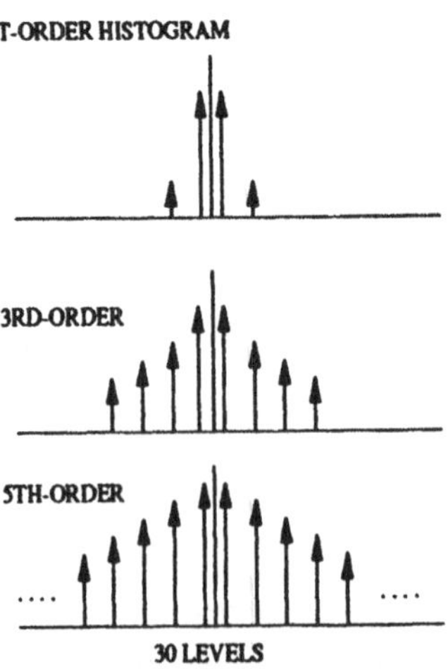

Figure 3 Quantizer histogram for various orders of pure differentiation.

4.3.6.2 Butterworth high-pass response.

The problem encountered in the previous noise-shaping function was the large high-frequency noise-shaping gain for large n, which causes the idling waveform at the comparator input to be very large, resulting in low comparator gain and hence instability. We can modify the pure differentiating response by introducing poles into $H(z)$, as shown below

$$H(z) = \frac{(z-1)^n}{D(z)} \tag{4}$$

where $H(z)$ is an nth-order polynomial with a leading coefficient of one.

The purpose of adding $D(z)$ is to flatten the high-frequency portion of $H(z)$, as shown in Figure 4. $H(z)$ is now simply an nth-order highpass function. With the correct choice of $D(z)$, a Butterworth alignment of the poles can be obtained, resulting in a maximally flat high-frequency region of $H(z)$. With a correct choice of cutoff frequency, it is possible to meet both the 3 dB gain rule as well as the causality rule as outlined earlier. Note that to meet both requirements, there are no degrees of freedom; once a Butterworth topology is chosen, there is one and only one choice of cutoff frequency that meets both conditions.

The Butterworth alignment of the NTF is often a good choice and is commonly used in commercial products. One reason for this is that the poles are relatively low-Q, and therefore the Butterworth alignment tends to be less susceptible to oscillations caused by input signals that are at the same frequency as the poles.

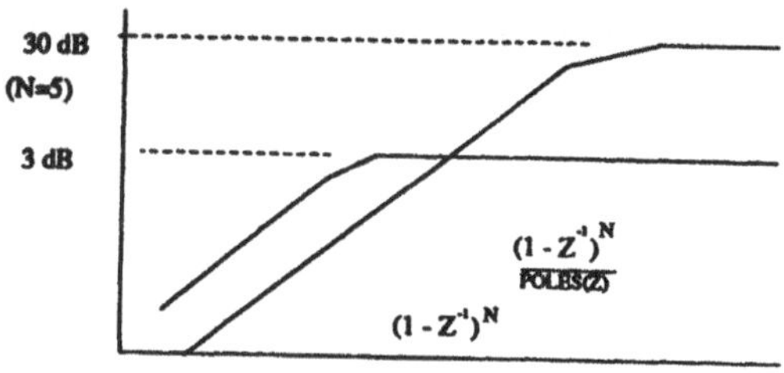

Figure 4 Modifying the NFT to reduce the high-frequency gain.

4.3.6.3 Complex zeros on the unit circle (inverse Chebychev)

The Butterworth high-pass response of the previous section can be modified to move the real stopband zeros at the $(1+j\,0)$ point out on the unit circle to produce nulls in the noise-shaping transfer function at frequencies other than DC. Compared with the Butterworth alignment, improved signal-to-noise ratio can be obtained, as the filter exhibits greater attenuation over the desired signal passband range.

For a given order, the designer must decide how many complex zero pairs will be moved to non-zero frequencies on the unit circle. For example, a fifth-order system could have three real zeros and one complex pair, or it could have one real zero and two complex pairs. Generally, using the largest number of complex pairs possible for a given order results in the best noise-shaping characteristic, although there may be circuit-related reasons for choosing otherwise.

The same iterative design methodology outlined previously can be used here as well, the only difference being that the commercial IIR design program should be instructed to design an inverse Chebychev filter instead of a Butterworth filter, and that the stopband edge should be entered as the edge of the desired signal passband. Note that the design program will result in a filter that is equiripple in the stopband. This is not the optimum solution in terms of the total integrated noise over the audio band, and adjusting the complex zero locations may yield several dB of improvement in the SNR. It is also advisable to do this optimization in conjunction with the known frequency response of the decimation filter, as the noise from the noise-shaper rises at a very fast rate at frequencies just beyond the last complex pair frequency, which is typically quite close to the decimator cutoff frequency.

4.3.7 Comparison of loop topologies

There are as many architectures for building delta-sigma modulators as there are ways to build a lowpass filter. In this section we will focus on the most common loop topologies.

All delta-sigma feedback topologies may be characterized by two transfer functions; the signal transfer function (STF) and the noise transfer function (NTF). The NTF determines to what extent the quantization noise is reduced in a given bandwidth and hence determines the overall SNR of the converter. Depending on the chosen architecture, it may or may not be possible to independently specify the STF. Normally the desired STF is flat over the band of interest.

From a circuit design standpoint, it is desirable to use integrators as the fundamental active building block. This allows switched-capacitor circuits to be designed in a parasitic-insensitive manner, with the integrator virtual ground receiving the charge stored on various capacitors in the circuit. Moreover, as we shall show shortly, loop filters built with integrators result in high-quality NTF notches.

In general it is possible to reduce all single-quantizer loops into the universal architecture shown in Figure 5. From this equivalent topology we easily see that the NTF and STF are

$$H(z) = \frac{1}{1 + L(z)} \tag{5}$$

and

$$G(z) = \frac{L_0(z)}{1 + L_1(z)} \tag{6}$$

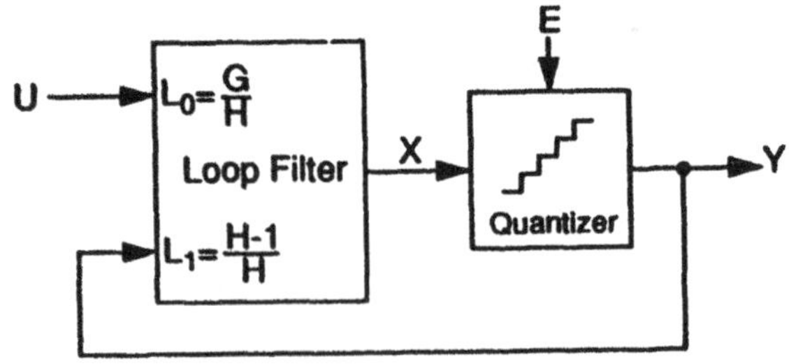

Figure 5 A universal delta-sigma structure.

Given a particular topology, the loop filters $L_0(z)$ and $L_1(z)$ can be expressed as functions of the loop parameters. Matching these expressions to (5) and (6) yields the desired parameters. For loops of high order, the algebra may become very messy. In some cases non-linear equation solvers may be used to solve for the coefficients without having to solve the set of equations. Alternatively, symbolic equation solvers may be used.

In practice, $L_1(z)$ has high gain in the band of interest and is thus responsible for attenuating the quantization noise. Specifically, the poles of $L_1(z)$ are the zeros of the NTF. Also note that both the NTF and the STF generally share the same poles (the roots of $1+L_1(z)=0$), unless pole-zero cancellation occurs by judicious choice of $L_0(z)$.

In the previous chapter the signal-dependent gain of the loop quantizer was analyzed, and from this result we can observe that (5) and (6) should be taken as approximations only. Any attempt to use high-Q poles (poles very close to the unit circle) for either the NTF or the STF will result in potential instability, or at the very least large variations in response shape as a function of the input signal.

The following circuit topologies have been successfully used in delta-sigma modulator design. This list is by no means exhaustive, and many other loops are possible, especially if one allows the use of continuous-time circuits.

1) Chain of integrators with weighted feed-forward summation.
2) Chain of integrators with feed-forward summation and local resonator feedbacks.
3) Chain of integrators with distributed feedback.
4) Chain of integrators with distributed feedback and distributed feed-forward input paths.
5) Chain of integrators with distributed feedback, distributed feed-forward input paths and local resonator feedbacks.
6) Error feedback only (normally only used for digital implementations of delta-sigma modulators).

We will now give the basic design equations for each type of loop, as well as the advantages and disadvantages for each type.

4.3.7.1 Chain of integrators with weighted feed-forward summation

For the topology shown in Figure 6, the loop filters are identical

$$L_0(z)=L_1(z)=L(z)=\frac{a_1}{z-1}+\frac{a_2}{(z-1)^2}+\frac{a_3}{(z-1)^3}+\ldots \tag{7}$$

Note that as a result of this equivalence, once the loop filter is set for optimum noise-shaping, the STF is fixed. Specifically,

$$G(z)=1-H(z) \tag{8}$$

The poles of $L_1(z)$ are restricted to DC ($1+j\,0$ in the z plane) by the nature of the filter topology. Since the poles of $L_1(z)$ translate to the zeros of the NTF, the NTF must have all its zeros at DC. Inverse Chebychev NTFs, which have stopband zeros at non-zero frequencies, can not be implemented with this architecture. Butterworth high-pass filters can be implemented and are often used in practice. A drawback of a Butterworth NTF is that it produces a STF that invariably contains peaking at high frequencies. Thus, it is wise to either add a pre-filter to the input to prevent inputs in this range from overloading the modulator or to modify the NTF family such that maximally-flat STFs result from (8).

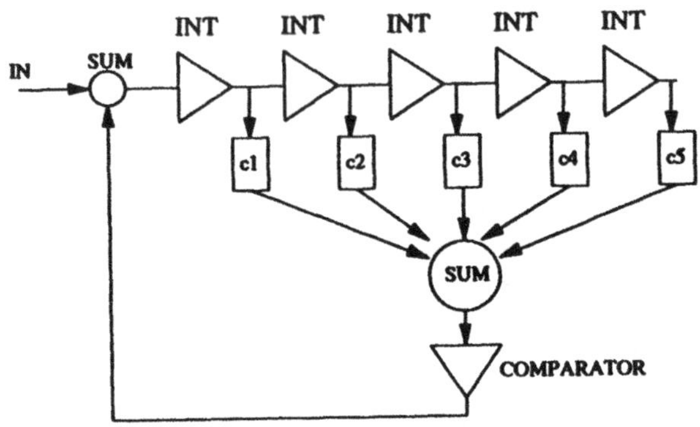

Figure 6 Chain of integrators with weighted feed-forward summation.

4.3.7.2 *Chain of integrators with feed-forward summation and local resonator feedbacks*

By adding a small negative feedback term around pairs of integrators in the loop filter as shown in Figure 7, it is possible to move the open-loop poles (which become the NTF zeros when the loop is closed) away from DC along the unit circle. This causes the frequencies of infinite loop gain (and hence infinite noise attenuation) to be shifted away from DC to finite positive frequencies. The equation for a pair of integrators with feedback is

$$R(z) = \frac{z}{z^2 + (2-\alpha)z + 1} \tag{9}$$

The poles have a radius of one and an angle θ given by;

$$\theta = \mathrm{acos}\left(1 - \frac{\alpha}{2}\right) \tag{10}$$

The equations above assume negative feedback.

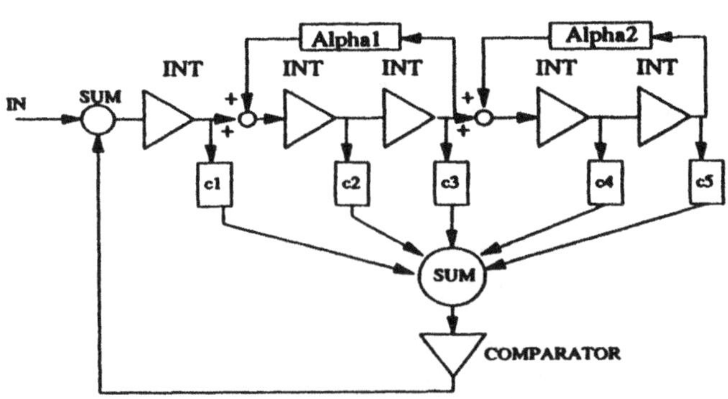

Figure 7 Chain of integrators with feed-forward summation and resonator feedbacks.

Equations (9) and (10) were derived assuming that one of the integrators has a 1-sample delay while the other does not. This can be accomplished in switched-capacitor circuits by correctly phasing the switches that form the negative feedback network.

A slightly less effective resonator can be built by using feedback around pairs of integrators, but allowing both integrators to have a z^{-1} delay term in the numerator. In this case, the poles move on a straight line from the $1, j\,0$ point away from the real axis, and therefore do not exhibit infinite gain at the resonance frequency. For small angles of θ, however, the gain at resonance is quite large, and in most cases is equally effective at removing quantization noise as the single-delay pair of integrators. The switchedcapacitor implementation for the dual-delay integrator pair is somewhat simpler, and does not require double op-amp settling, as does the single-delay integrator pair. By allowing complex NTF zeros, it is possible to implement the inverse Chebychev NTF described earlier. To solve for the coefficients, the transfer function of the loop filter $L_1(z)$ is first written, and then the NTF is derived from

$$NFT(Z) = \frac{1}{1+G(Z)} \tag{11}$$

By equating this expression with the desired NTF, it is possible to solve for the coefficients. This assumes that the designer has chosen the number of resonators in $L_1(z)$ equal to the number of unit-circle zero pairs in the NTF.

4.3.7.3 *Chain of integrators with distributed feedback*

The topology shown in Figure 8 is an "inverted" form of the previous topologies, and may in fact be graphically derived from the feedforward topology by invoking the flow-graph inversion rule. The loop filters for this topology are

$$L_0(z) = \frac{1}{(z-1)^n} \tag{12}$$

and

$$L_1(z) = \frac{a_1}{(z-1)^n} + \frac{a_2}{(z-1)^{n-1}} + \frac{a_3}{(z-1)^{n-2}} + \dots \tag{13}$$

All NTF zeros are required to be at DC. Also as before, the NTF determines the $L_1(z)$ loop filter uniquely and this in turn specifies the STF. In particular, if the NTF is of the form

$$H(z) = \frac{(z-1)^n}{D(z)} \tag{14}$$

then the STF is

$$G(z) = H(z)L_0(z) = \frac{1}{D(z)} \tag{15}$$

Therefore, if we design the NTF to be a classical high-pass filter such as a Butterworth filter, then the STF will be a low-pass Butterworth filter with the same poles. For example, a modulator designed with a Butterworth NTF with a high-pass corner frequency of 200 kHz would have a signal frequency response of a low-pass Butterworth filter with a 200 kHz cutoff. We conclude that it is possible to achieve nearly flat passband response with this topology by using a classical NTF filter shape such as Butterworth. This is in fact something of an advantage over the previous loops, as the STF does not contain significant peaking. In fact, the STF itself is a lowpass filter, which should result in improved stability when driven by large transient signals with significant out-of-band energy.

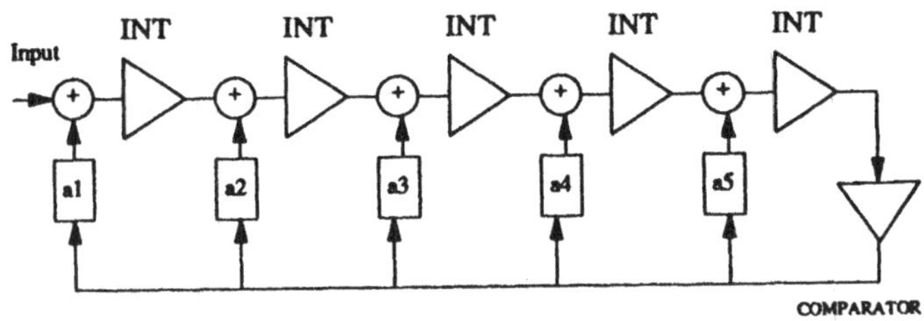

Figure 8 Chain of integrators with distributed feedback.

One drawback of this topology is that the integrator outputs contain significant amounts of the input signal as well as filtered quantization noise. This can be seen by considering what happens when a DC signal is applied to the input. Since each integrator has infinite gain at DC, the sum of the two input paths to each integrator must be zero to prevent any DC content from appearing at the integrator input. One of these paths is the 1-bit feedback signal multiplied by the feed-in coefficient. The other integrator input path is the output of the previous integrator in the loop. The previous integrator output must therefore contain a DC component to counteract the weighted 1-bit feedback.

Each integrator output thus contains a combination of filtered quantization noise and a low-frequency component equal to the input signal. In simulations it is apparent that the signal component is larger than the quantization noise component by a considerable amount. When designing switched-capacitor circuits, the output swings must be scaled by adjusting capacitor ratios to stay within the available supply range. Larger integrator swings therefore result in larger integrator feedback capacitors to keep the swings within the allowable range. Distributed-feedback circuits thus tend to be larger and more powerhungry than circuits that use the feed-forward topology.

4.3.7.4 Chain of integrators with distributed feedback, distributed feedforward inputs

The previous topology can be modified to allow a path from the input node to each integrator's summing junction, as shown in Figure 9. The addition of input feed-forward paths allows a certain degree of independence in specifying the NTF and STF. The loop filters for this topology are given by

$$L_0(z) = \frac{b_1}{(z-1)^n} + \frac{b_2}{(z-1)^{n-1}} + \frac{b3}{(z-1)^{n-3}} + \dots \quad (16)$$

and (13). Thus, if the NTF is given by (14), the STF is

$$G(z) = \frac{b_1 + b_2(z-1) + b_3(z-1)^2 + \ldots}{D(z)} \tag{17}$$

The numerator of $G(z)$ is arbitrary, except that it has an order which is one less than the denominator order. The zeros of $G(z)$ may be placed in such as way as to cancel some of the poles, thus allowing the STF to have a lower roll-off rate than could otherwise be obtained. For example, if the NTF was designed to be a fifth-order Butterworth high-pass function, the STF with no pre-filter would be constrained to be a fifth-order Butterworth low-pass. With the addition of the feed-forward coefficients, one might choose to cancel all the poles except for the real pole, resulting in an STF with a single-pole rolloff characteristic.

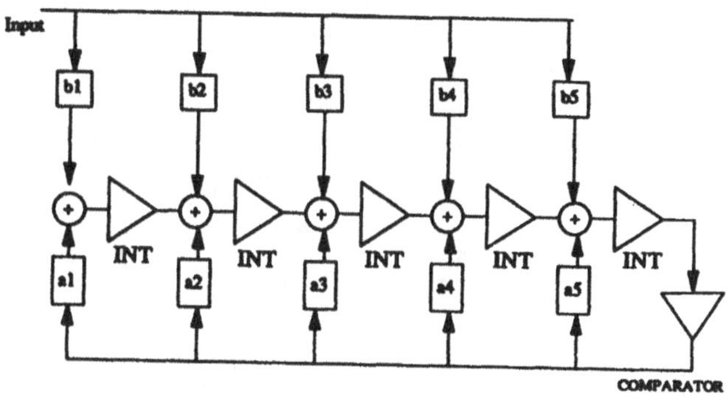

Figure 9 Chain of integrators with distributed feedback, distributed feed-forward branches.

This topology may also be modified to include local resonator feedback paths around pairs of integrators, as shown in Figure 10. Again, this allows the placement of zeros on the unit circle at finite positive frequencies, allowing the NTF magnitude response to exhibit one or more notches in its frequency response.

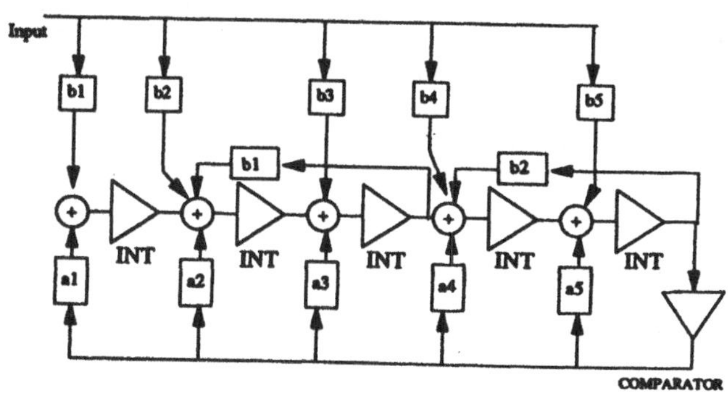

Figure 10 Chain of integrators with distributed feed-forward, distributed feedback and local resonators.

4.3.7.5 Error-feedback only

The topology of Figure 11 is often used in all-digital implementations of delta-sigma loops. It contains a single FIR (all-zero) feedback filter $F(z)$. While it may seem at first that this topology is fundamentally different than those we have discussed so far, it is in fact just another special case. The loop filters for this topology are given by

$$L_0(z) = \frac{1}{1 - F(z)} \tag{18}$$

$$L_1(z) = \frac{-F(z)}{1 - F(z)} \tag{19}$$

with the result that the NTF and STF are simply $H(z)=1-F(z)$ and $G(z)=1$.

Equation (19) indicates that a first-order modulator may be obtained with only a single delay ($F(z)=z^{-1}$). A second-order modulator may be obtained with a simple 2-tap filter with weightings of 2 and -1 for the first and second tap, respectively. It is interesting to note that the signal response is flat and delay-free regardless of the loop filter used.

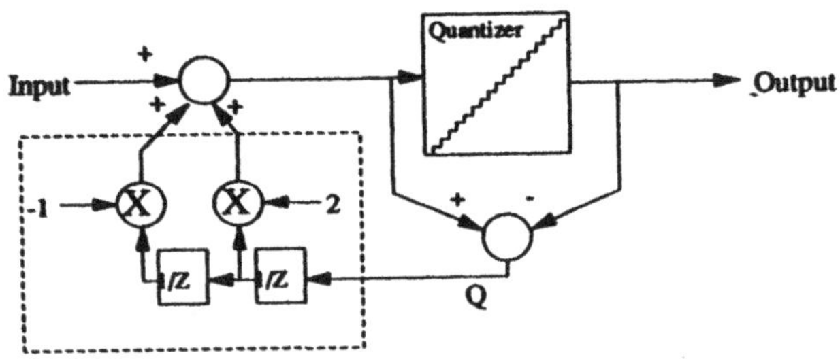

Figure 11 Error feedback only.

This topology is never used in ADCs, as analog FIR loop filters require accurate sampleand-hold circuits, which are more difficult to design than integrators. Also, any mismatches that occur in generating the error signal by subtracting quantizer input and output will seriously affect the performance of the loop. For this reason, only digital implementations of delta-sigma converters use this topology. In a digital implementation, this topology has some definite advantages. The difference between quantizer input and output can be done by simply taking the LSB's of a digital number; no hardware is required! Also, the feedback FIR filter is very easily accomplished, especially when the coefficients are simple powers of 2. An equivalent loop filter designed with conventional integrators requires several 3-input adders, as each integrator must sum its input signal as well as its own feedback signal to generate a new output sample.

4.3.8 Loop filter design example

In this section a simple fourth-order example will be given, with the following design parameters.

1) NTF filter type: fourth-order Butterworth highpass.

2) Clock rate: 1 MHz.

The "cookbook" design procedure given in section 4.3.4 was used to design this modulator. A commercial IIR filter design package (FDAS) was used to find the cutoff frequency of a fourth-order Butterworth filter whose impulse response begins with a 1 and whose passband gain is 3 dB. Since filter design packages normally design filters with unity-gain passbands, a filter with an initial impulse value of 0.707 was found by experimenting with various cutoff frequencies of a fourth-order Butterworth filter, and then the resulting polynomial was scaled by 1/.707, resulting in a 3 dB passband gain and an initial impulse value of 1. Note that this search for the correct high-pass corner is not completely random, as there exists a simple inverse relationship between the high-pass cutoff frequency and the value of the initial impulse response. If the initial impulse value for a given cutoff frequency is too low, for example, the high-pass corner frequency needs to be reduced to increase it. A successive-approximation type of search pattern usually gives the correct result after only a few iterations. The resulting NTF is

$$H(z)=\frac{(z-1)^4}{(z^2-1.5526z+0.609)(z^2-1.753z+0.817)} \tag{20}$$

Since this is a Butterworth and not a Chebychev design, all the NTF zeros are placed at the $1, j0$ point (DC). The cutoff frequency (-3 dB point) required to achieve the correct 3 dB of passband gain after impulse scaling was 50 kHz.

If we assume a single loop filter $L(z)$, we can solve for $L(z)$ in terms of the desired NTF using

$$L(z)=\frac{1-H(z)}{H(z)} \tag{21}$$

Substituting (20) into (21), we find

$$L(z)=\frac{(z^2-1.5526z+0.609)(z^2-1.753z+0.817)-(z-1)^4}{(z-1)^4} \tag{22}$$

Note that the $(z-1)^4$ term in the denominator implies a structure that uses four cascaded integrators for the loop filter so the structure shown in Figure 6 could be used.

To derive the tap weights from each integrator, we must do a partial-fraction expansion on (22). To avoid messy algebra, we can use a nonlinear equation solver (the author has used EUREKA by Borland), and equate (22) to the following expression

$$L(z)=+\frac{a_1}{z-1}+\frac{a_2}{(z-1)^2}+\frac{a_3}{(z-1)^3}+\frac{a_4}{(z-1)^4} \tag{23}$$

Alternatively, one can observe that (23) is linear in $a_{1...4}$ and thus form a system of linear equations by evaluating (23) at four points in the z-plane and equating these to the values of (22) evaluated at those same four points. The solution can then be found by invoking a numerical linear-equation solver.

As a third alternative, one could derive the symbolic equations by hand or by using a symbolic equation solver.

The equation solver finds values given in Table 1. The rather large ratio of these coefficients can be reduced by adding gain to each integrator stage, as is commonly done in switched-capacitor circuits to avoid large capacitor ratios.

Coefficient	a_1	a_2	a_3	a_4
Value	0.694025	0.230824	0.042539	0.0036151

Table 1 Coefficient values for the example modulator.

It was noted earlier that an arbitrary gain placed in front of the comparator does not change the operation of the actual circuit, since the sign of the comparator input voltage is not affected by gain scaling. This implies that $a_{1...4}$ may be scaled by an arbitrary factor with no effect on the operation of the actual circuit, and therefore it is only necessary to observe the correct ratios between the coefficients.

At this point in the design, it is necessary to use a non-linear difference equation simulator to establish the SNR, stable DC range, and other important characteristics of the loop. In addition to the loop itself, it is usually desirable to simulate the decimation filter at the same time. It is then possible to simulate the SNR of the loop in the time domain, using a running RMS sum of the filtered bit stream. Commercial software simulators such as MATLAB may be used for this purpose, with the advantage that a graphical description of the loop using simple building blocks such as integrators and comparators may be used. Otherwise, custom C programs may easily be written to simulate the loop. The loop designed here was simulated with the following results.

SNR: 85 dB over a 5 kHz bandwidth.

Stable DC range: -.7 to +.7 Volts (comparator feedback levels = 1).

4.3.9 Non-linear global stabilization techniques

In the previous section, two strategies for insuring global stability were discussed; the state-variable clamping technique and the integrator reset technique. In this section we will delve more deeply into the practical implications of these techniques.

All high-order modulators become unstable for inputs that exceed certain bounds. This may also occur after power-on since op-amp integrators with arbitrary initial states may place the modulator in an unstable region of its state-space.

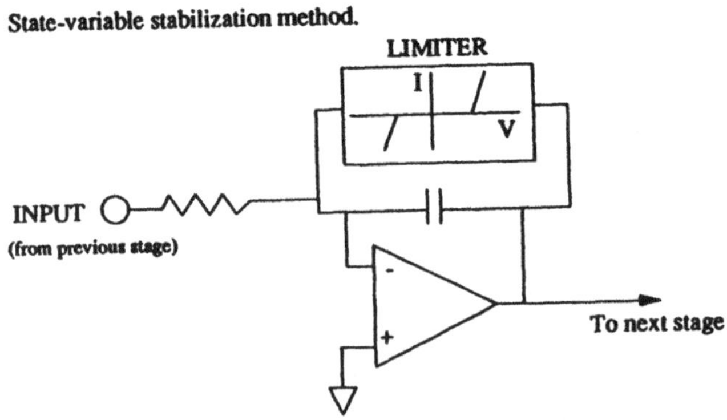

Figure 12 Non-linear limits on integrator state.

The state-variable clamping technique is shown in Figure 12. A set of nonlinear elements is connected across each integrating capacitor to prevent large values from appearing at the integrator outputs. Note that placing a voltage limiter on the op-amp output rather than around the capacitor will not work, as the inverting input of the op-amp will cease to be a virtual ground, and extra charge from the previous stage can still accumulate on the capacitor.

One question that arises using this technique is "how close do the limits have to be placed to the normal *p-p* operating range of the integrators?". The answer depends entirely on the NTF; high-*Q* poles are troublesome. For typical fourth or fifth-order circuits, experiments suggest that the non-linear elements should turn on at about 20 to 50% higher than the peak-to-peak integrator swings. In a multi-bit modulator, the effective step size is much smaller, and non-linear limits may therefore be much larger than the normal integrator swing while still insuring stability. This difference between multi-bit and 1-bit stabilization may be explained by noting that instability is triggered by overloading the quantizer, and in a multi-bit modulator the integrator swings can become large compared to the step-size of the quantizer while still staying within the saturation limits of the quantizer. In many cases the op-amp supply rails themselves may serve to limit the state variables to the stable region.

When building switched-cap modulators in a CMOS process, the integrator limiter approach is less attractive. A simple MOS device does not have a sharp enough *V-I* characteristic to make a good limiter, especially if the integrator swings are only ±1 Volt or so. This results in the requirement for complex active limiters around each op-amp. For this reason, state-variable limiting is used more often in multi-bit delta-sigma converters, where the limiting characteristic need not be so sharp.

In single-bit designs, the integrator reset approach is the preferred solution to this problem. Usually the reset circuitry simply takes the form of a MOS switch connected across the integrating capacitor of each op-amp. The duration of the reset pulse is not critical as long as the integrating capacitor is fully discharged.

There are two common ways of sensing instability. One is to look for integrator states above a certain value using a comparator, and use this to trigger the reset circuitry. The other method is to look for long strings of ones or zeros in the digital bit-stream. A normally-operating modulator usually has a maximum "run-length" anywhere from 6 to 10 bits in a row, whereas oscillations often produce run-lengths of up to 100 cycles. A threshold of 12 to 32 same-valued bits is usually sufficient to detect oscillation.

In many cases the overloading signal will persist at the input for many thousands of clock cycles. A sequence of events then occurs as follows;

a) Modulator overload is detected, and the integrators are reset with a short pulse.

b) The modulator starts running again, and as the overloading signal is still present, the output pulse density ramps towards a high value with a time constant determined by the design of the loop filter. At a certain point, oscillations begin once again. Loops that are designed with high-bandwidth signal-transfer functions (STF) will begin to oscillate very quickly after coming out of the reset state, whereas loops with a "slow" STF will take some time to achieve an output pulse density which is high enough to trigger the start of oscillations.

c) The oscillation detection circuitry goes into action, and once again resets the integrators.

This sequence of events happens repeatedly for the entire duration of the overloading input event. The frequency at which these repeated reset events occur is determined by the roll-off characteristic of the STF. If the frequency of reset events is much higher than the cutoff frequency of the decimation filter, then the noise introduced by the reset sequence will be partially filtered by the decimation filter. Simulations show that the DC value of the bit-stream during these repeated reset events is larger than the normal full-scale bit-stream density, and therefore a decimator that is designed with saturation logic to prevent the output signal from exceeding full-scale may completely hide the increase in noise that occurs during signal overload and repeated reset events. This results in "well-behaved" decimator outputs during modulator overload. On the other hand, modulators that are designed to have low-pass STF characteristics may have a frequency of reset events that is lower than the cutoff frequency of the decimation filter, causing large amounts of noise to appear at the decimator output. Such a modulator may also exhibit a reduction in average bit-stream densities during sustained overload, causing an inverted "cusp" to appear at the peaks of the decimator output signal

4.3.10 Practical measures for preventing idle tones

Idling tones are undesirable tones that appear in the baseband spectrum of the sigma-delta converter. One of the advantages of higher-order sigma-delta loops is that they tend to produce fewer idling tones in the baseband, at least when simulated on the computer. The actual circuit usually has idling tones that are far in excess of what has been predicted by simulation. These excess idling tones can usually be traced to large tones in the high-frequency region of the 1-bit signal folding down into the baseband by a variety of nonlinear practical circuit effects. Two of the most important mechanisms are listed below.

a) Interfering clock signals on the reference voltage, especially at odd multiples of $Fs/2$. The reference pin is a multiplicative input to the modulator, and hence directly translates signals from the region around $Fs/2$ to the region around DC. The reference pin is extremely sensitive to this effect. Often only a few micro-volts of interference is enough to degrade idle-tone performance.

b) Non-linearity in the input op-amp. Since there are multiple large tones at high frequencies, the difference frequency between pairs of tones may fold down into the baseband. If the requirement for low signal distortion is not very severe, the unwary designer will often use an op-amp with poor linearity. This design approach may lead to serious idle-tone problems. Most successful designs insist on op-amp linearity that is at least the equal of the desired SNR of the converter, even if this figure is overkill as far as signal distortion is concerned.

To minimize problem a), an external large reference bypass capacitor is often used to eliminate noise on the reference pin. Problem b) can only be addressed by careful op-amp design.

These high-frequency idling patterns may be largely eliminated by adding a "dither" signal, often produced by a random noise generator, to the quantizer input. Discrete-time simulation shows that as the amplitude of the dither signal is increased, the energy of the idle tone is spread out in frequency and eventually becomes indistinguishable from the surrounding shaped quantization noise. Unfortunately, the amount of the dither signal required to effectively eliminate the idle tone is often as large as the normal quantizer input signal itself. If a dither signal of this magnitude is added to the comparator input in a single-bit system, the available signal range before instability occurs is substantially reduced. When using a multi-bit quantizer, though, the dither signal need only be slightly larger than 1 LSB of the quantizer to insure that the quantization error is uncorrelated, and hence the available signal range may only be reduced by a small amount. In this case, dither is a useful and recommended addition to the circuit design. In the 1-bit quantizer case, the authors experience is that the loss in signal input range precludes the use of dither, although this result may vary with the exact loop filter design.

4.3.11 Practical implementation of a stereo delta-sigma 18-bit ADC IC

Detailed circuit design techniques for single-bit high-order loops is not markedly different than those used for 2nd-order loops, covered in detail in chapter 4.1 of this book. The only additional hardware is normally the integrator reset or non-linear state-variable limit circuit to insure global stability. Since most high-order loops are used in designs where high SNRs are desired, individual building blocks such as op-amps must normally achieve very high performance in areas such as noise, distortion, slew-rate and bandwidth.

In this section we will describe the design of a commercially available dual, 20 kHz bandwidth, 18-bit delta-sigma ADC with an oversampling ratio of 64 and a noise floor that is 104 dB below full scale. The converter is implemented on two chips: a 10 Volt 3μm CMOS modulator chip, and a 5 Volt CMOS decimator chip. Only the modulator design will be discussed here. The overall block diagram of the ADC is shown in Figure 12.

4.3.12 Noise-shaping modulator IC

The techniques described earlier were used to design a fifth-order noise-shaping filter with an inverse Chebychev characteristic for the NTF. The loop filter employs two resonator sections and one integrator, resulting in the NTF shown in Figure 13.

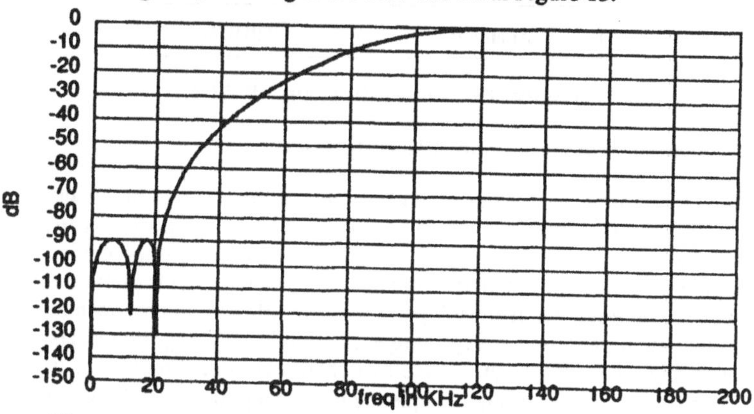

Figure 13 Noise-shaping transfer function of the ADC.

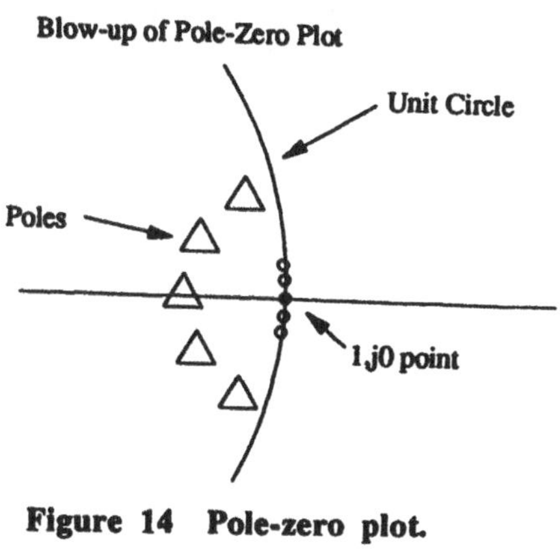

Figure 14 Pole-zero plot.

The plot in Figure 13 does not include the normalization of the transfer function to make the first value of the impulse response equal to 1, which would increase the magnitude response by 3 dB at all frequencies. The corresponding pole-zero diagram is shown in Figure 14.

A block diagram of the circuit is shown in Figure 15. The modulator chip contains a dual output 3.0 V reference, clocking and interface circuitry and two fifth-order, one-bit, fullydifferential switched-capacitor delta-sigma modulators. Each modulator contains five amplifiers, a comparator, switches, capacitors and associated non-linear stabilization circuitry.

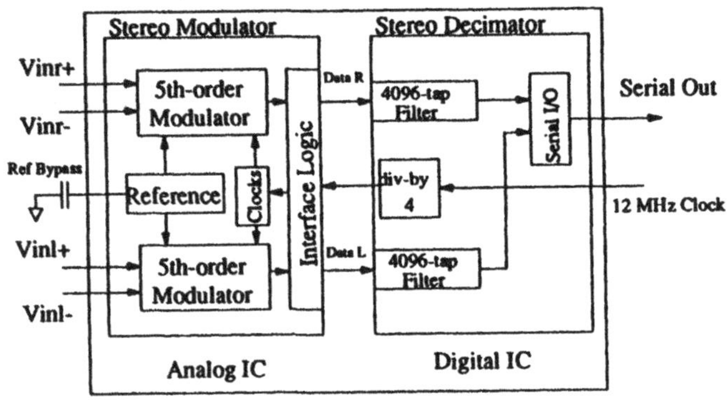

Figure 15 Block diagram of ADC.

4.3.13 Switched-capacitor loop filter design

The architecture chosen for this design corresponds to the inverted architecture shown in Figure 11. This design was chosen to avoid having to use an extra active summing amplifier at the comparator input (a requirement of the non-inverted topologies of Figures 6 and 7). Figure 16 shows the detailed switching arrangement used for both the signal input paths as well as the reference feedback path. A fully-differential architecture was used to increase signal swing and power supply rejection while reducing the effects of substrate coupling and digital feedthrough. Since switched-capacitor circuits are true sampling devices, even small amounts of high-frequency noise coupling through the supplies or any other path will fold down and cause noise and/or distortion in the audio band. The switches are sized to meet the requirement of 18-bit settling in one-half the clock period, and the *R-C* bandwidths of the switch/capacitor combination may be 30 MHz or more. There is therefore little intrinsic filtering of high-frequency noise that may cause foldovers.

On phase 2 of the non-overlapping clock, the differential input is sampled onto capacitors C_1 and C_2. To avoid distortion due to signal-dependant charge injection, the grounded phase 2 switch is opened slightly before the input phase 2 switch. This prevents the signaldependant channel charge of the input switch from appearing on the input capacitor. On phase 1, the input capacitors are connected to the op-amp input terminals, and the stored charge is transferred. This charge is summed with the reference feedback charge, the polarity of which is determined by the comparator decision.

The reference switching scheme is also shown in Figure 16. The signals "X" and "$\overline{X}$" are the true and complement outputs of the comparator. Instead of using both a positive and a negative reference, the feedback sense of the reference is inverted by swapping phase 1 and phase 2 clocks on the reference input sampling switches. In one case, the capacitor is shorted in phase 2 and in phase one it is connected between V_{ref} and the integrator input, delivering a positive packet of charge to the integrator. In the other case, the capacitor is connected between V_{ref} and ground on phase 2, and between ground and the integrator input on phase 1, delivering a negative charge packet to the integrator. Depending on the decision of the comparator, these charge packets are steered on phase 1 into either one side or the other side of the differential integrator. This scheme has the important advantage that the current drawn from the reference does not depend on the signal. Were this not true, some non-linear version of the signal would appear on the reference and cause distortion, as the reference is a multiplicative input to the system. An external large capacitor is used to reduce noise on the reference voltage.

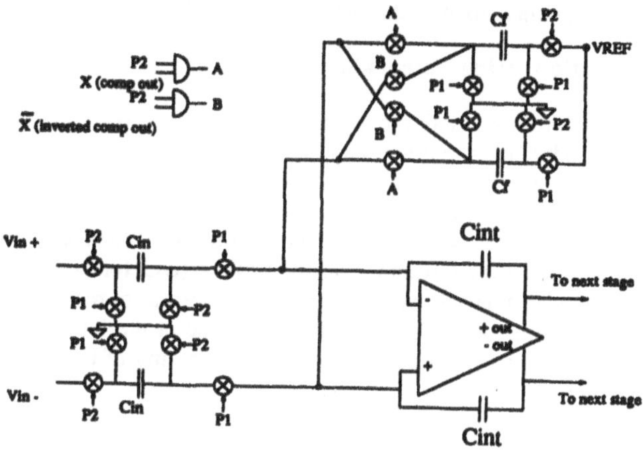

Figure 16 Details of switched-cap implementation.

The differential architecture is fully-pipelined, with each integrator having an effective delay of one sample. This avoids the double-settling problem, where two op-amps are connected in series during the settling phase of the circuit. This would require much faster settling times from each op-amp.

The notches in the NTF are implemented by feeding back around pairs of integrators, as shown in Figure 10. As described previously, this feedback moves the open-loop poles of the feedback filter along the unit circle to frequencies in the audio band. It is interesting to note that in the open-loop configuration, these sections would in theory oscillate with constant amplitude at their resonant frequency. Fortunately, the application of negative feedback to the system turns these open-loop poles into zeros on the unit circle, where they serve to reduce in-band quantization noise.

4.3.14 Circuit noise considerations

Careful attention was paid to all noise sources in the circuit. The theoretical quantization noise of the loop (as simulated using a difference-equation discrete-time simulator) is at the -112 dB level. One noise source that is frequently overlooked is the digital truncation noise that occurs at the decimator output when the accumulator is truncated to 18 bits.

This noise is by definition at the 18-bit level (110 dB). This truncation is properly dithered by the analog circuit noise so as to remove any distortion effects that might otherwise occur.

The capacitors were sized from *kT/C* noise considerations, with the first-stage capacitors being the dominant noise source and hence the largest capacitors. These capacitors must be quite large to give 18-bit noise performance, and this results in large op-amps with high bias currents to meet the 18-bit settling requirement. The total noise of the circuit is the power sum of the theoretical loop quantization noise, the *kT/C* noise, the digital truncation noise and the op-amp thermal noise.

The theoretical performance of the loop with no analog noise sources was -112 dB. It is quite typical to over-design the loop in this way, because it is surprising how fast all of the noise sources that result from practical circuit effects can add up. Since many high-resolution delta-sigma converter IC's are dominated by capacitor area, it makes sense to insure that *kT/C* noise dominates all the rest of the noise sources by a large margin. It is often cheaper in terms of area to increase the order of the loop filter than it is to decrease the *kT/C* noise by 3 dB, which doubles the capacitor area.

4.3.15 Stabilization using integrator reset

Higher order modulators are stable over a limited range of one's density at the output. Instability in this modulator is sensed digitally, by counting the number of consecutive ones or zeros in the modulator bit stream. A sufficiently long string of either ones or zeros indicates modulator instability and triggers circuitry which resets the state in the integrators to put the modulator into a stable operating condition. When a signal continuously overloads the modulator for an extended period of time (DC, for example), the modulator goes into a mode where it detects overload, resets, comes out of reset, and after some time goes into overload again. As long as the average value of the 1-bit stream remains high enough, the digital clipper in the decimator will prevent any of this noisy behavior from appearing at the output.

4.3.16 Op-amp design

The performance of the input amplifier in higher order loops determines overall converter performance. Therefore, to achieve the stated performance, the following features were designed into our front-end amplifier

- Fully Differential, including a common mode loop.
- Settling to 18-bit accuracy (23 μV) in one-half clock cycle (120 nsec).
- Gain linearity of 0.001%, or 10μV peak non-linearity over the entire 3V peak output range.
- Input referred thermal noise contribution below the modulator noise floor of -105 dB full scale.

The question of whether or not the op-amp must settle to the full resolution of the converter is still unsettled. In this design, a conservative approach was taken to avoid risk, and full settling was a design goal of the op-amp (at least according to SPICE; measurements to confirm this settling performance are not easily done!). Arguments can be made that less than full settling can be tolerated if the "non-linear" portion of the settling waveform is less than the resolution of the converter. This non-linear settling behavior is a result of op-amp slewing at the beginning of the settling period, the effects of which are still present at the end of the settling period.

A simplified circuit diagram of the amplifier is shown in Figure 17. A single-stage foldedcascode architecture was chosen for both the differential and common mode paths. This topology has an inherently high bandwidth which enables the amplifier to meet the settling time specifications. Compensation for both the common mode and differential mode paths is provided using the switched-capacitor circuit capacitances surrounding the amplifier.

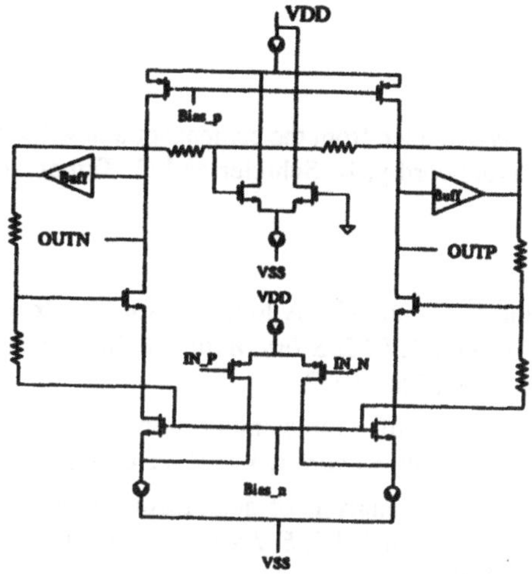

Figure 17 Folded-cascode op-amp design.

The common mode output voltage is sensed using PMOS source followers on each output driving a resistive and capacitive summing network at the output. The summed voltage is compared to ground through an NMOS differential pair, and the difference current fed back to the amplifier to control the common mode current.

The amplifier linearity is limited by impact ionization current in the NMOS output devices. This impact ionization current causes an effective non-linear resistive load on the amplifier, and this will cause distortion when reflected through the finite gain of the amplifier to the input terminals. This problem is corrected by placing an NMOS transistor between the output and the NMOS current source and driving its gate with half the difference between the output and the minus analog supply. This has the effect of maintaining a low drain-source voltage across all NMOS signal transistors. Since impact ionization is a strong function of the drain-to-source voltage, this circuit dramatically reduces impact ionization. Without the correction circuitry, the third harmonic distortion products were only 88 dB below the fundamental signal. The correction circuitry improves the distortion performance by almost 20 dB.

4.3.17 Results and comments

The actual silicon results were within about 2 dB of the original noise analysis (103 dB achieved vs. 105 dB expected). Most of this difference can be traced to digital-analog interference; when the decimator and modulator are physically separated, performance improves. One must recall that the *RC* bandwidth of the sampling switches is large, and therefore it is very difficult to achieve very high resolution in a noisy digital environment.

Distortion was measured at 1 kHz and found to be approximately 98 dB below the fundamental. Frequency response was flat within a few hundredths of a dB.

Considerable time was spent on PC-board layout and decoupling techniques to achieve this performance. In particular, the reference pin is quite sensitive to pickup. Any signal on the reference has the potential to fold down shaped noise, as the reference pin is a multiplicative input to the system. Even the small amount of signal picked up on the bond wire to the internal reference pad was found to be significant. One must be prepared to spend considerable amount of lab time tracking down such subtle effects if the highest performance levels are to be reached.

Acknowledgements

Some of this material is extracted from the contents of a new IEEE Press book, *Delta-Sigma Converters*, eds. S. Norsworthy, R. Schreier and G. Temes, to be published in 1994 or 1995.

References

[1] W. L. Lee and C. G. Sodini, "A topology for higher order interpolative coders", *Proceedings of the 1987 IEEE International Symposium on Circuits and Systems*, Vol. 4, pp. 459-462, May 1987.

[2] J. C. Candy, "A use of double integration in sigma-delta modulation", *IEEE Transactions on Communications*, Vol. 33, No. 3, pp. 249-258, March 1985.

[3] R. M. Gray, "Spectral analysis of quantization noise in a single-loop sigma-delta modulator with dc input", *IEEE Trans.on Communications*, Vol. 37, No. 6, pp. 588-599, June 1989.

[4] U. S. Patent # 4,509,03, "Enhanced delta modulation encoder", Robert W. Harris. Filed Dec. 1, 1982. Assigned to Gould Inc.

[5] G. C. Temes, R. H. Walden and T. Catalepe, "Architectures for high-order multibit sigma-delta modulators", *Proceedings of the 1990 IEEE International Symposium on Circuits and Systems*, Vol. 2, pp. 895 - 898.

[6] M. O. J. Hawksford, "Chaos, oversampling, and noise-shaping in digital-to-analog conversion", *Journal of the Audio Eng. Society*, Vol. 37, No. 12, Dec. 1989.

[7] R. W. Adams, "Companded predictive delta modulation: a low-cost conversion technique for digital recording", *Journal of the Audio Engineering Society*, Vol. 32, pp. 106-113, March 1984.

[8] R. W. Adams, "Design and Implementation of an audio 18-bit analog-to-digital converter using oversampling techniques", *Journal of the Audio Engineering Society*, Vol. 34, pp. 153-166, March 1986.

[9] R. W. Adams, P. F. Ferguson, A. Ganesan, S. Vincelette, A. Volpe and R. Libert, "Theory and practical implementation of a fifth-order sigma-delta A/D converter", *Journal of the Audio Engineering Society*, Vol. 39, pp. 515-528, July 1991.

[10] D. Welland, "A stereo 16-bit delta sigma A/D converter for digital audio", *85th Convention of the Audio Engineering Society*, November 1988.

[11] P. F. Ferguson Jr., A. Ganesan, and R. W. Adams, "One bit higher order sigma-delta A/D converters", *Proceedings of the 1990 IEEE International Symposium on Circuits and Systems*, pp. 890-893, May 1990.

Tutorial 5

DSP in mobile telephony

Organised by: Yrjö Neuvo, *Nokia Headquarters*

Chapter 5.1

Mobile Telephony

Timo Ali-Vehmas and Yrjö Neuvo

5.1.1 Introduction

During the last one and half decades the world has become familiar with not only a new technology but also a phenomenon, Mobile Telephony. Mobile telephony is most often understood as cellular telephony but it actually includes a much wider range of technologies and services. During these years mobile telephony has developed from clumsy and heavy manually operated vehicle mounted radio telephones, to comprehensive systems that can be used by anybody almost anywhere and small but sophisticated devices that can connect people to people. Not even the sky is the limit for the mobile technology to develop, as mobile satellite systems will come to wide service during next 15 years.

Mobile telephony has its early roots equally and independently in North America, North Europe and in Japan. In these regions different systems were developed to fulfil local needs and under local regulative atmospheres. Today, enhanced versions of these early cellular networks form the backbone of the mobile telephony in the world.

North American AMPS (Advanced Mobile Phone System) is in largest use, not only in the USA and Canada, but also in South America and in several countries in the Pacific rim. Most subscribers in Europe are in TACS (Total Access Communications System) networks. TACS and its derivative E-TACS have their roots in AMPS. Largest TACS networks can be found in the United Kingdom and in Italy.

Another national network that has a strong position on its home market is the NTT network in Japan. The japanese market has been quite different for many years because of the strong monopoly of the NTT. Only recently Japan has opened its markets and now there is also the so called J-TACS network operational in Japan.

The only first generation system providing international roaming is the Nordic Mobile Telephone (NMT) which has been developed in the Nordic countries and later has been adopted by several other countries (NMT has 450 MHz and 900 MHz versions). Most of these NMT operators have implemented automatic roaming between the networks.

In addition to these major systems several other systems have been developed; Netz C in Germany, Portugal and South Africa, Radiocom 2000 in France and so on.

Particularly in Europe there was a real patchwork of analog networks during the 1980's. Since it was obvious from the beginning that none of the first generation systems could have been accepted to become a pan European system, the CEPT initiated a new development that is now known as GSM, Global System for Mobile Communication. This process required about a decade from the start to commercial operation 1991 and several thousands of man-years of work. During that long period several test systems were developed to confirm the selected technologies. Also the standardisation process was transferred from CEPT to ETSI to widen the technological and financial basis of the process.

The large number of organizations involved in the development work was not the only challenge for GSM to overcome. The technological challenges were extremely demanding: all digital transmission, vocoder based speech source coding and requirements for signalling that materialised in the form of massive software and ASIC development projects in several companies all around the world. Later similar work started in the USA and Japan respectively; Telecommunications Industry Association (TIA) in the USA and RCR in Japan.

The starting point for digital mobile telephony in each of these regions was quite different and therefore it is quite obvious that different technology and market approaches were selected. In the USA there was no need for a brand new system, on the contrary, it was highly desirable to create a system that could coexist with the mighty AMPS network. The D-AMPS system, Digital and Dual mode AMPS was developed and documented as Interim Standard IS-54, which already has several revisions. However, it must be noted, that no other mobile telephone system standard except GSM is but an air interface specification. Only in the GSM system also the network interfaces have open standards for anybody to provide network elements.

The highly competing community of the telecommunications industry in the US could not be satisfied with a single standard. Two other major developments emerged. Narrow band AMPS, N-AMPS, introduced by Motorola, is a straight forward solution to provide a three fold capacity increase without any new principal technology. The other new candidate is technologically much more interesting since it introduced a totally new multiple access method for cellular, the Code Division Multiple Access method (CDMA). Earlier it had been used only in military or other specialised application.

In Japan the new digital standard has been developed primarily based on research activities by NTT. This standard, called RCR STD 27, has also its roots in national grounds. Many of the principal elements of the system are similar to either D-AMPS or to GSM but are modified and adjusted according to geographical, technological and political traditions and conditions of the country. The PDC (Personal Digital Communincations) system is also based on a TDMA air interface.

Mobile telephony is actually much more than just cellular. The other technology offered for the general public is cordless telephony. Cordless telephones originally were just extensions to fixed PSTN, having radio access instead of physical wire. Since the early days of analogue cordless telephones (CT0) the development has gone through CT1 and CT2 to more advanced digital technologies such as DECT in Europe and Personal Handy Phone (PHP) in Japan, which actually in many respects closes the gap to cellular services by advanced features such as authentication, ciphering and also data services in the future.

The other section of Mobile Telephony, which in many ways is comparable to previous systems as technology but has very different users' requirements is Private Mobile Radio, PMR. Without going into detail it shall be recognised that from a technological (like DSP) point of view, there is much in common in the new digital truncked PMR systems and the current digital cellular systems. The most recent is the digital PMR specification in Europe, TETRA (Trans-European Truncked Radio).

The last and also the least developed sector of mobile telephony is mobile satellite communication. Currently there are some special systems for special needs. None of the current satellite telephone systems can be considered as personal telephony. In the future this may well be different if some of the currently planned personal satellite communications systems, such as Inmarsat P21, Iridium, Globalstar etc. are launched. However, the limited number of subscribers in any of the systems will limit the commercial potential of any such system and therefore mobile satellite telephony will be rather a complementary system to mainstream terrestrial mobile telephone systems.

In the future the integration of all these different mobile systems is evident. However, the air interfaces may in many cases be different for different systems because of different spectrum and service requirements. In some other cases these different systems will be more overlapping each other and for the customers of mobile telephony there will be many options to select. This development will be supported by the developing fixed networks and particularly concepts such as Intelligent Networking. Also it must be recognised that mobile communications networks often are based on state-of-the-art technologies and therefore provide at least the same amount of intelligence as any of their fixed counterparts. The integration of both mobile and fixed networks will finally form the future environment of telecommunication.

Regardless of any research and specification activities, the development of mobile telephony would have been quite different than experienced if thousands of enthusiastic engineers in hundreds of companies had not been able to materialise all their new ideas into physical designs of products. The development of size and performance of mobile communications equipment has been so rapid that it can only be compared to the fast decrease of the prices of this same equipment particularly on the markets which are open for competition. The customers, e.g. end-users of these systems do not consider standards and technologies to be important but features and services the systems can provide and total cost of the ownership.

When analysing the reasons for this extraordinary progress it can be seen quite easily that two technological key elements in this process have been Digital Signal Processing and Advanced embedded software and processing. Digital processing of physical signals as understood from a theoretical point of view is one step and its practical implementation as hardware and software is the other.

5.1.2 Usage

Today there are over ten million users in public cellular networks and another similar amount of users of cordless telephones and of other mobile telephone systems. In the early days most of the users were using their car phones but today more than 90% of the new subscribers select a hand portable or pocket phone when they buy a new mobile telephone.

Also the profile of the users today is quite different from what it was only a few years ago. Still in many markets and societies the mobile telephone is considered as a tool for a business man or a status symbol for wealthy people. In some areas where the penetration is about to exceed 10% this is quite different. For instance in Scandinavia quite ordinary people are using mobile telephones to widen the opportunities in their daily life. In Scandinavia the total cost of ownership is about the same or sometimes even below that of the fixed telephone and therefore the mobile telephones are used today by all groups of the society. In the Nordic countries mobile telephony is already becoming personal telephony in the office, at home and on the road.

Tailored tariffs for various groups of people have been the key promoter to this wide acceptance of mobile telephones. There are different tariff structures for low use subscribers such as elderly people or people having their vacation, that typically want to be reachable but do not call during office hours in the town centres. These tariffs are so low that any fixed service cannot compete. Also mobile centrex services are available covering the office or the town or the whole country depending on the needs of the subscribers. It has been recognised that once a person gets used to his/her personal communication tool, e.g. cellular telephone, he/she wants to use it anywhere regardless whether fixed telephone services are available in that location or not.

These type of behavioural patterns become even more dominating when very personalised additional services are available only via advanced mobile systems. Such services include messaging and other data services as well as advanced supplementary services such as call diverting and voice mail.

While many users today are quite happy if they have wide national coverage available, there is a larger and larger segment of subscribers that truly appreciate international roaming, particularly in Europe, where there are only national networks and at the same time European wide roaming of people is increasing because of, for instance, the development of the European Union.

The next very important step in this process is the extension of user's service profile from speech to data and further to modest multimedia during the next few years. This will be more visible for ordinary people than any high technology innovations deep in the networks. Multimedia will also require more than ever multitechnologies in the mobile terminals, both in ordinary telephones and in personal and portable digital assistants.

5.1.3 Evolutions and revolutions

When forecasting the future it is often useful first to backcast. The mobile telephony is right now living its first revolution, the digitalization. This revolution has taken different paths in Europe, Japan and in the USA and for good reasons. By analysing these reasons and their consequences we may start to draft the future.

It can be forecast that the next revolution requires quite significant forces, i.e. customers' needs and innovations and technical solutions to materialise. Also there must be very wide political will to influence positively for the revolution. On the other hand there will most likely be very strong evolution of the current systems and networks in any case before and parallel to any major revolution.

However, there will be developments which are part of the evolution, but which we can call revolutionary break throughs. Development of services will migrate towards multimedia and the best guess for the next revolution is actually evolutional integration of several existing digital networks through interworking. This will not slow down the development of new services and introduction of new technologies. On the contrary, it will make the transition much easier and less risky.

Whatever new name or names will be used for this new generation of mobile telephony will be in the technologies we already know by names even today. Whether this will lead to global mobile telephony the answer is again the word: interworking. All global networks will interwork. The only question is: how. In the current European thinking the role of smart cards is very essential. One possibility to provide truly global roaming is to rely on the SIM roaming as will be implemented in GSM and PCN.

5.1.4 Basic concepts

When considering the role of digital signal processing in mobile telephony, the basic concepts are generic. None of the first generation analog systems have any feature that require any DSP at all. Even the modems and voice processors typically have been implemented using analog processing. Respectively, all second generation digital mobile telephone systems have DSP used in well defined sections of the system.

1. Source signal processing (audio [and video] source coding and decoding, pre-, and post-processing of those signals).
2. Modulation and demodulation and detection of the digital transmission.

3. Measuring and optimising the employment of radio resources (power control, frequency control and timing, also adaptive antennas).

The first section is what all the students in universities typically think when talking about DSP and its applications. This is an important section also because the major part of the processing complexity lies there. The source coding algorithms used in mobile telephones fall into two categories: waveform coders, such as Adaptive PCM (ADPCM is used typically in cordless telephones) and vocoders, such as CELP and VSELP or RPE-LPC (used in cellular phones). In vocoders Linear Prediction is the core of the processing. When coding video signal, then Discrete Cosine Transform (DCT) is an essential trick.

On top of the coding itself there are few a many pre- and post-filtering algorithms used to enhance the (perceived) quality of the signal. All kinds of linear and also unlinear filter techniques are used.

Independently from specifications of mobile telephone systems, several kinds of accessories and additional functions have been introduced as well. One of the most promising such features is voice recognition, which independently of the methods and algorithms used is heavily relying on DSP.

The second section actually is the area that has given the characteristical name for all second generation systems, i.e. digital systems. Modulation, demodulation and detection are actually communications theory terms. In this area, however, most of the mathematical procedures are executed in the form of digital signal processing. Such procedures include several types of convolutions, correlations and filtrations. Some of the algorithms even carry names of famous experts of this field, such as Andrew Viterbi.

In radio communications many of the digital processing algorithms are adaptive. The need for adaptation is obvious due to the fast changing radio channel. Very robust algorithms are needed since they must work regardless of the wild behaviour of input signals.

The third section of digital signal processing used in mobile telephony can be described also using terms of control theory rather than communications theory. Automatic control functions for frequency source stability as well as for signal strength indication and gain setting in reception and power adjustment in transmission are typically variants of first and second order control loops, executed by the digital signal processing units of the mobile telephones.

Status of these control circuits, output of some special measurement processses and also status of detection algorithms of the receiver are used as basic information to initiate and complete the essential function of a mobile telephone network, the handover.

It is important to recognize that digital signal processing is used widely both in the terminals and in the network side. The aim of this chapter is to give an understanding on how DSP enters in mobile telephony by concentrating largely on the speech processing used in mobile telephony and on circuit structures for DSP. It is quite apparent that mobile terminals require a very high processing power accompanied by low power consumption both in standby and during operation.

Chapter 5.2

Digital Signal Processing in Mobile Telephony

Nick Courtis

Abstract

The digital signal processing elements of a typical digital cellular mobile phone are presented. We discuss in broad terms how digits are extracted from the speech waveform and the life of these digits until they are reconstituted as speech at a mobile phone receiver.

5.2.1 Introduction

Digital mobile phones are at long last appearing in the market, competitively positioned against their analog counterparts.

Nokia Mobile Phones recently announced the launch of a family of phones for the three main global digital cellular systems, i.e. GSM (Pan European), DAMPS or Digital AMPS (North America) and JDC (Japanese Digital System). This tutorial chapter and expanded conference presentation are addressed to the engineer who wishes to gain an appreciation of the complex digital signal processing activities that occur at the heart of such a mobile phone, and some architectural design considerations that apply in digital cellular telephony.

	GSM	JDC	DAMPS
Speech Coding (kbit/s) (Full Rate Channel)	13.0	6.7	7.95
Speech+Channel Coding (kbit/s)	22.8	11.2	13.0
No of Slots/Carrier (Full Rate)	8	3	3
Modulation Scheme	GMSK	$\pi/4$ Shifted QPSK	$\pi/4$ Shifted QPSK
Modulation Rate (kSymbols/s)	270.8 (1bit per symbol)	21 (2bits/Symbol)	24.3 (2bits/symbol)
Carrier Spacing (kHz)	200	25	30

Table 1 The main features and differences between the systems.

The principal reasons for going toward an all digital cellular system can be seen from the last row of Table 1. Typically in an analog network, one speech channel occupies on both transmit and receive side the full 25 or 30 kHz of bandwidth that separates the carriers. In

JDC and DAMPS, three speech channels are supported on one such carrier for the full rate systems. When half rate speech coders are introduced, there will be 6 slots per channel and hence 6 speech channels per carrier. In GSM, there are 8 channels per carrier, which roughly equates to 25 kHz per channel. Not as spectrally efficient as JDC or DAMPS, and only a slight improvement over the AMPS network (North American Analog), but the ability to withstand higher carrier signal to interferer ratios means that cell sizes can be smaller and the network as a whole can support many more users. Other benefits of a digital system include security, better speech quality, and enhanced services - such as the ability to connect to ISDN. Introduction of half rate into GSM will again mean the doubling of capacity.

The benefits of CDMA have not been discussed here, and that will be left to other proponents, but suffice to say that it is being heralded as offering yet more efficient spectral usage.

Obviously, the prime consideration in choosing the data rate on the air interface and the modulation scheme is to take account of the radio channel. We shall not dwell much on this topic, and leave that for the previous tutorial chapter. Some introduction is necessary to explain why the standards are how they are, what the mobile phone must do as defined by the standards, and what strategies are best employed by the receiver to recover the data sent from the base station. In the main, the mobile phone designer has little scope for creativity in the choice of algorithms used in the transmit section; these are largely laid down of course in the system specification, and has to reserve his flair and expertise for the receiver.

5.2.2 The adverse radio channel

The mobile radio channel presents a very severe and hostile environment to our data bits as they journey from transmitter to receiver.

Reflections from obstacles can be categorized into a variety of different effects. Multiple signals arrive at the antenna at different times which causes the data to become smeared or convolved with the varying impulse response of the changing radio channel. In a typical urban environment, the delay between the primary signal and subsequent reflections can be up to 5 μs, and in a rural hilly background, possibly up to 40 μs. The higher the data rate, the more chance there is of symbols running into and becoming mixed up with neighbouring symbols, causing *Inter Symbol Interference* (ISI). This is one motivation for making the data rate on the air interface as low as possible to avoid the requirement for excessively complicated receivers, particularly important for the mobile phone designer and cost of the phones. From Table 1, it can be seen that GSM has an appreciably higher data rate (1 symbol = 5 μs) than the other two systems, and therefore data is more prone to intersymbol interference; consideration of this will be given later.

If the delay between one signal arrival and the next is of the order of a wavelength or so, fading results when the carrier signal phases add constructively or destructively at the receiver, and this is known as Rayleigh fading. The depth and spacing of these fades depend upon the carrier frequency, the spacing at 900 MHz being just a few centimetres and typically occur at intervals of tenths of a second. The actual variations are governed by a statistical distribution, the Rayleigh probability density function. A vehicle moving relative to the basestation produces frequency (Doppler) shifts to the radio signal. This contributes random frequency modulations, which in turn effects the characteristics of the Rayleigh fading.

Slow fading, or shadowing, sometimes occurs when the signal is absorbed or partially blocked by obstacles such as buildings or hills. From the perspective of a moving vehicle, such obstacles can produce a shadow between transmitter and receiver that appears and disappears and which may be from many metres to several hundreds of metres across.

Then there is plain old path loss, which in free space is proportional to $1/d^2$, but in the real world of mobile radio with terrain and atmospheric absorption tends to be more severe and losses are inversely proportional to higher powers of *d*.

5.2.3 The transmitter

A highly simplified block diagram of the transmitter part of a digital mobile phone is shown in Figure 1. In the following sections, each component is explained in more detail.

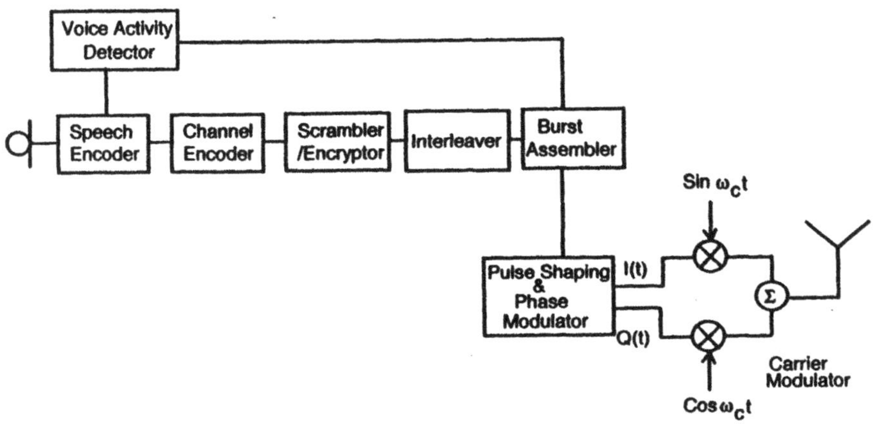

Figure 1 Mobile phone transmitter.

5.2.3.1 Speech coding

Fundamental to a digital system is the ability to express the speech waveform, i.e. the information we wish to convey, as a series of digits. The fewer the number of bits per second that can be used, the more channels that can be allocated per unit bandwidth of available spectrum. This consequently allows the data rate on the air interface to be kept reasonably low, avoiding excessive intersymbol interference.

In all three systems the speech is first bandpass limited to the range 300 - 3300 Hz, i.e. telephone voiceband. It is then sampled at 8 kHz with a resolution of 13 bits to produce our digital signal. The digital speech signal is analyzed using a technique known as Linear Prediction which attempts to model the mechanism whereby speech is produced in the human articulatory system. In this model, speech is viewed as the result of passing an excitation through the vocal tract. The excitation is produced either by the opening and closing of the vocal chords, giving rise to the vowel sounds, or else at some constriction forward of the vocal chords, which creates the so-called unvoiced sounds such as '*sh*'. The speech signal is then

$$S(z) = U(z) / A(z) \tag{1}$$

where $U(z)$ is the excitation, and $1/A(z)$ is the vocal tract response. In this model, the vocal tract response is an all pole construct; standing waves or resonances are set up in the vocal tract in response to an excitation, whose frequencies and bandwidths at a particular time are determined by the shape of the vocal tract and the placement of the so-called articulators (tongue, teeth, etc.). This model relates well to the production of vowel sounds, but is limited in its ability to adequately represent nasals and some fricatives where zeroes appear in the spectrum because of reflected wavefronts within the vocal apparatus causing cancellations.

Sweeping this objection aside, in the sampled time domain the speech waveform is expressed as

$$s_n = u_n + \sum_{i=0}^{p-1} a_i \cdot s_{n-i} \qquad (2)$$

where u_n is the excitation sequence and s_n is the output speech. The current sample s_n can be then viewed as the current excitation u_n plus a weighted sum of the previous p output samples. It is the estimation of these weights $\{a_i, i = 0, p-1\}$ that is the Linear Prediction analysis. The number of pole pairs, $p/2$, is equivalent to the number of resonances in the speech waveform that we are trying to represent and in the bandwidth of interest 0-3.3 kHz this is typically 4, and p is usually chosen as 8 or 10. In most coders, the linear prediction coefficients are calculated every 20 ms which is a compromise between on the one hand the dynamics of the vocal tract and the rate at which it can change shape, and on the other the amount of computation required and resultant coded data rate. To prove the rule, an exception is the recently announced half rate coder for JDC which uses a 40 ms frame interval. The coefficients are then transformed to another domain such as Log Area Ratios or Line Spectral pairs which are more robust to the effects of channel errors and then quantized for transmission. Linear prediction analysis has been applied to Speech Coding since it was first invented in 1970 [1]. The difficulty since and still, is how to adequately represent the excitation waveform, and what information about it to send to the receiver, and still manage to significantly reduce its associated data rate.

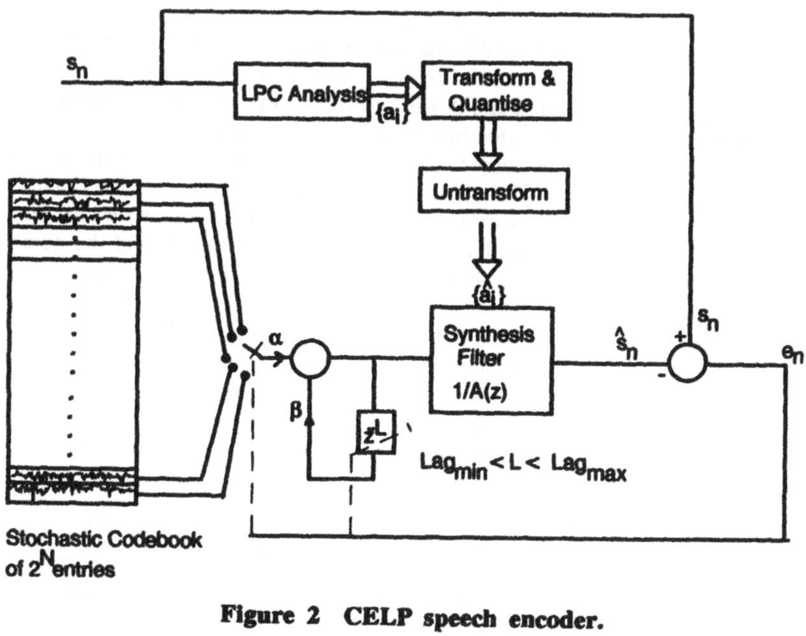

Figure 2 CELP speech encoder.

Modern speech coders as used in digital cellular systems employ analysis by synthesis, in that a variety of parameters are selectively combined to form an excitation waveform, and this waveform is made to excite a synthesis filter as illustrated in Figure 2. The combination of parameters that in some sense minimize the error term e_n are quantized and sent. GSM full rate apart, all other coders are based on Code Excited Linear Prediction (CELP), the idea for which was formed by the observation that residual waveforms after passing the speech signal through an inverse filter (the all zero filter $A(z)$) are found to be reasonably well modelled as samples from a Gaussian white noise source [2]. The codebook is populated with samples from such a source, and one by one are successively used to synthesize an excitation waveform by passing them through the long term predictor filter, a subframe at a time, where a subframe is typically 5 ms. The index of the codeword and gain α which minimizes the error term $\Sigma(s_n - \hat{s}_n)^2$ over the subframe is sent to the receiver, together with the coefficients and gain β of the long term predictor. The long term predictor itself synthesizes a waveform that attempts the best fit to any periodic components in the original speech waveform, and this best fit is determined by exhaustively trying all allowable values of lag (from 2.5 to 18 ms) and again choosing that which minimizes the error term. The codebook should be reasonably large (2^8 to 2^{10}) to afford adequate speech quality and many structures have been proposed to render a feasible computational burden for today's digital signal processing chips in having to explore the likelihood of each different entry for every new subframe. Efficient codebook structures have led to the VSELP coder specification used in DAMPS and JDC and the recently announced JDC half rate coder known as PSI CELP.

5.2.3.2 *Channel coding*

The coded data words emanating from the speech coder are segregated into two classes. Those deemed to cause very poor speech reproduction if corrupted are afforded some error protection for their onward journey, and the remainder are sent unprotected. A CRC is calculated for these more perceptually important bits and this together with those bits are convolved with two or more standardized code sequences. Typically two are used and for every data bit input, two are output and this is known as a rate 1/2 coder. The outputs are gathered together with the unprotected bits, thereby increasing the bit rate as indicated in Table 1. This powerful coding technique adds redundancy to the data in a highly structured manner, enabling it to be optimally recovered at the receiver after errors in the channel have been introduced. The ability to recover from bit errors in the data depends on the length of the sequences that the data is convolved with (the constraint length) and the number of sequences employed. As the bit rates of the speech coders decrease, so the more significant the information that is conveyed by each bit. The JDC half rate Speech Coder encodes the speech at 3.45 kbit/s and the convolutional coder uses a constraint length of 8 as compared with 6 for the full rate. The computational complexity of the optimal convolutional decoder rises exponentially with the constraint length.

The decoder has the ability to correct the most errors if the data presented to them is random. So, at least in JDC, the data at the transmitter is convolved with a pseudo random sequence, which also serves to encrypt the data. The pseudo random sequence employed is calculated by the phone using a key programmed into it and an authentication number sent by the network at registration. Convolutional decoding is not the only reason for randomizing the data; receivers generally have a better chance if the power spectrum they are presented with is flat. To reduce burst errors caused by fading in the radio channel, the data is buffered and individual bits selectively put out over successive transmission slots and sent along with data obtained from previous speech frames and frames yet to be encoded. This process is known as Interleaving.

5.2.3.3 *Burst assembling*

A relatively simple operation. When on a traffic channel for instance, control messages need to be inserted into the same burst as the encoded speech data, together with other information like frame synchronization data and guard bits. In GSM and DAMPS, a specially constructed training sequence is also added into the burst to allow the receiver to estimate the radio channel impulse response and unravel the intersymbol interference. In JDC, the system designers deemed it to be unnecessary to have to deal with channel dispersion. In practice for the DAMPS system, this is largely true as well, but the system designers here envisaged some rare scenarios where echoes off distant, highly reflective objects might occur.

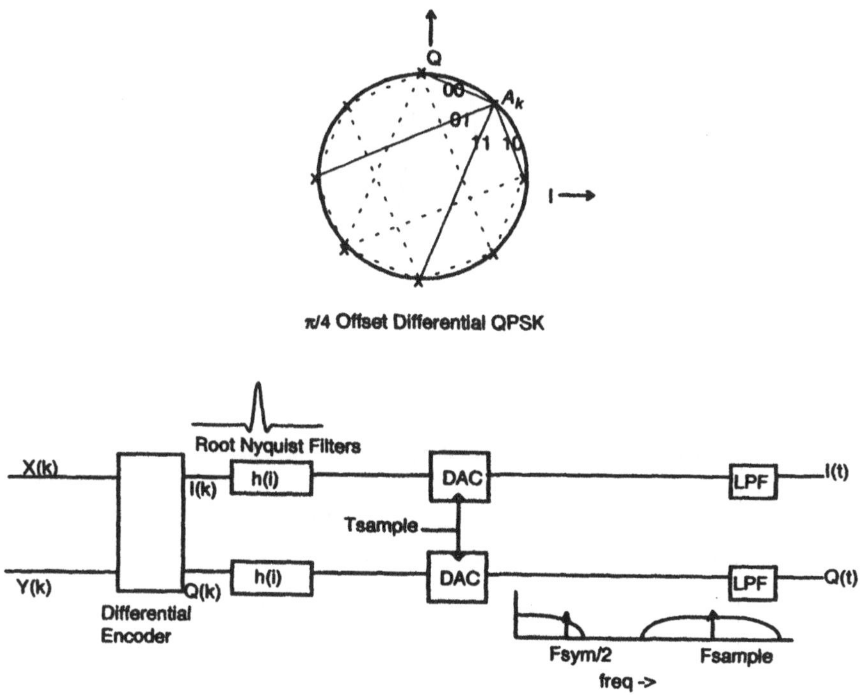

Figure 3 Pulse shaping and phase modulation.

5.2.3.4 *Modulation*

Our data bits must now be converted into a bandlimited analog signal suitable for transmission over the radio interface. The data bits are first converted into a series of real and imaginary pulses, and this series is known as the alphabet $\{A_k\}$, where with reference to Figure 3, $A_k = I_k + jQ_k$. The mapping of the data bits into the alphabet is the Modulation scheme, and that illustrated in Figure 3 is $\pi/4$ offset differential QPSK as used in JDC and DAMPS. By way of example, suppose we have just transmitted the pulse $A_k = \sqrt{2} + j\sqrt{2}$, and the next two data bits to be sent are **00**. The phase of the current A_k is advanced by π/4 and the next pulse to be transmitted is $A_{k+1} = 0 + j$. If the next two bits had instead been one of the pair **01**, **10**, or **11** the phase would have changed by either 3π/4, -π/4, or -3π/4 respectively. Two bits, or one symbol, are encoded by a change of phase which determines the value of the complex pulse to be emitted at the symbol rate $F_{sym} = \frac{1}{T_{sym}}$. It is of course not possible to just transmit and recover this pulse without bandlimiting it first. To minimize bit errors, the filter should ideally obey the Nyquist Criteria, which loosely states that the impulse response of the filter must be zero at all symbol intervals kT_{sym} other than at the time, $t = 0$, at which it was sent. Two such pulses are illustrated in Figure 4.

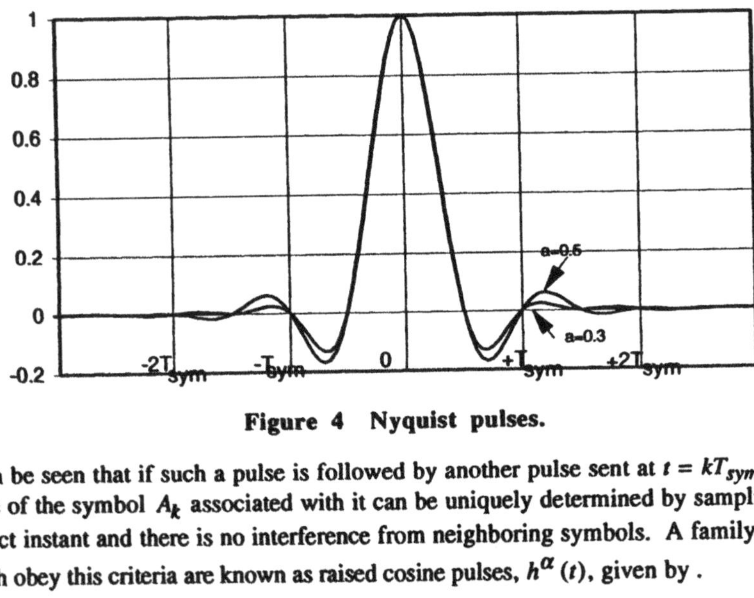

Figure 4 Nyquist pulses.

It can be seen that if such a pulse is followed by another pulse sent at $t = kT_{sym}$ then the value of the symbol A_k associated with it can be uniquely determined by sampling at the correct instant and there is no interference from neighboring symbols. A family of filters which obey this criteria are known as raised cosine pulses, $h^{\alpha}(t)$, given by .

$$h^{\alpha}(t) = \frac{\sin(\pi t / T_{sym})}{\pi t / T_{sym}} \cdot \frac{\cos(\alpha \pi t / T_{sym})}{\left[1 - (2\alpha t / T_{sym})^2\right]} \quad (3)$$

For $\alpha = 0$, the pulse becomes a sync pulse; the ideal but unrealisable bandlimited pulse, whose frequency response is a brick wall filter with cut off at $F_{sym}/2$. As α increases to a maximum of one, the bandwidth occupied increases, but the effective length of the impulse response decreases making the filter easier to implement. For JDC, α is set at 0.5, and for DAMPS at 0.35.

The Nyquist criteria must apply from the moment the information is bandlimited at the transmitter right through to when it is sampled for its value at the receiver. The filter is applied in two halves, half the contribution at the transmitter and half at the receiver. In the presence of noise, the optimum pulse detector at the receiver is a matched filter - its response is the same as the filter to originally shape the pulse, except time reversed, and for these reasons, the filter $\sqrt{h^{\alpha}(t)}$, or Root Nyquist filter is employed at both transmitter and receiver.

In a digital mobile phone, it is convenient to implement the Root Nyquist filters digitally. The symbol stream $\{A_k\}$ is convolved digitally with the impulse response of the filter, and the resulting bit stream is clocked out of the D/A converters at a sampling rate T_{sample}. The spectrum at the output of the D/A converters then repeats at $1/T_{sample}$. Low pass filters are applied to filter out any components above $F_{sample}/2$ and we are left with the desired baseband spectrum. The higher F_{sample} is above $F_{sym}/2$ the less stringent the requirements of the low pass filters, but the more power is consumed in clocking the relevant circuits with higher frequency clocks. Finally the I and Q signals are modulated onto the selected carrier and transmitted as indicated in Figure 1 to produce a signal:

$$x(t) = \cos(w_c t)\cdot \sum_{k=-\infty}^{\infty} \mathrm{Re}\{A_k\}\cdot g(t-kT_{sym}) + \sin(w_c t)\cdot \sum_{k=-\infty}^{\infty} \mathrm{Im}\{A_k\}\cdot g(t-kT_{sym}) \quad (4)$$

where $g(t) = \sqrt{h^{\alpha}(t)}$

5.2.4 Receiver architectures

At Nokia Mobile Phones, we have explored many different methods of recovering the transmitted symbols with a view to deriving the most compact, cost effective and power saving designs that operate within the system specifications. Two major classes of receiver have been considered; non-linear and linear.

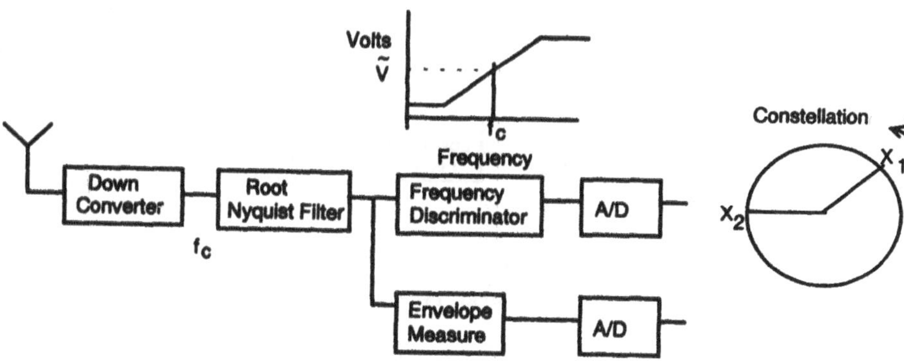

Figure 5 Non-linear receiver.

5.2.4.1 Non-linear receivers

Figure 5 illustrates the front-end of a non-linear receiver that utilizes standard components manufactured in millions for analog cellular phones (except for the root Nyquist filter). The incoming RF signal is down-converted to a convenient intermediate frequency (IF), f_c, where it is root Nyquist filtered. The signal is then applied to a frequency discriminator, which produces a voltage proportional to the frequency of the input signal. This voltage is sampled, to give a scaled, sampled representation of the instantaneous frequency. From the constellation diagram, we can see that over a symbol period, the phase changes by a certain amount in going from X_1 to X_2. If the instantaneous frequency samples are averaged over this symbol period, this gives a value for the change of phase between X_1 and X_2 plus a value for the average frequency due to the average voltage, $\tilde{V}$, on the A/D converter. Subtracting off the average frequency offset gives a number representing the phase change. Then if the averaging process was started at the correct instant, the phase change tells us the symbol that was sent during that period. There are several strategies for determining the correct instants at which to start the averaging, i.e. the optimum sampling point. A simple technique is to look for the instants at which abrupt changes in instantaneous frequency occur. Having recovered the phase changes, the amplitudes of the symbols can be restored from a separate measurement made on the signal envelope, and the symbols $\{A_k\}$ that were sent are resolved as soft (i.e. multi valued) decisions. Any long term frequency offset caused by the frequency f_c not being set to its nominal value will be manifested as continually increasing offsets of A_k from their idealized positions on the constellation. A way to evaluate this frequency error is to saturate the IF signal, which happens to be a by-product of standard frequency discriminators, and then to count the number of zero crossings that occur within a prescribed interval. The number and size of phase changes due to modulation are known from the recovered bit sequence, which enables the expected number of zero crossings in the interval to be calculated. The difference between observed and expected is a measure of the frequency error in f_c, which is adjusted accordingly.

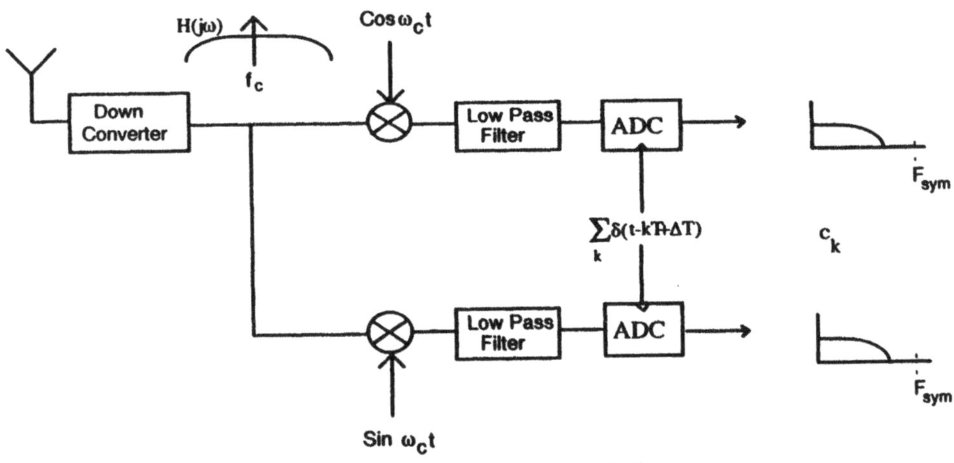

Figure 6 Linear receiver (RF).

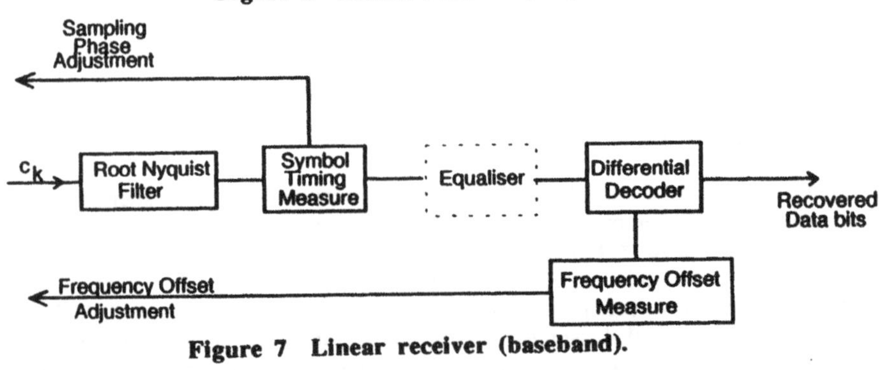

Figure 7 Linear receiver (baseband).

5.2.4.2 *Linear receivers*

Figure 6 illustrates one possible structure for a linear receiver. The RF signal received at the antenna is first down-converted to some convenient IF carrier, denoted f_c. Quadrature and in-phase signals at baseband are generated by multiplying the signal at IF by $\cos \omega_c t$ and $\sin \omega_c t$ respectively and anti-aliasing filters are applied to ensure no frequency components above F_{sym} are present prior to sampling. The two signals are then sampled at T, where $T = \frac{1}{2F_{sym}}$. The sampling process can alternatively be viewed as multiplying the signals by the pulse train $\sum_k \delta(t - kT + \Delta T)$, whose only non-zero values are at $t = kT + \Delta T$. ΔT is made to be an alterable fraction of the sampling interval which can be achieved by deriving the sampling clock from decimation of a faster integer multiple clock. By this means, the sampling phase can be adjusted. We now have our I and Q signals sampled at twice the symbol rate, which are then digitally filtered using a root nyquist filter.

5.2.4.3 *Bit recovery*

The times at which the IF signal is sampled should be adjusted until the sampling occurs at every symbol and half symbol instants. The general schemes for bit recovery in a linear receiver are then as depicted in Figure 7.

In digital cellular radios, the acquisition does not have to be fast, and simplicity is the watchword. One method known as *sampling derivative* can be employed where the timing is adjusted until the time derivative of the signal at the output of the root Nyquist filter becomes zero, which occurs when sampling at the maximum of the Nyquist pulse. The time derivative is measured by subtracting the value of the sample 1/2 symbol in advance of the nominal maximum sample position from the value of the sample 1/2 symbol behind.

It sometimes becomes necessary to combat the intersymbol interference introduced by the radio channel in order to achieve an acceptable bit error rate. The channel impulse response is estimated and the inverse filter or equalizer applied to the received data. It is also desirable that the equalizer be adaptive, i.e. is capable of changing its response in sympathy with a varying channel. The simplest structure for an adaptive equalizer is shown in Figure 8.

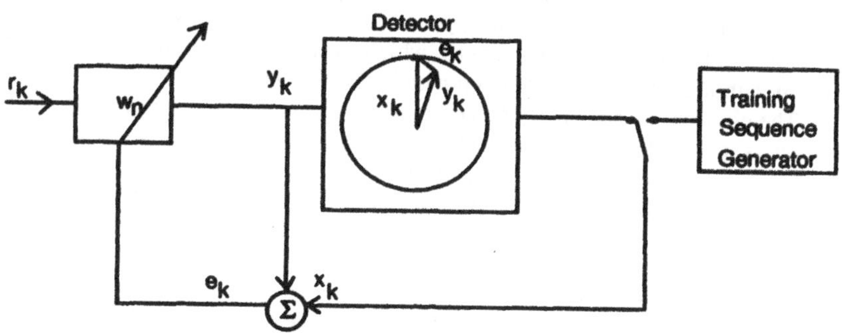

Figure 8 Adaptive linear equalizer.

The incoming complex I and Q signal r_k is filtered by the equalizer whose impulse response $\{w_n, n = 0, L-1\}$ is also complex. The output of the filter z_k is then fed into a decision device which selects as the received symbol x_k the nearest legitimate point on the true constellation, and the error is e_k as in Figure 8. The equalizer coefficients are then updated according to some algorithm that causes this filter to minimize the mean-square error, such as the stochastic gradient algorithm [3]. This type of equalizer is known as decision directed and the strategy assumes the equalizer is already making a reasonable estimate of the inverse channel response so that the nearest idealized point on the constellation to that produced by the output of the filter is indeed the correct one. This can be done by training switching in a generator of the same sequence as indicated.

The GSM system is designed for the receiver to use a Viterbi Equalizer to gain optimum performance in the presence of noise, and this is a non-linear equalizer. A specific training sequence whose autocorrelation is an impulse is sent during a traffic channel at the centre of each slot. At the receiver, the symbols in the center of the slot are correlated with a local copy of the training sequence which leaves as the output of the correlator a sequence of symbols which is an estimate of the (complex) channel finite impulse response or filter of M coefficients, where M is small (6 or less). After the channel filter coefficients have been determined, the received symbol sequence is compared with the all possible candidate sequences that could have emerged from such a filter, given that it was fed with ones or zeroes. The candidate sequence that most closely matches the received sequence is selected as the most likely sequence to have actually been transmitted and to have become convolved with the channel impulse response. Since the candidate symbol sequence is a linear combination of the original transmitted symbols, they are then easily recovered. The searching for the closest match between the received sequence and all the possible allowed sequences is done using the Viterbi algorithm [4] (in a sensible design that is; an exhaustive search would be prohibitive).

In JDC and DAMPS, the phase changes over a symbol interval now have to be detected and the original transmitted data is recovered as soft (multi valued) decisions. Their positions on the constellation are compared with the idealized, and if there is a continually increasing phase error this is due to there being a carrier frequency offset and this is then measured and adjusted for.

5.2.4.4 *Diversity*

One method to deal with fading is to make use of space diversity. At one point in space, the radio wave may be undergoing a fade, yet a few centimetres away the signal may be good and strong. This requires two antennas, which for a small handportable phone may be a problem, although this technique is common in JDC phones.

The choice then becomes whether to use two complete receiving paths and to combin the outputs (post detection) or to select one antenna, measure the signal strength and signal t noise ratio and swap to the other if the measurements indicate that the reception is inadequat (pre-detection). Suffice to say that post-detection has better performance that pre-detection but obviously costs more.

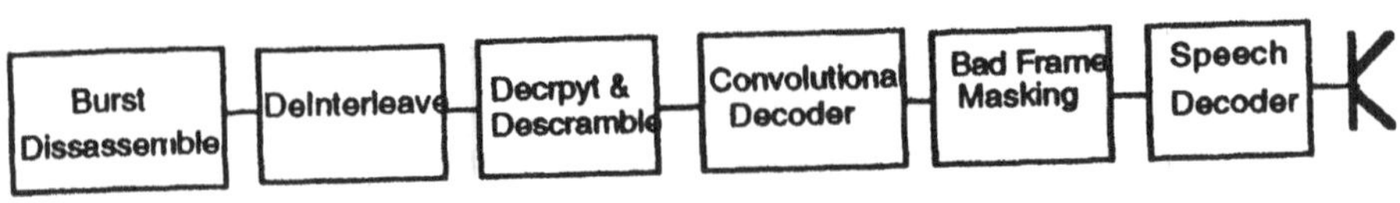

Figure 9 Receiver decoding section.

5.2.4.5 Decoding

After bit recovery, the data undergoes the reverse processes it went through in the transmitter, except it is now multi valued, i.e. it has a sign and a degree of confidence that is either a +1 or -1. The various stages in the decoding are shown in Figure 9. First the burst is disassembled, and the traffic channel data de-interleaved, after which it is decrypted and descrambled. The data is then sorted out into the two classes; protected and unprotected. The protected bits were convolutionally encoded, and now they must be decoded by generating all the possible sequences that could have been transmitted and selecting the one which matches closest the sequence of soft decisions that are the received data. Again it is recommended that this search be done using the Viterbi Algorithm, as in the Viterbi equalizer discussed previously. After decoding, the CRC bits are checked to determine if an error has been detected in the perceptually most significant bits for each frame. If there is an error, the bad frame masking strategy is bought into operation, which would typically cause the output speech to become gradually muted as more and more consecutive bad frames are received.

Hard decisions are made on the unprotected data and combined together with the decoded bits to form the input to the speech decoder. The speech is then synthesized in the speech decoder, which is the same as in the encoder, to produce the signal $\hat{s}_n$ as in Figure 2. An adaptive spectral postfilter may be applied that emphasizes the speech formats (vocal tract resonances) that were calculated in the Encoder by the linear prediction analysis. Finally, the 16 bit synthesized speech signal samples are clocked out of a D/A converter at 8 kHz and the speech is reconstituted.

5.2.5 Implementation issues

Small size, minimal power consumption, least cost and top grade performance all vie for number one priority on the mobile phone designers' list. Some, or parts of the algorithms are more suited to implementation on hardwired logic, and for flexibility (say, being able to reconfigure a phone for a different system) parts are better implemented on a programmable digital signal processor (DSP). A judicious partitioning is advocated. Fixed point realizations of the algorithms are essential to reduce the first three on the list, and the smaller the word lengths used the greater the benefits. Fixed point DSPs with word lengths of 16 bits are typically employed, being the lowest number of bits that can be used to adequately realize the dynamic range required within a speech coder without incurring severe processing penalties. This exercises the abilities of the designer in having to realize the speech coder in fixed point; to maintain the same quality as the original floating point specification, whilst at the same time minimizing cycles (cost and power) and program memory (cost and size). This is not so much of a challenge in GSM, where the work has been done a-priori and the precise arithmetic for implementing the speech coder is dictated (using 16 bit as the basic word length) to yield an implementation that must conform to a bit exact result.

Acknowledgments

The author wishes to acknowledge the received wisdom of many of his colleagues at Nokia Mobile Phones who have formulated and wrought the substance of this paper over many years of digital mobile phone developments since the original GSM concept.

References

[1] B. S Atal, "Speech analysis and synthesis by linear prediction of speech waveform," *J. Acoust Sor. Am.*, Vol. 47, 1970.

[2] M. R. Schroeder and B. S. Atal, "Code-excited linear prediction (CELP): high-quality speech at very low bit rates," in *Proc. 1985 Int. Conf. on Acoust. Speech and Sig Proc.*, pp. 25.1.1 - 25.1.4, 1985.

[3] B. Widrow and M. Hoff, "Adaptive Switching Circuits", in *IRE WESCON Conf Rec.*, pp. 96-104, 1960.

[4] A. J. Viterbi, "Convolutional codes and their performance in communication systems," *IEEE Trans. Commun. Technol.*, Vol. COM-19, pp. 751-772, Oct. 1971.

Chapter 5.3

High Speed DSP and Low Power Circuits for Future Mobile Phone

Masakazu Yamashina

5.3.1 Introduction

DSP speed has been increased year by year and GHz operation could be achieved at year 2005 [1]. Power dissipation has been thought to be increased proportional to operating frequency. DSP operating energy, however, has been reduced while keeping speed improvement, according to progress on device technology, as shown in Figure 1. Here, operating energy means power dissipation divided by operating frequency per transistor. Scaling device, which reduces load capacitances and a supply voltage, reduces power. Innovations on architectures and circuits, which decrease active transistor counts and load capacitances, also lower the power.

The reduction of operating energy will produce the future portable system which is capable of handling picture data in sub-mW power, and DSPs will play an important role in the system.

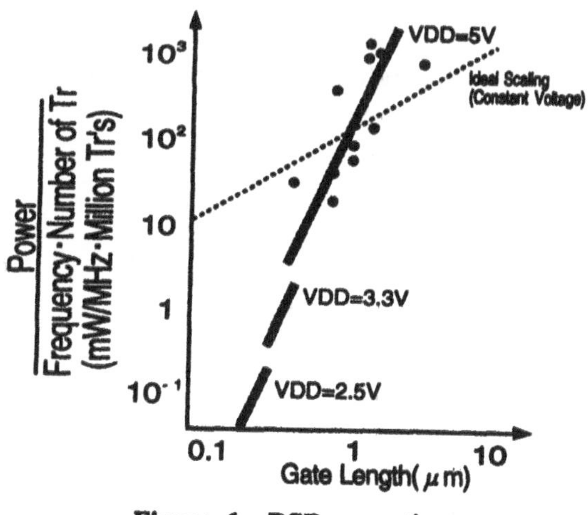

Figure 1 DSP operating energy.

This tutorial presents low power circuits for producing sub-mW power and GOPS-performance DSPs for future mobile systems.

Several approaches realizing mW and GOPS DSPs are discussed. The tutorial also discusses LSI circuits which can operate at over a GHz frequency on a low supply voltage.

5.3.2 DSPs for future mobile systems

Fast market progress of mobile equipment accelerates producing low power DSPs, which dissipate less than 100 mW and target to process voiceband signals. Portable computer and communication systems, which are capable of processing picture and video data, require GOPS performance and Mega-bit per second data-rate DSPs as shown in Figure 2. Figure 3 shows data compression rates for mobile equipments handling video data [2]. The compression rate of a mobile picture phone should be 1/100000, which is 1/100 compression rate of present teleconference systems, because of low transmission rate of wireless communication. New compression algorithm will help to reduce data to be processed by DSPs, however, GOPS performance will still be necessary. For achieving the specification, we can take two approaches: parallel processing and high-speed processing. These techniques are discussed in the next section.

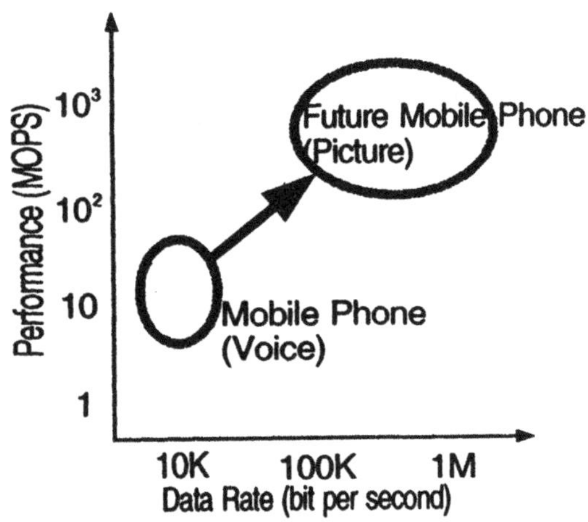

Figure 2 Performance requirement for future mobile DSPs.

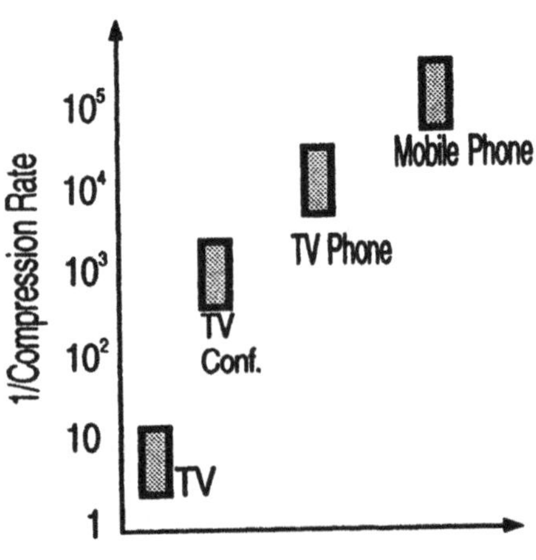

Figure 3 Data compression rate of mobile picture phone.

5.3.3 Parallel or high speed?

5.3.3.1 Is parallel processing practical for low-power DSPs?

Parallel processing architecture is a candidate for producing high-performace and low-power DSPs, as shown in Figure 4. Low supply voltage and parallel processing architecture is useful for low power DSPs [3]. Even though not reducing supply voltage, parallel processors dissipate lower power compared to high speed processors, as shown in Figure 5. This result is obtained from the fact that higher speed operation requires larger input power and input power does not increase speed proportionally because of parasitic load capacitance.

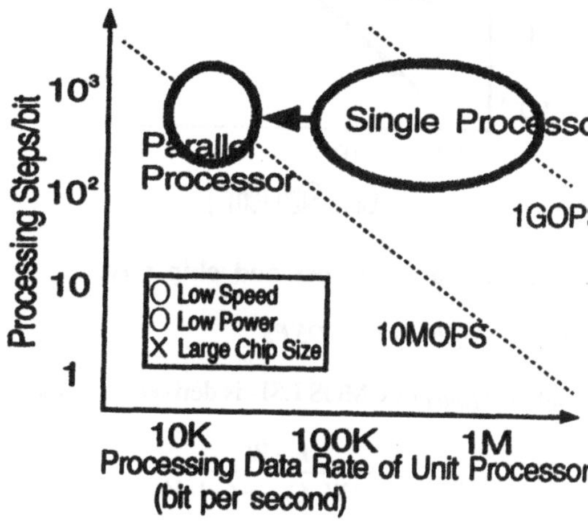

Figure 4 Processing data rate of an unit processor.

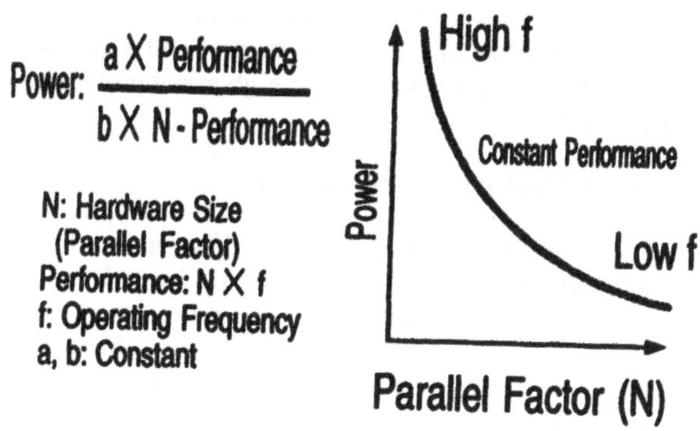

Figure 5 Low speed and parallel operation reduce power.

Parallel processing, however, has two disadvantages. One is a lack of architecture scalability. Performance is not improved proportional to the number of parallel processing units because of the lack of parallel processing capability of each job [4].

Another problem is a cost caused by a large chip size. Figure 6 shows LSI costs increased by increasing chip size. Cost of a chip smaller than 0.1 cm^2, is dominated by packaging and testing costs [5]. Therefore, parallel processing architecture is useful for processing specific algorithm such as DCT and filtering, and is practical for small chips.

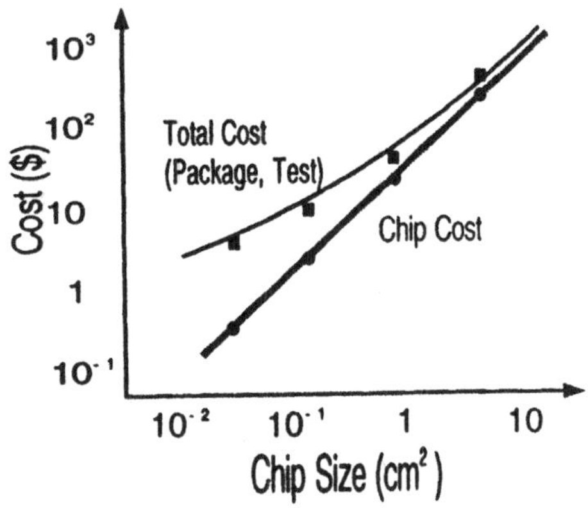

Figure 6 LSI cost and chip size.

5.3.3.2 *How to produce high speed DSPs*

Maximum operating frequency f_{max} of CMOS LSIs is derived from the following equation

$$f_{max} = \frac{I_0 \cdot W}{a \cdot (W \cdot C_{go} + C_1) \cdot V} \tag{1}$$

being W the gate width, I_0 the transistor current per unit gate width, a the number of gates along the critical path, C_{go}: the gate capacitance per unit gate width, C_1 the interconnection capacitances, and V the supply voltage. Equation (1) shows that (a) reduction of the critical path gate number, and (b) increase of a transistor current and a gate width to drive load capacitances in a small delay time, are essential for high speed processing. In this section, these techniques are discussed.

(i) Reduction of the critical path gate number

Reduction of the critical path gate number is the most effective way to reduce critical path delay time because each gate delay time is determined by device technology.

An implementation technique to reduce the number of gates along the critical path is pipeline architecture. Pipeline processors having the same performance as non-pipeline processors have dissipate lower power compared to non-pipeline processors, and eliminate overhead of a chip size which parallel processors have. High speed pipeline operation, however, doesn't increase performance proportional to speed improvement achieved by pipeline architecture, because of pipeline hazards. Pipeline hazards occur when data or instruction streams are disturbed. For example, a double-stage pipeline multiply-accumulator has twice throughput compared to a non-pipeline one, when data to be processed are input continuously. However, for single input datum, the throughput isn't improved by pipeline operation. Therefore pipeline architecture can be applied only to specific applications such as filtering operation. This is the same disadvantage as a parallel architecture has.

Obstacles to increase pipeline processors speed are a register delay time and clock skews, (pipeline overhead). A register delay time and a clock skew is about 500 ps and 300 ps, respectively, which cannot be neglected for high speed processors operating at over 100 MHz. A circuit technique to make pipeline processors operate at a faster speed is a wave pipeline [6], which has no pipeline registers that are usually put among pipeline stages. Eliminating pipeline registers by wave pipeline reduces not only delay time but also power dissipation of clock drivers and pipeline registers. Many logic design techniques have been also proposed to reduce delay time and power [1, 7]. Most of them utilize parallel operation and tend to increase hardware and power.

(ii) Increase of transistor current

There are two approaches to increase transistor current: the device approach and the circuit approach. Device technology plays an important role to increase the transistor current per unit gate width I_O, that is derived from equation

$$I_o = \frac{\mu \cdot C_o \cdot (V - V_t)^2}{2L} \tag{2}$$

where μ is the mobility, C_O the gate capacitance per unit gate area, V_t the threshold voltage and L the gate length. μ is determined by semiconductor materials, C_O is determined by gate oxide thickness and V_t is also controlled by process parameters. Reduction of L and V_t increases the transistor current.

A circuit technique to increase a transistor current is the BiCMOS circuit shown in Figure 7, which utilizes large current amplification of bipolar transistors. BiCMOS circuits have disadvantages caused by built-in potential of bipolar transistors and are not suitable for full-swing operation and low-supply voltage operation. Not full-swing operation increases short circuit currents. The limitation to lower supply voltage limits power reduction.

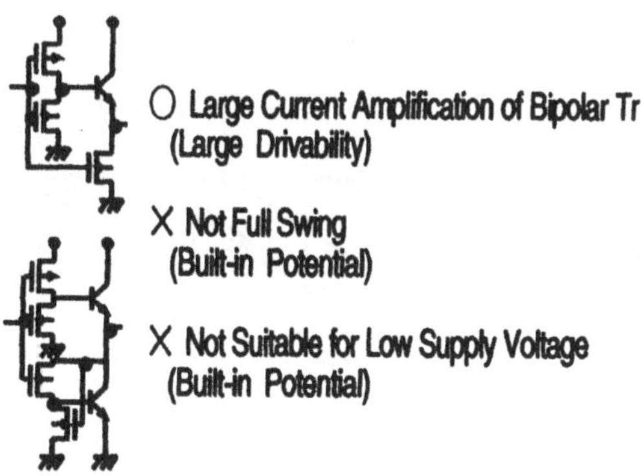

Figure 7 BiCMOS circuit.

5.3.3.3 Is a developed high speed DSP practical for mobile systems?

Figure 8 is a microphotograph of a 300 MHz DSP [8]. The DSP is capable of processing video data and especially suitable for compressing video data of teleconference systems. The number of transistors is 1.3 million and power is over 10 W at a 3.3 V supply voltage.

This large power results from circuit and architecture designs specialised for high-speed operation, which are large gate width to increase transistor current, many sense amplifiers of on-chip multiple memories, large fanouts caused by a lot of pipeline registers, and so on.

Figure 8 300 MHz DSP.

5.3.4 Low power circuits for high speed DSPs

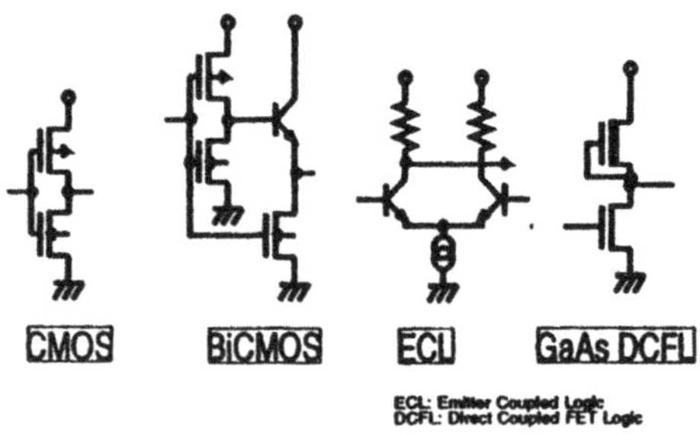

(a) circuit configurations

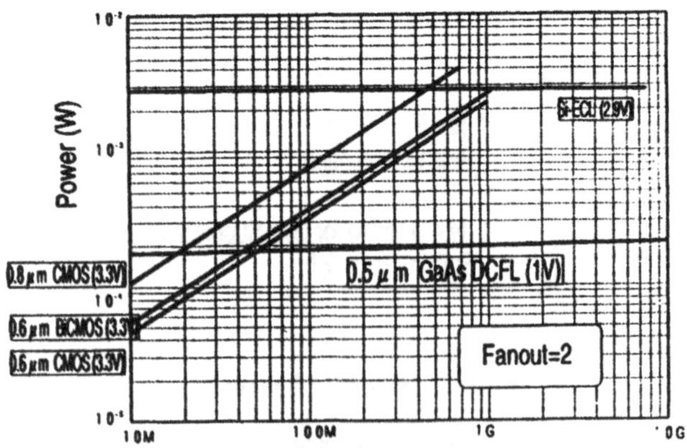

(b) power dissipation and operating frequency

Figure 9 Comparison of low power circuits.

Low power circuits operating over GHz are needed in order to produce GOPS and mW DSPs. Figure 9a shows several LSI circuits and 9b their respective performance. ECL circuits can operate at over GHz, however, dissipate large power and reduction of its supply voltage is limited by built-in potential. GaAs DCFL (Direct Coupled FET Logic) circuits could operate at high frequency with low power. However, it is hard to integrate a lot of GaAs devices on the chip, as shown in Figure 10, because of instability of materials. CMOS circuits operating at higher frequency dissipate larger power.

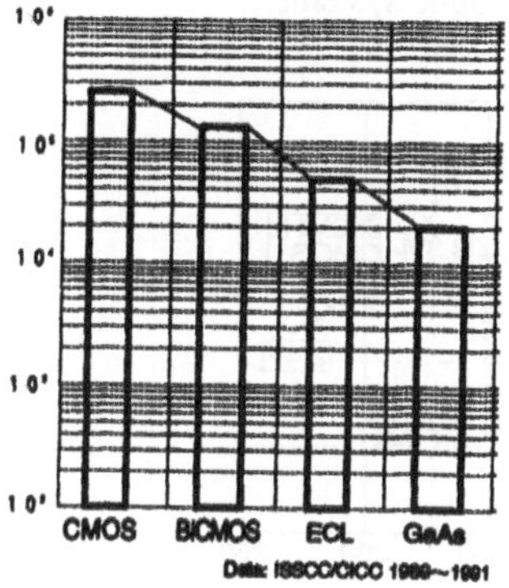

(a) the number of gates

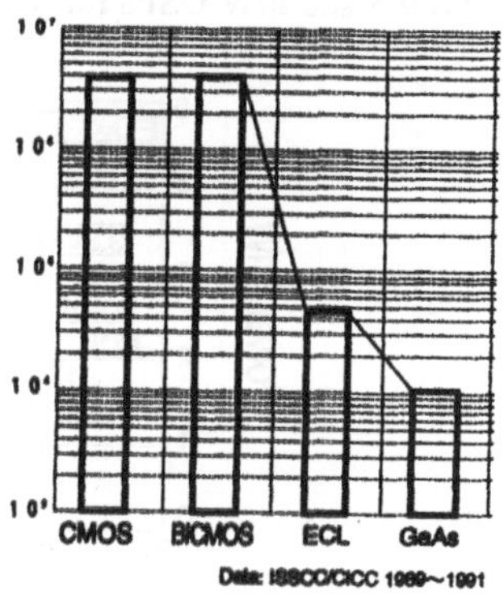

(b) memory capacity (bit)

Figure 10 Integration density of circuits.

What we need is a high speed operating circuit with small power dissipation. The target can be achieved by reducing supply voltage without speed loss. The circuit which is able to do this is a current-mode circuit using MOS transistors (MCML: MOS Current Mode Logic), as shown in Figure 11. Current mode circuits consist of differentially operating transistors and constant current sources, such as ECL. Current-mode circuits can realize high speed operation with a small signal voltage on low supply voltage.

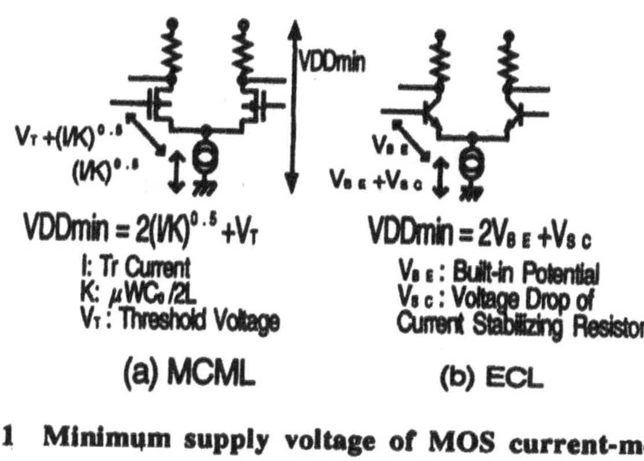

Figure 11 Minimum supply voltage of MOS current-mode circuits.

The delay time of current-mode circuits is given by $\frac{C \cdot dV}{I}$, where dV is the signal voltage swing, I is the constant driving current and C is the load capacitance. The switching speed of current-mode circuits can be increased by reducing *dV*. The power dissipation of current-mode circuits is given by $V \cdot I$, where V is the supply voltage. The current-mode circuits power is constant even by increasing operating frequency and can be reduced by lowering V. Current-mode circuits using MOS transistors can reduce supply voltage, as shown in Figure 11, because there is no built-in potential. Figure 12 shows characteristics of power and delay time. Normalized by the MCML circuit power-delay product, the power-delay product of CMOS, CML and ECL circuits are 3.8, 1.2, and 1.5, respectively. MCML circuits can be integrated on the CMOS chip. MCML/CMOS circuits will be useful for producing GOPS and mW DSPs for future mobile systems.

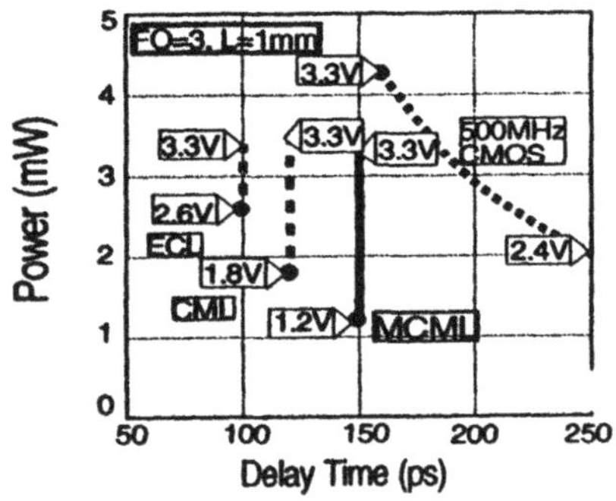

Figure 12 Power and delay characteristics of MCML circuits.

5.3.5 Summary

This tutorial describes future mobile DSPs and technologies achieving GOPS and mW performance. Comparison of parallel processing and high speed processing makes clear that parallel processing exactly reduces power, however, has chip size overhead compared to pipeline architecture. High speed processing increases power dissipation by circuit design specialised for high speed operation. Proposed MCML circuits have good characteristics reducing power at GHz operating conditions. Furthermore, MCML circuits can be integrated on a CMOS chip. MCML/CMOS circuits will be useful for producing GOPS and mW DSPs for future mobile systems.

References

[1] M. Yamashina, *et al.*, "A 200 MHz 16bit super high speed signal processor", *IEEE Journal of Solid-State Circuits*, Vol. 24, No. 6, pp. 1668-1674, Dec. 1989.

[2] Nikkei Business Publications, Inc., *Nikkei new media -Mobile computing*, 1993.

[3] A. P. Chandrakasan, *et al.*, "Low-power CMOS digital design", *IEEE Journal of Solid-State Circuits*, Vol. 27, No. 4, pp. 473-484, Apr. 1992.

[4] N. P. Jouppi, "The nonuniform distribution of instruction-level and machine parallelism and its effect on performance", *IEEE Trans. Comput.*, Vol. C-38, pp. 1645-1658, Dec. 1989.

[5] J. L. Henessy and D. A. Patterson, *Computer Architecture. A Quantitative Approach*, Morgan Kaufmann Publishers Inc., 1990.

[6] M. Izumikawa, *et al.*, "A 400 MHz 300 mW 8 kb CMOS SRAM macro with a current sensing scheme", *Proceedings of IEEE 1994 CICC.*

[7] K. Suzuki, *et al.*, "A 2.4 ns 16 b 0.5 um CMOS arithmetic logic unit for microprogrammable video signal processor", *Proceedings of IEE 1993 CICC*, pp. 12.4.1-12.4.4.

[8] T. Inoue, *et al.*, " A 300 MHz 16 b BiCMOS video signal processor", *ISSCC Digest of Technical Papers*, pp. 36-37, Feb. 1993.

Tutorial 6

Chaos from basic concepts to engineering applications

Organised by: Martin Hasler, *Swiss Federal Institute of Technology*

Chapter 6.1

Basic Concepts of Nonlinear Dynamics and Chaos

M. Peter Kennedy

6.1.1 Electrical and electronic circuits as dynamical systems

A *system* is something having parts which may perceived as an single entity. A *dynamical system* is one which changes with time; what changes is the *state* of the system. Mathematically, a dynamical system consists of a space of states (called the *state space* or *phase space*) and a rule, called the *dynamic*, for determining which state corresponds at a given future time to a given present state. A *deterministic* dynamical system is one whose state at any time is *completely determined* by its initial state and dynamic. In this chapter, we consider only deterministic dynamical systems.

A deterministic dynamical system may have a *continuous* or *discrete* state space and a *continuous-time* or *discrete-time* dynamic.

A *lumped*[1] circuit containing resistive elements (resistors, voltage and current sources) and *energy-storage elements* (capacitors and/or inductors) may be modeled as a *continuous-time* deterministic dynamical system in $\mathbf{R}^n$. The evolution of the state of the circuit is described by a system of ordinary differential equations called state equations.

Discrete-time deterministic dynamical systems occur in electrical engineering as models of switched-capacitor and digital filters, sampled phase-locked loops, and sigma-delta modulators. Discrete-time dynamical systems also arise when analyzing the stability of steady-state solutions of continuous-time systems. The evolution of a discrete-time dynamical system is described by a system of difference equations.

6.1.1.1 Continuous-time dynamical systems

Theorem 1: Existence and uniqueness of solution for a differential equation

Consider a continuous-time deterministic dynamical system defined by a system of ordinary differential equations of the form

$$\dot{\mathbf{X}}(t) = \mathbf{F}(\mathbf{X}(t), t) \tag{1}$$

where $\mathbf{X}(t) \in \mathbf{R}^n$ is called the *state*, $\dot{\mathbf{X}}(t)$ denotes the derivative of $\mathbf{X}(t)$ with respect to time, $\mathbf{X}(t_0) = \mathbf{X}_0$ is called the *initial condition*, and the map $\mathbf{F}(\cdot,\cdot) : \mathbf{R}^n \times \mathbf{R}_+ \rightarrow \mathbf{R}^n$ is (i) *continuous almost everywhere* on $\mathbf{R}^n \times \mathbf{R}_+$ and (ii) *globally Lipschitz*.

Then, for each $(\mathbf{X}_0, t_0) \in \mathbf{R}^n \times \mathbf{R}_+$, there exists a *continuous* function $\phi(\cdot; \mathbf{X}_0, t_0) : \mathbf{R}_+ \rightarrow \mathbf{R}^n$ such that

$$\phi(t_0; \mathbf{X}_0, t_0) = \mathbf{X}_0$$

[1] A *lumped* circuit is one whose physical dimensions are small compared to the wavelengths of its voltage and current waveforms [1].

and

$$\dot{\phi}(t; \mathbf{X}_0, t_0) = \mathbf{F}(\phi(t; \mathbf{X}_0, t_0), t) \tag{2}$$

Furthermore, this function is ***unique*** [2].

The function $\phi(\cdot; \mathbf{X}_0, t_0)$ is called the *solution* or *trajectory* through $(\mathbf{X}_0, t_0)$ of the differential equation (1).

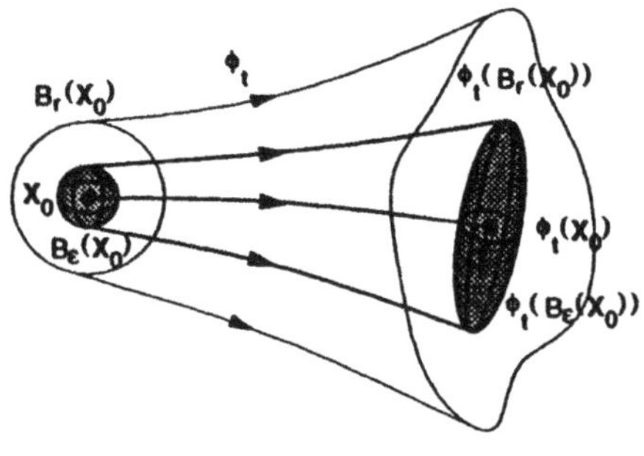

Figure 1 The vector field F of an autonomous continuous-time dynamical system generates a flow ϕ which maps a point $\mathbf{X}_0$ in the state space to its image $\phi_t(\mathbf{X}_0)$ t seconds later. A volume of state space $B_r(\mathbf{X}_0)$ evolves under the flow into a region $\phi_t(B_r(\mathbf{X}_0))$. Sufficiently close to the trajectory $\phi_t(\mathbf{X}_0)$, the linearized flow maps a sphere of radius ϵ into an ellipsoid.

6.1.1.2 Autonomous continuous-time dynamical systems

If the vector field of a continuous-time deterministic dynamical system depends only on the state and is *independent* of time t, then the system is said to be *autonomous* and may be written as

$$\dot{\mathbf{X}}(t) = \mathbf{F}(\mathbf{X}(t))$$

or simply

$$\dot{\mathbf{X}} = \mathbf{F}(\mathbf{X}) \tag{3}$$

If, in addition, the vector field $\mathbf{F}(\cdot) : \mathbb{R}^n \rightarrow \mathbb{R}^n$ is *Lipschitz*[2] then there is a ***unique*** continuous function $\phi(\cdot, \mathbf{X}_0) : \mathbb{R}_+ \rightarrow \mathbb{R}^n$ (called the *trajectory* through $\mathbf{X}_0$) which satisfies

$$\dot{\phi}(t, \mathbf{X}_0) = \mathbf{F}(\phi(t, \mathbf{X}_0)), \quad \phi(t_0, \mathbf{X}_0) = \mathbf{X}_0 \tag{4}$$

Since the vector field is independent of time, we choose $t_0 \equiv 0$. For shorthand, we denote the flow by ϕ and the map $\phi(t, \cdot) : \mathbb{R}^n \rightarrow \mathbb{R}^n$ by ϕ_t.

The *t-advance map* ϕ_t takes a state $\mathbf{X}_0 \in \mathbb{R}^n$ to state $\mathbf{X}(t) = \phi_t(\mathbf{X}_0)$ t seconds later. In particular, ϕ_0 is the identity mapping. Furthermore, $\phi_{t+s} = \phi_t \phi_s$ since the state $\mathbf{Y} = \phi_s(\mathbf{X})$ to which $\mathbf{X}$ evolves after time s, evolves after an additional time t into the

[2]There exists a finite $k \in \mathbb{R}$ such that $||\mathbf{F}(\mathbf{X}) - \mathbf{F}(\mathbf{X}')|| \leq k||\mathbf{X} - \mathbf{X}'||, \ \forall \mathbf{X}, \mathbf{X}' \in \mathbb{R}^n$.

same state $\mathbf{Z}$ as that to which $\mathbf{X}$ evolves after time $t+s$

$$\mathbf{Z} = \phi_t(\mathbf{Y}) = \phi_t(\phi_s(\mathbf{X})) = \phi_{t+s}(\mathbf{X})$$

A bundle of trajectories emanating from a ball $B_r(\mathbf{X}_0)$ of radius r centered at $\mathbf{X}_0$ is mapped by the flow into some region $\phi_t(B_r(\mathbf{X}_0))$ after t seconds (see Figure 1). Consider a short segment of the trajectory $\phi_t(\mathbf{X}_0)$ along which the flow is differentiable with respect to $\mathbf{X}$: in a sufficiently small neighbourhood of this trajectory, the flow is almost linear, so the ball $B_\epsilon(\mathbf{X}_0)$ of radius ϵ about $\mathbf{X}_0$ evolves into an ellipsoid $\phi_t(B_\epsilon(\mathbf{X}_0))$, as shown.

An important consequence of Lipschitz continuity in an autonomous vector field and the resulting uniqueness of solution of (3) is that a trajectory of the dynamical system cannot go through the same point twice in two different directions. In particular, no two trajectories may *cross each other*; this is called the *noncrossing property* [3].

6.1.1.3 Non-autonomous dynamical systems

A non-autonomous n-dimensional continuous-time dynamical system may be transformed to an $(n+1)$-dimensional *autonomous* system by appending time as an additional state variable and writing

$$\begin{aligned} \dot{\mathbf{X}}(t) &= \mathbf{F}(\mathbf{X}(t), X_{n+1}(t)) \\ \dot{X}_{n+1}(t) &= 1 \end{aligned} \tag{5}$$

6.1.1.4 Discrete-time dynamical system

Consider a *discrete-time deterministic dynamical system* defined by a system of difference equations of the form

$$\mathbf{X}(k+1) = \mathbf{G}(\mathbf{X}(k), k) \tag{6}$$

where $\mathbf{X}(k) \in \mathbb{R}^n$ is called the *state*, $\mathbf{X}(k_0) = \mathbf{X}_0$ is the *initial condition*, and $\mathbf{G}(\cdot,\cdot) : \mathbb{R}^n \times \mathbb{Z}_+ \to \mathbb{R}^n$ maps the *current* state $\mathbf{X}(k)$ into the *next* state $\mathbf{X}(k+1)$, where $k_0 \in \mathbb{Z}_+$.

By analogy with the continuous-time case, there exists a function $\phi(\cdot, \mathbf{X}_0, k_0) : \mathbb{Z}_+ \to \mathbb{R}^n$ such that

$$\phi(k_0; \mathbf{X}_0, k_0) = \mathbf{X}_0$$

and

$$\phi(k+1; \mathbf{X}_0, k_0) = \mathbf{G}(\phi(k; \mathbf{X}_0, k_0), k)$$

The function $\phi(\cdot; \mathbf{X}_0, k_0) : \mathbb{Z}_+ \to \mathbb{R}^n$ is called the *solution* or *trajectory* through $(\mathbf{X}_0, k_0)$ of the difference equation (6).

The image $\{\phi(k; \mathbf{X}_0, k_0) \in \mathbb{R}^n | k \in \mathbb{Z}_+\}$ in $\mathbb{R}^n$ of the trajectory through $(\mathbf{X}_0, k_0)$ is called an *orbit* through $(\mathbf{X}_0, k_0)$.

If the map $\mathbf{G}(\cdot,\cdot)$ of a discrete-time dynamical system depends only on the state $\mathbf{X}(k)$ and is *independent* of k, then the system is said to be *autonomous* and may be written more simply as

$$\mathbf{X}_{k+1} = \mathbf{G}(\mathbf{X}_k) \tag{7}$$

where $\mathbf{X}_k$ is shorthand for $\mathbf{X}(k)$ and the initial iterate k_0 is chosen, without loss of generality, to be zero. Using this notation, $\mathbf{X}_k$ is the image of $\mathbf{X}_0$ after k iterations of the map $\mathbf{G}(\cdot) : \mathbb{R}^n \to \mathbb{R}^n$.

Example: nonlinear parallel RLC circuit

Consider the parallel RLC circuit shown in Figure 2. This circuit contains a linear inductor L, a linear capacitor C_2, and a nonlinear resistor N_R', whose continuous piecewise-linear

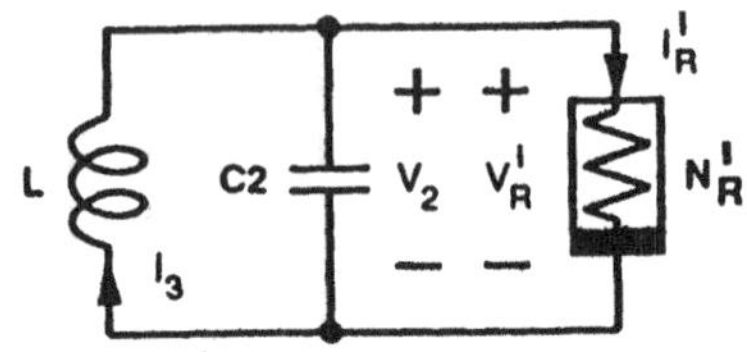

Figure 2 Parallel RLC circuit whose nonlinear resistor N_R' has a DP characteristic as shown in Figure 3. By Kirchhoff's voltage law, $V_R' = V_2$.

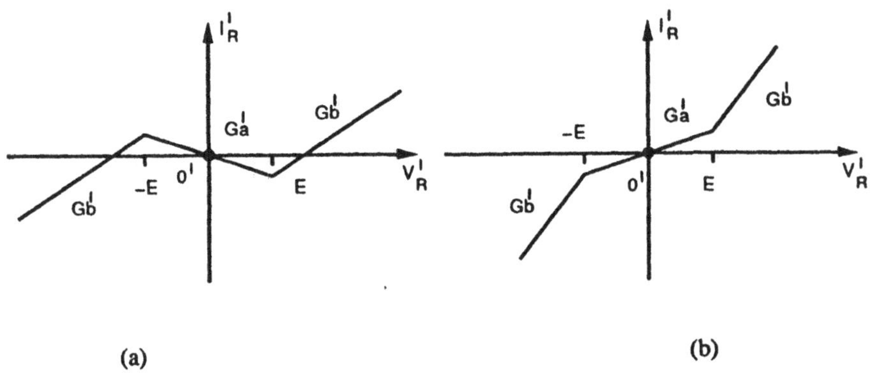

Figure 3 DP characteristic of N_R' in Figure 2 when (a) $G_a' < 0$, and (b) $G_a' > 0$.

driving-point (DP) characteristic (shown in Figure 3) has slope G_a' for $|V_R'| \leq E$ and slope G_b' for $|V_R'| > E$. The DP characteristic of N_R' may be written explicitly

$$I_R'(V_R') = G_b' V_R' + \frac{1}{2}(G_a' - G_b')(|V_R' + E| - |V_R' - E|)$$

This circuit may be described by a pair of ordinary differential equations and is therefore a *second-order continuous-time* dynamical system. Choosing I_3 and V_2 as *state variables*, we write

$$\begin{aligned} \frac{dI_3}{dt} &= -\frac{1}{L}V_2 \\ \frac{dV_2}{dt} &= \frac{1}{C_2}I_3 - \frac{1}{C_2}I_R'(V_2) \end{aligned}$$

with $I_3(0) = I_{30}$ and $V_2(0) = V_{20}$.

We illustrate the vector field by drawing vectors at uniformly-spaced points in the two-dimensional *state space* defined by (I_3, V_2). Starting from a given initial condition (I_{30}, V_{20}), a solution curve in state space is the locus of points plotted out by the state as it moves through the vector field, following the direction of the arrow at every point. Figure 4 shows typical vector fields and trajectories of the circuit.

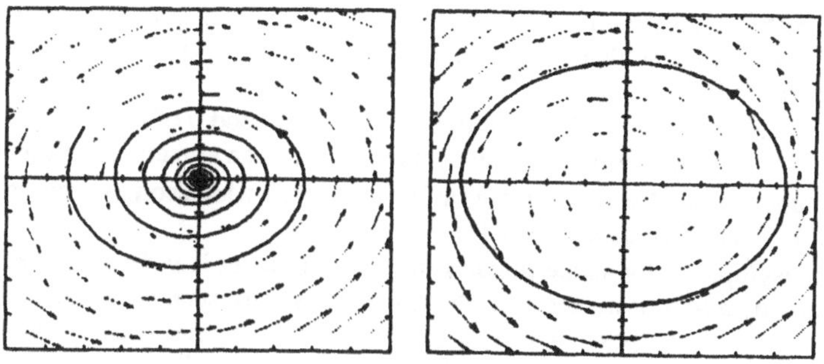

(a) (b)

Figure 4 Vector fields for the nonlinear RLC circuit in Figure 2. $L = 18$mH, $C_2 = 100$nF, $E = 0.47$V. (a) $G'_a = 242.424\mu$S, $G'_b = 1045.455\mu$S: all trajectories converge to the origin. (b) $G'_a = -257.576\mu$S, $G'_b = 545.455\mu$S: the unique steady-state solution is a limit cycle. Horizontal axis: I_3, 400μA/div; vertical axis: V_2, 200mV/div.

If L, C_2, and G'_b are positive, the steady-state behaviour of the circuit depends on the sign of G'_a. When $G'_a > 0$, the circuit is dissipative everywhere and all trajectories collapse towards the origin. The unique steady-state solution of the circuit is the stable DC equilibrium condition $I_3 = V_2 = 0$.

If $G'_a < 0$, N'_R looks like a negative resistor close to the origin and injects energy into the circuit, pushing trajectories *away*. Further out, where the characteristic has positive slope, trajectories are pulled in by the dissipative vector field. The resulting balance of forces produces a periodic steady-state orbit called a *limit cycle*, which is approached asymptotically by all initial conditions of this circuit.

This limit cycle is termed *attracting* because nearby trajectories move *towards* it and is *structurally stable* in the sense that, for almost all values of G'_a, a small change in the parameters of the circuit has little effect on it. In the special case when $G'_a \equiv 0$, a perturbation of G'_a causes the steady-state behaviour to change from an equilibrium point to a limit cycle; this is called a *bifurcation*.

In the following subsections, we consider in detail steady-state behaviours, stability, structural stability, and bifurcations.

6.1.2 Classification and uniqueness of steady-state behaviours

A trajectory of a dynamical system from an initial state $\mathbf{X}_0$ settles, possibly after some transient, onto a set of points called a *limit set*. The ω-limit set corresponds to the asymptotic behaviour of the system as $t \to +\infty$ and is called the *steady-state* response. We use the idea of *nonwandering states* to determine when the system has reached steady-state.

A state $\mathbf{X}$ of a dynamical system is called *nonwandering* under the flow ϕ_t if, for every neighbourhood $B_\epsilon(\mathbf{X})$ of $\mathbf{X}$ and for every $T > 0$, there is a time $t > T$ such that $\phi_t(B_\epsilon(\mathbf{X})) \cap B_\epsilon(\mathbf{X}) \neq \emptyset$. Thus, a nonwandering state $\mathbf{X}$ lies on or near trajectories which come back within a specified distance of themselves [4].

Wandering points correspond to transient behaviour, while *steady-state* or asymptotic behaviour corresponds to orbits of *nonwandering* states.

A point $\mathbf{X}_\omega$ is an *ω-limit point* of $\mathbf{X}_0$ if and only if $\lim_{k\to+\infty} \phi_{t_k}(\mathbf{X}_0) = \mathbf{X}_\omega$ for some sequence $\{t_k | k \in \mathbb{Z}\}$ such that $t_k \to +\infty$. The set $L(\mathbf{X}_0)$ of ω-limit points of $\mathbf{X}_0$ is called the *ω-limit set* of $\mathbf{X}_0$.[3]

A limit set L is *attracting* if there exists an neighbourhood U of L such that $L(\mathbf{X}_0) = L$ for all $\mathbf{X}_0 \in U$. Thus, nearby trajectories converge towards an attracting limit set as $t \to \infty$.

An attracting set $\mathcal{A}$ which contains at least one orbit which comes arbitrarily close to every point in $\mathcal{A}$ is called an *attractor* [4].

In an asymptotically stable *linear* system the limit set is *independent of the initial condition* and *unique* so it makes sense to talk of *the* steady-state behaviour. By contrast, a nonlinear system may possess several different limit sets and therefore may exhibit a variety of steady-state behaviours, depending on the initial condition.

The set of all points in the state space which converge to a particular limit set L is called the *basin of attraction* of L.

Since non-attracting limit sets cannot be observed in physical systems, the asymptotic or *steady-state* behaviour of a real electronic circuit corresponds to motion on an *attracting* limit set.

6.1.2.1 Equilibrium point

The simplest steady-state behaviour of a dynamical system is an equilibrium point. An *equilibrium point* or *stationary point* of (3) is a state $\mathbf{X}_Q$ at which the vector field is zero. Thus $\mathbf{F}(\mathbf{X}_Q) = 0$ and $\phi_t(\mathbf{X}_Q) = \mathbf{X}_Q$; a trajectory starting from an equilibrium point remains indefinitely at that point.

In state space, the limit set consists of a single nonwandering point $\mathbf{X}_Q$. A point is a zero-dimensional object. Thus, an equilibrium point is said to have *dimension* zero.

In the time domain, an equilibrium point of an electronic circuit is simply a DC solution or operating point.

An equilibrium point or *fixed point* of a discrete-time dynamical system is a point $\mathbf{X}_Q$ which satisfies

$$\mathbf{G}(\mathbf{X}_Q) = \mathbf{X}_Q$$

Example: nonlinear parallel RLC circuit

The nonlinear RLC circuit shown in Figure 2 has just one equilibrium point $(I_{3Q}, V_{2Q}) = (0,0)$. When G'_a is positive, a trajectory originating at any point in the state space converges to this attracting DC steady-state (as shown in Figure 4a). The basin of attraction of the origin is the entire state space.

Since all trajectories, and not just those which start close to it, converge to the origin, this equilibrium point is said to be a *global* attractor.

When $G'_a < 0$, the circuit possesses two steady-state solutions: the equilibrium point at the origin, and the limit cycle Γ. The equilibrium point is unstable in this case. All trajectories, except that which starts *at* the origin, are attracted to Γ.

---

[3]The set of points to which trajectories converge from $\mathbf{X}_0$ as $t \to -\infty$ is called the α-limit set of $\mathbf{X}_0$. Since we consider only positive time, by *limit set* we mean the ω-limit set.

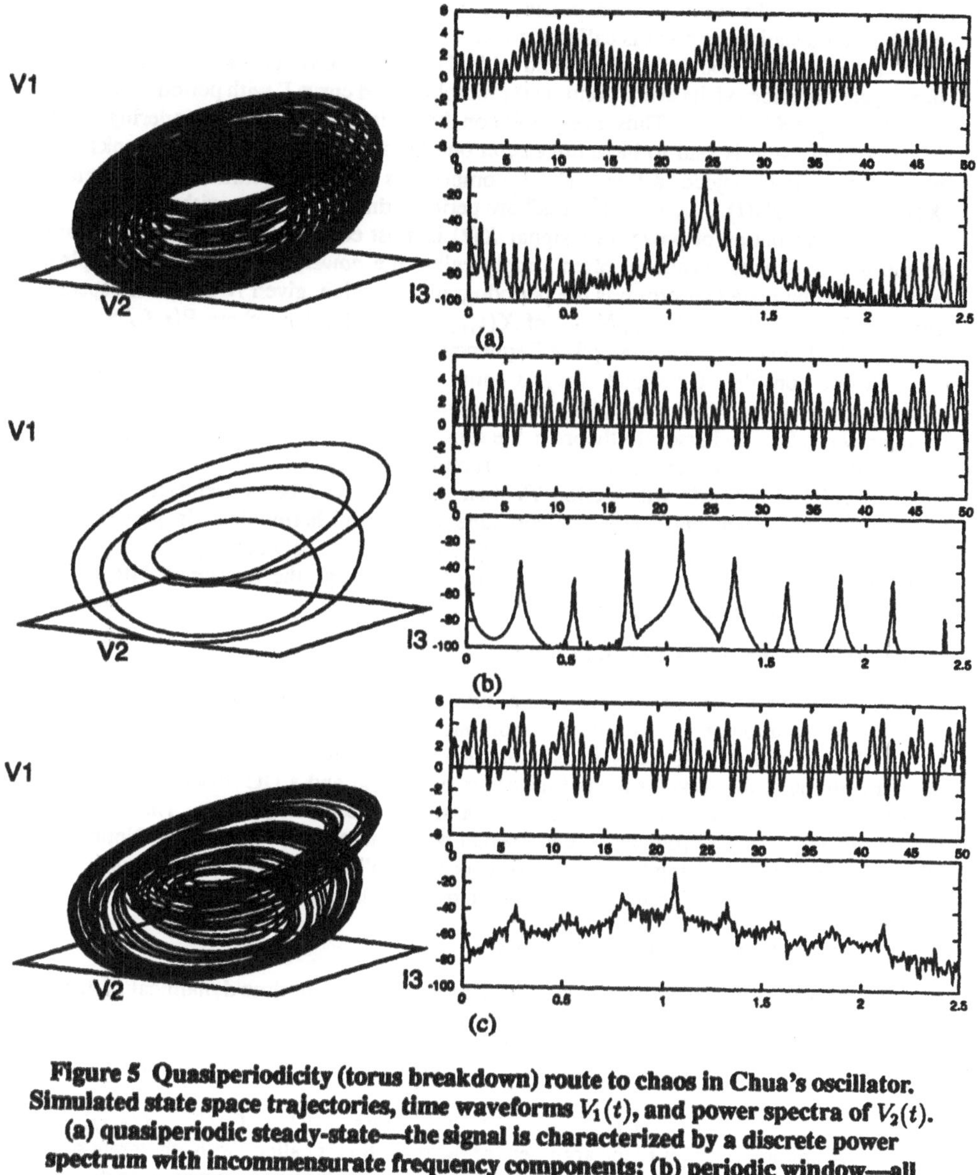

Figure 5 Quasiperiodicity (torus breakdown) route to chaos in Chua's oscillator. Simulated state space trajectories, time waveforms $V_1(t)$, and power spectra of $V_2(t)$. (a) quasiperiodic steady-state—the signal is characterized by a discrete power spectrum with incommensurate frequency components; (b) periodic window—all spikes in the power spectrum are harmonically related to the fundamental frequency; (c) chaotic steady-state following breakdown of the torus—the waveform has a broadband power spectrum. Time plots: horizontal axis—t (ms); vertical axis—V_1 (V). Power spectra: horizontal axis—frequency (kHz); vertical axis—power (mean squared amplitude) of $V_2(t)$ (dB).

6.1.2.2 Periodic steady-state

A state $\mathbf{X}$ is called *periodic* if there exists $T > 0$ such that $\phi_T(\mathbf{X}) = \mathbf{X}$. A periodic orbit which is not a stationary point is called a *cycle*.

A *limit cycle* Γ is an isolated periodic orbit of a dynamical system (see Figure 5b). The limit cycle trajectory visits every point on the simple closed curve Γ with period T. Indeed, $\phi_t(\mathbf{X}) = \phi_{t+T}(\mathbf{X}) \forall \mathbf{X} \in \Gamma$. Thus, every point on the limit cycle Γ is a nonwandering point.

A limit cycle is said to have *dimension* one because a small piece of it looks like a one-dimensional object: a line. The n components $X_i(t)$ of a limit cycle trajectory $\mathbf{X}(t) = (X_1(t), X_2(t), \ldots, X_n(t))^T$ in $\mathbb{R}^n$ are periodic time waveforms with period T.

The distribution of power in a signal $X(t)$ is most commonly quantified by means of the *power density spectrum*, often simply called the *power spectrum*. The simplest estimator of the power spectrum is the periodogram [5] which, given N uniformly spaced samples $X(k/f_s), k = 0, 1, \ldots, N-1$ of $X(t)$, yields $N/2+1$ numbers $P(nf_s/N), n = 0, 1, \ldots, N/2$, where f_s is the sampling frequency.

If one considers the signal $X(t)$ as being composed of sinusoidal components at discrete frequencies, then $P(nf_s/N)$ is an estimate of the power in the component at frequency nf_s/N. By Parseval's theorem, the sum of the power in each of these components equals the mean squared amplitude of the N samples of $X(t)$ [5].

If $X(t)$ is *periodic* with period T, then its power will be concentrated in a DC component, a fundamental frequency component $1/T$, and harmonics. In practice, the discrete nature of the sampling process causes power to "leak" between adjacent frequency components; this leakage may be reduced by "windowing" the measured data before calculating the periodogram [5].

Example: periodic steady-state solution

Figure 5b shows a state-space orbit, time waveform, and power spectrum of a periodic steady-state solution of a third-order autonomous continuous-time dynamical system.

The orbit in state-space is an asymmetric closed curve consisting of four loops. In the time domain, the waveform has four crests per period and a DC offset. In the power spectrum, the DC offset manifests itself as a spike at zero frequency. The period of approximately 270 Hz produces a fundamental component at that frequency. Notice that the fourth harmonic (arising from "four crests per period") has the largest magnitude. This power spectrum is reminiscent of subharmonic mode locking in a forced oscillator.

6.1.2.3 Subharmonic periodic steady-state

A subharmonic periodic solution or *period-K orbit* of a discrete-time dynamical system is a set of K points $\{\mathbf{X}_1, \mathbf{X}_2, \ldots, \mathbf{X}_K\}$ which satisfy

$$\mathbf{X}_2 = \mathbf{G}(\mathbf{X}_1), \quad \mathbf{X}_3 = \mathbf{G}(\mathbf{X}_2), \quad \ldots, \quad \mathbf{X}_K = \mathbf{G}(\mathbf{X}_{K-1}), \quad \mathbf{X}_1 = \mathbf{G}(\mathbf{X}_K)$$

More compactly, we may write $\mathbf{X}_i = \mathbf{G}^{(K)}(\mathbf{X}_i)$ where $\mathbf{G}^{(K)} = \mathbf{G}(\mathbf{G}(\ldots(\mathbf{G}(\cdot))\ldots))$ denotes $\mathbf{G}$ applied K times to the argument of the map; this is called the K-th iterate of $\mathbf{G}$.

Subharmonic periodic solutions occur in systems which contain two or more competing frequencies, such as forced oscillators or sampled-data circuits. Subharmonic solutions also arise following period-doubling bifurcations (see section 6.1.5).

6.1.2.4 Quasiperiodic steady-state

The next most complicated form of steady-state behaviour is called *quasiperiodicity*. In state space, this corresponds to a *torus* (see Figure 5a). While a small piece of a limit cycle in $\mathbb{R}^3$ looks like a line, a small section of two-torus looks like a plane; a two-torus has dimension two.

A quasiperiodic function is one which may be expressed as a countable sum of periodic functions with incommensurate frequencies, i.e. frequencies which are not rationally related. For example, $X(t) = \sin(t) + \sin(2\pi t)$ is a quasiperiodic signal. In the time domain, a quasiperiodic signal may look like an amplitude- or phase-modulated waveform.

While the Fourier spectrum of a periodic signal consists of a discrete set of spikes at integer multiples of a fundamental frequency, that of a quasiperiodic solution comprises a discrete set of spikes at *incommensurate* frequencies, as shown in Figure 5a.

In principle, a quasiperiodic signal may be distinguished from a periodic one by determining whether or not the frequency spikes in the Fourier spectrum are harmonically related. In practice, it is impossible to determine whether a measured number is rational or irrational; therefore, any spectrum which appears to be quasiperiodic may simply be periodic with an extremely long period.

A two-torus in a three-dimensional state space looks like a doughnut. Quasiperiodic behaviour on a higher dimensional torus is more difficult to visualize in state space but appears in the power spectrum as a set of discrete components at incommensurate frequencies. A K-torus has dimension K.

Quasiperiodic behaviour occurs in discrete-time systems where two incommensurate frequencies are present. A periodically-forced or discrete-time dynamical system has a frequency associated with the period of the forcing or sampling interval of the system; if a second frequency is introduced which is not rationally related to the period of the forcing or the sampling interval then quasiperiodicity may occur.

Example: Discrete torus

Consider a map from the circle S^1 onto itself. In polar coordinates, a point on the circle is parametrized by an angle θ. Assume that θ has been normalized so that one complete revolution of the circle corresponds to a change in θ of 1. The state of this system is determined by the normalized angle θ and the dynamics by

$$\theta_{k+1} = (\theta_k + \Omega) \bmod 1$$

If Ω is a rational number (of the form J/K), then the steady-state solution is a period-K (subharmonic) orbit. If Ω is irrational, we obtain quasiperiodic behaviour.

6.1.2.5 Chaotic steady-state

DC equilibrium, periodic, and quasiperiodic steady-state behaviours have been correctly identified and classified since the pioneering days of electronics in the 1920s. By contrast, the existence of more exotic steady-state behaviours in electronic circuits has only been acknowledged in the past thirty years. While the notion of chaotic behaviour in dynamical systems has existed in the mathematics literature since the turn of the century, unusual behaviours in the physical sciences as recently as the 1960s were described as "strange." Today, we classify as *chaos* low-dimensional nonwandering motion in a deterministic dynamical system which exhibits both "randomness" and "order" [3].

From a experimentalist's point of view, chaos may be defined as *bounded steady-state behaviour which is not an equilibrium point, not periodic, and not quasiperiodic* [6].

Chaos is characterized by repeated stretching and folding of a bundle of trajectories in state space. Two trajectories started from almost identical initial conditions diverge and soon become uncorrelated; this is called *sensitive dependence on initial conditions* and gives rise to long-term unpredictability.

In the time domain, a chaotic trajectory is neither periodic nor quasiperiodic but looks "random." This "randomness" manifests itself in the frequency domain as a broad "noise-like" Fourier spectrum, as shown in Figure 5c.

While an equilibrium point, a limit cycle, and a K-torus each have integer *dimension*, the repeated stretching and folding of trajectories in a chaotic steady-state gives the limit set a more complicated structure which, for three-dimensional continuous-time circuits, is something more than a surface but not quite a volume. A set which has non-integer dimension is called a *fractal*.

6.1.3 Stability of steady-state trajectories

Consider once more the nonlinear RLC circuit shown in Figure 2. If G_a' is negative, this circuit settles to a periodic steady-state from almost every initial condition. However, a trajectory started from the origin will, in principle, remain indefinitely at the origin since this is an equilibrium point. The circuit has *two* possible steady-state solutions. Experimentally, only the limit cycle will be observed. Why?

If trajectories starting from states close to a limit set *converge* to that steady-state, the limit set is called an *attracting limit set*. If, in addition, the attracting limit set contains at least one trajectory which comes arbitrarily close to every point in the set, then it is an *attractor*. If nearby points *diverge* from the limit set, it is called a *repellor*.

In the nonlinear RLC circuit with $G_a' < 0$, the equilibrium point is a repellor and the limit cycle is an attractor.

6.1.3.1 Stability of equilibrium points

Qualitatively, an equilibrium point is said to be *stable* if trajectories starting close to it remain nearby for all future time and unstable otherwise. Stability is a *local* concept, dealing with trajectories in a small neighbourhood of the equilibrium point.

To analyze the behaviour of the vector field in the vicinity of an equilibrium point $\mathbf{X_Q}$, we write $\mathbf{X} = \mathbf{X_Q} + \mathbf{x}$ and substitute into (3) to obtain

$$\begin{aligned} \dot{\mathbf{X}}_\mathbf{Q} + \dot{\mathbf{x}} &= \mathbf{F}(\mathbf{X_Q} + \mathbf{x}) \\ \mathbf{F}(\mathbf{X_Q}) + \dot{\mathbf{x}} &\approx \mathbf{F}(\mathbf{X_Q}) + D_\mathbf{X}\mathbf{F}(\mathbf{X_Q})\mathbf{x} \end{aligned}$$

where we have kept just the first two terms of the Taylor series expansion of $\mathbf{F}(\mathbf{X})$ about $\mathbf{X_Q}$. The *Jacobian matrix* $D_\mathbf{X}\mathbf{F}(\mathbf{X})$ is the matrix of partial derivatives of $\mathbf{F}(\mathbf{X})$ with respect to $\mathbf{X}$

$$D_\mathbf{X}\mathbf{F}(\mathbf{X}) = \begin{bmatrix} \frac{\partial F_1(\mathbf{X})}{\partial X_1} & \frac{\partial F_1(\mathbf{X})}{\partial X_2} & \cdots & \frac{\partial F_1(\mathbf{X})}{\partial X_n} \\ \frac{\partial F_2(\mathbf{X})}{\partial X_1} & \frac{\partial F_2(\mathbf{X})}{\partial X_2} & \cdots & \frac{\partial F_2(\mathbf{X})}{\partial X_n} \\ \vdots & \vdots & \ddots & \vdots \\ \frac{\partial F_n(\mathbf{X})}{\partial X_1} & \frac{\partial F_n(\mathbf{X})}{\partial X_2} & \cdots & \frac{\partial F_n(\mathbf{X})}{\partial X_n} \end{bmatrix}$$

Subtracting $\mathbf{F}(\mathbf{X_Q})$ from both sides of (8), we obtain the *linear* system

$$\dot{\mathbf{x}} = D_\mathbf{X}\mathbf{F}(\mathbf{X_Q})\mathbf{x} \tag{8}$$

where the Jacobian matrix is evaluated at $\mathbf{X_Q}$. This *linearization* describes the behaviour of the circuit in the vicinity of $\mathbf{X_Q}$; we call this the *local* behaviour.

Note that the linearization is simply the *small-signal equivalent circuit* at the operating point $\mathbf{X_Q}$. In general, the local behaviour of a circuit depends explicitly on the operating point $\mathbf{X_Q}$. For example, a pn-junction diode exhibits a small incremental resistance under forward bias but a large small-signal resistance under reverse bias.

6.1.3.2 Eigenvalues

If $\mathbf{X_Q}$ is an *equilibrium point* of (3), a complete description of its *stability* is contained in the eigenvalues of the linearization of (3) about $\mathbf{X_Q}$. These are defined as the roots λ of the *characteristic equation*

$$\det(\lambda\mathbf{I} - D_{\mathbf{X}}\mathbf{F}(\mathbf{X_Q})) = 0 \tag{9}$$

where $\mathbf{I}$ is the identity matrix.

If the real parts of all of the eigenvalues of $D_{\mathbf{X}}\mathbf{F}(\mathbf{X_Q})$ are strictly negative, then the equilibrium point $\mathbf{X_Q}$ is *asymptotically stable* and is called a *sink* because all nearby trajectories converge towards it.

If *any* of the eigenvalues has a positive real part, the equilibrium point is *unstable*; if *all* of the eigenvalues have positive real parts, the equilibrium point is called a *source*. An equilibrium point which has eigenvalues with both negative and positive real parts is called a *saddle*. A saddle is unstable.

An equilibrium point is said to be *hyperbolic* if all of the eigenvalues of $D_{\mathbf{X}}\mathbf{F}(\mathbf{X_Q})$ have *non-zero* real parts. All hyperbolic equilibrium points are either *unstable* or *asymptotically stable.*

6.1.3.3 Discrete-time systems

The stability of a fixed point $\mathbf{X_Q}$ of a discrete-time dynamical system

$$\mathbf{X}_{k+1} = \mathbf{G}(\mathbf{X}_k)$$

is determined by the eigenvalues of the linearization $D_{\mathbf{X}}\mathbf{G}(\mathbf{X_Q})$ of the vector field $\mathbf{G}$, evaluated at $\mathbf{X_Q}$.

The equilibrium point is classified as *stable* if all of the eigenvalues of $D_{\mathbf{X}}\mathbf{G}(\mathbf{X_Q})$ are strictly less than unity in modulus, and *unstable* if any has modulus greater than unity.

6.1.3.4 Eigenvectors, eigenspaces, stable and unstable manifolds

Associated with each distinct eigenvalue λ of the Jacobian matrix $D_{\mathbf{X}}\mathbf{F}(\mathbf{X_Q})$ is an eigenvector $\vec{v}$ defined by

$$D_{\mathbf{X}}\mathbf{F}(\mathbf{X_Q})\vec{v} = \lambda\vec{v}$$

A real eigenvalue γ has a real eigenvector $\vec{\eta}$. Complex eigenvalues of a real matrix occur in pairs of the form $\sigma \pm j\omega$. The real and imaginary parts of the associated eigenvectors $\vec{\eta_r} \pm j\vec{\eta_c}$ span a plane called a *complex eigenplane.*

The n_s-dimensional subspace of $\mathbb{R}^n$ associated with the stable eigenvalues of the Jacobian matrix is called the *stable eigenspace*, denoted $E^s(\mathbf{X_Q})$. The n_u-dimensional subspace corresponding to the unstable eigenvalues is called the *unstable eigenspace,* denoted $E^u(\mathbf{X_Q})$.

The analogs of the stable and unstable eigenspaces for a general nonlinear system are called the local stable and unstable *manifolds*[4] $W^s(\mathbf{X_Q})$ and $W^u(\mathbf{X_Q})$.

---

[4]An m-dimensional *manifold* is a geometrical object every small section of which looks like $\mathbb{R}^m$. More precisely, M is an m-dimensional manifold if, for every $x \in M$, there exists an open neighbourhood U of x and a smooth invertible map which takes U to some open neighbourhood of $\mathbb{R}^m$. For example, a limit cycle of a continuous-time dynamical

The stable manifold $W^s(X_Q)$ is defined as the set of all states from which trajectories remain in the manifold and converge under the flow to X_Q. The unstable manifold $W^u(X_Q)$ is defined as the set of all states from which trajectories remain in the manifold and diverge under the flow from X_Q.

By definition, the stable and unstable manifolds are *invariant* under the flow (if $X \in W^s$, then $\phi_t(X) \in W^s$). Futhermore, the n_s- and n_u-dimensional *tangent* spaces to W^s and W^u at X_Q are E^s and E^u (as shown in Figure 6). In the special case of a linear or affine vector field **F**, the stable and unstable manifolds are simply the eigenspaces E^s and E^u themselves.

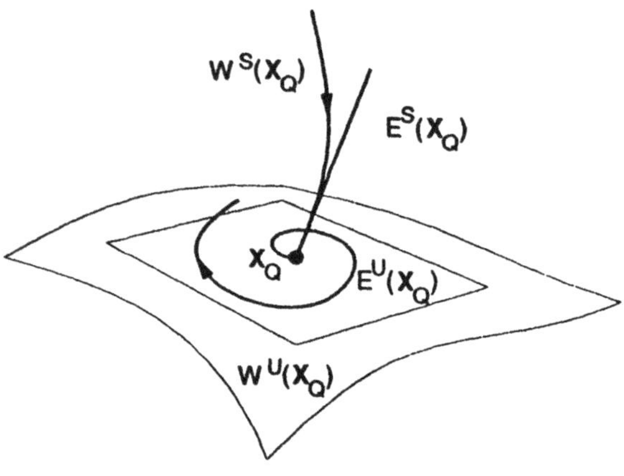

Figure 6 Stable and unstable manifolds $W^s(X_Q)$ and $W^u(X_Q)$ of an equilibrium point X_Q. The stable and unstable eigenspaces $E^s(X_Q)$ and $E^u(X_Q)$ derived from the linearization of the vector field at X_Q are tangent to the corresponding manifolds W^s and W^u at X_Q. A trajectory approaching the equilibrium point along the stable manifold is tangential to $E^s(X_Q)$ at X_Q; a trajectory leaving X_Q along the unstable manifold is tangential to $E^u(X_Q)$ at X_Q.

Chaos is associated with two characteristic connections of the stable and unstable manifolds. A *homoclinic* orbit (shown in Figure 7a) joins an isolated equilibrium point X_Q to itself along its stable and unstable manifolds. A *heteroclinic* orbit (figure 7b) joins two distinct equilibrium points X_{Q_1} and X_{Q_2} together along the unstable manifold of one and the stable manifold of the other.

6.1.3.5 Stability of limit cycles

While the stability of an equilibrium point may be determined by considering the eigenvalues of the linearization of the vector field near the point, how does one study the stability of a limit cycle, torus, or chaotic steady-state trajectory?

The idea introduced by Poincaré is to convert a continuous-time dynamical system into an equivalent discrete-time dynamical system by taking a transverse slice through the

---

system is a one-dimensional manifold.

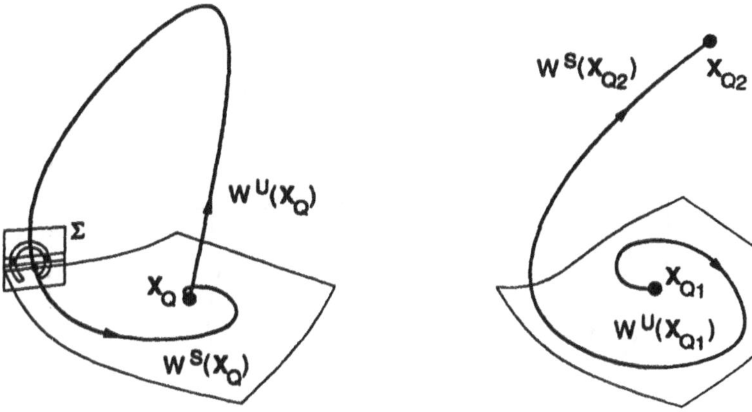

Figure 7 (a) A homoclinic orbit joins an isolated equilibrium point X_Q to itself along its stable and unstable manifolds. (b) A heteroclinic orbit joins two distinct equilibrium points X_{Q_1} and X_{Q_2} together along the unstable manifold of one and the stable manifold of the other.

flow. Intersections of trajectories with this so-called *Poincaré section* define a *Poincaré map* from the section to itself. Since the limit cycle is a fixed point $\mathbf{X}_Q$ of the associated discrete-time dynamical system, its stability may be determined by examining the eigenvalues of the linearization of the Poincaré map at $\mathbf{X}_Q$.

6.1.3.6 Poincaré sections

A *Poincaré section* of an n-dimensional autonomous continuous-time dynamical system is an $(n-1)$-dimensional hyperplane Σ in the state space which is intersected *transversally*[5] by the flow.

Let Γ be a closed orbit of the flow of a smooth vector field $\mathbf{F}$, and let $\mathbf{X}_Q$ be a point of intersection of Γ with Σ. If T is the period of Γ and $\mathbf{X} \in \Sigma$ is sufficiently close to $\mathbf{X}_Q$, then the trajectory $\phi_t(\mathbf{X})$ through $\mathbf{X}$ will return to Σ after a time $\tau(\mathbf{X}) \approx T$ and intersect the hyperplane at a point $\phi_{\tau(\mathbf{X})}(\mathbf{X})$, as shown in Figure 8.

This construction implicitly defines a function (called a *Poincaré map* or *first return map*) $\mathbf{G} : U \to \Sigma$

$$\mathbf{G}(\mathbf{X}) = \phi_{\tau(\mathbf{X})}(\mathbf{X})$$

where U is a small region of Σ close to $\mathbf{X}_Q$. The corresponding discrete-time dynamical system

$$\mathbf{X}_{k+1} = \mathbf{G}(\mathbf{X}_k)$$

[5]A *transverse* intersection of manifolds in $\mathbb{R}^n$ is an intersection of manifolds such that, from any point in the intersection, all directions in $\mathbb{R}^n$ can be generated by linear combinations of vectors tangent to the manifolds.

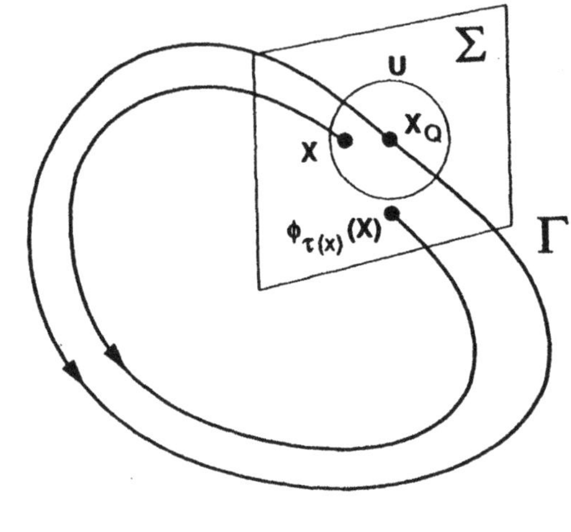

Figure 8 A transverse Poincaré section Σ through the flow of a dynamical system induces a discrete Poincaré map from a neighbourhood U of the point of intersection X_Q to Σ.

has a fixed point at X_Q.

The stability of the limit cycle is determined by the eigenvalues of the linearization $D_X G(X_Q)$ of G at X_Q. If all of the eigenvalues of $D_X G(X_Q)$ have modulus less than unity, the limit cycle is *asymptotically stable*; if any has modulus greater than unity, the limit cycle is *unstable*.

Note that the stability of the limit cycle is independent of the position and orientation of the Poincaré plane, provided that the intersection is chosen *transverse* to the flow. For a non-autonomous system with periodic forcing, a natural choice for the hyperplane is at a fixed phase θ_0 of the forcing.

In the Poincaré section, a limit cycle looks like a fixed point. A period-K subharmonic of a non-autonomous system with periodic forcing appears as a period-K orbit of the corresponding map.

The Poincaré section of a quasiperiodic attractor consisting of two incommensurate frequencies looks like a closed curve—a transverse cut through a two-torus.

The Poincaré section of a chaotic attractor has fractal structure.

6.1.4 Horseshoes and chaos

Chaotic behaviour is characterized by *sensitive dependence on initial conditions*. This phrase emphasizes the fact that small differences in initial conditions are persistently magnified by the dynamics of the system so that trajectories starting from nearby initial conditions reach totally different states in a finite time.

Trajectories of the nonlinear *RLC* circuit in Figure 2 which originate near the equilibrium point are initially stretched apart exponentially by the locally negative resistance in the case $G'_a < 0$. Eventually, however, they are squeezed together onto a limit cycle, so the stretching is not persistent. This is a consequence of the non-crossing property and eventual passivity.

Although perhaps locally active, every physical resistor is *eventually passive* meaning that, for a large enough voltage across its terminals, it dissipates power. This in turn limits the maximum values of the voltages and currents in the circuit giving a *bounded* steady-state solution. All physical systems are bounded so how can small differences be magnified *persistently* in a real circuit?

6.1.4.1 Chaos in the sense of Shil'nikov

Consider a flow ϕ in $\mathbb{R}^3$ which has an equilibrium point at the origin with a real eigenvalue $\gamma > 0$ and a pair of complex conjugate eigenvalues $\sigma \pm j\omega$ with $\sigma < 0$ and $\omega \neq 0$. Assume that the flow has a homoclinic orbit Γ through the origin.

One may define a Poincaré map for this system by taking a transverse section through the homoclinic orbit, as shown in Figure 7a.

Theorem 2 Shil'nikov

If $|\sigma/\gamma| < 1$, the flow ϕ can be perturbed to ϕ' such that ϕ' has a homoclinic orbit Γ' near Γ and the Poincaré map of ϕ' defined in a neighbourhood of Γ' has a countable number of horseshoes in its discrete dynamics.

The characteristic horseshoe shape in the Poincaré map stretches and folds trajectories repeatedly (see Figure 9). The resulting dynamics exhibit extreme sensitivity to initial conditions [4].

The presence of horseshoes in the flow of a continuous-time system which satisfies the assumptions of Shil'nikov's theorem implies the existence of a countable number of unstable periodic orbits of arbitrarily long period as well as an uncountable number of complicated bounded nonperiodic *chaotic* solutions [4].

Horseshoes

The action of the Smale horseshoe map is to take the unit square (figure 9a), stretch it, fold it into a horseshoe shape (figure 9b), and lay it down on itself (figure 9c). Under the action of this map, only four regions of the unit square are returned to the square.

Successive iterations of the horseshoe map return smaller and smaller regions of square to itself, as shown in Figures 9d – 9f. If the map is iterated *ad infinitum*, the unit square is ultimately mapped onto a set of points. These points form an invariant (fractal) limit set L which contains a countable set of periodic orbits of arbitrarily long periods, an uncountable set of bounded non-periodic orbits, and at least one orbit which comes arbitrarily close to every point in L.

The properties of the map still hold if the horseshoe is distorted by a perturbation of small size but arbitrary shape. Thus, the dynamical behaviour of the horseshoe map is *structurally stable* (see section 6.1.5).

While the invariant limit set of a horseshoe map consists of non-wandering points, it is not attracting. Therefore, the existence of a horseshoe in the flow of a third-order system does not *imply* that the system will exhibit chaotic steady-state behaviour. However if a typical trajectory in the Poincaré map remains in a neighbourhood of the invariant set, then the system may exhibit chaos. Thus, while Shil'nikov's theorem is a strong *indicator* of chaos, it is not a definitive proof that a system is chaotic.

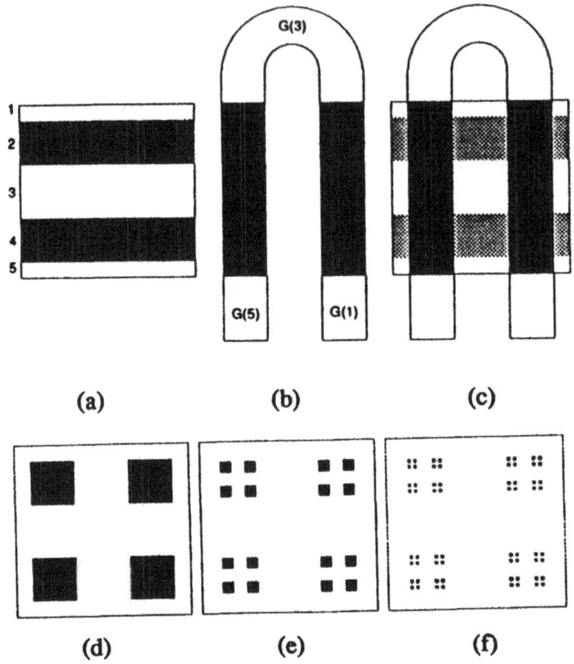

Figure 9 The Smale horsehoe map stretches the unit square (a), folds it into a horseshoe (b), and lays it back on itself (c), so that only points lying in bands 2 and 4 of (a) are mapped into the square. At the next iteration, only those points in (G(2)∪G(4))∩(2∪4) (d) are mapped back to the square. Repeated iterations of the map (d)–(f) remove all points from the square except an invariant (fractal) set of fixed points.

Example: chaos in a piecewise-linear system

Although we have stated them for the case $\sigma < 0, \gamma > 0$, Shil'nikov's theorem and the corollary also apply when the equilibrium point at the origin has an *unstable* pair of complex conjugate eigenvalues and a *stable* real eigenvalue. In that case, it is somewhat easier to visualize the stretching and folding of trajectories close to a homoclinic orbit.

Consider the trajectory in a three-region piecewise-linear vector field shown in Figure 10. We assume that the equilibrium point P_- has a stable real eigenvalue γ_1 (whose eigenvector is $E^r(P_-)$) and an unstable complex conjugate pair of eigenvalues $\sigma_1 \pm j\omega_1$, the real and imaginary parts of whose eigenvectors span the plane $E^c(P_-)$ [1], as shown. A trajectory originating from a point X_0 on $E^c(P_-)$ spirals away from the equilibrium point along $E^c(P_-)$ until it enters the D_0 region, where it is folded back into D_{-1}. Upon re-entering D_{-1}, the trajectory is pulled towards P_- roughly in the direction of the real eigenvector $E^r(P_-)$, as shown.

Now imagine what would happen if the trajectory entering D_{-1} from D_0 were in *precisely* the direction $E^r(P_-)$. Such a trajectory would follow $E^r(P_-)$ towards P_-,

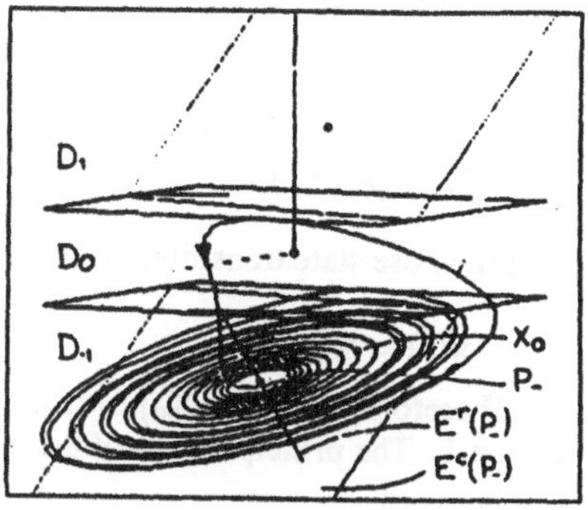

Figure 10 Stretching and folding mechanism of chaos generation in Chua's circuit [7]. A trajectory spirals away from the equilibrium point P_- along the eigenplane $E^c(P_-)$ until it enters the D_0 region, where it is folded back into D_{-1} and returns to the unstable eigenplane $E^c(P_-)$ close to P_-.

reaching the equilibrium point asymptotically as $t \to \infty$. Similarly, if we were to follow this trajectory *backwards* in time though D_0 and back onto $E^c(P_-)$ in D_{-1}, it would then spiral towards P_-, reaching it asymptotically as $t \to -\infty$. The closed curve thus formed would be a homoclinic orbit, reaching the same equilibrium point P_- asymptotically in forward and reverse time.

While the homoclinic orbit itself is not structurally stable, and therefore cannot be observed experimentally, horseshoes *are* structurally stable. A flow ϕ which satisfies the assumptions of Shil'nikov's theorem contains a countable infinity of horseshoes; for sufficiently small perturbations ϕ' of the flow, finitely many of the horseshoes will persist. Thus, both the original flow and the perturbed flow exhibit chaos in the sense of Shil'nikov.

In Figure 10, we see that a trajectory lying close to a homoclinic orbit exhibits similar qualitative behaviour: it spirals away from P_- along the unstable complex plane $E^c(P_-)$, is folded in D_0, re-enters D_{-1} above $E^c(P_-)$, and is pulled back towards $E^c(P_-)$ only to be spun away from P_- once more.

Thus, two trajectories starting from distinct initial states close to P_- on $E^c(P_-)$ are stretched apart exponentially along the unstable eigenplane before being folded in D_1 and reinjected close to P_-; this gives rise to sensitive dependence on initial conditions. The recurrent stretching and folding continues *ad infinitum*, producing a chaotic steady-state solution.

6.1.4.2 Lyapunov exponents

The notion of sensitive dependence on initial conditions may be made more precise through the introduction of Lyapunov exponents (LEs). Lyapunov exponents quantify the average exponential rates of separation of trajectories along the flow.

The flow in a neighbourhood of asymptotically stable trajectory is contracting so the LEs are zero or negative. Sensitive dependence on initial conditions results from a positive LE.

To determine the stability of an equilibrium point, we considered the eigenvalues of the linearization of the vector field in the vicinity of the equilibrium trajectory. This idea can be generalized to any trajectory of the flow.

The local behaviour of the vector field along a trajectory $\phi_t(\mathbf{X}_0)$ of an autonomous continuous-time dynamical system (3) is governed by the linearized dynamics

$$\begin{aligned}\dot{\mathbf{x}} &= D_{\mathbf{X}}\mathbf{F}(\mathbf{X})\mathbf{x}, \quad \mathbf{x}(0) = \mathbf{x}_0 \\ &= D_{\mathbf{X}}\mathbf{F}(\phi_t(\mathbf{X}_0))\mathbf{x}\end{aligned}$$

This is a linear time-varying system whose state transition matrix $\Phi_t(\mathbf{X}_0)$ maps a point $\mathbf{x}_0$ into $\mathbf{x}(t)$. Thus,

$$\mathbf{x}(t) = \Phi_t(\mathbf{X}_0)\mathbf{x}_0$$

Note that Φ_t is a linear operator. Therefore, a ball $B_\epsilon(\mathbf{X}_0)$ of radius ϵ about $\mathbf{X}_0$ is mapped into an *ellipsoid* as shown in Figure 1. The principal axes of the ellipsoid are determined by the singular values of Φ_t.

The singular values $\sigma_1(t), \sigma_2(t), \ldots, \sigma_n(t)$ of Φ_t are defined as the *square roots* of the eigenvalues of $\Phi_t^H\Phi_t$, where Φ_t^H is the complex conjugate transpose of Φ_t. The singular values are ordered so that $\sigma_1(t) > \sigma_2(t) > \ldots > \sigma_n(t)$.

In particular, a ball of radius ε is mapped by the linearized flow into an ellipsoid (see Figure 1), the maximum and minimum radii of which are bounded by $\sigma_1(t)\varepsilon$ and $\sigma_n(t)\varepsilon$ respectively.

The stability of a steady-state orbit is governed by the average local rates of expansion and contraction of volumes of state space close to the orbit. The *Lyapunov exponents* (LEs) λ_i are defined by

$$\lambda_i = \lim_{t\to\infty} \frac{1}{t} \ln \sigma_i(t)$$

whenever this limit exists. The LEs quantify the *average* exponential rates of separation of trajectories along the flow.

The LEs are a property of a steady-state trajectory. Any transient effect is averaged out by taking the limit as $t \to \infty$. Furthermore, the LEs are global quantities of an attracting set which depend on the local stability properties of a trajectory within the set.

The set $\{\lambda_i, i = 1, 2, \ldots, n\}$ is called the *Lyapunov spectrum*. An attractor has the property that the sum of its LEs is negative.

6.1.4.3 Lyapunov exponents of steady-state solutions

All of the eigenvalues of a stable equilibrium point have negative real parts and therefore the largest Lyapunov exponent of an attracting equilibrium point is negative.

Trajectories close to a stable limit cycle converge onto the limit cycle. Therefore, the largest LE of a periodic steady-state is zero (corresponding to motion along the limit cycle [6]) and all of its other LEs are negative.

A quasiperiodic K-torus has K zero LEs since the flow is locally neither contracting nor expanding along the surface of the K-torus.

On average, a chaotic trajectory is *locally unstable* and therefore has a *positive* LE; this produces sensitive dependence on initial conditions. Nevertheless, in the case of a chaotic attractor, this locally unstable chaotic trajectory belongs to an *attracting* limit set to which nearby trajectories converge.

The steady-state behaviour of a four-dimensional continuous-time dynamical system which has two positive, one zero, and one negative LE is called *hyperchaos*.

The Lyapunov spectrum may be used to identify attractors, as summarized in table 1.

Steady-state	Limit set	Spectrum	LEs	Dimension
DC	fixed point	spike at DC	$0 > \lambda_1 \geq \cdots \geq \lambda_n$	0
Periodic	closed curve	fundamental plus integer harmonics	$\lambda_1 = 0$, $0 > \lambda_2 \geq \cdots \geq \lambda_n$	1
Quasiperiodic	K-torus	incommensurate frequencies	$\lambda_1 = \ldots = \lambda_K = 0$, $0 > \lambda_{K+1} \geq \cdots \geq \lambda_n$	K
Chaotic	fractal	broad spectrum	$\lambda_1 > 0$, $\sum_{i=1}^{n} \lambda_i < 0$	non-integer

Table 1 Classification of steady-state behaviours according to their limit sets, power spectra, LEs, and dimension.

6.1.5 Structural stability and bifurcations

Structural stability refers to the sensitivity of a phenomenon to small changes in the parameter of a system. A structurally stable vector field $\mathbf{F}$ is one for which sufficiently close vector fields $\mathbf{F}'$ have *equivalent*[6] dynamics [3].

The behaviour of a typical circuit depends on a set of parameters one or more of which may be varied in order to optimize some performance criteria. In particular, one may think of a one-parameter family of systems

$$\dot{\mathbf{X}} = \mathbf{F}_\mu(\mathbf{X}) \tag{10}$$

whose vector field is parametrized by a *control parameter* μ. A value μ_0 of equation (10) for which the flow of (10) is not structurally stable is a *bifurcation value* of μ [4].

The dynamics in the state space may be qualitatively very different from one value of μ to another. In the nonlinear RLC circuit example, the steady-state solution is a limit cycle if the control parameter G_a' is negative and an equilibrium point if G_a' is positive. If G_a' is identically equal to zero, trajectories starting from $I_{30} = 0, V_{20} < E$ yield sinusoidal solutions. These sinusoidal solutions are not structurally stable because the slightest perturbation of G_a' will cause the oscillation to decay to zero or converge to the limit cycle, depending on whether G_a' is made slightly larger or smaller than zero. If we think of this circuit as being parametrized by G_a', then its vector field is not structurally

[6] *Equivalent* means that there exists a continuous invertible function h which transforms $\mathbf{F}$ into $\mathbf{F}'$.

stable at $G'_a \equiv 0$. We say that the equilibrium point undergoes a *bifurcation* (from stability to instability) as the value of the *bifurcation parameter* G'_a is reduced through the bifurcation point $G'_a = 0$.

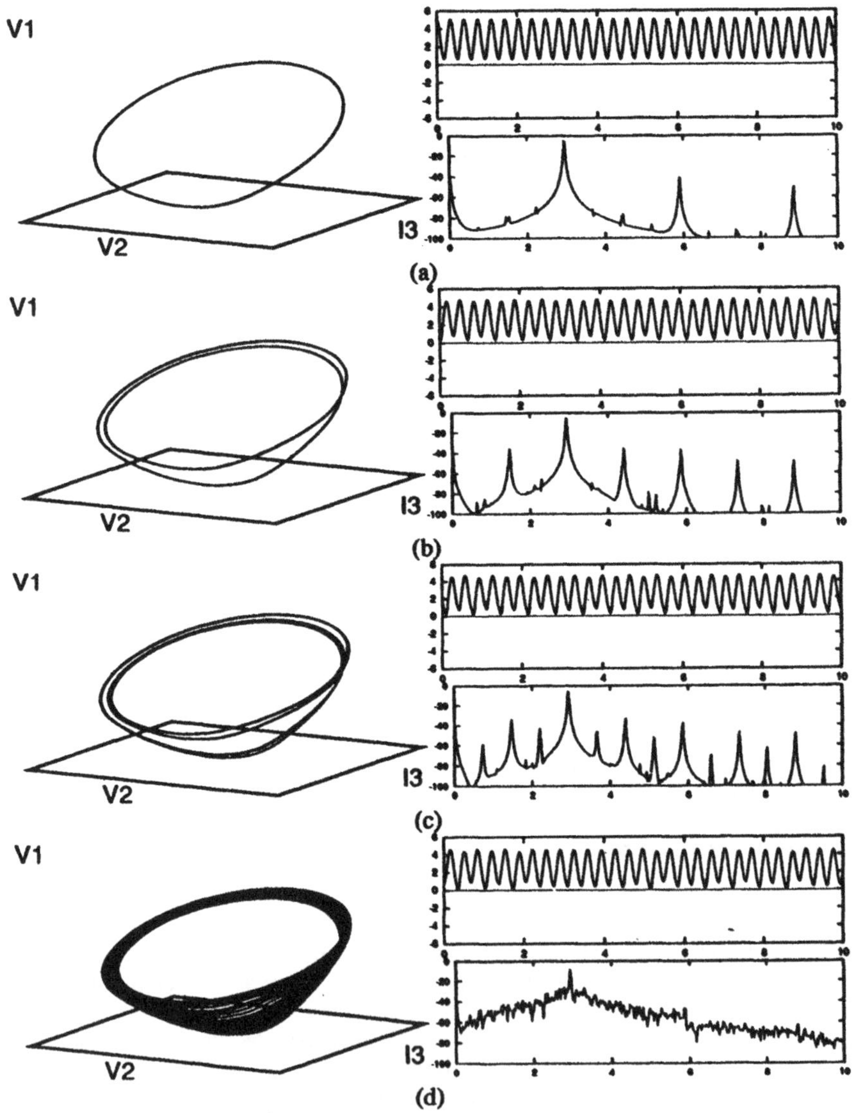

Figure 11 Period-doubling route to chaos in Chua's oscillator. Simulated state space trajectories, time waveforms $V_1(t)$, and power spectra of $V_2(t)$. (a) periodic steady-state; (b) period-two; (c) period-four; (d) Spiral attractor. Time plots: horizontal axis—t (ms); vertical axis—V_1 (V). Power spectra: horizontal axis—frequency (kHz); vertical axis—power (mean squared amplitude of $V_2(t)$) (dB).

6.1.5.1 Bifurcation types

In this section, we consider three types of local bifurcation: the *Hopf bifurcation*, the *saddle-node* bifurcation, and the *period-doubling* bifurcation [3]. These bifurcations are called *local* because they may be understood by linearizing the system close to an equilibrium point or limit cycle.

(i) Hopf bifurcation

A *Hopf bifurcation* occurs in a continuous-time dynamical system (3) when a simple pair of complex conjugate eigenvalues of the linearization $D_{\mathbf{X}}\mathbf{F}(\mathbf{X}_Q)$ of the vector field at an equilibrium point $\mathbf{X}_Q$ crosses the imaginary axis.

Typically, the equilibrium point changes stability from stable to unstable and a stable limit cycle is born. The bifurcation at $G_a'' \equiv 0$ in the nonlinear RLC circuit is Hopf-like.[7]

When an equilibrium point undergoes a Hopf bifurcation, a limit cycle is born. When a limit cycle undergoes a Hopf bifurcation, motion on a two-torus results.

(ii) Saddle-node bifurcation

A *saddle-node* bifurcation occurs when a stable and an unstable equilibrium point merge and disappear; this typically manifests itself as the abrupt disappearance of an attractor.

A common example of a saddle-node bifurcation in electronic circuits is switching between equilibrium states in a Schmitt trigger. At the threshold for switching, a stable equilibrium point corresponding to the "high" saturated state merges with the high-gain region's unstable saddle-type equilibrium point and disappears. After a switching transient, the trajectory settles to the other stable equilibrium point, which corresponds to the "low" state.

A saddle-node bifurcation may also manifest itself as a switch between periodic attractors of different size, between a periodic attractor and a chaotic attractor, or between a limit cycle at one frequency and a limit cycle at another frequency.

(iii) Period-doubling bifurcation

A *period-doubling* bifurcation occurs in a discrete-time dynamical system (7) when a real eigenvalue of the linearization $D_{\mathbf{X}}\mathbf{G}(\mathbf{X}_Q)$ of the map $\mathbf{G}$ at an equilibrium point crosses the unit circle at -1 [4].

In a continuous-time system, a period-doubling bifurcation occurs only from a periodic solution (an equilibrium point of the Poincaré map). At the bifurcation point, a periodic orbit with period T changes smoothly into one with period $2T$, as shown in Figures 11a and 11b.

6.1.5.2 Routes to chaos

Each of the three local bifurcations may give rise to a distinct route to chaos, and all three have been reported in electronic circuits. These routes are important because it is often difficult to conclude from experimental data alone whether irregular behaviour is due to measurement noise or to underlying chaotic dynamics. If, upon adjusting a control parameter, one of the three prototype routes is observed, this indicates that the dynamics might be chaotic.

---

[7]Note that the Hopf bifurcation theorem is proven for sufficiently smooth systems and does not strictly apply to piecewise-linear systems. However, a *physical* implementation of a piecewise-linear characteristic, such as that of N_R is always smooth.

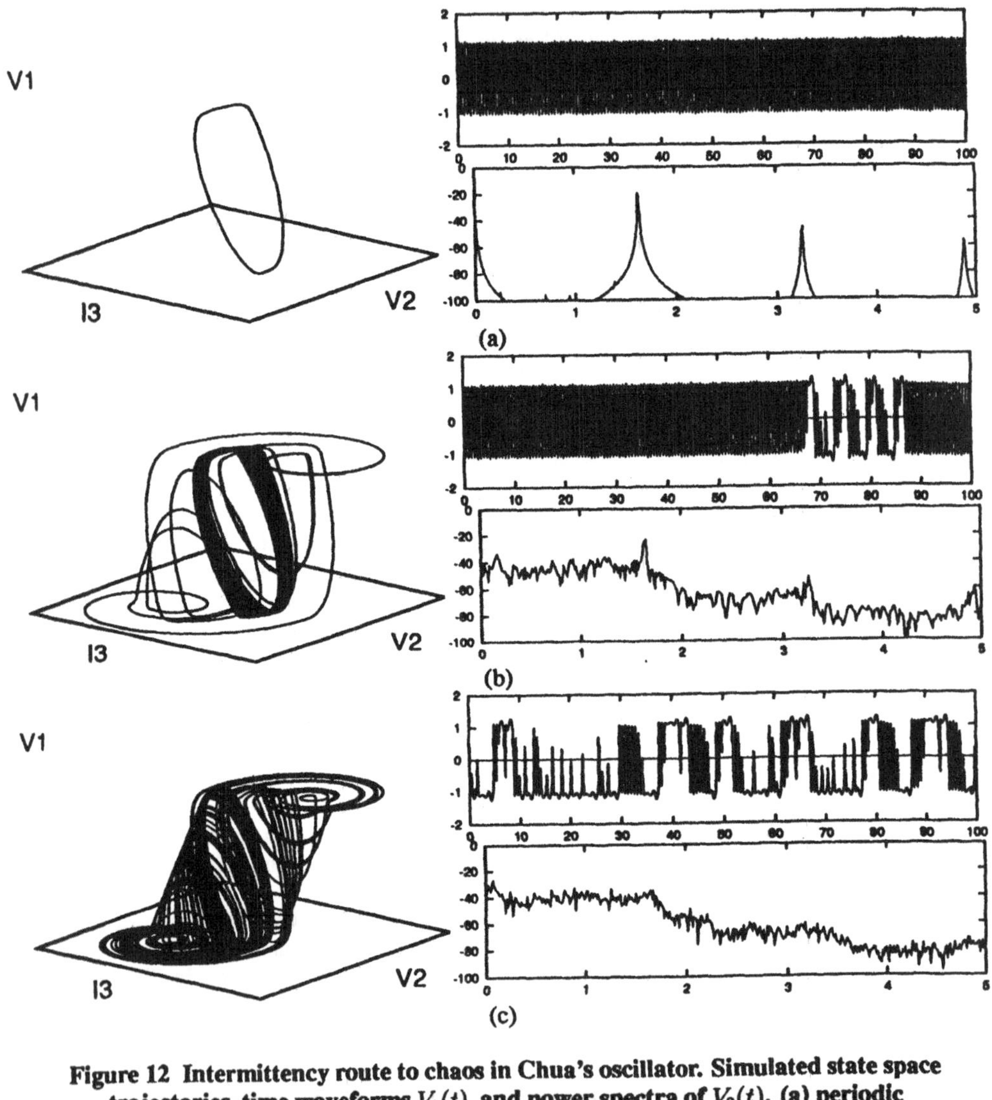

Figure 12 Intermittency route to chaos in Chua's oscillator. Simulated state space trajectories, time waveforms $V_1(t)$, and power spectra of $V_2(t)$. (a) periodic steady-state—the signal is characterized by a discrete power spectrum with energy at integer multiples of the fundamental frequency; (b) onset of intermittency—the time signal contains long regular "laminar" phases and occasional "bursts" of irregular motion—in the frequency domain, intermittency manifests itself as a raising of the noise floor; (c) fully-developed chaos—laminar phases are infrequent and the power spectrum is broad. Time plots: horizontal axis—t (ms); vertical axis—V_1 (V). Power spectra: horizontal axis—frequency (kHz); vertical axis—power (mean squared amplitude of $V_2(t)$) (dB).

6.1.5.3 Period-doubling route to chaos

The period-doubling route to chaos is characterized by a cascade of period-doubling bifurcations. Each period-doubling transforms a limit cycle into one at half the frequency, spreading the energy of the system over a wider range of frequencies. An infinite cascade of such doublings results in a chaotic trajectory of infinite period and a broad frequency spectrum which contains energy at all frequencies. Figure 11 is a set of snapshots of the period-doubling route to chaos in Chua's oscillator.

The period-doubling route to chaos is readily identified from a state-space plot, time series, power spectrum, or a Poincaré map.

6.1.5.4 Intermittency route to chaos

The route to chaos caused by saddle-node bifurcations comes in different forms, the common feature of which is a direct transition from regular motion to chaos. The most common type is the intermittency route and results from a single saddle-node bifurcation. This is a *route* and not just a jump because straight after the bifurcation, the trajectory is characterized by long intervals of almost regular motion (called *laminar phases*) and short bursts of irregular motion. The period of the oscillations is approximately equal to that of the system just before the bifurcation. This is shown in Figure 12.

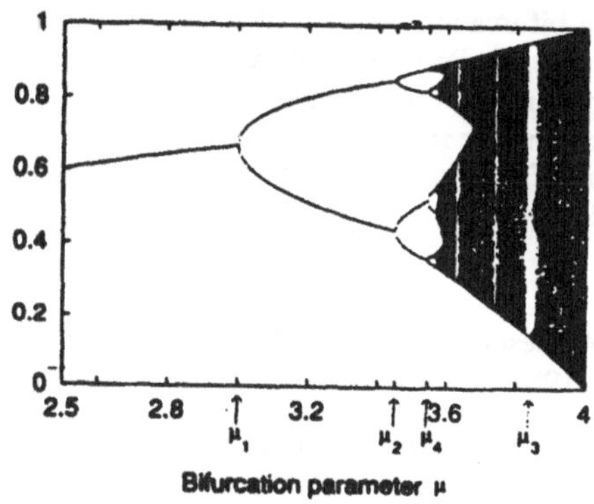

Figure 13 Bifurcation diagram for the logistic map: $X_{k+1} = \mu X_k(1 - X_k)$. The first period-doubling bifurcation occurs at $\mu = \mu_1$, the second at μ_2, and the third at μ_4. μ_3 corresponds to a period-three window. When $\mu = 4$, the entire interval (0,1) is visited by a chaotic orbit $\{X_k, k = 0, 1, \ldots\}$.

At the parameter passes through the critical value μ_c at the bifurcation point into the chaotic region, the laminar phases become shorter and the bursts become more frequent, until the regular intervals disappear altogether. The scaling law for the average interval of the laminar phases depends on $|\mu - \mu_c|$ so chaos is not fully developed until some distance from the bifurcation point [3].

Intermittency is best characterized in the time domain since its scaling law governs on the length of laminar phases.

Another type of bifurcation to chaos associated with saddle-nodes is the direct transition from a regular attractor (fixed point or limit cycle) to a coexisting chaotic one, without the phenomenon of intermittency.

6.1.5.5 Quasiperiodic (torus breakdown) route to chaos

The quasiperiodic route to chaos results from a sequence of Hopf bifurcations. Starting from a fixed point, the three-torus generated after three Hopf bifurcations is not stable in the sense that there exists an arbitrarily small perturbation of the system (in terms of parameters) for which the three-torus gives way to chaos.

A quasiperiodic-periodic-chaotic sequence corresponding to torus breakdown in Chua's oscillator is shown in Figure 5.

Quasiperiodicity is difficult to detect from a time series; it is more readily identified by means of a power spectrum or Poincaré map.

6.1.5.6 Bifurcation diagrams

While state-space, time- and frequency-domain measurements are useful for characterizing steady-state behaviours, nonlinear dynamics offers several other tools for summarizing qualitative information concerning bifurcations.

A *bifurcation diagram* is a plot of the attracting sets of a system versus a control parameter. Typically, one chooses a state variable and plots this against a single control parameter. In discrete systems, one simply plots successive values of a state variable. In the continuous-time case, some type of discretization is needed, typically by means of a Poincaré section.

Figure 13 shows a bifurcation diagram of the logistic map $X_{k+1} = \mu X_k(1 - X_k)$ for $\mu \in [2.5, 4]$ and $X_k \in [0, 1]$. Period-doubling from period-one to period-two occurs at μ_2; the next two doublings in the period-doubling cascade occur at μ_2 and μ_4 respectively. A periodic *window* in the chaotic region is indicated by μ_3. The map becomes chaotic by the period-doubling route if μ is increased from μ_3 and by the intermittency route if μ is reduced out of the window [8].

References

[1] L. O. Chua, C. A. Desoer, and E. S. Kuh, *Linear and Nonlinear Circuits*, McGraw-Hill, New York, 1987.

[2] F. M. Callier and C. A. Desoer, *Linear System Theory*, Springer-Verlag, New York, 1991.

[3] J. M. T. Thompson and H. B. Stewart, *Nonlinear Dynamics and Chaos*, Wiley, New York, 1986.

[4] J. Guckenheimer and P. Holmes, *Nonlinear Oscillations, Dynamical Systems, and Bifurcations of Vector Fields*, Springer-Verlag, New York, 1983.

[5] W. H. Press, B. P. Flannery, S. A. Teukolsky, and W. T. Vetterling, *Numerical Recipes in C*, Cambridge University Press, Cambridge, 1988.

[6] T. S. Parker and L. O. Chua, *Practical Numerical Algorithms for Chaotic Systems*, Springer-Verlag, New York, 1989.

[7] M. P. Kennedy. "Three Steps to Chaos—Part I: Evolution," *IEEE Trans. Circuits and Systems—I: Fundamental Theory and Applications*, Vol. 40, No. 10, pp. 640–656, October 1993; "Three Steps to Chaos—Part II: A Chua's Circuit Primer," *IEEE Trans. Circuits and Systems—I: Fundamental Theory and Applications*, Vol. 40, No. 10, pp. 657–674, October 1993.

[8] C. W. Wu and N. F. Rul'kov, "Studying Chaos via 1-D Maps—A Tutorial," *IEEE Trans. Circuits Syst.—I:Fundamental Theory and Applications*, Vol. 40, No. 10, pp. 707–721, October 1993.

Chapter 6.2

Synchronization Principles and Applications

Martin Hasler

6.2.1 Introduction

In this chapter, we consider a set-up where a *master system* drives a *slave system* in order to impose its wave forms. This situation is depicted schematically in Figure 1. Both systems should be thought of as being dynamic and nonlinear. In general, the slave system closely resembles the master system, except that it has an additional control input which is either driven directly with the transmitted signal $y(t)$, as shown in Figure 1, or it is driven by some error signal.

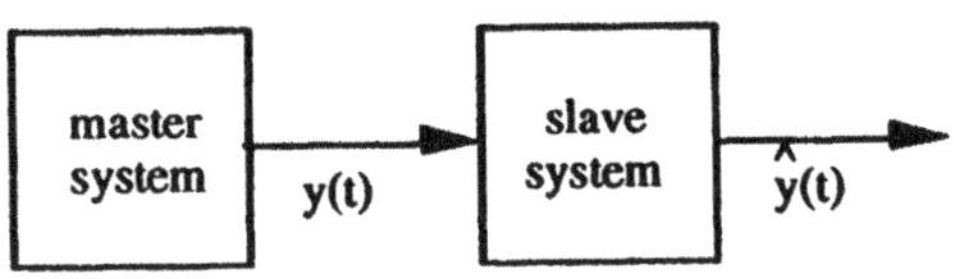

Figure 1 Master-slave set-up.

The transmitted signal $y(t)$ may have the effect to force the output $\hat{y}(t)$ of the slave system to copy its wave form. Since the signal $y(t)$ represents the only interaction between the two systems, the initial state of the two systems is not coordinated. Since the time evolution of a dynamic system depends on its initial state, we cannot expect $\hat{y}(t)$ to be identical to $y(t)$. Only at time $t \to \infty$ the influence of the initial state can be expected to fade away. This justifies the following definition.

Definition

The slave system *synchronizes* with the master system if

$$|\hat{y}(t) - y(t)| \xrightarrow[t\to\infty]{} 0 \tag{1}$$

for any combination of initial states of the master and the slave system.

Usually, the notion of *synchronization* is used for periodic signals, e.g. a quartz oscillator synchronizes a watch. This example suggests a more general definition of synchronization

$$|\hat{y}(t) - Ky(t)| \xrightarrow[t\to\infty]{} 0 \quad \text{for some } K \tag{2}$$

Still more general is the condition, which expresses the essential of the intuitive notion of synchronization

$$|\hat{y}(t) - f(y(t))| \xrightarrow[t\to\infty]{} 0 \quad \text{for a suitable function } f \tag{3}$$

We shall not need this level of generality. Definition (1) is sufficient for our purposes.

In this chapter we consider the rather peculiar case where both the master and the slave system exhibit chaotic behaviour. Thus, $y(t)$ has completely irregular behaviour, yet, the above definition makes sense even in this case. It means that, asymptotically, $\hat{y}(t)$ copies the irregular behaviour of $y(t)$.

It is surprising that the synchronization of chaotic systems is at all possible. In fact, a chaotic system has sensitive dependence on initial conditions. This means that even if the two systems in Figure 1 are perfectly identical, and if they are started at almost identical states, after some time, the two time evolutions become completely uncorrelated and (1) is far from being satisfied. This is the case as long as the two systems do not interact. But, surprisingly, the interaction may be able to force synchronization.

We shall show three methods to synchronize chaotic systems, namely

a) by decomposition into subsystems

b) by linear feedback

c) by the inverse system

Synchronization can be used in transmission systems. A useful, information carrying signal is transmitted hidden in a chaotic signal. Again, we shall present different methods used for hiding the information, namely

a) chaotic masking

b) chaotic switching

c) direct chaotic modulation

The discussion will reveal that the design of such transmission systems is very delicate because of the conflicting design objectives *robustness* and *security*.

Part of this tutorial has been inspired by the introductory text [1].

6.2.2 Syncronization by decomposition into subsystems

The idea of synchronization by decomposition into subsystems has first been proposed in [2]. In this section, we start by discussing it in terms of systems and state equations. This presentation has been inspired by [3]. We then discuss a circuit realization of the systems and the resulting conditions for synchronization.

Suppose a nonlinear dynamical system is described by state equations of the form

$$\frac{dx_1}{dt} = f_1(x, y_1) \qquad \cdots \qquad \frac{dx_n}{dt} = f_n(x, y_1)$$

$$\frac{dy_1}{dt} = g_1(x_1, y) \qquad \cdots \qquad \frac{dy_m}{dt} = g_m(x_1, y) \tag{4}$$

where $x = (x_1, \ldots, x_n)$ and $y = (y_1, \ldots, y_m)$. The system can be decomposed into two subsystems that interact only through the signals x_1 and y_1 (Figure 2). To this decomposition corresponds the simple separation of the equations in (4) into two groups, the first n equations that are the state equations for the variables x_i and the following m equations, the state equations for the variables y_i.

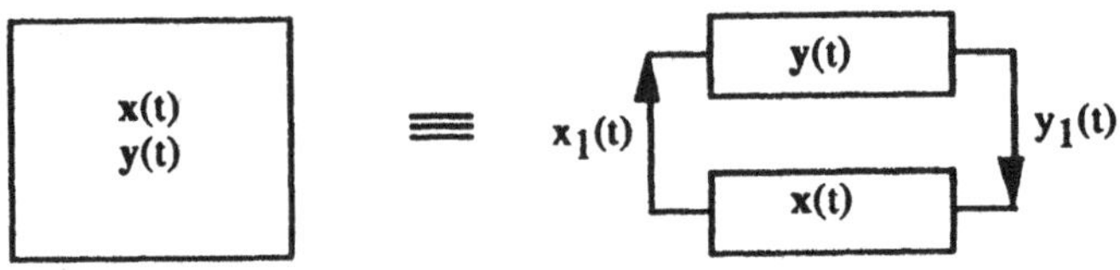

Figure 2 Decomposition of the system.

If we now cut the interaction by the signal y_1, we obtain the two subsystems connected in cascade, as shown in Figure 3.

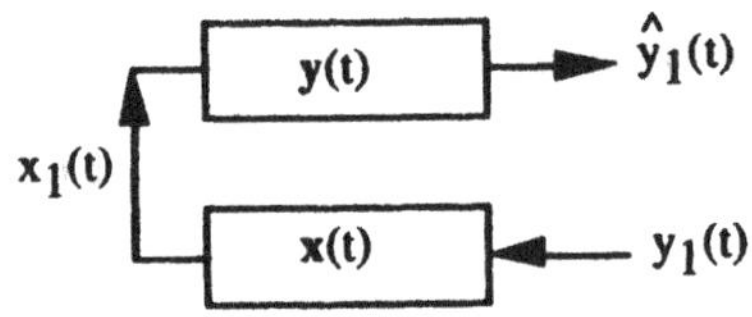

Figure 3 Decomposition cancelling the y-interaction.

This corresponds to a separation of the system of state equation (4) into two parts. Indeed, the lower system is described by

$$\frac{dx_1}{dt} = f_1(x, y_1) \qquad \cdots \qquad \frac{dx_n}{dt} = f_n(x, y_1) \tag{5}$$

and the upper system is described by

$$\frac{d\hat{y}_1}{dt} = g_1(x_1, \hat{y}) \qquad \cdots \qquad \frac{dy_m}{dt} = g_m(x_1, \hat{y}) \tag{6}$$

where $\hat{y} = (\hat{y}_1, y_2, \ldots, y_m)$.

We now would like to synchronize the system of Figure 3 with the system of Figure 2. For this purpose, we transmit the signal $y_1(t)$, as represented in Figure 4. Here, $\hat{x} = (\hat{x}_1, \ldots, \hat{x}_n)$ and $\hat{y} = (\hat{y}_1, \ldots, \hat{y}_m)$

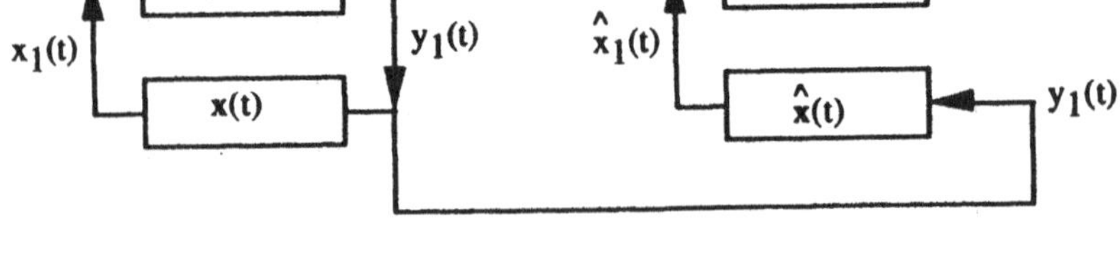

Figure 4 Master-slave set-up for synchronization by decomposition into subsystems.

Both the systems in Figure 4 started exactly at the same initial conditions $x(0) = \hat{x}(0)$, $y(0) = \hat{y}(0)$, then clearly the time evolution of the state variables in both systems would be identical, i.e. the two systems would be perfectly synchronized at all times. However, in most practical situations, we have no control over the initial conditions and therefore synchronization may or may not take place in the master-slave system of Figure 4.

Let us now look at one of the numerous possible realizations of this synchronization scheme. As a system with chaotic behaviour, we take Chua's circuit (Figure 5)

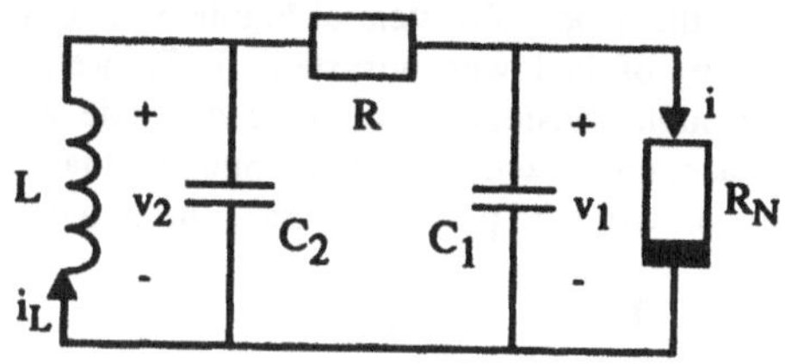

Figure 5 **Chua's circuit.**

described by the state equations

$$C_1 \frac{dv_1}{dt} = \frac{1}{R}(v_2 - v_1) - g(v_1) \tag{7}$$

$$C_2 \frac{dv_2}{dt} = -\frac{1}{R}(v_2 - v_1) + i_L \tag{8}$$

$$L \frac{di_L}{dt} = -v_2 \tag{9}$$

with the nonlinear resistor characteristic of Figure 6.

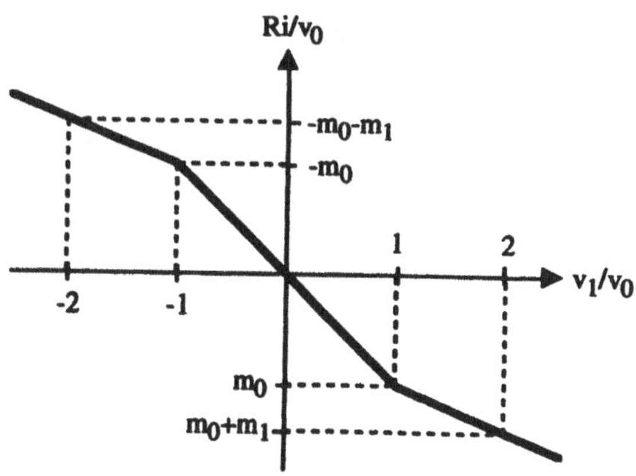

Figure 6 **Characteristic of the nonlinear resistor in Chua's circuit.**

We choose the parameters to be $R = 1730\ \Omega$, $L = 18$ mH, $C_1 = 10$ nF, $C_2 = 100$ nF, v_0=1V, $m_0 = -0.44$, $m_1 = -0.23$, $r = 150\ \text{k}\Omega$.

The roles of the state variables x and y in Figure 2 are played by $x = (v_2)$, and $y = (v_1, i_L)$. We can consider the synchronization of the two subsystems in Figure 3 separately. The lower subsystem, driven by the complete circuit through transmission of the signal $y_1 = v_1$ is shown in Figure 7.

The slave subcircuit in Figure 7 is linear and its elements have positive values. Thus, it is globally asymptotically stable, which implies that its state variables i_L and v_2 synchronize with the corresponding state variables of the master circuit, in the sense of the definition in section 6.2.1.

Let us now consider the upper subsystem of Figure 3. In our case, it would be driven by the state variable $x_1 = v_2$ of the lower subsystem. In Figure 8, we let the master circuit drive this subcircuit, through transmission of the same state variable. If in this set-up synchronization takes place, then, except for very singular situations, in the original set-up of Figure 4 synchronization will take place, too, since the two driving signals converge to each other.

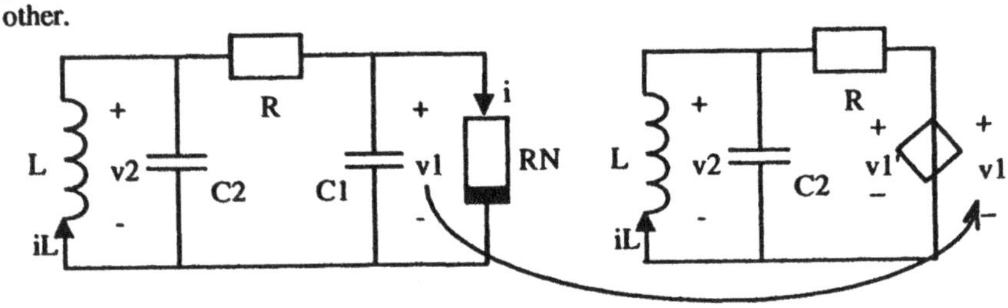

Figure 7 Subcircuit driven by the "master" circuit.

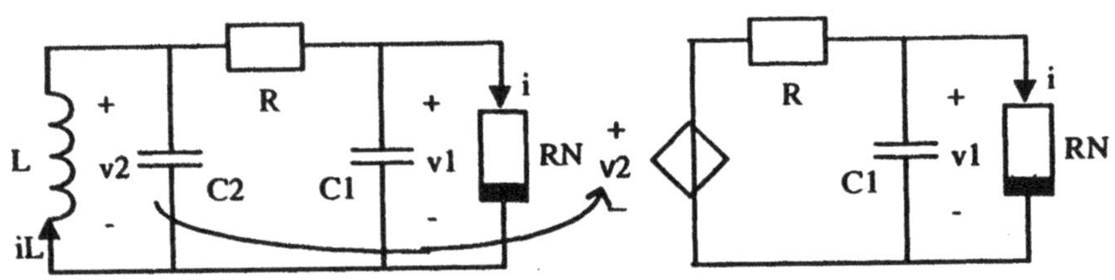

Figure 8 Second subcircuit driven by the master circuit.

In Figure 8, synchronization is a much more delicate issue than in Figure 7, and we cannot prove it from first principles. The reason is that the slave subcircuit is neither linear nor are its solutions globally asymptotically stable. More precisely, it is a piecewise linear circuit with three linear regions, and with our choice of parameters, the inner region is an unstable linear circuit, whereas the outer regions correspond to asymptotically stable linear circuits. It is thus intuitive that state variables v_1 of the master and the slave circuit drift apart whenever one of them is in the inner region and that they converge to each other whenever both are in the same outer region. It is not difficult to show that synchronization takes place if in the average over time, the state variable v_1 of the master circuit remains a sufficiently long proportion of the time in the outer regions. For a precise statement, we refer to [4]. There remains to show that almost all solutions of the master circuit satisfy this condition, which is highly nontrivial.

While we cannot show synchronization from first principles, we could observe from simulations that, with our choice of parameters, the typical solution of Chua's circuit remains indeed a sufficiently large fraction of the time in the outer linear regions. The expected synchronization takes place, which was confirmed by simulation and by laboratory experiment. For more details, we refer to [4].

6.2.3 Synchronization by linear feedback

The point of view of this method is typical for automatic control. We consider two identical systems as master and slave, compare their outputs and use the difference to control the slave system (Figure 9). This approach has been advocated in [5, 6] under the topic *control of chaos*.

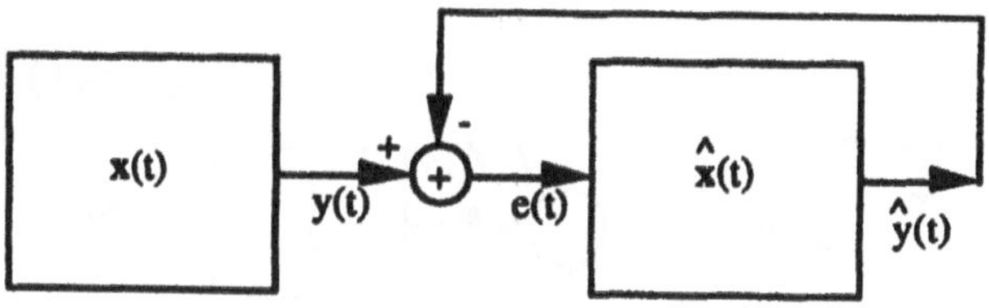

Figure 9 Master - slave set-up for synchronization by linear feedback.

If the error signal controls the state variables, the state equations of the whole system of Figure 9 become

$$\frac{dx}{dt} = f(x) \tag{10}$$

$$y(t) = c^T x(t) \tag{11}$$

$$\frac{d\hat{x}}{dt} = f(\hat{x}) + ke(t) \tag{12}$$

$$\hat{y}(t) = c^T \hat{x}(t) \tag{13}$$

$$e(t) = y(t) - \hat{y}(t) \tag{14}$$

Again, if both the master and the slave system started from exactly the same initial conditions then at all times, $x(t) = \hat{x}(t)$, $y(t) = \hat{y}(t)$, $e(t) \equiv 0$. With no constraint on the initial conditions, however, the slave system may or may not synchronize with the master system. In some instances, synchronization can be proved by Lyapunov functions [5, 7].

This method of synchronization can also be applied successfully to Chua's circuit. In Figure 10 Chua's circuit is represented as a block diagram. This corresponds to rewriting the state equations (7) - (9) in the form

$$\frac{dx(\tau)}{d\tau} = A(x_1)x(\tau) + u(x_1) \tag{15}$$

where

$$A(x_1) = \begin{pmatrix} -(1+m(x_1))\alpha & \alpha & 0 \\ 1 & -1 & 1 \\ 0 & -\beta & 0 \end{pmatrix}, \quad u(x_1) = \begin{pmatrix} \eta(x_1)(m_0 - m_1) \\ 0 \\ 0 \end{pmatrix} \tag{16}$$

and

$$m(x_1) = \begin{cases} m_0 & if\ |x_1| \le 1 \\ m_1 & if\ |x_1| > 1 \end{cases}, \quad \eta(x_1) = \begin{cases} 0 & if\ |x_1| \le 1 \\ 1 & if\ |x_1| > 1 \end{cases} \tag{17}$$

with the normalized variables $x_1 = v_1/v_0$, $x_2 = v_2/v_0$, $x_3 = R i_L/v_0$, $\tau = t/(RC_2)$, and the parameters $\alpha = C_2/C_1$, $\beta = R^2 C_2/L$.

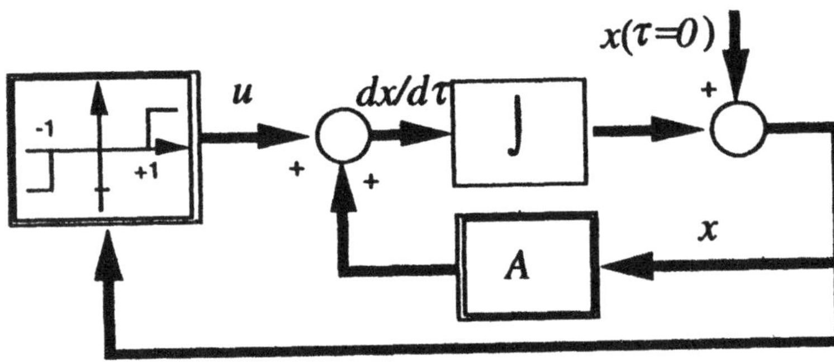

Figure 10 Chua's circuit as a block diagram.

The whole master - slave system is represented in Figure 11. The slave system is described by the state equations

$$\frac{d\hat{x}}{d\tau} = A(\hat{x}_1) + \hat{u}(x_1) + k\,c^T(x - \hat{x}) \tag{18}$$

Subtracting (8) from (5), we get for the synchronization error on the state variables $\Delta x\ (\tau) = x(\tau)$ - $\hat{x}(\tau)$, omitting the argument x_1

$$\frac{d\Delta x}{d\tau} = \left(\hat{A} - kc^T\right)\Delta x + u - \hat{u} \tag{19}$$

Note that the right hand side of (19) can be interpreted as a homogeneous part and an excitation part. The excitation is given by the difference between the two piecewise constant functions u and $\hat{u}$. Note that this term vanishes when the master and the slave system are in the same linear region.

In the case where the output signal is the first state variable, i.e. $y(t) = x_1(t)$, synchronization can be proved by using a Lyapunov function of the form

$$V(\Delta x) = \frac{\xi_1}{2}\Delta x_1^2 + \frac{\xi_2}{2}\Delta x_2^2 + \frac{\xi_3}{2}\Delta x_3^2 \tag{20}$$

With different output signals, we were also able to achieve synchronization by simulation and in laboratory mounted circuits. For more details, we refer to [7].

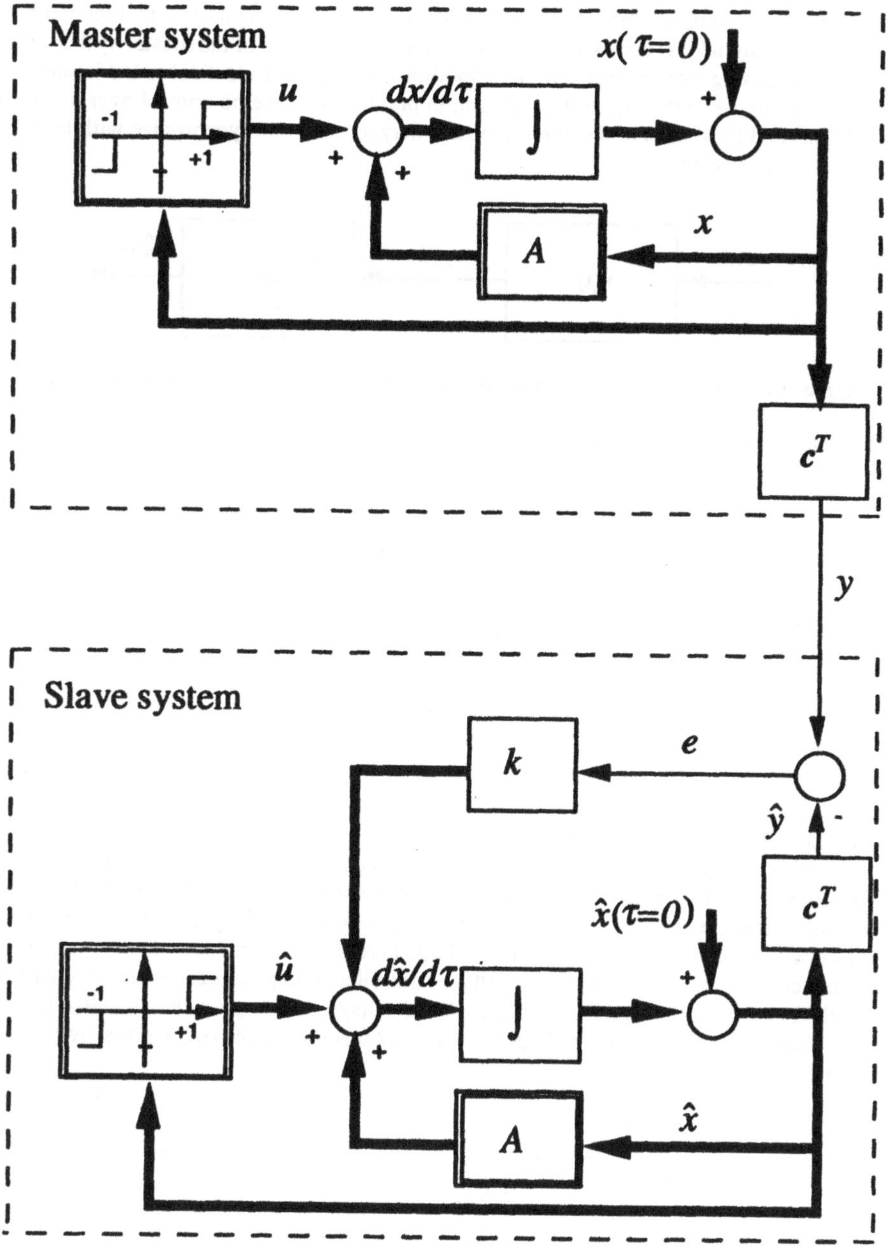

Figure 11 Synchronization of Chua's circuits by linear feedback.

6.2.4 Synchronization by the inverse system

For this method the set-up is somewhat different. Instead of using, as the basic building block, an autonomous system with chaotic behaviour, we start with a master system that is excited by an input signal $s(t)$ and produces an output signal $y(t)$. The signal $s(t)$ can be thought of as having a rather regular wave form, whereas the signal $y(t)$ is intended to be chaotic. The signal $y(t)$ drives a slave system that is the inverse of the master system in the sense that, starting from identical initial conditions, its output is identical to the input of the master system, i.e. $s(t)$ (Figure 12). Again, in practice we have no control over the initial conditions and the output of the inverse system may, or may not, synchronize with the input of the original system.

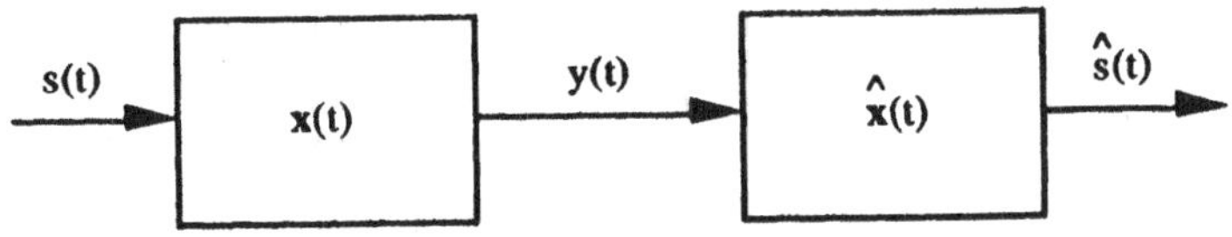

Figure 12 Master-slave set-up for synchronization by the inverse system.

The state equations for the system of Figure 12 have the form

$$\frac{dx}{dt} = f(x,s) \tag{21}$$

$$y = g(x,s) \tag{22}$$

$$\frac{d\hat{x}}{dt} = \hat{f}(\hat{x},y) \tag{23}$$

$$\hat{s} = h(\hat{x},y) \tag{24}$$

and the condition for the inverse system is that the solution of (21) - (24) with initial conditions $x(0)=\hat{x}(0)$ satisfies $\hat{s}(t)=s(t)$ for all $t \geq 0$. We expect in this case also $\hat{x}(t)=x(t)$ to hold for all $t \geq 0$.

At first sight, it might seem a formidable task to find an inverse for a nonlinear dynamical system. In the context of circuits, however, there are evident candidates for inverse systems. Consider a nonlinear dynamical 1-port excited by an independent current source with current $i(t)$. If we take the voltage across the current source and inject it into another copy of the same 1-port by a voltage controlled voltage source, we usually will find exactly the same current $i(t)$ flowing through the voltage source, provided the initial currents in the inductors and the initial voltages across the capacitors in the two 1-ports are the same (Figure 13).

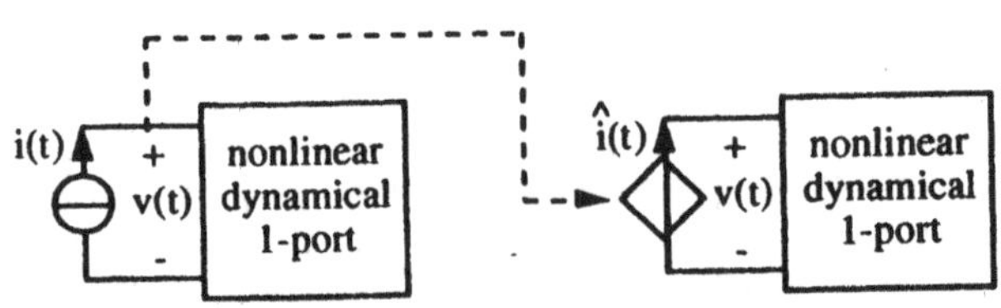

Figure 13 Realization of the inverse system by circuits.

This method of synchronization has been successfully applied to Chua's circuit and to Saito's circuit [8, 9]. In [8] it was possible to prove synchronization rigorously, because the nonlinear resistor and the linear part of the slave circuit have their voltage imposed by the controlled source (Figure 14). Therefore, $\hat{i}_1(t) = i_1(t)$ at all times and $\left|\hat{i}_2(t) - i_2(t)\right| \xrightarrow[t\to\infty]{} 0$ since i_2 is the port current of a linear passive 1-port. In other words, the inverse system is globally asymptotically stable. This implies that $\hat{i}(t)$ synchronizes with $i(t)$. In the set-up of [9], neither the voltage nor the current of the nonlinear element is imposed, and therefore the proof of synchronization is not that easy. Nevertheless, we have confirmed synchronization by computer simulation.

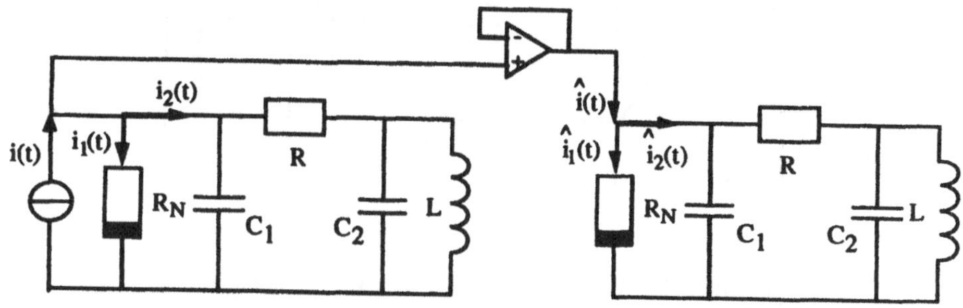

Figure 14 Synchronization of Chua's circuit by the inverse system.

6.2.5 Transmission with a chaotic carrier signal

The main application of synchronization of chaotic systems that has been proposed so far is to hide an information carrying signal in a chaotic carrier signal. This can be viewed as *analog cryptography*.

The master system is the transmitter and the slave system is the receiver. The cryptographic key is the parameter set of the chaotic system. In analog circuit realizations, of course, the parameters of the transmitter and the receiver cannot be completely identical. Therefore, the synchronization phenomenon must have a certain robustness against parameter variations. This question has not been addressed so far in this chapter, but the mere fact that laboratory experiments of synchronization were successful show that in the tested circuits synchronization had a minimum of robustness.

On the other hand, synchronization should not be too robust, because otherwise an intruder could synchronize onto the transmitted signal without precise knowledge of the parameter values. This shows that the design of analog cryptographic systems is delicate.

The information carrying signal may be analog or digital. In the sequel we will show methods for both cases. The advantages and disadvantages of this choice are the usual issues of speed and precision.

At this point of the development, of course, analog cryptography cannot yet compete with the highly developed methods of digital cryptography.

6.2.6 Chaotic masking

In this method [10, 11] an analog information carrying signal $s(t)$ is added to the output $y(t)$ of the chaotic system in the transmitter. The receiver tries to synchronize with the component $y(t)$ of the transmitted signal $s(t) + y(t)$. If synchronization takes place, the information signal can be retrieved by subtraction (Figure 15). As a first idea, one would expect that synchronization can only take place if the signal $s(t)$ is small with respect to $y(t)$.

Apparently, this is not confirmed by numerical experiments [12]. On the other hand, $s(t)$ should be small in order to remain inaccessible to an intruder without knowledge of the chaotic system parameters. Therefore, it can be expected that the method is sensitive to channel noise. Indeed, additive noise cannot be distinguished from $s(t)$ by the set-up of Figure 15. Therefore, it has to be eliminated at a later stage. This is a difficult if not impossible task if the amplitude of $s(t)$ is not large with respect to the noise level.

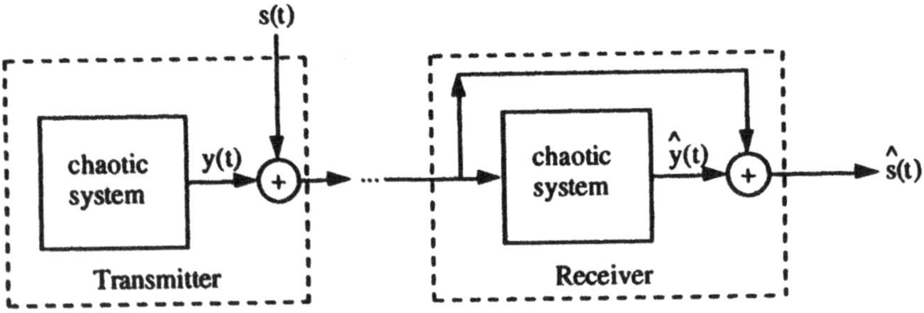

Figure 15 Transmission using chaotic masking.

6.2.7 Chaotic switching

In this method the information signal $s(t)$ is supposed to be binary. It controls a switch whose action changes the parameter values of the chaotic system. Thus, according to the value of $s(t)$ at any given instant t, the chaotic system has either the parameter vector $\boldsymbol{p}$ or the parameter vector $\boldsymbol{p'}$. The output $y(t)$ of the chaotic system is transmitted to two copies of the chaotic system, one with the parameter vector $\boldsymbol{p}$ and the other with the parameter vector $\boldsymbol{p'}$ (Figure 16).

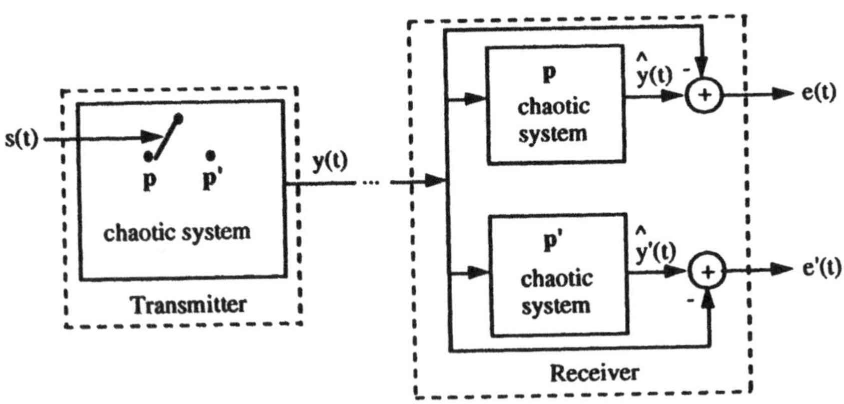

Figure 16 Transmission using chaotic switching.

If the momentary position of the switch in the transmitter is on position $\boldsymbol{p}$, then the system with parameter vector $\boldsymbol{p}$ in the receiver will synchronize, whereas the system with parameter vector $\boldsymbol{p'}$ will desynchronize. Thus the error signal $e(t)$ will converge to zero, whereas $e'(t)$ will have an irregular wave form with a distinctly non zero amplitude. If the switch in the transmitter is on position $\boldsymbol{p'}$, then we have the opposite situation, $e'(t)$ will converge to zero and $e(t)$ will be of non zero amplitude. Consequently, the signal $s(t)$ can be retrieved from the error signals $e(t)$ and $e'(t)$. Clearly, one has to leave the switch in the transmitter a certain time in the same position in order to be able to observe the convergence of the corresponding error signal to zero.

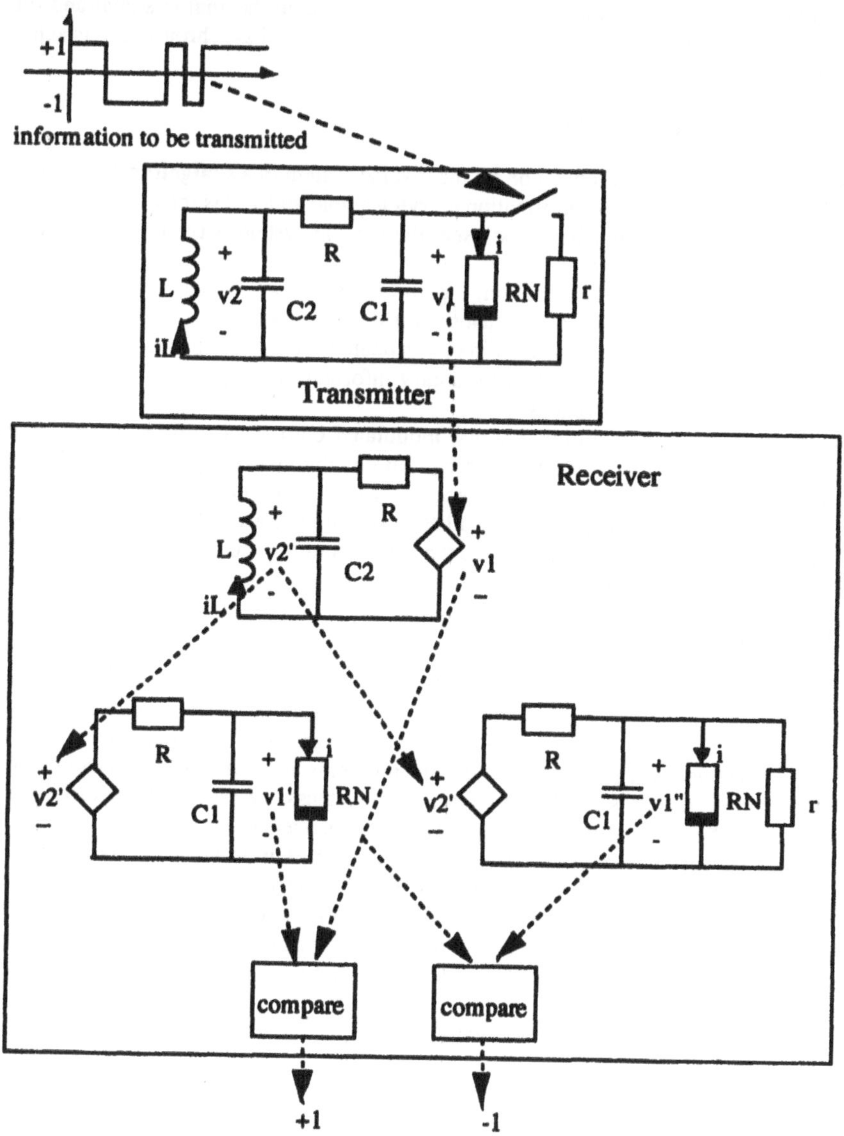

Figure 17

Chaotic switching has been realized with Chua's circuit, switching a linear resistor in parallel with the nonlinear resistor and performing the synchronization with decomposition into subcircuits [4]. The complete system is represented in Figure 17. In [7], the same parameter is switched, but synchronization is performed by linear error feedback. Finally, in [13], the capacitors and the inductor are switched between two values whereas synchronization is performed by decomposition into subcircuits.

While chaotic switching is expected to be more robust against noise than chaotic masking, its information transmission rate is lower, because on the one hand, the binary signal has a lower information content per unit of time than the analog signal and on the other hand, for each bit that is transmitted, one has to wait until synchronization is achieved in the receiver.

6.2.8 Direct chaotic modulation

This method uses the synchronization by the inverse system in a straightforward way. The signal $s(t)$ in Figure 12 is the information carrying signal and $y(t)$ is the transmitted signal. Thus, no additional circuitry has to be used, the chaotic system is the transmitter and the inverse system is the receiver.

If we look at a circuit realization, e.g. in Figure 14, we can see that $s(t)$ drives the chaotic circuit and thus modulates the chaotic signal in some way.

The information can be injected directly in analog form, as proposed in [8], or $s(t)$ can itself be an analog signal modulated by binary information, as proposed in [9], with the obvious advantages and drawbacks.

The transmission of a digital signal modulated onto $s(t)$ can be expected to reach higher bitrates than with chaotic switching. In chaotic switching, whenever the signal changes its value, one has to wait for synchronization since the initial conditions in the transmitter and the receiver subsystem that has to synchronize are different. In direct chaotic modulation, the receiver continuously tracks the transmitter and thus the states of the two chaotic systems are never very different.

6.2.9 Conclusion

We have presented various methods to synchronize chaotic systems and various methods to use this synchronization for transmitting information hidden in a chaotic signal. Thus, there are today several methods available for "analog cryptography". However, none of them is fully developed yet. We have mentioned some problems associated with this technique and others remain to be addressed. In particular, the robustness against parameter variations is a delicate issue, because on the one hand, robustness is needed because the sender and the receiver can never have perfectly matching parameters, but on the other hand, too much robustness would allow an unauthorized receiver to synchronize and extract the information without previous knowledge of the "encryption key", the parameter values. Other problems that should be studied systematically are the sensitivity against channel noise and channel characteristics. Independently of these questions, that are crucial for the applicability of analog cryptography, it is fascinating to see how the rich dynamics of chaotic systems may find their way into engineering applications.

References

[1] M. Ogorzalek, "Taming chaos - part I", *IEEE Trans. Circuits Syst.*, Vol.40, pp.693-699, 1993.

[2] L. M. Pecora, T. L. Carroll, "Synchronization in Chaotic Systems", *Phys. Rev. Letters*, Vol. 64, pp. 821-824, 1990.

[3] A. Tesi, A. De Angeli, R. Genesio "On the System Decomposition for Synchronizing Chaos", Technical Report, RT 12/93, Università di Firenze.

[4] H. Dedieu, M. P. Kennedy, M. Hasler, "Chaos shift keying: modulation and demodulation of a chaotic carrier using self-synchronizing Chua's circuits", *IEEE Trans. Circ. Syst., Part II*, pp. 634-642, Oct. 1993.

[5] G. Chen, X. Dong, "Controlling Chua's Circuit", *Journal of Circuits, Systems and Computers*, Vol. 3, pp. 139-149, 1993.

[6] G. Chen, X. Dong, "From chaos to order: perspectives and methodologies in controlling nonlinear dynamical systems", *Intern. Journal on Bifurc. and Chaos*, Vol. 3, pp. 1343-1389, 1993.

[7] M. Hasler, H. Dedieu, J. Schweizer, M. P. Kennedy, "Synchronization of Chaotic Signals", *Proc. Workshop on Nonlinear Dynamics of Electronics Systems*, Dresden, July 1993, to appear, World Scientific.

[8] K. Sean Halle, Chai Wah Wu, Makoto Itoh and L. O. Chua, "Spread Spectrum Communication Through Modulation of Chaos", *Int. Journal of Bifurcation and Chaos*, Vol. 3, pp. 469-477, 1993.

[9] M. Hasler, H. Dedieu, M. P. Kennedy, J. Schweizer, "Secure Communication via Chua's Circuit", *Proc. NOLTA'93 Workshops*, pp.87-92, Hawaii, USA, Dec. 1993.

[10] A. V. Oppenheim, G. W. Wornell, S. H. Isabelle, K. M. Cuomo, "Signal Processing in the Context of Chaotic Signals", *Proc. IEEE ICCASP'92*, pp.IV-117-IV-120.

[11] Lj. Kocarev, K. S. Halle, K. Eckert, L. O. Chua and U. Parlitz, "Experimental Demonstration of Secure Communications via Chaotic Synchronization", *International Journal of Bifurcation and Chaos*, Vol. 2, pp. 709-713, Sep. 1992.

[12] R. Lozi, private communication.

[13] U. Parlitz, L. O. Chua, Lj. Kocarev, K. S. Halle, A. Shang, "Transmission of digital signals by chaotic synchronization", *International Journal of Bifurcation and Chaos*, Vol. 2, pp. 973-977, 1993.

Chapter 6.3

Control Concepts and Applications

Maciej J. Ogorzalek

Abstract

We describe various control concepts developed for systems operating in a chaotic regime. The aim of these control schemes is to achieve a desired type of dynamic behaviour ("goal") such as a fixed point or a periodic orbit. The control concepts described here include: (i) system parameter variation; (ii) chaotic oscillation absorber; (iii) entrainment; (iv) linear feedback control; (v) time-delayed feedback; (vi) methods for stabilising unstable periodic orbits: occasional proportional feedback and sampled input. Advantages and drawbacks of these methods are described.

6.3.1 Introduction

As we have seen from the previous chapters chaotic oscillations, although commonly encountered in electronic circuits and systems, represent rather a peculiar type of behaviour commonly considered as a disastrous one in most applications and thus unwanted. It is clear that we do not want an amplifier, a phase-locked loop or a digital filter to produce chaotic outputs. Thus, a most common goal of control for chaotic electronic circuits is suppression of oscillations of this kind.

In contrast to the classical control problems where stabilisation has a clear meaning, stabilisation/suppression of chaos can be understood in many ways. Some authors define the control problem as stabilisation of the equilibrium points, others put as a goal elimination of multiple basins of attraction, and in some cases the goal can be in stabilising periodic orbits [32-34]. Considering the possibilities of influencing the dynamics of a chaotic circuit one can distinguish four basic approaches

- Variation of an existing accessible system parameter.
- Change in the system design - modification of its internal structure.
- Injection of an external signal(s).
- Introduction of a controller (classical PI, PID, linear or nonlinear, neural, stochastic etc.) - this means changing the overall structure without changing the internal design/dynamics of the system.

Depending on the application so-called "low energy" and "high energy" controls are considered, the first notion meaning that very small control signal/parameter changes are admitted. Thus, due to very rich dynamic phenomena encountered in typical chaotic systems, there exist a large variety of approaches to controlling such systems.

Below we present a selection of methods developed for controlling chaos in various aspects - starting from the most primitive concepts like parameter variation, through classical controller applications (open- and closed-loop control), to quite sophisticated ones like stabilisation of unstable periodic orbits embedded within the strange attractor.

6.3.2 Simple techniques for suppressing chaotic oscillations

6.3.2.1 Parameter variation technique

The simplest way of suppressing chaotic oscillations is to change the system parameters (redesign !!!) in such a way as to produce the desired kind of behaviour. It is quite interesting to analyse the influence of parameter variations on the asymptotic behaviour of the system - this can be done using a standard tool used in analysis of chaotic systems - the bifurcation diagrams.

As an example let us analyse one of the bifurcation diagrams obtained for the RC-ladder chaos generator [29, 30]. A variety of dynamic behaviours can be observed for appropriate choices of circuit parameters. The bifurcation diagram shown in Figure 1 has been constructed for fixed values of all circuit parameters except one of the capacitances.

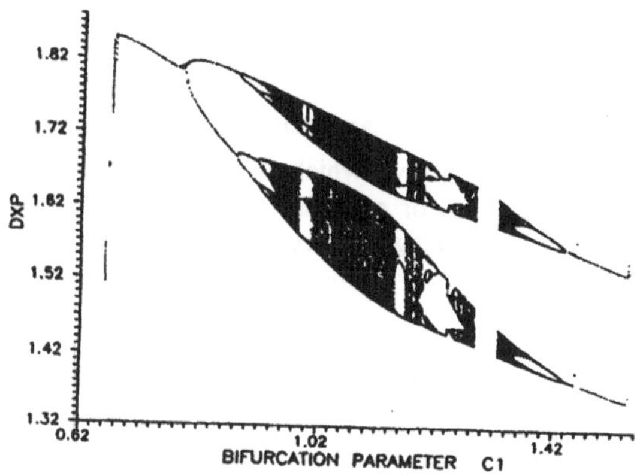

Figure 1 Typical one-parameter bifurcation diagram showing dependence of steady state observed in the system on the variations of one of the parameters. For each distinct value of the bifurcation (variable, controlled) parameter the number of corresponding points in the diagram determines the steady-state behaviour, e.g. one point - period-one orbit, five points - period-five orbit, etc.

Using a bifurcation diagram one can choose the appropriate value of the variable parameter for any prescribed type of behaviour: stable fixed points, period-1 orbits, period-2 or any higher-period orbits. This method, although intuitively simple, has some major drawbacks - the main one being the requirement of large parameter variations ("large energy control"). This requirement cannot be met in many physical systems where the construction parameters are either fixed or can be changed in very small ranges.

6.3.2.2 "Shock absorber" concept

This simple technique is being used in a variety of applications. For motivation let us consider the movement of a car wheel in motion. Due to uneven surface of the road and changes in the direction of the movement the wheels and the suspension are subject to unpredictable aperiodic vibrations. Such vibrations could be very dangerous.

A common method for suppressing such vibrations is via installing shock absorbers, i.e. adding supplementary dynamical systems whose role will be to quench the oscillations. Thus we modify the original chaotic system (add the absorber without major changes in the design or construction) in order to change its dynamics in such a way that a new stable orbit appears in a neighbourhood of the original attractor. In an electronic system the absorber can be as simple as an additional shunt capacitor or an *LC* tank circuit.

Kapitaniak, Kocarev and Chua [22] proposed such a "chaotic oscillation absorber" for Chua's circuit [5, 26] - it is a parallel *RLC* circuit coupled with original Chua's circuit via the resistor R_x. The equations describing dynamics of this augmented system can be given in a dimensionless form

$$\begin{aligned}
\dot{x} &= \alpha[y - x - g(x)] \\
\dot{y} &= x - y + z + \varepsilon(y^1 - y) \\
\dot{z} &= -\beta y \\
\dot{y}^1 &= \alpha^1\left[-\gamma^1 y^1 + z^1 + \varepsilon(y - y^1)\right] \\
\dot{z}^1 &= -\beta^1 y^1
\end{aligned} \tag{1}$$

In terms of circuit equations we have an additional set of equations for the "absorber" (y^1, z^1) and a small term $[\varepsilon(y^1 - y)]$ via which the original equations of Chua's circuit are modified. Depending on the choice of ε various stable asymptotic states can be obtained in the originally chaotic circuit. Looking at the equations alone one could identify a feedback action of the absorber subsystem. However this is not apparent from the physical implementation.

6.3.2.3 Properties of the simplest methods for suppression of chaotic oscillations

- Extremely simple principle of operation.
- Usually some kind of redesign of the system is required (either in terms of large changes of parameters or building additional subsystems). In many practical situations such changes in the design are not allowed.
- In many cases the goal behaviour has to be chosen by trial-and-error. The bifurcation behaviour in real physical systems is difficult to learn.
- Implementation of these methods is usually quite easy. Neither access to internal variables nor complicated controlling devices are needed.

6.3.3 External perturbation techniques

6.3.3.1 "Entrainment" - Open loop control

Let us briefly describe another control concept developed for systems with complicated dynamics and developed in a series of papers [17-20, 25].

Let us assume that we have an experimental system and its model

$$\frac{dx}{dt} = F(x) \tag{2}$$

where $x \in R^n$, $F(x)$ - differentiable, solutions exist $\forall t \geq 0$. The system possesses attracting limit sets for trajectories.

Control problem: Find suitable perturbation of the problem such that the solutions $x(t)$ will be entrained to arbitrarily chosen "goal dynamics" $g(t)$ i.e.

$$\lim_{t\to\infty} |x(t) - g(t)| = 0 \tag{3}$$

Solution: Modify the equations by adding the control term $E(g,\dot{g}) = \dot{g} - F(g)$:

$$\frac{dx}{dt} = F(x) + E(g,\dot{g})S(t) \tag{4}$$

where $S(t) = 0$ for $t<0$, $S(t)=1$ for $t \geq 0$. Control can be achieved for the initial conditions in some set $BE(g)$ called the basin of entrainment [$x_0 \in BE(g)$].

6.3.3.2 *Properties of the entrainment method*

- No feedback.
- Applicable for entrainment and also elimination of multiple basins of attraction (in systems with many attractors).
- A good model of the system dynamics is necessary.
- Generally there is no *a priori* knowledge of the basins of entrainment. ($BE(g_0)$ exists for all systems when the goal dynamics is a fixed point $g(t) = g_0$).
- Any particular solution of the autonomous system cannot be the goal of the control (for $E(g,\dot{g}) = 0$ - no control).
- In real applications, the implementation can be rather problematic.

6.3.3.3 *Weak periodic perturbation*

Interesting results have been reported by Breiman and Goldhirsch [1] who studied effects of adding a small periodic driving signal in a system behaving in a chaotic way. They discovered that external sinusoidal perturbation of small amplitude and appropriately chosen frequency can eliminate chaotic oscillations in a model of the dynamics of a Josephson junction and cause the system to operate in some stable periodic mode. Unfortunately there is little theory behind this approach and the goal behaviours can be learned by trial-and-error only. Some hope for further understanding and applications can be based on using theoretical results known from the theory of synchronisation [1, 32].

6.3.4 Control engineering approach

Several attempts have been made to use known methods belonging to a "control engineer's toolkit". For example the use of PI and PID controllers for chaotic circuits has been considered by Hartley and Mossayebi [15].

Applications of stochastic control techniques, Lyapunov-type methods, robust controllers and many other methodologies including intelligent control and neural controllers have been described in literature [4, 11, 14].

Below as an example, we describe application of a standard technique - linear feedback control - to chaos stabilisation.

6.3.4.1 Linear feedback control

Interesting applications of the linear feedback methodology to controlling chaotic dynamical systems have been developed by Pyragas [41] and Chen and Dong [2, 3]. This technique has been successfully applied to controlling chaos in discrete-time and continuous-time systems. Among the applications in electrical circuits let us mention Duffing's oscillator and Chua's circuit. The block diagram of the control scheme is shown in Figure 2.

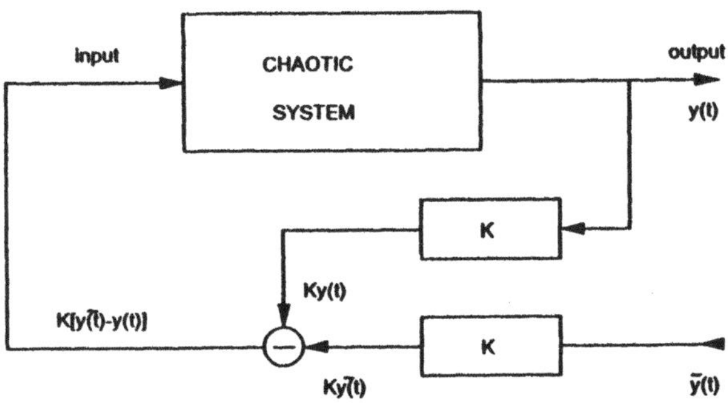

Figure 2 Block diagram of the linear feedback control considered by Pyragas and Chen and Dong.

Let $(\tilde{x}(t), \tilde{y}(t), \tilde{z}(t))$ be any solution of a chaotic circuit existing for a given parameter set (e.g. an unstable periodic orbit), the goal is to control the actual system trajectory such that for any given $\varepsilon > 0$, there exists $T_\varepsilon > t_0$ for which

$$|x(t) - \tilde{x}(t)| \le \varepsilon,\ |y(t) - \tilde{y}(t)| \le \varepsilon,\ |z(t) - \tilde{z}(t)| \le \varepsilon, \quad \text{for all } t \ge T_\varepsilon \tag{5}$$

The chaotic trajectory $(x(t), y(t), z(t))$ of the circuit can be driven to reach this desired solution by a linear feedback control of the form (analytical proof of validity of this result has been obtained eg. for the Duffing system and Chua's circuit).

$$\begin{bmatrix} u_1 \\ u_2 \\ u_3 \end{bmatrix} = -K \begin{bmatrix} x - \tilde{x} \\ y - \tilde{y} \\ z - \tilde{z} \end{bmatrix} \tag{6}$$

Using the described method Pyragas and Chen and Dong were able to stabilise the chaotic trajectories directing them towards chosen fixed points and periodic orbits.

6.3.4.2 Properties of the linear feedback method

- Any solution of the original system can be a goal of the control (fixed point, unstable periodic orbit etc.).
- In real applications lack of knowledge of these solutions might be a major difficulty - one has to generate the goal behaviour.

- Feedback control method. Very simple structure of controller.
- Access to system parameters not required. Immune to small parameter variations.
- Might be difficult to apply in real systems (interactions of many system variables needed).

6.3.4.3 Time - delay feedback control (Pyragas method)

Pyragas in his recent work [41] proposed an interesting feedback structure (Figure 3) employing a delayed copy of the output signal. He obtained very promising results in control of many different chaotic systems. Despite the lack of mathematical rigor this method is being successfully used in several applications.

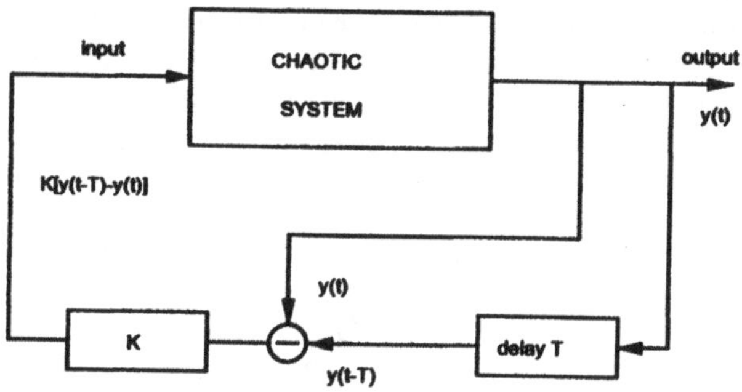

Figure 3 Block diagram of the Pyragas time-delay control scheme.

An interesting application of this technique is described by Mayer-Kress *et al.* [27]. Pyragas' control scheme has been used for tuning chaotic Chua's circuits to generate musical tones and signals.

6.3.4.4 Properties of the delay feedback control method

- Any solution of the original system can be a goal of the control (fixed point, unstable periodic orbit etc.). The knowledge of these solutions beforehand is not required.
- There is no *a priori* knowledge of the goal - this is chosen by trial-and-error.
- This is a self-control (no external signals injected) feedback method.
- Access to system parameters is not required. Immune to small parameter variations.
- Might be difficult to apply in real systems - variable delay element required.

6.3.5 Control in terms of stabilising unstable periodic orbits. Ott-Grebogi-Yorke approach (OGY)

This method of controlling chaos is based on the key property of chaotic attractors namely the existence of a dense set of unstable periodic orbits in the attractor [36, 37].

Let us assume that the dynamics of the system are described by a k-dimensional map

$$\xi_{n+1} = f(\xi_n, p) \tag{7}$$

This map in the case of continuous-time systems can be constructed eg. by introducing a transversal surface of section for system trajectories, p is some accessible system parameter which can be changed in some small neighbourhood of its nominal value p^*. For the value of p^* there is a periodic orbit within the attractor around which we would like to stabilise the system $-\xi_F$ is the chosen fixed point of the map f corresponding to this orbit. The position of this orbit changes smoothly with p parameter changes, and there are little changes in the local system behavior for small variations of p. In a close vicinity of this fixed point with good accuracy we can assume that the dynamics are linear and can be expressed approximately by

$$\xi_{n+1} - \xi_F = M(\xi_n - \xi_F) \tag{8}$$

The elements of the matrix M can be calculated using the measured chaotic time series and analysing its behaviour in the neighbourhood of the fixed point. Further the eigenvalues λ_s, λ_u and eigenvectors e_s, e_u of this matrix can be found. These eigenvectors determine the stable and unstable directions in the small neighbourhood of the fixed point. Denoting by f_s, f_u the contravariant eigenvectors ($f_s \cdot e_s = f_u \cdot e_u = 1, f_s \cdot e_u = f_u \cdot e_s = 0$) we can find the linear approximation valid for small $|p_n - p^*|$

$$\xi_{n+1} = p_n g + [\lambda_u e_u f_u + \lambda_s e_s f_s][\xi_n - p_n g] \tag{9}$$

where $$g = \left.\frac{\partial \xi_F(p)}{\partial p}\right|_{p=p^*}.$$

Because ξ_{n+1} should fall on the stable manifold of ξ_F so choose p_n such that $f_u \xi_{n+1} = 0$

$$p_n = \frac{\lambda_u \xi_u f_u}{(\lambda_u - 1) g f_u} \tag{10}$$

Figure 4 schematically explains the action of the OGY algorithm.

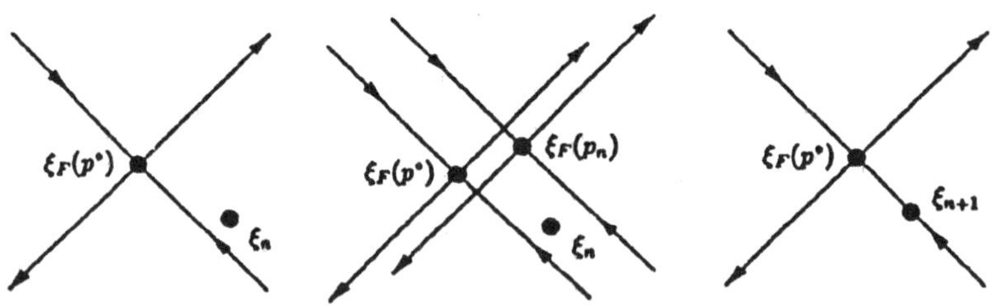

Figure 4 Schematic explanation of the OGY algorithm - a linear perturbation is applied in such a way that the successive iterate falls onto the stable manifold of the fixed point.

6.3.5.1 *Properties of the OGY technique:*

- This is a feedback control method.
- No model of dynamics is required. One can use either full information from the process or use delay co-ordinate embedding techniques using the single variable experimental time series (see Dressler and Nitsche [11]).

- Any accessible variable (controllable) system parameter can be used for applying perturbation (control parameter).
- In the absence of noise and error in characterisation of the system control is achieved with arbitrarilly small parameter perturbations. In practice noise and inevitable errors in signal processing may cause the control not to be so effective.
- Before settling into desired periodic mode the trajectory exhibits chaotic transients the length of which depends on the actual starting point (initial condition).

We have carried out an extensive study of application of the OGY technique to controlling chaos in Chua's circuit [5, 26]. Using an application-specific software package [7, 8] we were able to find some of the unstable periodic orbits embedded in the double scroll chaotic attractor and use them further as control goals.

Figure 5 shows the results of stabilisation of period-one and period-two unstable periodic orbits. Before the control is achieved the trajectories exhibit chaotic transients.

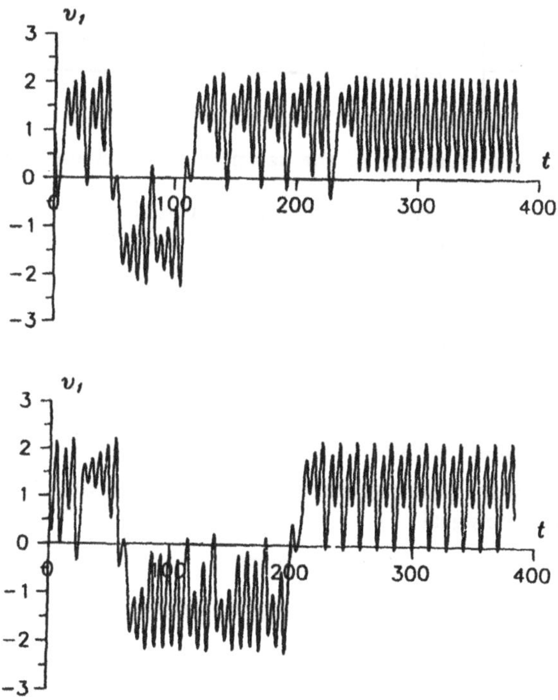

Figure 5 Two results of stabilisation of unstable periodic orbits via OGY method - period-one orbit and period-two orbit.

When applying the OGY method to control chaos in a real physical circuit the main problem encountered is the error introduced by inevitable noise of the circuit elements, A/D and D/A conversion of signals (quantification), rounding operations in the computer calculations etc.

The method was found to be very sensitive to the noise level - very small control signals sometimes are hidden within the noise and control is impossible.

6.3.5.2 Sampled input waveform method

A very simple, robust and effective method of chaos control in terms of stabilisation of an unstable periodic orbit has been proposed by Dedieu and Ogorzalek [9]. A sampled version of the output signal corresponding to a chosen unstable periodic trajectory, uncovered from a measured time series is applied in the chaotic system causing the system to follow this desired orbit.

The block diagram of this control scheme is shown in Figure 6.

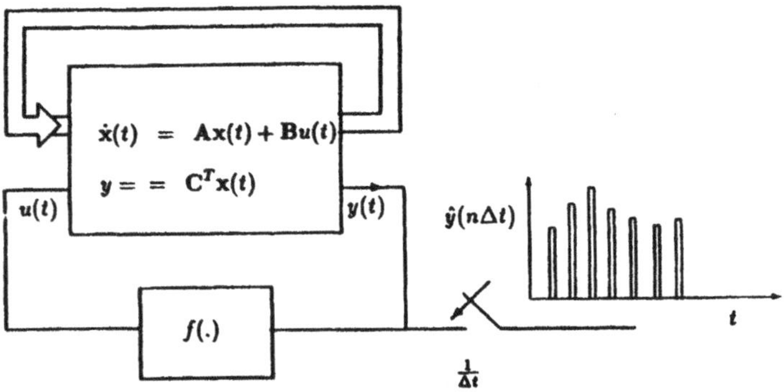

Figure 6 Block diagram of the sampled input chaos control system.

In real systems this sampled version of the unstable periodic orbit can be programmed into a programmable waveform generator and used as the forcing signal.

6.3.5.3 Properties of the sampled input control method

- Any solution of the original system can be a goal of the control (fixed point, unstable periodic orbit). The goal orbit can be pre-selected.
- Extremely simple structure.
- Immune to parameter variations, noise, scaling and quantisation - robust operation.
- Generator of the goal signal required. Reduced memory requirements due to sampling of the signal.
- Access to system parameters not required.

6.3.5.4 Electronic chaos controllers- OPF (occasional proportional feedback)

So far, the most promising control technique for real applications is a "one-dimensional" version of the OGY method, so-called Occasional Proportional Feedback. This method has been successfully implemented in a continuous-time analog electronic circuit and used in a variety of applications ranging from stabilisation of chaos in laboratory circuits [16, 21] to stabilisation of chaotic behaviour in lasers [6, 38, 42].

6.3.5.5 Properties of the OPF method

- Applicable in real chaotic systems (also higher-dimensional ones) - analog feedback controller.

- The accessible goal trajectories have to be determined by trial-and-error. Thus applicability of the control strategy limited to systems in which the goal is supression of chaos without more strict requirements.
- Access to single system parameter required.

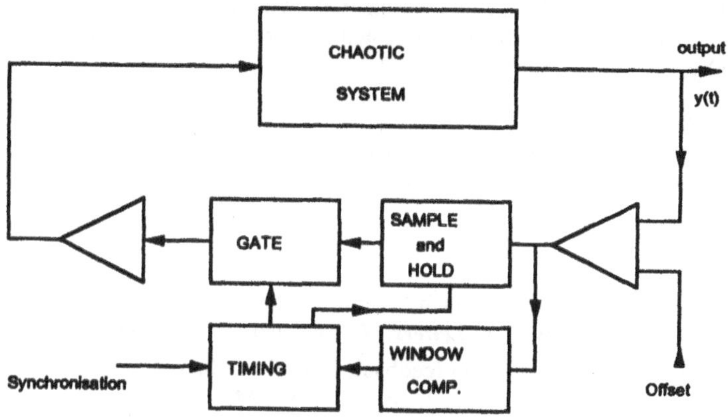

Figure 7 Block diagram of the electronic OPF chaos controller.

6.3.6 Controlling systems towards chaotic states- Synchronisation as a control problem

It should be pointed out that synchronisation and control problems of chaotic systems have common points. In particular, synchronisation can be considered as a particular type of control problem in which the goal of control is to track (follow) the desired (input) chaotic trajectory. It is only very recently that such a control problem has been recognised in control engineering.

The linear coupling technique and the linear feedback approach to controlling chaos can be applied for obtaining any chosen goal - no matter if it is chaotic, periodic or constant in time.

Using the approach described by Kocarev *et al.* [23] and Kocarev and Ogorzalek [24] we can even think of synchronising/controlling chaotic systems to chaotic trajectories being solutions of a qualitatively different chaotic system. We believe that this kind of chaotic synchronisation - control to a chaotic goal - could lead to new developments and possibly new applications.

6.3.7 Discussion

The control problems existing in the domain of chaotic systems are far from being fully identified, to say nothing about their solutions. Due to extreme richness of the phenomena one can treat every problem as a new challenge for scientists and engineers. Among many problems to be solved let us mention here the basic ones:

How can the methods already developed be used in real applications?

What are the limitations in terms of convergence, initial conditions etc. of these methods?

What are the limitations in terms of system complexity, possibilities of implementations?

There are certainly many others.

A host of possibly new application areas open up thanks to these new developments in various aspects of controlling chaos - these include neural signal processing, biology and medicine [13, 28] and many others. In this chapter we have presented only a very small portion of existing results believing that this will give the reader an idea of the subject and encourage further studies. For the interested reader we have included a comprehensive bibliography.

Acknowledgement

The author would like to acknowledge the support received from the University of Mining and Metallurgy (Grant 11.120.15).

References

[1] Y. Breiman and I. Goldhirsch, "Taming Chaotic Dynamics with Weak Periodic Perturbation", *Phys. Rev. Letters*, Vol. 66, No. 20, pp. 2545-2548, 1991.

[2] G. Chen and X. Dong, "On feedback control of chaotic dynamic systems", *Int. J. Bifurcation and Chaos*, Vol. 2, pp. 407-411, 1992.

[3] G. Chen and X. Dong, "Controlling Chua's circuit", *J. Circuits, Systems and Computers*, Vol.3, No. 1, pp. 139-148, 1993.

[4] G. Chen and X. Dong, "Control of chaos - A survey", *Proc. IEEE Conf. Decision and Control*, San Antonio, Texas, December 1993.

[5] L. O. Chua, "Global unfolding of Chua's circuit", *IEICE Trans. Fund. Elect. Comm. Comp. Sci.*, (Tokyo, Japan), Vol. E76-A, No. 5, pp. 704-734, May 1993.

[6] E. Corcoran, "Kicking Chaos out of Lasers", *Scientific American*, pp. 19 November 1992.

[7] A. Dabrowski, Z. Galias and M. J. Ogorzalek, "A Study of Identification and Control in a Real Implementation of Chua's Circuit", *System Structure and Control - 2nd IFAC Workshop*, pp. 278-281, Prague 1992.

[8] A. Dabrowski, Z. Galias and M. J. Ogorzalek, "On-line identification and control of chaos in a real Chua's circuit", To appear in *Kybernetika*, Czech Academy of Sciences, 1994.

[9] H. Dedieu and M. J. Ogorzalek, "Controlling chaos in Chua's circuit via sampled inputs", *Int. J. Bifurcation and Chaos*, Vol. 4, No. 2, 1994 (in press).

[10] W. L. Ditto, S. N. Rauseo and M. L. Spano, "Experimental Control of Chaos", *Phys. Rev. Letters*, Vol. 65, No. 26, pp. 3211-3214, 1990.

[11] U. Dressler and G. Nitsche, "Controlling Chaos Using Time Delay Coordinates", *Phys. Rev. Letters*, Vol. 68, No. 1, pp. 1-49, 1992.

[12] T. B. Fowler, "Application of Stochastic Control Techniques to Chaotic Nonlinear Systems", *IEEE Trans. Automatic Control*, Vol. AC-34, pp. 201-205, 1989.

[13] A. Garfinkel, M. L. Spano, W. L. Ditto and J. N. Weiss, "Controlling Cardiac Chaos", *Science*, Vol. 257, pp. 1230-1235, 1992.

[14] R. Genesio and A. Tesi, "Distortion Control of Chaotic Systems: The Chua's circuit", *J. Circuits Syst. Comput.*, Vol. 3., pp. 151-171, 1993.

[15] T. T. Hartley and F. Mossayebi, "Control of Chua's circuit", *J. Circuits Syst. Comput.*, Vol. 3., No. 2, pp. 173-194, 1993.

[16] E. R. Hunt, "Stabilizing High-Period Orbits in a Chaotic System: The Diode Resonator", Phys. Rev. Letters, Vol. 67, No. 15, pp. 1953-1955, 1991.

[17] A. Hübler, "Adaptive control of chaotic systems", *Helvetica Physica Acta*, Vol. 62, pp. 343-346, 1989.

[18] A. Hübler and E. Lüscher, "Resonant stimulation and control of nonlinear oscillators", *Naturwissenschaft*, Vol. 76, pp. 76, 1989.

[19] E. A. Jackson, "Control of dynamic flows with attractors", *Physical Review A*, Vol. 44, No. 8, pp.4839-4853, 1991.

[20] E. A. Jackson, "On the control of complex dynamic systems", *Physica D*, Vol. D 50, pp. 341-366, 1991.

[21] G. A. Johnson, T. E. Tigner and E. R. Hunt, "Controlling chaos in Chua's Circuit", *Circuits Syst. Comput.*, Vol. 3., No. 1, pp. 109-117, 1993.

[22] T. Kapitaniak, L. Kocarev and L. O. Chua, "Controlling chaos without feedback and control signals", *Int. J. Bifurcation Chaos*, Vol. 3, No. 2, pp. 459-468, 1993.

[23] L. Kocarev, A. Shang and L. O. Chua, "Transition in Dynamical Regime by Driving: A Method of Control and Sysnchronisation of Chaos", *Int. J. Bifurcation and Chaos*, Vol. 3, No. 2, pp. 479-483, 1993.

[24] L. Kocarev and M. J. Ogorzalek, "Mutual synchronisation between different chaotic systems", *Proc. 1993 International Symposium on Nonlinear Theory and Applications*, Vol. 3, pp. 835-840, 1993.

[25] E. Lüscher and A. Hübler, " Resonant stimulation of complex systems", *Helvetica Physica Acta*, Vol. 62, pp. 543, 1989.

[26] R. N. Madan, (ed.), *Chua's Circuit; A Paradigm for Chaos*. World Scientific, 1993.

[27] G. Mayer-Kress, I. Choi, N. Weber, R. Bargar and A. Hubler, "Musical signals from Chua's circuit", *IEEE Trans. Circ. Systems Part II*, Vol. 40, No. 10, pp. 688-695, 1993.

[28] J. S. Nicolis, "Chaotic dynamics in biological information processing: A heuristic outline", *Chaos in Biological Systems*, eds. H. Degn, A.V. Holden and L.F. Olsen, Plenum Press, pp. 221-232, New York 1987.

[29] M. J. Ogorzalek, "Remerging bifurcation sequences in an autonomous electronic circuit", *Proc. 1989 European Conference on Circuit Theory and Design*, IEE Publication 308, pp. 142- 146, Brighton 1989.

[30] M. J. Ogorzalek, "Order and chaos in RC- ladder network with nonlinear feedback", *IEEE Trans. Circuits and Systems,* Vol. CAS-36, No. 9, pp. 1221-1230.

[31] M. J. Ogorzalek, "Controlling chaos in Chua's circuit", *Proc. 1993 Int. Symp. Nonlinear Theory and Applications - Workshops*, Honolulu, pp. 47-52, 1993.

[32] M. J. Ogorzalek,"Taming Chaos: Part I - Synchronisation", *IEEE Trans. Circuits Syst.*, Vol. CAS-40, No. 10, pp. 693-699, 1993.

[33] M. J. Ogorzalek, "Taming Chaos: Part II - Control", *IEEE Trans. Circuits Syst.*, Vol. CAS-40, No. 10, pp. 700-706, 1993.

[34] M. J. Ogorzalek and Z. Galias, "Characterisation of chaos in Chua's oscillator in terms of unstable periodic orbits", *J. Circuits Syst. and Computers*, Vol. 3, No. 2, pp. 411-429, 1993 .

[35] M. J. Ogorzalek, Z. Galias and L. O. Chua, "Exploring chaos in Chua's circuit via unstable periodic orbits", *Proc. 1993 IEEE International Symposium on Circuit and Systems,* Chicago, Vol. 4, pp. 2608-2611, 1993.

[36] E. Ott, C. Grebogi and J. A. Yorke, "Controlling Chaos", *Phys. Rev. Letters*, Vol. 64, No. 11, pp. 1196-1199, 1990.

[37] E. Ott, C. Grebogi and J. A. Yorke, "Controlling Chaotic Dynamical Systems", in *Chaos: Soviet-American Perspectives on Nonlinear Science*, D.K.Campbell ed., American Institute of Physics, New York 1990.

[38] I. Peterson, "Ribbon of Chaos: Researchers develop a Lab Technique for snatching Order out of Chaos", *Science News*, Vol. 139, No. 4, pp. 60-61, 1991.

[39] L. M. Pecora and T. L. Carroll,"Driving systems with chaotic signals", *Physical Review A*, Vol. 44, No. 4, pp. 2374-2383, 1991.

[40] L. M. Pecora and T. L. Carroll, "Pseudoperiodic driving: Eliminating multiple domains of attraction using chaos", *Phys. Rev. Letters*, Vol. 67, No. 8, pp. 945-948, 1991.

[41] K. Pyragas, "Continuous control of chaos by self-controlling feedback", *Physics Letters A*, Vol. A170, pp. 421-428, 1992.

[42] R. Roy, T. W. Murphy Jr., T. D. Maier, Z. Gills and E. R. Hunt, "Dynamical Control of a Chaotic Laser: Experimental Stabilization of a Globally Coupled System", *Phys. Rev. Letters*, Vol. 68, No. 9, pp. 1259-1262, 1990.

[43] I. B. Schwartz and I. Triandaf, "Tracking unstable orbits in experiments", *Phys. Rev. A*, Vol. 46, No. 12, pp. 7439-7444, 1992.

[44] S. Sinha, "An Efficient Control Algorithm for Nonlinear Systems", *Physics Letters A*, Vol. 156, No. 9, pp. 475-478, 1991.

[45] H. Wang and E. H. Abed, "Bifurcation Control of Chaotic Dynamical Systems", *Proc. 2nd IFAC Symposium NOLCOS*, pp. 57-62, Bordeaux-France, 1992.

Chapter 6.4

Fractals in Signal Processing

Arnaud Jacquin

6.4.1 Introduction

This tutorial aims at presenting a very brief history of *fractals*, describing their characteristics and their relation with *chaos*, as well as pointing to where and why fractals and chaotic systems are commonly found in nature, implying that they should be good candidates for modelling different types of real world signals. It then describes some specific applications of fractals in *signal processing*, more specifically in the areas of *image modelling*, *synthesis*, and *compression*[1].

These application areas are losing their boundaries more and more, as one witnesses the merging of image synthesis and computer animation - which traditionally derive from *computer graphics*, and digital image and video compression - which derive from *digital signal processing*, specifically *information theory*. To a large extent, this merging is due to the need to find ever more powerful models for real world digital signals in order to transmit these signals efficiently. We will show examples involving differents types of signals, and correspondingly, different types of fractal models, and will try to describe the advantages and shortcomings of each of these models. The reader should be warned that our choice was to present several frameworks and examples which prevented mathematical rigor. This tutorial is mostly aimed at engineers who are interested in an overview of fractals and chaos, rather than in mathematical detail. However, references to articles and books which give the details of the various frameworks are indicated in each section.

The general problem statement which underlies the use of *fractals* for signal modelling and compression - the main purpose of the last and largest section of this paper - is the following. Given any original *object* or *signal*, for example a discrete monochrome image specified by an array of 8-bit pixel values, how can a computer construct a fractal object - the encoded object - which is both visually close to the original one, and has a digital representation which requires fewer bits than the original. Note that this latter requirement is not part of the notion of modelling *per se* but is essential to the efficiency of the modelling procedure for *compression* or *coding*. For each type of object/signal to model, this problem will be presented in the following format. Firstly, we describe the objects under study, the class of *iterated systems* which can generate them, and the generation process. Secondly, we describe several approaches to solving what is commonly referred to as the *inverse problem*, which consists of devising a procedure for *controlling* the generation process in such a way that it produces fractal models of original real world objects. While

---

[1] A recent issue of the Proceedings of the IEEE [39] includes a special section on fractals in electrical engineering. Fractal techniques are also being used in other fields of image processing, such as *image segmentation* [69], *scene analysis* [70, 76] and *texture coding* [50, 73]. These applications are not covered here.

the first part is mathematically simple and merely represents several instances of a very general theory of iterated contractive transformations in a metric space, the inverse problem can only be dealt with on a per-framework basis and does require creativity on the part of the designer of a fractal modelling system.

6.4.2 Fractals and chaos

6.4.2.1 What is a fractal object ?

Texts on *fractal geometry* abound. We refer the reader with a strong mathematical interest in the topic to [6, 17, 21, 22, 53, 68, 75]. Although it is difficult to give an all-encompassing definition since fractals can be objects of different types, it is usually agreed that (deterministic) fractals arise from the **repeated iteration** of a transformation[2]. In other words, fractals are mathematical objects which can in general be written as

$$A = \lim_{n \to 0} \tau^n (A_0) \tag{1}$$

where A_0 denotes an *initial object*, and

$$\tau^n = \underbrace{\tau \circ \tau \circ \ldots \tau}_{n \text{ times}} \tag{2}$$

denotes the iteration of τ n times. They are typically generated by computing and displaying a sequence of iterates

$$A_0, A_1, A_2, \ldots, \text{where} \quad A_n = \tau(A_{n-1}) \tag{3}$$

From equation (1), it is clear that fractals satisfy the *invariance equation*

$$A = \tau(A) \tag{4}$$

which confers to them a property which we generically refer to as *self-transformability*, and which leads to the well-known properties of fractals to be rugged objects with an infinite amount of detail; objects which can be found again and again in magnified pieces of themselves, however small.

This very simple formulation can lead to objects which have "pathological"[3] mathematical properties, as well as tremendous visual complexity. It is quite sufficient to generate the "mathematical monsters" such as the Cantor dust, Sierpinski triangle, von Koch curve and snowflake, space-filling Peano curve, Menger sponge, etc. [53, 68], which kept a number of mathematicians at the beginning of the twentieth century puzzled for a number of years, but which are by now considered to be rather tame examples of fractal objects and are all well-understood. Their status of "monsters" caused them to fall into near-oblivion until they were "rediscovered" by the mathematician Benoit Mandelbrot in the 1960's, which caused a shift of attitude towards them.

---

[2] Note that where matemathicians see "iteration", engineers might be more likely to see a "feedback process". The idea is the same.

[3] Such as nowhere differentiability for some continuous fractal curves and surfaces, or infinite length or area. "Pathological" is meant here as opposed to the regularity of the curves and surfaces of Cartesian geometry and differential calculus.

One can assume that this shift of attitude was due to the advent of the computer and display devices which made these objects *visual*, and often strikingly so, as opposed to being characterized only by their construction and mathematical properties, or rather lack thereof. This was also due to the realization, initially by Mandelbrot but soon followed by others, that their complexity was *typical* of objects found in nature (cf. section 6.4.2.2) as opposed to being *aberrant*. An illustration of this idea can be found in a paper by Mandelbrot entitled "How long is the coast of Britain ?..." [51]. The argument is the following. If one set out to measure the coastline of a geographic region such as the west coast of Great Britain with yardsticks of ever decreasing size, the measurements would show that the length associated with each yardstick keeps increasing as the size of the yardstick decreases, eventually reaching infinity[4] .The impracticality of describing most natural objects in terms of straight lines - the basis of differential calculus and engineering - and the hint at the possibility to find accurate fractal models for them both contribute to the fascination exercised by fractals.

(i) Two simple examples of self-similar fractals: the von Koch curve and snowflake

The fractal curve known as the *von Koch curve* can be generated in the following way (see Figure 1)

1. Take as initial object a line segment of unit length S_0= [0,1].
2. To produce the first iterate S_1 , distort the segment by introducing a "bump" in the middle of S_0 in the form of two sides of an equilateral triangle with side of length one third of the length of S_0.
3. Repeat this introduction of triangular bumps to each line segment until convergence.

One can see that this construction can be equivalently formulated in the framework of equation (1), with a transformation τ consisting of the union of four similarities which map S_0 to S_1.The construction of the snowflake is very similar and is left as a simple exercise to the reader. The iterative construction of both objects is illustrated in Figure 1.

6.4.2.2 Fractals and chaos in the real world

(i) Fractal objects in nature

Many objects found in nature such as: i) weeds, ferns, vegetables, and trees, ii) clouds and mist, iii) mountain ranges, coastlines, terrain, rocks, aggregates, and ice crystals, iv) galaxies, etc. exhibit typical fractal characteristics such as self-transformability, which manifests itself as "the object looks the same at many different scales", and which can be seen for example in the photographs of Figure 2. This observation gives weight to the assumption that the "geometry of nature" is a *fractal* geometry which arises from iterative

---

[4] A series of such measurements was studied by L. F. Richardson who showed that the approximate length $L(\eta)$ of the west coast of Great Britain, as measured with a yardstick of size η, satisfies an empirical law of the form $L(\eta) \propto \eta^{-\alpha}$ [72]. In [51, 53] Mandelbrot interprets the quantity $D = 1 + \alpha$ as the *fractal dimension* of a curve. According to the measurements, the west coast of Great Britain has a fractal dimension approximately equal to 1.25. It is easy to see that a curve with a fractal dimension strictly greater than one has infinite length. This feature is characteristic of fractal curves.

processes; a notion which was popularized by Mandelbrot [53]. Yet it would be naive to believe that the iterative processes at work in nature are as simple as equation (1). Rather, it is reasonable to assume that they are influenced by external forces and perturbations. The growth of two trees from two identical seeds, one in a protected (ideal) environment, the other in an environment plagued with acid rain, poor soil, aridity, strong winds, etc. illustrates this concept. Rather than trying to find perfect deterministic fractals in nature (the naive view) one should realize that real world objects are more likely to be the result of an iterated process of the type

$$A_n = \tau(A_{n-1}) + \varepsilon_n \tag{5}$$

where ε_n denotes a random perturbation[5].

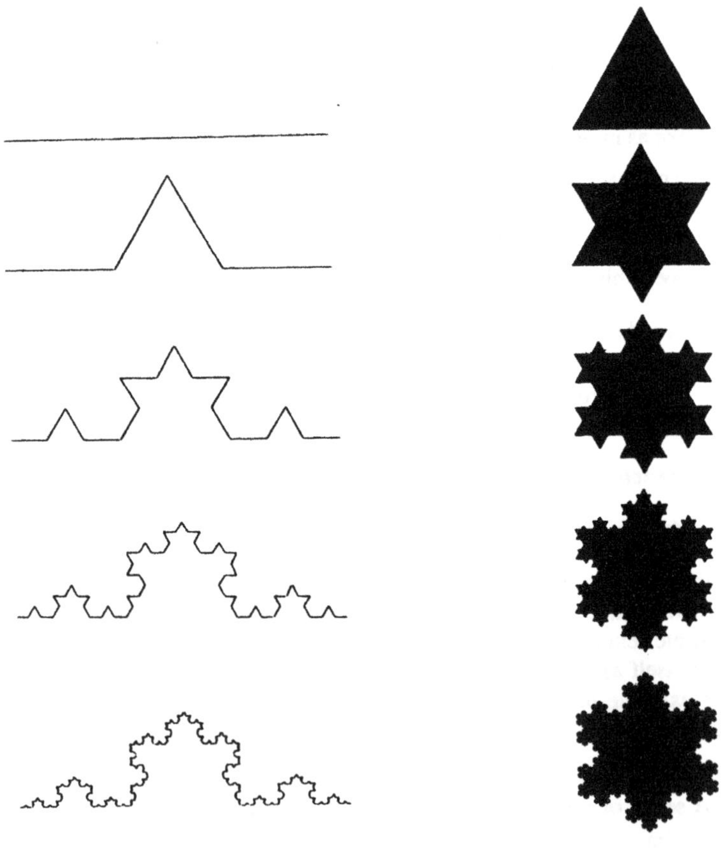

Figure 1 Construction of the von Koch curve (left) and snowflake (right).

[5] This type of process can lead to what Mandelbrot refers to as *statistical self-similarity* [52, 53].

Figure 2 Fractals in nature (a) "Georgia sky" (b) "Brazilian landscape".
© A. Jacquin

(ii) Chaotic dynamical systems

Fractals are closely related to chaotic systems. In this section, we introduce the mathematical notion of *chaos* by describing and analyzing a simple chaotic dynamical system, and briefly describe the relationship between fractals and chaos. For an introduction to chaos we refer the reader to [28] and to the texts [6, 13, 30, 68, 77] for a comprehensive treatment of chaotic dynamical systems and fractals.

A *dynamical system* consists of a transformation or *map* s defined from a metric space of points into itself - we consider an example where the space is the unit interval $I = [0,1]$ endowed with the Euclidean metric. The *orbit* of a point $x_0 \in I$ is the sequence of iterates

$$x_0, x_1, x_2, \ldots \text{ where } \quad x_n = s(x_{n-1}) \tag{6}$$

The orbit of x_0 is said to be *periodic* if there exists an integer P such that[6]

$$\forall n \in N, \qquad x_{n+P} = x_n \tag{7}$$

and it is then written $\overline{x_0 x_1 \ldots x_{P-1}}$.

We consider the *dyadic* map s defined by:

$$s(x) = \begin{cases} 2x & \text{if } 0 \le x < \frac{1}{2} \\ 2x - 1 & \text{if } \frac{1}{2} \le x \le 1 \end{cases} \tag{8}$$

sometimes referred to as the "stretch-cut-and-paste" transformation because of the analogy with an idealized process of kneading dough[7]. The analysis of the dynamical system $\{I, s\}$ can be more easily carried-out when representing the real numbers in I by their *binary expansions*, whereby a point x_0 in I is written

$$0 \cdot a_1 a_2 a_3 \ldots, \quad \text{with } a_k \in \{0,1\} \quad \text{for all } k \tag{9}$$

to denote that

$$x_0 = a_1 \cdot 2^{-1} + a_2 \cdot 2^{-2} + \ldots \tag{10}$$

One can easily see that the transformation s is equivalent to the *left-shift operator* S which acts on the binary expansions of points in I in the following way

$$S(0 \cdot a_1 a_2 a_3 \ldots) = 0 \cdot a_2 a_3 a_4 \ldots \tag{11}$$

This dynamical system is said to be chaotic because it can be shown[8] that it exhibits the three characteristics, which we intuitively describe below.

---

[6] The smallest non-zero integer which satisfies equation (7) is called the *period* of the orbit.

[7] A piece of dough is modeled by the line segment [0,1]. The transformation s can be seen as "stretching" the dough to twice its initial length, then "cutting" it into two equal-length pieces, and finally "pasting" the pieces together. The "kneading" comes from the repetition of the above sequence of steps.

[8] The proofs of these properties can be found in [19, 68]. They make extensive use of the following lemma. Let x and y be two real numbers in $I = [0,1]$, whose binary expansions agree up to the kth position, i.e. $x = 0 \cdot a_1 \cdot a_2 \ldots$, and $y = 0 \cdot a_1 \cdot a_2 \ldots a_k \cdot b_{k+1} \cdot b_{k+2} \ldots$, then $|x-y| \le 2^{-k}$.

Sensitivity to initial conditions

Let x_0 and y_0 be two "arbitrarily close" real numbers in I. There exists a strictly positive threshold T (consider for example $T = 1/2$) and a (large) number of iterations N such that x_N and y_N differ by the threshold T. This property is also known as the "butterfly effect" because of the chaotic weather models developed by Lorenz [46, 47]. This is the property which explains the impossibility to successfully perform long-term predictions with chaotic dynamical systems.

Mixing

The points in an arbitrarily small subinterval of [0,1] of initial values eventually "spread" over the whole interval [0,1].

Existence of dense periodic points:

There exists a set of points with periodic orbits for S which is *dense* in [0,1].

Dynamical systems theory is closely related to fractal geometry. One can show that fractals - attractors of Iterated Function Systems (cf. section 6.4.3.1) in particular - have a naturally associated dynamical system which is chaotic. Fractals are *attractors* of dynamical systems; the place where chaotic dynamics occur. For details on the relation between chaos and fractals, we refer the reader to [6, 13, 30, 68, 77].

(iii) Chaos in the real world

The list of the areas of science and engineering where chaotic dynamics are observable is long: Astronomy, Biology (eg. models of population growth [54]), Chemistry (eg. diffusion-limited aggregation), Climatology [46, 47], Economics (eg. stock market prices [53]), Electrical Engineering [39], Fluid Dynamics (eg. turbulent flows), Geology, Medicine, Physics, Physiology and Psychiatry, Seismology (eg. earthquake prediction). What is often of most interest in each of these areas is the extremely difficult task of trying to accurately model the underlying systems which give rise to the chaotic behavior observed - attempts which have been met with varying degrees of success so far.

6.4.3 Modelling real world objects with fractals

The artificial, "machine-world" quality of virtually all computer-generated images and animation before 1980 illustrates the impact that fractal modelling tools have had in computer graphics in the past fifteen years. Because of their inherent ability to look like typical objects found in nature, fractal models have been extensively used for image synthesis to model and render objects such as plants, trees, terrain, clouds, landscapes, seascapes, etc. [8, 25, 31, 32, 53, 66-68, 71]. Some of the natural scenes generated from fractal models such as the "fractal forgeries" of Voss [67, 68], Musgrave *et al.* [68], Peitgen *et al.* [67, 68], have a strikingly natural quality.

However, the examples of synthetic images given above remain limited by the fact that the fractal techniques used only allow a limited range of objects to be modelled and "forged". Besides, they do so by generating fractal scenes which, although they do look like they could be real natural scenes, are not fractal versions of an underlying **original**, and are therefore not candidates for *image compression*. In the following sections, we will show how other types of fractal forgeries obtained either as attractors of Iterated Function Systems or of Recurrent IFSs, or, in the framework of section 6.4.3.3, as fixed points of contractive image transformations defined piecewise, are explicitely modelled from original real-world signals.

Why is this interesting to engineers? The modelling of signals such as images is interesting if it can lead to very high compression. With the current state-of-the-art digital communication and coding technology, the transmission of simple "head-and-shoulder" scenes for video teleconferencing/telephony remains a challenge at bit rates under 64 kbps[9]. It seems that classical image compression techniques will not be sufficient, and different types of model-based approaches are being considered [1, 24, 42, 59].

6.4.3.1 Iterated function systems theory

(i) Fractals as attractors of iterated function systems

An Iterated Function System in R^P is entirely specified by a set of k affine transformations $w_1,\ldots,w_k$ in a metric space (R^P, d). A transformation w, which operates on (compact) sets of points in R^P, is defined by

$$\text{For every } B \subset R^P,\ w(B) = \bigcup_{1 \le i \le k} w_i(B) \tag{12}$$

When the k affine transformations w_i are *contractive*, i.e.

$$\forall (x,y) \in R^P,\quad d(w_i(x), w_i(y)) < s_i \cdot d(x,y) \tag{13}$$

for some real number s_i smaller than one[10], it can be shown that w is also a contractive transformation in the metric space of compact sets in (R^P, d) endowed with the Hausdorff metric [6] associated with d. The *contraction mapping principle* [20] due to Banach, ensures that w has a unique *fixed point* A - a set of points in (R^P, d), which is called the *attractor* of the iterated function system. The set A satisfies the invariance equation

$$A = w(A) \tag{14}$$

which means that A is exactly made of the union of k affinely transformed copies of itself. It is said to be a *globally self-affine fractal*[11]. The famous 2D Black Spleenwort fern of Barnsley [4, 6] is an example of such an object; it is the self-affine attractor of a set of four two-dimensional affine transformations. It is easy to show that any iterated sequence of the type

$$A_0, w(A_0), w^2(A_0), \ldots \tag{15}$$

---

[9] Even though many approaches have been proposed and many systems have been developed, the image quality of these systems is well below that of analog television signals.

[10] The smallest s_i for which equation (13) holds is called the *contractivity* factor of the transformation w_i.

[11] If the transformations w_i are contractive **similarities** - i.e. each w_i contracts the p dimensions of R^P with the same factor, A is said to be a (globally) *self*-similar fractal.

converges to the attractor A regardless of the choice of the initial set A_0. This property provides a simple procedure for the *generation* or *synthesis* of the attractor.

The above results are due to the work of Hutchinson in the context of self-similar fractals [33]. They were later extended by Barnsley [3, 6] to encompass the self-affine case. Barnsley further extended these results by introducing the notion of IFS *with condensation set*[12], whereby the IFS is of the form

$$C, w_1, \ldots, w_k \tag{16}$$

where C is the condensation set associated to the constant set transformation $w_0(B) = C$, and where w is now defined by

$$w(B) = \bigcup_{0 \le i \le k} w_i(B) \tag{17}$$

(ii) The inverse problem of modelling an object as the attractor of an IFS

Let us now suppose that we start with an original object A_{orig} which seems to be approximately globally self-affine, and therefore a good candidate for an attempt at modelling it as the attractor of an IFS. For the sake of illustration, let this original object be a digitized binary image of a real Black Spleenwort fern for example. The *inverse problem* consists in trying to "recover" a set of affine transformations under which the original object is approximately invariant. This can be expressed as finding a set of k (resp. $k+1$) transformations $\{w_i\}_{1 \le i \le k}$ (resp. $\{w_i\}_{0 \le i \le k}$) which satisfy

$$\bigcup_{1 \le i \le k} w_i(A_{orig}) \quad \text{(resp. } \bigcup_{0 \le i \le k} w_i(A_{orig})\text{)} \quad \text{is close to } A_{orig} \tag{18}$$

The union of the transformed copies in equation (18) is called a collage of A_{orig} [4, 6]. The goal of the modelling is to find the "best" possible - in the sense of equation (18) - collage of A_{orig} with as few affine transformations as possible, in the shortest amount of time. Note that although the notion of "closeness" should be in the sense of the Hausdorff metric, it was interpreted in the sense of "visual closeness" as defined by a human operator in the examples of natural scenes described at the end of this section.

The *collage theorem* of Barnsley *et al.* [4, 6] states that the fixed point A of the transformation

$$w = \bigcup_{1 \le i \le k} w_i \quad \text{(resp. } \bigcup_{0 \le i \le k} w_i\text{)} \tag{19}$$

an exactly globally self-affine fractal which in practice can be synthesized by iterating w a large number of times on any initial object A_0 is also close to A_{orig}, i.e.

$$w^n(A_o) \approx A_{orig}\,, \quad \text{when } n \text{ is large} \tag{20}$$

Therefore, the information contained in the parameters of the transformation w is sufficient to enable anyone with a knowledge if the iterative reconstruction procedure to synthesize an

---

[12] Fractal plants and trees, for example, are objects which arise as attractors of such systems.

object which is close to the original. The set of transformations $\{w_i\}_{1 \le i \le k}$ can thus be considered as a *fractal code* - lossy in general - for A_{orig}.

The result of an important *stability theorem*, which is referred to in [6] as *continuous dependence of attractors on transformation parameters* can be informally stated as follows. Small errors in the parameters of w - such as the ones that occur by quantizing the parameters of the individual w_i's for transmission on a digital channel - will induce equally small changes in the attractor. This property is a consequence of the contractivity of the transformation w. It ensures that two different decoders of the attractor of w, which could for example consist of the same reconstruction algorithm implemented on computers with different floating point precision, will produce two visually indistinguishable fractals.

(iii) Synthetic natural scenes

The images in Figure 3 referred to as "Monterey Coast", "Black Forest", "Sunflower Field", and "Andes Girl", [6, 7, 8], were all synthesized with this type of fractal modelling technique. More specifically, they were obtained from the following three-step procedure

1. Real world color images were segmented by a human operator who would strive to identify various self-affine objects.
2. Each object was modelled as the attractor of an IFS, using an interactive software package[13].
3. The final images were synthesized by successively reconstructing, rendering, and overlaying the pieces in a specific order - from background to foreground - using another software package described in [71].

The fact that this procedure, when viewed as an image coding and decoding scheme[14], can be seen as "compressing" images with ratios of 1000:1 and higher provoked great, and partially unfulfilled, excitement in the signal processing community. The major hurdles with the approach described above are clear

i) natural scenes are in general not globally self-affine (the way a fern is), and scene segmentation - a difficult problem by itself - is one essential feature,

ii) the scheme fundamentally involves a human operator whose function cannot easily be automated.

(iv) Piece-wise self-transformability, recurrent IFS

Foreseeing that the above procedure could only be automated with great difficulty, if it could be automated at all, lead the author to try and bypass the segmentation step by capturing the notion of *piecewise self-transformability* in the transformations themselves, albeit by considering slightly more general transformations. Instead of segmenting *a priori* an object A_{orig} into several pieces and modelling each of the pieces separately, one could look for a set of contractive transformations which operate on an object restricted to pieces of its support, i.e. of the general form

---

[13] In practice, these objects were not as obviously self-affine as ferns. The number of transformations as well as the transformations themselves involved a great deal of trial-and-error.

[14] Albeit only semi-automated and very lossy.

Figure 3 Natural scenes synthesized with iterated function systems (a)"Monterey coast" (b)"Black Forest"(c)"Sunflower field"(d)"Andes girl".
©Georgia Tech. Research Corp.

$$w(B) = \bigcup_{1 \le i \le k} w_i(B]_{D_i}) \tag{21}$$

and which would leave the object approximately invariant, i.e. such that

$$\bigcup_{1 \le i \le k} w_i(A_{orig}]_{D_i}) = A_{orig} \tag{22}$$

This is a considerable relaxation of the self-affinity requirement, which in the context of IFS theory is a *global* one, and the only reasonable one when the objects to model are not self-affine, such the image of a human face. When working in a space of digital images, the pieces D_i can be for example squares of pixels of size 8×8 which form a partition of the image support. This idea can be applied to the modelling of contours of real-world objects as well as to the coding of digital images, two examples that are developed in the next section.

6.4.3.2 Recurrent IFS theory for contour modelling

(i) Fractal curves as attractors of recurrent IFSs

This particular theory of *Recurrent Iterated Function Systems*[15] (RIFS) was developed in order to be used as a synthesizer of planar, continuous fractal curves which interpolate a set of data points. It was independently developed by Hutchinson [33], and can also be seen as an extension of the theory of *Fractal Interpolation Functions* of Barnsley [5].

Let $\{z_i\}_{i=0}^{N}$ be a set of $N+1$ *interpolation nodes* in $\boldsymbol{R}^2$. Let C denote the space of continuous parametrized curves with compact support [0,1][16], which interpolate the set of nodes, i.e.

$$\begin{aligned} &Z \in C \ \text{ if } \ Z(0) = z_0, \ \ Z(1) = z_1 \text{ , and} \\ &\exists (t_1, \ldots, t_{N-1}) \in \boldsymbol{R}^{N-1} \ \text{ such that } \ Z(t_i) = z_i \text{ for every } i \in \{1, \ldots, N-1\} \end{aligned} \tag{23}$$

Note that $Z(t)$ is a *closed* curve if $z_0 = z_N$. Let $Z]_{[t_i, t_j]}(t)$ denote the restriction of the parametrized curve $Z(t)$ to the segment $[t_i, t_j]$.

To each ordered pair of consecutive *range nodes* (z_i, z_{i+1}), is associated an ordered pair of *domain nodes* $(z_{d_i}, z_{d_{i+1}})$, also among the set of interpolation points, and a contractive two-dimensional affine transformation w_i, such that

$$w_i(z_{d_i}) = z_i, \ \text{ and } \ w_i(z_{d_{i+1}}) = z_{i+1} \tag{24}$$

which we refer to as a *node mapping condition*. Let T denote the transformation which operates on a continuous interpolating curve Z in the following way

---

15 The term "recurrent" was coined by Barnsley *et al.* [6, 10]. Its selection will probably not seem obvious to the reader, whom we refer to [10] for justification of that choice.

16 For the sake of simplicity, we use the same notation for a curve $Z(\cdot)$ - a parametric vector function, and its graph - the set of points $\{z(t), t \in [0,1]\}$.

$$T(Z) = \bigcup_{0 \le i \le N} w_i(Z|_{[t_{d_i}, t_{d_{i+1}}]}) \tag{25}$$

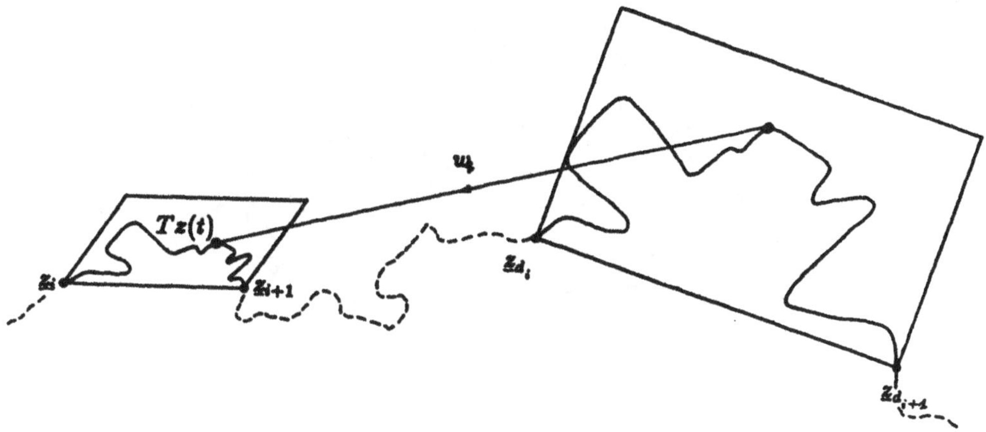

Figure 4 Action of the transformation T on an interpolating curve.

The action of this mapping is illustrated in Figure 4. It is easy to see that the transformed curve $T(Z)$ is also a continuous interpolating curve. It can also be shown [9] that the mapping T is contractive in the space of interpolating curves endowed with an appropriately defined Hausdorff metric and therefore has a unique fixed point A - *the attractor* of the RIFS $\left\{\{z_i\}_{i=0}^{N}, T\right\}$ - which is a continuous fractal curve that interpolates the set of points $\{z_i\}_{i=0}^{N}$. The invariance equation

$$A = \bigcup_{0 \le i \le N} w_i(A|_{[t_{d_i}, t_{d_{i+1}}]}) \tag{26}$$

shows that the curve A is the union of N affinely distorted copies of pieces of itself. This type of fractal curve is therefore said to be *piecewise self-affine*. It can be generated by plotting the iterates of the sequence

$$Z_0, T(Z_0), T^2(Z_0), \ldots \tag{27}$$

where Z_0 is, for example, the linear interpolation of the nodes. An example of a recurrent IFS structure and its associated attractor are shown in Figure 5.

(ii) Contour modelling

The inverse problem can in this context be viewed as a *contour modelling* problem, whereby given a set of digital object contours extracted from image data, one would look for a model in the form of the attractor of a recurrent IFS. A version of the collage theorem can be shown to hold. It states that given an original contour and a set of interpolation points on that curve, the attractor of a recurrent IFS which leaves the original curve approximately invariant will be visually close to the original.

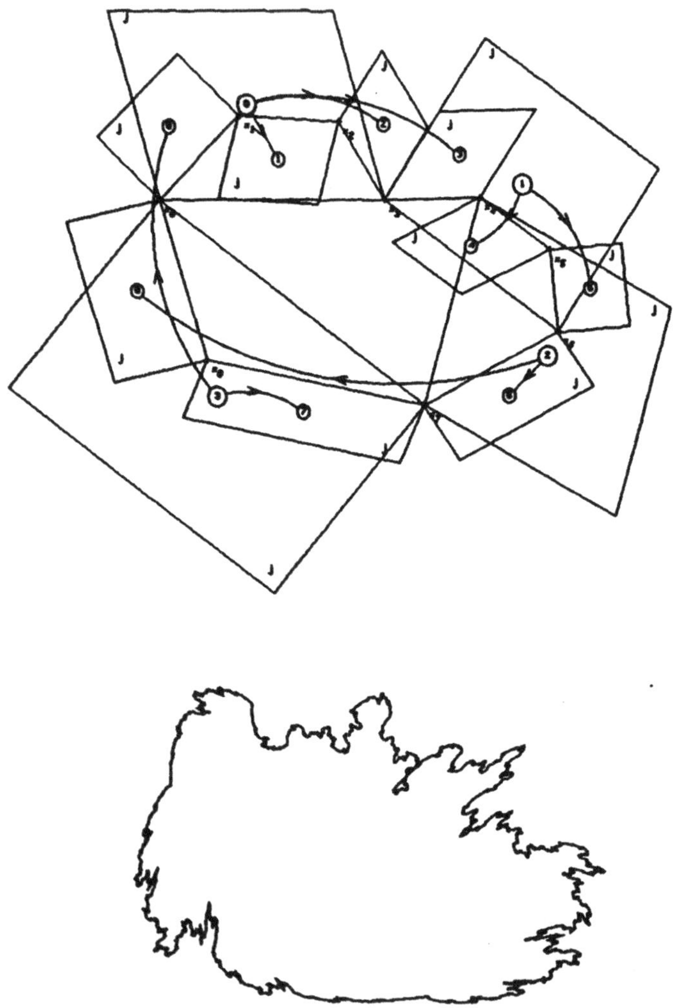

Figure 5 Example of a recurrent IFS structure and its fractal attractor.

Examples of contours extracted from the image "Cumulus" of Figure 6a are shown in Figure 6b. These contours were extracted by first applying a median edge-preserving filter and quantizing the image to eight gray levels, producing uniform gray plateaus whose outer boundaries was traced using a left-most looking rule [9]. A collage of the cloud boundaries was constructed using interactive software described in [9, 35]. The final set of interpolation points chosen on the boundaries is shown in Figure 6c, with two node mapping conditions indicated in this figure by arrows from domain to range nodes. The collage, made of transformed pieces of the original boundaries is shown in Figure 6d. Figure 7 shows a reconstruction sequence of the RIFS structure resulting from the collage procedure. The top-left frame shows the initial graph; a linear interpolation of the nodes. At the fifth iteration, the sequence has converged to the attractor shown in the lower-right frame.

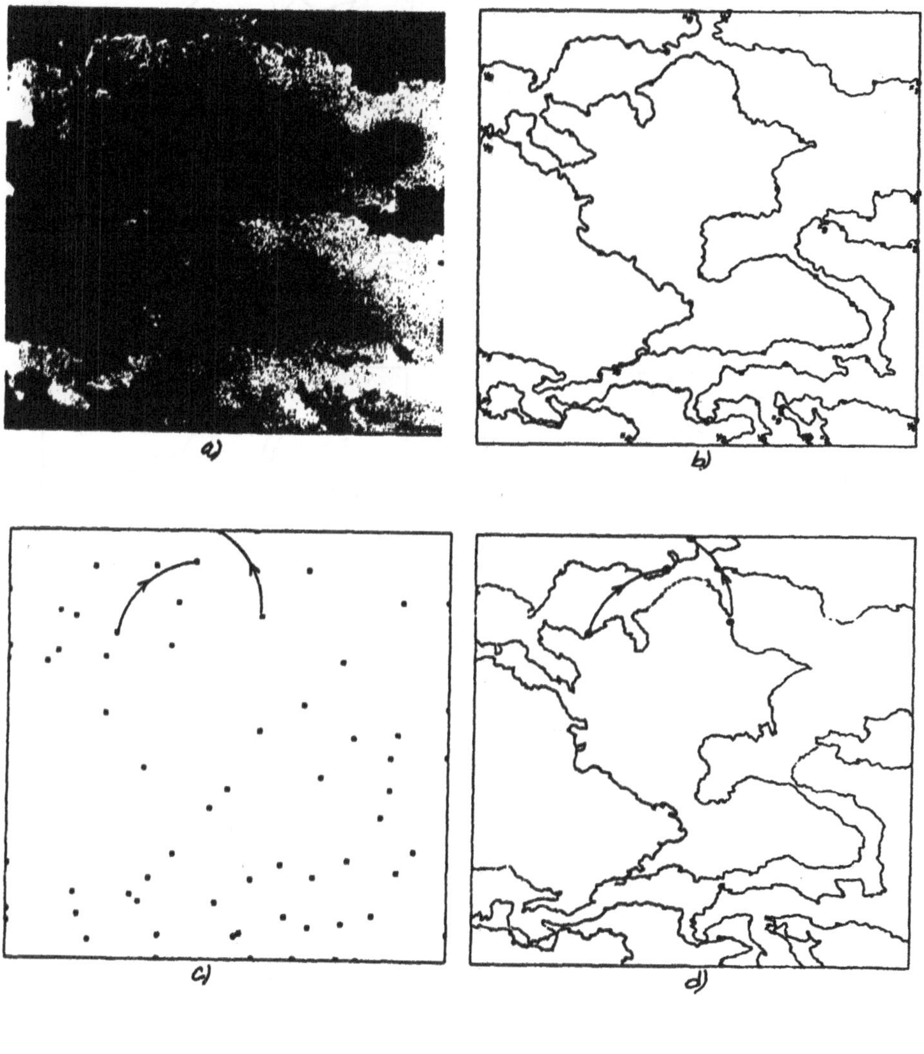

Figure 6 Modelling contours extracted fron "Cloud" image recurrent IFSs (a)"Cloud" image (b)Extracted contours (c)interpolation points (d) collage.

The parameters of the model for the cloud boundaries considered here consisted of approximately 70 nodes, 60 affine transformations, amounting to about 500 bytes of data. Note that this modelling approach still has the undesirable characteristic of requiring human interaction in the selection of nodes and transformation. The author is not aware of any successful attempts to fully automate this procedure, although a similar procedure was implemented by Mazel *et al.* for the modelling of speech signals and seismic data [55].

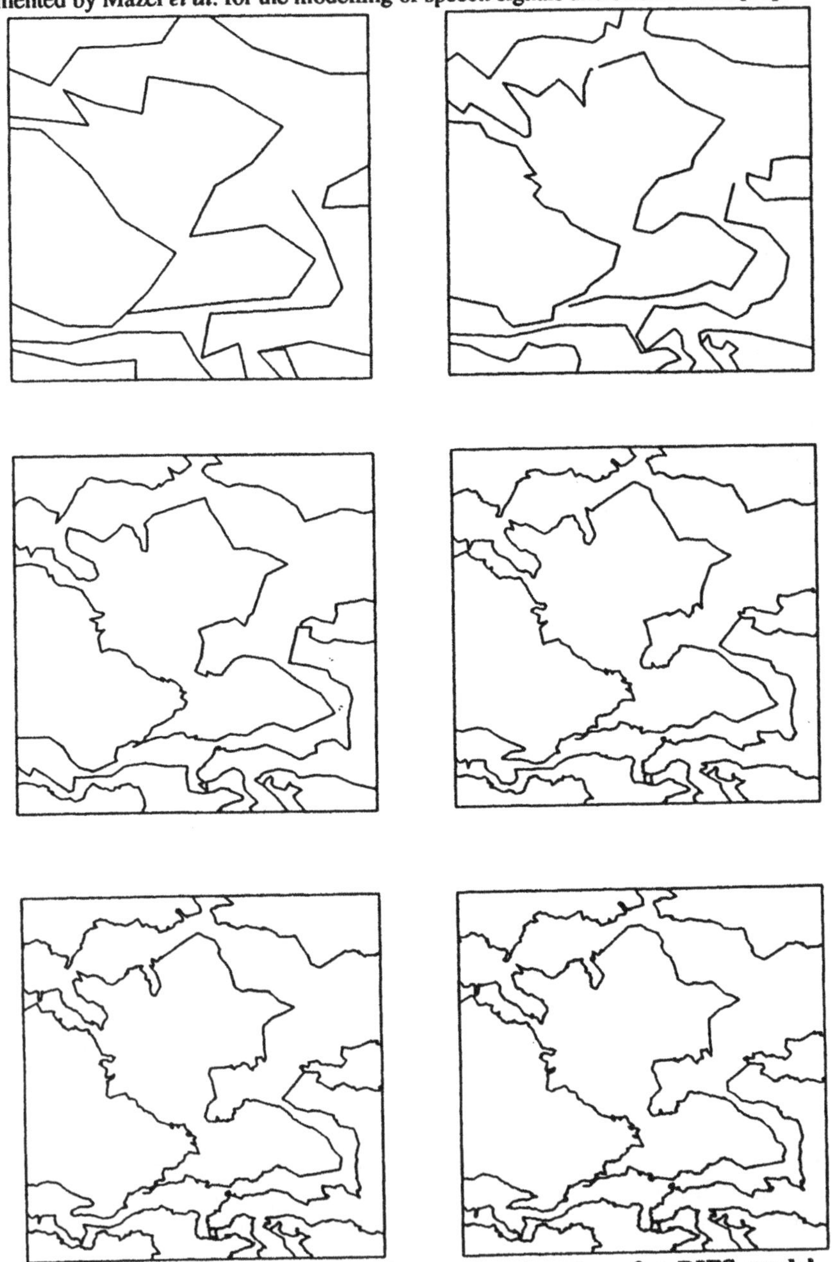

Figure 7 Six iterations converging to the attractor of a RIFS model for "Cloud" contours.

6.4.3.3 *Fractal block coding of digital images*

In this section, we look at the modelling and encoding of monochrome 8-bit digital images. *Fractal Block Coding* (FBC) is a modelling approach to image coding based, once again, on a theory of iterated contractive transformations defined piecewise. This work was initially published by the author in [35, 36], and is fully described in [35, 37]. The two main characteristics of the approach are that

i) it relies on the assumption that image redundancy can be efficiently captured and exploited through piecewise *self-transformability* on a blockwise basis,

ii) it approximates an original image by a *fractal image* obtained from a finite number of iterations of an image transformation defined blockwise called a *fractal code.*

It is truly an image coding technique since it can be used to encode any digital image, as opposed to only images of a specific type, and is implementable as a computer algorithm which requires no human intervention. We now summarize this coding technique.

Let (M, d) denote a metric space of digital images, where d is for example the familiar ***root-mean-square*** error metric. The encoding of an original image μ_{orig} consists in the automatic construction of a contractive image transformation τ defined piecewise, which leaves μ_{orig} approximately invariant, i.e. such that

$$d(\mu_{orig}, \tau(\mu_{orig})) \text{ is as "close to zero" as possible.} \tag{28}$$

The transformation τ is chosen to be of the general form

$$\tau(\mu) = \sum_{0 \le i < N} (\tau\mu)]_{R_i} = \sum_{0 \le i < N} \tau_i(\mu]_{D_i}) \tag{29}$$

where

$$P = \{R_i\}_{0 \le i < N} \tag{30}$$

denotes a non-overlapping partition of the image support into N ***range cells*** - usually, but not necessarily, square blocks of possibly different sizes. The symbol $\mu]_{R_i}$ denotes the restriction of the image μ to the cell R_i; we call it the image block over R_i and write

$$\mu = \sum_{0 \le i < N} \mu]_{R_i} \tag{31}$$

simply to indicate that an image is the union of its restrictions to the cells of the partition. The symbol τ_i denotes an elemental block transformation from a domain cell D_i[17] to the range cell R_i. For clarity, τ_i is written as the composition of two transformations S_i and T_i

$$\tau_i = T_i \circ S_i \tag{32}$$

---

[17] These cells have to be larger than the range cells so as to ensure contractivity. A common choice is to consider square blocks of twice the size of the range blocks.

where S_i and T_i are the so-called *geometric* and *massic* parts of τ_i, respectively. In the case of square blocks, S_i *contracts* domain blocks the the size of range block R_i, while T_i *processes* an image block without altering its square support.

Designing a fractal block coding system consists in defining *a priori* the following

i) a procedure for constructing a "good" image dependent partition, which can consist for example of a quadtree structure or any other type of tiling,

ii) classes of discrete block transformations and a procedure for searching these classes.

The construction of a fractal code τ for μ_{orig} consists in finding, for each value of the index i, the "best pair" of domain and block transformation (D_i, T_i), i.e. the pair such that the distortion

$$d(\mu]_{R_i},\ T_i \circ S_i(\mu]_{D_i}) \text{ is minimal.} \tag{33}$$

The *fractal code* consists of the description of the image partition along with the quantized parameters of the block transformations. Its structure is illustrated in Figure 8 in which the arrows indicate block transformations from domain block to range block.

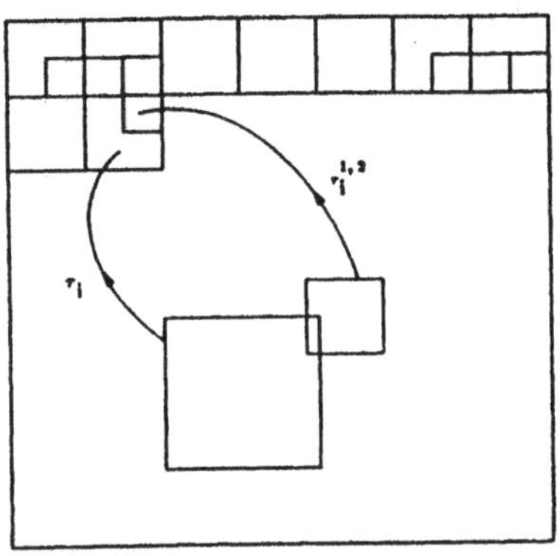

Figure 8 Structure of a fractal image block code.

(i) Decoding/reconstruction

As always with fractal techniques, the reconstruction of a decoded image is easily achieved through iteration. Here, the fractal block code τ is iterated on an initial image, such as a black square (or any other image) until convergence to a stable image - in practice after about ten iterations. This reconstruction procedure is illustrated in Figure 9 for a fractal code of the familiar digital image "Lena", with the initial image "peppers".

(ii) Extensions

Over the past five years, a large number of extensions of fractal image block coding have been proposed by various researchers in the signal processing research community [2, 11, 12, 14-16, 23, 27, 34, 41, 43-45, 48, 49, 56-58, 61-65, 74, 78], which shows that the field remains a very active research area. These extensions broadly address the following issues

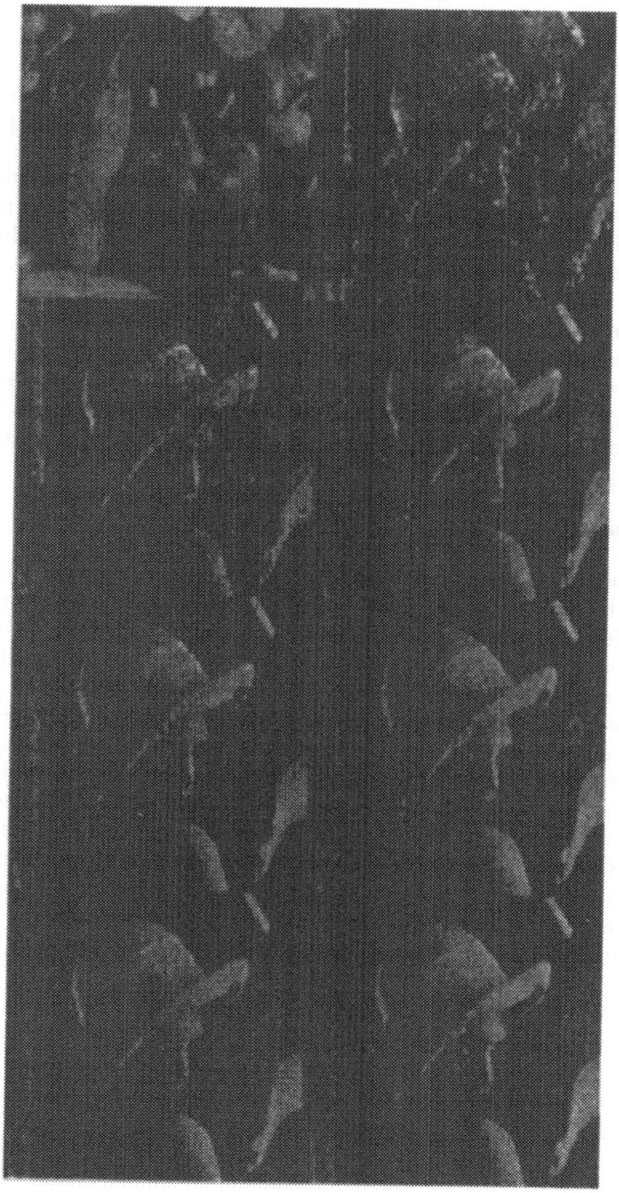

Figure 9 First eight iterations of "peppers-to-lena" decoding sequence.

i) the influence of the type of image partition, pool of block transformations, and optimization of the parameters defining these transformations,

ii) the reduction of the complexity of the encoding process - the search for best matching blocks in the image to encode,

iii) the comparison and possible merging of fractal block coding with more traditional block-based image coding techniques,

iv) the extension of the theory to the three-dimensional case for fractal block coding of sequences of images. A summary of these extensions can be found in [38].

Unfortunately, the compression ratios achieved with block fractal coding techniques turn out to be quite similar to those obtainable with other more classical image coding techniques, in the order of 30:1[18], which seems modest compared with the 1000:1 ratios claimed for the pictures described in section 6.4.3.1(ii). This is because, in order to make it tractable, the inverse problem had to be broken up into elementary subproblems which can be efficiently dealt with automatically. In particular, the image segmentation with human operator described in such section has been replaced by an image partitioning into fairly small blocks, which is of course reminiscent of other block-based image coding techniques such as block-based DCT [18, 40, 60], or vector quantization (VQ) [26, 29]. By doing so however, the possibility to model very large image segments by only very few transformations - as was done in the images of section 6.4.3.1 - has been lost.

6.4.4 Conclusion

In this tutorial paper, we briefly introduced the notions of fractal objects - objects with an intricate geometry which look the same at many different scales, and chaotic systems - systems for which long term prediction is impossible due to the property of sensitivity to initial conditions. We then described applications of fractals in signal processing, specifically in the modelling and compression of images.

Three basic types of deterministic fractal modelling techniques were described. The first two, *iterated function systems* (IFS) theory and *recurrent* IFS theory, could potentially achieve extremely high compression ratios - provided that they could be automated. The third, *fractal block coding*, is a true image compression technique but leads, to this date, to more modest compression ratios, comparable to those obtained with standard image compression techniques. Even though fractal block coding of images seems to offer a promise of both understanding and exploiting the iterative processes commonly found in nature which give rise to fractal patterns, the actual process of modelling natural scenes with both a very compact (i.e. very low bit rate) fractal model and a great degree of fidelity seems to be eluding us, still.

References

[1] K. Aizawa, H. Harashima, T. Saito, "Model-based analysis synthesis image coding (MBASIC) system for a person's face", *Signal Processing: Image Communication*, Vol. 1, No. 2, pp. 139-152, October 1989.

---

[18] A rather severe breakdown seems to occur fairly systematically at coding rates below 0.2 *bpp*.

[2] Z. Baharav, D. Malah, and E. Karnin, "Hierarchical interpretation of fractal image coding and its application to fast decoding", *Proc. Digital Signal Processing Conference*, Cyprus, 1993.

[3] M. F. Barnsley and S. Demko, "Iterated function systems and the global construction of fractals", *Proc. of Royal Soc.*, London, Vol. A399, pp. 243-275, 1985.

[4] M. F. Barnsley, V. Ervin, D. Hardin, J. Lancaster, "Solution of an inverse problem for fractals and other sets", *Proc. of the National Academy of Sciences*, Vol. 83, pp. 1975-1977, 1986.

[5] M. F. Barnsley , "Fractal functions and interpolation", Constructive Approximation, Springer-Verlag, Vol. 2, pp. 303-329, 1986.

[6] M. F. Barnsley, *Fractals everywhere* (Second edition), New York Academic, 1988.

[7] M. F. Barnsley, A. Sloan, "A better way to compress images", *BYTE Magazine*, January 1988.

[8] M. F. Barnsley, A. Jacquin, F. Malassenet, L. Reuter, "Harnessing chaos for image synthesis", *Proc. SIGGRAPH'88* , pp. 131-140, 1988.

[9] M. F. Barnsley, A. Jacquin, "Application of recurrent iterated function systems to images", *Proc. SPIE*, Vol. 1001, pp. 122-131, 1988.

[10] M. F. Barnsley, J. H. Elton, and D. P. Hardin, "Recurrent iterated function systems", *Constructive Approximation*, pp. 3-31, Berlin, Germany: Springer-Verlag, 1989.

[11] K. U. Barthel, T. Voyé, and P. Noll, "Improved fractal image coding", *Picture Coding Symposium*, Lausanne, 1993.

[12] J. M. Beaumont, "Image data compression using fractal techniques", *BT Technology Journal*, Vol. 9, No. 4, pp. 93-109, October 1991.

[13] K-H. Becker, M. Dörfler, *Dynamical systems and fractals*, Cambridge University Press, 1989.

[14] T. J. Bedford, F. M. Dekking, M. S. Keane, "Fractal image coding techniques and contraction operators", Delft University of Technology, Internal Report, 1992.

[15] T. J. Bedford, F. M. Dekking, M. Breeuwer, M. S. Keane, D. van Schooneveld, "Fractal coding of monochrome images", Delft University of Technology, Internal Report, 1992.

[16] J-M. Chassery, F. Davoine, et E. Bertin, "Compression fractale par partitionnement de Delaunay", *Quatorzieme colloque GRETSI*, Juan-Les-Pins, Septembre 1993.

[17] G. Cherbit, *Fractals, dimensions non entieres et applications*, Paris, Masson, 1987.

[18] R. J. Clarke, *Transform coding of images*, Academic Press, London, 1985.

[19] R. Devaney, *Introduction to chaotic dynamical systems*, Benjamin-Cummings, 1986.

[20] N. Dunford and J. T. Schwartz, *Linear operators*, New York: Wiley, 1966.

[21] K. Falconer, *The geometry of fractal sets*, Cambridge University Press, 1985.

[22] K. Falconer, *Fractal geometry, mathematical foundations and applications*, J. Wiley & Sons, 1990.

[23] Y. Fisher, E. W. Jacobs, R. D. Boss, "Iterated transform image compression", NOSC TR-1408, Naval Ocean Systems Center, San Diego, CA, 1991.

[24] R. Forchheimer, T. Kronander, "Image Coding-From Waveforms to Animation", *IEEE Trans. on Acoustics, Speech and Signal Processing*, Vol. 37, No. 12, pp. 2008-2023, December 1989.

[25] A. Fournier, D. Fussell, L. Carpenter, "Computer rendering of stochastic models", *Com. of the ACM*, Vol. 25, pp. 371-384, 1982.

[26] A. Gersho and R. M. Gray, *Vector quantization and signal compression*, Kluwer Academic Publishers, 1992.

[27] M. Gharavi-Alkhansari and T. S. Huang, "A fractal-based image block-coding algorithm", *Proc. ICASSP'93*, pp. 345-348.

[28] J. Gleick, *Chaos, the making of a new science*, Vicking, New York, 1987.

[29] R. M. Gray, "Vector quantization", *IEEE ASSP Mag.*, April 1984.

[30] D. Gulick, *Encounters with chaos*, McGraw-Hill, 1992.

[31] J. C. Hart, D. J. Sandin and L. H. Kauffman, "Ray tracing deterministic 3-D fractals", *Computer Graphics*, Vol. 23, No.3, pp. 91-100, 1989.

[32] J. C. Hart and F. K. Musgrave (Co-Chairs), *Fractal modelling in 3D computer graphics and imaging*, in *SIGGRAPH'91*, course notes, 1991.

[33] J. E. Hutchinson, "Fractals and self-similarity", *Indiana University Mathematics Journal*, Vol. 30, No. 5, 1981.

[34] E. W. Jacobs, Y. Fisher and R. D. Boss, "Image compression: A study of the iterated transform method", *Signal Processing*, Vol. 29, No. 3, pp. 251-263, December 1992.

[35] A. E. Jacquin, "A fractal theory of iterated Markov operators with applications to digital image coding", PhD Dissertation, Georgia Tech, 1989.

[36] A. E. Jacquin, "A novel fractal block-coding technique for digital images", *Proc. ICASSP'90*, pp. 2225-2228.

[37] A. E. Jacquin, "Image coding based on a fractal theory of iterated contractive image transformations", *IEEE Trans. on Image Processing*, Vol. 1, No. 1, pp. 18-30, January 1992.

[38] A. E. Jacquin, "Fractal Image Coding: A Review", *Proceedings of the IEEE*, Vol. 81, No. 10, pp. 1451-1465, October 1993.

[39] D. L. Jaggard (Editor), "Special section on fractals in electrical engineering", *Proceedings of the IEEE*, Vol. 81, No. 10, pp. 1423-1533, October 1993.

[40] N. S. Jayant, P. Noll, *Digital coding of waveforms: principles and applications to speech and video*, Prentice-Hall, 1984.

[41] T. Laurencot, A. Jacquin, "Hybrid image block coders incorporating fractal coding and vector quantization, with a robust classification scheme", AT&T Technical Memorandum, February 1992.

[42] M. Kunt, A. Ikonomopoulos, M. Kocher, "Second-generation image coding techniques", *Proc. IEEE*, Vol. 73, No. 4, pp. 549-574, April 1985.

[43] S. Lepsoy, "Attractor image compression-fast algorithms and comparisons to related techniques", PhD Thesis, The Norwegian Institute of Technology.

[44] S. Lepsoy, G. E. Oien, T. A. Ramstad, "Attractor image compression with a fast non-iterative decoding algorithm", *Proc. ICASSP'93*, pp. 337-340.

[45] H. Li, M. Novak, R. Forchheimer, "A fractal-based image sequence compression scheme", Preprint, 1993.

[46] E. N. Lorenz, "Deterministic non-periodic flow", *Journal of Atmospheric Science*, Vol. 20, pp. 130-141, 1963.

[47] E. N. Lorenz, "Dimension of weather and climate attractors", *Nature*, Vol. 353, pp. 241-241, 1991.

[48] L. Lundheim, "An approach to fractal coding of one-dimensional signals", *Proc. KOMPRESJON'89* Oslo, 1989.

[49] L. Lundheim, "Fractal signal modelling for source coding", PhD Thesis, The Norwegian Institute of Technology, September 1992.

[50] F. J. Malassenet, "Texture coding using a pyramid decomposition", *Proc. ICASSP'93*, Vol. V, pp. 353-356.

[51] B. Mandelbrot, "How long is the coast of Britain ?-- Statistical self-similarity and fractional dimension", *Science*, Vol. 155, pp. 636-638, 1967.

[52] B. Mandelbrot, *Les objects fractals* , (Deuxieme edition), Flammarion, 1984.

[53] B. Mandelbrot, *The Fractal Geometry of Nature*, San Francisco, CA: Freeman, 1982.

[54] R. M. May, "Simple mathematical models with very complicated dynamics", *Nature*, Vol. 261, pp. 459-467, 1976.

[55] D. S. Mazel, M. H. Hayes, "Using Iterated Function Systems to Model Discrete Sequences", *IEEE Trans. on SP*, Vol. 40, No. 7, pp. 1724-1734, July 1992.

[56] D. M. Monro, F. Dudbridge, "Fractal approximation of image blocks", *Proc. ICASSP'92*, Vol. III, pp. 485-488.

[57] D. M. Monro, F. Dudbridge, "Fractal block coding of images", *Electronics Letters*, Vol. 28, pp. 1053-1055, May 1992.

[58] D. M. Monro, "Generalized fractal transforms: Complexity issues", *Proc. Data Compression Conference*, IEEE Computer Society Press, pp. 254-261, March-April 1993.

[59] H. G. Musmann, M. Hötter, J. Ostermann, "Object-oriented analysis-synthesis coding of moving images", *Signal Processing: Image Communication*, Vol. 1, No. 2, pp. 117-138, October 1989.

[60] A. N. Netravali and B. G. Haskell, *Digital pictures: representation and compression*, New York: Plenum, 1989.

[61] G. E. Øien, S. Lepsøy, T. A. Ramstad, "An inner product space approach to image coding by contractive transformations", *Proc. ICASSP'91*, pp. 2773-2776.

[62] G. E. Øien, S. Lepsøy, T. A. Ramstad, "Reducing the complexity of a fractal-based image coder", *EUSIPCO'92*.

[63] G. E. Øien, A. Sollid, T. A. Ramstad, "Hybrid image compression combining transform coding and attractor coding", *NOBIM'92*.

[64] G. E. Øien, "L2-Optimal attractor image coding with fast decoder convergence", PhD Thesis, The Norwegian Institute of Technology, 1993.

[65] G. E. Øien, Z. Baharav, S. Lepsøy, E. Karnin, "A new improved collage theorem with applications to multiresolution fractal image coding", *Proc. ICASSP'94*, 1994.

[66] H-O. Peitgen, P. H. Richter, *The beauty of fractals*, Berlin: Springer, 1986.

[67] H-O. Peitgen, D. Saupe, *The science of fractal images*, Springer-Verlag, 1988.

[68] H-O. Peitgen, H. Jurgens, D. Saupe, *Chaos and fractals, new frontiers of science*, Springer-Verlag, 1992.

[69] A. P. Pentland, "Fractal-based descriptions of natural scenes", *IEEE PAMI*, Vol. 6, No. 6, 1984.

[70] A. P. Pentland, "Fractal surface models for communications about terrain", *SPIE Visual Com. and Image Proc. II*, Vol. 845, 1987.

[71] L. Hodges-Reuter, "Rendering and magnification of fractals using iterated function systems", PhD. Dissertation, Georgia Tech, 1987.

[72] L. F. Richardson, in *General Systems Yearbook*, 6, 139, 1961.

[73] R. Rinaldo, A. Zakhor, "Fractal approximation of images", *Proc. Data Compression Conference*, pp. 451, March-April 1993.

[74] D. van Schooneveld, "Fractal coding of monochrome images", Delft University of Technology, Internal Report, December 1991.

[75] M. Schroder, *Fractals, chaos, power laws*, W. H. Freeman and Co., New York, 1991.

[76] M. C. Stein, "Fractal image models and object detection", *SPIE Visual Com. and Image Proc. II*, Vol. 845, 1987.

[77] A. A. Tsonis, *Chaos, from theory to applications*, Plenum Press, New York, 1992.

[78] G. Vines, and M. H. Hayes, III, "Adaptive IFS image coding with proximity maps", Proc. ICASSP'93, pp. 349-352.

Tutorial 7

The fundamentals of analog micropower design

Organised by: Eric Vittoz, *Swiss Centre for Electronics and Microtechnology*

Chapter 7

The Fundamentals of Analog Micropower Design

Eric A. Vittoz

7.1 Introduction

Micropower IC techniques were originally developed more than 20 years ago for electronic watches, in which the power is limited to below 1 microwatt. Since that time, low-voltage very low-current circuits have found applications in various kinds of small-size battery-operated systems, such as pocket calculators, implanted or portable medical devices, hearing instruments and paging receivers. Another interesting niche is that of data acquisition systems powered by light or by a radio-frequency field. Most of these applications require low-voltage operation (1 to 3 Volts) and a current consumption typically below 100 microamperes, which are severe limitations, especially for analog sub-circuits. Some trade-off between dynamic range and speed must generally be accepted, but specific design techniques had to be developed.

More recently, micropower circuits have found new applications for the analog VLSI implementation of novel processing systems based on the neural paradigm.

The interest in low-power has dramatically increased in the last few years, mainly because of the growing need for portability in computer and telecommunication products. Other concomitant reasons are heat generated on-chip, the growing relative cost of power supplies, or even unacceptable network power required by very large systems. Low-voltage may be a requirement imposed by the process or by the system, or it may be an element of the strategy to reduce the power consumed by digital circuits. It has little influence on the power consumption of analog circuits, but it complicates their implementation since the corresponding reduction of maximum signal amplitude must be compensated by lowering the noise floor.

This chapter describes the fundamental device and circuit aspects of low-voltage very low-current integrated circuits. The discussion is limited to CMOS which is unquestionably the best technology for micropower, and the emphasis is put on analog sub-circuits. Many of these techniques can help to meet the new requirements for more general low-power VLSI.

Most of the material presented is extracted from [1], whereas the presentation itself is an excerpt of [2].

7.2 Devices for low-voltage and very low-current

Many of the problems and solutions encountered in micropower circuits are directly related to the properties of the transistor itself, which must therefore be properly understood and modelled down to very low currents.

The basic static behaviour of a MOS transistor is best characterized by a symmetrical model [3,4,1] in which the drain current I_D is decomposed into a forward component I_F and a reverse component I_R

$$I_D = I_F - I_R = I_S[\,f(V_S, V_G) - f(V_D, V_G)] \tag{1}$$

where source voltage V_S, drain voltage V_D and gate voltage V_G are all referred to the local substrate. $f(V, V_G)$ is a monotonically decreasing function of V which tends to a square law for $f \gg 1$ (corresponding component of current in strong inversion) and to an exponential for $f \ll 1$ (weak inversion). It can be approximated by

$$f(V, V_G) = \ln^2\{\,1 + \exp\frac{V_P - V}{2U_T}\,\} \tag{2}$$

where $U_T = kT/q$ and V_P is called the pinch-off voltage, approximately given by

$$V_P = \frac{V_G - V_{T0}}{n} \tag{3}$$

V_{T0} is the threshold voltage and $n<2$ is a slope factor which is only weakly dependent on V_G (tends to 1 for V_G very large).

The third and last basic parameter of the transistor is the specific current I_S in (1). It is given by

$$I_S = 2n\mu C_{ox} U_T^2 \frac{W}{L} = 2n\beta U_T^2 \tag{4}$$

and can alternatively be replaced by the usual transfer parameter β.

This specific current, which characterizes the level for which the component is in moderate inversion, typically ranges from 10 to 200 nA for minimum size devices. The overall mode of operation of the transistor depends on the respective levels of both I_F and I_R, as illustrated in Figure 1.

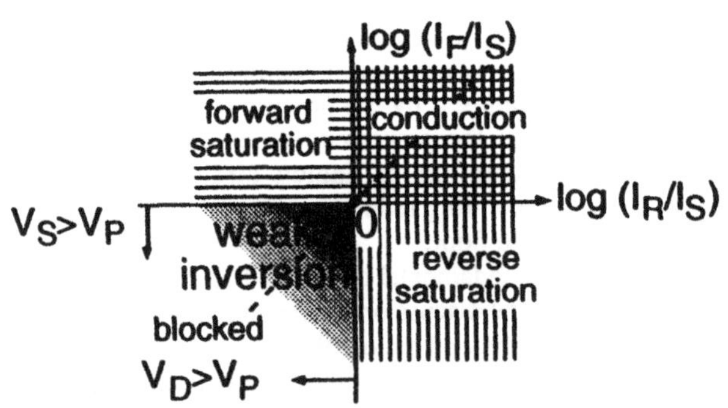

Figure 1 Modes of operation of a MOS transistor.

If both I_F and I_R are in weak inversion (V_S and $V_D > V_P$), then the whole transistor is said to operate in the weak inversion mode and behaves, in many respects, similarly to a bipolar transistor without any DC control current. Combining (1) to (3) gives the corresponding drain current

$$I_D = I_S\, e^{\frac{V_G - V_{T0}}{nU_T}} \left\{ e^{-\frac{V_S}{U_T}} - e^{-\frac{V_D}{U_T}} \right\} \tag{5}$$

Weak inversion provides the minimum possible value of drain to source saturation voltage (V_{DSsat}= 4 to $6U_T$), which helps to keep peak-to-peak signal amplitudes close to the (battery) supply voltage V_B.

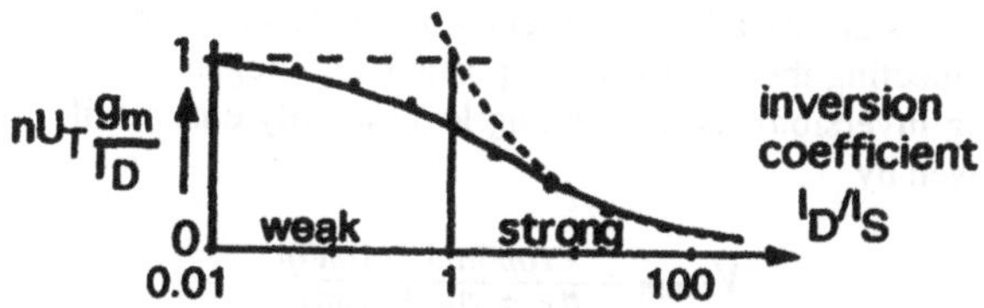

Figure 2 Transconductance-to-current ratio; solid line obtained from the model; experimental points for comparison.

The gate to drain transconductance g_m in saturation is obtained by differentiating $I_D=I_F$ given by (1) to (3). The result is represented in Figure 2 as a function of the *inversion coefficient* I_D/I_S. It can be seen that weak inversion provides the maximum value of g_m/I_D. Thus, if the current is limited, weak inversion provides the maximum value of transconductance g_m and of bandwidth g_m/C for a given value of capacitive load C, and the minimum value of input noise resistance $R_N \cong 1/g_m$. This maximum value of g_m/I_D also results in a maximum value of the intrinsic voltage gain, which helps to simplify the architecture of amplifiers and in minimum input offset in a pair.

Weak inversion is not applicable to very high frequencies, since

$$f_{Tmax,weak} = \frac{g_m(I_D=I_S)}{2\pi(C_G+C_D)} < \frac{\mu U_T}{\pi L^2} \tag{6}$$

cannot exceed a few hundred MHz for L=1µm.

Other problems with weak inversion are the large relative noise content of the drain current and the degradation of drain current mismatch due to threshold mismatch. Both effects result from the maximum value of g_m/I_D. Thus, using weak inversion to counter the low value of supply voltage V_B by maximizing the signal swing results in very inaccurate and noisy currents. The precision of currents at low-voltage can be drastically improved by operating the MOS transistor as a lateral bipolar [5], or by using true bipolar transistors in a BiCMOS process.

No particular requirements are imposed on capacitors for the realization of micropower circuits. The lowering of the values of functional capacitors to help reduce power consumption is bounded by the acceptable kT/C noise level. On the contrary, the realization of usable resistors is made difficult by the need for high values. Linear resistive networks operating in current mode can be implemented in a very dense manner by using the transistor as a pseudo-conductance [6]. Some processes intended for micropower provide lateral diodes in the polysilicon layer [7]. These floating diodes can be exploited to implement high-impedance biasing schemes or on-chip voltage multipliers.

Parasitic capacitors and noise impose a trade-off between bandwidth and current drain. However, an absolute limit on minimum operating current is given by leakage currents. Leakage is mostly due to generation of electron-hole pairs within space charge regions, including those underneath the channel of a transistor. Latch-up can be prevented by limiting the current available from the power source below the minimum value of hold current, without affecting normal operation at the microampere level.

7.3 Circuit limitations and solutions

Reducing the supply voltage V_B does usually not help to reduce the power consumption of analog circuits. However, several problems must be solved when low-voltage is imposed by the system or by the source of energy.

Analog switches are usually implemented by means of complementary pairs made conducting by connecting their gates to the positive and negative power rails. However, conduction (strong inversion) for any analog level is only ensured if V_B is larger than a minimum value given by

$$V_{Bmin} = \frac{V_{TOp}n_n + V_{TOn}n_p}{n_p + n_n + n_p n_n} \tag{7}$$

which is typically around 3V for maximum threshold values of the order of 0.7 V. This very severe limitation to low-voltage operation of analog switches can be circumvented by various solutions: lower values of V_{TO} may be required from the process, but it would increase the off-current of the switch due to the weak inversion behaviour. Circuits may be architectured to avoid the need for switches conducting in the whole voltage range. Another solution amounts to provide on-chip multiplication of the clock voltage [8].

The effect of the charge injected from switches is worse at low supply voltages, since it corresponds to a relatively larger fraction of the signal amplitude for the same on-conductance g of switches (same channel charge $q=gL^2/\mu$).

Low-voltage operation precludes the use of any transistor stacking. In particular, cascodes must be implemented to bias the drain voltage of the common source device just at the verge of saturation. A well-known solution is illustrated in Figure 3a [1,9]. If the three transistors are operated in strong inversion, they can be sized to impose $V_{D1}=V_{P1}$ independently of the value of current I_0, for example by choosing $\beta_1=\beta_2=4\beta_3$.

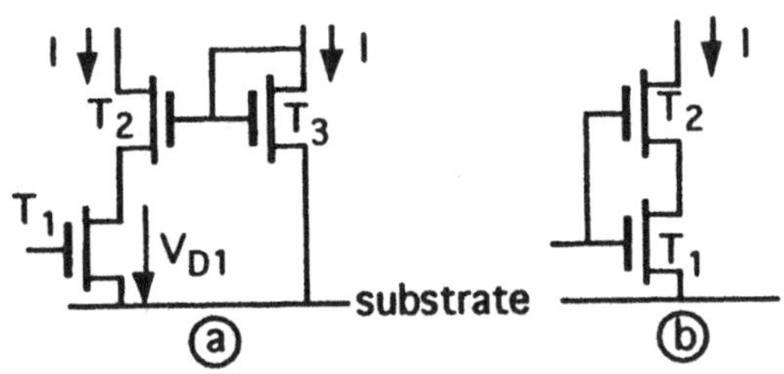

Figure 3 Low-voltage cascode (a) for strong inversion, (b) self-cascode for weak inversion.

This scheme must be modified to be applicable to weak inversion [1]. An alternative is given by the compact solution illustrated in Figure 3b [10]. From (5), it can be shown that this combination of two transistors with Early voltages V_{E1} and V_{E2} provides an output conductance

$$g_o = \frac{I}{V_{E2}} \frac{1 + FU_T/V_{E1}}{1 + F} \tag{8}$$

which approaches that of a real cascode if the factor

$$F = \frac{\beta_2}{\beta_1} e^{\frac{V_{T01} - V_{T02}}{nU_T}} \tag{9}$$

is large. This can be obtained by using channel width $W_2 \gg W_1$, with W_1 as small as possible. V_{T01} is then increased by narrow channel effect, which further increases F.

The properties of basic configurations such as current mirrors, differential pairs and voltage gain cells strongly depend on the inversion coefficient (I_F/I_S) of each of their transistors. Special solutions are needed to implement current and voltage references that avoid high-value resistors. Micropower analog blocks exploit the features of weak inversion in architectures that are as simple as possible.

As an example, the optimization of operational amplifiers for low current and low voltage can be illustrated by the very simple single stage OTA shown in Figure 4.

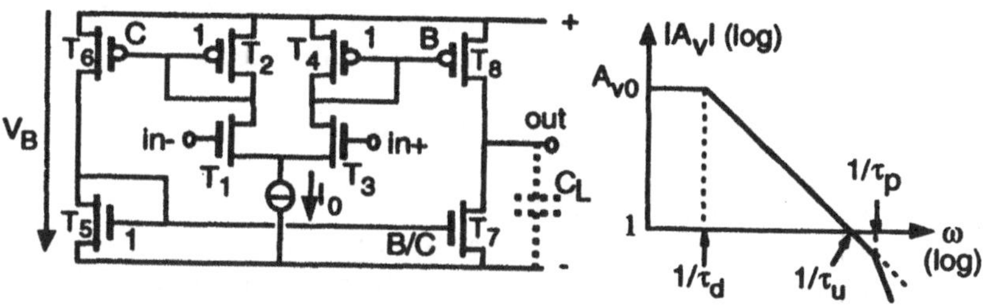

Figure 4 Single stage OTA and its frequency behaviour.

The overall transconductance g_m of this amplifier is that of input transistors T_1-T_3 multiplied by mirror ratio B. Loaded by C_L, the circuit behaves essentially as an integrator with time constant

$$\tau_u = C_L/g_m \tag{10}$$

If all other poles are at a much higher frequency, their effect may be approximated by a single equivalent pole with time constant τ_p.

Ignoring the finite DC gain, the transfer function of the amplifier may then be expressed by

$$A_v = \frac{1}{s\tau_u (1 + s\tau_p)} \tag{11}$$

The OTA always operates in closed loop, with a certain amount of negative voltage feedback $\beta \leq 1$. The settling time for small perturbations can then be approximated [11] by

$$T_s = (2\tau_p + \tau_u/\beta) ln(1/\varepsilon) \tag{12}$$

where ε is the relative settling error after time T_s. Settling time T_s also increases with τ_p and settling becomes oscillatory when this term dominates (small phase margin). Using (11), the total output noise V_N^2 in closed loop due to an equivalent input white noise resistance R_N can be calculated analytically [1]

$$V_N^2 = \frac{kTR_N}{\beta\tau_u} = \frac{\gamma}{\beta}\frac{kT}{C_L} \tag{13}$$

Surprisingly, this result does not depend on τ_p. Thus, capacitors located elsewhere than at the output node indirectly increase the power by reducing the speed and degrading stability, but they have no effect on reducing the noise. By definition, $\gamma = g_m R_N$ is a factor of merit which should be as small as possible. The smallest value is obtained when the noise is limited to that of the input pair, by operating T_1 and T_3 in weak inversion and all other transistors deeply in strong inversion. Then:

$$\gamma = \gamma_{min} = nB \tag{14}$$

The price to pay is a reduced maximum signal amplitude due to the large saturation voltage V_P of T_7 and T_8. It also increases τ_p by reducing $g_{m2(4)}$. If these drawbacks are avoided by operating all the transistors in weak inversion, then γ is degraded to

$$\gamma = \gamma_{max} = n(2B+1+B/C) \tag{15}$$

which is 3 to 4 times γ_{min} if $B \geq 1$ and $C=1$.

As shown by (13), noise is increased above the theoretical minimum kT/C_L by a factor γ/β. It might seem interesting to increase the power efficiency of the amplifier (percentage of supply current delivered to the load C_L) by selecting a large value for mirror ratio B. But this would increase γ and hence the noise proportionally. Relation (13) also shows that any reduction of the feedback factor β below unity (due for example to parasitic capacitors at the input node) increases the noise proportionally.

Except for small possible variations of γ, the noise power expressed by (13) is not directly dependent on the bias current. However, reducing the noise by increasing C_L requires an increase in current to maintain the small-signal settling time T_s and the slew rate BI_0/C_L.

Power efficiency and slew rate can be drastically increased by using class-AB amplifiers, which can provide to the load a current much larger than their standby bias current. A very simple and efficient solution intended for switched capacitor circuits is illustrated in Figure 5 [12].

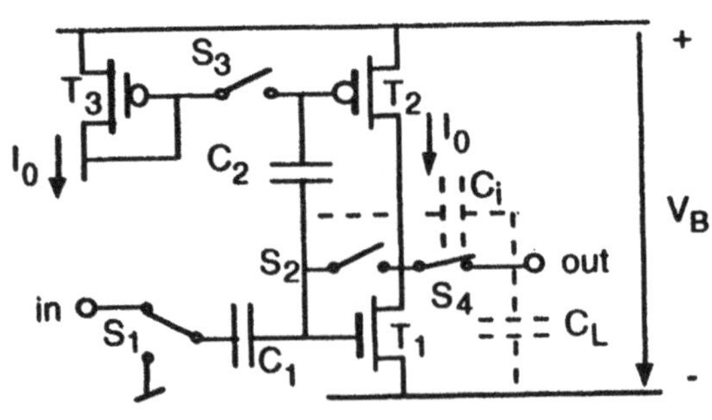

Figure 5 Low-power class-AB transconductance amplifier.

This circuit operates in two phases: an amplifying phase (position of switches shown in the figure) and a biasing phase (opposite position of switches). During the biasing phase, T_3-T_2 are configured as a mirror which imposes bias current I_0 through diode-connected transistor T_1. At equilibrium, the gate voltage of T_1 reaches an internal virtual ground level (no current available from the output node). Switch S_1 is grounded, thus capacitor C_1 is charged to adapt the external virtual ground level to the desired value. In the amplifying phase, T_1 and T_2 are configured as a CMOS inverter and the two transistors are driven in parallel by any voltage difference between input and ground.

This solution is an excellent illustration as to how simplicity can bring high performance when power is limited. In weak inversion, this circuit provides the maximum possible value of transconductance (and thus minimum equivalent input noise resistance R_N) for current I_0, and a maximum value of small-signal DC gain equal to the intrinsic gain of a single isolated transistor. Gain may be further increased beyond 100 dB by cascoding T_1 and T_2; the maximum peak-to-peak signal amplitude is then only 400 to 600 mV below V_B. The auto-zeroing properties of the switching scheme [4,13] result in a good low-frequency PSRR and eliminate the $1/f$ noise. The noise figure of merit γ is almost as low as that of a single transistor. This circuit provides class AB operation, since a current much larger than bias I_0 is supplied to the load C_L for large input amplitudes; power efficiency is then excellent, since most of the current absorbed from the power supply is delivered to the load. Last but not least, this amplifier can be transformed into a stray- and offset-insensitive SC-integrator by just adding the integrating capacitor C_i. Capacitor C_1 then becomes a SC-resistor which can be duplicated with S_1 if several inputs are needed.

7.4 Conclusion

For fundamental reasons, very low power operation of analog circuits limits their bandwidth and/or the achievable signal-to noise ratio [14]. Low supply voltage affects the maximum signal amplitude, degrades the noise content and the precision of currents, complicates the realization of analog switches and prohibits stacks of transistors.

Low current requirements limit the use of resistors, reduce speed and bandwidth and increase the noise voltage spectral densities. However, the possibility of operating transistors in weak inversion helps to counter these limitations. The very low value of saturation voltage permits peak-to-peak amplitudes close to the supply voltage. The maximum value of transconductance-to-current ratio provides a large DC gain per stage, as well as minimum input-referred noise for a given current. The well controlled exponential characteristics permit interesting analog sub circuits.

Other favorable factors are the absence of thermal gradients on the chip, an easy elimination of latch-up, and reduced PSRR requirements. Special design techniques include single stage transconductance amplifiers, class-AB amplifiers to eliminate slew rate limitations, and the use of special devices, such as the lateral diode in the polysilicon gate layer and compatible bipolar transistors.

Micropower analog circuits are limited in their applications to restitutive tasks for which high precision and/or high local speed are needed. On the contrary, micropower analog VLSI is expected to be very efficient for the implementation of biologically inspired perception systems using massively parallel collective computation [15].

References

[1] E. Vittoz, "Micropower Techniques," in J. Franca and Y. Tsividis, Eds., *Design of Analog-Digital VLSI Circuits for Telecommunications and Signal Processing*, Prentice Hall, 1993.

[2] E. Vittoz, "Low-power low-voltage limitations and prospects in analog design," in Dig. of *Advances in Analog Circuits Design*, Eindhoven, March 29-31, 1994.

[3] H. Oguey and S. Cserveny, "MOS modelling at low current density," *Summer course on Process and Device Modelling*, KU-Leuven, 1983.

[4] C. Enz, *High Precision CMOS Micropower Amplifiers*, Ph.D. Thesis No 802, EPFL, Lausanne, 1989.

[5] E. Vittoz, "MOS transistors operated in the lateral bipolar mode and their application in CMOS technology," *IEEE J. Solid-State Circuits*, Vol. SC-18, pp. 273-279, June 1983.

[6] E. Vittoz and X. Arreguit, "Linear networks based on transistors," *Electronics Letters*, Vol. 29, pp. 297-299, Feb. 4, 1993.

[7] M. Dutoit and F. Sollberger, "Lateral polysilicon p-n diodes," *J. Electrochem. Soc.*, Vol. 125, pp. 1648-1651, Oct. 1978.

[8] F. Krummenacher *et al.*, "Higher sampling rate in SC circuits by on-chip clock-voltage multiplication," in *Dig. ESSCIRC '83*, pp. 240-241, Lausanne 1983.

[9] T. C. Choi *et al.*, "High-frequency CMOS switched-capacitor filters for communications applications," *IEEE J. Solid-State Circuits*, Vol. SC-18, pp. 652-664, Dec. 1983.

[10] R. C. Schober *et al.*, "Ultra low power circuit techniques for focal plane electronics applications," *JPL Task Plan 80.3295*, 1993.

[11] E .Vittoz, "Microwatt switched capacitor circuit design," *Electrocomponents Science and Technology*, Vol. 9, pp. 263-273, 1982.

[12] F. Krummenacher, E. Vittoz and M. Degrauwe, "Class AB CMOS amplifiers for micropower SC filters," *Electronics Letters*, Vol. 17, pp. 433-435, 25th June 1981.

[13] E. Vittoz, "Dynamic analog techniques," in J. Franca and Y. Tsividis, Eds., *Design of Analog-Digital VLSI Circuits for Telecommunications and Signal Processing*, Prentice Hall, 1993.

[14] E. Vittoz, "Low power design: ways to approach the limits," in *ISSCC '94 Dig. Tech. Papers*, pp. 14-18, Feb. 1994.

[15] E. Vittoz, "Analog VLSI signal processing: why, where and how?," to be published in *J. VLSI Signal Processing*, Kluwer, 1994.

Tutorial 8

Low-power digital design

Organised by: Jan Rabaey, *University of California*

Chapter 8

Low Power Digital Design

Jan M. Rabaey

8.1 Introduction

Most of the research and development efforts in the area of digital electronics have been oriented towards increasing the speed and the complexity of single chip digital systems. This has resulted in powerful design techniques, which enabled the development of personal workstations, sophisticated computer graphics, and multi-media capabilities such as real-time speech recognition and real-time video. While focusing the attention on speed and area, power consumption has long been ignored.

This picture is, however, undergoing some essential changes. Power consumption of individual components is reaching the limits of what can be dealt with by economic packaging technologies, resulting in reduced device reliability. This is illustrated by some of the recent micro-processor designs, which achieve impressive clocking speeds (up to 300 MHz) at the expense of very large power dissipation (larger than 30 Watts for CMOS, 110 Watt for ECL) [1, 2]. Until now, this power consumption has not been of great concern, since large packages, cooling fins and fans have been capable of dissipating the generated heat. However, as the density and size of the chips and systems continues to increase, the difficulty in providing adequate cooling might either add significant cost to the system or provide a limit on the amount of functionality that can be provided.

Dealing with power is, therefore, rapidly becoming one of the most important issues in digital system design. This situation is aggravated by the increasing demand for portable systems in the areas of communications, general purpose computing and consumer electronics. On the horizon are portable communication systems, which will support speech recognition, graphics and full motion video, while communicating with a high speed network over a high bandwidth wireless link [3]. Improvements in battery technology are easily offset by the increasing complexity of those applications: it is projected that only a 30% improvement in battery performance will be obtained over the next 5 years [4]. To guarantee a reasonable battery operation time, a dramatic reduction of the power consumption is essential.

Such a reduction is not easily achieved and requires optimizations at all levels of the design hierarchy: technology, circuit, logic, architecture and algorithm. The global picture of how power can be reduced is readily understood: the dynamic power consumed by a circuit equals $C\,V_{dd}^2 f$, with f the clocking frequency of the circuit, V_{dd} the supply voltage and C the effective capacitance, which is the product of the physical load capacitance and the switching activity. Reducing the consumption can be accomplished by reducing any of those three factors. Lowering the supply voltage is the most attractive choice, however, due to the quadratic dependence. While clearly beneficial from a power point of view, this also results in a loss of speed, which has to be compensated by other means (such as other logic structures, architectures or improved algorithms). Besides reducing the supply voltage, power reduction is possible through a reduction of the effective capacitance: the physical capacitance can be minimized through the utilization of certain circuit styles or by utilizing better interconnect or clocking schemes. Furthermore, the switching activity can be minimized by using optimized logic structures to reduce hazards (for instance, wallace multipliers versus carry save multiplication), by avoiding excessive multiplexing of the hardware and by shutting down unused functions.

The paper first presents a detailed analysis of the sources of power dissipation in digital CMOS circuits. The discussion starts with an analysis of dynamic dissipation, as this is the dominant source of consumption in contemporary circuits. In particular, the dependence of dynamic consumption on signal statistics is analyzed. Other components of the power budget include dynamic hazards ("glitches") and short-circuit currents. The impact of these contributions is discussed and techniques to minimize their impact are proposed.

The second part is devoted to a short description of power minimization approaches. To be effective, power has to be optimized at all levels of the design hierarchy, being the technology, circuit, architectural and algorithmic levels. Such a design methodology requires innovations in the areas of circuit design, cell libraries, architectures and design tools (both analysis and synthesis). It is our experience that combining optimizations at all those levels easily results in orders of magnitude of power reduction.

8.2 Sources of power dissipation in digital circuits

8.2.1 Dynamic power dissipation

In CMOS circuits, power is mainly consumed during the switching of gates. The static power consumption for most gate structures (besides pseudo-NMOS) is limited to leakage. The power consumed during the charging and discharging of a capacitor is given by the well known

$$P_{dyn} = C_L V_{DD}^2 f_{0\to 1} \tag{1}$$

with $f_{0\to 1}$ the frequency of energy consuming transitions (or $0\to 1$ transitions for static CMOS). It is easily realized that this expression also holds for more complex gates as the nature of the energy consumption remains identical: charging and discharging capacitors. Minimizing power consumption then boils down to reducing one or a number of factors of equation (1). The V_{DD} factor is, obviously, the most influential due to quadratic dependence.

Computing the dissipation of a complex gate is complicated by the $f_{0\to 1}$-factor, also called the *switching activity*. While this factor is easily computed for an inverter, it turns out to be far more complex when higher order gates or circuits are concerned. One essential concern is that the switching activity of a network is a function of the nature and the statistics of the input signals: if the input signals remain unchanged, no switching will happen and the dynamic power consumption is nihil! On the other hand, fast changing signals will provoke plenty of switching and hence dissipation. Other factors influencing the activity are the circuit style (e.g. dynamic versus static), the function to be implemented and the overall network topology. These factors can be accounted for by introducing a slight modification in equation (1)

$$P_{dyn} = C_L V_{DD}^2 f_{0\to 1} = C_L V_{DD}^2 P_{0\to 1} f \tag{2}$$

with f the average event rate of the inputs and $P_{0\to1}$ the probability that an input transition results in a 0→1 (or power consuming) event.

A plot of the output transition probability (and hence also the average power consumption) of a 2 input NOR gate in function of input probabilities P_A and P_B is shown in Figure 1. This picture clearly demonstrates the impact of input probabilities on the dissipation. Ignoring signal statistics in power analysis results in substantial errors.

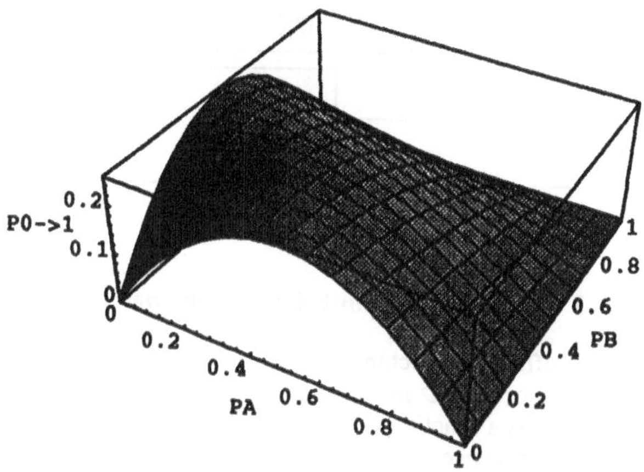

Figure 1 Transition probability (or equivalent energy dissipation of 2 input NOR gate in function of the input probabilities P_A and P_B .

The evaluation of the switching activity becomes more evolved, when complex logic networks have to be analyzed. When a signal propagates through a number of logic layers, its signal and transition probabilities become 'colored'. For these networks, advanced circuit analysis or extensive simulation over a large number of input vectors is the only means to produce a reasonable estimation of the activity factor.

8.2.2 Glitching in static CMOS circuits

When analyzing the transition probabilities of complex multi-stage logic networks in the previous section, we have ignored the fact that the gates have a non-zero propagation delay. In reality, the finite propagation delay from one logic block to the next can cause spurious transitions (called glitches, critical races or dynamic hazards) to occur: a node can have exhibit multiple transitions in a single clock cycle before settling to the correct logic level. For example, consider the cascade of two NOR gates, as shown in Figure 2.

Assume that each gate has a unit delay and that inputs A, B and C arrive at the same time. Since there is a finite propagation delay time through the first NOR gate, the second NOR gate initially evaluates with the previous value of X. When the correct value from the first gate finally ripples through, the second NOR gate evaluates again.

The first evaluation was redundant, consuming power without performing any 'useful' function. For example, consider the transition from **101** to **000** for inputs ABC (Figure 3). The output should remain **0**, but makes an extra transition to **1** before settling to the correct value of zero, dissipating some additional energy.

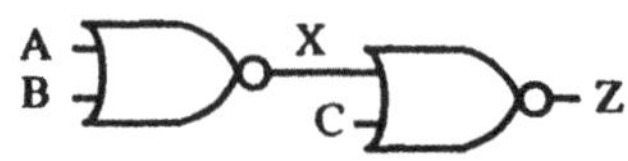

Figure 2 Glitching in static CMOS circuits.

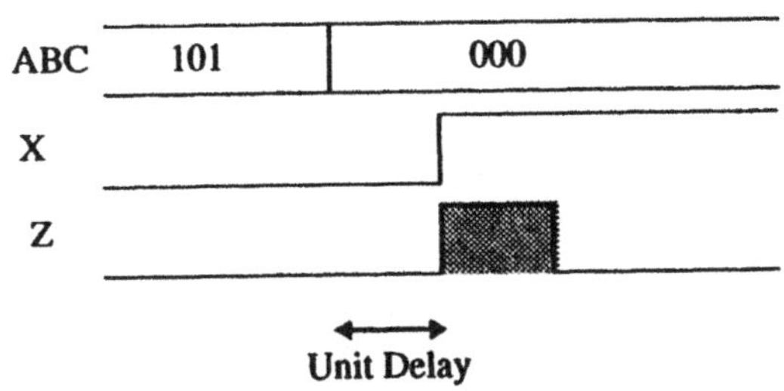

Figure 3 Glitching in multi-level logic networks.

A typical example of the effect of glitching is shown in Figure 4, which displays the simulated response of a chain of NAND-gates for all inputs going simultaneously from 0 to 1. The output bits are shown as a function of time. A logical analysis of the circuit shows the correct outputs to be a sequence of alternating 0's and 1's. As can be observed, all signals initially evolve in the direction of the GND rail. This trend is later reversed for half of the signals, as it takes some time for the correct outputs to propagate down the chain. Actually, the farther down the chain the worse it gets. Although the glitches are only partial (i.e. not from rail to rail), they contribute significantly to the power dissipation. In later chapters, it will be shown that long chains of gates often occur in important logic structures, such as adders and multipliers. The dissipation caused by these spurious transitions can reach up to 25% of the total dissipation for some of those circuits.

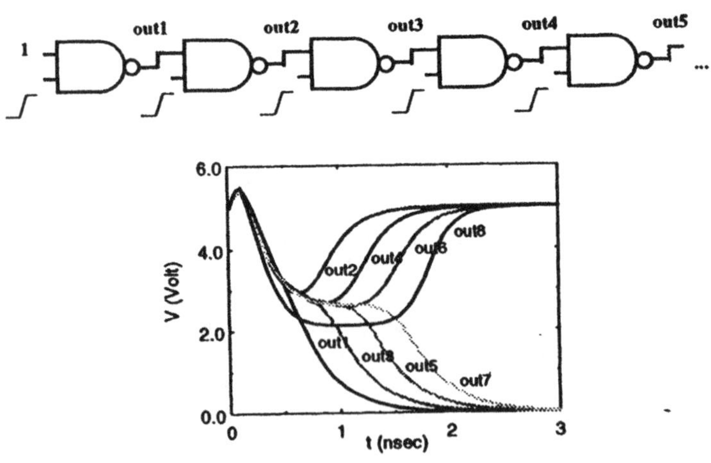

Figure 4 Glitching in a chain of NAND gates.

The occurrence of glitching in a circuit is mainly due to a mismatch in the path lengths in the network. If all input signals of a gate change simultaneously, no glitching occurs. On the other hand, if input signals change on different time-instances (for instance, as a result of different path lengths with respect to the primary inputs of the network), a dynamic hazard might develop. This statement can be illustrated with the example of Figure 5. Assume that all operators F have the same unit-delay. The first network (a) will suffer from glitching as a result of the wide disparity between the arrival times of the input signals for a gate. For instance, for gate F_3, one input settles at time 0, while the second one only arrives at time 2. In contrast, redesigning the network such that all arrival times are identical, can dramatically reduce the number of transitions. This observation can be translated in the following rule:

> *To eliminate the impact of dynamic hazards on the power dissipation of a static CMOS circuit, it is sufficient to equalize the arrival times of all signals at the inputs of each gate.*

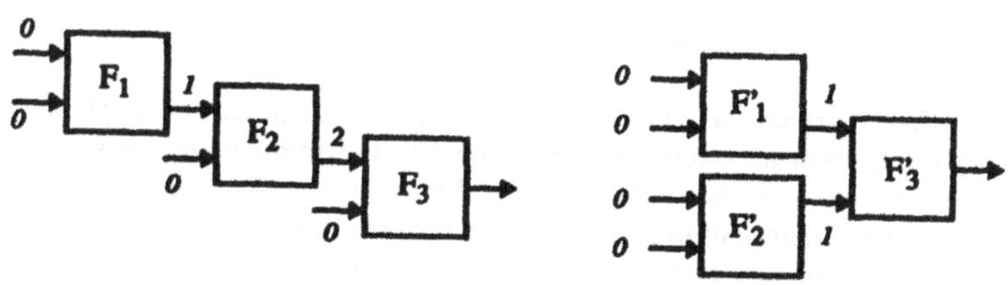

(a) Network sensitive to glitching (b) Glitch-Free network

Figure 5 Glitching is influenced by matching of signal paths lengths. The annotated numbers point to the signal arrival times.

Strictly adhering to this rule is hard in practice. Glitching can be virtually eliminated, though, by making the path lengths to the inputs of a gate approximately the same length. The attentive reader will observe that this condition will be beneficial for the performance of the structure as well.

Glitching is not an issue in dynamic logic, as the correct operation of such a circuit requires that each node undergoes at most one transition per clock cycle.

8.2.3 Short-circuit currents in static CMOS circuits

As discussed in section 8.3, another source of power dissipation in a static CMOS circuit is the flow of current from V_{DD} to GND during switching, when both NMOS and PMOS are conducting simultaneously. Such a path never exists in dynamic circuits, as precharge and evaluate transistors should never be on simultaneously (this would lead to malfunction). Short-Circuit currents are, therefore, a problem solely encountered in static designs. The total amount of energy, dissipated in the short-circuit current, is a function of the on-time of the transistors and the operation modes of the devices.

Consider a static CMOS inverter with a 0→1 transition at the input. Assume first that the load capacitance is very large, such that the output fall time is significantly larger than the input rise times. Under those circumstances, the input will move through the transient region before the output starts to change. As the source-drain voltage of the PMOS device is essentially 0 during that period, the device shuts off without ever delivering any current. The short-circuit current is therefore essentially zero. Consider now the reverse case, where the output capacitance is very small and the output fall time is substantially smaller than the input rise time. The drain-source voltage of the PMOS device is equal to V_{DD} during most of the transition period, guaranteeing the maximal short-circuit current (equal to the saturation current of the PMOS) during most of the transient period. This is clearly the worse case.

A detailed analysis of static CMOS gates revealed that the short circuit current contribution to power dissipation is minimized by adhering to the following rule [11]:

> *The power dissipation due to short circuit currents is minimized by matching the rise/fall times of the input and output signals. At the overall circuit level, this means that rise/fall times of all signals should be kept constant within a range.*

Making the input and output rise times of a gate identical is not the optimum solution for that particular gate on its own, but keeps the overall short circuit current within bounds. When the rise/fall times of inputs and outputs are equalized, most power dissipation is associated to the dynamic power and only a minor fraction (< 20%) is devoted to short circuit currents.

8.3 Low power CMOS design

Assuming that short circuit current, glitching and leakage can be kept in bounds by employing the appropriate circuit design approaches, the dominant power source in CMOS is the dynamic power consumption, expressed by equation (2). Suppose further that the throughput or performance of the circuit is to be kept constant, which means that f is a fixed parameter. Under those circumstances, the power of a design can be reduced by two parameters: the supply voltage V_{DD} and the *effective capacitance* C_{eff}, which is the product of the physical capacitance C and the switching activity $P_{0 \to 1}$. Each of those factors will be discussed in sequence.

8.3.1 Reducing the supply voltage

This is, obviously, the most effective way of reducing the power consumption, as the dissipation is proportional to V_{DD}^2. This is confirmed in Figure 6, which plots the measured power-delay product (which equals $C_{eff}V_{DD}^2$) of two experimental circuits.

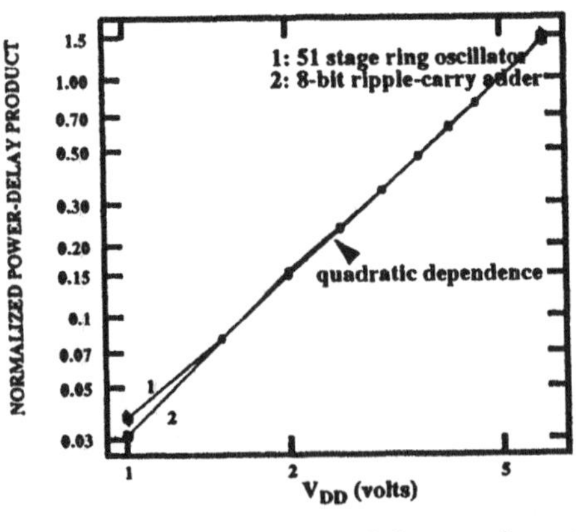

Figure 6 Power-Delay product of 2 experimental circuits as a function of V_{DD}.

Therefore, it is only necessary to reduce the supply voltage for a *quadratic* improvement in the power-delay product of a logic family [5]. It is interesting to observe that the PDP (which is considered as one of the quality measures of a logic style) continues to improve when dropping the supply voltage.

Unfortunately, this simple solution to low power design comes at a cost. As shown in Figure 7, the effect of reducing V_{DD} on the delay is shown for a variety of different logic circuits, that range in size from 56 to 44,000 transistors spanning a variety of functions, all exhibit essentially the same dependence. Clearly, we pay a speed penalty for a V_{DD} reduction, with the delays drastically increasing as VDD approaches the sum of the threshold voltages of the devices. This is apparent in equation (3), which approximates the propagation delay of a static CMOS gate.

$$t_p \approx \left(\frac{C_L V_{DD}}{2}\right)\left[\frac{1}{k_n(V_{DD}-V_{Tn})^2}+\frac{1}{k_p(V_{DD}-|V_{Tp}|)^2}\right] \tag{3}$$

We see that the power-delay product improves as delays increase (through reduction of the supply voltage), and, therefore, it is desirable to operate at the slowest possible speed. Since the objective is to reduce power consumption while maintaining the overall system throughput, compensation for the increased delays at low voltages is required.

A first approach is to adopt a technology with *lower threshold voltages*. Reducing the thresholds of the devices moves the curves in Figure 7 to the left, which means that the performance penalty for lowering the supply voltage is reduced. In general, the threshold voltages are lower bounded by the amount of allowable sub-threshold leakage current: if the V_T becomes too low, even setting the gate-source voltage to zero does not completely turn off the device (Figure 8). This raises the minimum clock-frequency in dynamic circuits (due to the increased leakage on the dynamic nodes) and causes stand-by currents and reduced noise margins in static designs. Thresholds are, therefore, never smaller than 0.5 - 0.6 Volt in standard processes (most often, they are even substantially larger and go up to 1 Volt) and lowering the supply voltage below 1.2-2 Volt (2 times V_T) is excessively expensive in terms of lost performance.

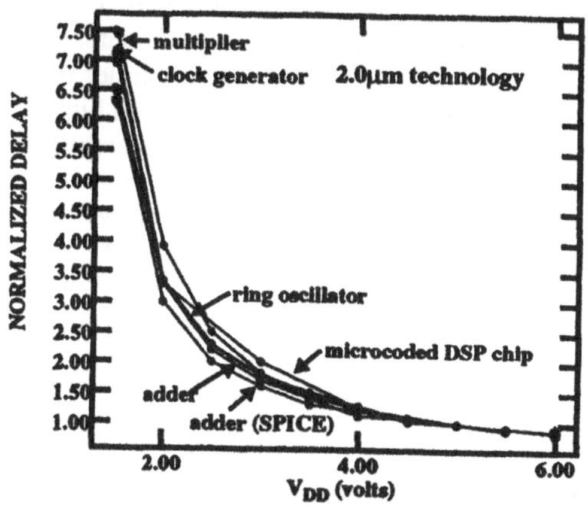

Figure 7 Normalized propagation delay in function of supply voltage for a number of experimental circuits.

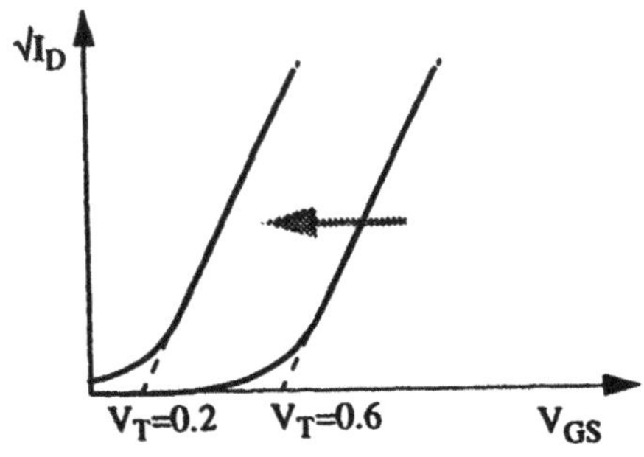

Figure 8 Decreasing the threshold increases the subthreshold current at V_{GS} =0.

It has been suggested that this lower bound on the thresholds is artificial [9]. The idea that the leakage current in a static CMOS circuit has to be zero is a pre-conception. Leakage current causes a degradation of the noise margins, as the logic levels are no longer equal to the supply rails (similar to the operation of a pseudo-NMOS gate). As long as the noise margins are within range, this is not a compelling issue. The leakage currents, of course, cause an increase in static power dissipation. This is offset by the drop in supply voltage, which is enabled by the reduced thresholds at no cost in performance. Remember that the static power consumption is equal to $I_{static}V_{DD}$. Liu and Svenson demonstrated that for a 0.25 mm CMOS process, the following circuit configurations obtain the same performance: (3 V Supply, 0.7 Volt V_T) and (0.5 Volt Supply, 0.1 Volt V_T) The power consumption of the latter is, however, 45 times smaller! The same holds for dynamic circuits. The minimum allowed threshold voltages in dynamic circuits are severely limited by the minimum operating frequency. The potential power savings are not as impressive and are limited to approximately 8 when maintaining the performance level (for a supply voltage of 1 Volt and a minimum threshold of 0.3 Volt). This demonstrates that *static circuits are more amenable to supply scaling than dynamic ones.*

Changing the process parameters is not an option for most of the digital circuit designers. The loss in performance due to the supply reduction, hence, has to be compensated for by other means. There are ample opportunities to do so at the circuit or architectural design levels. For instance, it is well known that high performance does not always rely on the highest possible clock frequency. Parallel architectures, which combine multiple slow computational elements, can achieve exactly the same goal. This might result in an increase in area, but the reduction in power might be well worth it (as the cost of silicon is gradually being reduced with the scaling of the technology).

To illustrate how architectural techniques can be used to compensate for reduced speeds, a simple 8-bit datapath consisting of an adder and a comparator is analyzed assuming a 2 mm technology [5]. As shown in Figure 9, inputs A and B are added, and the result compared to input C. Assuming the worst-case delay through the adder, comparator and latch is approximately 25 ns at a supply voltage of 5 V, the system in the best case can be clocked with a clock period of $T = 25$ ns.

When required to run at this maximum possible throughput, it is clear that the operating voltage cannot be reduced any further since no extra delay can be tolerated, hence yielding no reduction in power. We will use this as the reference datapath for our study and present power improvement numbers with respect to this reference. The power for the reference datapath is given by

$$P_{ref} = C_{ref} V_{ref}^2 f_{ref} \tag{4}$$

where C_{ref} is the total effective capacitance being switched per clock cycle. The effective capacitance was determined by averaging the energy over a sequence of input patterns with a uniform distribution.

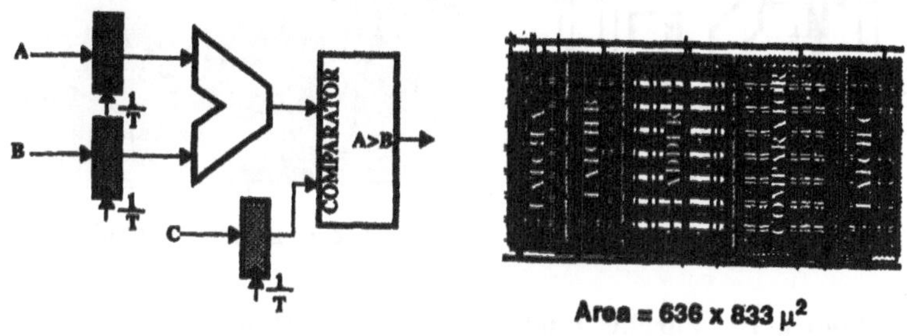

Figure 9 A simple datapath with corresponding layout.

One way to maintain throughput while reducing the supply voltage is to utilize a parallel architecture. As shown in Figure 10, two identical adder-comparator data paths are used, allowing each unit to work at half the original rate while maintaining the original throughput. Since the speed requirements for the adder, comparator, and latch have decreased from 25 ns to 50 ns, the voltage can be dropped from 5 V to 2.9 V (the voltage at which the delay doubled, from Figure 7).

While the datapath capacitance has increased by a factor of 2, the operating frequency has correspondingly decreased by a factor of 2. Unfortunately, there is also a slight increase in the total effective capacitance, due to the extra routing. This results in an increased capacitance by a factor of 2.15. The power for the parallel datapath is, thus, given by

$$P_{par} = C_{par} V_{par}^2 f_{par} = (2.15C_{ref})(0.58V_{ref})^2 \left(\frac{f_{ref}}{2}\right) \approx 0.36P_{ref} \tag{5}$$

The presented approach trades-off area for power (the resulting area is approximately 3.4 larger than the original design). This is obviously only possible when the design is not area constrained. Furthermore, parallelism has the overhead of extra routing (which might cause extra power), and careful optimization must be performed to minimize this overhead.

Parallelism is not the only way to compensate for the loss in performance. Other architectural approaches, such as the use of pipelining, can just as well achieve the same goal.

The most important message in the above analysis is that, if power dissipation is the prime concern, dropping the supply voltage is the most effective means to achieve that goal. The subsequent loss in performance can be compensated, if necessary, at a loss of area. Within certain bounds, this is acceptable as area is not any more the compelling issue it used to be as a result of the dramatic increase in integration levels of the last decade.

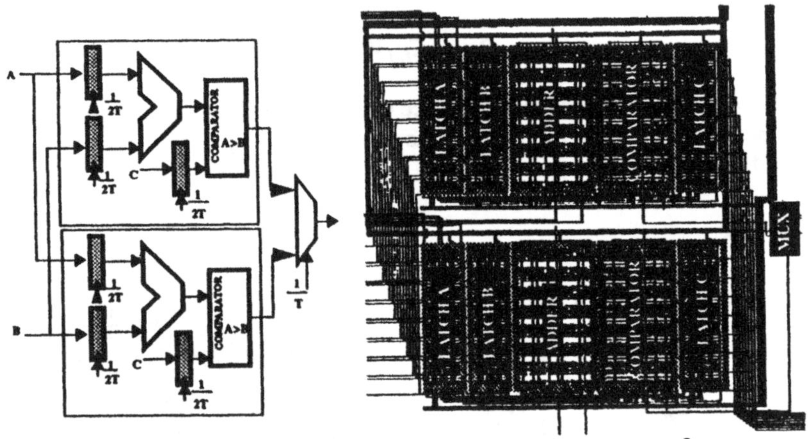

Figure 10 Parallel implementation of the simple datapath.

8.3.2 Reducing the effective capacitance

When the supply voltage is lower bounded by external constraints (as is often the case in real world designs) or when the performance degradation due to the lowering supply voltage is intolerable, the only means of reducing the dissipation is by lowering the effective capacitance. This can be achieved by addressing both of its components: the physical capacitance and the switching activity.

Lowering the physical capacitance is an overall worthwhile goal, which can also help to improve the performance of the circuit. For instance, some circuit cycles such as Complementary Pass Transistor Logic (CPL) come with a substantially reduced capacitance and may result in lower power operation. A CPL adder was reported [10] to reduce the dissipation with 30% compared to a conventional adder (for a 4 Volt supply). This reduction, which even becomes more significant for lower supply voltages, is mainly due to the decrease in input capacitance.

As most of the capacitance in a logic combinational circuit is due to transistor capacitances (gate and diffusion), it makes sense to keep those to a minimum when designing for low power. This means that transistors should be kept minimal size, whenever possible or reasonable. This definitely will affect the performance of the circuit, but this can be offset by using logic or architectural speed-up techniques, as was briefly discussed in the previous section. The only instances where transistors should be sized up is when the load capacitance is dominated by extrinsic capacitances (such as fanout or wiring capacitance). Again, this is contrary to common design practices used in cell libraries, where transistors are generally made large to accommodate for a range of loading and performance requirements.

To illustrate this point, let us analyze the simple case of a static inverter driving a load capacitance, consisting of an intrinsic (C_{int}) and an extrinsic component (C_{ext}) (Figure 11a). While the former represents the diffusion capacitances, the latter stands for wiring capacitance and fanout. It is assumed that the ratio between PMOS and NMOS transistors is constant. The factor N stands for the inverter sizing factor, where N is equal to 1 for an inverter constructed of minimum size devices. It is easily realized that the intrinsic capacitance of the scaled device is proportional to N (or $C_{int\,(scaled)}' = NC_{int}$). Figure 11b plots the normalized energy (per transition) in function of the scaling factor N the ratio between the extrinsic and intrinsic capacitance as a parameter: $\alpha = C_{ext} / C_{int}$. The speed of all possible versions is kept constant by appropriately adjusting the supply voltage: larger values of N normally mean lower values of the supply voltage.

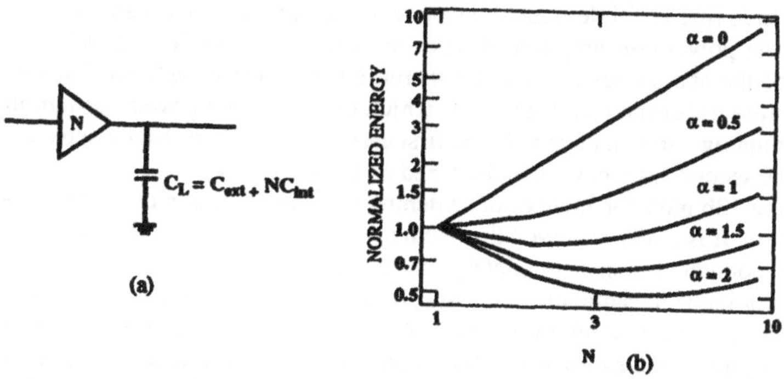

Figure 11 Normalized energy of a MOS inverter with load capacitance as a function of the inverter size N and the ratio between the extrinsic and intrinsic capacitance α (=C_{ext} / C_{int}).

When $\alpha = 0$ (or the load capacitance is zero), the lowest energy consumption is obtained when using minimum size devices. Only when the extrinsic capacitances dominate ($\alpha > 1$) does it make sense to widen the devices. Notice that this analysis assumes a sustained performance.

Another approach to reducing the physical capacitance is to avoid the extensive sharing of resources. Figure 12 shows a common bus architecture, where a single shared bus is connected to a number of drivers and receivers. This configuration results in a large bus capacitance, due to the large number of devices connected to the bus as well as the long wire-length of the bus (as it probably will stretch over a large part of the chip). From a power perspective, it is beneficial to avoid the extensive sharing and use a number of dedicated point-to-point buses instead. This probably will have a negative effect on the chip area, but reduces the physical capacitance being switched per transition. Once again, this can amount to *trading off area versus power.*

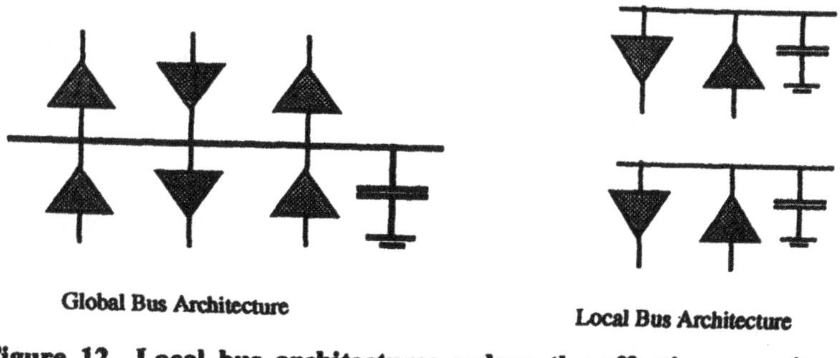

Figure 12 Local bus architectures reduce the effective capacitance.

While reducing the physical capacitance helps, another way to reduce the dissipation is to switch that capacitor less often, or, in other words, *reduce the switching activity* . A good example in this class is the choice between static and dynamic circuits.

Due to the periodic precharging, dynamic circuits tend (on the average) to exhibit a larger switching activity than their static counter-parts (although one should be aware that this depends upon the logical function). This is further aggravated by the necessary connection with the clock, which performs a transition every cycle.

Multiplexing multiple operations on a single hardware unit can have a detrimental effect on the power consumption. Besides increasing the physical capacitance (as in the example of the shared bus), it can also increase the switching activity. This is illustrated with a simple experiment in Figure 13, which compares the power consumption of two counters running simultaneously. In the first case, both counters run on separate hardware, while in the second case they are multiplexed on the same unit.

Figure 13b plots the number of switching events in function of the skew between the two counters. It is easily observed that the non-multiplexed case is always superior, except when both counters run in a completely synchronous fashion.This is caused by the fact that multiplexing tends to randomize the data signals presented to the operational unit, which results in an increased switching activity. When power consumption is a concern, it is often beneficial to avoid the excessive re-use of resources. Observe that CMOS hardware units consume only negligible amounts of power, when idle. Providing dedicated, specialized operators, therefore, only presents an extra cost in area, while being generally beneficial in speed and power.

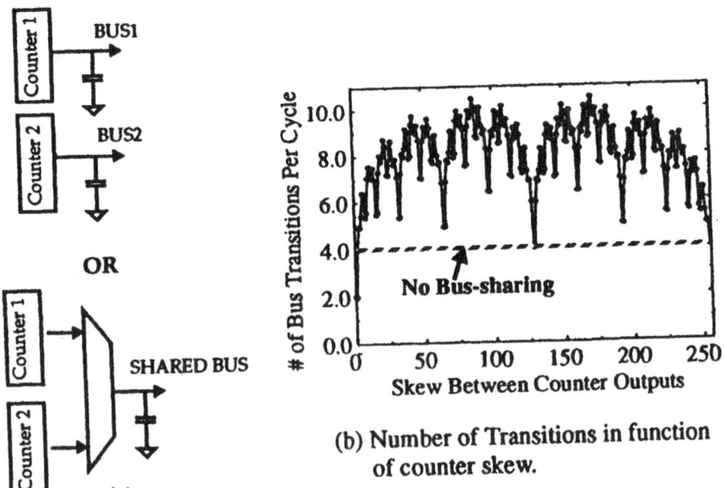

(b) Number of Transitions in function of counter skew.

Figure 13 Multiplexing increases the switching activity.

Taking all the aforementioned factors together, power savings of one or more orders of magnitude are possible. These can be complemented by system level techniques, such as the power-down and as the clock speed reduction approaches, which are commonly used in portable devices such laptop computers and video cameras.

8.4 A case study

The impact of a number of the aforementioned power minimization techniques is illustrated with a simple, but real example. Figure 14 shows two implementations of a Quadrature Mirror Filter (QMF), implemented in 2 mm CMOS technology [8]. The first implementation was automatically implemented using a high level synthesis tool and uses a time multiplexed processor architecture. It achieves a sample rate of 2.5 MHz when running at 5 V. The second design instance has an identical performance, but uses a hand crafted optimized fully parallel architecture. Its power dissipation is, however, 140 times lower than the first implementation (2.5 mW versus 350 mW), while its area is only marginally larger (15.9 mm^2 versus 13.6 mm^2).

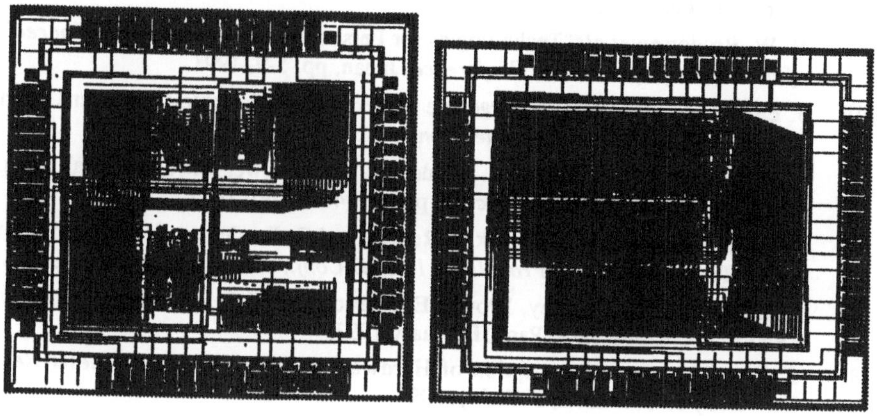

(a) Multiplexed Version (b) Parallel Version

Figure 14 Two implementations of a QMF structure.

This improvement can be attributed to a number of factors. First of all, the second implementation is deeply pipelined. As a result, it can maintain the same performance level, while running at a substantially reduced supply voltage. Reducing V_{DD} from 5 V to 1.2 V results in a power reduction by a factor of 17.4. Secondly, the fully parallel version has the advantage that the physical capacitance on the chip is dramatically reduced: multiplexed, high capacitance buses are replaced by local point to point connections, the controller overhead is eliminated and the programmable shifters are replaced by simple wires. Finally, the parallel implementation also reduces the switching activity. Data propagating through the hardware modules, registers and busses is highly auto-correlated. Consequently, the number of bits switching at each transition is minimized. The measured effective capacitance for the multiplexed and the parallel version equals 5,171 and 643 pF, respectively. This translates into an additional gain of 8.

8.5 Summary

Minimizing power dissipation has become a prime goal in the design of advanced digital circuits. To be effective, optimization has to occur at all levels of the design hierarchy. It has been demonstrated that by doing so, dissipation can be reduced by orders of magnitude. This requires an in-depth understanding of the nature of power dissipation in digital CMOS circuits. To realize a low power design methodology, which crosses all boundaries of the design hierarchy, an extensive set of design tools (for both analysis and synthesis) is a necessity, almost none of which are available today [6, 7]. Low power design, therefore, presents a challenge to a number of communities, ranging from circuit, system, algorithm to CAD tool designers.

References

[1] D. Dobberpuhl *et al.*, "A 200 MHz Microprocessor implemented in CMOS Technology", *IEEE Journal of Solid-State Circuits*, pp. 1585-1598, Nov. 1992.

[2] N. Jouppi *et al.*, "A 300 MHz 115W 32b Bipolar ECL Microprocessor with on-chip Caches", *Proc. IEEE International Solid-State Circuit Conference*, pp. 84-85, San Fransisco, Feb. 1993.

[3] R. W. Brodersen *et al*, "Technologies for Personal Communications", *Proceedings of the VLSI Symposium 91 Conference*, Japan, pp. 5-9, 1991.

[4] J. Eager, "Advances in Rechargeable Batteries Pace Portable Computer Growth", *Silicon Valley Personal Computer Conference*, pp. 693-697, 1991.

[5] A. Chandrakasan, S. Sheng, R. Brodersen, "Low-power CMOS Digital Design", *IEEE Journal of Solid-State Circuits*, pp. 473-484, April 1992.

[6] A. Chandrakasan, M. Potkonjak, J. Rabaey and R. Brodersen, "Optimizing Power Using Transformations", *Proc. IEEE ICCAD Conf.*, Santa Clara, November 1992.

[7] P. Landman and J.Rabaey, "Power Estimation for High Level Synthesis", *Proc. 1993 EDAC Conference*, Paris, February 1993.

[8] S. Wu, "A generic Description of Hardware Libraries", *Master Thesis Report*, University of California, Berkeley, 1994.

[9] D. Liu and C. Svensson, "Trading Speed for Low Power by Choice of Supply and Threshold Voltages", *IEEE Journal of Solid-State Circuits*, pp 10-17, Jan. 1993.

[10] K. Yano *et al.*, "A 3.8 nsec CMOS 16x16 multiplier using Complementary Pass Transistor Logic", *IEEE Journal of Solid-State Circuits*, pp 388-395, April 1990.

[11] H.J.M. Veendrick, "Short-Circuit Dissipation of Static CMOS Circuitry and Its Impact on the Design of Buffer Circuts", *IEEE Journal of Solid-State Circuits*, Vol. SC-19, pp. 468-473, August 1984.

Tutorial 9

Taking the mystery out of RF design

Organised by: Barrie Gilbert, *Analog Devices*
Chris Toumazou, *Imperial College*

Chapter 9.1

Taking the Mystery out of RF Design

Barrie Gilbert

9.1.1 Introduction

Profound changes have occurred in the field of communications during the past two decades, directly as a result of the dramatic advances in VLSI, in particular, microprocessors and memory technologies. Most obvious is the rapidly growing dependence on digital signal processing (DSP) to make up for deficiencies in the analog path in order to push data rates and transmission accuracy ever higher.

Examples include modems which can reliably handle 19.2 kB/s data in an under-3 kHz phone-line bandwidth, and cellular phone systems, such as GSM, which can tolerate quite poor radio-system performance through the use of clever modulation techniques followed by DSP baseband filtering and data recovery algorithms. These embody "mysteries" of their own - what might be called "system mysteries". But our objective here is something more basic, namely, taking (at least some of) the mystery out of small-scale analog circuits that ought to exhibit simple behavior but often do not. It is this propensity for perversity, at least in the minds of many new graduates, that has only strengthened the conviction that it's little more than common sense to use DSP wherever possible, and that analog - particularly RF analog - is best left to the veterans of WW-II.

How did this situation arise? The explanation is not hard to find. The strong - and deserved - emphasis on the superlative benefits of digital electronics, in both industry and academia, to the neglect of classical topics in electronic design, has led to a situation where many students are ill-equipped to deal with the more 'physical' aspects of RF systems. Digital systems largely require an understanding of how to manipulate logical, dimensionless variables (even simpler than that - nothing but binary variables) on any suitable 'engine'. In an important sense, all digital systems are merely hardware incarnations of software. The fact that they utilize electronics is incidental: it's entirely a matter of convenience; silicon logic cells are tiny, cheap, fast and reliable.

Classical analog circuits, on the other hand, manipulate 'real', that is, physical variables, such as voltages and currents, existing in (and, at high frequencies, around) physically real components, such as inductors and capacitors, the performance of which depends strongly on the particular materials used, even on detailed aspects of the quality of these materials. Furthermore, even the most carefully-designed analog circuits do things at high frequency that are sometimes anomalous, mysterious or downright baffling. The same 20 GHz transistor used as a logic inverter, with perhaps only time-delay as a parameter of interest, has the potential for becoming a many-horned monster when biased up as a low-noise Class-A amplifier.

It is the physical reality of RF circuits that leads me to call this type of electronics 'Newtonian'. Whereas digital electronics is largely about the management of 'data' (or, at a higher level of abstraction, about 'information', which at the top level ultimately means being concerned with 'knowledge objects'), Newtonian electronics, by contrast, is about the management of 'signals', having dimensional attributes, existing in devices that are subject to, and limited by, the laws of physics (and, very often, statistics, in the guise of that universal leveler, thermal noise). This kind of signal leaks and leaps from wanted places to unwanted ones, often by routes that simply aren't visible to the innocent eye, such as a few millimeters of solid ground plane, or empty space. Somewhat like quantum particles, analog signals defy being counted like their brilliant but boring binary brothers.

It has been this multi-dimensional individualism of analog circuits (and, to some minds, their eternally enigmatic nature) that has resulted in their falling out of favour. Analog circuits, we hear, 'don't stay calibrated', or 'drift', or 'are noisy', all of which may be true, especially if poorly designed with little regard for robustness and manufacturability. Analog circuits are also more limited in their domain of application; you have probably noticed that there is no genuine equivalent to the microprocessor in the world of Newtonian electronics. (Ironically, the nearest thing is a DSP engine, which claims to be analog in nature, but which lacks a soul). Stated another way: analog circuits aren't usually very programmable, certainly not in the same way and with the broad range of applications and sure-fire certainty of digital circuits. Nor are they "user-friendly".

In view of these limitations, it's not surprising to see a strong push to digitize wayward signals, and turn them into benign logical variables, at the earliest possible point in a receiver (or at the latest point in a transmitter), with the objective of eliminating as much as possible of the poorly-understood analog processing sections. The ultimate realization of this brilliant strategy (or 'nightmare', take your pick) is to put data converters (an ADC and/or a DAC) right at the antenna, and 'do everything else in DSP'. The problem is that this is not only impractical at high carrier frequencies, but it also makes needless demands on the converters, and will often result in higher power consumption, at those more moderate frequencies (say, below 100 MHz) where the technology permits the use of converters of borderline consumption.

Of course, once the bit-stream has been acquired, one now has to burn more megaflops (and watts) to DSP the raw data. Furthermore, while it may be possible to attain adequate resolution to meet minimal intermodulation requirements at moderate to high signal levels, it is impossible (I dare to say: watch me go down in flames) to achieve the same sort of dynamic range (typically 80 dB for a cellular phone receiver) in the ADC alone, due partly to the size of the LSB (often more than a millivolt) and partly due to the difficulty of achieving the same sort of noise performance from the small-geometry, low-biased comparator cells typically used in a flash converter compared to what can be achieved in an optimized amplifier.

The dynamic range requirement might be met by the availability of a VERY good programmable-gain pre-amplifier: such products will appear on the scene in time. And certainly, one cannot afford to be naive about these trends; they are a fact of modern electronics and will continue to shape the future of communications systems, usually for the best of reasons.

Nevertheless, there remains much to be done in the Newtonian realm, a reality that is gradually dawning on both new and mature IC companies alike, as they scramble to corner the design talent that is in short supply. Those who understand the "mysteries of RF" are nowadays accorded the same reverence as that enjoyed by alchemists in medieval times - unusual people having special connections with the infinite not available to the rest of us mortals.

All of which may be a slight exaggeration. But anyone whose job it is to seek out and hire skilled analog designers, particularly with experience in the UHF and microwave arena, knows only too well that first-class monolithic RF skills are quite rare. This may in part be due to the over-inflated predictions on the part of the digital community: we often hear that "the future is digital", leading teachers and students alike to put unrelenting emphasis on the DSP content of communications courses.

That fact is, the future "is" one in which a thorough understanding of complex mixed-signal systems, implemented in advanced integrated-circuit form, will be even more critical to serving the marketplace. And the least well-understood area of any modern RF-based communications system will continue to be the "front end" (both send and receive), operating at somewhere between 900 MHz and 2.5 GHz in the near future, and possibly as high as many tens of gigahertz (in local-area services) as bandwidth demands continue to escalate.

9.1.2 The mystery of terminology

One of the many challenges facing the newcomer to the world of RF systems is understanding the often obscure and confusing terminology. While every discipline is blighted by an excess of three-letter-acronyms, at least in the world of data-related signal processing these are just straightforward abbreviations, while in RF they often refer to dimensional attributes of signals. Thus, we have numerous variants of the decibel, including dBc, dBr, dBm, dBV, dBW and even dBx. In each case, these have no single meaning. Sometimes (like dB, dBr and dBc) they refer to ratios; at other times (like dBm and dBW) they refer to power levels; at still others (dbV, dBμV) to voltages. But it's even more detailed.

Thus, dBm means "decibels above 1 mW", but unless the impedance is specified this reference to power can leave one guessing about the equivalent voltage. In the RF context, the impedance is assumed to be 50 Ω (by practical convenience, as much as anything else), so 1 mW corresponds to the square-root of 1 mW times 50 Ω, which evaluates to 223.6 mV RMS. Now, while the power in a signal ought to be its most important attribute in defining the dynamic range of a receiver that will only be true if an amplifier is power-responding, that is, somehow makes use of both the voltage present at its input port and the current flowing into that port. When the ratio of these two quantities - that is, the input impedance - is the same as that of the source, maximum power transfer occurs.

The problem with the popular usage of "dBm" in connection with signal level is that input ports of such things as amplifiers and mixers are very often NOT matched, and only the voltage part of the signal is used. In such cases, the specification of signal level should be in "dBV", which means "decibels above 1 V" (or sometimes dBμV). This is now unambiguous in the context of bridging-style (high-input-impedance) components. In spite of its more accurate description of signal level, dBV is rarely used in an RF context.

A further problem with the usage of "dBm" - expressly meaning 223.6 mV RMS - is that the system impedance is often not 50 Ω. Thus, we see instances of some receiver's IF input signal being, say, -95 dBm to -15 dBm, meaning a sinusoidal amplitude of from 5.62 μV to 56.2 mV, while the preceding band-pass SAW filter may have an impedance of as high as 400 Ω.

This particular term - dBm - is abused in other ways. Thus, in connection with the slope of a logarithmically responding detector (such as the "Received Signal Strength Indicator", or RSSI, function) one sometimes sees this expressed as so-many mV/dBm, which of course is completely incorrect. dBm is an absolute measure of power, while dB is a relative measure, and the correct one for a slope. On the other hand, the intercept of the transfer characteristic of a log-amp is properly expressed in dBm, or even more correctly as dBV, as explained above.

Anomalies of this sort - and there are many other similar examples - stand between the designer of a new RF system and the designer of the components with which the system architecture is to be implemented. In many cases, they cause needless confusion and are one of the reasons why RF systems seem so frustratingly difficult to "break into" on the part of the neophyte designer. The idealist might argue that it ought to be possible to gradually displace the imprecise terminology that plagues the RF world by a more exact set of terms, but that is about as likely as changing the names of the days and the months to rid the world of Hellenism.

9.1.3 Mysteries of measurement

Another area of mystery arises in connection with how to define, predict and later measure, certain aspects of RF component performance. Simple examples include noise characterization, which of course defines the lower limit of the dynamic range, and such metrics as "1 dB compression point" and "third-order intercept" important parameter for an amplifier or mixer, in defining the upper end of the dynamic range. Out of these we can define the "spurious-free dynamic range". RF abounds with many terms like this, and confusion can sometimes arise because they often do not correspond to some definite quantity but depend critically on the measurement technique employed (the Heisenberg principle of RF circuits).

For example, consider "1 dB compression point", which we'll call P1dB. This is conventionally defined as "the signal level at which the gain is 1 dB lower than for very small signals". But how is it to be measured? Our choices are several, including the use of (1) a true RMS-responding AC voltmeter (like a bolometer); (2) a mean-responding, but RMS-calibrated, voltmeter (like a diode detector); (3) a spectrum analyzer showing the fundamental frequency of the input. In fact, each of these will yield a different result for P1dB. But there is more uncertainty.

First, it's often not clear whether the figure quoted is referred to the input or the output of the component. One might say that it doesn't matter much, because if one knows the gain, the signal level at the alternative port is related simply by that gain. But is it? If an amplifier has a gain of 10 dB and a measured output-referred P1dB of 3 dBm, what then is the input-referred P1dB? It's not -7 dBm, because the gain is no longer 10 dB. In this case, the reader should have no trouble in finding the answer, but even some handbooks fail to explain how to make the needed correction.

Second, P1dB is a strong function of many parameters, including the test frequency, the supply voltage and temperature. Providing a single figure which is supposed to characterize such components well enough to make a selection from several alternatives may lead to disappointment, as the chosen component is found to behave 'mysteriously' in the user's system.

Most RF components - particularly monolithic ICs - are poorly specified. There are several reasons for this state of affairs. They are often very cheap components; they have to be, to find their way into aggressively priced modern equipment. This leaves little room for expensive testing. Any measurements made on the packaged component are costly, but this is even more true when characterization at high frequencies is involved. Just finding ways of handling the component in a mass-production line to allow meaningful measurement at, say, 900 MHz, is a big challenge. That aside, the cost of test equipment tied up in a manufacturing line, which may be producing hundreds of quite different components, is prohibitive.

But above and beyond all those reasons, it is the very 'Newtonian nature of the RF component that makes it hard to characterize fully. There are just too many dimensions to explore in a test context. One strives to avoid the need for such tests by the use of good design practices, but these can never guarantee that an occasional part may exhibit some unique quirkiness under certain rare conditions.

There are many cases where 'mystery' arises because of the persistent use of performance metric which may have worked well for an older style of component, but which is the wrong way to characterize a modern component. An example of this is 'third order intercept', which we'll call 3OI. This is another metric that pertains to the linearity of an amplifier or mixer, and it is usually measured by applying two equal-amplitude test signals at almost the same frequency, and then looking for the third-order cross-products.

A simpler definition will be used here, to make a particular point. We can define 3OI as that signal level at which, in response to a sinusoidal stimulus, the third-harmonic of the (distorted) output would be equal in amplitude to the fundamental component in the output spectrum. This is determined by measuring the fundamental and the third-harmonic at some moderate input level and constructing projections through these two points. The line through the fundamental is given a slope of 1 and that through the third-harmonic is given a slope of 3. Where these lines meet is the 3OI, which corresponds either to a certain input level (in dBm, although more accurately, in dBV, as noted earlier) or to a certain output level.

In fact, one does not have to carry out geometrical constructions to find the 3OI. Thus, if D_O is the input level (in dBm or dBV) at which the test is made, and D is the difference (in dB) between the fundamental and third-harmonic outputs, then the input-referred 3OI is just $D/2+D_O$ (in dBm or dBV, respectively).

The problem with 3OI is that it makes a very basic assumption about the nature of the non-linearity. It assumes that the output y can be expressed as a cubic function of the input x: $y = ax + bx^3$, where a is (roughly) the simple numerical gain and b controls the magnitude of the distortion. When one inserts $A\sin\omega t$ for x, and expands the polynomial, the basis for the observation that the third-order component in the output spectrum increases at a "dB rate of 3" becomes apparent.

Unfortunately, this naive supposition about the form of the non-linearity is not always justified. A simple thought experiment will make this clear. Suppose one has an amplifier which is very linear for moderate signals but whose gain drops abruptly for larger inputs. That is, we are postulating a limiter, with a gain of G up to some (symmetrical) input voltage $\pm E$, whereafter the incremental gain drops to zero. Now, this function does not lend itself to polynomial expansion. Furthermore, if we carried out the 3OI test described above, it would come through with shining colours, since there would be (in principle) ZERO third-order distortion at ALL inputs of amplitude less than E. This would produce a 3OI which would be "infinite", or at least, misleadingly large in practice.

But would it be "misleading"? After all, that IS the definition of 3OI, and the ideal limiter WOULD have a clean bill of health in this department. This is more than an academic question. It turns out that in the design of receiver front ends one can choose to use design techniques which result in very flat "gain-vs-instantaneous-input" performance, which actually is beneficial, although the 3OI "rating" may be a little unbelievable.

The point here is that 3OI has served us well for several decades, but new HF circuit techniques, not possible before the present-day availability of high-f_T monolithic transistors, are making it possible to achieve performance that defies the existing taxonomy. The dilemma often arises in drafting a data sheet for such products: should they be couched in the words of the old metrics - to "give them a familiar look" - or should one try to "educate the customer"?

Here's a minor example of metrical mania, not expressly from the RF world, but certainly an analog dilemma. For years, IC manufacturers have been specifying op-amp offset voltages as "1 mV typ", or whatever the number may be. But what is this number really referring to? One is justified in assuming that the "typical offset" had better be ZERO, not skewed one way or the other. What OUGHT to be meant (that is, specified) is the standard deviation of the offset. But no-one is willing to do that, because (a) "all the other manufacturers say typ", and (b) "customers would be ringing us up night and day". Both true.

The same problem is already being felt in the burgeoning world of RF microcircuits. Expect that these bumbling attempts at saying everything in the same erroneous way as everybody else to lead to a further enhancing of the mystery.

9.1.4 Mysterious behaviour

This is by far the biggest category in the RF experience. Circuits at 900 MHz - or even at 90 MHz - just don't seem to do what they're supposed to do. The gain is too low, the noise is too high, the phase jitter's excessive..... the list is endless, but we would be remiss not to mention the most insidious of all unwelcome behavior, the conditional oscillation.

The one thing that can be said is: there is always a reason. James Bryant will address this topic with some hard-won practical advice in chapter 9.5.

Chapter 9.2

Direct Digital Synthesis

Peter Saul

9.2.1 Introduction

Direct Digital Synthesis (DDS) is conceptually one of the simplest forms of frequency synthesis, and is, unlike phase locked loop (PLL) techniques, very accurately described as synthesis, in that the output waveform is built up from mathematically derived points in amplitude and phase. This chapter will describe the mathematical basis of the DDS technique, and will show how this can be related to practical devices.

The basic technique has been known [1] and used in audio and sub-audio frequency research for many years, but only recently has it become practical to build R.F. DDS devices [2-4]. The limiting factors were primarily digital to analog converter speeds and resolutions, and read only memory (ROM) access times. These remain the causes of current performance limits, and are the subject of continuing research. However, DDS devices are now commercially available from several manufacturers and are in some cases coming down to commodity price ranges.

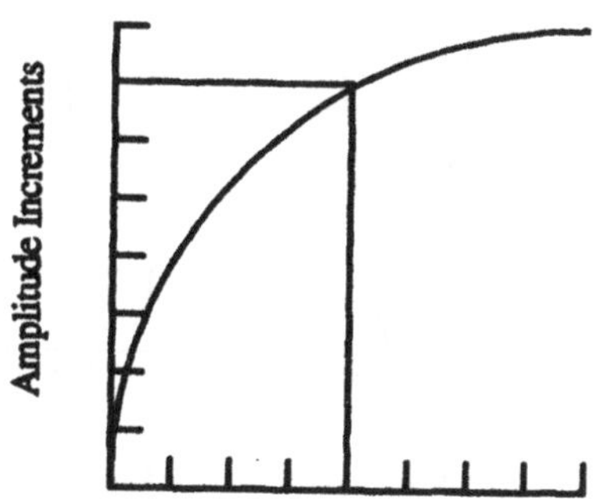

Figure 1 Time and amplitude quantization.

Figure 1 shows the concept of time and amplitude quantization of a waveform.

In direct digital synthesis, the output need not be sinusoidal, although conventionally it is for R.F. purposes. Any point on the sinusoid can be constructed by calculation of the amplitude required at a point in time and phase. The coarse increments on the axes indicate that the quantization is a finite process, on both axes.

This brings out the limitation in the DDS technique; finite resolution in quantization leads to inaccurate representation of the sinusoid and hence spurious outputs.

Before describing the practical methods of implementation of DDS, it is worth considering some of the implications of finite sampling intervals.

9.2.2 The mathematics of DDS

Figure 2 illustrates a mathematically accurate sine wave. The frequency has been set such that 4 cycles occur in the range of interest. The limitations to the accuracy are only numerical in drawing. However, there is quantization in time applied to this sinewave, along the y axis, since the points are calculated at finite intervals. In this example, a total of 4096 points are used for 4 cycles of the waveform, i.e. 1024 points per cycle. On the scales used, this gives the impression of continuity; this will be returned to later.

The sinewave equation is

$$q_i := \frac{2^m}{2} \sin\left(i.\frac{\pi}{F}\right) \quad (1)$$

where *Frequency* = 4 and $m = 3$

$$F := \frac{n+1}{(Frequency.2)} \quad (2)$$

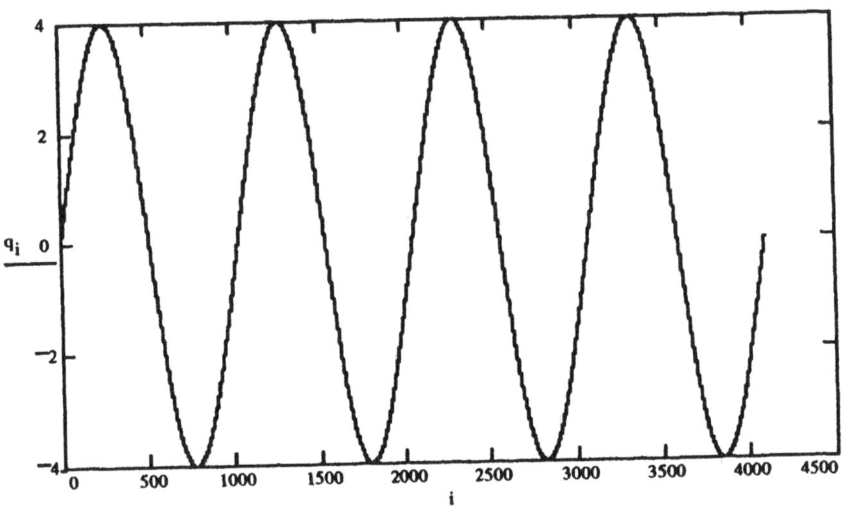

Figure 2 A sinewave sampled at 1024 points per cycle.

In amplitude terms, Figure 2 has "perfect" resolution. The practical DDS is much more limited. In general, high speed DACs are limited to 8 or 10 bits of resolution; 256 or 1024 levels. For the purposes of illustration, this will be reduced here to 3 bits, or 8 levels. Figure 3 shows quantization in amplitude to this resolution. The identity of the sinewave has been all but lost in this coarsely quantized view.

The discrete Fourier transform of this waveform is shown in Figure 4. The vertical scale is in dB relative to the zero frequency (DC) term of the output

$$fdB_j := 20log\left(\frac{|f_j|}{|f_0|}\right) \quad (3)$$

where j is a time axis increment. Units of frequency are not here employed; they could be Hz, MHz or any other unit; the analysis is a mathematical treatment at this stage.

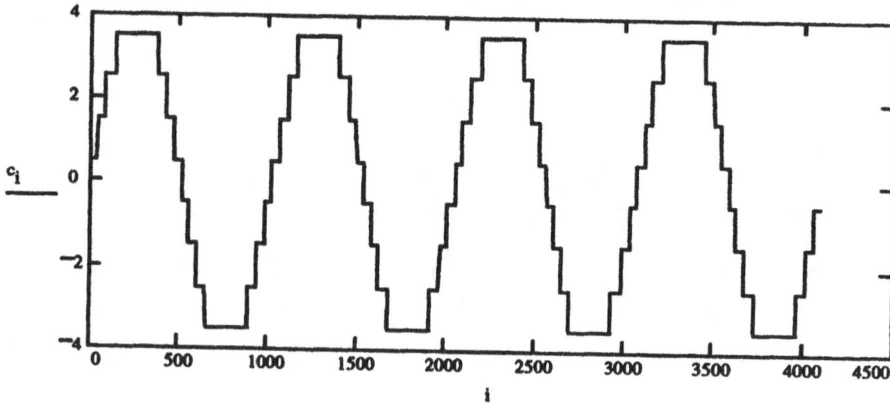

Figure 3 Three bit quantization.

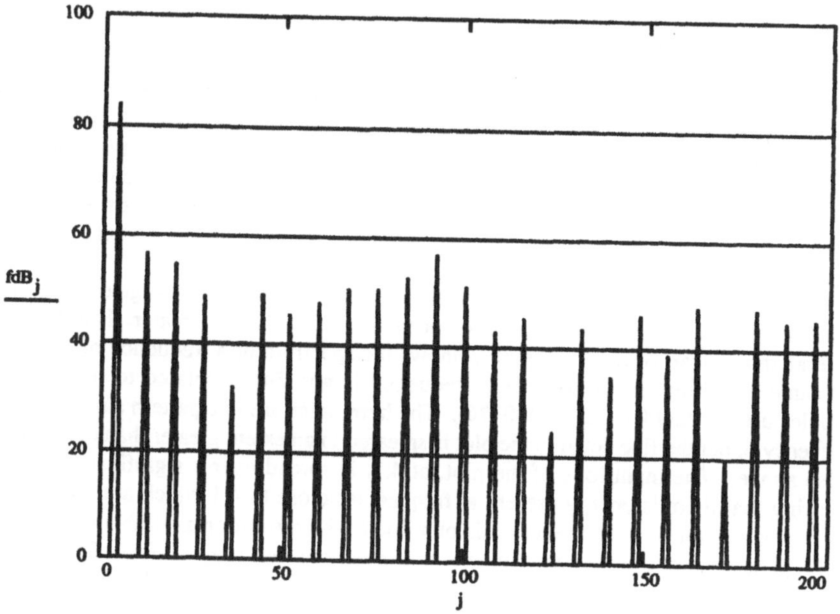

Figure 4 Spectrum of a digitized (m = 3 bits) sinewave.

It can be seen that the spectrum spreads up to very high relative frequencies due to the mathematically precise edge speeds involved. The fundamental, at a frequency of 4, is seen to the left, while the graph has been arbitrarily truncated at a limit of 200. In this example, since the signal is sampled synchronously by the clock at 1024 point per cycle, the output spurious signals are only in the form of harmonics. It can be seen that the largest of these is at 26 dB below the level of the wanted signal. In the general case, spurs in DDS devices are said to be "-6 dB per bit" below the wanted output. This is a simplification, but is well justified by the real measurements. The sources of this will now be examined.

The first consideration is random errors in the system, most particularly in the DAC. This can be simulated mathematically by the addition of random errors up to ±0.5 bit to the levels of Figure 3, to give Figure 5. Of course, 3-bit DACs would have much better performance than this; the diagram is for illustration related to what might happen in 8-bit or more DACs.

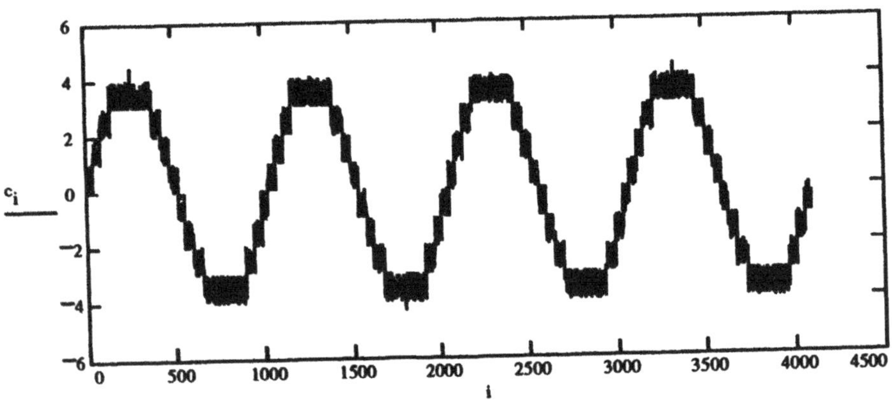

Figure 5 Random errors up to ± 0.5 bits.

The average output with time is clearly still the same, but the error content spreads this out. The spectrum is shown in Figure 6. The overall background noise level is increased, but the pattern of the output is basically unchanged. This is primarily because the signal and the sampling interval are synchronous. This illustrates a significant point concerning DDS devices, that if the output frequency is sub-harmonically related to the clock frequency, then the output is relatively free of spurious lines. Even low resolution DACs would be perfectly adequate if this were always the case.

Non-harmonically related frequencies are more difficult to investigate using Fourier transform techniques, since the resolution needed is progressively greater. However, some generalizations can be made. Figure 7 shows the spectrum of a frequency of "4.1" under the same conditions as above. This frequency is not simply related to the number of samples used (1024/4.1 = 249.7560976). The harmonics are in a pattern similar to that of frequency 4, but the fine detail in the plot is smeared to an extent greater than the addition of noise above. The main lobe of the plot, at 4.1, is spread over a significant bandwidth, inferring that there is energy present at frequencies close to 4.1 which are non-harmonic spurs; this cannot be confirmed in this plot due to the resolution available from practical FFTs. The apparent loss of amplitude is not important, but is an artifact of the choice of reference level and range. Even such a small transition as from 4 to 4.1 produces a radical change in the character of the output spectrum. This is clearly seen in the output of practical DDS devices when examined in the frequency domain.

It can be concluded from the above that the output spectrum is therefore not a simple function, and that it is determined by the exact relationship, if any, between the sample interval (4096 steps in the above) and the output frequency, 4 (or 4.1 in the above). These results must be considered when designing a practical DDS. The other key issue is the ratio between the sample rate and the frequency. In the above example, 1024 steps per cycle had a negligible effect on the 3 bit quantized signal. The effect of this oversampling was seen in Figure 4, where, because the "DAC" of the DDS is perfect, the background noise levels are zero. In Figure 6, the addition of errors has raised the background noise very significantly.

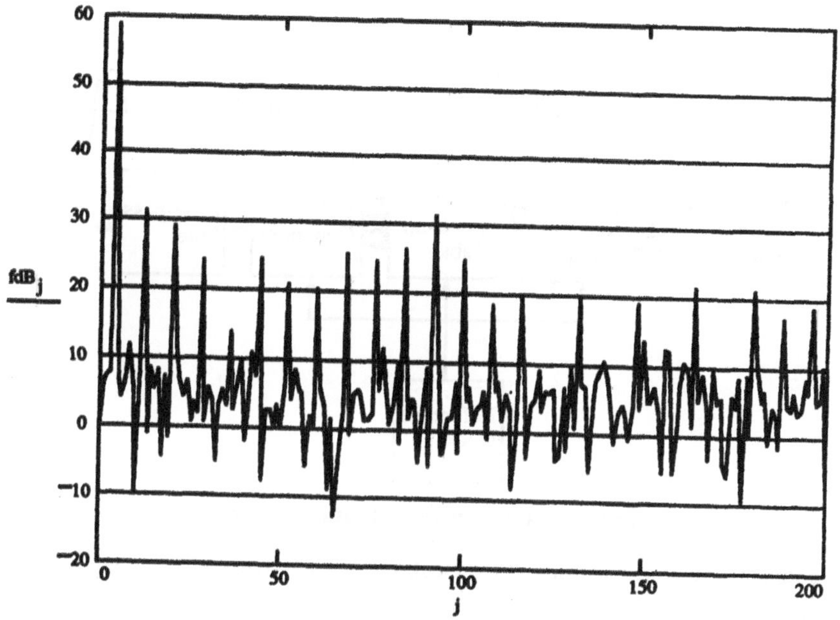

Figure 6 Spectrum of noisy reconstruction.

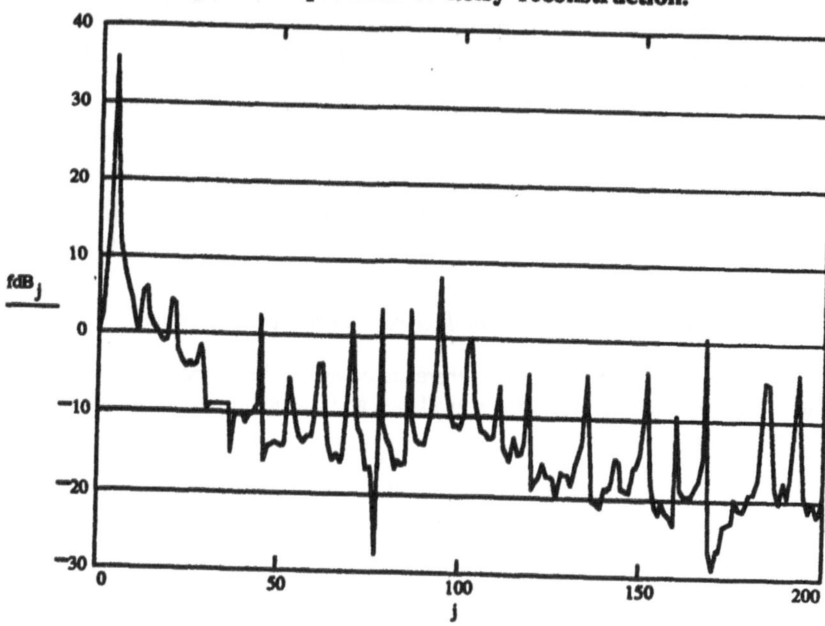

Figure 7 Spectrum of frequency "4.1".

9.2.3 DDS systems

The block diagram of a DDS is shown in Figure 8.

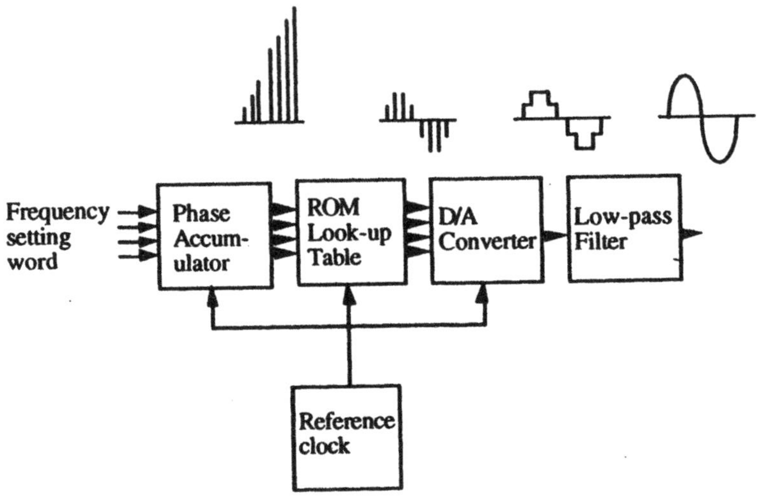

Figure 8 Basic direct digital synthesizer.

A digital accumulator, operating at the clock rate, successively adds the input frequency setting word to the accumulated total. The output from the accumulator is this rolling total, with of course periodic roll over as the maximum count in the accumulator is reached. This represents, in digital form, a saw tooth wave form with a mean frequency derived from the clock modified by the accumulator and the frequency setting word. Figure 8 includes analog representations of the wave forms at points in the DDS. The sawtooth is converted to a sampled sine wave by the ROM. A digital to analog converter (DAC) then converts the output to real analog form, although in a stepped manner due to the finite sampling intervals. A low pass filter then attenuates the steps. The length of the accumulator is determined by

$$Accumulator\ length = \frac{Clock\ Frequency}{Frequency\ Increment} \tag{4}$$

and the output frequency is

$$Output\ Frequency = Input\ data\ word\ x\ Frequency\ Increment \tag{5}$$

Unlike the conventional PLL synthesizer, there are no real limits on the size of the frequency increment; 1 Hz increments are routinely available, and fractional Hz increments are equally feasible and available.

There is no feedback loop in the basic design, so there are no conflicting design parameters to be reconciled as there are in a PLL. With the same synthesizer, it is also possible to have very large increments, and there can also be frequency offset, and phase, frequency and amplitude modulation, or combinations of these. Programming of the output frequency is very simple; most fast devices have a parallel binary or BCD programming input which, in accordance with the equations above amounts to the 'channel number' if expressed in terms of the frequency increment.

As can be seen from Figure 8, the DDS is essentially an all-digital device, with no variable oscillators or tuning involved. The clock oscillator source will usually be a phase locked loop circuit, but at a single, fixed frequency. This can be made very spurious free and of low noise. The only analog parts in the DDS are in the digital to analog converter.

These are usually defined by the manufacturer and are either not alterable by the user, or only to a limited degree, e.g. for final setting of the output current or voltage. Fast DACs tend to have current-mode outputs, often driving into a 50 Ω load impedance.

DDS circuits and concepts have advanced rapidly in recent years, building to the point where an output bandwidth of 350 MHz - 500 MHz is possible with production devices. It is of course possible to up-convert this frequency range using multiplying, mixing or PLL techniques. Lower frequency DDSs are available, typically as digital output parts, to be used with external DACs of 10, 12 or more bits. The digital chip is clearly much easier to realize than its corresponding DAC and is available from several manufacturers in different forms; there is as yet very little functional standardization.

9.2.3.1 *The spurious content*

The spurious output content of DDS systems has been described in general terms above. Where there is synchronism between the clock and output frequencies, the output is particularly spur free. However, where this is not the case, the spurs rise to a level which can be predicted from the number of bits in the output, the output frequency and the clock frequency. The amplitude quantization noise, which is due to the finite number of quantization levels available = - $(6.02N_a + 1.76)$ dB relative to the wanted output [5],where N_a = No. bits of amplitude quantization.

In general, the provision of more bits in the accumulator/ROM is much easier than more accurate bits in the DAC, so in general amplitude quantization is due to finite bit count in the DAC.

Phase truncation also contributes a term = $-6.02N_\phi + 3.922$ dB, where N_ϕ = No. bits of phase quantization. This is slightly worse than the amplitude term for a given N, but more time points are easily arranged, so it would not be sensible to use fewer bits of phase resolution than DAC resolution; usually one extra bit of phase is added. More phase resolution tends to be avoided since the amplitude or the DAC terms then dominate, and nothing is added. In extreme cases, ROM access time considerations dictate that minimum phase points are used.

The major limiting factor in the DDS in practice is the DAC. Spurs occur at $\pm mF_{clk} \pm nF_{DDS}$ and have a maximum amplitude of $-6N$ dB below the wanted signal, where N is the DAC linearity in bits. It is possible to have a worse spur levels, if for example the ROM access time is too long, or if the DAC does not achieve full accuracy in the clock period. In addition, DAC inaccuracies and 'glitch' add to the spur energy. This latter is the reason why some DDS manufacturers recommend DACs which have fewer than the ideal number of bits, but are proportionately faster; the final criterion is output spur level.

As with all sampled data phenomena, DDS outputs exhibit an amplitude degradation based on the *(sin x)/x* function.

$$A_{Degradation} = \frac{\sin(\pi F_{DDS}/F_{clk})}{\pi F_{DDS}/F_{clk}} \tag{6}$$

This is plotted in Figure 9. Figure 10 shows a more detailed plot of the first lobe of the response. The figures assumed are frequency in MHz and a sample rate of 1 GHz. Others frequencies can be easily had by scaling.

Specific points on this curve which are of interest are

A_{10} = - 0.001

A_{100} = - 0.143 all results in dB below the low frequency level.

A_{300} = - 1.326

A_{445} = - 3.041 i.e. ~-3 dB @0.445 of the clock

A_{500} = - 3.922

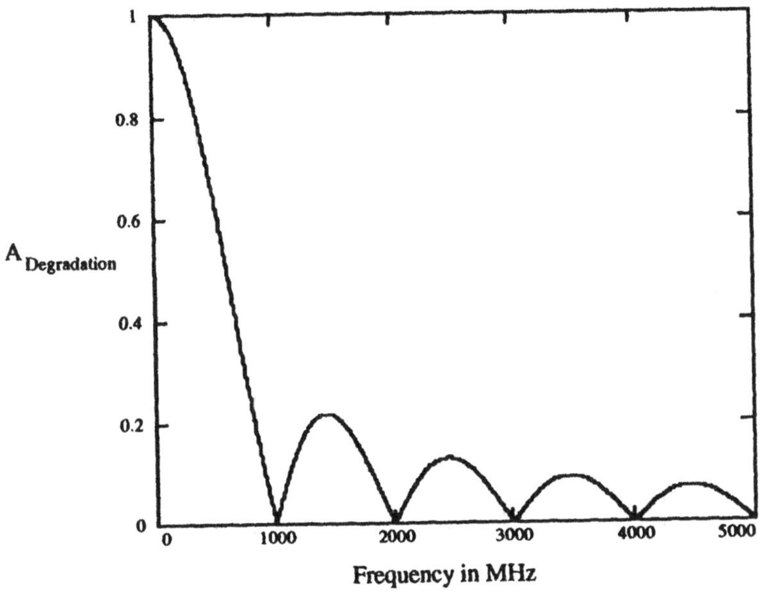

Figure 9 Sampling roll-off.

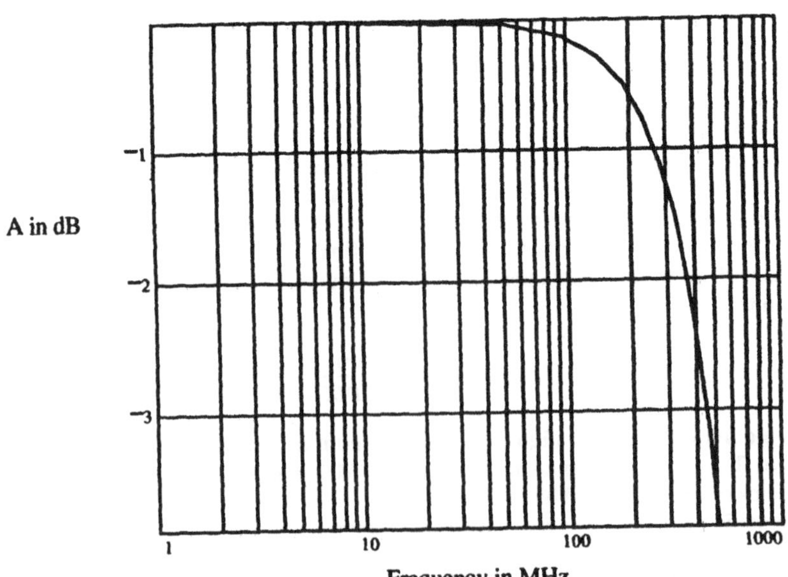

Figure 10 Amplitude roll-off due to sampling.

This analysis assumes that there is no further roll-off. In real systems, some roll off due to low pass filter effects is inevitable.

Most DDS designs use the first response, since the amplitude is the greatest, higher order responses are illustrated in Figure 11. However, it is possible to use the first image response, for a higher output frequency, or to use harmonics. In most cases, however, it would be more efficient to use separate means for achieving these frequencies, such as external mixers or multipliers.

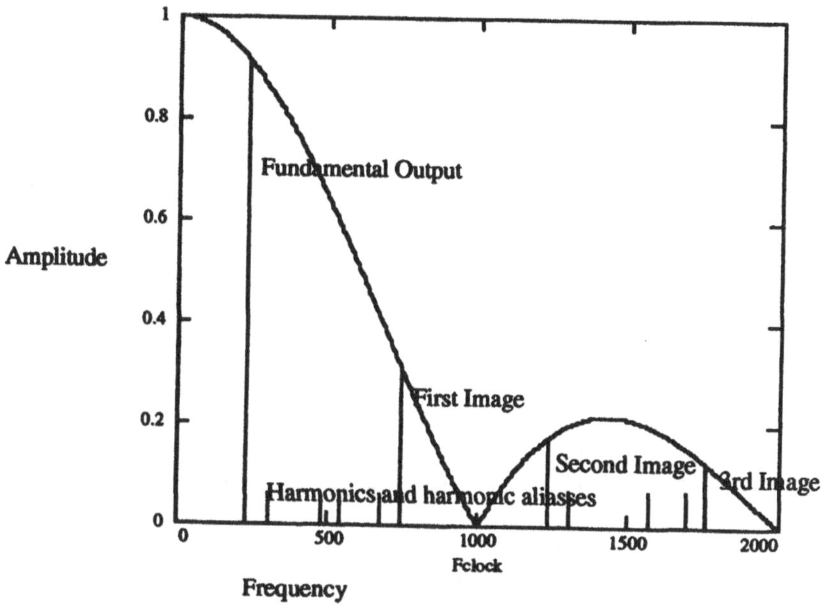

Figure 11 Signal, harmonics and aliases.

If all the above comments tend to show DDS as a spur prone technique, there is at least one mitigating factor in this respect. The spur content is improved by having many clock cycles per output cycle, in the ratio *(F_{clk}/F_{DDS})* or $20Log_{10}(F_{clk}/F_{DDS})$ dB. This is closely linked to "oversampling" in the DAC, and indeed may well prove an attractive means of increasing DDS performance by reducing spur levels in the future.

A typical fast DDS uses an eight-bit output data word to a DAC which achieves full accuracy, but no more, in one clock period. Where there is no oversampling, this theoretically results in 49.8 dB suppression of spurs; practical devices are - 45 dBc to - 47 dBc. This performance is close to the theoretical limit, but is not adequate in some applications, e.g. for local oscillators in radio. However, it is adequate where the signal to be emitted is a jammer etc., where some spurs are acceptable provided the frequency agility is adequate. In this area, the DDS is very much better that the PLL, since the former has no loop setting or stability problems like the PLL. In addition, the DDS can switch very quickly between frequencies, again without stability or overshoot difficulties.

The facility for very small frequency increments in the DDS can also be translated in frequency by using PLLs. Conventionally, a PLL loop bandwidth is limited to less that the required frequency increment. Fast frequency changes need loop speed-up devices and are notoriously difficult to design. Where a DDS can be substituted for the reference in the loop, the PLL can be made very broad band and hence fast in response, without compromising the fine increment characteristics of the DDS. This is illustrated in Figure 12.

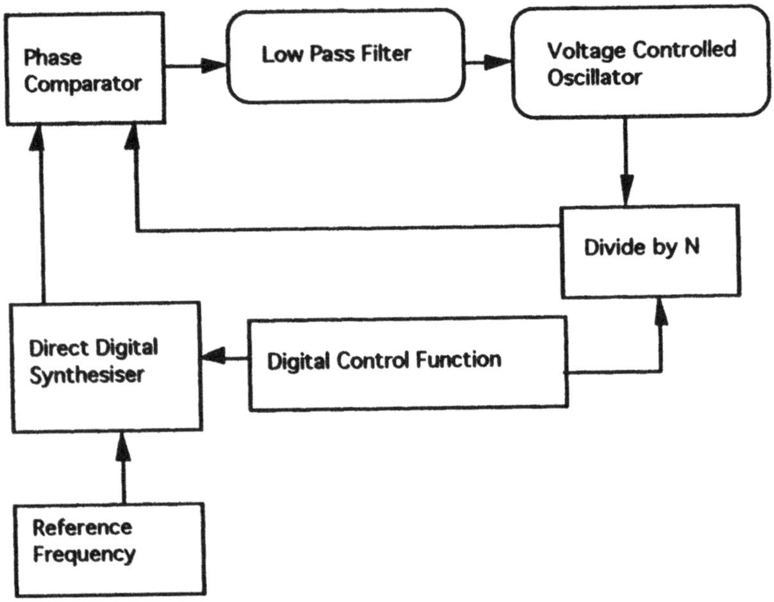

Figure 12 Phase locked loop DDS.

This figure shows a PLL multiplying the output of a DDS through the action of the loop. This does retain a wide operating frequency range, provided the VCO design is adequate, and can be quite fast in operation, although, since a feedback loop is present, the response must be much slower than a simple DDS. In addition, phase and frequency related spurs which occur inside the loop are multiplied by 6.02 dB/octave. Careful design of loop parameters can alleviate this problem in circuits where this is the preferred solution.

Figure 13 shows an alternative structure with a mixer upconconversion. This is more restricted in frequency range, and has the other PLL disadvantages, but preserves the DDS spur levels inside the loop bandwidth, and attenuates the spurs outside the loop.

Until recently, application areas for DDS devices have been primarily in radar and EW, possibly extending to ECM and ESM. Facilities such as 'chirps' for radar are easily programmed in, with full software control, unlike alternatives such as SAW techniques, where the form of the chirp must be designed into the device. Indeed, it is possible to conceive of a DDS that could offer full phase, frequency and amplitude control, and could change any of these parameters almost arbitrarily quickly. Frequency hopping in less than 20 ns has been demonstrated, with full phase coherence.

It has been suggested that DDS techniques might be applicable to lower cost receivers, i.e. for the domestic market, where the major advantage of DDS for the set manufacturer is the complete absence of "setting-up" and tuning components. This concept may well be seen in the next generation of domestic broadcast radio receivers.

In Europe and the US, DDS is seen as a mainly military technique, especially in EW. Radio transceivers will probably use DDS-PLL methods; at least one military radio has already done so.

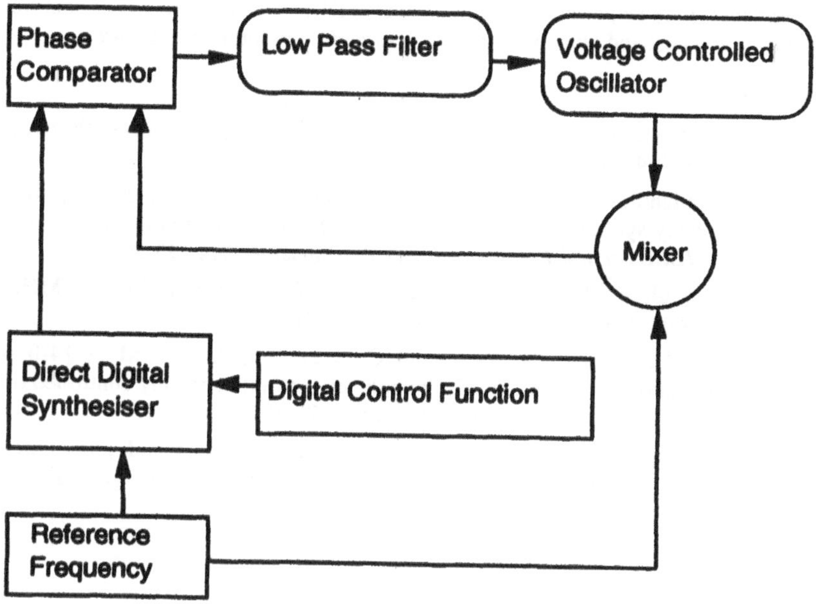

Figure 13 Offset phase locked loop DDS.

The practical improvements needed in DDS are

1) Integration of complete devices.

   This has been carried out on one design i.e. GEC-Plessey SP2002. Very fast devices are unlikely to work as well with multiple-chip arrangements since the parallelism of the data highways is difficult to maintain.

2) Increase in bit-count.

   This will lead to lower spurious levels. For use in radio receivers, a local oscillator spur level of below - 90 dB is desirable, i.e. equivalent to a 15 - 16 bit DAC. This is currently only possible up to about 1 MHz. The digital part of the system is of course relatively straightforward, i.e. the DAC is the limitation. Alternative DAC techniques need urgently to be sought.

3) Addition of digital or analog filtering techniques, again for spurious reduction. It is possible that digital or even analog FIR or IIR filters could be designed into the system.

4) Replacement of the ROM structure. While of lesser importance than the above, ROM access time is, after the DAC settling time and accuracy, the limiting factor. Hence, replacement of the ROM by a digital filter looks very attractive.

9.2.4 Conclusions

This chapter has described the theoretical background to direct digital synthesis, and has illustrated it with plots of ideal and noisy systems. DDS techniques are already in use in radio systems, primarily in combination DDS/PLL synthesizers, but, with further refinement it may be possible to build dedicated DDS only synthesizers or communications quality at low cost and with minimal user intervention in setting up and use.

References

[1] V. Mannassewitch, *Frequency synthesis and design,* New York: Wiley, pp. 37-49 and pp. 494-501, 1980.

[2] H. T. Nicholas and H. Samueli, "An analysis of the output spectrum of direct digital frequency synthesisers in the presence of phase-accumulator truncation," in *Proc. 41st. IEEE Annual Frequency Control Symposium,* pp. 495-502.

[3] P. H. Saul, W. Barber, D. G. Taylor and T. Ward, "Single chip 500 MHz function generator," *IEE Proceedings,* Part G, Vol. 138, April 1991.

[4] F. Lu, H. Samueli, J. Yuan and C. Svensson, "A 700-MHz 24-b pipelined accumulator in 1.2 μm CMOS for application as a numerically controlled oscillator," *IEEE J. Solid-State Circuits,* Vol. 28, pp. 878-886, Aug. 1993.

[5] J. Tsimbinos, "The use of direct digital synthesisers and phase locked loop frequency multipliers for generating local oscillator signals in digital radios," *JIREE.*

Chapter 9.3

RF Design in GaAs and BiCMOS

Peter Mole and Ken Searle

9.3.1 Introduction

As in all good mysteries, in RF design things are never quite what they seem. Surprises are common when we assume the elements that make up a design to be simple, but if we expand these elements to include what are sometimes called parasitics, then design and reality become closer.

This tutorial uses integrated circuit design in BiCMOS and GaAs as an example, but the principles, and the clues that unveil the mystery, are common to all RF design.

9.3.2 RF design

9.3.2.1 What is RF ?

Radio frequencies are that part of the electromagnetic spectrum extending from a few kHz to many GHz. Frequencies above 300 MHz are also known as microwave. Recent interest, however, centres on the market for cellular and satellite communications. These frequencies extend from 866 MHz for CT2 cordless phones to 12 GHz TV satellite downlinks. The most popular frequencies for new products, around 2 GHz, is high enough to expose a few mysteries, and will therefore be used for examples given.

9.3.2.2 The 'unexpected' parasitic component

When first looking at a circuit schematic of a UHF amplifier, several years ago, I was surprised to see a component I did not recognise. Reading the article with care, this new component turned out to be no more than a length of line printed onto a PCB and behaved like an inductor. The length and width of this line and the PCB material and thickness controlled the inductance. This was surely an easy and low cost way of constructing an inductor, but it naturally leads to the question; when can a length of interconnect on a PCB be considered as an electrical short circuit and when must it be regarded as a component in its own right?

The consequences of treating an inductor as a 'mere piece of wire' in a circuit can be severe if the frequency of operation of the circuit is sufficiently high that the inductance offers an impedance comparable to the local impedance of the circuit. The problem is not restricted to interconnect on a PCB; the use of high value decoupling capacitors with long internal interconnect can often lead to the capacitor behaving as an inductor at high frequencies. This is because the interconnect impedance, which is in series with the physical capacitor, dominates the component impedance. If care is not taken in the choice of construction of decoupling capacitors, they may not be doing the intended job at the frequency of interest.

To acquire an understanding of the magnitude of these parasitics on a PCB, consider an isolated signal track of width $2a$ running centrally in a gap of width $2b$ in a ground plane on a PCB. Such a configuration will present a parasitic capacitance between signal track and ground due to electrostatic field set up both in the PCB dielectric and in the air above the PCB. Equally, any current flowing in the signal track will set up a magnetic field in the gap between the tracks which ensures that the signal path has an inductance. The capacitance and inductance can be calculated by solving the appropriate field equations [1]. In Figure 1, the field equations have been solved for such a signal path (known as a coplanar waveguide) and its characteristic impedance Z_0 has been calculated.

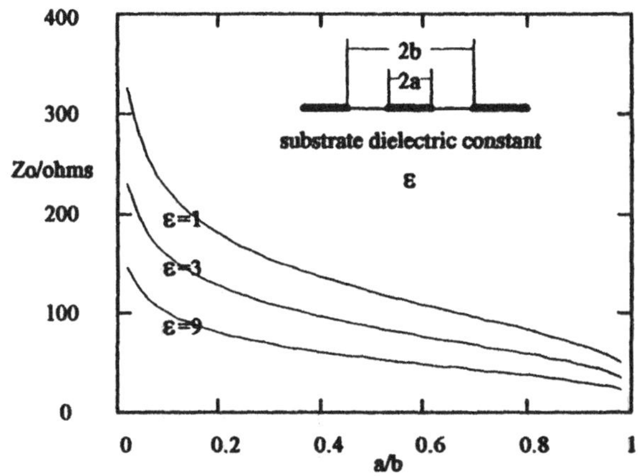

Figure 1 Characteristic impedance of a coplanar waveguide.

Where the wavelength at the signal frequency is large compared to the length of the signal line, then from standard transmission line theory it can be shown that

- the inductive series impedance of the line is $Z_0 \cdot \frac{2\pi}{\lambda}$ (1)

- the capacitive shunting impedance of the line is $Z_0 \cdot \frac{\lambda}{2\pi l}$ (2)

where l is the length of the track and λ the wavelength of the signal in the coplanar line. At higher frequencies the above simplifications no longer hold and the full transmission line equations must be considered.

From Figure 1, it can be seen that for reasonable dimensions of line the impedance does not vary greatly and is in the range 30 - 300 Ohms. Even if the geometry of interest is changed (e.g. we consider track to track or track to back-plane parasitics) we find a similar result.

Equations (1) and (2) now offer very simple interpretation with three characteristic regions

1. If we are operating at very low frequencies, l/λ is small and the series inductance and the shunt capacitance can be neglected. The track behaves like a short circuit.

2. As the frequency of the signal increases, the effect of both the shunting and series impedance gets more severe. If the connection is at a high impedance point the

shunting capacitance is more important, if at a low impedance point then the inductance is the dominant parasitic. However, if a successful design is to be completed as λ gets closer to l, the local circuit impedances must be close to Z_0 if the parasitics are not to dominate. Thus as the frequency rises the circuit designer becomes advised to operate at impedances of 50 - 100 Ohms.

3. At frequencies where l approaches 0.1λ, then a full transmission line design is essential.

The above argument has been presented for a single signal line and its ground return. If, however, two signal lines run close to one another, it could also be applied to estimate the crosstalk capacitance. Thus, coupling between signal lines will also depend on l/λ. This is most significant if an output and input signal run close to one another. To avoid a problem, signal lines should be physically separated by a grounded path to ensure that each couples electrostatically to a local ground in preference to the other.

The above argument leads to the conclusion that the effect of interconnect parasitics depends on the local circuit impedance levels and the length of the interconnect. With little freedom remaining to alter impedance levels, circuits need to be made physically smaller if higher and higher frequencies of operation are to be reached without embedding the circuit in a transmission line environment. Integrated circuitry provides the perfect means to reduce interconnect distances and simplify the RF design - at least whilst the signal remains on the chip. At first sight the reduction of size of the circuit looks as if it might cause problems with increased coupling between input and output. Usually, however, the reduced length, width and separation from ground of the input and output tracks more than compensate for the reduced separation, and even crosstalk problems are reduced with integration.

9.3.2.3 *Matching circuit impedances*

The efficient transfer of power between components in the signal path is essential if a good power gain is to be achieved. To transfer power efficiently between the stages it is essential that the impedances of the circuits are conjugately matched. To investigate what this means, consider the circuit in Figure 2.

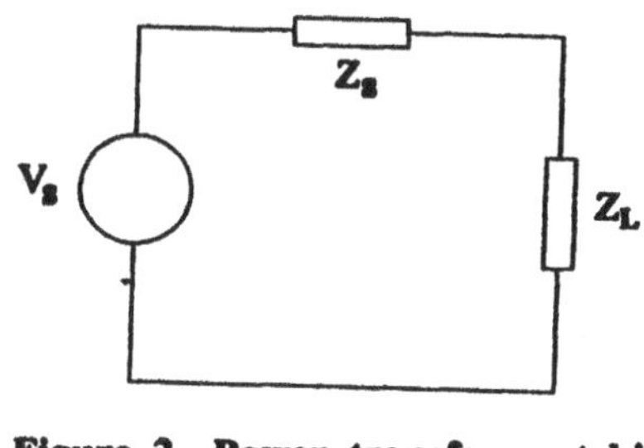

Figure 2 Power transfer matching.

The current flowing is given by

$$i = \frac{v_s}{Z_s + Z_L}$$

Only the real part of Z_L can dissipate power so the power dissipated is

$$P_L = \frac{v_s^2 \cdot \mathrm{Re}(Z_L)}{(Z_s + Z_L)^2} = \frac{v_s^2 \cdot \mathrm{Re}(Z_L)}{[\mathrm{Re}(Z_s + Z_L)]^2 + [\mathrm{Im}(Z_s + Z_L)]^2}$$

With a little inspection it can be seen that this is a maximum when $\mathrm{Im}(Z_S+Z_L) = 0$ and $\mathrm{Re}(Z_S) = \mathrm{Re}(Z_L)$.

The first requirement indicates that the imaginary part of the impedance, although unable to dissipate power itself, can affect the power transferred. Thus if two components are joined by a wire with a significant parasitic series inductance, power transfer can be degraded.

The second requirement indicates that efficient power transfer will occur when the real parts of the source and load impedances are equal. Thus to perfectly match two components, the complex parts of their impedances must be conjugate and the real parts of their impedance must be transformed so they are equal.

Figure 3 illustrates the principles of matching with lumped components for a source that is connected to a load through a wire which appears mainly inductive. Consider what happens when a capacitance is coupled across the load. The first step of the analysis is to replace this parallel *RC* combination with its equivalent series network as shown in Figure 3. It is then simple to see that matching would be achieved when both

$$\omega L_p = \frac{1}{\omega C\left(1 + \frac{1}{\omega^2 \tau^2}\right)} \quad \text{(where } \tau = R_L C\text{)} \tag{3}$$

and

$$R_s = \frac{R_L}{1 + \omega^2 \tau^2} \tag{4}$$

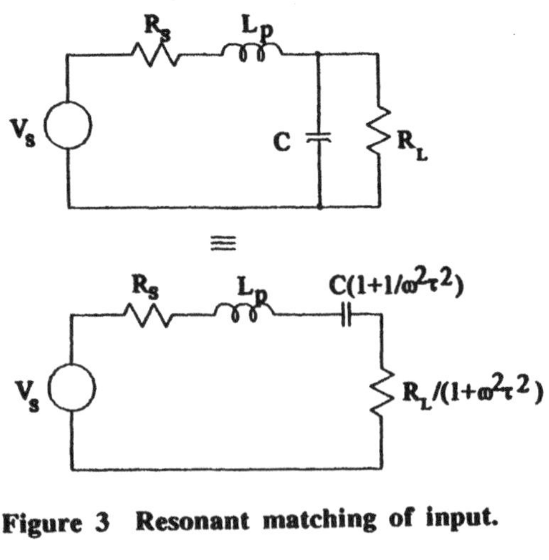

Figure 3 Resonant matching of input.

These cannot in general be achieved with a single capacitor, but they serve to illustrate the principles. Thus at one frequency equation (3) indicates that, by selection of the value of the capacitance, the imaginary components of the impedances cancel and the effects of parasitic components introduced by wiring can be eliminated. The band of frequencies over which conjugate matching is adequately achieved is determined by the Q of the resonant circuit. Thus the larger R_s is in comparison to L_p the wider the band over which matching can be achieved.

Equation (4) illustrates that a frequency dependent transformation of the real part of impedance has also been achieved with this circuit. Thus matching of the real parts of impedances need not be achieved with conventional transformers but can be obtained by introducing controlled L and C into the interconnect path. Again it can be simply shown that if R_L is very much larger than R_s then the Q of the circuit will be increased and the matching circuit will be more narrow band and more sensitive to component tolerances. It is worth noting that, although matching at a single frequency can always be achieved in theory by adding suitable capacitances and inductances, the value of reactance needed may not always be practical. For example, it is not practical to match the capacitive load at a collector of a common emitter gain stage on chip as the capacitance requires too large a value shunt inductor for practical integration.

The idea of resonant input matching can be extended to cross-talk capacitance or feedback capacitance; this can be 'neutralised' at one given frequency by shunting the capacitance with an inductance. However, such a technique in a feedback path needs to be undertaken with great care as not only must it be designed to improve the performance at the desired frequency, it must not lead to an instability at any other frequency.

In the above argument, the concept of matching is presented purely from a low frequency perspective. As the frequency increases, it becomes important to allow for the time it takes for a signal to propagate along a wire. Consider what happens when a generator with a 50 Ohm source impedance is connected to a capacitive load via several meters of 50 Ohm cable. When the signal is first launched, the generator can have no idea of the load at the end of the cable. It is not until the signal has had time to travel the length of the cable that the capacitor can have any effect. The capacitor cannot dissipate any power, so the generator cannot supply any power. Thus all power that is sent down the cable is reflected back along the wire and returned to the generator. The RF circuit designer usually thinks in terms of the launching, transmission and reflection of waves by a circuit rather than input impedances and conjugate matching. Thus a perfectly matched circuit will not reflect any power, all will be transmitted to the load.

This approach has given rise to the *s*-parameter concept. If a wave of unit magnitude is launched at the input port (port 1) of a component, then the wave reflected at the input is s_{11}, whilst the wave transmitted from the output port (port 2) is s_{21}. Since the concept of *s* (or scattering) parameters deals with waves, the *s*-parameters have both magnitude and phase. The *s*-parameter concept can be generalised to a system with as many input/output ports as required. If a reference wave is input at port *m* of a system and the wave output is monitored at port *n*, then s_{nm} is the resulting output wave. The power output is given by s_{nm}^2. Thus it can be seen that if a source is perfectly matched into port 1 of a component, no power is reflected and s_{11} will be zero. If the input port presents a purely capacitive input then all the power is reflected and s_{11} is unity with a phase which will vary with frequency.

It should be noted that strictly the *s*-parameter concept is a small signal concept. A small-signal analysis of circuit performance can be made if the *s*-parameters of a device are known. Further information is needed if the DC bias operation or large signal behaviour is to be analysed. However, since the *s*-parameters of a circuit are usually the easiest properties to measure at high frequencies, it is usual to characterise RF circuits in terms of their *s*-parameter performance.

For a detailed discussion of *s*-parameters and their representation on a Smith chart the reader is referred to [2] or similar text.

9.3.2.4 Packaging

Many packaging options exist for RF circuits, from simple 4 lead stripline packages to 40 or more pin multi-chip modules. There are also options to avoid bond wires by using solder-bump or tab bonds. However, focusing on the commercial market and inexpensive plastic packages will emphasise many of the package effects.

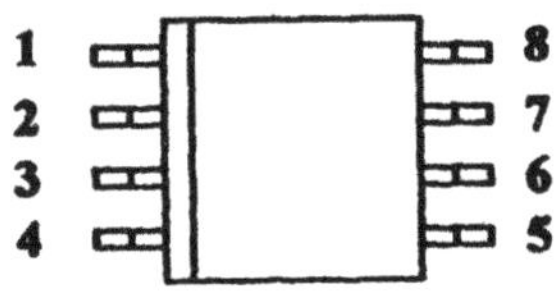

Figure 4 SO8 package.

The bond wire and leadframe present a number of coupling problems that limit isolation and gain as well as dynamic range if for example digital noise is coupled to a sensitive amplifier input. This coupling is both capacitive and inductive as shown in the table below for a 8 pin Small Outline (SO8) package. Inter pin capacitance in femtofarads, for pin 1 grounded and all other pins open circuits, is shown in shaded italic while the lower half of the table shows self and mutual inductance in nanohenries.

pin #	1	2	3	4	5	6	7	8
1	4.89							
2	1.54	4.04	*82*	*26*	*4*	*4*	*7*	*6*
3	0.92	1.44	4.04	*72*	*71*	*4*	*5*	*8*
4	0.67	0.92	1.54	4.89	*10*	*5*	*9*	*5*
5	0.31	0.29	0.29	0.31	4.89	*72*	*27*	*10*
6	0.28	0.26	0.26	0.29	1.54	4.04	*73*	*18*
7	0.29	0.27	0.26	0.29	0.92	1.44	4.04	*58*
8	0.30	0.28	0.27	0.31	0.66	0.92	1.54	4.89

Self inductance is often all that is considered but from the above it can be seen that, even for this small package, mutual inductance is around 25% of self inductance. In larger packages this ratio can exceed 50%. Capacitive coupling across the package is very low due to the lead frame (die attach) being grounded through pin 1. If this pin is left floating, then the capacitance between 3 and 7, for example, would increase from 5 fF to 67 fF.

Which is the more important depends upon the impedances, capacitance being more significant for high impedance pins whereas inductance can couple ground currents to low impedance inputs.

Bond inductance is often reduced by adding parallel bonds, but unless the spacing is large mutual inductance prevents the inductance being halved by the additional bond. Ground inductance is a particular problem since the chip ground exists on the top surface of the die, but may also be connected resistively or capacitively, depending on the die fixing method, to the die attach plate. This plate is not connected externally except in special plastic packages.

The equivalent circuit of a typical package, essentially a low pass filter, is shown in Figure 5. The inductance and capacitance cause a deviation from the normal 50 Ohm characteristic impedance of the signal path, and insert a reactive impedance in the ground and supply path.

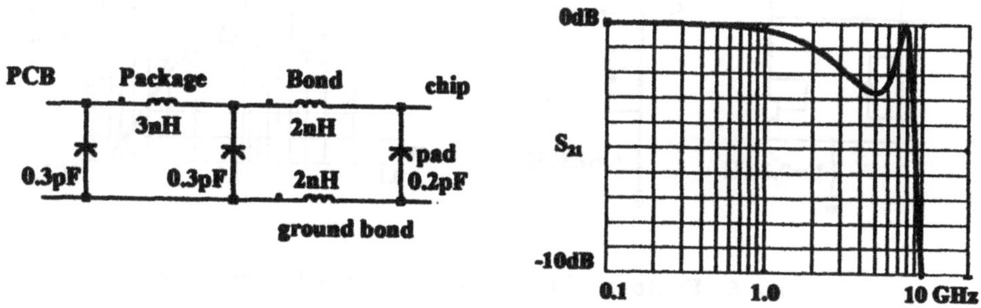

Figure 5 Package pin model.

Many of these effects are avoided by ceramic or metal packages; some using screened and matched 50 Ohm transmission lines for the signal paths and wide low impedance lines for the power supplies. In ceramic packages grounding and thermal characteristics may also be improved by including vias from the die attach plate to the external package surface.

9.3.2.5 Feedback

What's the difference between an amplifier and an oscillator? In most designer's experience, 'not very much'! In low frequency designs, for example using op-amps, where the amplifier characteristic is dominated by the feedback path, the gain and feedback elements are easily recognised and well defined. However, in RF design there are probably more significant feedback paths than nodes in the circuit, all interacting with one another. This would appear to be a recipe for instability but essentially it just restricts the range of impedances usable in the amplifier and what source and load impedances may be used.

Up to 2 GHz, broadband feedback may be applied in the normal way over perhaps two stages to set gain and port impedances, but above this frequency, phase shift in the loop limits feedback to a single stage.

If there is insufficient gain in the circuit to support feedback then circuits may be designed without feedback, but the circuit characteristics then become much more dependent on processing variations in the active element and the yield, or the number of circuits meeting specification is reduced. Given the process limits, yield may be optimised using design centering simulations with Monte Carlo analysis techniques.

Unexpected feedback will often cause unwanted peaks in the gain characteristics even if it is insufficient to cause instability. A frequent source of this is a common impedance in the ground line, usually a bond wire or package pin. The example in Figure 4 illustrates this effect with two similar feedback amplifiers. In the first the ground is common to both stages and is typically also connected to the substrate, this is then connected to the external ground via a bond wire. In the second case the ground for each stage is connected through its own bond wire to the external ground. The gain plot clearly shows the effect of the common impedance in the first amplifier, producing a large peak in the gain response, verging on oscillation.

The significance of this example is further emphasised by the fact that the inductance used is only 0.3 nH or approximately 0.5 mm of bond wire.

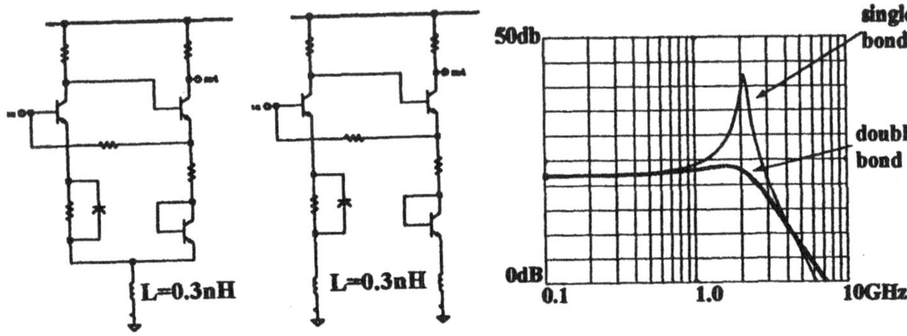

Figure 6 Feedback due to common ground impedance.

9.3.2.6 *Balanced or single-ended*

Antennas, switches and filters, typically connected with coax or microstrip, are termed single ended since the signal return current flows in the ground path. Single ended is also the simplest monolithic configuration with fewer devices, less DC power and a lower noise figure. However, as discussed above, a good ground reference is difficult to maintain through the IC interface and a balanced configuration becomes attractive.

A balanced configuration not only produces very little signal current in the ground but also rejects signals which are seen as common mode, for example from ground, power supply or electromagnetic coupling from other circuits. This same mechanism also provides increased isolation between ports in balanced mixers. These advantages may be extended to include balanced filters, particularly SAW devices which are by default balanced structures.

Two single ended amplifiers may also be used in a quasi-balanced configuration using quadrature couplers as shown in Figure 16. This gives improved matching and gain flatness, higher compression point and 3rd order intercept, and some degree of redundancy; should one of the amplifiers fail, gain is only reduced by 3 dB.

9.3.2.7 *Measurement and testing*

Complete RF ICs and MCMs (multi-chip modules) are usually designed to have an interface impedance around 50 Ohms that matches with standard test equipment, for example network analysers, spectrum analysers and oscilloscopes. Probing of internal nodes is generally more difficult, both because of the small dimensions, a track on a silicon wafer could be only 2 microns across and 2 microns from another track, and the additional capacitive load caused by the probe. However, for probing DC or large signals of frequencies up to 500 MHz, special active probes are available with capacitance as low as 20 fF and tip radius less than 1 micron. At high frequencies the poor definition of the ground path makes the accuracy of this method of measurement doubtful and a coplanar probe must be employed.

Coplanar probes, where signal and ground contact are both made at the probe tip, are accurate at frequencies in excess of 20 GHz, but provision must be made in the original design to site adjacent signal and ground test pads.

RF testing on-wafer is essential to avoid packaging non-functioning die or to reliably remove the effect of the package from a measurement. Again, coplanar probes are used where the probe tip signal to ground pitch is conveniently close to the normal bond pad spacing of around 200 micron.

9.3.2.8 Integration of RF circuitry

In the above, many of the basics of RF design concept have been discussed. The essential problem of parasitics introduced by interconnect track lengths indicates that a move to integrated or at least hybrid circuitry will be essential to achieve RF performance without needing to develop a transmission line circuit. Integrated circuitry also offers the potential of achieving low cost circuits through high volume manufacture once the ability to reliably design circuits including their interface off-chip has been established. Both high performance silicon and gallium arsenide process offer the capability of integrating RF circuitry on a single chip.

9.3.3 Designing RF circuitry in a silicon BiCMOS technology

9.3.3.1 What a BiCMOS technology has to offer

A modern silicon BiCMOS technology such as the Northern Telecom BATMOS process [3] offers the RF designer a large range of components to integrate. Typical features of an advanced BiCMOS technology of interest to an RF designer are

- A fast bipolar transistor (f_{max} 10 - 20 GHz).
- Complementary MOS devices.
- Resistors with a low parasitic capacitance.
- A linear capacitor with low parasitic capacitance to ground.
- 2 or more levels of metal to enable simple wiring and good grounding.
- The capability to make low value inductors (< 1 nH).

With the above features, the capability of the bipolar transistor enables useful RF circuitry up to frequencies of 1 to 2 GHz. The MOS devices on a technology with channel lengths just under 1 μm will have an intrinsic f_T of about 2 GHz. Thus in the signal path, bipolar transistors are usually preferred. However, the MOS devices can prove useful in bias circuitry and as passive loads. Most importantly, they provide the most efficient way to provide support logic and similar circuitry that operates at lower speeds.

Capacitors in silicon technology inevitably have one plate which has a parasitic shunting capacitance to ground. When the capacitance is between two polysilicon plates the lower plate is separated by about 1μm of oxide from the substrate. The parasitic capacitance is then typically about 5% of the capacitor value and it becomes practical to use the capacitor to AC couple stages without adding a significant load capacitance. Technologies which do not offer poly-poly capacitors have a much higher parasitic value to ground for a given capacitance and AC coupling of signal paths is no longer efficient.

Resistors will also introduce significant parasitic capacitance to ground if they are made by diffusion into the substrate. This will be made worse where tight matching control of resistors is needed, as the first defence of the designer against random fluctuations of resistor geometry is to make the resistors large in area. This will introduce prohibitive stray capacitance. In a technology suitable for RF design, resistors made in polysilicon will be well characterised. These will then offer a low parasitic capacitance. For typical RF circuitry resistors in the range 10-10,000 Ohms are required; for convenience of resistor geometry this is best achieved if polysilicon is available in two resistivities (about 50 and 1000 Ohm/square).

To date, silicon technology has not had the basic speed to tackle designs which operate at 10 GHz and above. Thus, there has not been the reason to develop a range of couplers and such devices that will be discussed later in the section on GaAs. However, as the capability silicon technology advances into the GHz regime inductors are being used. Typically inductors of values up to 10 nH can be 'wound' on a silicon substrate and these will have a Q of about 5. The Q is usually limited by resistive losses in the metallisation which is sufficiently thin that the physical thickness rather than the 'skin depth' limits the resistance up to frequencies in the low GHz range. The large physical size of the inductors coupled with their low Q means that they are used sparingly.

9.3.3.2 Design with bipolar transistors at RF frequencies

When designing on chip at frequencies about 1 to 2 GHz, the design procedure for silicon follows that used at lower frequencies. There already exists a range of CAD tools available for circuit simulation. These support non-linear modelling of the bipolar transistor using a Gummel-Poon based model as used in the SPICE [4] simulator. The parameters for such a model are provided by the manufacturers of BiCMOS technology and there is no reason why SPICE and its derivatives cannot be used to design RF circuits. Dedicated microwave design tools often make it easier to represent the results in *s*-parameter format and have models for the microwave elements such as lossy transmission lines and couplers. Currently silicon circuitry does not operate at frequencies where such microwave structures can be efficiently used, so the motivation to use a dedicated microwave simulator is not strong. There is, however, a stronger motivation to use non-linear simulators with a harmonic balance capability which offer an easier route to simulate intermodulation and non-linear noise mixing. Such simulators can provide a significant time saving over using SPICE where the lengthy route of transient simulation and Fourier transformation must be used.

As circuits are designed without use of special microwave components, RF circuit design techniques are very similar to those at lower frequency. Two general points of note are worth making though

1. Feedback can only be applied locally as the phase shift over each stage is significant and very component sensitive.
2. The bipolar devices will normally be operating well above the frequency at which their current gain starts to decrease ($f\beta$).

The second point is worth considering a little further as it has some consequences that are initially a little unexpected.

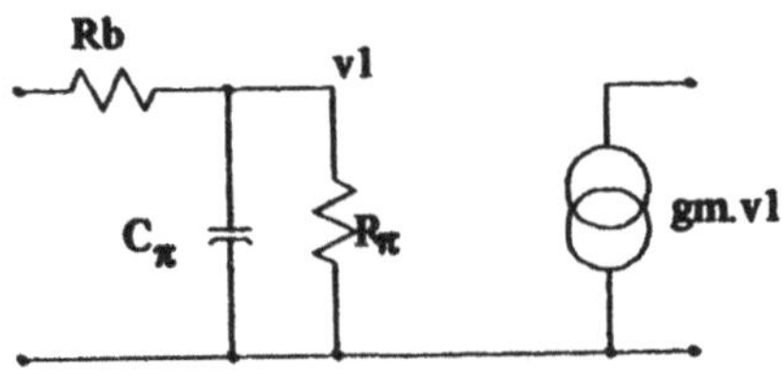

Figure 7 A simple equivalent circuit for a bipolar transistor.

Referring to Figure 7, once the frequency f_β is passed, the input impedance of the device is dominated by the base to emitter capacitance C_π and thus the input current is capacitive. The output current is controlled by g_m. The current gain of the transistor can simply be written as

$$h_{fe} = -\frac{j g_m}{\omega C_\pi} = -j\frac{\omega_T}{\omega}$$
$$\omega_T = 2\pi f_T \qquad (5)$$

The phase shift in the base current has the simple property of transforming resistances in the base to reactances when viewed from the emitter and vice-versa. This is illustrated in the two circuits in Figure 8.

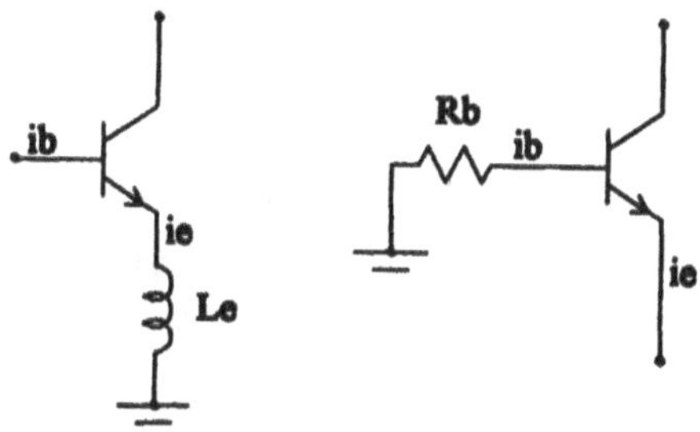

Figure 8 Impedance transformations through a bipolar transistor.

In the case of the emitter inductor, the voltage drop across the inductor is

$$V_L = i_e\, j\omega L_e = i_b\left[-j\frac{\omega_T}{\omega}\right] j\omega L_e = i_b\, \omega_T L_e \qquad (6)$$

This voltage drop when viewed at the base terminal is therefore equivalent to that of a resistor in the base of value $\omega_T L_e$.

In the case of the resistance in the base the voltage drop across the base resistor is

$$V_R = i_b R_b = i_e\left[j\frac{\omega}{\omega_T}\right] R_b = i_e\, j\,\omega\frac{R_b}{\omega_T} \qquad (7)$$

Thus, the resistor in the base appears to behave like an inductor of value R_b/ω_T when viewed from the emitter.

These two simple results can have some useful consequences. The second example provides a means of providing a small signal active inductor with very little power cost. Such an inductor can be used to improve the gain bandwidth of a small signal gain stage. It also gives an indication of the problems that can occur when an emitter follower drives into a capacitive load as the output inductance will form a series resonance with the load capacitance.

The first example can prove useful if it is important to increase the input impedance of a bipolar transistor. It has the advantage over adding extra series base resistance that it does not degrade the voltage noise contribution of the device. These are just two examples of impedance transformations that are inherent in the high frequency operation of a bipolar transistor. It is left to the reader to investigate others such as capacitive loading of the base input which can appear like a negative resistance at the emitter.

Further points in the design of a bipolar RF circuit are brought out in the example of an LNA first gain stage below.

9.3.3.3 The design of a bipolar LNA gain stage

Figure 9 shows a typical bipolar input stage for an LNA omitting the details of biasing for simplicity. The major considerations in the design of such a stage are

1. Matching to a source impedance, R_s
2. Ensuring low noise contribution from the main transistor
3. Ensuring enough gain bandwidth to minimise noise from later stages.

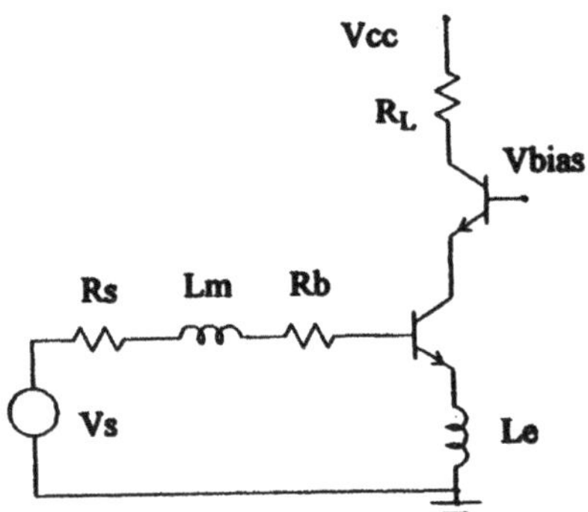

Figure 9 Input stage of an LNA.

To match the source impedance we firstly note that the input impedance of the transistor comprises the base resistance R_b shown explicitly in Figure 9 and the emitter-base capacitance C_π. The emitter inductor appears as a resistance in the base circuit (as discussed above) thus for conjugate matching we require:

$$\omega L_m = \frac{1}{\omega C_\pi} \quad \text{and} \quad R_s = R_b + \omega_T L_e \tag{8}$$

Thus over a finite band we can match the input impedance by choice of L_e and L_m. Part of the matching inductance can be formed by the bond wire to the base, while the emitter inductance may be formed by bonding the emitter to the external ground plane. In this case the problems of common impedance on the signal input ground will also be reduced.

To minimise the noise contribution of the transistor we must first note that if $R_b = R_s$ then its thermal noise will equal that of the source resistor and the noise figure of the amplifier can never be better than 3 dB. Thus the first goal is to choose Tr_1 to have a low R_b (cf R_s) which usually requires a reasonably large transistor. The bias current of the transistor is now selected so that the balance between input current shot noise at high currents, and second stage noise contribution at low currents is found. This is best done by simulation as the optimum current can be found whilst taking into consideration bias circuit contributions to the noise. If the noise figure and current consumption are not acceptable then a transistor with a different R_b needs to be selected and matching by impedance transformation needs to be considered.

To maintain the gain bandwidth product it should be noted that Tr_1 acts as a current amplifier whilst Tr_2 acts as a common base, voltage gain stage. The ideal power gain of the circuit into the load resistor can therefore be written as

$$P_{gain} = \left(\frac{f_T}{F}\right)^2 \cdot \frac{4R_L}{R_s} \qquad (9)$$

It is therefore important to maintain an f_T for Tr_1 as high as possible. This requires both a high operating current and also the elimination of Miller feedback via the collector-base capacitance. The latter has been achieved by the use of the common base, cascode stage. It is also important to increase R_L as much as possible. This, however, is limited by the shunting impedance of the output of Tr_2. Again using a cascode transistor, which need not be as large as Tr_1 as it does not require a low base resistance, minimises the collector capacitances which shunt R_L and cause gain reduction at high frequencies. The gain performance could be further improved if the collector capacitances could be cancelled by resonating with an inductor. The inductance values required are, however, too large to implement in integrated form.

Whilst the above points cover many of the details of the design of an LNA using bipolar devices, there are other considerations. The dynamic range of the LNA will also be influenced by its compression point (CP1), the point where the input is sufficiently large that non-linear device behaviour distorts the gain. This will be controlled by the bias current of Tr_1 and must be taken into consideration in the design trade-offs.

An alternative means of defining the input matching whilst still maintaining a low R_b is to use shunt feedback as shown in Figure 10. Whilst the gain of the amplifier is high, R_f can be kept high and its noise contribution will remain small. This approach provides a more broad band match.

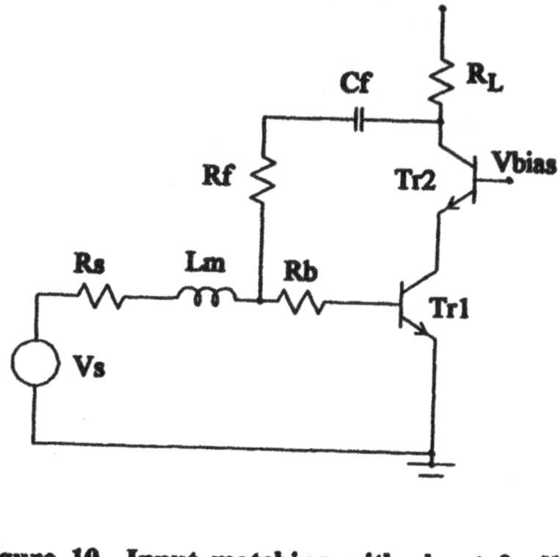

Figure 10 Input matching with shunt feedback.

In addition to the above comments on the design of the on chip components of an LNA it is essential to remember the problems referred to earlier about interactions with the package. In particular, the problems with common ground and supply impedances caused by shared bond wires. Careful consideration should always be given to the ground path for the signal input and to on chip power supply decoupling capacitors which minimise the high frequency power supply currents which flow in common impedances.

9.3.4 GaAs Design

9.3.4.1 Design route

Many RF designs begin as a hybrid using discrete components. This allows a design to be fabricated and adjusted relatively easily and quickly. If the circuit is part of a system this also allows early system evaluation. As the production volume increases, the circuit is often replaced by a monolithic circuit to reduce cost, or the physical styling and weight of the equipment may require the circuit to occupy less space making a monolithic circuit essential.

As a substrate, the characteristics of GaAs are similar to alumina, which often allows easy translation of a hybrid design to GaAs. Unfortunately, it is not so easy to adjust a monolithic circuit and each iteration is both costly and time consuming. Understanding and simulation of the circuit must therefore be accurate and complete before fabrication.

9.3.4.2 Where has GaAs been?

GaAs discrete devices were used in the late sixties for microwave amplification but most research and investment was aimed at the potential large volume digital applications. The first GaAs integrated circuit appeared in 1974 using a configuration called Buffered FET Logic (BFL). These early circuits used crude material technology and liquid phase epitaxy. By the late seventies rapid advances in technology and the replacement of epitaxy by ion implantation produced the first GaAs LSI circuit of just over 1000 gates in Schottky Diode FET Logic (SDFL), achieving an order of magnitude improvement in power consumption. The need for level shifting between gates was removed in the early eighties by replacing the depletion mode FETs with enhancement mode FETs giving Direct Coupled FET Logic (DCFL). Today's digital GaAs circuits have up to 350 k gates and are twice as fast as a BiCMOS IC for the same power requirement.

Digital circuits were expected to lead to increased volume and lower costs but today many vendors have abandoned GaAs and more than 75% of remaining foundry capacity is devoted to analogue circuits of which 35% is devoted to telecom products. Although wafer size has now reached 4 inch, silicon wafers are now 8 inch.

9.3.4.3 Transistors

The advantage of GaAs over hybrid substrates is that transistors may be fabricated in it. Furthermore, if the main current carrier in these transistors is electrons, then the mobility, a measure of electron speed and f_t, is around 5 times faster than in silicon. The three most common transistors in GaAs are:

- MESFET : Schottky gate FET.
- HEMT or TEGFET or MODFET : Similar to MESFET but with channel conduction taking place in an undoped layer giving even higher mobility, particularly at low temperature.

- HBT : Bipolar transistor with highly doped emitter and base giving very low base resistance.

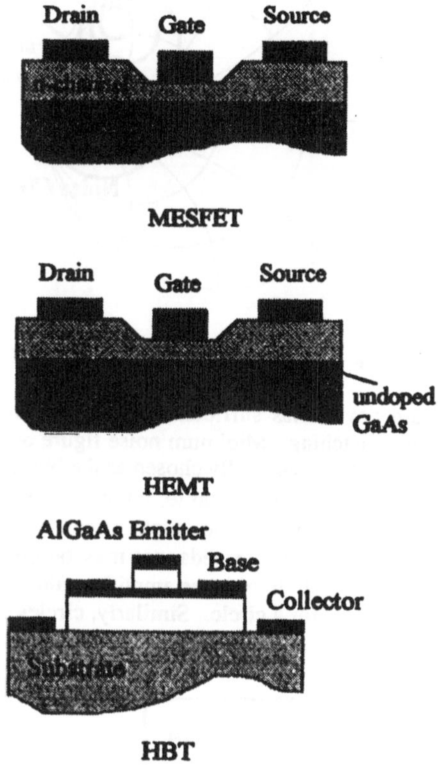

Figure 11

Of these the MESFET is the most mature and easiest to fabricate reliably and is therefore the active element in most GaAs monolithic circuits. A useful feature of the MESFET is the ability to customise its characteristics to suit a particular application by adjusting the gate recess depth and the implant profile. A typical profile for a small signal FET would give a high transconductance g_m, and maintain the high g_m down to low drain current values. A power device profile will be designed to give increased drain current while maintaining a high breakdown voltage. In addition the usual design freedoms of the MESFET are the gate length, which determines the f_t, and the gate width that determines the g_m and maximum drain current.

9.3.4.4 Typical design

Restricting the circuit to operate in a linear mode (small signal) removes the need to model the FET non-linear characteristics, and the circuit may be designed using a 2 port *s*-parameter representation of the FET. These parameters are often supplied for a range of bias conditions. Of particular interest is the Low Noise Amplifier (LNA) because there is an additional challenge to minimise the noise figure as well as maximising the power gain.

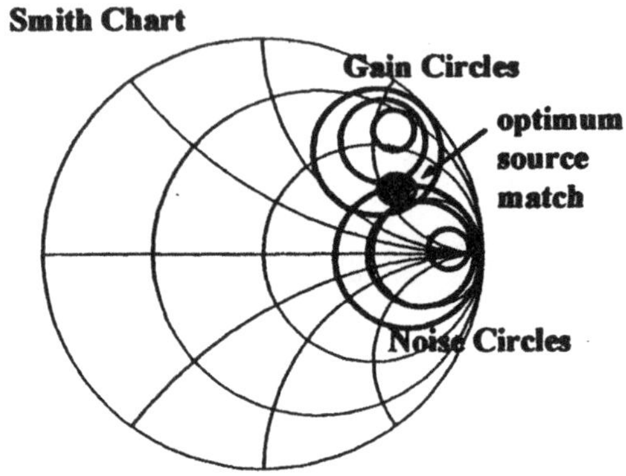

Figure 12 Source matching for LNA.

A 300 micron gate width device has sufficient transconductance and reasonable input impedance to allow simple matching. Minimum noise figure occurs at around 20% of the open channel current I_{dss}, so this is generally chosen as the bias point. Firstly it is useful to ensure that the device is unconditionally stable, which normally FETs are not. This is usually corrected by adding resistance to the output circuit or by adding inductance from source to ground. As a graphical aid, impedances may be plotted on a Smith chart. If source (input) impedances that result in the same amplifier gain are plotted on this chart, then conveniently they are found to form a circle. Similarly, circles of constant stability factor and noise figure may be plotted.

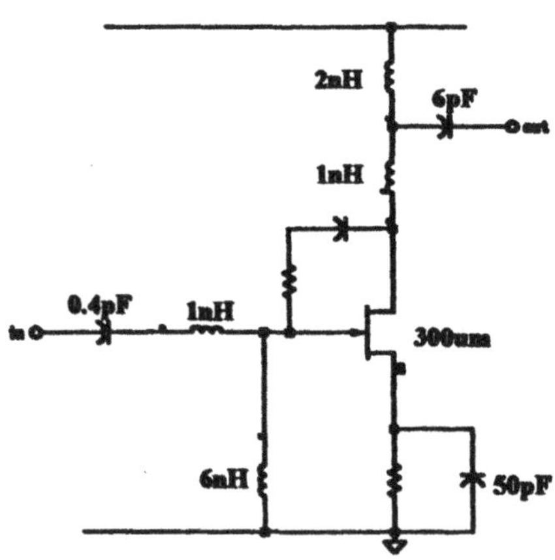

Figure 13 Typical GaAs LNA.

Typical noise and gain circles are shown in Figure 12, where it can be seen that a compromise impedance must be found between that for optimum noise and that for optimum gain. A matching circuit must now be designed from this impedance to the external impedance, usually 50 Ohms. For narrow band circuits the match can be achieved with simple *LC* networks as shown in Figure 13. A similar design is done for the output network. At 2 GHz these matching networks are implemented as lumped spiral inductance and nitride dielectric capacitors. At higher frequencies they may be implemented as transmission lines rather than *LC* networks with similar results. Bias may be simply arranged by a resistor in the source.

Figure 13 also illustrates the use of low loss reactive components for gain and impedance control, this avoids the use of resistors which dissipate signal power and limit dynamic range.

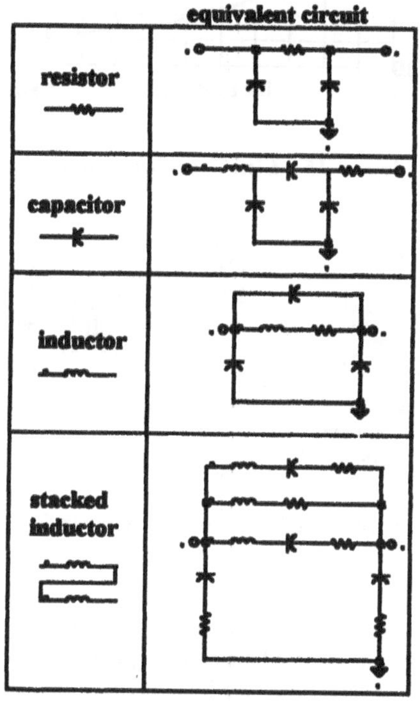

Figure 14 Equivalent circuits.

Although shown as lumped element components, *R*, *C* and *L* in reality should include their parasitics as shown in Figure 14. This considerably increases the complexity of the design but still does not include perhaps the most important, and the most difficult parasitic to assess, that of coupling. Coupling can be one of the most critical issues, particularly when significant gain also exists between the coupled elements. Figure 15 shows the magnitude of the coupling for two 1 mm long parallel lines on a 200 micron GaAs substrate.

This coupling can, however, be an advantage, allowing us to fabricate couplers and transformers on GaAs. For example, Figure 16 shows the use of the Lange quadrature coupler where reflected power is dissipated in the load resistors. This configuration avoids the additional constraints of matching once we have optimised the amplifier gain and efficiency. Another configuration that combines FETs and transmission lines is the distributed or travelling wave amplifier. This technique, originally used in the 1940s, absorbs the FET capacitances in a series of artificial transmission line sections allowing very wide band operation.

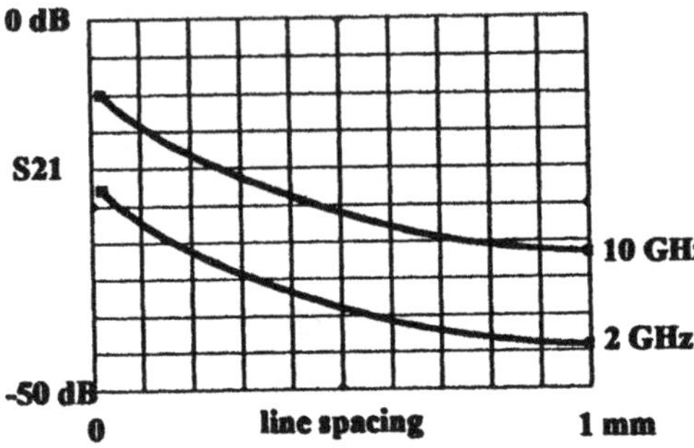

Figure 15 Line coupling on GaAs.

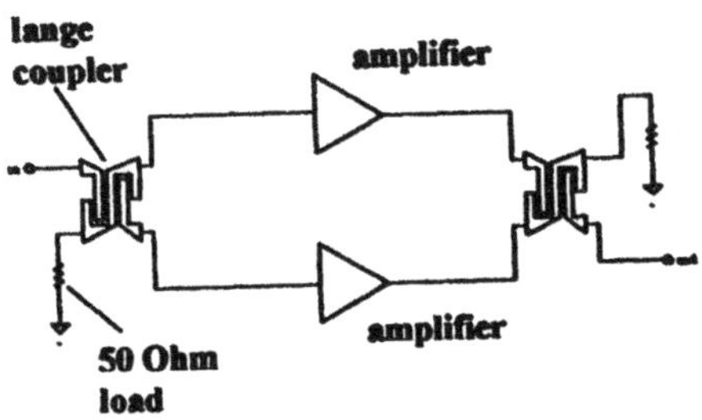

Figure 16 Balanced amplifier configuration.

9.3.4.5 *Oscillators*

The FET may be induced to oscillate in the normal way by coupling the input and output ports by series or parallel elements. This coupling may also be achieved using low cost dielectric resonators giving very good stability and much reduced noise. However, use of FETs in low phase noise oscillators is limited by the 1/*f* or flicker noise, the GaAs MESFET being significantly worse than bipolar devices.

9.3.4.6 *Construction*

Many of the features mentioned, particularly couplers, are limited by die size to frequencies greater than 10 GHz. The same circuit techniques may be used at lower frequencies by fabricating couplers off chip, typically on ceramic as part of the package.

The balance between reducing chip area, and hence material cost, and simplifying the design task by increasing chip area and reducing interface parasitics is a constant challenge to the RF designer. Any permutation between totally discrete and totally monolithic is possible and frequently the RF designer must consider all aspects from system architecture to IC design to achieve the required cost, size, and time to market.

9.3.4.7 *CAD tools*

Today it is possible to model and simulate most aspects of RF design. The problem has become one of deciding which effects and parasitics to include in the simulation to remain within the constraints of memory and simulation time. A number of two and three dimension field simulators are available to model abstract geometries, while non-linear simulators, using harmonic balance, can now model such effects as frequency translation of noise all the way through to full circuit and system effects on the error rate of complete digital radio systems.

It is, however, still necessary to exercise considerable care when using these tools, particularly in the area of FET models. Unlike bipolar modelling that is extensive, mature and restricted to one model, every GaAs vendor has their favourite model, usually a modified Statz, which generally only covers the saturated operation area with any accuracy. Operation with low drain to source voltage, forward gate bias or with drain and source reversed as in switch operation, has to be approached with an in depth understanding of the model implementation.

References

[1] C. P. Wen, *IEEE Trans MTT-16*, pp 1087-1090, 1969.

[2] *HP Application Note 154*, Available from Hewlett Packard.

[3] R. Hadaway *et al*, *Microelectronics Engineering*, Vol. 15 pp 513-516, 1991.

[4] Antognetti and Massobrio, *Semiconductor Device Modeling with Spice*, McGraw-Hill, 1987.

Further reading

[1] R. S. Pengelly, *Microwave Field-Effect Transistors-Theory, Design and Applications*, John Wiley & Sons Inc., 1986.

[2] G. D. Vendelin, A. M. Pavio, U. L. Rohde, *Microwave Circuit Design Using Linear and Nonlinear Techniques*, John Wiley & Sons Inc., 1990.

Chapter 9.4

Introduction to Spread Spectrum Techniques

Peter E. Chadwick

9.4.1 Introduction

Spread Spectrum techniques have become very popular in recent years. Originally developed and used for military purposes, the popularity of these techniques is mainly due to the release in the US by the FCC of various ISM (Industrial, Scientific and Medical Bands) for free, unlicensed use, using Spread Spectrum techniques. The decision to allow spread spectrum techniques was taken on the basis of freedom from interference to other services using those bands. This may however, not prove achievable in practice, however, and only time will tell.

9.4.2 What is spread spectrum ?

Spread Spectrum (SS) Modulation systems are ones in which the bandwidth of the radiated signal exceeds the information modulation bandwidth because of the addition of a spreading signal. This is done to provide anti-jamming features, to reduce the radiated power density of a transmission so that interception or the probability of causing interference is reduced, or to provide a high degree of security. In general, spread spectrum systems are used for the transfer of digitally modulated signals, although this is by no means a *sine qua non*; several of the tactical military systems are able to transmit analog signals in spread spectrum form.

9.4.2.1 Types of spread spectrum

There are a number of forms of spread spectrum in general use: these include Fast Frequency Hopping (FFH), Slow Frequency Hopping (SFH or FH), Direct Sequence (DS), and Chirp. All have their own advantages and disadvantages, and some are in very widespread use.

(i) Frequency hopping systems

In these systems, the frequency of operation of the radio system varies over the band, hopping between various defined channels according to a predetermined, and generally pseudo-random pattern.

FAST frequency hoppers are generally defined as those in which each bit of a digitized transmission is sent on one or more channels, while SLOW frequency hoppers send at least two bits on each hop. The main advantage of the Fast Frequency Hopper (FFH) is that because each bit is sent on more than one frequency, the effects of multipath fading (which is frequency dependent) are mitigated. However, especially at high bit rates, the technological requirements for implementation are by no means negligible.

Furthermore, FFH techniques start to blend into the Direct Sequence (DS) techniques in terms of jamming susceptability (Figure 1).

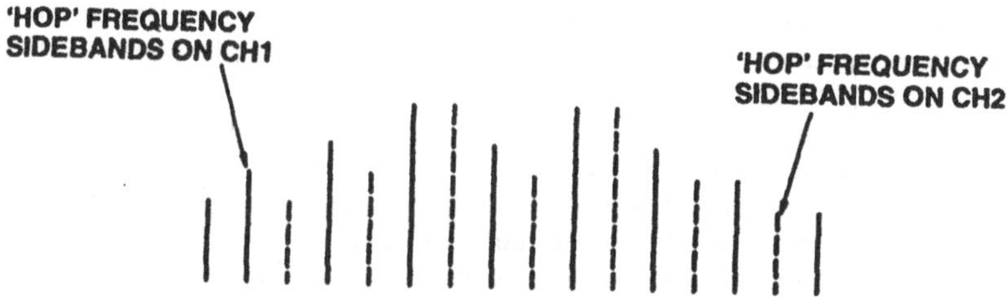

Figure 1 Fast frequency hoping.

SLOW frequency hoppers exist in very large numbers, both in the military and commercial fields. Hop rates of between 10 and 400 hops per second are common, and military systems such as the Racal JAGUAR-V (JAmming GUArded Radio), and SINGCARS (SINGle ChAnnel Radio System), both of which are tactical VHF systems operating in the 30 - 88 MHz band, and HAVE-QUICK, a tactical airborne equipment operating in the 225 - 400 MHz band, are widely deployed. Commercially, there is a multiplicity of products available for the US 900 MHz and the 2.4 GHz ISM band: the GEC-Plessey Semiconductors DE6003 is an example of the latter variety (Figure 2).

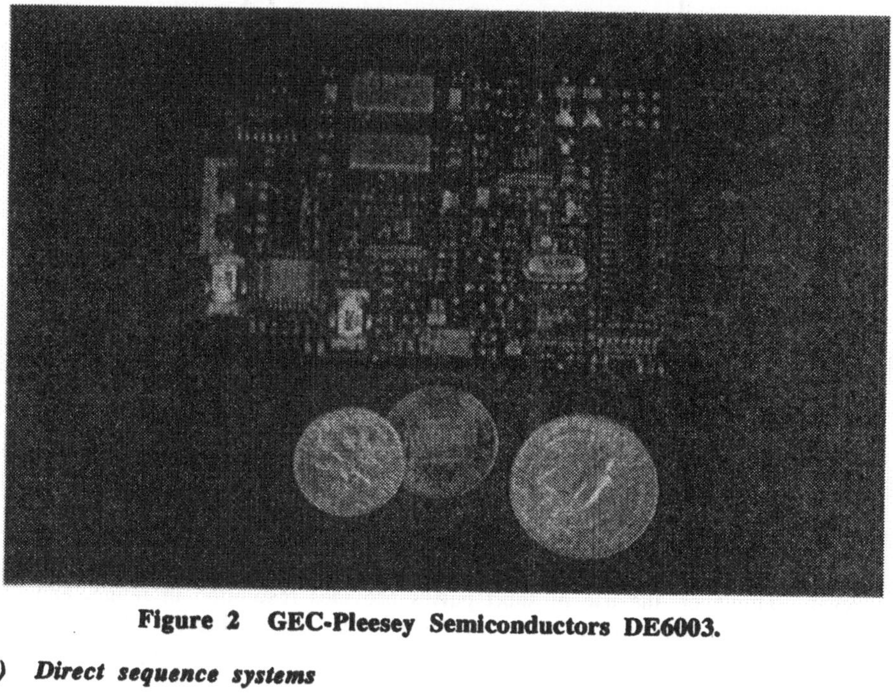

Figure 2 GEC-Pleesey Semiconductors DE6003.

(ii) Direct sequence systems

In these systems, the RF signal, (modulated in the simplest form by the information) is then modulated by the Spreading Sequence. This modulation is usually of BPSK (Bi-Phase

Shift Keying, as shown in Figure 3) or QPSK (Quadrature Phase Shift Keying), or a variant thereof. The Spreading Sequence is a pseudo random digital bit stream of clock rate (usually referred to as the Chip rate) much greater than the information rate, and the choice of the Spreading Sequence is important if the maximum performance of the spread spectrum system is to be obtained. In practice, the spreading sequence and the digital modulation are usually combined by such techniques as modulo 2 addition, and the resulting bit stream is used to modulate the carrier. In the receiver, the use of the same bit stream, suitably synchronized, modulating the local oscillator of the receiver, despreads the received signal to one in which, in the simplest case, the original modulated signal is left at the receiver IF; the more usual practical case leaves the modulated signal to be demodulated in the usual way. Where the spreading signal or code does not match the transmitted code, then the resultant output is still noise, and the lack of correlation between successive chip bits means that the receiver merely sees noise. As the degree of correlation increases, the level of signal rises; lack of correlation degrades received signal to noise ratio. Obviously, by using different codes, a number of stations are able to occupy the same radio frequency simultaneously, and this is known as CDMA (Code Division Multiple Access). The choice of codes is obviously important, as for minimum interference, the codes chosen need to have minimum auto-correlation between them, while to achieve synchronisation readily, it is important to have as little correlation between parts of the same sequence as possible. The ratio of the Chip Rate to the information rate gives, to a first order, the ratio of the spectrum width over which the signal is spread, and thus the Spreading Gain.

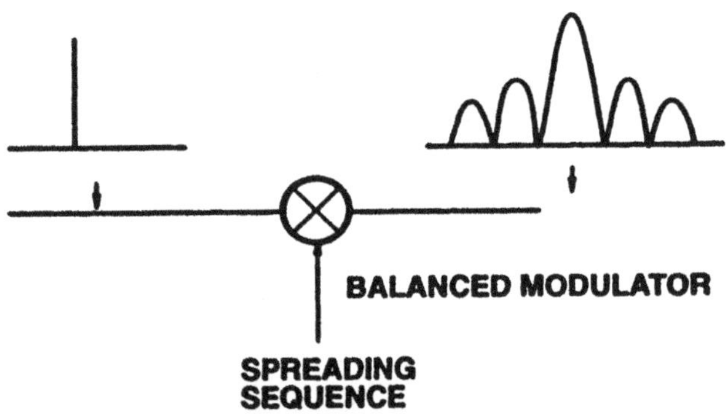

Figure 3 BPSK direct sequence generator.

(iii) Chirp systems

These are systems in which the frequency transmitted varies during the bit period in (generally) a linear manner. They are not used very much for communications, but see maximum use in radar systems, where they can provide pulse compression. In a radar, the sensitivity or detection probability is dependent upon the amount of reflected energy from a target; the longer the transmitted pulse incident upon a target, the greater the returned energy. However, the use of short pulses provides resolution in range, and these incompatible requirements can be resolved by chirping the radar frequency during the pulse. In the receiver, the received signal is fed via a dispersive delay line, usually a SAW filter, in which the delay of signals through the filter is dependent upon frequency. By arranging the filter so that the speed of the chirp or frequency sweep matches the dispersion of the filter, a very narrow pulse of energy appears at the output of the filter. In the presence of wideband jamming, the integration of the wanted energy by this method improves the signal to jam ratio, while the jamming signals do not correlate, and are thus rejected. Chirp receivers can use other forms of correlation than SAW filters, and reference can be made to various texts on radar. Chirp spread spectrum is not to be confused in this respect with chirp modulation, in which such possibilities as chirping the frequency up for a '1' and down for a '0' exist, or in triple chirp modulation [1].

9.4.2.2 Spreading gain

Spreading gain is an important concept in any spread spectrum system, for without it, the whole *raison d'etre* of spread spectrum is in doubt. In terms of frequency hopping (FH) systems, the existence of spreading gain is somewhat dependent upon the outlook adopted in analysing the system, but in Direct Sequence (DS) systems, the concept is fairly straightforward.

Starting with DS system, it has been shown how spreading may be achieved by modulating the signal with BPSK at the chip rate, and despread by modulating in the receiver chain. An interfering signal at the receiver will be spread by spreading sequence, and the amount of power in the receiver bandwidth after de spreading or correlating will fall by at least the ratio of the post correlation bandwidth to the pre-correlation bandwidth. Where the interferer is wide band signal, it will become even wider, eg. a 1 MHz wide interferer in a system with a 5 MHz spread will become spread over 6 MHz, thus reducing its interfering properties even further. The ratio of pre to post correlator bandwidth is the **Processing Gain**; it is not completely realisable in practice because of losses in implementation.

In the FH system, the concept is somewhat more difficult. For any given total interfering power, then the interference or jamming is reduced if the total power is spread over all the channels, and the concept of processing gain still applies in the same way. If a number of channels are jammed, then the the concept still applies, but it is not so straightforward in that by avoiding the jammed channels (or even stopping hopping and staying on a clear channel) the interference disappears, and under those circumstances, it could be argued that the processing gain becomes infinite!

9.4.3 Why spread spectrum ?

With the ever increasing demand for RF spectrum, it appears *prima facie* unacceptable to use more spectrum than the minimum possibly required to transmit a signal. However, this is not necessarily the case, especially when it becomes necessary to look at 'overlaying' new systems into bands which are already allocated to, and in use by, other services. If we consider the situation where a 1000 Watt signal is spread over a bandwidth of 1 MHz, then the power density in Watts/Hz is 1 mW/Hz, and in the 3 KHz bandwidth occupied by say an analog SSB channel, this only represents a total power of 3 Watts, and if the SSB signal is transmitted at 100 Watts, then all things being equal, the resulting signal to noise ratio is of the order of 15.5 dB. In the case of frequency hopping, if the spread spectrum transmitter hops at a rate of 200 hops/sec over a range of 1000 channels, it will take at least 5 seconds to visit all channels, and assuming each channel gets visited equally often, will see interference appearing on any particular channel only for about 5 ms in 5 seconds. It has been suggested that spread spectrum systems allow more stations per MHz than other systems, but this is fallacious. Taking the case of a frequency hopping system for example, in which 1000 channels are available, then when 1000 stations are attempting to use the channels, there is no advantage over 1 channel per station. In the case of direct sequence, the situation is slightly different, in that the total number of signals is limited by the processing gain as well as the number of codes available. If it is assumed that there are 1025 codes available, in which the cross and auto correlation functions are zero, then 1024 unwanted stations will raise the noise floor by 29 dB. If the spreading gain is 30 dB, then the available theoretical signal to noise ratio becomes 1 dB, assuming that all stations are received at the same level. So if a 25 KHz wide channel is assumed, with a spread of approximately 26 MHz, the signal to noise ratio achieved is 1 dB; yet if the channels were contiguous in a non-spread system, each would be independent and to a first order, suffer no interference. These cases serve to illustrate that the simpler forms of spread spectrum do not increase spectrum utilisation, while in the case of CDMA, requiring some control of signal levels to even achieve useability.

9.4.3.1 Applications

Spread spectrum systems have already been mentioned in terms of specific military and commercial equipments. The major civil uses currently are local area communications links, such as wireless LAN's, point-of-sale communications, warehouse stock control, etc, and digital cellular radio. Navigation is an important user, although the systems used (Global Positioning System, Glonass) are primarily military in design and original purpose. The navigation use is derived from the fact that if a DSSS signal is received and correlated, the time difference between the receiver's master clock and the transmitters master clock is measurable. This gives ranging information, and if more than one transmitter's range can be determined, then position can be derived. If three transmitters are used, then there are sets of hyperbolas of constant time difference betwen the arrival of signals from any two of them, and these hyperbolas may be plotted on a map. This is the well known 'hyperbolic' navigational system that has been in use for over fifty years. Obviously, the resolution achievable will be dependent upon the chip rate, being higher with higher chip rates. Probably the best known of these systems is the NavStar or GPS satellite navigation system. Another system, which *prima facie* has no excuse for existence is GEOLOC, which is in the 2 MHz region, with a main modulation lobe 700 KHz wide, giving limited range resolution.

9.4.4 The near-far problem, jamming and interference

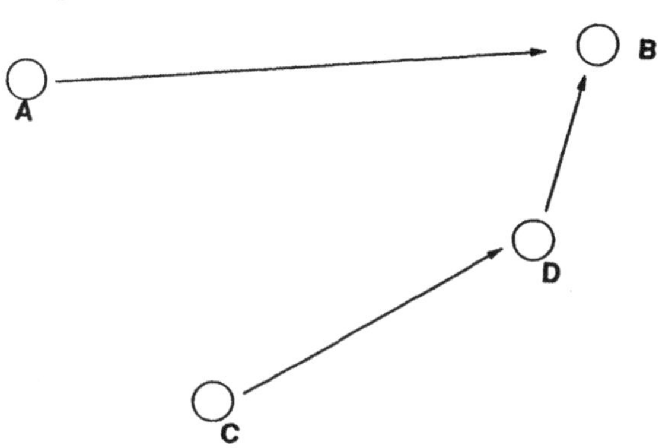

Figure 4 Near far problem.

It has been shown earlier that the fundamental requirement of a spread spectrum system is spreading gain. Indeed, without spreading gain, the system will inevitably fail in the presence of interference. The major question is 'how much interference?' Consider the situation where two systems exist close together (Figure 4). Station *A* transmits to station *B*, while station *C*, physically closer to *B* than *A* is, transmits to *D*. This means that station *B* receives *A*'s signals at a lower level than station *C*'s signals. Indeed, the signals from *C* could well be totally unwanted.

9.4.4.1 Direct sequence systems

In the case of a DSSS system, the received signal to noise ratio will be degraded by the presence of *C*, and when the difference in signal levels exceeds the spreading gain, the system will fail entirely. In practice of course, failure will occur as soon as the (S+N)/N ratio drops below some predetermined level: this may well be determined by the bit error ratio in a digital system. This problem limits the immediate application of DSSS systems in an uncoordinated manner, and in the cellular radio application, a lot of effort is involved in providing power control to maintain the received signals at the cell site receiver identical in level. If a system has a 20 dB (S+N)/N requirement, and 20 dB of spreading gain, then if the unwanted signal is equal in level to the wanted, there will be no fade margin available. In practice, this would happen at a lower level of interference because of the inevitable inefficiencies of the system. It is this problem which provides the biggest challenge to DS systems in networks which are orthogonal. Work has been done in attempting to minimise the problems, and in certain limited cases, improvements can be achieved. These cases have been particularly analyzed where the unwanted signals are DS signals of known characteristics - modulation, chip rate and spreading sequence. Under these circumstances, the system can be arranged to subtract the unwanted signal from the wanted one, but of course, the difference in levels has still to be such that the receiver has adequate instantaneous dynamic range, and can provide the necessary S(+N)/N ratio on the weaker, wanted signal [2, 3].

9.4.4.2 Frequency hopping systems

In the frequency hopper, the near-far effects are dependent upon the hop rate: the fast hopper can suffer in a way similar to the DS system if bit rates are high enough. The slow hopper will suffer only when the 'near'signal is high enough to cause gain compression in the receiver, or map enough receiver phase noise on to the channel to degrade the (S+N)/N, or if the transmitter phase noise is high enough to raise the r_X noise floor, or if the overall rejection of unwanted signals (receiver selectivity) is inadequate.

In the fast hopper, the situation is different. Because the receiver is changing frequency, it is effectively seeing an FM signal as the local oscillator input, and this signal has sidebands spaced at the hop rate. The number and absolute level of these sidebands will be dependent upon the modulation index (β), i.e. the distance in frequency between hops and the hop rate. These sidebands will have exactly the same effect on the rejection of unwanted signals as phase noise does: the reciprocal mixing performance of the receiver will be reduced. This phenomena limits the performance of fast hopping systems, and as stated earlier, the fast hop system gradually merges into a DS system as the frequency of the hop increases.

9.4.4.3 Jamming

The resistance of a communications circuit to deliberate jamming is of prime importance in military systems. However, in the commercial world, although deliberate jamming to produce disruption of communications is unlikely, the use of non-regulated bands produces a similar scenario in real operation, and the effects of high jam resistance in a system are precisely those which maximize spectrum efficiency by reducing the physical distance between stations using the same radio frequency without mutual interference. In this characteristic, slow frequency hopping systems using FM as their prime modulation method have the advantage of being able to use the Capture Effect of FM to minimise interference, although this can be a two edged sword as interference levels rise to the capture effect knee. In the DS system, an interfering signal is spread in spectrum to at least that produced by the chipping rate: wideband signals are spread more than narrow band signals, which thus reduces the power per unit bandwidth and thus the effects of the jamming. Therefore, it can be seen that the most effective jammer of a DS system is a single carrier, while in a frequency hop system, the most effective jammer (leaving aside 'follower jammers') is a broadband signal. In the fast frequency hop case, it has been shown that the effects of reciprocal mixing are not negligible, and the single carrier will cause a degree of jamming dependent upon the effective of the receiver LO.

Narrow band signals of known frequency can be hardware filtered if their existence and level can be pre-determined, but where a number of such carriers exist, this becomes impractical. It is this susceptability to jamming by a single carrier of level exceeding the wanted by somewhat less than the spreading gain of the system which leaves an undesirable vulnerability in a public telecommunications system, such as digital cellular radio.

(i) Mutual jamming by DS and FH systems

In the case of the unregulated bands which are currently the focus of so much commercial interest, consideration needs to be given to the effects of mutual jamming by disparate orthogonal systems. *Prima facie*, it would appear that the DS system will be interfered with more destructively than will the slow FH system, but this has been shown to not necessarily be the case [4]. However, practical DS systems are limited in their attainable spreading gains, and this reduces the available jamming rejection. The major advantage of the DS system in such frequency bands is the relative insensitivity to multipath propagation, although the use of suitable coding schemes can improve the FH system performance such that it is overall more attractive. Again, the use of modulation schemes for FH systems such that the benefits of capture effect can be exploited is highly desirable.

(ii) Jamming of narrow band systems by FH and DS systems

As one of the primary reasons for the use of FH and DS systems is the Low Probability of Intercept in the military environment, and the low probability of causing interference in the commercial environment, this may appear at first be trivial. However, some consideration requires to be given where Spread Spectrum (SS) systems are to be overlaid onto frequencies in use by existing services. The FCC Part 15 rules, which historically opened the door to unlicensed SS systems addresses the problem by making a series of arbitrary assumptions. For FH systems, the maximum time for which the system could dwell on a channel was set at 400 ms, which had been found empirically to be a value which in the majority of cases, is not long enough to open the squelch on a narrow band FM communications radio. Additionally, transmissions using FH must not exceed a width of 1 MHz when measured at points -20 dB relative to the frequency of maximum power. On DS systems, the problem is one of the power received in the subject receivers passband, and if, for example, a 1 Watt DS system is spread over 10 MHz with an average power of 0.1 μW/Hz, then the maximum power that can be received in a 15 KHz wide passband receiver is 1.5 mW. With a path loss of 60 dB between receiver and transmitter, the noise floor of the victim receiver is still at around -59 dBm, which may be totally unacceptable: when the attenuation between thereceiver and the DS transmitter is say 100 dB, the victim receiver sees -99 dBm, which may cause negligible loss of service. In the case of an FH system, the interference is a function of the time spent on the victim receivers channel.

9.4.5 Specialised equipment requirements

In terms of RF parameters, the basic requirements of SS systems differ little from those of conventional equipments. Probably the major requirement is where high processing gains are required, the ultimate (S+N)/N ratio of the receiver is important if the processing gain is to be realised. This can have an effect on the allowable phase noise performance in systems where a direct mapping of phase noise on to the received signal takes place. DS transmitters require carrier suppression if they are not to have the problem that an appreciable part of the output power is to be wasted, and this is not always easy to achieve at high frequencies. Typically, double balanced modulators are used, and cascading them can be useful. [4] is the preliminary data sheet for an integrated circuit designed and developed as an experimental approach to this application, although not commercially available.

Regulatory requirements often require that the radiated spectral power density falls within certain limits, and where the carrier in say a BPSK system is inadequately suppressed, the available spectral power density may well be very low in order to meet the overall power allowed.

Frequency synthesizers for FH systems require careful consideration. Practical PLL type synthesizers are currently limited to loop bandwidths of the order of a few hundred KHz at best, and are non-negligible designs to achieve that. This suggests that hop and settling times of much less than 20 - 30 μseconds are unlikely to be readily achieved, so any hop scheme requiring faster hopping necessitate the use of more complex systems, such as Direct Digital Synthesis. It is also highly desirable to switch the transmitter off while hopping, to minimize the amount of general interference caused, but this requirement itself leads to the necessity for careful shaping of the amplitude modulation envelope impressed upon the transmitter output if so-called 'key clicks' are to be avoided.

DS systems require a degree of linearity in both the transmitter PA and receiver IF stages. Although the use of systems such as OQPSK can reduce the amplitude excursions, and thus the linearity requirements, the PA stage cannot be run in saturation, and so consideration needs to be given to the implications of this. Intermodulation in the PA stage will effectively 'broaden' the transmitted signal, and it will become necessary to consider the out of band radiation produced. Although filtering of the signal can be used, the effects of differential group delay on the edge of the filter passband cannot be ignored, as this is effectively the same as multipath propagation distortion introduced at the transmitting end. For the same reason, the transmitter needs to be inherently capable of instantaneous wide band operation, and this can produce difficulties in situations where the antenna tuning has to be considered. In this respect, slow FH systems have an advantage, as if necessary, antenna tuning can be carried out on each hop, or, with the use of a priori knowledge, suitable components switched into circuit.

9.4.6 Coding sequences

The requirements for coding sequences were briefly touched on in section 2. To reiterate, the DS system requires codes which have minimum auto and cross correlations, while the FH system requires a hopping sequence which provides low cross correlation to other sequences in use by orthogonal systems. In order to facilitate synchronization, FH sequences have been used in which certain channels are visited more frequently than the average, but this has certain difficulties in that those channels become more vulnerable to jamming. Adaptive systems have been proposed in which jammed channels are avoided, although the definition of what is a jammed channel is not as easy as would at first sight appear. It is not proposed to examine the spreading sequence choice in any great detail, but given the requirements for low auto and cross correlation in DS systems, the choice of code is very important in maximizing the capability of the DS system. To avoid granularity in the spectrum, short codes should not be used, while it is also desirable that the code is one in which the search time to achieve synchronization (given knowledge of the code) is minimized [5]. It is also desirable that the spreading sequence is one which has a high degree of overall symmetry (low DC bias): long strings of 0's or 1's produces an offset on the balanced modulator when BPFSK is used such that the carrier suppression suffers, and the spread spectrum signal then starts to have discrete lines appearing (granularity).

In FH schemes that are intended to operate in non-benign environments, such as the ISM bands or in ECM situations, consideration needs to be given to error detection and correction codes. Schemes have been proposed [6] in which the basic data is coded using a concatenated code, and then split into 5 segments. These segments are then transmitted on 5 separate channels, such that reception of the blocks on any three channels allows the re-constitution of the original data. However, this method, although extremely robust, has the disadvantage of producing an extremely high overhead - in this particular case, one of approximately three times. Data coding is not a consideration peculiar to spread spectrum systems *per se*, but the choice of FH channel selection codes is dependent upon a number of factors. However, careful consideration is required to optimise the system such that the loss in throughput caused by the overhead is not greater than the loss occasioned by using ARQ (Automatic Repeat reQuest), and merely repeating the data as necessary. This affects the length of the packets used at any given Bit Error Ratio.

9.4.7 Conclusion

The use of Spread Spectrum Radio systems is increasing rapidly, in a wide and diverse number of applications. Certain advantages and disadvantages accrue to the two major schemes -Frequency Hopping and Direct Sequence, both in terms of robustness and probability of interception, but it appears that both will grow in use in particular application areas where communication without cables is required.

References

[1] T. Wysocki, jr. and R. Weber, "Triple Chirp narrow band modulation", *IEE Conference on Personal Mobile Communications*, Brighton, ISBN 0 85296 607 5 ISSN 0537-9989, pp. 90 *et seq*, 1993.

[2] L. A. Rusch and H. V. Poor, "Narrowband interference suppression in spread spectrum communications via multiuser detection techniques", *Ibid.*

[3] U. Mitra and H.V. Poorr, "Analysis of an adaptive decorrelating detector for synchronous CDMA channels", *Ibid.*

[4] SL801 Data Sheet.

[5] Robert C. Dixon, *Spread Spectrum Systems*, Wiley , ISBN 0-471-88309-3, 1984.

[6] E. Geiger, "An RF Data Transport Protocol. The RF Adaption Sublayer and RF Physical Layer Specifications for Slow Hopping Spread Spectrum Radio LANs", IEEE 802.11 Committee paper, DOC.IEEE.P802.11-93/104, July 1993.

Chapter 9.5

Taking the Mystery out of RF Design: Obey the Law!

James Bryant

9.5.1 Introduction

Practical RF design, the successful construction of a circuit which has been analyzed to death in its theoretical design phase, has a reputation of being a black art, which can only be acquired by the highly talented, and then only at the price of infinite study, and possibly the sacrifice of a virgin or two. In fact RF circuitry obeys very simple laws, the ones we learned in the nursery: Ohm's law; Kirchoff's law, Lenz's law and Faraday's laws. The problem, however, lies in Murphy's law.

RF CIRCUITRY OBEYS BASIC LAWS

Ohm's law
Kirchoff's law
Faraday's laws
Lenz's law
&
Murphy's law

Figure 1 RF circuitry obeys basic laws.

Murphy's law is the subject of many engineering jokes, but, in its simplest form "If Anything Can Go Wrong - It Will!" it states the simple truth that physical laws do not cease to operate just because we have overlooked or ignored them.

MURPHY'S LAW

Whatever can go wrong, will go wrong.
[Failsafes don't]
[Interchangeable parts aren't]
[After working well for a week it will fail when you call your boss to see it]

The basic principle behind Murphy's law is that all physical laws always apply - when ignored or overlooked they do not stop working.

Figure 2 Murphy's law.

At high frequency there are plenty of effects which are easy to overlook, but if we adopt a systematic approach to circuit layout it is possible to consider all likely causes of circuit malfunction without wasting much time.

In this chapter we shall consider some simple issues which are likely to affect the success of RF (and, indeed, LF precision analog) designs, namely resistance (including skin effect and the non-ideal characteristics of resistors), capacitance (including the behavior of real capacitors), inductance (both self inductance and mutual inductance), noise, and the effect of current routing.

ISSUES AFFECTING SUCCESSFUL RF LAYOUT

Resistance
Capacitance
Inductance (self- and mutual)
Noise
Grounding and decoupling

Figure 3 Issues affecting successful RF layout.

9.5.2 Resistors

As an applications engineer I shall be relieved when room-temperature superconductors are finally invented, as too many engineers suppose that they are already available, and that copper is one of them. The assumption that any two points connected by copper are at the same potential completely overlooks the fact that copper has a definite resistivity and that the resistance of copper tracks on PCBs is frequently large enough to affect analog and RF circuitry, although it is rarely important to digital circuits.

CONDUCTORS ARE NOT *SUPER*CONDUCTORS

Consider 10 cm of 1 mm PC track

Standard track thickness is 0.038 mm
ρ for copper is $1.724 \times 10^{-6}\ \Omega$ cm @ 25°C
∴ PCB sheet resistance is 0.45 mΩ/sq
Resistance of the track is 45 mΩ
THIS IS ENOUGH TO MATTER!

Figure 4 Conductors are not superconductors.

The diagram in Figure 4 shows the effect of copper resistance at DC and LF At HF matters are complicated by "skin effect". Inductive effects cause HF currents to flow only in the surface of conductors. The skin depth (defined as the depth at which the current density has dropped to 1/e of its value at the surface) at a frequency f is $1/\sqrt{\mu\sigma\pi f}$ where μ is the permettivity of the conductor, and σ is its conductivity in Ohm-metres. $\mu = 4\pi \cdot 10^{-7}$

henry/metre except for magnetic materials, where $\mu = 4\mu_r\pi\cdot10^{-7}$ henry/metre (μ_r is the relative permettivity). For the purposes of resistance calculation, in cases where the skin depth is less than one fifth the conductor thickness, we can assume that all the HF current flows in a layer the thickness of the skin depth, and is uniformly distributed.

SKIN EFFECT

At high frequencies inductive effects cause currents to flow only in the surface of conductors.

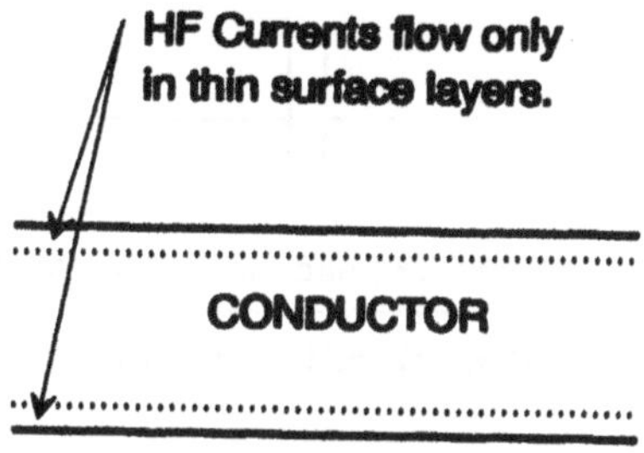

Skin depth at frequency f in a conductor of resistivity ρ Ohm-metre and permettivity μ henry/metre is

$$\sqrt{\frac{\rho}{\mu\pi f}}$$

In copper the skin depth is $6.61/\sqrt{f}$ cm and

the skin resistance is $2.6\cdot10^{-7}\sqrt{f}$ Ω/sq

(Remember that current *may* flow in both sides of a PCB [this is discussed later] and that the skin resistance formula is only valid if the skin depth is less than the conductor thickness).

Figure 5 Skin effect.

Skin effect has the effect of increasing the resistance of conductors at quite modest frequencies and must be considered when deciding if the resistance of wires or PC tracks will affect a circuit's performance. It also affects the behavior of resistors at HF.

Resistors are not the simple components that one would wish. In addition to resistance they have inductance, capacitance, thermo-electric effects and noise. Each type of resistor is different: wirewound and spirally cut film resistors are inductive, solid carbon resistors are less inductive, but noisy (and part of the noise may be a function of the current in the resistor), all have capacitance. At HF it is probably unnecessary to consider temperature coefficients and thermoelectric effects, but these are important in high precision LF applications.

These parasitic effects can be dealt with in one of two ways: use only components whose HF behavior has been characterized by their manufacturer (and remember that changing manufacturers may change the nature of the parasitics), or characterize resistors yourself at the frequency at which they will be used (do, however, remember that many resistors in RF circuitry are decoupled or bypassed and their HF performance is unimportant - this should be verified by simple calculation, though, and never just assumed). HF parasitics must never be ignored.

EQUIVALENT CIRCUIT OF A "SIMPLE" RESISTOR

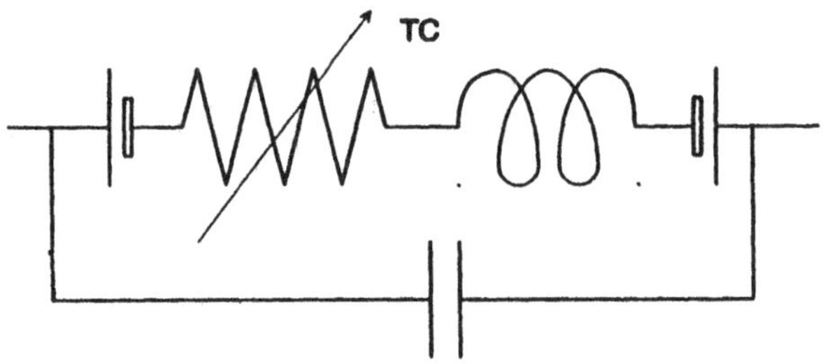

At HF the temperature coefficient (*TC*) and thermocouples may usually be ignored.

Figure 6 Equivalent circuit of a 'simple' resistor.

JOHNSON NOISE

All resistors have Johnson noise with a spectral density of

$$\sqrt{4kTR}\ \mathrm{V}/\sqrt{\mathrm{Hz}}$$

Where *T* is the absolute temperature, *R* is the resistance, and *k* is Boltzmann's constant ($1.38\cdot10^{-23}$ J/K).

(It is possible to reduce *T* or *R*, but *NOT* *k* because Boltzmann is dead.)

Figure 7 Johnson noise.

All resistors have Johnson noise. A resistor of *R* Ohms has noise spectral density of $\sqrt{4kTR}\ \mathrm{V}/\sqrt{\mathrm{Hz}}$ where *k* is Boltzmann's constant and *T* is the absolute temperature. This noise is intrinsic to resistors (but not to capacitors or inductors) and must be considered in all low-noise designs. Most modern resistors, with the exception of carbon rod types, have very little more noise than their basic Johnson noise, but carbon rod resistors are convenient at HF because of their low inductance and when they are used their noise specification may need to be considered, although many applications of such resistors are in loads where noise is not important.

9.5.3 Capacitors

Capacitors are even more complex than resistors. As the diagram in Figure 8 shows, a complete model has at least three resistances, an inductor and *three* capacitances.

Fortunately at HF the leakage and dielectric absorption (provided it is not too great) are rarely important and the model may be simplified to two capacitances, a resistor and an inductance. In many cases the small shunt capacitance can also be ignored. The temperature coefficient of a capacitor is very important if it is being used in a tuned circuit.

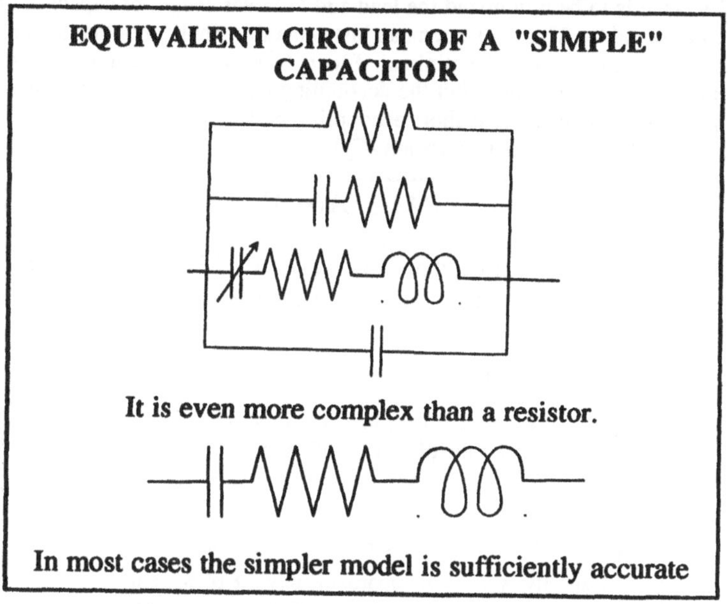

Figure 8 Equivalent circuit of a 'simple' capacitor.

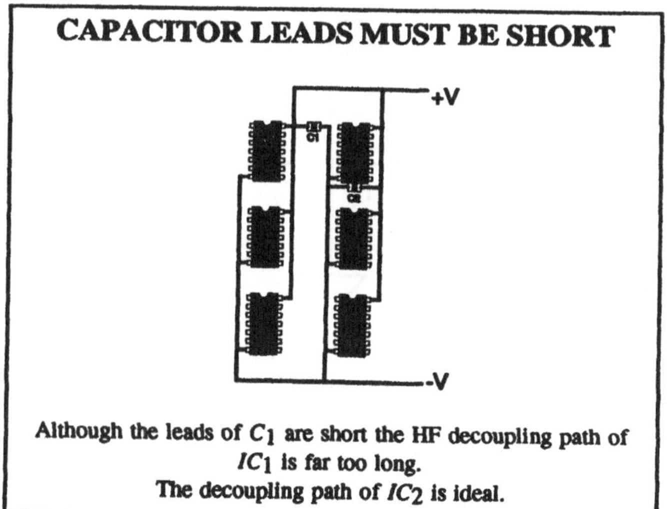

Figure 9 Capacitor leads must be short.

Where capacitors are used for coupling and decoupling, their resistance and inductance are important, but their variation with temperature rarely matters. Ceramic types have low resistance and inductance, but the high capacity ones have lousy temperature coefficients

(NPO types are fine, but tend to be larger), foil capacitors generally have high inductance, but are stable. (There are low inductance foil types, but it may be inadvisable to use them where there is any risk of the low inductance type being changed for a cheaper high inductance type after the circuit has gone to production).

If inductance is to be minimized the lead and PC track length of capacitors must be kept as small as possible. This does not mean just generally "short", but that the inductance in the actual circuit function must be minimal. Figure 9 shows both a common mistake (the leads of the capacitor C_1 are short, but the decoupling path for IC_1 is very long); and the correct way to decouple an IC (the short decoupling path C_2 provides for IC_2 - the path would be even shorter if C_2 were underneath IC_2).

CAPACITANCE

Wherever two conductors are separated by a dielectric (including air or a vacuum) there is capacitance.

For a parallel plate capacitor $C = .0885 \varepsilon_r \frac{A}{d}$ pF

where A is the plate area in sq. cm
d is the plate separation in cm
& ε_r is the dielectric constant

Epoxy PCB material is often 1.5 mm thick and $\varepsilon_r = 4.7$
Capacity is therefore approximately 2.8 pF/sq. cm

Figure 10 Capacitance.

FARADAY SHIELDS

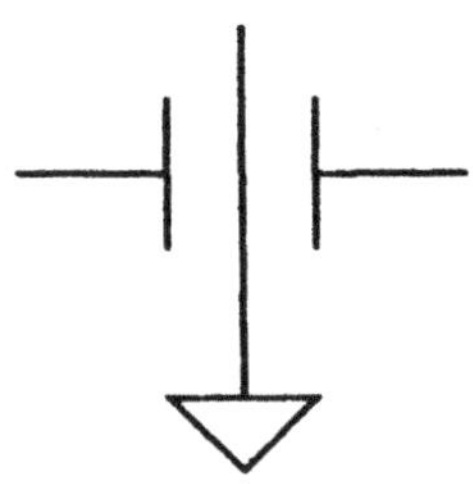

Capacitively coupled noise can be very effectively shielded by a grounded conductive shield, known as a Faraday shield.
But it must be grounded or it increases the problem.
For this reason coil and quartz crystal cans should always be grounded.

Figure 11 Faraday shields.

Good RF design must incorporate stray capacitance. Wherever two conductors are separated by a dielectric there is capacitance. The formulae for parallel wires, concentric spheres and cylinders, and other more exotic structures may be found in any good physics book but the commonest structure, found on all PCBs, is the parallel plate capacitor.

When stray capacitance appears as parasitic capacity to ground it can be minimized by careful layout and routing, and incorporated into the design. Where stray capacity couples a signal where it is not wanted the effect may be minimized by design but often must be cured by the use of a Faraday shield.

A Faraday shield consists of a *grounded* conductor placed between the signal source and its destination. It is surprising how often designers omit to ground a Faraday shield, which almost invariably makes the problem worse. It is good practice to ground any substantial conductor which would otherwise float: crystal cans, coil cans, metal semiconductor packages, and structural metal.

9.5.4 Inductors

INDUCTANCE

Any conductor has some inductance
A straight wire of length *L* and radius *R* (both mm & *L*>>*R*)
has inductance $0.2L\left[\ln\left(\frac{2L}{R}\right)-.75\right]$ nH

A strip of conductor of length *L*, width *W* and thickness *H* (mm)
has inductance $0.2L\left[\ln\left(\frac{2L}{W+H}\right)+0.2235\left(\frac{W+H}{L}\right)+0.5\right]$ nH

1 cm of thin wire or PC track is somewhere between 7 and 10 nH

Figure 12 Inductance.

INDUCTANCE

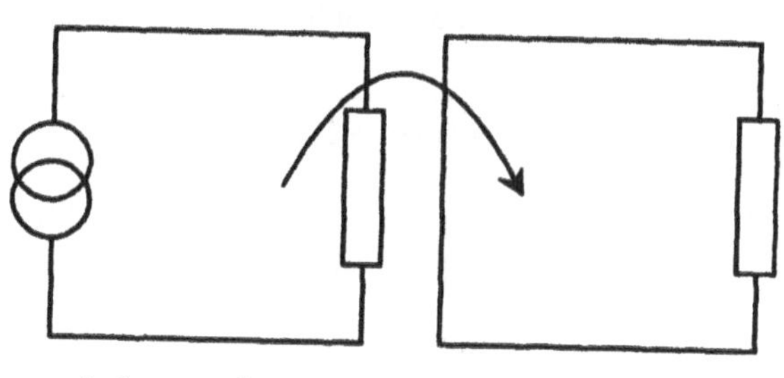

A loop of conductor has inductance -
two adjacent loops have mutual inductance.

Figure 13 Inductance.

Any length of conductor has inductance and at HF, this matters. In free space a 1 cm length of conductor has inductance of 7-10 nH (depending on diameter), which represents an impedance of 4-6 Ω at 100 MHz. This may be large enough to be troublesome, but badly routed conductors can be far worse as they form, in effect, single turn coils with quite substantial inductance.

If two such coils are close to each other we must consider their *mutual* inductance as well as their self inductance. An HF current in one will induce an EMF in the other. Defining the problem, of course, at once suggests cures: reducing the area of the coils by more careful layout, and increasing their separation. Both will reduce mutual inductance and reducing area reduces self inductance.

INDUCTANCE

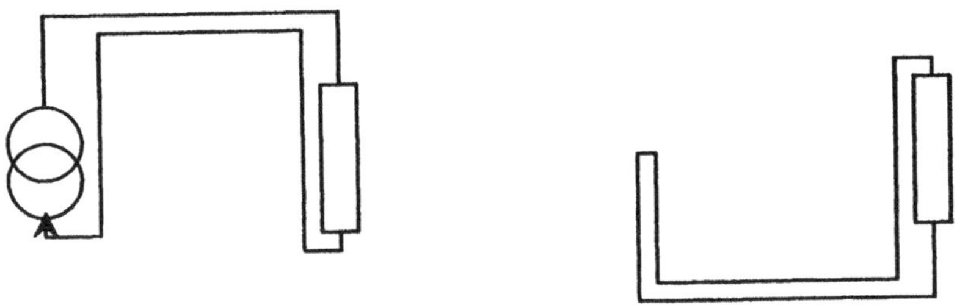

Inductance is reduced by reducing loop area - mutual inductance is reduced by reducing loop area and increasing separation.

Since the magnetic fields around coils are dipole fields they attenuate with the *cube* of the distance - so increasing separation is a very effective way of reducing mutual inductance.

Figure 14 Inductance.

It is possible to reduce inductive coupling by means of shields. At LF shields of μ-metal are necessary (and expensive, heavy and vulnerable to shock, which causes loss of permettivity) but at HF a continuous Faraday shield (mesh will not work well here) blocks magnetic fields too, provided that the skin depth at the frequency of interest is much less than the thickness of the shield.

MAGNETIC SHIELDS

At LF magnetic shielding requires μ -Metal which is heavy, expensive and vulnerable to shock.

At HF a conductor provides effective magnetic shielding provided the skin depth is less than the conductor thickness.

PC foil is an effective magnetic shield above 10-20 MHz.

Figure 15 Magnetic shields.

Magnetic fields are dipole fields, and therefore the field strength diminishes with the *cube* of the distance. This means that quite modest separation increases attenuation a lot. In many cases physical distance is all that is necessary to reduce magnetic coupling to acceptable levels.

9.5.5 Grounds

KIRCHOFF'S LAW

The net current at any point in a circuit is zero.

OR

What flows in flows out again.

OR

Current flows in circles.

THEREFORE

All signals are differential.

AND

Ground impedance matters.

Figure 16 Kirchoff's law.

Kirchoff's law suggests that return currents in ground are as important as signal currents in signal leads. We find here another example of the "superconductor assumption" - too many engineers believe that all points marked with a ground symbol on the circuit diagram are at the same potential. In practice ground conductors have resistance and inductance - and potential differences.

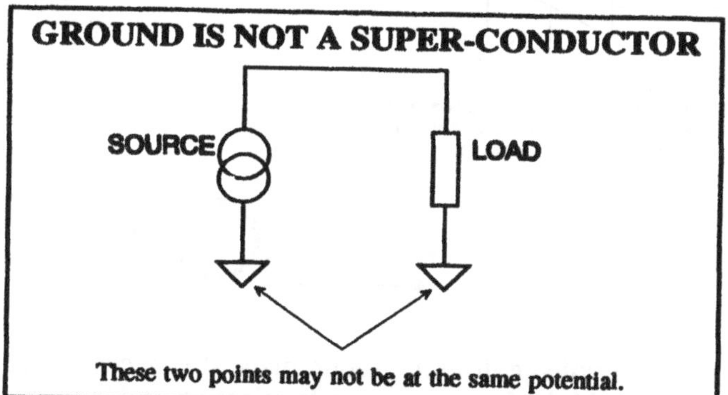

Figure 17 Ground is not a super-conductor.

GROUND PLANE

The philosophy of ground plane is that one side (or layer) of a PCB consists of a single layer of continuous metal.
It is hoped that all points on the plane are at the same potential.
In practice it may be necessary to configure ground currents by means of breaks in the plane, or careful placement of sub-systems.
Nevertheless ground plane is generally the most effective ground technique for RF and high precision analog designs.

Figure 18 Ground plane.

The best ground arrangement on a PCB is a "ground plane" - a layer of continuous copper, generally on one side of the board, but occasionally, in multi-layer boards, internal to it. A ground plane has minimal resistance and inductance, but its impedance may still be too great at high currents or high frequencies. Sometimes a break in a ground plane can configure currents so that they do not interfere with each other, sometimes physical separation of different subsystems is sufficient.

It is often easy to deduce where currents flow in a ground plane, but in complex systems it may be difficult. It is possible to measure differential voltages of as little as 5 μV on a ground plane. At DC and LF this is done by using an instrumentation amplifier with a gain of 1,000 to drive an oscilloscope working at 5 mV/cm. The sensitivity at the input terminals of the inamp is 5 μV/cm, there will be some noise present on the oscilloscope trace, but it is quite possible to measure ground voltages of the order of 1 μV with such simple equipment. It is important to allow a path for the bias current of the inamp, but its common-mode rejection is so good that this bias path is not critical.

GROUND PLANE

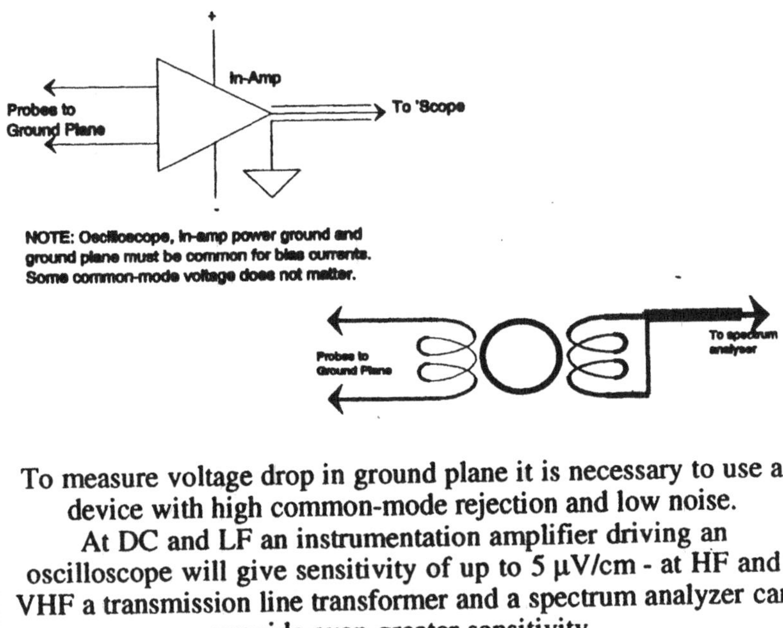

To measure voltage drop in ground plane it is necessary to use a device with high common-mode rejection and low noise. At DC and LF an instrumentation amplifier driving an oscilloscope will give sensitivity of up to 5 μV/cm - at HF and VHF a transmission line transformer and a spectrum analyzer can provide even greater sensitivity.

Figure 19 Ground plane.

The upper frequency of most inamps is 25-50 kHz (the AD830 is an exception - it works up to 50 MHz at low gains, but not at X1,000). Above LF a better technique is to use a broadband transmission line transformer to remove common-mode signals. Such a transformer has little or no voltage gain, so the signal is best displayed on a spectrum analyzer, with μV sensitivity, rather than on an oscilloscope which only has 5 mV or so.

If HF signal conductors run over a ground plane they form a microstrip transmission line. The characteristic impedance of the line is defined by its dimensions and it must be correctly terminated if it is not to behave reactively. This feature of conductors running over ground plane may be exploited to simplify HF layout: if the conductor strip is the correct width there will be no impedance mismatch at a connector, alternatively microstrip of various impedances may be used to build impedance matching circuitry which uses nothing but PCB. Glass fiber/epoxy PCB is somewhat lossy at VHF and UHF, but the losses may be tolerable if microstrip runs are short - if not, low loss (and expensive) PCB material such as teflon must be used.

MICROSTRIP TRANSMISSION LINE

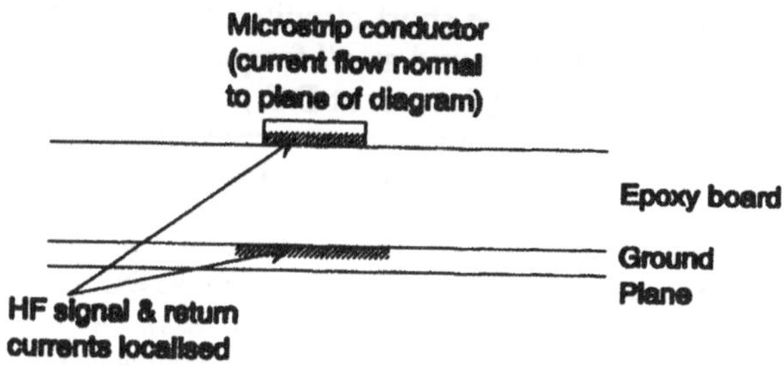

When a conductor runs over a ground plane it forms a microstrip transmission line.

The characteristic impedance is $\frac{377H}{W\sqrt{\varepsilon_r}}$ Ω (note that units of *H* and *R* are unimportant)

The transmission line determines where both the signal *and return* currents flow.

Figure 20 Microstrip transmission line.

It is important to realize that current flow in a microstrip transmission line is constrained by inductive effects. The signal current flows only on the side of the conductor next to the ground plane (its skin depth is calculated in the normal way) and the return current flows *only directly beneath the signal conductor*, not in the entire ground plane (skin effect naturally limits this current, too, to one side of the ground plane). This is helpful in separating ground currents, but increases the resistance of the circuit.

MICROSTRIP TRANSMISSION LINE

The ground under a microstrip line must not be broken or the HF ground current will be forced to flow around the edge of the break, increasing impedance.

The best way of making a crossover of a microstrip line is to pass one of the signals to a wire link, which crosses the ground plane on the opposite side from the microstrip. The return currents are thus completely isolated.

Figure 21 Microstrip transmission line.

It is clear that breaks in the ground plane under a conductor carrying HF will force the return current to flow around the break, increasing impedance. Even worse, if the break is made to allow two HF circuits to cross, the two signals will interact. Such breaks should be avoided if at all possible. The best way to enable two HF conductors on a ground plane to cross without interaction is to keep the ground plane continuous and use a wire ***on the other side of the ground plane*** to carry one of the signals over the other. If the skin depth is much less than the ground plane thickness the interaction of ground currents will be negligible.

9.5.6 Decoupling

DECOUPLING

Supplies must be short-circuited to each other
and to ground at *all* frequencies.
(But not at DC.)

Figure 22 Decoupling.

The final issue we must consider is decoupling. The power supplies of HF circuits must be short-circuited together and to ground at all frequencies above DC. (DC short-circuits are undesirable for reasons which I shall not bother to discuss). At low frequencies the impedance of supply lines is (or should be) low and so decoupling can be accomplished by relatively few electrolytic capacitors which will not generally need to be very close to the parts of the circuit they are decoupling, and so may be shared among several (the exception to this is where a component draws a large LF current, when a local, dedicated, electrolytic should be used).

At HF we cannot ignore the impedance of supply leads (as we have already seen in Figure 9) and ICs must be individually decoupled with low inductance capacitors having short leads and PC tracks. Even 2-3 mm of extra lead/track length may make the difference between the success and failure of a circuit layout.

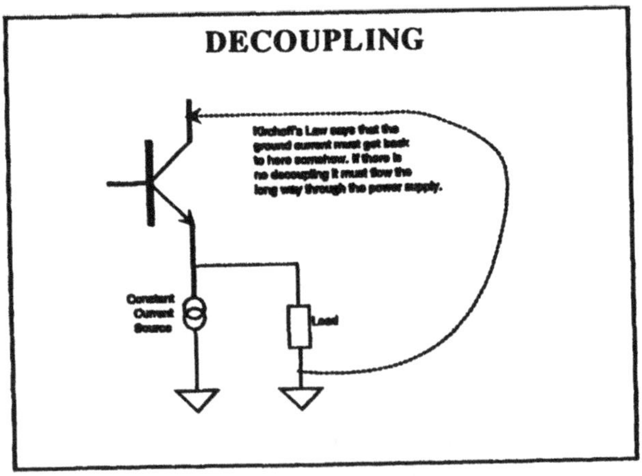

Figure 23 Decoupling.

Where the HF currents of a circuit are mostly internal (as is the case with many ADCs) it is sufficient that we short-circuit its supplies at HF so that it sees its supplies as stiff voltage sources at all frequencies. When it is driving a load with HF, the decoupling must be arranged to ensure that the total loop in which the load current flows is as small as possible. Figure 23 shows an emitter follower without supply decoupling - the HF current in the load must flow through the power supply to return to the output stage (remember that

Kirchoff's law says, in effect, that currents *must* flow in circles). Figure 24 shows the same circuit with proper supply decoupling.

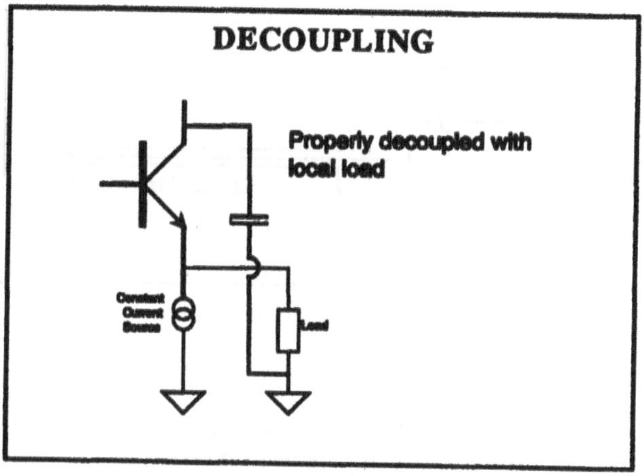

Figure 24 Decoupling.

This principle is easy enough to apply if the load is adjacent to the circuit driving it. Where the load must be remote it is much more difficult, but there are solutions: transformer isolation and the use of a transmission line. If the signal contains no DC or LF components it may be isolated with a transformer close to the driver. Such an arrangement is shown in Figure 25. (The nature of the connection from the transformer to the load may present its own problems - but not of supply decoupling).

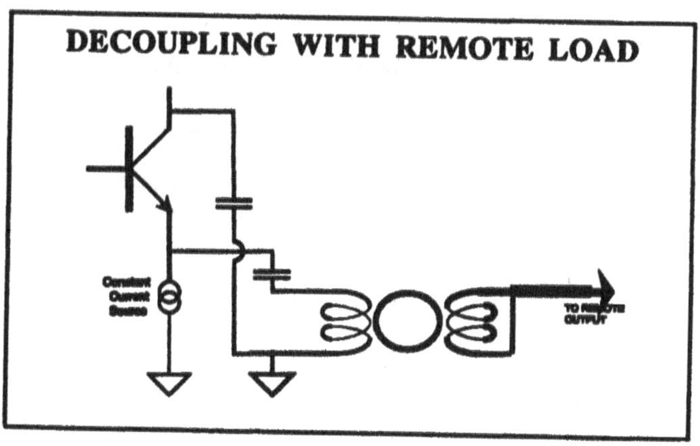

Figure 25 Decoupling with remote load.

A correctly terminated transmission line constrains HF signal currents so that, to the supply decoupling capacitors, the load appears to be adjacent to the driver. Even if the line is not precisely terminated, it will constrain the majority of the return current and is frequently sufficient to prevent ground current problems.

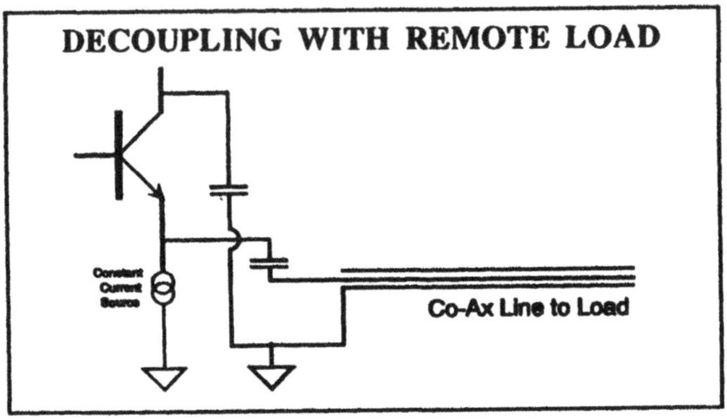

Figure 26 Decoupling with remote load.

9.5.7 Conclusion

It is not possible in a short paper to discuss all the intricacies of successful RF construction, but we have seen that the basic principle is to remember all the laws of nature which apply and consider their effects on the design.

SUCCESSFUL RF DESIGNS

Pay attention to:
Resistance
Capacitance
Inductance
Decoupling
Ground
&
Separating sensitive circuits from noisy ones

Figure 27 Successful RF designs.

In addition to the considerations of resistance, skin effect, capacitance, inductance and ground current, it is important to configure systems so that sensitive circuitry is separated from noise sources and so that the noise coupling mechanisms we have described (common resistance/inductance, stray capacitance, and mutual inductance) have minimal opportunity to degrade system performance. ("Noise" in this context means a signal we want [or which *somebody* wants] in a place where we don't want it, not natural noise like thermal, shot or popcorn noise.) The general rule is to have a signal path which is roughly linear, so that outputs are physically separated from inputs and logic and high level external signals only appear where they are needed.

Thoughtful layout is important, but in many cases screening may be necessary as well.

A final consideration is the power supply Switching power supplies are ubiquitous today, because of their low cost, high efficiency and reliability, and small size. But they can be a major source of HF noise, both broadband and at frequencies harmonically related to their switching frequency. This noise can couple into sensitive circuitry by all the means we have discussed, and extreme care is necessary to prevent switching supplies from ruining system performance. In addition to screening and decoupling, it is wise wherever possible to site switching power supplies as far as possible from sensitive parts of the system. At some time during development it is also interesting (and frightening, and helpful) to replace the switching supply with a battery and observe the difference in noise.

SWITCHING POWER SUPPLIES

Generate noise at every frequency under the Sun
(and some interstellar ones as well).

Every mode of noise transmission is present.

If you must use them you should filter, screen,
keep them far away from sensitive circuits,
and still worry!

Figure 28 Switching power supplies.

OBEY THE LAW

Unexpected behavior of RF circuitry is almost always due to the designer overlooking one of the basic laws of electronics. Remember and obey Ohm, Faraday, Lenz, Maxwell, Kirchoff **and MURPHY.**

"Murphy always was an optimist" - Mrs. Murphy.

Figure 29 Obey the law.

Chapter 9.6

RF Receivers for Cellular Telephone Base Stations

J. David Rhodes

Abstract

In most countries there are two cellular telephone operations in each geographic region with the designated transmit and receive bands shared between the two operators. There are no guard bands between the frequencies allocated to each operator and in the UK and USA in particular, the frequency allocations are interleaved. Two distinct problems arise from having the competing operators base stations either co-located or remotely located. These problems are investigated and solutions proposed and evaluated. The solutions involve receivers designed as a cascade of filters and amplifiers in order to minimise noise figure for sensitivity and to increase the dynamic range by attenuating the other strong signals.

9.6.1 UK 900 MHz

In the UK in the 900 MHz cellular system the two operators Cellnet and Vodaphone share 35 MHz transmit and 35 MHz receive to cover the TACS, ETACS and GSM systems. The receive band in the base stations is 880 MHz to 915 MHz with the corresponding transmit band from 915 MHz to 960 MHz. In Figure 1 we have the block diagram for a typical receive multicoupler for each antenna feed in the base station. The initial filter after the antenna is a cover filter with a passband of 35 MHz designed to provide low loss but with high selectivity on the high frequency side to provide an attenuation of 30 dB. This filter is constructed as a 10th degree machined combline device with additional cross-coupling to provide the finite transmission zeros. The measured response is shown in Figure 2.

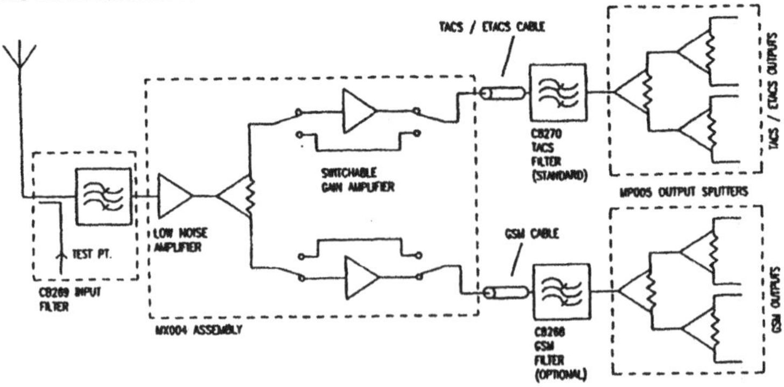

Figure 1 Typical receive multicoupler for each antenna.

The amplifier consists of a low noise front end which is constructed using an isolator and a single ended FET. The signal is then fed to two paths each of which consist of high intercept balanced BJT amplifiers both having switchable gains dependent upon the location of the base station. These amplifiers which are manufactured using normal pick and place techniques are shown in Figure 3.

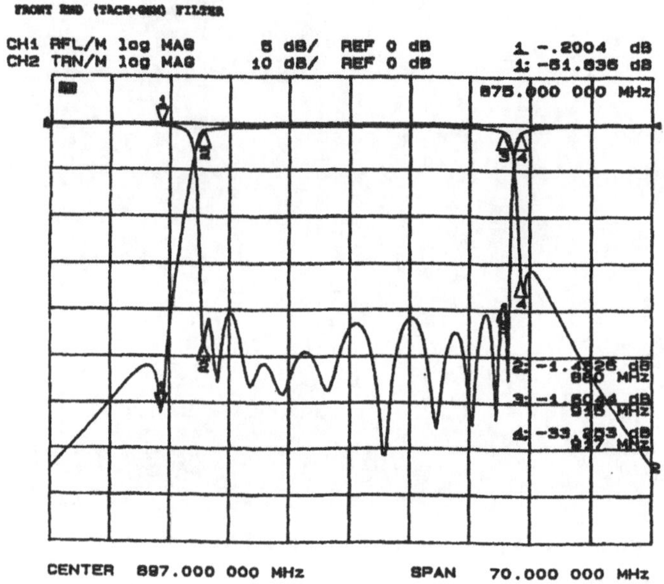

Figure 2 Measured response of cover filter.

Figure 3 Low noise front end amplifiers.

For the TACS and ETACS channels the following filter is required to pass a 25 MHz band but must provide sharp attenuation on the high side to reject signals from possible GSM mobiles transmitting at high power in the band 905→910 MHz. This filter has asymmetric poles on the high frequency side similar to the device shown in Figure 4 and with a measured response is shown in Figure 5.

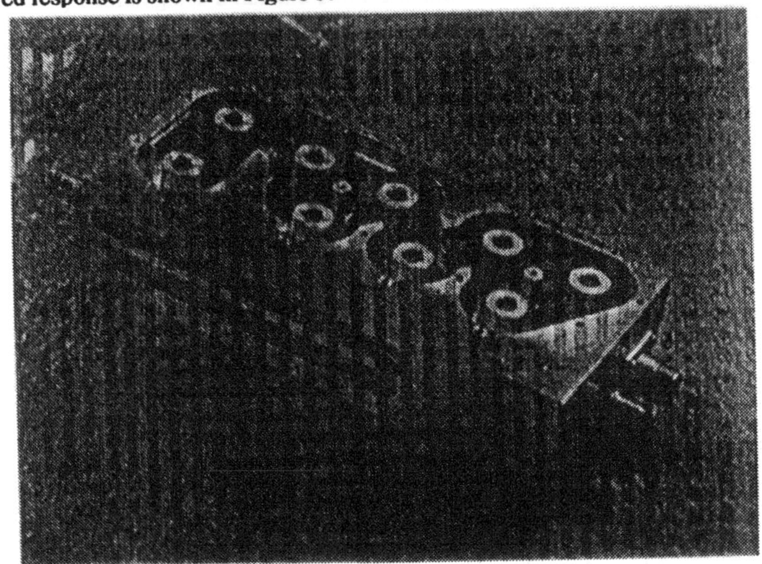

Figure 4 Filter for TACS and ETACS channels.

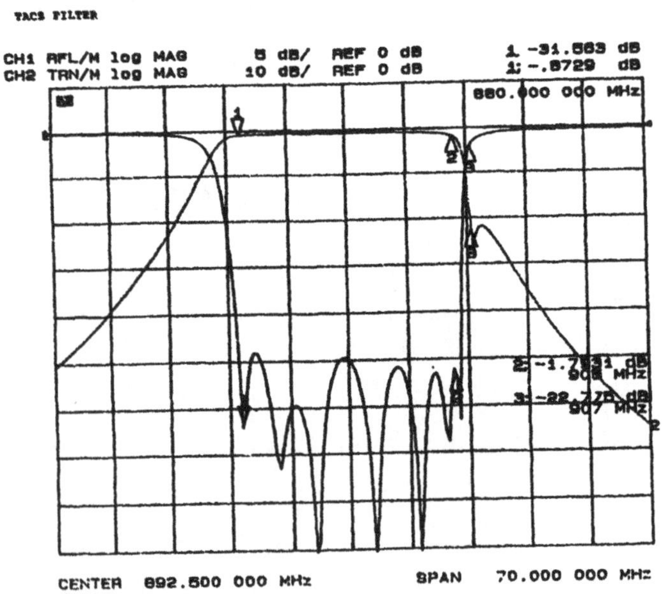

Figure 5 Measured response of the filter in Figure 4.

The GSM channels require filtering on both sides of the band. On the low frequency side this is to reject the other operator's GSM mobiles and on the high side to provide further attenuation to the other operator's base station ETACS transmissions. The response of a 5th degree device with symmetrical cross-coupling is shown in Figure 6.

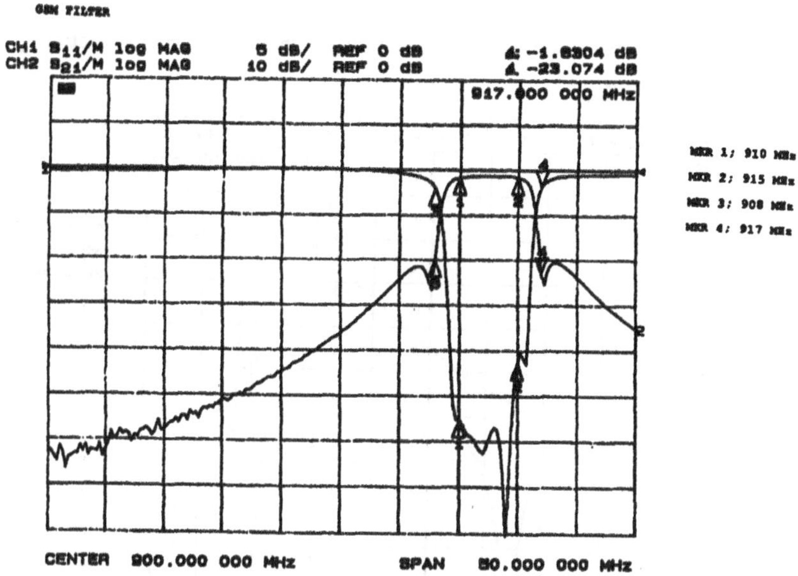

Figure 6 Response of a 5th degree device with symmetrical cross-coupling.

Finally, the overall response for the TAC/ETACS output are shown in Figure 7 and for the GSM outputs in Figure 8. The power divided outputs are then fed into individual receivers before down conversion.

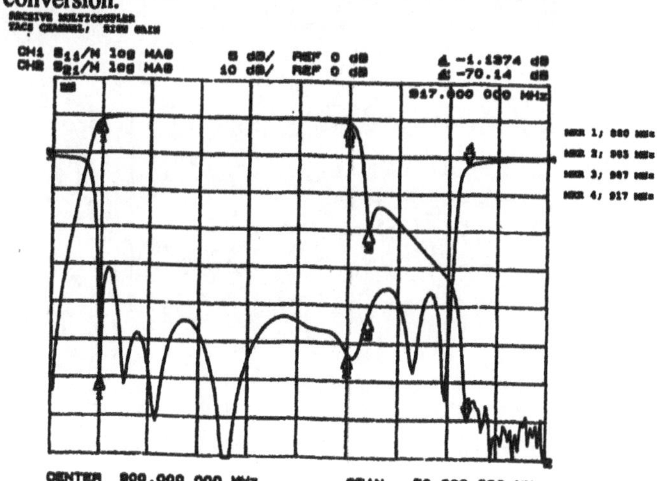

Figure 7 Overall response for the TAC/ETACS output.

A typical base station has six antennas and a complete multi-coupler receiver unit as shown in Figure 9. Also, in the cases where a common antenna is used for both transmit and receive, an additional triplexer is required after the antenna to provide additional attenuation of the transmit signals over the receive filters, as shown in Figure 10. In this case the TACS/ETACS and GSM transmitters use separate inputs to further reduce the 3rd ordered passive intermodulation products which would appear in the receive band. These channels also have to withstand several kWs of peak power due to the multi-carrier operation.

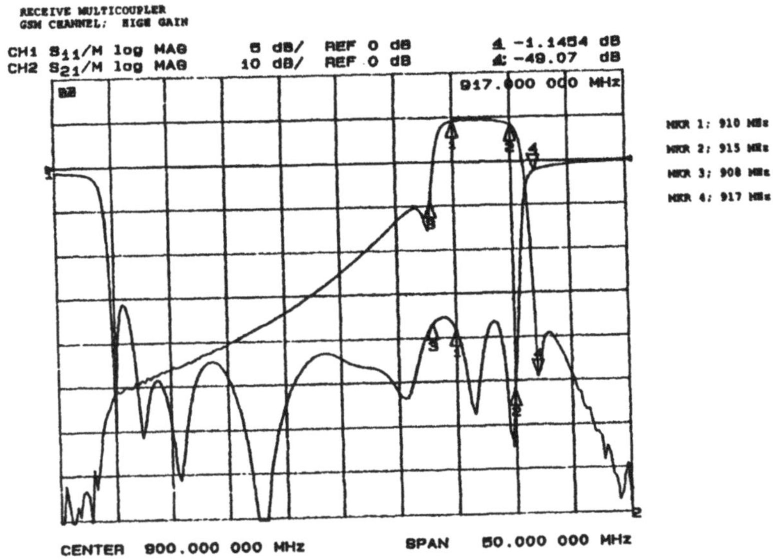

Figure 8 Overall response for the GSM output .

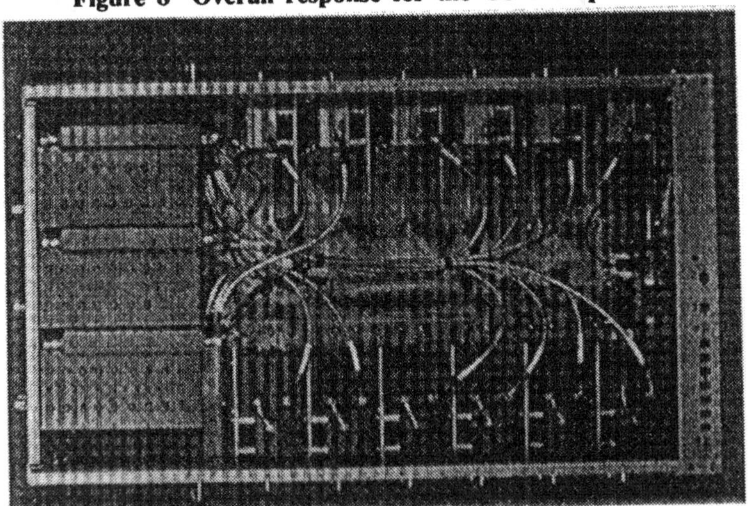

Figure 9 A typical base station.

For most of the rest of Europe where there was no 900 MHz analogue system, the introduction of GSM requires a simpler receiver. However, some base stations are required to cover large cell areas and hence antenna mounted receivers may be necessary. Figure 11 shows a typical block diagram for the antenna mounted unit and the subsequent receiver in the base station and Figure 12 shows the units themselves. The receive channel is 880 MHz to 915 MHz and a high level of attenuation has to be maintained over the 925 MHz to 960 MHz transmit band and a plot of a typical unit is shown in Figure 13.

Figure 10 Additional triplexer.

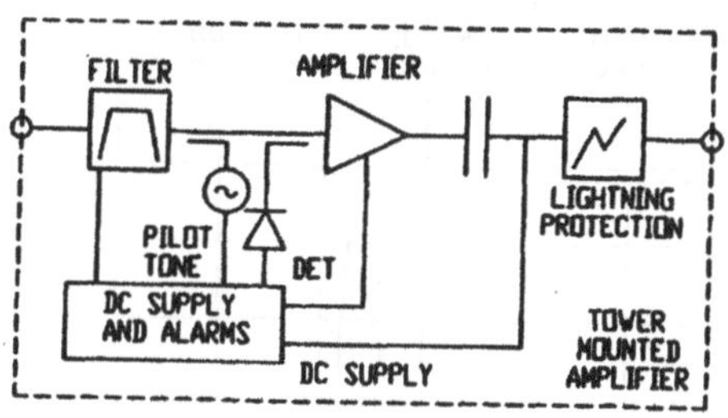

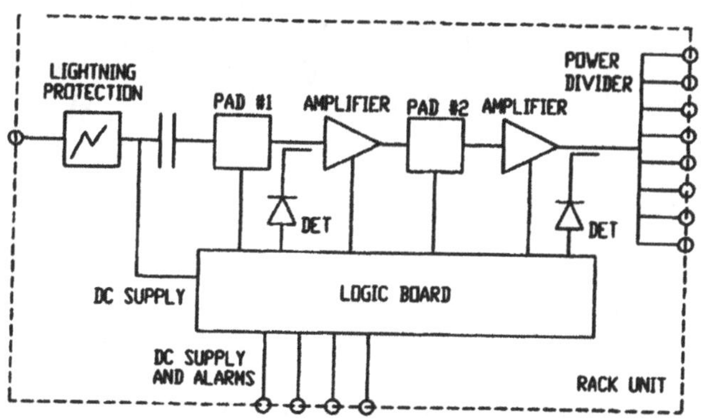

Figure 11 Typical block diagram for the antenna mounted unit and the subsequent receiver in the base station.

Figure 12 Typical units.

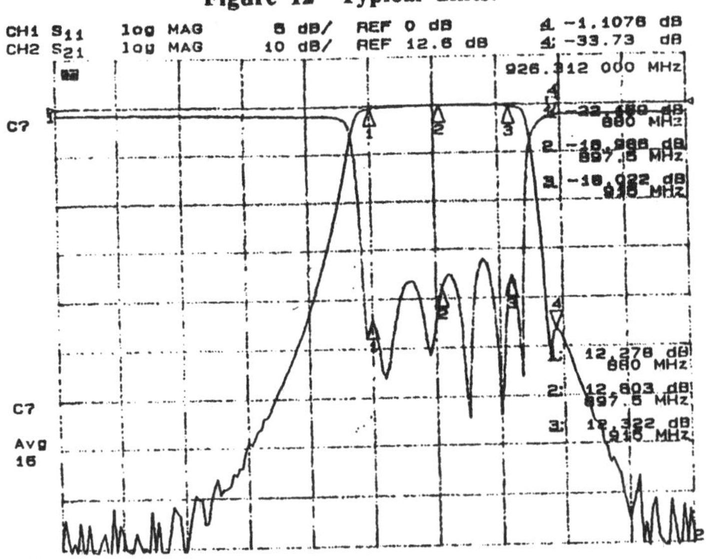

Figure 13 Plot of the typical unit in Figure 12.

Similarly, the new PCN/PCS systems at 1800 MHz and 1900 MHz are new systems which predominantly use GSM. A typical mast mounted transmit receive diplexer, with receiver module is shown in Figure 14, with the block diagram and diplexer performance shown in Figures 15 and 16 respectively.

Figure 14 Mast mounted transmit receive diplexer, with receiver module.

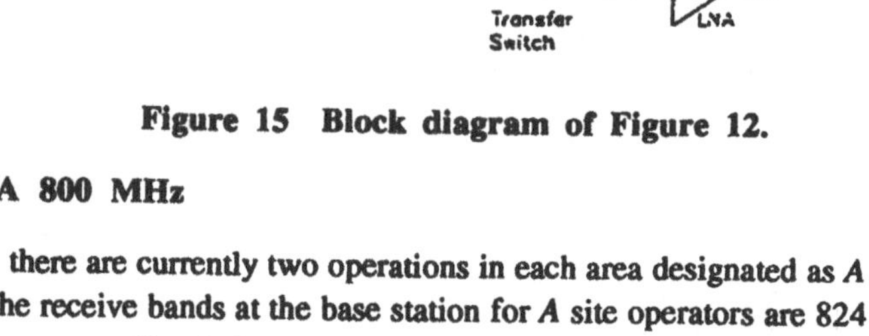

Figure 15 Block diagram of Figure 12.

9.6.2 USA 800 MHz

In the USA there are currently two operations in each area designated as *A* site and *B* site systems. The receive bands at the base station for *A* site operators are 824 MHz→ 846.5 MHz. The corresponding *B* site operators are 835 MHz→ 845 MHz and 846.5 MHz to 849 MHz. The corresponding transmit channels are 45 MHz higher in frequency. The situation is further complicated by the air phase system at 849→ 851 MHz and the special mobile system at 851→ 856 MHz. The latter problem also gives rise to the need for selective filtering in the transmit band. For example, the *B* site transmits over 880 MHz to 894 MHz and at least 30 dB of attenuation is required with 2 MHz of bandedge to reduce spurious emissions. This, coupled with the need to withstand several kWs of peak power demands a high performance from the filter.

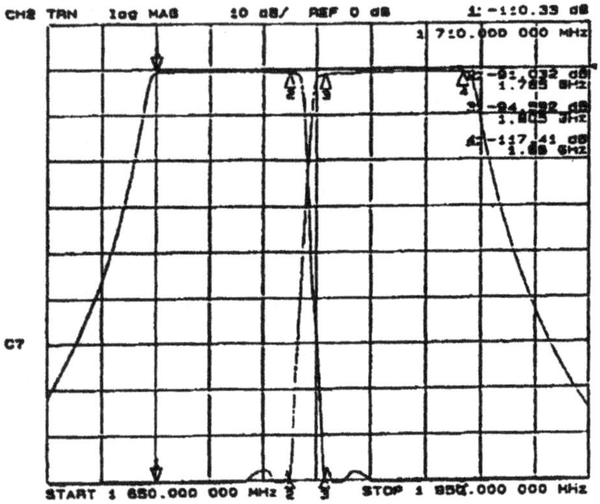

Figure 16 Diplexer performance.

On the receive side an *A* site receive filter may be constructed from the parallel connection of a combline 8th degree filter with additional cross-coupling and a 6th degree dielectric filter operating in the TEois mode again with cross-coupling. A typical performance is shown in Figure 17.

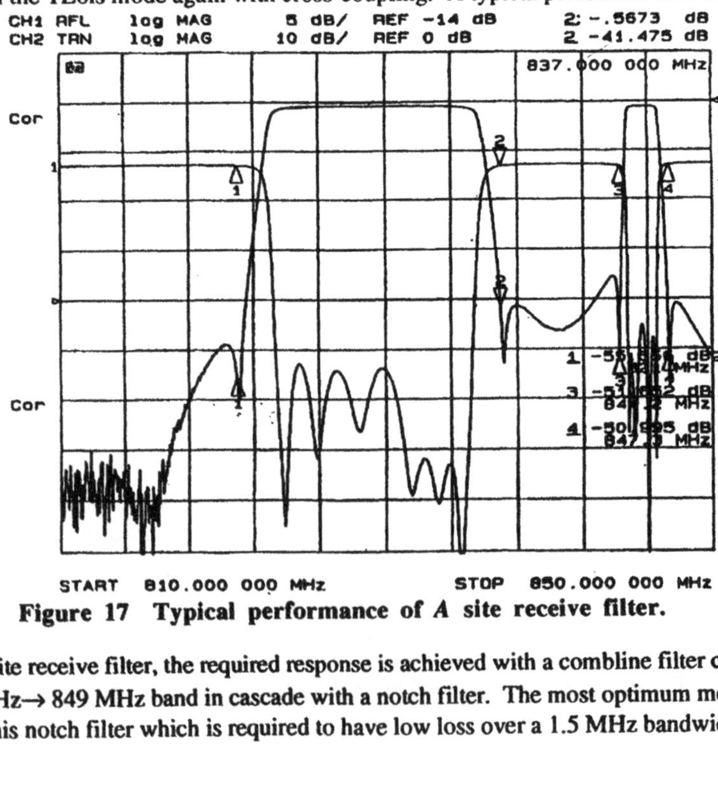

Figure 17 Typical performance of *A* site receive filter.

For the *B* site receive filter, the required response is achieved with a combline filter covering the 835 MHz→ 849 MHz band in cascade with a notch filter. The most optimum method of realising this notch filter which is required to have low loss over a 1.5 MHz bandwidth with

20 dB attenuation over a 1.1 MHz bandwidth is to use a 'Hybrid Notch Filter'. This is basically a conventional bandpass filter connected to a 3 dB hybrid. The hybrid may be the integrated into the filter and the resulting transmission characteristic is the return loss characteristic of the basic bandpass filter. A typical response for a 6th degree dielectric resonator filter is shown in Figure 18.

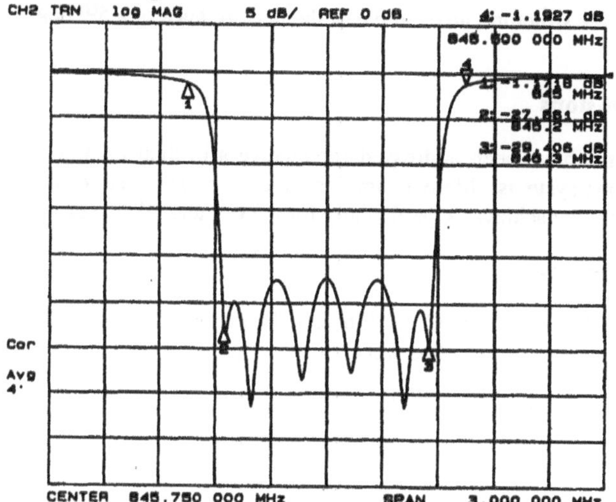

Figure 18 Typical response for a 6th degree dielectric resonator filter.

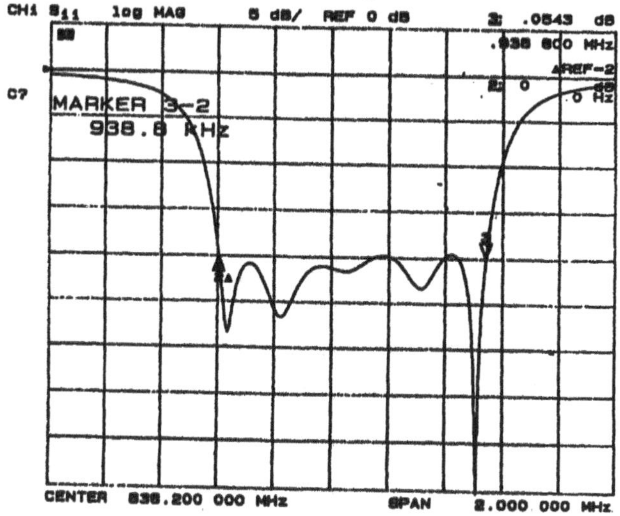

Figure 19 Performance of a 5th degree filter.

In order to reduce size it is necessary to integrate the filter amplifier chain and provide high selectivities with lower Q resonators. Using circulators a variable Q reflection mode filter may be constructed. The design is relatively complicated and for the quasi equiripple lowpass prototype response the transfer characteristic may be shown to be

$$|S_{12}(j\omega)|^2 = \frac{\alpha^2(1-\omega^2)\left[1-T_{n+1}^2(\alpha\omega)\right]}{(1-\alpha^2\omega^2)T_{n+1}^2 - T_{n+1}^2(\alpha\omega)}$$

Where $\alpha > 1$ and $T_n(x)$ is the n-th degree Chebyshev polynomial. Transforming to a bandpass combline structure, an example of the performance of a 5th degree filter is shown in Figure 19.

9.6.3 Conclusions

Several problems have been addressed for receivers in dual operator environments for cellular telephone systems. Many examples have been presented to illustrate the types of performances that are achievable particularly for the UK and USA systems.

Acknowledgements

The author would like to thank C. I. Mobbs, T. H. Fripp, M. Perkins, D. Collins and P. D. Sleigh for their contributions to this work.

Tutorial 10

Current-mode analog signal processing

Part 1

Switched-Currents

Organised by: Nick Battersby, *Northern Telecom*
Gordon Roberts, *McGill University*
Chris Toumazou, *Imperial College*

Chapter 10.1

Switched-Current Circuits and Systems

Chris Toumazou and Geir Saether

10.1.1 Introduction

In a competitive electronics industry, particularly in consumer electronics, there is a desperate need to keep processing costs and hence production costs to a minimum. On chip analog interface circuits are generally a costly part of an IC not only because of additional design time and effort, but also the cost of inclusion of analog components on primarily a digital process technology.

The role of analog integrated circuits is therefore changing, no longer are complex systems entirely analog, rather they typically consist of a core of digital signal processing and computation circuits, buffered from an analog external environment by a layer of analog interface circuits. Complete analog systems will still continue to be required in some applications, mainly those in which the frequency of operation is too high for digital implementation, low complexity does not justify a digital implementation or in very low power applications. In recent years, the quest for ever smaller and cheaper electronic systems has led manufacturers to integrate entire systems onto a single chip. It is now becoming common to find that a single mixed analog and digital (mixed-mode) integrated circuit contains both a digital signal processor and all the analog interface circuits required to interact with its external analog transducers and sensors.

This tutorial is aimed at engineers who are interested in an overview of just some of the research openings in Switched-Currents. The tutorial is light in mathematical content, and geared towards circuit design and not analysis.

An important building block for front end processing is the analog to digital converter (ADC) and an exciting development in this area over the past few years has been the Sigma-Delta (ΣΔ) modulation technique [1] (see Chapter 4 of this book). This technique is an attempt to trade precision in the analog amplitude domain for precision in the digital time domain, by the well known oversampling technique. Now while such a technique places less stringent demands on analog performance the realisation of an over-sampling converter in both standard digital CMOS and GaAs MESFET process technologies still requires the use of analog components such as an integrators. An integrator generally requires the use of a linear floating capacitor and so a second layer of polysilicon is required to implement this capacitor and this adds to the overall cost of the technology. The marriage between switched-currents and oversampling converters would appear to be ideal. In one technology analogue precision is traded for digital precision, from a circuits perspective, and in the other analogue process technology is traded for digital process technology.

In this chapter we review some of the research at Imperial College aimed at developing suitable circuit techniques to implement good quality switched-current cells and high speed ΣΔ modulators on a pure digital process technology. Some basic techniques are presented in both Silicon and GaAs technologies, and in both cases performance is optimised for speed. Simulations presented for the silicon circuits employ MOSFETs from the 0.8 μm Northern Telecom BiCMOS technology, since the technology was made available for our research work.

10.1.2 The switched-current approach

Whilst mixed-mode integrated circuits are advantageous from both economic and systems design viewpoints, combining both analog and digital circuits on a single chip makes the circuit design and simulation process considerably more complex. As a typical mixed-mode integrated circuit contains primarily digital circuits, it is natural that the processing technology be tailored to optimise digital performance. This has perhaps been the thrust for "Current-Mode Processing". Unfortunately **current-mode** has become a marketing word, but we must not let that hinder innovative research. In our experience "a piece of silicon is a piece of silicon", there are many times when we have, for example, decomposed and redrawn a switched-current schematic which to the end designer can almost be viewed as a classical voltage-mode circuit. It is what you do with the circuit that makes it current-mode or voltage-mode, it is how you load and drive it, not necessarily its internal architecture. The output impedance of a current-mode circuit should be high, but if you want high voltage gain from a high frequency voltage mode circuit you still require a high output impedance! We still await practical demonstrators of good so called current-mode performance, but what clearly seems to be the potential of current-mode is high speed and simple circuits.

Traditionally, the switched-capacitor technique has been employed extensively in the analog interface portion of mixed-mode designs such as data converters. The second layer of polysilicon required is not required in pure DSP and may even become unavailable as process dimensions shrink to the deep submicron range. This trend towards submicron processes is also leading to reduced power supply voltages, which in turn makes the realisation of wide dynamic range, high gain, high speed operational amplifiers more difficult. The difficulties faced by switched-capacitor (SC) techniques, and other "voltage-mode" analog interface circuits, in coping with the advance of digital processing and technology has revived interest in "current-mode" techniques [2], and in particular an alternative current-mode analog sampled data technique known as switched-currents (SI) [3]. Unlike SC techniques SI circuits do not require linear floating capacitors or operational amplifiers, and this makes the technique most eminently suited to exploiting pure digital technology for ΣΔ modulator implementation, for example. Furthermore, the SI technique, as the name implies, operates with current samples and so voltage excursions are limited, leading to the potential for high speed, low voltage operation. Signal to noise ratio may be a problem due to low voltage excursions, but if the structure is placed within an enviroment that can tolerate this performance then the approach is acceptable.

10.1.3 Switched-current ΣΔ modulator building blocks in CMOS technology

Our work is aimed at the feasibility of realising fairly high performance SI ΣΔ converters on pure digital CMOS process technology for both audio and video applications. Although, some very encouraging work has already been reported (by Roberts and Crawley in Chapter 14 [3] and Roberts in Chapter 10.3 of this book). Our work is aimed at exploiting more recently reported circuit techniques for the building blocks rather than developing specific switched-current data converter architectures. An excellent treatment of various data converter architectures can also be found in chapter 10.3 of this book. The main building blocks for the modulator are the integrator, comparator and digital to analog converter (DAC) and so now consider each in turn. The main reason for choosing the ΣΔ modulator architecture is to demonstrate the requirements placed upon key SI building blocks in a realistic mixed-mode systems enviroment, and how the system specs are decomposed into individual circuits architectures.

10.1.3.1 SI memory cell

The basis of the SI integrator is the memory cell. The idea on which the switched-current memory cell is based is that an MOS transistor requires no gate current in order to maintain a constant flow of current between its drain and source terminals. The first application of this idea seems to date back to 1972 and was that of storing the current generated by a photodiode [Matsuzaki and Kondo, " Information Holding apparatus" UK patent 1359105, 10 July 1974, Filed 6 July 1972]. However, its use as a discrete time analog signal processing technique is more recent.

The basic element of the technique is the so-called switched-current memory [Hughes *et al.* All references to inventors of various types of cell can be found in reference [3]] or current copier cell, shown in Figure 1, which functions as a simple current track-and-hold element.

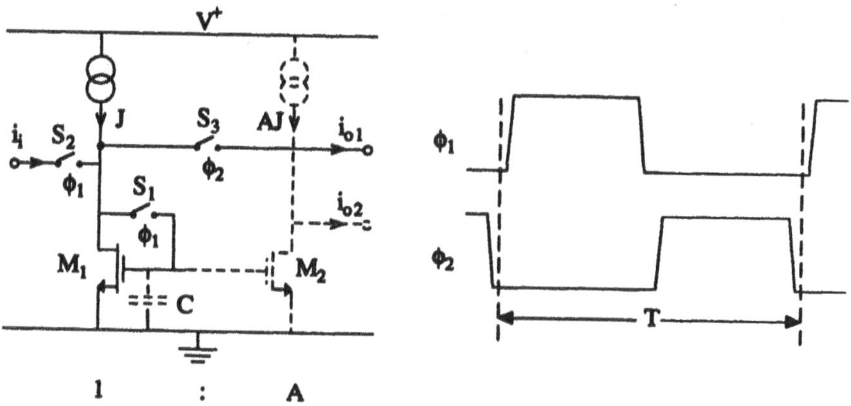

(a) Circuit structure (b) Clock waveforms

Figure 1 Basic switched-current memory cell.

During phase ϕ_1 the input current i_i adds to bias J, transistor M_1 is connected as a diode and so its gate-source capacitance charges up to the V_{gs} due to i_i. During phase ϕ_2, M_1's gate source capacitance stores the value of V_{gs} and so maintains a drain current output equal to the original input. The cell is essentially acting as a half wave discontinuous current-mirror, or a simple current sample-and-hold analog delay. What has been effectively realised is a half delay $z^{-1/2}$, a cascade of two memories gives a full delay, and with feedback a difference equation is formed to give integration following the classical z transformation.

Much work can be found in the literature [3] on cell performance optimisation in particular minimising errors which in a sense is equivalent to maximising voltage gain in an opamp used in SC applications. Note that in the memory cell the function of the gate capacitance is to store charge and not transfer charge as in SC techniques. Hence the capacitor can be non-linear; for this reason single poly technology can be employed.

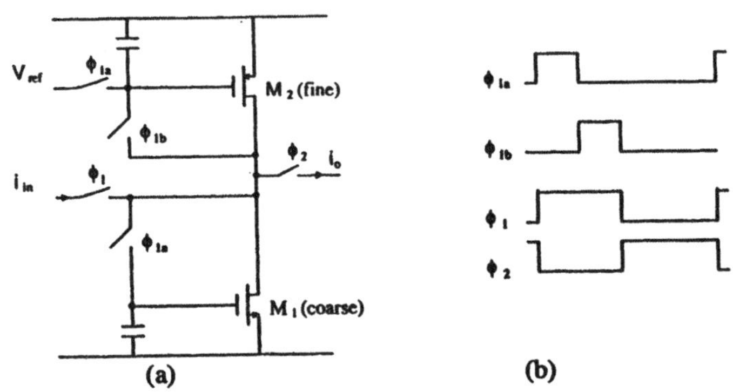

Figure 2 (a) S^2I memory cell [4] (b) clock waveforms.

Figure 2 shows the recently introduced high performance so-called S^2I (two-step) memory cell and the clock waveforms, respectively, proposed in [4]. The basic idea is that during ϕ_{1a} the coarse memory M_1 is connected as a diode and samples the input whilst fine-memory M_2 provides the bias. During ϕ_{1b} the gate switch of the coarse cell M_1 is open and the transistor holds its gate source voltage on the parasitic gate capacitor C_{gs}. The gate is now isolated and consequently M_1's drain-source current continues to flow during ϕ_{1b} i.e. it is memorised via the non linear C_{gs}.

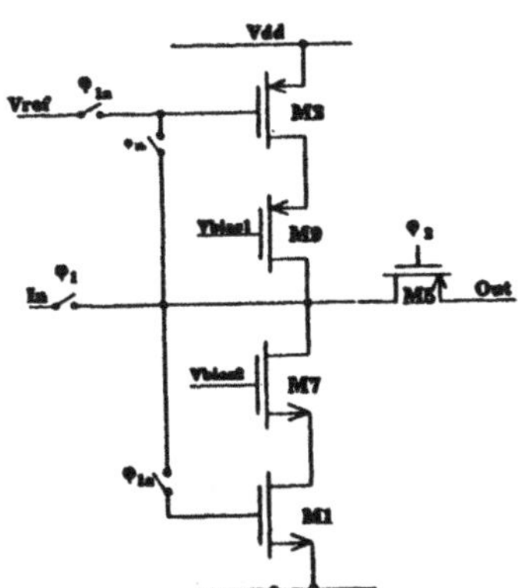

Figure 3 Cascoded S^2I memory cell.

Fine memory cell M_2 is now connected as a diode and will sample the difference between the input current and stored input, namely the error current. During the output phase ϕ_2, the error created in M_1 is subtracted by the same error memorised in M_2 and therefore the output current is ideally free of error. Note that the charge injection from the fine memory switch is treated as an offset and can be automatically cancelled when the cell is used in a delay cell or an integrator. Techniques have been recently proposed which allow an increase in the number of fine cells to further reduce errors [5].

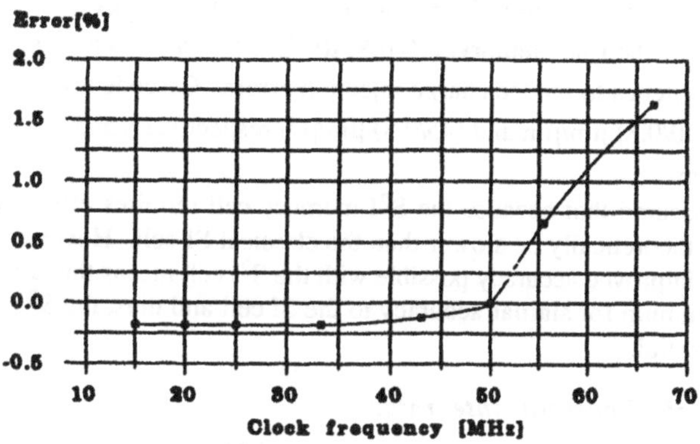

Figure 4 Memory current error versus sampling frequency.

To reduce the effects of channel length modulation a cascoded version of the memory cell is shown in Figure 3. Simulated performance of the cell based upon parameters of the Northern Telecom 0.8 μm BaTMOS technology and HSPICE is shown in Figure 4. The curve shows % error in cell current versus clock frequency. A maximum error in output current of -0.18% is achieved over an input signal range of ±50 μA. The error remains constant up to a clock frequency of 50 MHz beyond which the error begins to rise quite rapidly. Trade-offs have to be made between speed, accuracy, dynamic range, noise etc., in much the same way as SC circuits [6]. When designing the cascoded cell various procedures can be employed. In our work the following design procedure was adopted for optimising cell performance [15], the optimisation is in an approximate priority order.

(1) Choose V_{ref} close to $V_{dd}/2$ = Bias voltage.

(2) Choose L_1, L_2 close to minimum size to obtain fastest settling.

(3) Set the appropriate bias current by designing M_2 and adjust M_1's width to satisfy the condition $g_{m1}=g_{m2}$.

(4) Design the switch such that $1/R_{on}C_{gs}$ is not the dominant pole. Use minimum length and an aspect ratio of about 4.

(5) Choose $L_7=L_9$ and $L_1=L_2$ for balance.

(6) Adjust the widths of cascode transistors M_7 and M_9 to minimise the V_{ds} variations M_1 and M_2, while adjusting V_{B1} and V_{B2} to ensure the devices are saturated for $I_{in}=\pm I_{in\ max}$.

(7) Choose L% min and $(W/L)_5 = 25$, to get a good trade-off between on-resistance and stray capacitance.

(8) Finally adjust width of M_1 and M_2 to get critically damped behaviour giving close to optimum settling performance.

Probably the most attractive performance potential for the SI memory is that of speed at low voltage and several detailed chapters almost entirely devoted to enhanced SI memory performance can be found in [3]. In memory cell design transistor sizing is a very important procedure and many judicious decisions have to be made (see chapters 10.2 and 10.3). In our simulations close to minimum length devices were used to maximise speed, typically (15/0.8) µm/µm for the fine memory, (10/0.8) µm/µm for the coarse memory and (3/0.8) µm/µm for the switches to reduce charge injection errors. Cascode transistors M_7 and M_9 in Figure 3 are (20/0.8) µm/µm and (45/0.8) µm/µm respectively, and output transistor M_5 is (25/0.8) µm/µm.

One may assume that because the S²I memory cell requires effectively four clock phases, it would theoretically be slower than the classical SI cell. However, this is not the case because the improved accuracy possible with the S²I cell means that dimensions can be kept close to minimum for similar accuracy to the SI cell and thus, the S²I cell could have higher speed potential.

10.1.3.2 Switched-current integrator

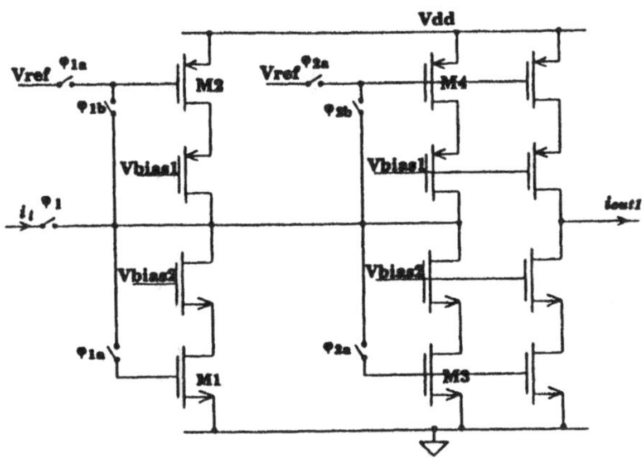

Figure 5 Lossless integrator.

A lossless integrator can be easily realised by introducing a feedback loop around a delay cell as shown in Figure 5. Essentially two memory cells are cascaded to form a delay and the output of the second is fed back to the input of the first. The integrator output i_{outI} is formed by copying the current in the M_3/M_4 branch and this gives a non-inverting output with a forward Euler transformation of

$$H_n(z) = \frac{i_{outI}(z)}{I_{in}(z)} = \frac{\alpha z^{-1}}{(1-z^{-1})} \qquad (1)$$

and by using an inverting output (copying M_1/M_2 current) the transfer becomes

$$H_i(z) = \frac{i_{outI}(z)}{I_{in}(z)} = \frac{\alpha z^{-1/2}}{(1-z^{-1})} \qquad (2)$$

which is the midpoint transformation of a continuous time non-inverting lossless integrator.

10.1.4 CMOS current-comparator/quantiser [7]

The current comparator employed in the design is based upon the recently reported design in [7] and is shown in Figure 6.

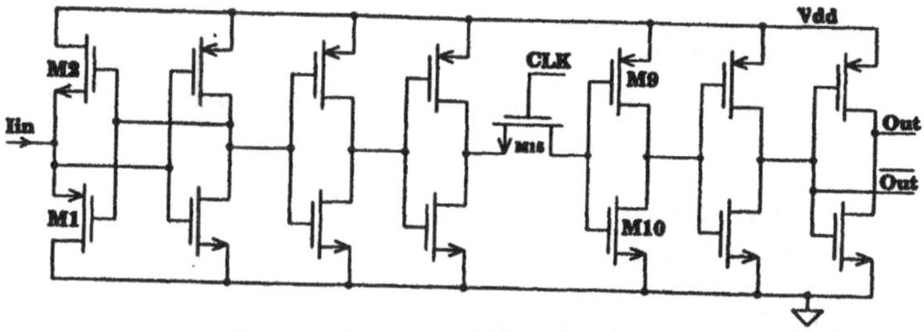

Figure 6 Current-comparator.

The input stage is a low impedance class B source follower which in addition to providing the necessary low impedance load for the integrator, it provides a feedback path from the first inverter setting the saturation level of the comparator. The connection largely decreases the response time of the comparator. The comparator will actually be latched in practice, as shown, to ensure that the comparator delay in the feedback balances the forward delay of the integrator allowing a stable current into the integrator during its sampling phase. The hold function of the comparator output is realised by M_9, M_{10} and M_{15} which form an inverter with a switched input. When the CLK is low (i.e. open switch) C_{gs9} and C_{gs10} will preserve their charge memorising the logic level. The delay here is $z^{-1/2}$ and so coupled with the transfer, $\frac{\alpha z^{-1/2}}{(1-z^{-1})}$, of the inverting integrator will give the required full delay.

10.1.5 Single bit current-mode D/A converter

A single bit DAC is formed by switching a positive and negative current-source in accordance to a control voltage V_{ctrl} as shown in Figure 7.

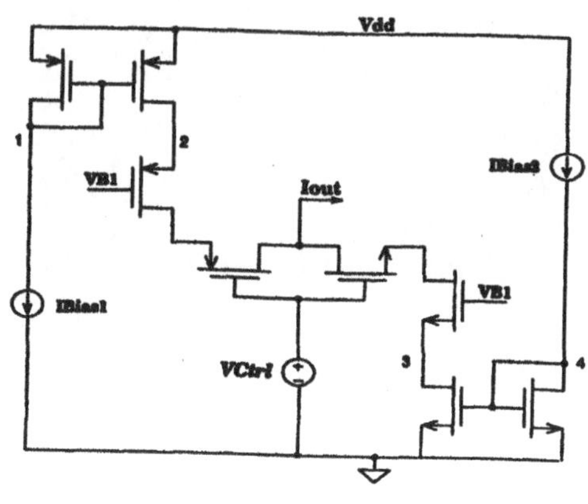

Figure 7 1-bit current-mode D/A converter.

The output current amplitude is made adjustable off chip, so that it can be tweaked to match the varying input current levels to the modulator. Figure 7 shows the basic design comprising two controlled current sources feeding the drains of an inverter. When V_{ctrl} is $+ve$, I_{bias1} sets the output level, and vice-versa for I_{bias2}.

10.1.6 CMOS SI ΣΔ modulator

The following specification for the SI modulator was chosen since it relates to the performance limitations of the previously described cells.

- Clock Frequency 50 MHz
- Base-band 500 KHz
- SNR 70-80 dB
- Power Consumption 4-5 mW
- Power supply 0-3.3 Volts

The architecture chosen is the two stage modulator design in Figure 8a. More details about this type of architecture can be found in Chapter 4.

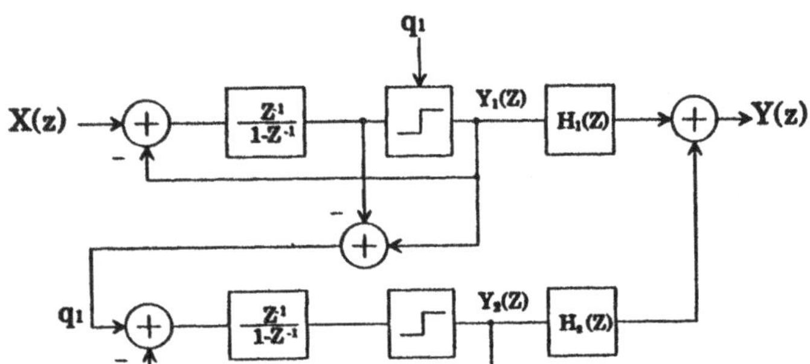

Figure 8a Two-stage ΣΔ modulator.

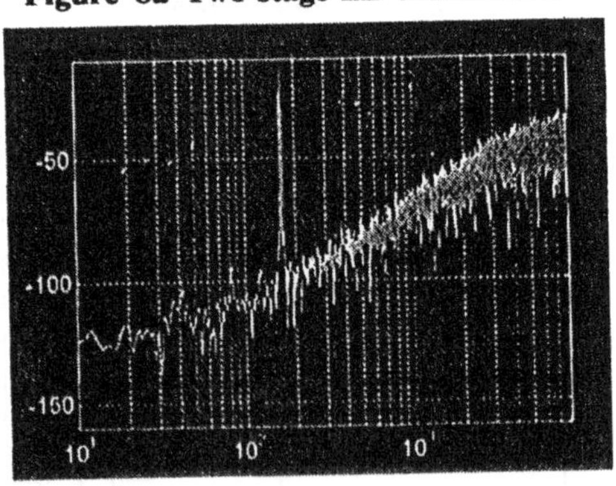

Figure 8b Spectrum from output of non-ideal modulator.

The main impetus for using a multistage architecture is that various stages of fine modulation can be performed on the quantisation error of the first stage essentially improving the SNR while maintaining single loop stability. The quantisation error out of the first stage q_1 is fed forward into a second modulator whose output is then summed to the output of the first modulator to cancel the error. Stage two will of course introduce its own quantisation error noise, but this noise is generally white noise from the previous stage and so does not contain much signal information. In much the same way as the S^2I switched-current memory cell the architecture can be viewed as one coarse stage and one fine.

Obviously the number of fine stages can be increased, but there will exist an optimum number beyond which the process is counter-productive. A full analysis of the modulator loop of Figure 1 yields the following output

$$Y(z) = X(z) \cdot z^{-1/2} + q_2 (1 - z^{-1})^2 \tag{3}$$

which gives a ideal SNR for the modulator of

$$\mathrm{SNR_{dB}} = 20\log\left(\frac{in_{rms}}{n_{0(output)}}\right) = \frac{\frac{\Delta/2}{\sqrt{2}}}{\frac{\Delta}{\sqrt{12}}\frac{\pi^2}{\sqrt{5}}\left(\frac{2}{\mathrm{OSR}}\right)^{5/2}} = \frac{\sqrt{30}}{2\pi^2}\left(\frac{2}{\mathrm{OSR}}\right)^{-5/2} \tag{4}$$

The output is essentially a delayed version of the input signal distorted by the high pass filtered quantisation error from the second stage. The function corresponds to the output of a conventional second order modulator. Assuming 100 times oversampling the model of equation (4) predicts an SNR of 73.8 dB. Note that in practice the integrators will introduce a loss factor α, equivalent to the finite gain loss of an opamp based integrator.

The main reason for choosing the architecture of Figure 8 to demonstrate the feasibility of an current-mode implementation is two-fold:

(i) The structure is repetitive which implies many summing operations which can easily be implemented in switched-currents just by output wire connections.

(ii) The maximum signal level in the loop of a second order modulator is generally higher than in a first order loop, since the 2-stage design is still a single order loop. It therefore requires less dynamic range in the integrator.

10.1.6.1 SI ΣΔ modulator performance

To verify the modulator performance, a complete high level simulation was performed using MATLAB. All the previously discussed building blocks were simulated using HSPICE and non idealities were extracted in order to perform a high level non-ideal simulation. The non-idealities included were the loss factor of the integrator $\alpha = 0.9964$ (worst case) which can be related to finite opamp gain in a classical integrator, 10% hysteresis in the comparator (worst case) and 5% mismatch between the two outputs of the DACs. The PSD of the modulator is shown in Figure 8b. The main difference between ideal and nonideal results is the reduction in low frequency SNR basically due to the loss factor of the integrator which in turn relates to the memory cell error. This loss manifests itself as a factor multiplying z in the integration described in equation (3). Essentially the term $q_2(1-\alpha z^{-1})^2$ is no longer ideal due to the factor α. An integrator in the classical sense would have a DC gain of $1/(1-\alpha)$ which is about 280 for $\alpha = 0.9964$. This yields a total loss of 30 dB SNR compared to the ideal $\alpha = 1$ situation. The effect of the loss introduced in the integrator, through the term $(1 - \alpha z^{-1})^2$, on the modulator performance is shown in Figure 9, where noise attenuation is plotted as a function of loss factor-α and oversampling sampling frequency f_S. A similar kind of curve would be obtained with the gain of a classical opamp integrator. **This is a very important result since it relates accuracy requirements of the memory cell to modulator S/N performance.**

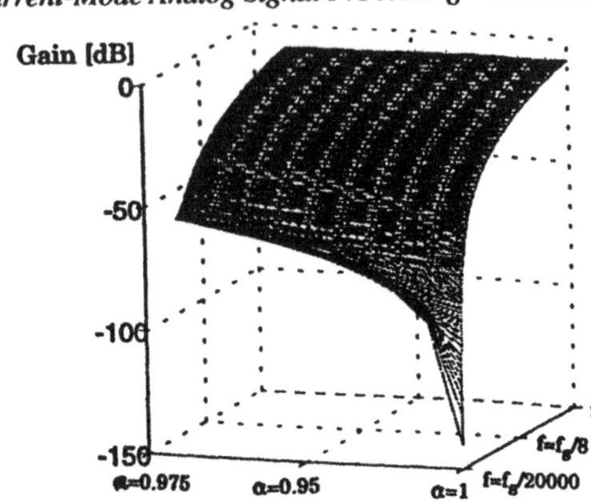

Figure 9 $(1 - \alpha z^{-1})^2$ **as a function of** α **and frequency.**

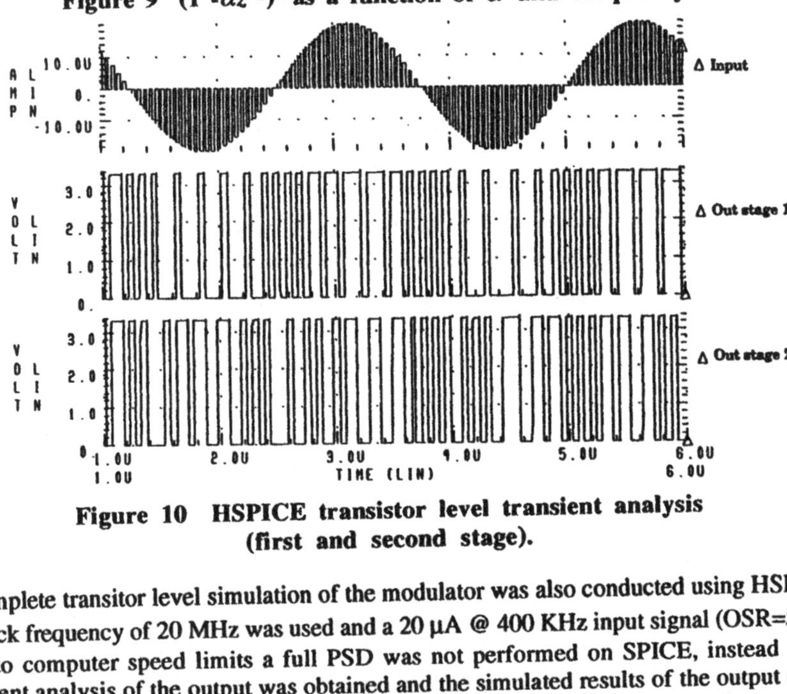

Figure 10 HSPICE transistor level transient analysis (first and second stage).

A complete transitor level simulation of the modulator was also conducted using HSPICE. A clock frequency of 20 MHz was used and a 20 μA @ 400 KHz input signal (OSR=50:1). Due to computer speed limits a full PSD was not performed on SPICE, instead a full transient analysis of the output was obtained and the simulated results of the output of the first and second stage of the modulator are shown in Figure 9. The results compare extremely well with those predicted from the non-ideal high level simulation. From simulation results the total power consumption of the modulator was found to be 4 mW and the rms SNR=73 dB. The OSR was set to 100 with a 50 MHz clock and the input current amplitude was 30 μA. Although the results are based upon simulations we are encouraged by the results of the full transistor level simulation on HSPICE since all the nonidealities of the cells have been considered.

10.1.7 Switched-current ΣΔ building blocks in GaAs technology

Traditionally circuits implemented in Gallium Arsenide technology have proved to be capable of higher operating speeds than their Silicon counterparts. The lower parasitics associated with the process combined with the higher electron mobility of the MESFET device gives GaAs technology a significant advantage over CMOS for very high frequency applications.

The trade-off between oversampling rate and resolution inherent in Sigma-Delta A/D conversion means, for some applications, the additional cost of the GaAs process may be justified by the increase in the modulator's operating speed. However, the high $1/f$ noise generated by the MESFETs becomes an important factor in the performance of the loop and may impair the resolution of a conventional low-pass modulator. For high-frequency band-pass applications no such problem exists.

An op-amp based design for a second-order $\Sigma\Delta$ modulator was presented in [8]. However, the design of a high speed high gain op-amp in GaAs is difficult due to the low device gain (g_m/g_o) caused by high frequency dispersion effects. The switched-current techniques proposed in [3] do not require op-amps and unlike switched-capacitor techniques are fully compatible with digital GaAs technology, since no high quality floating capacitor or complex switch control circuitry is required.

This section discusses the design of three building blocks required to realise a GaAs $\Sigma\Delta$ modulator based on switched-current techniques: integrator, 1-bit comparator and 1-bit D/A. The components are implemented as fully differential structures to reduce the effects of charge injection, offsets and power supply noise.

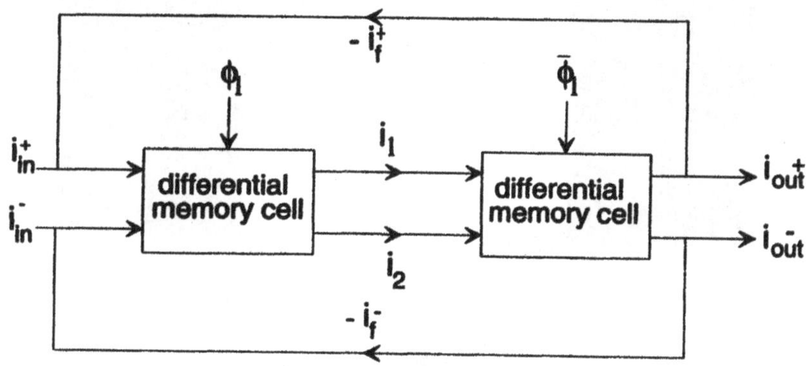

Figure 11 SI GaAs forward Euler integrator schematic.

10.1.7.1 GaAs switched-current integrator

The switched-current integrator is a forward Euler made from two differential SI memory cells as shown in Figure 11. The two cells are clocked on opposite clock phases to give a delay. A simplified diagram of the memory cell is shown in Figure 12.

It consists of two identical first generation SI memory cells presented in [9] and two continuous time non-inverting mirrors. The positive input current, i_{in}^+, is memorised in M_2 while the negative input, i_{in}^-, is stored in M_5 and duplicated in M_4. From M_4, i_{in}^- is mirrored to the high impedance drain of M_2 where it is combined with i_{in}^+ to give $i_{out}^+ = i_{in}^+ - i_{in}^-$. The widths of devices M_2-M_5 are chosen to be half those of M_1 and M_6 to give the correct integrator transfer function. In practice, due to the low output resistance of MESFETs, cascoding is required. Devices M_1-M_6 all require conventional cascoding while the devices in the mirrors need to be top cascoded since they act as current sources rather than sinks. A full treatment of the top cascoding technique is given in [10].

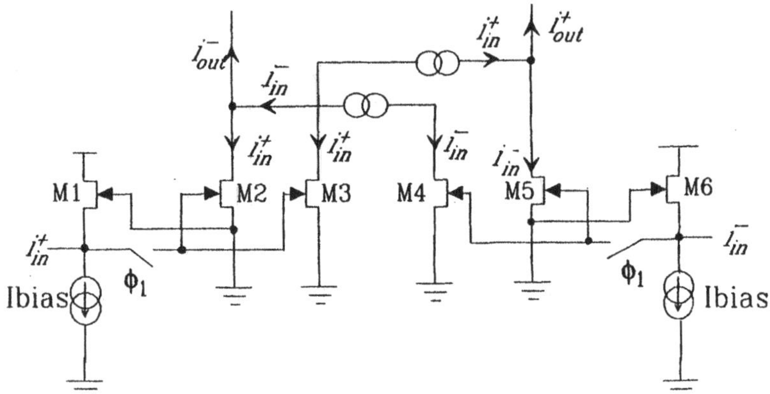

Figure 12 Simplified schematic of fully differential GaAs SI memory cell (current-mirror symbols are shown).

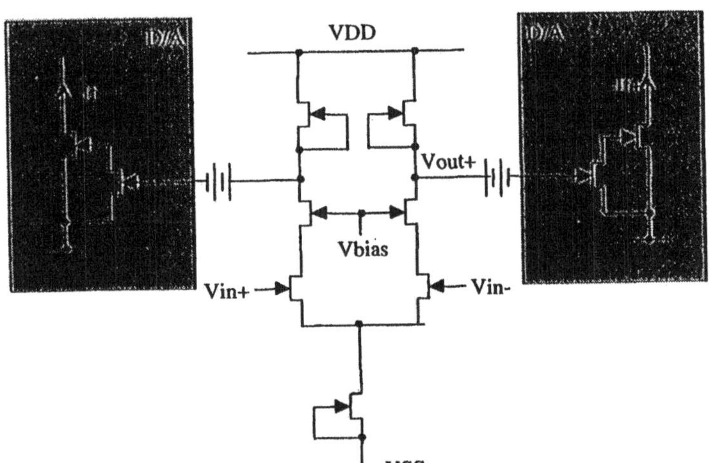

Figure 13 One-bit GaAs comparator and D/A circuit.

The practical integrator contains a coarse charge injection reduction method, dummy switch compensation. Here, a transistor, half the size of the main switch transistor, has its drain connected to its source and is inserted between the switch and the gate of the memory device. The dummy switch is clocked on the opposite phase to the main switch in order to absorb the charge flowing from the switch transistor's source.

10.1.7.2 One-bit GaAs voltage comparator

The comparator is a differential voltage amplifier [11] as illustrated in Figure 13. To give any reasonable gain the input devices require cascoding. The response to positive differential voltages is poor - the 'high' output can have a large range of values, but the performance is sufficient to provide a good current reference from the D/A covered in the next section. In a practical modulator the output of the comparator would be latched before the digital filtering stage. This would effectively provide a second stage of quantisation and produce a clean digital output.

10.1.7.3 One-bit GaAs D/A converter

The D/A converters which supply the current feedback references to the integrator are simply current sinks switched by the comparator output voltage as illustrated in Figure 13. Generally this voltage needs to be level-shifted so that the switching occurs about 0 V. Depending on the comparator output one of the sinks is switched on. This leads to a DC offset in the feedback references equal to half the sink's current, which can be simply removed by adjusting the bias current in the first memory cell in the integrator.

10.1.7.4 First order modulator

A first order ΣΔ modulator can be made from the building blocks presented above as illustrated in Figure 14. The conversion from the integrator current output to a voltage input for the comparator is performed by using a MESFET biased in its triode region as shown. This simple conversion is like that in [12] and is possible since the linearity of the conversion is not that important.

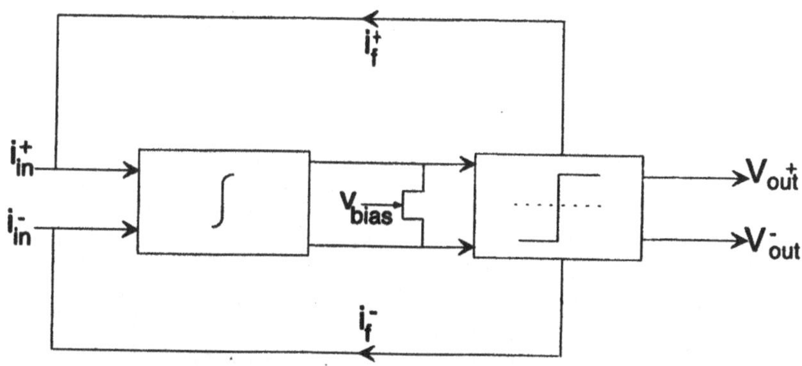

Figure 14 Schematic of first order GaAs ΣΔ modulator.

The first order modulator was simulated on HSPICE using parameters for a typical 1μm GaAs MESFET process with a threshold voltage of -1 V [13]. The modulator output for a 200 μA 10 MHz sinusoidal input with an oversampling rate of 1 GHz and a feedback reference of 150 μA is shown in Figure 15. The modulator does output more 'highs' when the input is around positive full scale and more 'lows' when negative full scale. Also when the input is near zero the output is rapidly changing states. The basic functionality of the modulator appears correct in the time domain and would benefit greatly from latching the output.

However, further simulation is required to provide sufficient data to estimate the power spectrum of the output to verify the correct performance of the noise-shaping in the frequency domain. The response does indicate the feasibility of the building blocks proposed above. Higher performance GaAs SI memory cells based upon second generation S^2I techniques have been recently introduced [14], and will form the basis of an IC version of the modulator currently being prepared for integration on the GEC Plessey F2 GaAs MESFET process.

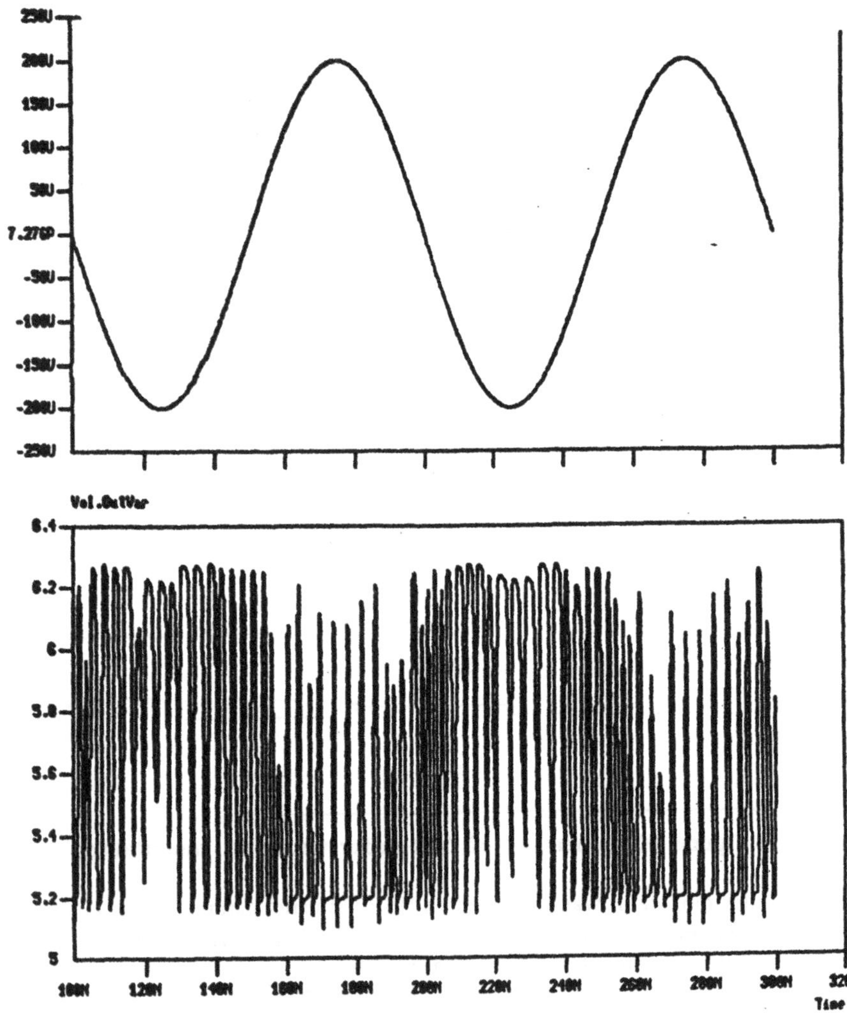

Figure 15 Output of first order GaAs ΣΔ modulator.

10.1.8 Enhanced CMOS memory cell development

In this section we propose an extension of the two-step S^2I technique to *n*-step (S^nI) which cancels higher order charge injection errors. Remembering that for high S/N ratios in a modulator the memory cell error (α) must be minimised. It will also be shown that this new scheme can be easily applied to class AB cells [5] and GaAs second generation cells [14].

10.1.8.1 n-step high precision memory cell (S^nI) scheme [5]

The idea behind the previously described S^2I scheme was that during the output phase ϕ_2, the error created in M_1 is subtracted by the same error memorised in M_2 and therefore the output current is free of error. The charge injection from the fine memory switch is treated as an offset and can be automatically cancelled when the cell is used in a delay cell or an integrator. It should be noted, however, that the charge injection in the fine cell is only to a first order signal independent and can be treated as an offset only if the switch is very small compared to the memory cell.

In some applications such as precision integrators and filtering, this first order cancellation may not be enough with the (α)-factor rather high. A multi-step error cancellation scheme [5] can be readily constructed based on the same principle as the two-step approach described earlier. Instead of using just one, we can use more fine memory cells to further access and cancel the residual error so that the output is an accurate representation of the input and the charge injection from the last fine cell is almost a pure offset even if a relatively small ratio of the memory cell to the switch is used. The number of fine stages can be determined by the accuracy requirement based on the applications. Figure 16 shows a proposed S^nI (*n*-step) memory cell and the clock waveforms. '*n*' represents the number of steps.

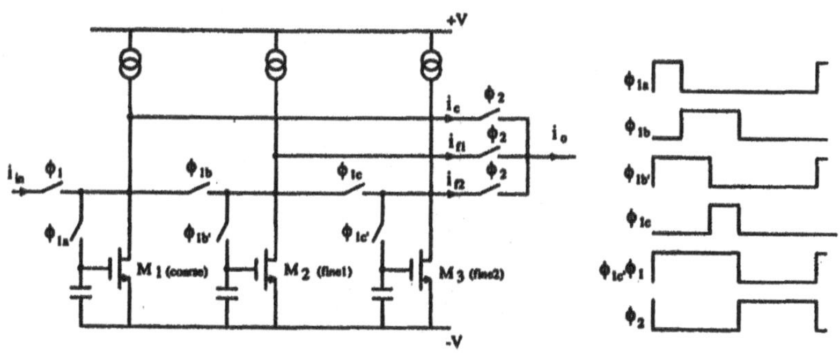

Figure 16 CMOS class A S^nI cell with three stages shown.

For simplicity, only three stages (one stage of coarse memory and two stages of fine memories) are shown. It can be seen that instead of using a p-channel MOSFET as the fine memory, the S^nI cell uses cascaded identical class A memory cells as the fine memories. In this way, an extension to *n* stages can be easily accomplished, which is difficult to achieve with conventional S^2I cells alone. The other advantage of using cascaded structure is that the requirement for the accurate g_m matching between the n- and the p-channel transistors is avoided. The operation of a three-stage S^nI cell is briefly described as below: The clock period ϕ_1 can be split into three sub-clocks. During ϕ_{1a}, the coarse memory M_1 sinks the input current. Then during ϕ_{1b}, the gate switch of M_1 is open while the input current is still kept on so that the first fine memory M_2 can access the error from the coarse memory M_1.

During ϕ_{1c}, the gate of the first fine memory is isolated while the error current from the coarse memory is still kept on so that the residual error from the first fine memory, which is now a second order signal dependent error, is accessed by the second fine memory M_3. Then during ϕ_2, the output current from the respective stage is

$$i_c = i_{in} - \delta_1 \qquad i_{f1} = \delta_1 - \delta_2 \qquad i_{f2} = \delta_2 - \delta_3 \tag{5}$$

where i_c, i_{f1} and i_{f2} are the output currents from the coarse, the first and the second fine memory, respectively, and δ_i represents the error from the i-th stage. The output current is the combination of the three:

$$i_o = i_c + i_{f1} + i_{f2} = i_{in} - \delta_3 \tag{6}$$

Here δ_3 is a second order error. This procedure can in principle be carried on in n successive fine memory stages until the accuracy requirement is finally reached. The output current with n stages is

$$i_o = i_{in} - \delta_{n+1} \tag{7}$$

and δ_{n+1} is an n-th order error.

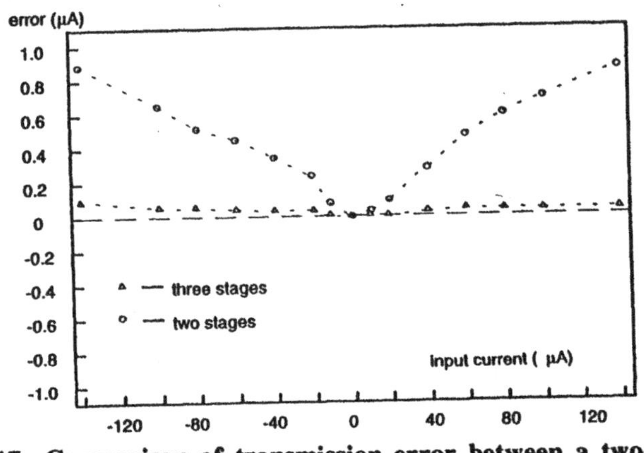

Figure 17 Comparison of transmission error between a two-and a three-stage S^{n}I cell.

Although this n-step error cancellation needs an initial clock to be split up into n sub-clocks, the speed does not necessarily decrease much because of the following reasons:

1) The error accessed in the S^{n}I cell is a lump error which includes charge injection, output-input conductance ratio and settling errors. So the error access carried out by the fine stages does not have to wait until the coarse stage is completely settled.

2) Since more fine memories are used, the aspect ratio of memory cell to switch can be relatively small so that the input capacitance is reduced.

Figure 17 shows the comparison of transmission errors between a two- and a three-stage S^{n}I cell obtained by using HSPICE simulations on a standard 1 µm digital CMOS process.

The aspect ratios of the cell FET's and the switch FET's are 20/1.2 and 2.4/1.2, respectively. It can be seen that an improvement of over a factor of 5 is achieved by adding one more fine stage. The percentage transmission error of the cell with two fine stages can be as small as 0.09% with the aspect ratio of the cell to the switch of about 8, as compared to a percentage error of 0.6% with only one fine stage used.

10.1.8.2 CMOS class AB cell

The principle of S^nI can be easily applied to the recently proposed CMOS class AB [3, 5] and second generation GaAs MESFET memory cell [3, 14] as well. A three stage CMOS class AB S^nI cell is shown in Figure 18. Based on the clock waveforms given, its operation can be evaluated as follows: during ϕ_{1a}, the coarse memory (composed of M_5 and M_6) samples the input current through M_3 and M_4. Then during ϕ_{1b}, the memorised current in the coarse memory which consists of the signal and error is fed back to the input so that the input current is subtracted and only error is accessed by the first fine memory (composed of M_7 and M_8). The same procedure is carried on during ϕ_{1c} when the residual error from the first fine memory is accessed by the second fine memory (composed of M_9 and M_{10}). Then during ϕ_2, the output current i_o is

$$i_o = i_{in} - \delta_3 \tag{8}$$

where δ_3 is the residual error from the second fine memory and is a second order error. The final error can be made minimum if enough number of fine memory stages are cascaded. The number of fine memory stages needed can be decided based on the accuracy and therefore speed requirements. The simulations indicate that the transmission error of a three-stage S^nI cell is low enough to meet the requirements for most of precision applications.

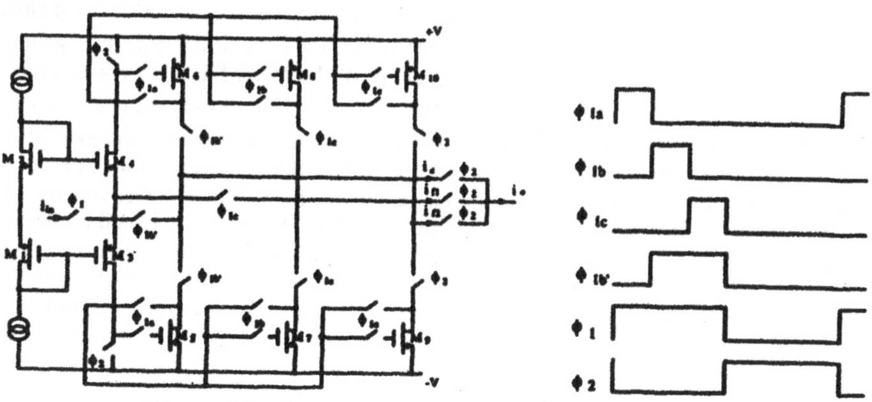

Figure 18 Class AB S^nI cell with three stages.

10.1.9 Enhanced GaAs memory cell development

10.1.9.1 GaAs SI memory cell

A first generation GaAs MESFET switched-current memory cell was proposed in [9]. The main problem with this first generation GaAs MESFET memory cell (Figure 12), used in the previous GaAs modulator design, is that it exhibits a degree of non-linearity under large signal conditions which will introduce offset, gain error and distortion components to the signals. The operation of the first generation cell is also sensitive to the device mismatch which is relatively large in GaAs MESFET technology. Therefore, a memory cell with better linearity and lower process sensitivity needs to be developed. Figure 19 shows a recently reported second generation GaAs MESFET SI memory [14]. Due to the Schottky diode gate, a GaAs MESFET cannot be diode connected as in the case of a CMOS second generation memory cell. This problem can be solved by using a diode chain and a current source to shift the input DC level down to a value which is low enough to avoid gate conductance while the difference between the input and gate voltage keeps the memory FET in saturation as shown in Figure 19. Since the diode-current source branch is only used as biasing, the current can be designed to be quite small.

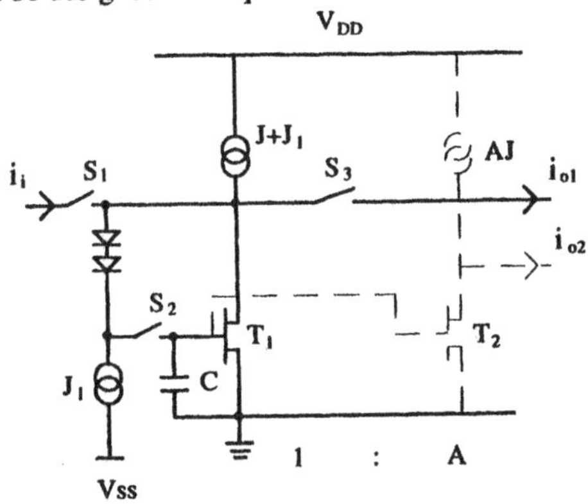

Figure 19 Second generation GaAs memory cell.

Furthermore, the two successive stages of cells in a delay cell or an integrator can share the same biasing chain since they operate in different clock phases. Hence the level shifting stage increases neither much power consumption nor much circuit complexity. During ϕ_1, switches S_1 and S_2 are closed and the gate voltage of the MESFET T_1 increases until it is sufficient to sink the sum of the input and bias currents. This Vgs also appears across the capacitor C, which is actually the gate capacitance of T_1. During ϕ_2, the gate of T_1 is isolated and its drain is connected to the output. The value of Vgs is now stored by C, enabling the output current to remain constant whilst the input current is free to vary. Thus the cell acts as a current-mode sample-and-hold element. Since the same cell transistor T_1 samples the input current as well as providing the output current, the effect of transistor mismatch is eliminated and the output is inherently linear if T_1 is ideal. It can be seen that C is only required to memorise the value of an applied voltage rather than to store and transfer an accurate amount of charge as in the case of switched-capacitor circuits, linearity requirements for the capacitor are not required. To obtain a scaled full period output, the current of T_1 can be mirrored and scaled using a current mirror stage (shown dashed in Figure 19).

10.1.9.2 Second generation two-step GaAS S^2I cell [14]

Although the techniques such as dummy switches and differential circuits have been used to reduce the charge injection, they can only partially cancel the errors. The S^2I CMOS memory cell, which uses a two-step cancellation scheme, has proved to be the most effective. However, the S^2I scheme used in CMOS cells cannot be duplicated directly in GaAs MESFET cells since p-channel MESFETs are not available. Here we propose a modified S^2I cell which suits a GaAs MESFET realisation. The cell and the clock waveforms are shown in Figure 20. Instead of using a p-channel transistor as a fine memory, the cell uses two identical n-channel memory cells T_1 and T_2, which function as coarse and fine memories respectively. During ϕ_{1a}, the coarse memory T_1 samples the input in the same way as a normal second generation cell. During ϕ_{1b}, the fine memory cell samples the difference between the input and the memorised input by T_1, which is the signal dependent error current. During output phase ϕ_2, the input current is disconnected and the output is formed by the difference in current between the two memories, which to a first order will be identical to the input current since the error has been subtracted.

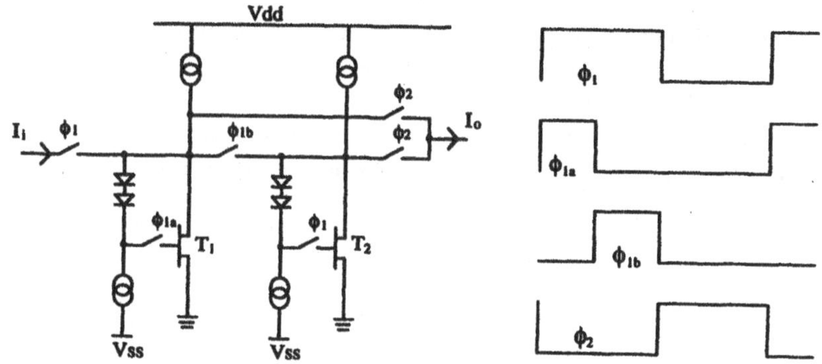

Figure 20 GaAS MESFET S^2I memory cell [14].

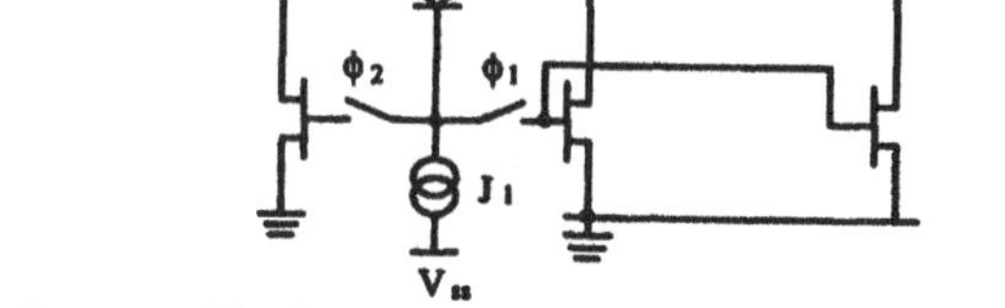

Figure 21 A generalised GaAs MESFET switched-current integrator.

A potential advantage of the modified S^2I memory cell over the conventional S^2I memory cell is that more fine cells can be cascaded to further cancel the residual error if the first order cancellation is not enough. The technique can similarly be extended to *n*-stages of fine memory similar to the CMOS version shown in Figure 16.

10.1.9.3 Generalised second generation GaAs integrator

A generalised switched-current integrator made from the superposition of forward and backward Euler and feed-forward inputs and realised in GaAs MESFET memory cells is shown in Figure 21. For simplicity, only general second generation cells are shown in the circuit. It can be seen that the level shifting stage is shared by the two cells. If the S^2I cells are used in the integrator, the level shifting stages for the coarse and fine cells can be shared respectively in the same way by the two cell stages. The transfer function of the integrator is given by

$$i_{out}(z) = \frac{A_1 z^{-1}}{(1 - Bz^{-1})} i_1(z) - \frac{A_2}{(1 - Bz^{-1})} i_2(z) - \frac{A_3(1 - z^{-1})}{(1 - Bz^{-1})} i_3(z) \quad (9)$$

where A_1, A_2, A_3 and B are scaling factors determined by transistor aspect ratios.

10.1.9.4 Simulation results

To verify the operation of the GaAs second generation cell of Figure 19, it was simulated using HSPICE on a commercial 1μm GaAs MESFET technology. To avoid the error introduced by the high output conductance of the GaAs MESFET, double cascodes are used for the SI and S^2I cell [3]. The transfer characteristics and the transmission errors of the modified S^2I cell and the second generation cell using dummy switches are shown in Figure 22 for an operating frequency of 50 MHz.

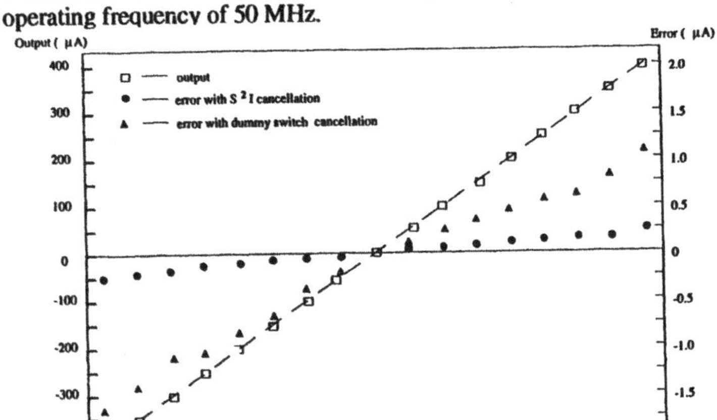

Figure 22 The transfer characteristics and transmission errors of the modified S^2I cell and the second generation cell using dummy switches.

It can be seen that by using the S^2I cell, the error is reduced by more than a factor of 4 compared to the SI cell with dummy switch compensation alone and the percentage transmission error in the S^2I cell is less than 0.05%. The input waveform and the output response of the integrator in Figure 21 with A_1=1, A_2=A_3=0 are shown in Figure 23. It can be seen that the integrator functions correctly at a clock rate of 1 GHz.

Since the operation of these GaAs MESFET SI cells is very similar to that of the CMOS counterpart, the design techniques developed for CMOS switched-current circuits [6] can in principle be duplicated in GaAs technology. The techniques presented show promise for the development of very high speed precision filters and date converters in pure digital GaAs process technology.

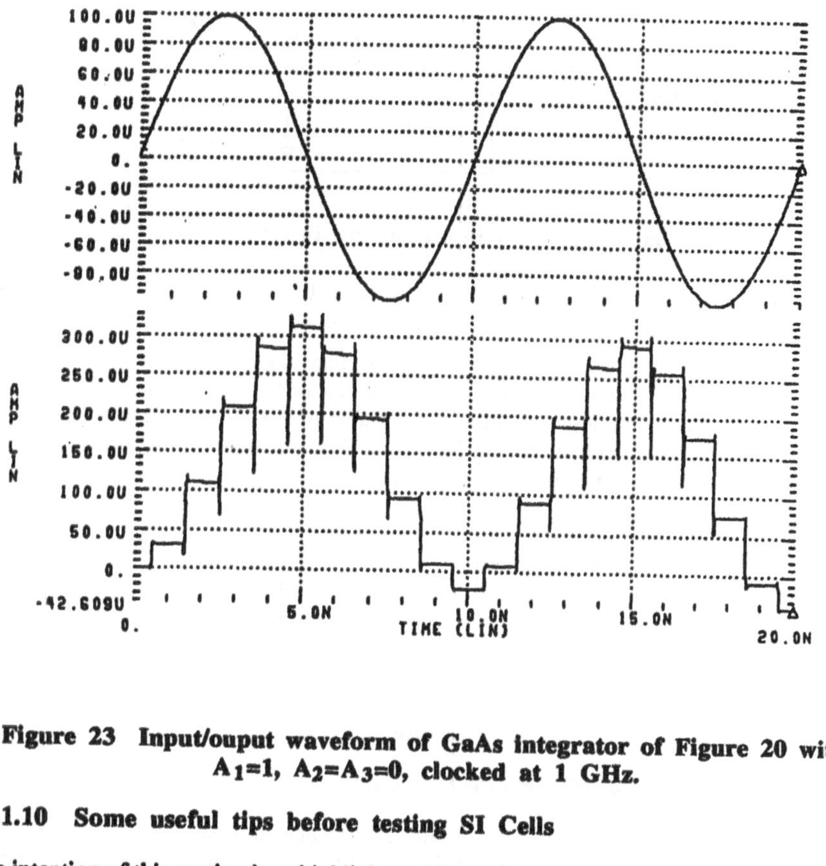

Figure 23 Input/ouput waveform of GaAs integrator of Figure 20 with A_1=1, A_2=A_3=0, clocked at 1 GHz.

10.1.10 Some useful tips before testing SI Cells

The intention of this section is to highlight problems related to interfacing and test of current mode chips in general and to suggest methods to overcome the difficulties.

10.1.10.1 Applying input signals

To get high frequency current-signals into an SI chip is not so straightforward. The problem is caused by the pole resulting from the parasitic input-pad capacitance in combination with a relatively high input impedance of the current cell.

Figure 24 above shows a current mode chip with a current input signal I_{in}. The input impedance and capacitance of the current mode cell are represented by the resistance $1/g_m$ and the capacitance C_i respectively, whereas the capacitance C_p is the sum of the parasitic pad and pin capacitances. It can then be shown that a resulting pole at the frequency f_{pi} is formed by the input network, where

$$f_{pi} = \frac{g_m}{2\pi\left(C_p + C_i\right)} \tag{10}$$

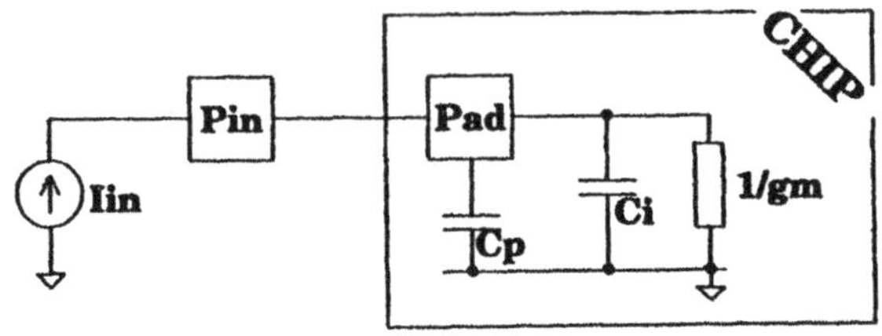

Figure 24 Applying a current input signal.

In the case of the ΣΔ modulator designed in section 10.1.6, $g_m \approx 400\ \mu S$, $C_i \approx 50$ fF and C_p is assumed to be approximately 5 pF, which results in a pole at $f_{pi} \approx 13$ MHz and a time constant of approximately 12.5 ns. The maximum simulated clock frequency of the current memory S^2I cell is 50 MHz which requires a settling time of 5 ns[1], hence the settling time of the input network is much larger than the settling time of the actual current memory cell. The practical effect is therefore that an individual S^2I cell cannot be tested over its full frequency range unless special precautions are taken.

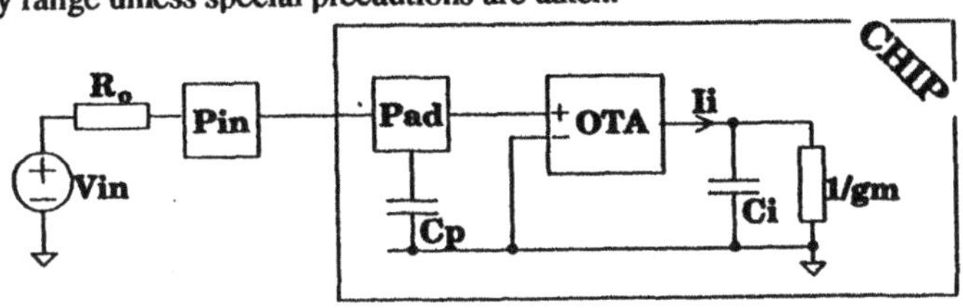

Figure 25 OTA used to interface current mode cell.

10.1.10.2 Cancellation of the effect of C_p

To cancel the effect of the parasitic capacitance C_p, the current-input can be converted to a voltage-input by employing an on-chip Operational Transconductance Amplifier (OTA). The OTA converts the input voltage to a corresponding current which is then used as the input signal to the current-mode circuit. An example system is shown in Figure 25. The pole at the input will now be determined by Cp together with the voltage source output impedance, and is located at frequency f_{pv} where

$$f_{pv} = \frac{1}{2\pi \cdot R_o\left(C_p + C_i\right)} \tag{11}$$

R_o is the output resistance of the voltage source. Assuming $R_o = 50\ \Omega$ and $C_p + C_i \approx 5$ pF implies a time constant of approximately 0.25 ns. This means that the settling time of the input network is far less than the settling time of the memory cell and hence the cell can be tested over its full frequency range.

[1] Remember that the settling time of the individual phases of the S^2I cell has to be 1/4 of 1/(max. clock frequency).

The on-chip OTA should be highly linear so that harmonic distortion does not corrupt the SNR of the SI -ΣΔ modulator. However, when testing a current memory cell on its own, the nonlinearity can be compensated for and is therefore of minor importance.

A very simple but sufficiently linear OTA was realised by making use of bipolar transistors available in the BaTMOS fabrication process used for the modulator design.

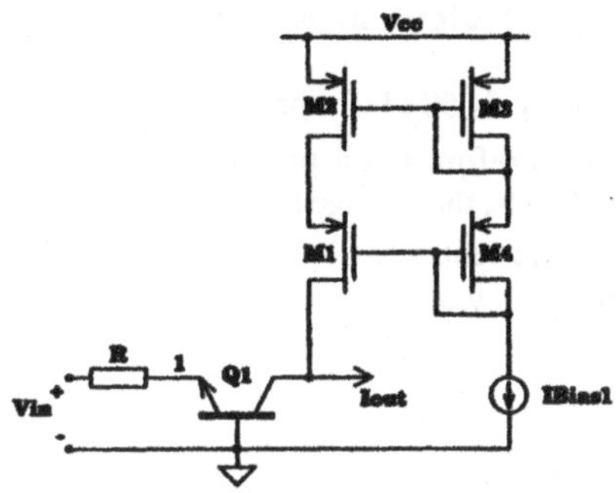

Figure 26 On chip OTA.

Figure 26 shows a transistor schematic of the OTA. The NPN transistor is operated in its triode region and biased with a 500 μA current resulting from DC-offsetting V_{in} to -5.9 V. The resulting transconductance g_m is then $g_m = \frac{I_c}{V_t} = \frac{500\,\text{mA}}{25.8\,\text{mV}} \approx 19\,\text{mS}$, which means that the voltage fluctuations at node 1 with a ±50 μA input signal is very low. The resulting non-linearity introduced by the resistor R is small. To choose $g_m \approx 19\,\text{mS}$ also means that the input impedance of the OTA becomes approximately 50 Ω which is important in order to gain a proper termination and avoid reflections in the input line. The bias current through Q_1 is sourced by a conventional improved MOS Wilson current source formed by M_1-M_4 and I_{Bias1}. I_{Bias1} may be off-chip but does not have to be trimable; an offset in I_{out} may be cancelled by adjusting the DC offset voltage of V_{in}. R was chosen to be 10 kΩ to provide a ±0.5 Volt input signal amplitude for a ±50 μA output current.

Simulations of the OTA were accomplished using the S^2I current memory cell as a load and the total harmonic distortion (THD) was measured. Results showed that the THD was limited to 20 m% for frequencies up to 200 kHz. For an input frequency of 20 MHz the THD had increased to 130 m%.

10.1.10.3 Measuring the output signal

The output stage of a current mode circuit is illustrated in Figure 27. The output current source of the transistor(s) is represented by I, while the finite output conductance of the transistor(s) is represented by g_{ds}. As we can see from Figure 27 this circuit will form a pole at the frequency f_{pio}, where

$$f_{pio} = \frac{1}{2\pi\left(R_L + \frac{1}{g_{ds}}\right)C_p} \tag{12}$$

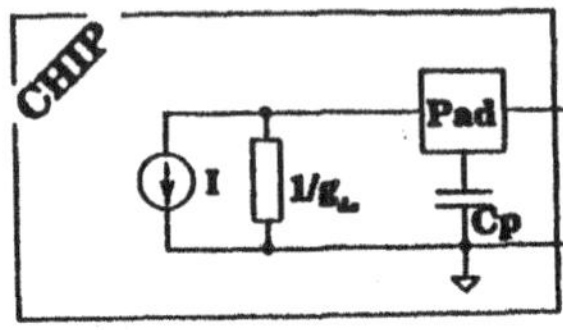

Figure 27 Output stage of current mode circuit.

Assuming the load resistor R_L is 50 Ω (for proper termination) it turns out that $f_{pio} \approx f_{pv}$ and the resulting pole is therefore not a problem. However, for current mode circuits operating with low signal levels, the voltage swing at the output may be very small when the current is dropped in a 50 Ω resistor. A solution is to use a current-conveyor/current-amplifier [2] or a high frequency transresistance amplifier maybe using a current-feedback opamp. A possible on-chip output stage which will amplify the current output signal is shown in Figure 28.

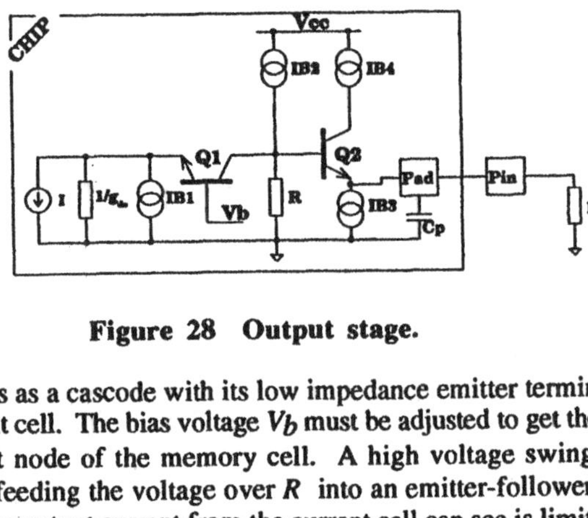

Figure 28 Output stage.

Transistor Q_1 functions as a cascode with its low impedance emitter terminal connected to the output of the current cell. The bias voltage V_b must be adjusted to get the correct voltage potential at the output node of the memory cell. A high voltage swing is obtained by choosing R large and feeding the voltage over R into an emitter-follower. However, the maximum resistance the output current from the current cell can see is limited by the small-signal input resistance $r_\pi = \dfrac{\beta}{g_m}$ of Q_2.

10.1.11 CAD tools and simulation results

10.1.11.1 About SI simulations

When entering the current mode domain one should remember a couple of things which are significantly different from voltage domain simulations.

- A signal from an ideal source cannot be cut off by simply applying a switch in its signal path. The signal has to be redirected to another drain.
- The output signal from a circuit has to be connected to a current drain with proper voltage biasing (i.e. an ideal voltage source). The magnitude of the signal can then be measured by monitoring the current through the drain component.

10.1.11.2 Simulation of S^2I cell

(i) Clock scheme and clock signals

The following three phase clock scheme was used for clocking the S^2I cell

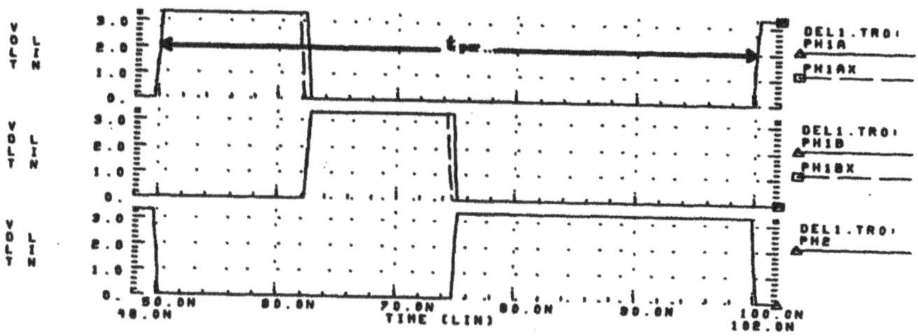

Figure 29 Clock scheme.

Phase φ_{1ax} (PH1AX in Figure 29) and φ_{1bx}, are shortened versions of phase φ_{1a} and φ_{1b} used to control the equivalent transistor switches in the S^2I cell of Figure 2 discussed earlier, to ensure that the currents are stored properly in the memory transistors before the next phase is entered. As can be seen from Figure 29, the φ_{1a} signal does not go low before , φ_{1ax} has reached 0 Volt. This is slightly restrictive, but for simplicity it should be acceptable. In addition to the clocks shown in the Figure, inverted versions of φ_{1a} and φ_{1b} were used to control the charge injection cancellation transistors.

Switched circuits often make use of many different clock signals for controlling switches. To be able to vary the clock frequency without changing tens of numbers, the SPICE PARAM OPTION should be used.

(ii) A one transistor settling problem

Unlike in the SC circuits, slew rate is not a major problem when using the SI technique due to the small voltage fluctuations in the circuit. Maximum operational frequency is therefore confined by the overall maximum settling-time of the SI circuit.

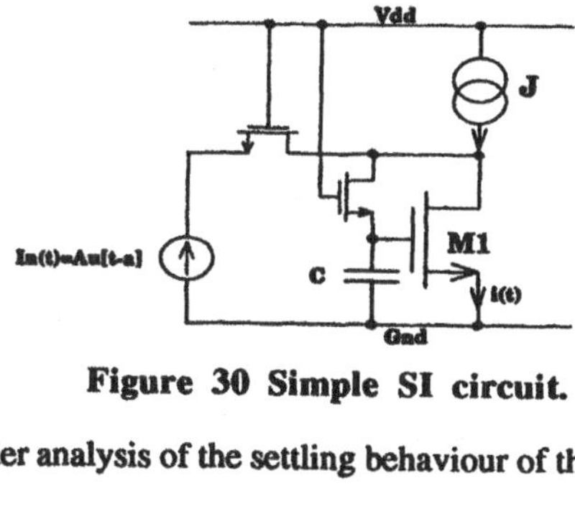

Figure 30 Simple SI circuit.

From a simple first order analysis of the settling behaviour of the simple SI circuit shown in Figure 30 we get

$$i(t) = A\left(1 - e^{\frac{g_m(t-a)}{C}}\right)u[t-a] + J \qquad (13)$$

where C is the parasitic gate-to-source capacitance of M_1 and g_m is the transconductance of the transistor. This gives a settling time-constant $\tau_s = C / g_m$. Since

$$C \propto W \quad \text{and} \quad g_m \propto \sqrt{W}$$

$$\Downarrow$$

$$\tau \propto \frac{W}{\sqrt{W}}$$

$$\Downarrow$$

$$\tau_s(W) > 0, \quad \forall W$$

The above indicates that the settling time is reduced to a minimum if W is minimised. However, this is not strictly true for high frequency systems where parasitic effects play an important role. For this kind of circuit a more detailed analysis reveals that the system is of higher order, and that the settling time is not a monotonous function of the transistor width.

The settling time of the circuit in Figure 30 was simulated by applying a 50 μA current step at its input. The results are shown in Figure 31 to 0.1 % settling..

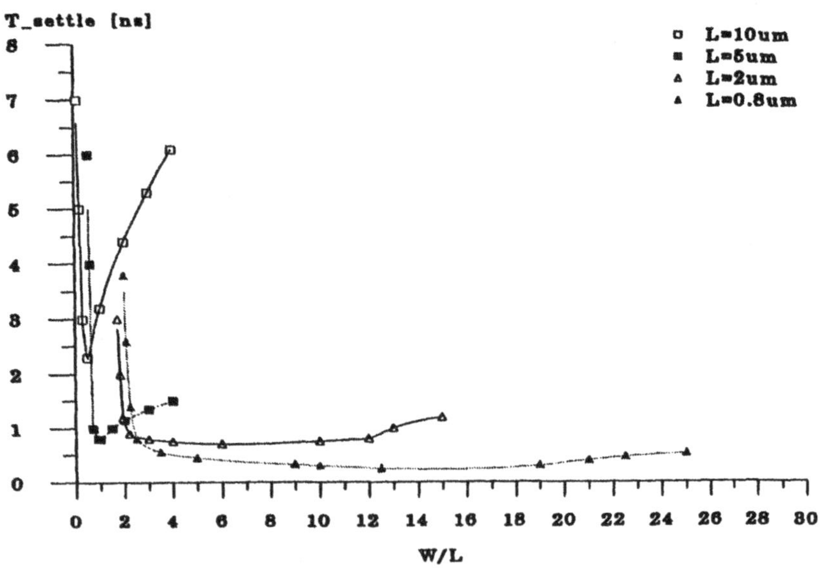

Figure 31 Settling time curves for basic cell.

10.1.12 Conclusions

We have hopefully shed some light on the realisation of high speed ΣΔ modulators employing SI techniques. Results are encouraging with respectable ΣΔ performance achieved in Silicon and GaAs process technology. Although the technique is aimed for use on pure digital processes, it needs to demonstrate equivalent performance to classical techniques using double-poly, without considerable increase in chip area. Full details of transistor dimensions have not been included in this chapter. In most cases minimum size devices have been employed to optimise speed, but obviously at the expense of accuracy.

In summary the principal advantages of using SI techniques particularly for $\Sigma\Delta$ modulators is the compatibility with the on-chip DSP and thus economical advantages of utilising a standard digital VLSI process. Summing nodes in multistage modulators are implemented simply by short circuits in the current domain and finally because we are dealing with current signals, the technique has high speed potential. A drawback of the technique is, however, the requirement for linear transconductors at the SI cells input to interface from generally a voltage mode environment. However, we are still at the tip of the iceberg and if one reminisces about the early stages of the SC techniques then clearly many SI circuits and systems innovations lie ahead. Hopefully this tutorial has given a brief overview of some of our research in the the SI area.

Acknowledgements

Part of this work (switched-current modulators etc.) is based on an IEE Colloquium presentation on Oversampling Techniques, March 1994 and also on collaboration work between Imperial College and the Norwegian Institute of Technology, Trondheim, Norway. Thanks are also due to my graduate student Mervin Adams and Dr Shuo Xiou for assistance with the work on Gallium Arsenide. The author would also like to gratefully acknowledge the financial support of the UK SERC (Grant numbers GR/H76944 and GR/J14530) and the Analog VLSI Group of Northern Telecom for both financial and technical support related to this research. Professor Tony Constantidides of the Dept. of EEE, Imperial College, is acknowlegded for very useful discussion related to the modulator design.

References

[1] J. C. Candy and G. C. Temes, *Oversampling Delta-Sigma data converters. Theory, design and simulation* , John Wiley & Sons Ltd; 1993.

[2] C. Toumazou, F. J. Lidgey and D. G. Haigh (Editors), *Analog IC design - the current-mode approach*, Peter Peregrinus Ltd., 1990 on behalf of the IEE.

[3] C. Toumazou, J. B. Hughes and N. C. Battersby (Editors), *Switched-Currents - An analog technique for digital technology*, Peter Peregrinus Ltd., 1993 on behalf of the IEE.

[4] J. B Hughes and K. Moulding, "S^2I: a switched-current technique for high performance", *Electronics Letters*, 29, pp. 1400-1401, 1993.

[5] C. Toumazou and S. Xiou "An N-Step charge injection cancellation scheme for switched current circuits", accepted for publication in *Electronics Letters*.

[6] P. Shah and C. Toumazou, "Trading speed for dynamic range in Switched-Current Circuits", *Proc. IEEE ISCAS*, London UK, June 1994.

[7] H Traff, "Novel approach to High Speed CMOS current comparators", *Electronics Letters*, Vol. 28, pp.310-12, Jan. 1992.

[8] K. T. Chan, "A GaAs Delta-Sigma Modulator for Oversampled A/D Converters", PhD Dissertation, UCLA 1991.

[9] C. Toumazou, N. C. Battersby and M. Punwani, "GaAs Switched-Current Techniques for Front-End Analog Signal Processing Applications", *IEEE Midwest Symposium on Circuits and Systems*, Washington D.C., August 1992.

[10] C. Toumazou and D. G. Haigh, " Design of GaAs operational amplifiers for analog sampled-data applications", *IEEE Trans. Circuits and Systems*, Vol. 37, pp. 922-935, July 1990.

[11] J. Mun (Editor), *GaAs Integrated Circuits*, BSP Professional Books, 1988.

[12] C. Toumazou and D. G. Haigh, "Cross-coupled GaAs MESFET circuits for potential MMIC applications", *Proc. IEEE International Symp. Circuits Systems*, June 1991.

[13] S. J. Daubert and D. Vallancourt, "A Transistor-Only Current-Mode ΣΔ Modulator", *IEEE Journal of Solid-State Circuits*, Vol. 27, No. 5, May 1992.

[14] S. Xiou and C. Toumazou "Second generation GaAS Switched-Current memory cells", accepted for publication in *Electronics Letters*, April 1994.

[15] Geir Edvard Saether, "High Speed Sigma Delta Modulator. The switched-current approach", Internal Project, Imperial College and Norwegian Institute of Technology.

Chapter 10.2

Analysis and Design of Speed and Dynamic Range in Switched Current Cells

Peter Shah and Chris Toumazou

10.2.1 Introduction

A new analogue sampled-data signal processing technique has evolved in later years which appears to be more attractive than the switched-capacitor technique for mixed-mode applications since it fully exploits digital CMOS process technology. The technique has been termed 'switched-currents' by virtue of its current-mode signal operation. It uses non-linear MOS capacitors to store charge and so does not require the 2nd layer of double poly required in SC circuits, and it does not use operational amplifiers. Over the past few years there has been a tremendous interest in this technique [2, 3, 6] resulting from the cost benefit of realising filtering and converter functions on digital technology. In situations however, where the cost benefit is not as imperative as high performance there is a need to design demonstrator circuits that compare in performance to equivalent SC techniques. The potential for higher speed and lower chip area than equivalent switched capacitor technique has yet to be fully demonstrated [3].

In this tutorial chapter we present an analysis which leads to an identification of the fundamental limits of speed, dynamic range, and power consumption in switched-current circuits. We show how the relationships are very inter-related and we define a figure of merit for the speed, dynamic range product. Once the key mechanisms that relate to the performance have been identified we describe an active compensation technique which assists in controlling the trade off between speed, dynamic range, and power consumption to suit a particular requirement.

The aim of this chapter is to present the reader with some analysis methods and design issues which may help in bringing good performance out of switched-current cells; and in some cases higher performance than equivalent switched capacitor circuits can offer.

10.2.2 Fundamental characteristics of switched current memory cells

All switched current circuits rely, in some form, on the basic current memory cell shown in Figure 1. We will therefore now examine the speed and dynamic range properties of this cell assuming first order behaviour. In the sample phase (ϕ_1 active) the transfer functions from i_{in} and i_n to v_{gs} are

$$H(s) \stackrel{\text{def}}{=} \frac{v_{gs}}{i_{in}} = \frac{v_{gs}}{i_n} = \frac{1}{g_m}\frac{1}{1+\frac{C_{gs}}{g_m}s} \tag{1}$$

The sample phase must be long enough for v_{gs} to settle properly. Thus, the duration of the sample phase should be a suitable number (e.g. 7–10) times the time constant C_{gs}/g_m. Consequently, this quantity can be used as a measure for the obtainable speed of the cell

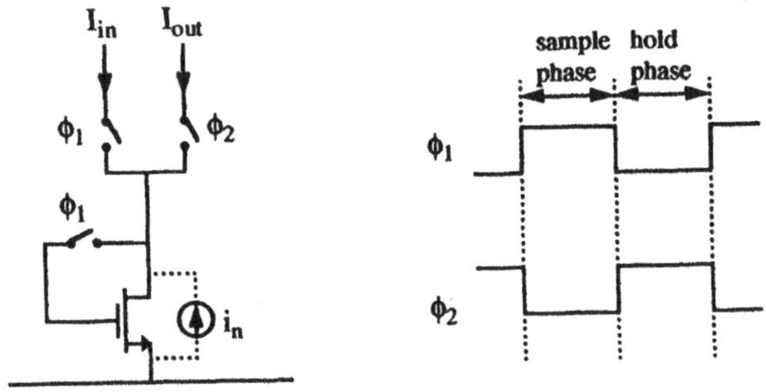

Figure 1 Basic current memory cell. i_n is the drain current noise of the MOS transistor.

and so we will define the cell's speed as

$$\omega_0 = g_m / C_{gs} \tag{2}$$

At the end of the sample phase (time t_1 in the figure) the voltage across the gate capacitor is stored together with a noise voltage and a voltage error due to charge injection from the switches. Assuming the cell has settled completely the statistical variance of the noise voltage is given by

$$V\{v_{gs}(t_1)\} = \int_0^\infty |H(j2\pi f)|^2 S_{i_n}(f)\, df$$

where $S_{i_n}(f)$ is the power spectral density of the i_n noise source. i_n contains both thermal and $1/f$ noise but we need only consider the thermal part because one of the inherent properties of switched current circuits is that they suppress low frequency noise. This is because i_n is sampled at the end of the ϕ_1 phase and during the ϕ_2 phase the resulting output noise current will therefore be the immediate value of i_n minus the stored one. Hence, if i_n does not change much during the ϕ_2 phase the noise is cancelled. Thus, slowly varying noise components, corresponding to the low frequency part of the noise spectrum, will be suppressed. This effect is known as *correlated double sampling* and can also be achieved in switched capacitor circuits [4] but at the cost of increased complexity and increased silicon area consumption.

When considering only thermal noise $S_{i_n}(f)$ is constant and [1]

$$S_{i_n}(f) = \frac{8}{3} kT g_m \tag{3}$$

where k is Boltzmann's constant, $k = 1.38 \cdot 10^{-23}$J/K, and T is the absolute temperature.

The integral then evaluates to

$$V\{v_{gs}(t_1)\} = \frac{1}{g_m^2}\frac{1}{4\frac{C_{gs}}{g_m}} \cdot \frac{8}{3}kTg_m = \frac{2}{3}\frac{kT}{C_{gs}} \tag{4}$$

The output noise during phase ϕ_2 consists of a sampled and a direct noise contribution. The sampled one is constant throughout the duration of the ϕ_2-phase and is due to the noise voltage stored on the gate. Thus, the variance of this current is

$$V\{i_{o,s}\} = g_m^2 V\{v_{gs}(t_1)\} = \frac{2}{3}g_m^2\frac{kT}{C_{gs}} \tag{5}$$

Apart from the sampled contribution there is also a direct one, namely that of the i_n-source during the output phase ϕ_2. Since i_n contains white noise it has, in principle, energy at infinitely high frequencies and consequently also infinite variance. However, in practice the cell's output will always be connected to some subsequent circuitry which limits the bandwidth and consequently yields a finite variance. To calculate the total noise variance it is therefore necessary to know the circuitry in which the memory cell is embedded. However, comparing (3) and (5) it is clear that the direct and sampled noise are linked by their dependence on g_m, i.e. lowering g_m reduces both types of noise. It is also obvious that the total noise variance cannot be lower than the sampled contribution and we will therefore define the *intrinsic dynamic range* of the cell as

$$DR \stackrel{\text{def}}{=} \frac{P_{signal,max}}{V\{i_{o,s}\}} \tag{6}$$

where $P_{signal,max}$ is the maximum signal power.

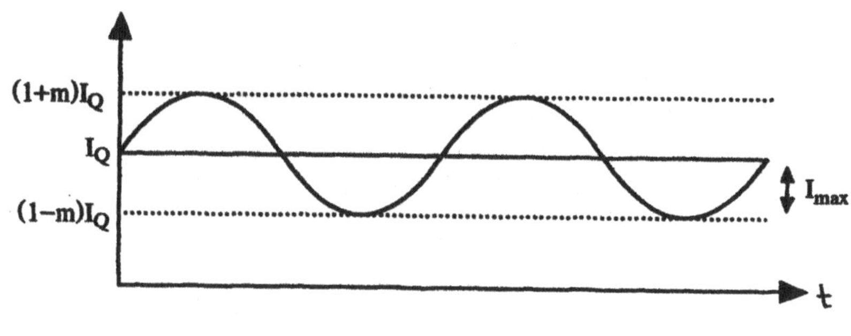

Figure 2 Signal swing around the quiescent current I_Q. I_{max} is the maximum signal amplitude and m is the socalled modulation index.

If we assume a sinusoidal signal the maximum signal power is given by the maximum amplitude I_{max}

$$P_{signal,max} = \tfrac{1}{2}I_{max}^2 \tag{7}$$

To ensure correct circuit operation a quiescent current I_Q is added to the signal. In a class A circuit I_Q serves to ensure that the resulting current never falls below zero. Thus, the *modulation index* m in Figure 2 must be less than 1. In class AB operation positive and negative going signals are handled by different circuit halves and m can be greater than 1.

In any case the maximum signal amplitude is

$$I_{max} = mI_Q \tag{8}$$

With these definitions and using (5) we can now find the intrinsic dynamic range of the cell

$$DR = \frac{3}{4}\frac{m^2 I_Q^2}{g_m^2 \frac{kT}{C_{gs}}} \tag{9}$$

Of course several other definitions could be contemplated which in one way or another include the direct noise. However (if they are sound), they will always end up with an expression of the same kind as above, except for some form of constant factor. To illustrate this point, appendix 10.2.A attempts an alternative definition which gives the dynamic range assuming that the cell's output is subsequently resampled (as would usually be the case in filters etc). The resulting expression differs from the above only by a factor of -3dB.

From (9) it is evident that to obtain maximum dynamic range g_m/C_{gs} should be as low as possible. Unfortunately (and not very surprisingly) this conflicts with a requirement for a high value of g_m/C_{gs} in (2) to achieve high speed. Since these two goals seem to be contradictory it is more informative to introduce a figure of merit for the cell, namely the *speed-dynamic range product*. Using (2) and (9) we get

$$DR \cdot \omega_0 = \frac{3}{4}\frac{m^2 I_Q^2}{g_m kT} \tag{10}$$

Clearly, to maximize this product g_m must be minimized and we proceed to do that now. g_m is given by

$$g_m = \frac{2I_Q}{(V_{GS} - V_T)_Q}$$

where $(V_{GS} - V_T)_Q$ is the quiescent value of the gate-source voltage less one threshold voltage. The larger this voltage is the lower g_m will be. However, there exists a maximum allowable value $(V_{GS} - V_T)_{max}$ of the voltage which is determined by the surrounding circuitry and, of course, ultimatively by the supply voltage. $(V_{GS} - V_T)_{max}$ occurs at the peak current through the cell, i.e. at $(1+m)I_Q$ (cf. Figure 2). Thus

$$\left.\begin{aligned} (V_{GS} - V_T)_{max} &= \sqrt{\frac{2(1+m)I_Q}{KW/L}} \\ (V_{GS} - V_T)_Q &= \sqrt{\frac{2I_Q}{KW/L}} \end{aligned}\right\} \Rightarrow \\ (V_{GS} - V_T)_Q = \frac{1}{\sqrt{1+m}}(V_{GS} - V_T)_{max} \tag{11}$$

With this result the transconductance becomes

$$g_m = \frac{2\sqrt{1+m}I_Q}{(V_{GS} - V_T)_{max}} \tag{12}$$

giving the following values for speed and dynamic range

$$\begin{aligned} \omega_0 &= \frac{2\sqrt{1+m}}{C_{gs}}\frac{I_Q}{(V_{GS} - V_T)_{max}} \\ DR &= \frac{3}{16}\frac{m^2}{1+m}\frac{(V_{GS} - V_T)_{max}^2}{kT/C_{gs}} \end{aligned} \tag{13}$$

and the figure of merit is

$$DR \cdot \omega_0 = \frac{3}{8} \frac{m^2}{\sqrt{1+m}} \frac{(V_{GS} - V_T)_{max} I_Q}{kT} \tag{14}$$

This shows that *the speed-dynamic range product is simply proportional to the quiescent supply power regardless of process parameters* [7] (except the threshold voltage, though). Thus, the speed-dynamic range product is a quite fundamental quantity and basically shows us that there are only two ways to gain overall improved circuit performance, namely to increase power consumption or to increase the modulation index m, i.e. entering into class AB operation[1]. However, the expression only concerns the *product* of dynamic range and speed and, so, it also shows that an increased dynamic range can be obtained at the cost of speed or vice versa — it is only their *product* which is constant.

Since the g_m is fixed by (12) the only way to trade off speed versus dynamic range is by altering the capacitance value C_{gs}. If a large dynamic range is required this means a large C_{gs} and, hence, a large silicon area. At the other end of the scale high speed operation necessitates very small capacitances, leading to huge charge injection errors and increased sensitivity to parasitic capacitances (especially the gate-drain capacitance). Also, the design rules of the process might not allow sufficiently low capacitance values, especially when considering that a specific aspect ratio is imposed by (12).

So, in the basic cell the speed-dynamic range tradeoff involves some penalties because of the need to alter the gate capacitance. However, the gate capacitance does not appear in the speed-dynamic range product and therefore the gate capacitance value should not determine the circuit behaviour in the same fundamental way as the supply power. And, indeed, by augmenting the basic cell with suitable circuit techniques it is possible to do the trade-off without needing excessively small or large capacitance values; this is the subject of the section 10.2.4. First, however, we will show how to use the derived results to design a basic cell for specified speed and dynamic range.

10.2.3 Designing for a specific speed and dynamic range — example

We will now use the fundamental relationship between speed, dynamic range, and power consumption to design a 'real' memory cell including biasing, etc.

The basic cell in Figure 1 works only for unipolar input signals so to be able to handle bipolar signals we must add a bias current source I_Q as shown in the left half of Figure 3. The bias current source must necessarily be implemented with transistors as shown to the right in the figure and therefore it also generates noise, symbolised by the i_{ns} noise current source. Furthermore, the V_{Bias} voltage which determines the quiescent current must be generated by some bias circuitry which also contributes noise. We will quantify this noise by a factor γ which is the excess noise generated by the bias circuitry relative to the noise generated by the bias current source transistor MS alone. Thus we will assume that the I_Q quiescent current source contributes a noise current with a power spectral density of

$$S_{i_{nq}} = (1+\gamma) S_{i_{ns}} \tag{15}$$

where $S_{i_{ns}}$ is the power spectral density of the i_{ns} source. Obviously γ should be kept

---

[1]Remember, though, that unlike class A circuits the actual power consumption in class AB circuits is signal-dependent and certainly greater than the quiescent value in (14). Thus, when $m \gg 1$ the figure of merit is only proportional to $\sqrt{m}$ times the power consumption at maximum input amplitude.

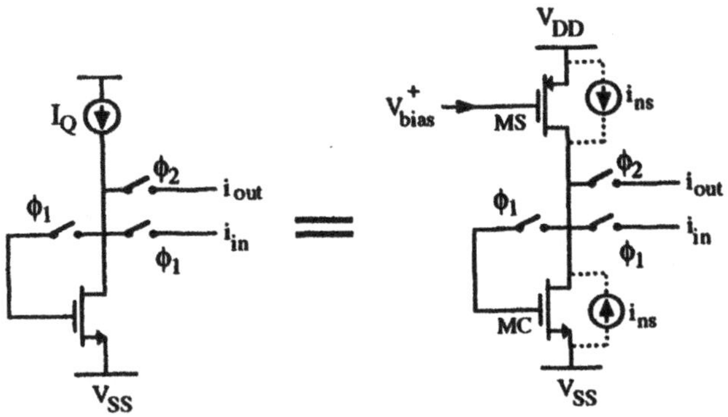

Figure 3 Basic current memory cell with biasing (left). The I_Q source must necessarily be implemented with transistors (right) which also add noise. i_{ns} and i_n are the drain current noises of the two transistors.

as small as possible, the ideal situation $\gamma = 0$ corresponding to a noise-free V_{Bias}. As a simple example to illustrate the expression Figure 4 shows a primitive biasing circuit where $\gamma = 1/a$ (this is easily derived by using (3) and $g_{ms} = g_{mb}/a$. In this case a small γ can be obtained by using a large a. However, this comes at the price of an increased power consumption in the biasing circuitry and a reasonable compromise must be found).

Because of the extra noise contribution from the biasing the results for dynamic range, speed, and figure of merit (13) and (14) must be modified. The calculations are carried out

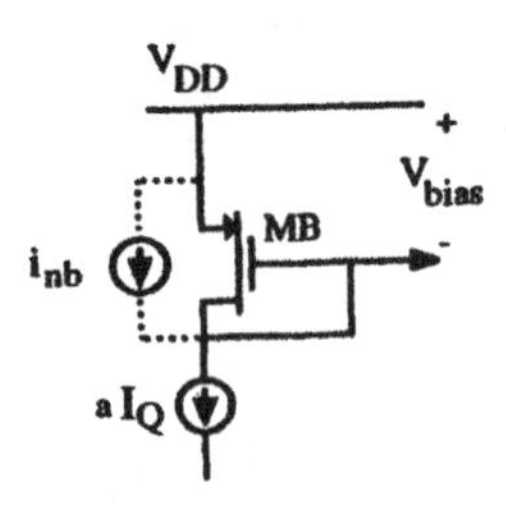

Figure 4 Simple biasing circuit.

in appendix 10.2.B. and they give the following results for speed and dynamic range

$$\begin{aligned} \omega_0 &= \frac{\beta-1}{\sqrt{\beta}-1}\frac{2\sqrt{1+m}}{C_{gs}}\frac{I_Q}{V_{DD}-V_{TN}-V_{saf}} \\ DR &= \frac{3}{16}\frac{1}{(1+\sqrt{\beta})^3}\frac{m^2}{1+m}\frac{(V_{DD}-V_{TN}-V_{saf}-V_{saf})^2}{kT/C_{gs}} \end{aligned} \quad (16)$$

and the figure of merit is

$$DR\cdot\omega_0 = \frac{3}{8}\frac{1}{(1+\sqrt{\beta})^2}\frac{m^2}{\sqrt{1+m}}\frac{(V_{DD}-V_{TN}-V_{saf})I_Q}{kT}$$

where

$$\beta = \frac{1+\gamma}{\sqrt{1+m}} \quad (17)$$

and V_{saf} is a 'safety margin' which can be used by the designer to cater for unmodelled effects and other design issues, etc. As an example, augmenting the memory cell with cascodes improves its accuracy without adding significant amounts of extra noise. However, it requires that the drain-source voltages of the cascode transistors are always greater than a certain minimum voltage to ensure they remain in saturation. This can easily be taken care of by setting V_{saf} equal to the sum of the desired cascode drain-source voltages and we can still use the above design equations.

The values for speed and dynamic range are attained at the following optimal gate-source voltages

$$\begin{aligned} (V_{GSc}-V_{TN})_Q &= \frac{1}{\sqrt{1+m}}\frac{\sqrt{\beta}-1}{\beta-1}(V_{DD}-V_{TN}-V_{saf}) \\ (V_{Bias}-|V_{TP}|) &= \sqrt{\beta}\frac{\sqrt{\beta}-1}{\beta-1}(V_{DD}-V_{TN}-V_{saf}) \end{aligned} \quad (18)$$

where $(V_{GSc}-V_{TN})_Q$ is the quiescent gate-source voltage of MC less one threshold voltage. The optimum is fairly flat and consequently both $(V_{GSc}-V_{TN})_Q$ and $(V_{Bias}-|V_{TP}|)$ can be lowered somewhat without any serious penalty in dynamic range or speed. Thus, it is possible to introduce the V_{saf} safety margin in the design without any serious noise penalty.

The optimal gate voltages must be established by some sort of bias circuitry. We will not go into detail on this, only give a simple example of how it can be done. The probably simplest imaginable biasing circuit is shown in Figure 5. The V_{Bias} voltage is

$$(V_{Bias}-|V_{TP}|) = \frac{1}{1+\sqrt{\frac{\mu_p}{\mu_n}\frac{W_{B1}/L_{B1}}{W_{B2}/L_{B2}}}}(V_{DD}-V_{TN}-|V_{TP}|)$$

where μ_p and μ_n are the mobilities of holes and electrons. The mobilities are usually fairly well controlled in modern IC processes and the factor in front of $(V_{DD}-V_{TN}-|V_{TP}|)$ is consequently reasonably immune to process variations. The circuit has the advantage of low noise because MB1 and MB2 small-signal wise act as two resistors in parallel, thus reducing the resistance value and thereby also the noise. If possible without consuming too much excess silicon area a capacitor C_B should be added as shown in the figure. This improves power supply rejection and, if it is large enough, it it also filters the noise on V_{Bias} so that the noise power is concentrated at low frequencies where it will be suppressed by the correlated double sampling feature of the memory cell. Without the C_B capacitor the γ

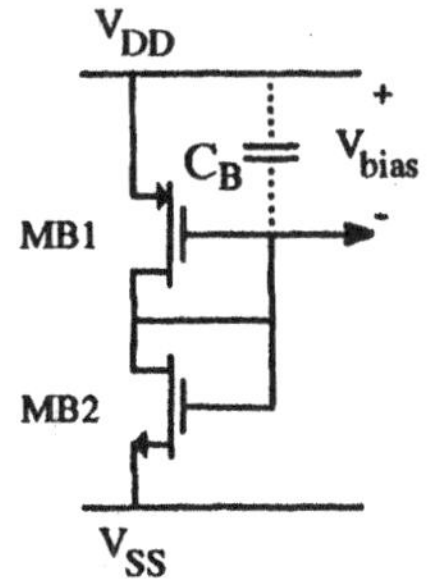

Figure 5 Biasing circuit.

factor can quite easily be shown to be

$$\gamma = \frac{1}{a}\frac{1}{1 + g_{mb2}/g_{mb1}} < \frac{1}{2a}$$

If we choose $a = 1$ then $\gamma < 0.5$. The precise value of γ is not terribly important as long as it is reasonably small. So we will just use a conservative estimate of $\gamma = 0.5$.

A disadvantage of the circuit is that $(V_{Bias} - |V_{TP}|)$ is not proportional to $(V_{DD} - V_{TN})$ but rather to $(V_{DD} - V_{TN} - |V_{TP}|)$. However, if V_{DD} is not too small the extra $|V_{TP}|$ can be used as a safety margin, i.e.

$$V_{saf} = |V_{TP}|$$

We are now in a position to design a cell which meets predetermined speed and dynamic range specifications. As an example we try to obtain

$$\begin{array}{ll} \text{Sampling frequency:} & f_s = 1\text{MHz} \\ \text{Dynamic range:} & DR(\text{DB}) = 80\text{dB} \end{array}$$

If we allow for a settling time of $N_{settl} = 8$ times the time constant and want the duration t_{ϕ_2} of the output phase to be equal to that of the input phase (t_{ϕ_1}) then the speed of the cell is given by

$$\begin{aligned} f_s &= 1/(t_{\phi_1} + t_{\phi_2}) \\ &= 1/(2t_{\phi_1}) \\ &= 1/(2N_{settl}/\omega_0) \end{aligned}$$

This gives $\omega_0 = 16 \cdot 10^6 s^{-1}$. Remembering that the dynamic range DR in (16) is a power quotient we have $DR(\text{DB}) = 10\log(DR)$ which gives $DR = 10^8$.

We use the biasing circuit of Figure 5 so $\gamma = 0.5$. Using a modulation index of $m = 0.7$ and designing for room temperature and using typical values for threshold values we now have the following design specification

$$\begin{aligned} DR &= 10^8 \\ \omega_0 &= 16 \cdot 10^6 s^{-1} \end{aligned}$$

	$\sqrt{V\{v_{GSc}\}}$	$V\{i_{o,s}\}$	DR	ω_0
With $1/f$-noise	27.8μVRMS	$8.5 \cdot 10^{-18}A^2$	80.3dB	$18.1 \cdot 10^6 s^{-1}$
Without $1/f$-noise	27.6μVRMS	$8.4 \cdot 10^{-18}A^2$	80.3dB	$18.1 \cdot 10^6 s^{-1}$

Table 1 Simulated speed ω_0 and gate noise RMS voltage and based on this the sampled noise current variance $V\{i_{o,s}\}$ and dynamic range DR.

$$
\begin{aligned}
m &= 0.7 \\
\gamma &= 0.5 \\
V_{DD} &= 5V \\
V_{TN} &= 0.8V \\
|V_{TP}| &= 0.9V \\
T &= 300K
\end{aligned}
$$

By (17) β is

$$\beta = 1.15$$

and by (16) and (18) we get

$$
\begin{aligned}
I_Q &= 61\mu A \\
C_{gs} &= 6.3pF \\
(V_{GSc} - V_{TN})_Q &= 1.21V \\
(V_{Bias} - |V_{TP}|) &= 1.70V
\end{aligned}
$$

This is straight-forward to translate into transistor dimensions which in the Northern Telecom 0.8μm BATMOS process become

$$
\begin{aligned}
W/L(MS) &= 52\mu m/26\mu m \\
W/L(MC) &= 75\mu m/64\mu m \\
W/L(MS1) &= 52\mu m/26\mu m \\
W/L(MS2) &= 44\mu m/56\mu m
\end{aligned}
$$

Unfortunately SPICE cannot do a proper discrete-time noise analysis so therefore the sampled noise was determined by doing a continuous-time noise analysis in the sample phase. The sampled noise current is then calculated by

$$
\begin{aligned}
V\{i_{o,s}\} &= \int_{f_l}^{f_h} S_{i_o}(f)\, df \\
&= g_{mc}^2 \int_{f_l}^{f_h} S_{v_{GSc}}(f)\, df
\end{aligned}
$$

where g_{mc} is the transconductance of MC and $S_{v_{GSc}}(f)$ is the power spectral density of MC's gate noise voltage. The dynamic range can then be found from (6) and (7). $\sqrt{S_{v_{gsC}}(f)}$ is plotted in Figure 6 both with $1/f$-noise and with the $1/f$-noise removed. In table 1 is summarized the simulation results. The total gate noise RMS voltage is measured in a

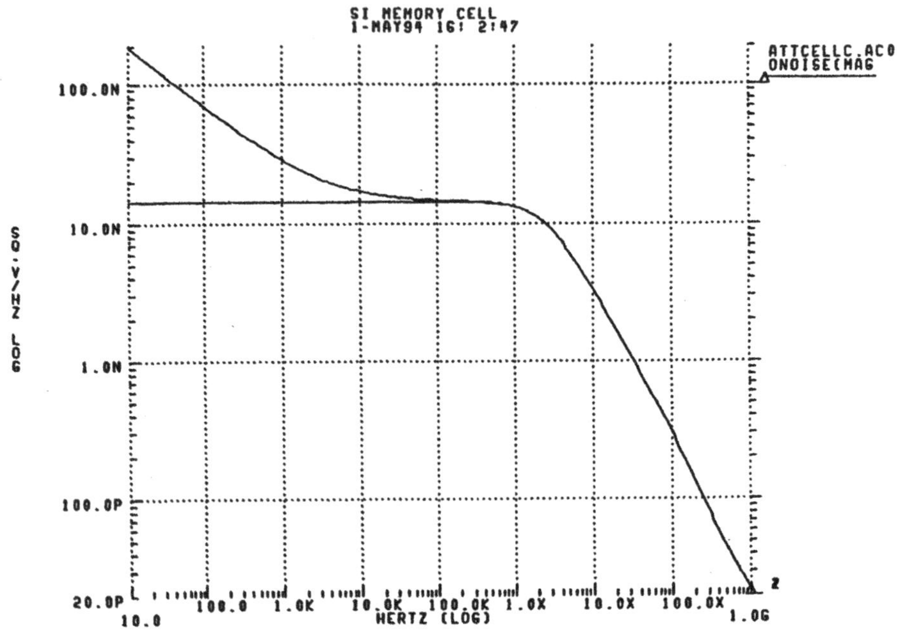

Figure 6 Simulated spectral density of the memory transistor gate noise voltage (in $V/\sqrt{Hz}$). The noise is both shown with $1/f$-noise and with $1/f$-noise artificially removed.

bandwidth from $f_l = 10Hz$ to $f_h = 1GHz$ and from this is calculated the equivalent sampled noise current variance $V\{i_{o,s}\}$ and the resulting dynamic range DR. Evidently, the effect of $1/f$-noise is completely negligible and the simulation results are in quite good agreement with the presented theory, although the resulting speed turns out to be a bit better than the design objective. This is probably due to that the target values of quiescent current and gate-source capacitance are not exactly attained in the simulation. Also, the theory is based on simple MOS models whereas the simulations used SPICE level 3 models.

10.2.4 Speed-dynamic range trade-off by feedback amplification or attenuation [7]

In Figure 7 is shown a modification of the basic cell with amplification or attenuation in the feedback path. The idea behind this is to change the apparent capacitance value or transconductance value (whichever way it is viewed) without changing their actual values. This enables us to achieve very large or very small equivalent capacitance values without physically needing extreme values.

The amplifier/attenuator can either be implemented as a voltage or a current amplifier/attenuator. In the current-mode case the amplifier/attenuator is assumed to have a low impedance input so that the effect of C_D is not significant. In the voltage-mode case the output impedance is assumed to be low, thereby eliminating the effect of C_{gs}.

When we to begin with assume the amplifier/attenuator noise-free ($i_{nA} = 0$, $v_{nA} = 0$)

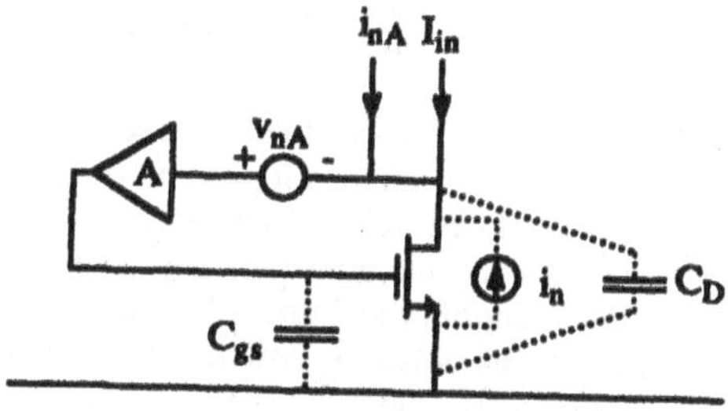

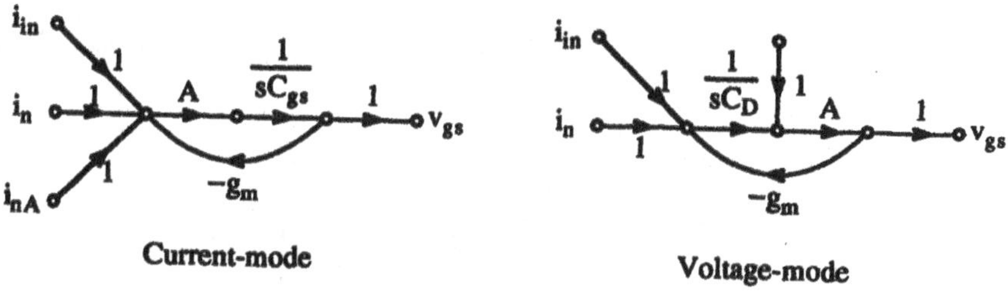

Figure 7 Current memory cell in sample phase with feedback amplification ($A > 1$) or attenuation ($A < 1$); both voltage and current amplification/attenuation can be used, leading to the two signal flow graphs. i_n is the drain current noise of the transistor and i_{nA} is the equivalent input noise of the amplifier/attenuator when A acts as a current amplifier or attenuator. v_{nA} is the equivalent input noise when a voltage amplifier/attenuator is used.

the speed and dynamic range in both the voltage and current mode cases are

$$DR = \frac{1}{A} \cdot \frac{3}{4} \frac{m^2 I_Q^2}{g_m^2 \frac{kT}{C_f}}$$
$$\omega_0 = A g_m / C_f \qquad (19)$$

where $C_f = C_{gs}$ in the current mode case and $C_f = C_D$ in the voltage mode case. So, compared to the basic cell, the speed goes up by a factor of A and the dynamic range goes down by the same factor. The speed-dynamic range product is thus still the same as in (10) but now *we can use the parameter A to do the trade off between dynamic range and speed independently of capacitance value.* In the voltage-mode case there is a catch, though — the voltage at the drain cannot exceed a certain maximum voltage V_{sup} which is determined by surrounding circuitry and ultimatively by the supply voltage. The problem occurs when $A < 1$. Here the voltage at the gate is restricted to AV_{sup} which will almost certainly be smaller than $(V_{GS} - V_T)_{max}$ in (13). This results in a poorer dynamic range and hence a poorer speed-dynamic range product. It is, however, possible to overcome the gate voltage restriction by an elegant two-step sampling procedure which will be explained in section 10.2.6. If $A > 1$ there is not a gate voltage restriction problem but now, of course, a fairly complex voltage amplifier is needed. By contrast the current-mode approach only needs simple current mirrors. For $A < 1$ the roles are reversed. Here the voltage-mode approach needs only a simple voltage divider whereas the current-mode one still needs current mirrors. Hence, the current-mode approach is best for amplification ($A > 1$) whereas the voltage-mode approach is better suited for attenuation ($A < 1$).

The concept of using feedback attenuation or amplification to tailor the speed-dynamic range trade-off appears to be new [7] and so does the current-mode approach for implementing it. The voltage-mode amplification approach is in fact also new, although similar techniques (operational amplifier and grounded gate feedback) have been used in the past for other purposes such as obtaining low input impedance, improved accuracy, etc [3][page 157–163]. It should be noted, though, that these techniques obtain voltage gain by dumping current into a high-impedance point and therefore do not implement the pure voltage controlled voltage source function of Figure 7. Thus, when using operational amplifier feedback [3][page 157] the behaviour will be determined by its gain-bandwidth product whereas the grounded gate approach [3][page 160] has no effect on the trade-off.

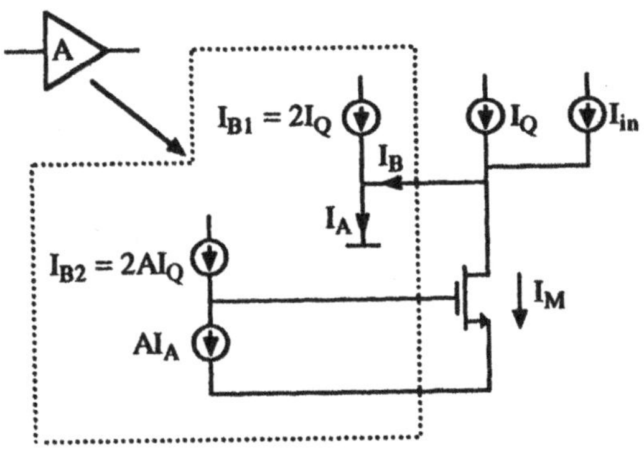

Figure 8 The current memory cell from Figure 7 with an abstract circuit representation of the amplifier/attenuator. I_M is the current through the memory transistor. At the beginning of the sample phase I_M equals the current sampled on the previous phase.

Until now we have assumed that the amplifier/attenuator is noise-free and does not consume supply power. In practice this of course does not hold true and we will therefore examine the effects on the speed-dynamic range trade-off and power consumption in the current-mode case. We will also assume the amplifier/attenuator operates in class A. Figure 8 shows the memory cell in the sample phase with an abstract representation of the amplifier/attenuator. The quiescent currents I_{B1} and I_{B2} serve to ensure that I_A is always positive, i.e. class A operation. Hence, their values are determined by the negative peak value of the feedback current I_B. Using Figure 2 we get

$$\left.\begin{array}{rcl} I_{in,min} &=& -mI_Q \\ I_{M,max} &=& (1+m)I_Q \end{array}\right\} \Rightarrow$$
$$I_{B,min} = I_{in,min} + I_Q - I_{M,max} = -2mI_Q$$

If a modulation index of m is used again, the value of I_{B1} becomes

$$I_{B1} = 2I_Q \tag{20}$$

and consequently

$$I_{B2} = AI_{B1} = 2AI_Q \tag{21}$$

Thus, the total supply current amounts to $(3 + 2A)I_Q$ which means $(3 + 2A)$ times larger power consumption than for the basic cell, this without any improvement in the speed-dynamic range product. In the amplification case, $A \gg 1$, the increased power consumption can be quite severe whereas in the attenuation case, $A \ll 1$, the supply power is "only" increased by a factor of 3.

Another problem is the noise of the amplifier/attenuator. In the amplifying case it can be shown not to be a severe problem, but in the attenuating case the requirement for minimum quiescent currents (20) and (21) in fact lead to a reduced dynamic range compared to the basic cell. And at the same time the speed decreases too.

So both concerning power consumption and excess noise it is desirable to reduce the quiescent current of the amplifier/attenuator. But, as explained, this cannot be done in the case of the cell of Figure 7. However, in the recently introduced S²I cell [5] it is possible. This will be explained below.

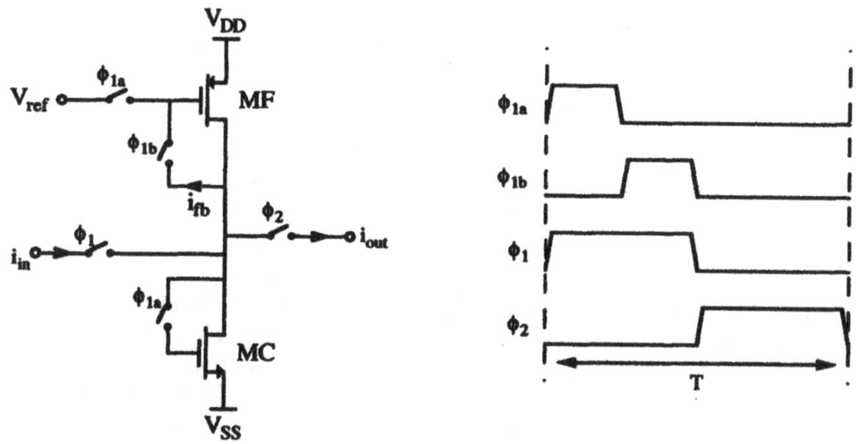

Figure 9 S²I cell with clock waveforms.

10.2.5 Using amplification or attenuation in connection with the S²I cell

The idea behind the S²I cell is to use a two-step approach to cancel charge injection errors etc.; the basic circuit is shown in Figure 9. At the end of phase ϕ_{1a} the input current is sampled on the coarse memory transistor MC. The sampled value will contain some errors due to charge injection, noise, etc. At the end of phase ϕ_{1b} these errors are stored, together with charge injection and noise, on the fine memory transistor MF. This way the errors of the coarse cell are cancelled by the fine cell during the output phase ϕ_2. The resulting errors during the output phase are then sampled noise on the fine memory, direct noise from both the fine and coarse memory, and charge injection on the fine memory. Because the fine memory only samples the errors the signal swing at its gate will be quite small. Consequently the charge injection will be nearly constant, i.e. it only leads to an offset. So, non-linearities due to deterministic phenomena such as charge injection, channel length modulation, etc. are cancelled by the S^2I cell *but noise is not.* Therefore the above speed-dynamic range and trade-off considerations are still valid. There is one important difference, though; the amplifier/attenuator quiescent current restrictions (20) and (21) can now be lifted because of the small signal swing on the fine memory transistor. Thus significantly lower quiescent currents can be used in the feedback amplifier/attenuator, thereby improving dynamic range and power consumption.

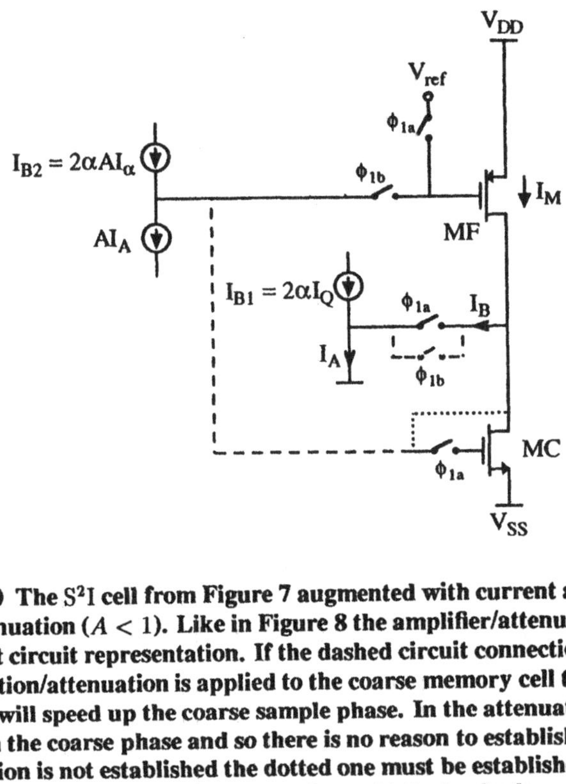

Figure 10 The S^2I cell from Figure 7 augmented with current amplification ($A > 1$) or attenuation ($A < 1$). Like in Figure 8 the amplifier/attenuator is shown as an abstract circuit representation. If the dashed circuit connection is established the amplification/attenuation is applied to the coarse memory cell too. In the amplifying case this will speed up the coarse sample phase. In the attenuating case it only slows down the coarse phase and so there is no reason to establish it. If the dashed connection is not established the dotted one must be established instead. I_M is the current through the fine memory transistor. At the beginning of the sample phase I_M equals the current sampled on the previous phase.

Figure 10 shows an S^2I cell augmented with feedback attenuation or amplification and from this figure we will now determine the minimum required quiescent currents I_{B1} and I_{B2}. If the maximum compensation current stored on the fine memory MF is $\pm I_{comp}$ we can define an error ratio factor α which is the ratio between I_{comp} and the maximum signal current amplitude, mI_Q

$$\alpha \stackrel{\text{def}}{=} \frac{I_{comp}}{mI_Q}$$

and using a similar argument as in the previous section the minimum values of I_{B1} and I_{B2} can be deduced

$$\begin{aligned} I_{B1} &= 2\alpha I_Q \\ I_{B2} &= 2\alpha A I_Q \end{aligned}$$

The total supply current is now only $(1 + \alpha(1 + A))I_Q$ and the noise from the amplifier/attenuator is also lowered. If α is small enough the amplifier/attenuator noise becomes insignificant in comparison with memory transistor noise and the power consumption is only slightly increased compared with the basic cell. Thus, the speed-dynamic range product (14) is restored.

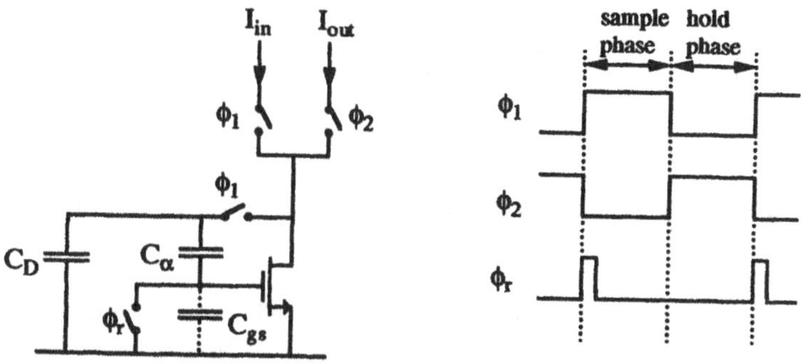

Figure 11 Current memory cell using voltage attenuation.

10.2.6 A two-step feedback voltage attenuation technique

As mentioned earlier, feedback voltage amplification is considerably more complex to implement than current amplification. On the other hand, voltage attenuation is extremely simple to implement, e.g. by using a capacitive divider as shown in Figure 11. The ϕ_r phase acts as a reset to provide a well controlled initial condition, thereby preventing the gate voltage from drifting away due to leakage currents etc. The transfer function from the drain current noise i_n to the gate-source voltage v_{gs} is

$$\begin{aligned} \frac{v_{gs}}{i_n} &= \frac{1}{g_m} \frac{1}{1 + s\left(C_{gs} + C_D\left(1 + \frac{C_{gs}}{C_a}\right)\right)/g_m} \\ &\approx \frac{1}{g_m} \frac{1}{1 + \left(1 + \frac{C_{gs}}{C_a}\right)\frac{C_D}{g_m}s} \end{aligned}$$

Thus, the attenuation ratio $1/A$ is

$$\frac{1}{A} \approx 1 + \frac{C_{gs}}{C_a} \tag{22}$$

On top of being simple the divider is also noise-free which is a marked advantage over the current feedback schemes.

Unfortunately the voltage division also leads to a severely restricted voltage swing

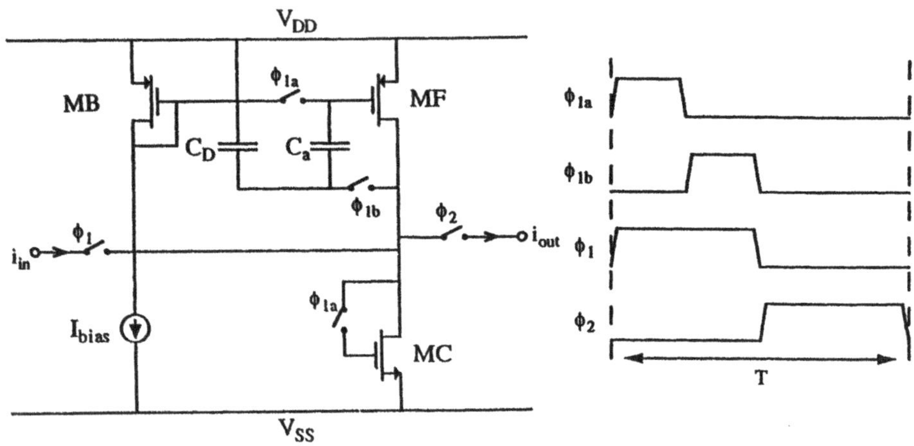

Figure 12 S^2I **current memory cell using voltage attenuation.**

at the transistor gate, giving a reduced dynamic range and hence a poorer speed-dynamic range product — $(V_{GS} - V_T)_{max}$ is, in effect, heavily reduced in (14). However, this problem can be overcome very elegantly by using the S^2I approach [7] because here only a small voltage swing is required on the gate of the fine memory transistor. The circuit is shown in Figure 12. Firstly the input current is sampled on the coarse memory transistor in phase ϕ_a without voltage attenuation. In this phase MF functions as a current source and the reset function of the ϕ_r switch in Figure 11 is, in effect, carried out by the topmost ϕ_{1a} switch. This generates approximately the correct gate voltage on MF and in the subsequent fine phase ϕ_{1b} only a small gate voltage change is required to cancel the errors from the coarse cell. Thus, the dynamic range problem of the voltage attenuator is solved and the voltage attenuator can now be activated during the fine phase.

Because the fine memory absorbs all the errors, including sampled noise, from the coarse memory the resulting dynamic range of the cell will be determined by the fine memory. Also, due to the attenuation, the fine memory will be significantly slower than the coarse one and the speed is therefore also determined by the fine memory. The speed-dynamic range product is consequently determined by the fine memory and the speed-dynamic range trade-off is controlled by the attenuator.

Apart from increasing dynamic range (at the cost of speed) the circuit also lowers the effect of charge injection because this is also attenuated by the voltage divider. Indeed, a similar circuit, shown in Figure 13, was published in [6] and [2, page 305] for exactly this purpose but its good dynamic range properties and usefulness for the trade-off of speed versus dynamic range have apparently not been realized before. Like the S^2I cell it uses a two-step approach. The coarse phase generates an approximately correct gate voltage and in the fine phase the voltage attenuator comes into play in the same way as in Figure 12. However, the circuit in Figure 13 only attenuates charge injection by the voltage division ratio whereas the new circuit in Figure 12 gives both error reduction due to the S^2I technique and to the voltage attenuation.

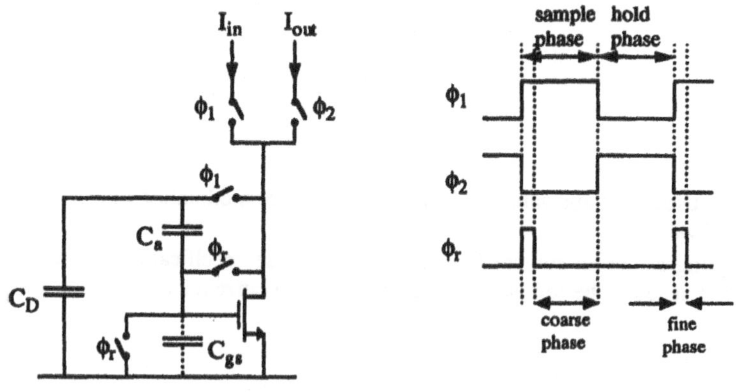

Figure 13 Current memory cell using two-step voltage attenuation.

The equations for speed and dynamic range are derived in appendix 10.2.C

$$\begin{aligned} \omega_0 &= A\frac{1-\sqrt{1+m}}{1-\sqrt[4]{1+m}}\frac{1}{C_D}\frac{2I_Q}{V_{DD}-V_{Tmax}-V_{saf}} \\ DR &= \frac{1}{A}\frac{3}{16}\frac{\sqrt[4]{1+m}}{\left(1+\frac{1}{\sqrt[4]{1+m}}\right)^3}\frac{m^2}{1+m}\frac{(V_{DD}-V_{Tmax}-V_{saf})^2}{kT/C_D} \end{aligned} \tag{23}$$

where $V_{Tmax} = \max(V_{TN}, |V_{TP}|)$ and V_{saf} is the 'safety margin' mentioned in section 10.2.3. The figure of merit is

$$DR\cdot\omega_0 = \frac{3}{8}\frac{1}{(1+\sqrt{\frac{1}{\sqrt[4]{1+m}}})^2}\frac{m^2}{\sqrt{1+m}}\frac{(V_{DD}-V_{Tmax}-V_{saf})_{max}I_Q}{kT}$$

and the optimal quiescent gate-source voltages are

$$\begin{aligned} (V_{GSc}-V_{TN})_Q &= \frac{1}{\sqrt{1+m}}\frac{\sqrt{\beta}-1}{\beta-1}(V_{DD}-V_{Tmax}-V_{saf}) \\ (V_{Bias}-|V_{TP}|) &= \sqrt{\beta}\frac{\sqrt{\beta}-1}{\beta-1}(V_{DD}-V_{Tmax}-V_{saf}) \end{aligned} \tag{24}$$

10.2.7 Design example

To illustrate some of the benefits of the presented compensation techniques we will now go through the design of a very high dynamic range memory cell using the voltage attenuation technique described above. We will aim for

Sampling frequency:	$f_s = 32\text{kHz}$
Dynamic range:	$DR(\text{DB}) = 100\text{dB}$

If we allow for a settling time of $N_{settl} = 6$ times the time constant this translates into the following design specification

$$DR = 10^{1}0$$

$$
\begin{aligned}
\omega_0 &= 384 \cdot 10^3 s^{-1} \\
m &= 0.7 \\
V_{DD} &= 5V \\
V_{TN} &= 0.8V \\
|V_{TP}| &= 0.9V \\
T &= 300K
\end{aligned}
$$

were we have used the same values as in section 10.2.3 for modulation index, temperature, and threshold values.

Designing a normal switched current cell for these specifications, as explained in section 10.2.3, leads to the following values of quiescent current and gate capacitance

$$
\begin{aligned}
I_Q &= 147\mu A \\
C_{gs} &= 631pF
\end{aligned}
$$

The quiescent current I_Q is within reasonable limits but the capacitance C_{gs} would require a vast amount of silicon area. Therefore, we will now take advantage of the extra degree of freedom offered by feedback attenuation to reduce the capacitor value. We will use the cell in Figure 12 with the following capacitance values

$$
\begin{aligned}
C_D &= 10pF \\
C_{gs,MF} &= 10pF \\
C_{gs,MC} &= 3pF
\end{aligned}
$$

We must unfortunately also design the cell to a higher speed than for the normal cell. This is because the fine memory must settle within the ϕ_{1b} phase whereas the memory transistor in the normal cell had the full length of the ϕ_1 phase to settle. If we use a simple four-phase clocking scheme as shown in Figure 12 the length of the ϕ_{1b} phase is half that of the ϕ_1 phase. Consequently, for the same sampling frequency the speed of the S^2I cell must be twice high as that of the normal cell. Hence

$$
\omega_0 = 768 \cdot 10^3 s^{-1}
$$

and by (23) and (24) we get

$$
\begin{aligned}
I_Q &= 263\mu A \\
1/A &= 48 \\
(V_{GSc} - V_{TN})_Q &= 1.24V \\
(V_{Bias} - |V_{TP}|) &= 1.41V
\end{aligned}
$$

So, using this circuit we have cut the total capacitor size from an unmanageable 631pF to a much more moderate 23pF while still retaining both dynamic range and speed. This means an area saving of a factor of 27 times!

There is also an apparent drawback, namely that the quiescent current has gone up by a factor of nearly 2. This is, however, solely due to the simple clocking scheme which necessitated the doubled value of ω_0. In this scheme the coarse cell has far more time to settle in than it needs so if we in stead use a slightly more sophisticated scheme, where the ϕ_{1a} phase is much shorter than the ϕ_{1b} phase, then the coarse cell still has ample time to settle and at the same time the fine cell has almost the full duration of the ϕ_1 phase. Thus it is no longer necessary to double ω_0 and the quiescent current will be around the same value as for the normal cell.

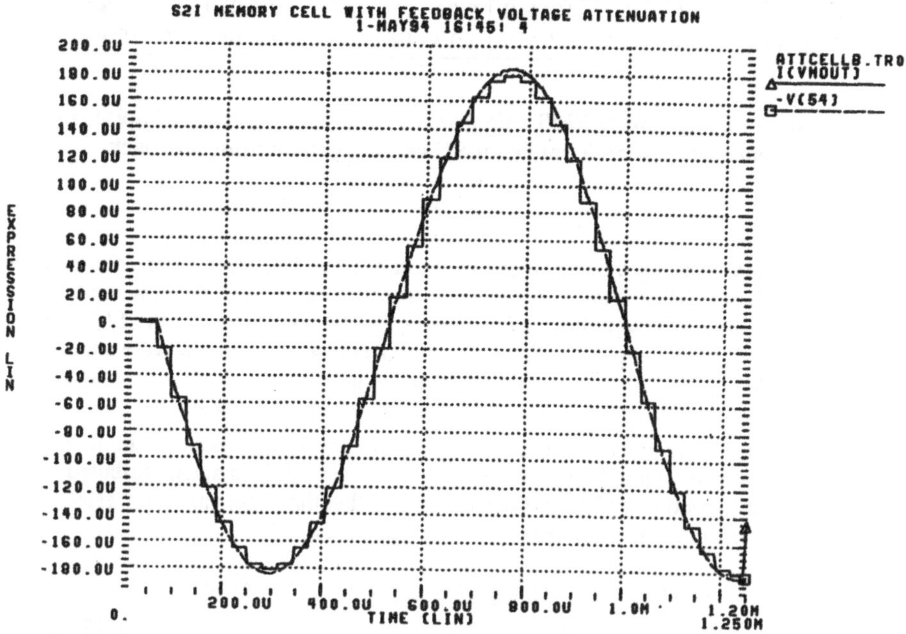

Figure 14 Sampled and held simulated response to a 1kHz sine input current.

The circuit was simulated in Northern Telecom's 0.8μm BATMOS process with the simple clocking scheme and the above derived values. The C_D capacitor was implemented as a MOS capacitor (i.e. a transistor, MD) whereas a linear capacitor was used for C_a (in the layout C_a can be quite elegantly implemented as a field oxide capacitor by simply placing a metal rectangle on top of MF's gate. Alternatively, to ensure process independent and reasonably precise attenuation, C_a can be implemented as a PMOS capacitor). The switches were implemented as minimum size NMOS transistors (denoted MS) and the component values used in the simulation were

$$\begin{aligned} W/L(\mathrm{MC}) &= 92\mu\mathrm{m}/25\mu\mathrm{m} \\ W/L(\mathrm{MF}) &= 268\mu\mathrm{m}/28\mu\mathrm{m} \\ W/L(\mathrm{MB}) &= 268\mu\mathrm{m}/28\mu\mathrm{m} \\ W/L(\mathrm{MD}) &= 71\mu\mathrm{m}/71\mu\mathrm{m} \\ W/L(\mathrm{MS}) &= 2.8\mu\mathrm{m}/0.8\mu\mathrm{m} \\ I_{bias} &= 263\mu\mathrm{A} \\ V_{DD} &= 5\mathrm{V} \end{aligned}$$

To verify the basic operation of the cell Figure 14 shows a simulated response to a 1kHz maximum amplitude sine input current. The sampled noise and resulting dynamic range was determined the same way as in section 10.2.3 and the simulation results are summarised in table 2. Figure 15 shows the gate noise spectral density on the fine transistor both with $1/f$-noise and with the $1/f$-noise removed. Both from the figure and the table it might look

	$\sqrt{V\{v_{gsF}\}}$	$V\{i_{o,s}\}$	DR	ω_0
With $1/f$-noise	5.10μVRMS	$2.36 \cdot 10^{-18}A^2$	98.6dB	$630 \cdot 10^3 s^{-1}$
Without $1/f$-noise	3.51μVRMS	$1.12 \cdot 10^{-18}A^2$	101.8dB	$630 \cdot 10^3 s^{-1}$

Table 2 Simulated gate noise RMS voltage and, based on this, sampled noise current variance and dynamic range for S^2I cell with attenuating feedback.

like the $1/f$-noise is a major contributor to the total noise. This is, however, not the case as Figure 16 shows. This figure also shows the spectral density but this time on a linear frequency axis and it is now quite clear that by far the largest amount of the $1/f$ noise is concentrated well below one third of the sampling frequency and is therefore suppressed by the correlated double sampling effect. The resulting dynamic range will consequently be close to the value in table 2 without $1/f$-noise, i.e. approximately 100dB.

Like for the basic cell in section 10.2.3 the theory and simulations are in quite good agreement. The speed, though, is 20% lower than desired. This is again attributed to the same reasons as in section 10.2.3, and in particular to that the target values for gate-source voltages etc. were not met very precisely in the simulated circuit.

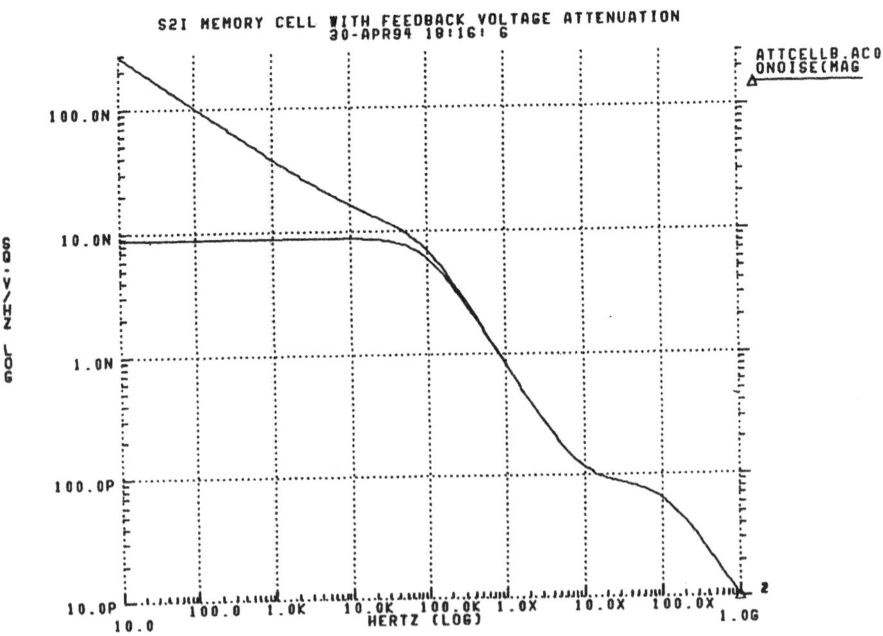

Figure 15 Simulated spectral density of the fine memory transistor gate noise voltage (in $V/\sqrt{Hz}$). The noise is both shown with $1/f$-noise and with $1/f$-noise artificially removed.

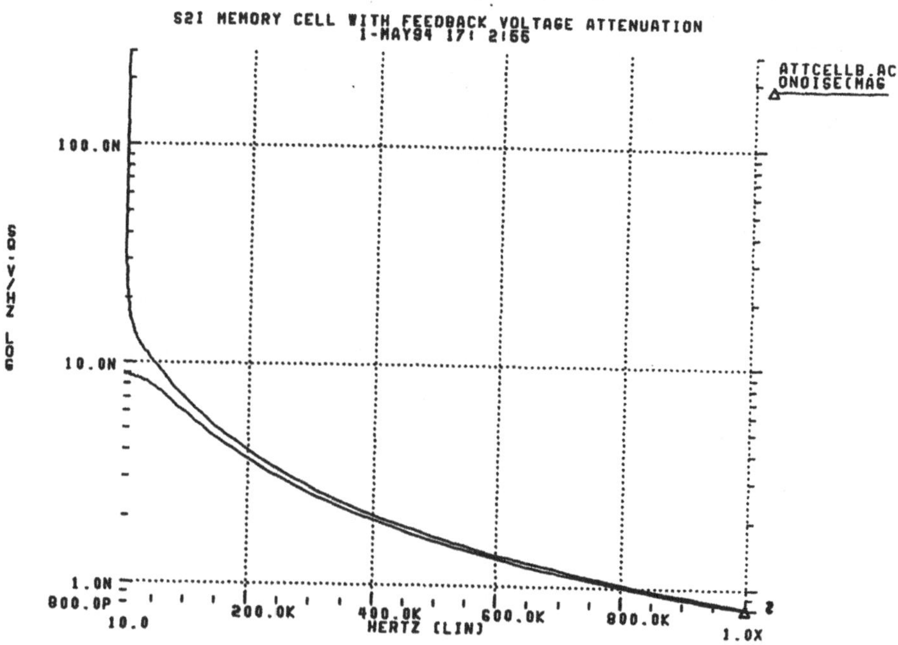

Figure 16 As Figure 15 but with linear frequency axis.

10.2.8 Conclusion

In this chapter we have demonstrated both analytically and via simulations that there exists a fundamental limit on the speed/dynamic range product of a switched-current memory cell and that this limit is directly proportional to the supply power.

This fundamental formulation of the relationship between speed, dynamic range, and power consumption has given clear insight to a mechanism for which one can be traded off for the other. Furthermore an active compensation technique has been described which provides a means of decoupling this trade-off from other circuit design issues. In fact, it has been shown that by employing the active attenuator technique for high dynamic range designs a significant saving in silicon area can be achieved, something which is not (in any obvious way) possible in switched capacitor technology. In the case of active amplification there is only a small penalty in silicon area and power consumption.

The results of this work gives insight and scope for the control and optimisation of speed, dynamic range and power consumption in switched-current circuits. We are currently integrating a high order, very wide dynamic range switched-current filter to demonstrate the compensation technique in practice.

Hopefully this chapter will inspire the switched-current novice to carry out similar analysis and reveal other powerful performance features. The switched-current technique is still immature and there is plenty of scope for design innovation in the future.

References

[1] Andrei Vladimirescu and Sally Liu. The Simulation of MOS Integrated Circuits using SPICE2 (memorandum no. UCB/ERL M80/7). Technical report, Electronics Research Laboratory, College of Engineering, University of California, Berkeley, 1980.

[2] C. Toumazou, F. J. Lidgey and D. G. Haigh. *Analogue IC design: the current-mode approach.* Peter Peregrinus (IEE), 1990.

[3] C. Toumazou, J. B. Hughes and N. C. Battersby (editors). *SWITCHED-CURRENTS: an analogue technique for digital technology.* Peter Peregrinus (IEE), 1993.

[4] R. Gregorian and G. C. Temes. *Analog MOS Integrated Circuits for Signal Processing.* Wiley-Interscience, 1986.

[5] John Hughes and Kenneth Moulding. "S²I: A Switched-Current Technique for High Performance," *Electronics Letters*, Vol. 29, No. 16, pp. 1400–1401, August 1993.

[6] S. J. Daubert, D. Vallancourt and Y. P. Tsividis. "Current Copier Cell," *Electronics Letters*, Vol. 24, pp. 1560–1562, December 1988.

[7] Peter Shah and Chris Toumazou. "Trading off speed versus dynamic range in switched current circuits," *To be presented at ISCAS'94*, London, June 1994.

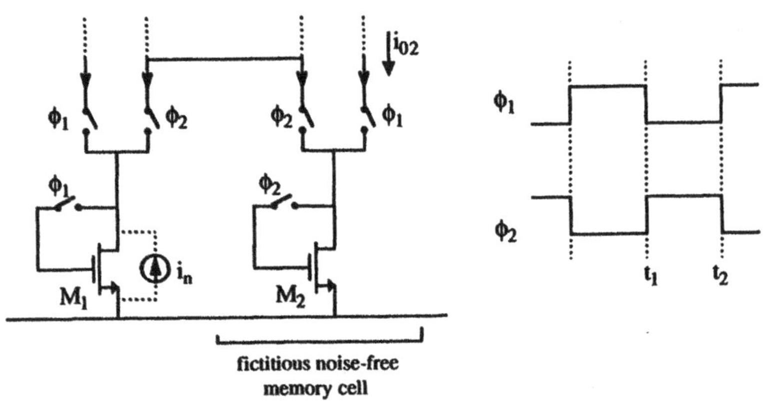

Figure 17 Memory cell in hold phase followed by a noise-free memory cell in sample-phase. i_n is the direct noise of the memory cell (which is simply the drain current noise of the transistor).

10.2.A Alternative definition for intrinsic dynamic range

In section 10.2.2 we defined the intrinsic dynamic range on the basis of the sampled noise contribution, disregarding the direct one. The direct contribution contained white noise and therefore had an infinite variance. However, as explained, the noise would be filtered by subsequent circuitry to obtain a finite variance. Indeed, in most cases the cell's output current will be resampled and memorized by another cell and in that case we are not really interested in the direct noise itself but in stead in what it contributes to the resulting error on the subsequent cell's memorized current. Thus, a reasonable measure for the intrinsic noise properties of a memory cell is the resulting sampled noise when its output current is assumed resampled by an identical but noise-free memory cell. In this appendix we will show that apart from a constant factor of 2 this definition leads to the same expression for intrinsic dynamic range as (10) in section 10.2.2.

In Figure 17 is shown the memory cell connected to the fictitious noise-free cell. At time t_1 a noise voltage is stored on M1's gate which, according to (4), has a variance of

$$V\{v_{gs1}(t_1)\} = \frac{2}{3}kT/C_{gs}$$

This gives rise to a constant output current during the ϕ_2 phase with a variance of

$$t_1 < t < t_2: \ V\{i_{o1}(t)\} = g_m^2 V\{v_{gs1}(t_1)\}$$

which, in turn, causes a gate voltage contribution on M2 with a variance of

$$V\{v_{gs2,1}(t_2)\} = V\{i_{o1}(t)\}/g_m^2 = V\{v_{gs1}(t_1)\}$$

During phase ϕ_2 the i_n-source also contributes to the gate noise voltage on M2. Like before the variance of this contribution is given by (4)

$$V\{v_{gs2,2}(t_2)\} = \frac{2}{3}kT/C_{gs}$$

and the total variance of the gate voltage is simply the sum of the two contributions[2]

$$\begin{aligned} V\{v_{gs2}(t_2)\} &= V\{v_{gs2,1}(t_2)\} + V\{v_{gs2,2}(t_2)\} \\ &= \frac{4}{3}kT/C_{gs} \end{aligned}$$

which then gives an output noise current variance of

$$V\{i_o(t_2)\} = g_m^2 V\{v_{gs2}(t_2)\} = g_m^2 \frac{4}{3}kT/C_{gs}$$

We can now find the intrinsic dynamic range, using the definition, (6), and equations (7) and (8)

$$DR = \frac{3}{8}\frac{m^2 I_Q^2}{g_m^2 \frac{kT}{C_{gs}}}$$

As evident when comparing with (9) the use of an alternative measure of dynamic range in this appendix does not alter the structure of the expression; it only gives rise to a constant factor of $\frac{1}{2}$.

[2]Strictly speaking this is not entirely true because $V\{v_{gs2,1}(t_2)\}$ and $V\{v_{gs2,2}(t_2)\}$ are correlated so that a covariance term should be included as well. This covariance term is, however, insignificant because we have assumed that i_n only contains white noise and that the cell has ample time to settle within the sample phase. When the $1/f$-noise part of i_n is considered the covariance term cannot be neglected and does in fact to some extent cancel the two variance terms, i.e. the resulting $1/f$-noise is suppressed (this is in fact just the well-known correlated double sampling property [3] of switched current circuits).

10.2.B Speed and dynamic range for memory cells with bias

To maximise the dynamic range of the cell in Figure 3 we must minimise the total noise current spectral density $S_{i_{nq}} + S_{i_n}$. Expressing $S_{i_{nq}}$ in terms of the bias conditions yields

$$\begin{aligned} S_{i_{nq}} &= (1+\gamma)\frac{8}{3}kTg_{ms} \\ &= (1+\gamma)\frac{8}{3}kT\frac{2I_Q}{V_{Bias} - |V_{TP}|} \end{aligned} \tag{25}$$

where V_{TP} is the PMOS threshold voltage. Likewise

$$S_{i_n} = \frac{8}{3}kTg_{mc}$$

which, using (12), can be expressed as

$$S_{i_n} = \frac{8}{3}kT\frac{2\sqrt{1+m}I_Q}{(V_{GSc} - V_{TN})_{max}} \tag{26}$$

where $(V_{GSc} - V_{TN})_{max}$ is the maximum gate-source voltage of MC less one NMOS threshold voltage, V_{TN}. Evidently, both V_{Bias} and V_{GSc} should be as large as possible but at the same time we must also ensure that both MS and MC remain in saturation at all times, Thus, during clock phase ϕ_1 we have to fulfil

$$(V_{Bias} - |V_{TP}|) + (V_{GSc} - V_{TN})_{max} + V_{TN} \leq V_{DD} - V_{saf}$$

where V_{saf} is a 'safety margin'. Ideally $V_{saf} = 0$ but in practice the might be unmodelled effects and other design issues, etc. which can be taken into account by using a $V_{saf} > 0$. The relation can now be used to find the largest possible value of $(V_{GSc} - V_{TN})_{max}$ at a given value of $(V_{Bias} - |V_{TP}|)$. Together with (25) and (26) we then get the noise power spectral density as a function of $(V_{GSc} - V_{TN})_{max}$

$$S_{i_{nq}} + S_{i_n} = \frac{16}{3}kT\frac{I_Q}{V_{DD} - V_{TN} - V_{saf}}\left(\frac{1+\gamma}{1 - \frac{(V_{GSc}-V_{TN})_{max}}{V_{DD}-V_{TN}-V_{saf}}} + \frac{\sqrt{1+m}}{\left(\frac{(V_{GSc}-V_{TN})_{max}}{V_{DD}-V_{TN}-V_{saf}}\right)}\right)$$

This expression can be rendered in the normalised form

$$S'_{ni} = \frac{\beta}{1 - V'_{GSc,max}} + \frac{1}{V'_{GSc,max}} \tag{27}$$

where

$$\begin{aligned} S_{i_{nq}} + S_{i_n} &= \frac{16}{3}kT\sqrt{1+m}\frac{I_Q}{V_{DD} - V_{TN} - V_{saf}}S'_{ni} \\ (V_{GSc} - V_{TN})_{max} &= (V_{DD} - V_{TN} - V_{saf})V'_{GSc,max} \\ \beta &= \frac{1+\gamma}{\sqrt{1+m}} \end{aligned} \tag{28}$$

Plotting the normalised noise S'_{ni} as a function of the normalised maximum memory transistor gate-source value $V'_{GSc,max}$, Figure 18, reveals an optimum. This optimum is

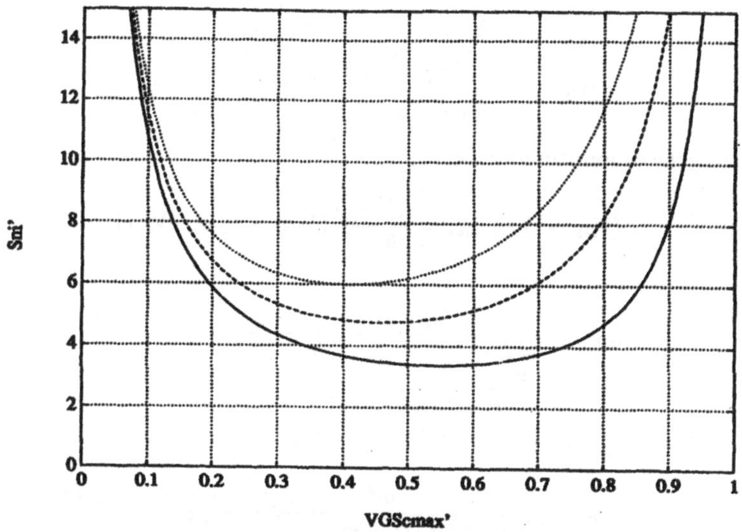

Figure 18 Normalised total current noise spectral density as a function of normalised maximum memory transistor gate-source voltage with $\beta = (1+\gamma)/(\sqrt{1+m})$ as a parameter. Solid curve: $\beta = 0.7$, dashed: $\beta = 1.4$, dotted: $\beta = 2.1$.

apparently quite flat and, hence, a somewhat lower value of $V'_{GSc,max}$ can be used without any serious degradation of noise performance. Thus, it is possible to introduce a 'safety margin' in the design without any serious noise penalty.

By differentiation the optimum can easily be derived to be

$$V'_{GSc,opt} = \frac{\sqrt{\beta}-1}{\beta-1}$$

At this value the normalised noise is

$$S'_{ni,opt} = (1+\sqrt{\beta})^2$$

For comparison, the noise in the basic cell of Figure 1 at the same gate-source voltage can be found by setting $\gamma = -1$ in (15). This gives $\beta = 0$ in (28) and inserting that in (27) yields

$$S'_{ni,bas} = \frac{1}{V'_{GSc,opt}} = \frac{\beta-1}{\sqrt{\beta}-1}$$

The cell in Figure 3 thus generates $S'_{ni,opt}/S'_{ni,bas}$ times more noise than the one in Figure 1 with

$$\frac{S'_{ni,opt}}{S'_{ni,bas}} = \sqrt{\beta}+1$$

Consequently, the dynamic range of the cell with biasing is a factor $S'_{ni,opt}/S'_{ni,bas}$ lower than that of the basic cell in Figure 1 and from (13)

$$\begin{aligned} \omega_0 &= \frac{2\sqrt{1+m}}{C_{gs}} \frac{I_Q}{(V_{GSc} - V_{TN})_{max}} \\ DR &\neq \frac{3}{16} \frac{1}{\sqrt{\beta}+1} \frac{m^2}{1+m} \frac{(V_{GSc} - V_{TN})^2_{max}}{kT/C_{gs}} \end{aligned}$$

and inserting the denormalised value

$$(V_{GSc} - V_{TN})_{max} = \frac{\sqrt{\beta}-1}{\beta-1}(V_{DD} - V_{TN} - V_{saf})$$

yields

$$\begin{aligned} \omega_0 &= \frac{\beta-1}{\sqrt{\beta}-1} \frac{2\sqrt{1+m}}{C_{gs}} \frac{I_Q}{V_{DD} - V_{TN} - V_{saf}} \\ DR &= \frac{3}{16} \frac{1}{(1+\sqrt{\beta})^3} \frac{m^2}{1+m} \frac{(V_{DD} - V_{TN} - V_{saf})^2}{kT/C_{gs}} \end{aligned}$$

The figure of merit is then

$$DR \cdot \omega_0 = \frac{3}{8} \frac{1}{(1+\sqrt{\beta})^2} \frac{m^2}{\sqrt{1+m}} \frac{(V_{DD} - V_{TN} - V_{saf})_{max} I_Q}{kT}$$

and this is attained at the following optimal gate-source voltages

$$\begin{aligned} (V_{GSc} - V_{TN})_Q &= \frac{1}{\sqrt{1+m}} \frac{\sqrt{\beta}-1}{\beta-1}(V_{DD} - V_{TN} - V_{saf}) \\ (V_{Bias} - |V_{TP}|) &= (1 - V'_{GSc,opt})(V_{DD} - V_{TN} - V_{saf}) \\ &= \sqrt{\beta}\frac{\sqrt{\beta}-1}{\beta-1}(V_{DD} - V_{TN} - V_{saf}) \end{aligned}$$

where $(V_{GSc} - V_{TN})_Q$ is the quiescent gate-source voltage of MC less one threshold voltage and is obtained from $(V_{GSc} - V_{TN})_{max}$ by using (11).

10.2.C Speed and dynamic range for S²I cells with amplifying or attenuating feedback

As mentioned the S²I cell has a number of advantages. One that has apparently not been realised previously is that it is almost immune to noise from the biasing circuitry. This is because the biasing noise is absorbed by the coarse memory during phase $\phi 1a$ and during the fine phase $\phi 1b$ the biasing circuitry is disconnected. Hence

$$\gamma = 0$$

in (15).

The optimum quiescent voltages for the S2I cell are obviously the same as for the normal cell except that in the normal cell in Figure 3 only MC is ever diode-connected whereas in the circuit in Figure 12 MC and MF are diode-connected during the ϕ_{1a} and ϕ_{1b} phases respectively. Hence, the V_{TN} in (18) should be replaced by V_{Tmax} where

$V_{Tmax} = \max(V_{TN}, |V_{TP}|)$. Thus, the optimal quiescent gate-source voltages are

$$\begin{aligned}(V_{GSc} - V_{TN})_Q &= \frac{1}{\sqrt{1+m}} \frac{\sqrt{\beta}-1}{\beta-1}(V_{DD} - V_{Tmax} - V_{saf}) \\ (V_{Bias} - |V_{TP}|) &= \sqrt{\beta}\frac{\sqrt{\beta}-1}{\beta-1}(V_{DD} - V_{Tmax} - V_{saf}) \qquad (29)\end{aligned}$$

The speed and dynamic range are now determined by the fine memory transistor MF which corresponds to the bias current transistor MS in the normal cell. Hence, the speed is a factor of g_{mf}/g_{mc} higher than for the normal cell at the same bias condition and the dynamic range is lower by the same factor. Thus

$$\begin{aligned}\omega_0(\mathrm{S^2I}) &= \frac{g_{mf}}{g_{mc}}\omega_0(\mathrm{normal}) \\ DR(\mathrm{S^2I}) &= \frac{g_{mc}}{g_{mf}}DR(\mathrm{normal})\end{aligned}$$

Using (29) we get

$$\frac{g_{mf}}{g_{mc}} = \frac{2I_Q/(V_{Bias} - |V_{TP}|)}{2I_Q/(V_{GSc} - V_{TN})_Q} = \frac{(V_{Bias} - |V_{TP}|)}{(V_{GSc} - V_{TN})_Q} = \frac{1}{\sqrt{\beta}}$$

Thus, with $\gamma = 0$ and using (16), (17), and (19)

$$\begin{aligned}\omega_0 &= A\frac{1-\sqrt{1+m}}{1-\sqrt[4]{1+m}}\frac{1}{C_D}\frac{2I_Q}{V_{DD} - V_{Tmax} - V_{saf}} \\ DR &= \frac{1}{A}\frac{3}{16}\frac{\sqrt[4]{1+m}}{\left(1+\frac{1}{\sqrt[4]{1+m}}\right)^3}\frac{m^2}{1+m}\frac{(V_{DD} - V_{Tmax} - V_{saf})^2}{kT/C_D}\end{aligned}$$

The figure of merit is then

$$DR \cdot \omega_0 = \frac{3}{8}\frac{1}{(1+\sqrt{\frac{1}{\sqrt[4]{1+m}}})^2}\frac{m^2}{\sqrt{1+m}}\frac{(V_{DD} - V_{Tmax} - V_{saf})_{max} I_Q}{kT}$$

Chapter 10.3

Switched-Current Data Converters

Gordon W. Roberts

10.3.1 Introduction

Switched-Current (SI) circuit techniques are now beginning to rival the performance capable from Switched-Capacitor (SC) circuit techniques in realizing many sampled-data analog circuit functions. This is beneficial to the VLSI community at large because SI circuit techniques provide an economical means to realize both analog and digital functions on the same IC using a low-cost digital CMOS process.

One important area of development underway in SI circuit research is in the creation of high-resolution data converter circuits for both high-speed and low-power applications. Since the introduction of the switched-capacitor technique, albeit referred to by various other names [1-3] , a wide range of data converter circuits have emerged. The intention here is not to re-cap the work which has already been written about extensively, such as in [4]. But, rather, the following will attempt to present to the novice the underlying principles of SI data conversion, from algorithm to circuit realization.

We shall begin by first describing the basic principles and some definitions of data conversion, followed by a description of the various data conversion algorithms, and then onto the implementation of these algorithms using SI circuit techniques.

10.3.2 Principles of data conversion

Data converters can be divided into two groups depending on the sampling rate of the converter: Nyquist-rate or oversampling converters. Nyquist-rate converters are defined as converters that sample a bandlimited continuous-time signal at its Nyquist frequency f_N, where f_N is twice the bandwidth of the analog signal. Oversampled converters are those converters that sample the continuous-time signal at a much higher rate than the Nyquist frequency f_N.

The following will describe the basic operation of these two classes of converter types from analog-to-digital (A/D) converters to digital-to-analog (D/A) converters.

10.3.2.1 Nyquist-rate converters

(i) Digital-to-analog conversion

Digital-to-analog conversion circuits, also called D/A converters or DACs, can be considered as decoding devices that accept digitally coded signals and provide analog outputs in the form of a voltage, current or charge. Figure 1 depicts the functional block diagram of the basic D/A converter. At the input to the D/A converter is a digital word D, consisting of a set of N binary bits. At the output an analog representation of this digital code, denoted A_{out}, is produced. Mathematically, A_{out} is related to the input according to

$$A_{out} = K_{ref} D \tag{1}$$

where K_{ref} is a reference level and D is the N-bit digital word. D can be represented as

$$D = \frac{b_1}{2^1} + \frac{b_2}{2^2} + \frac{b_3}{2^3} + \dots + \frac{b_N}{2^N} \tag{2}$$

with coefficients $b_1, b_2, \dots b_N$ taking on values of 1 or 0. Coefficient b_1 is regarded as the Most Significant Bit (MSB), as it has the largest effect on the output signal, and coefficient b_N is known as the Least Significant Bit (LSB) as it has the smallest effect.

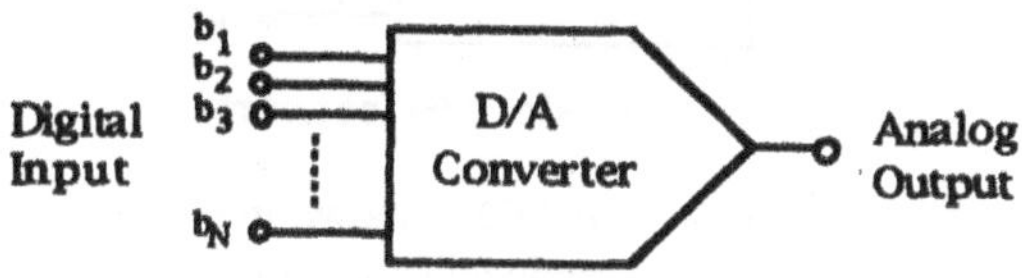

Figure 1 Basic D/A converter.

The ideal transfer characteristic of a D/A converter is shown in Figure 2. Here we illustrate the desired characteristic of a simple 3-bit converter, although a similar characteristic can be extended to any A/D converter, regardless of its bit length. Along the horizontal axis we indicate all possible combinations of the digital input word. For a 3-bit word length, this translates into 2^3 possible input code words. Correspondingly, along the vertical axis is the analog output voltage ranging from 0 to the full-scale (FS) value. For each input digital word, a single analog voltage level is produced, reflecting the one-to-one mapping of the D/A converter. In the ideal case, each analog output is separated from the others by one LSB, defined by

$$LSB = \frac{FS}{2^N} \tag{3}$$

In our 3-bit example shown in Figure 2 one LSB is equal to FS/8.

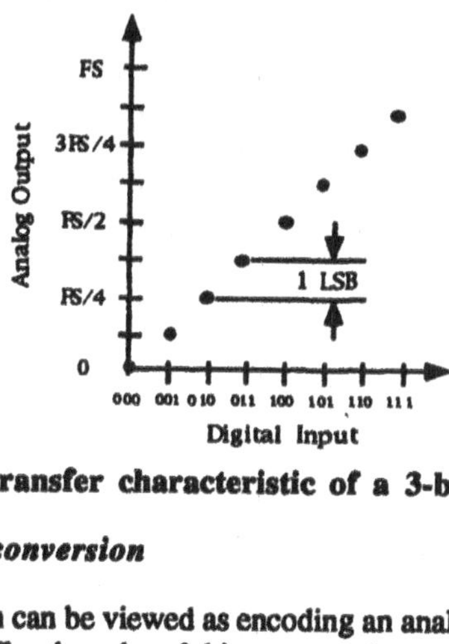

Figure 2 Ideal transfer characteristic of a 3-bit D/A converter.

(ii) Analog-to-digital conversion

Analog-to-digital conversion can be viewed as encoding an analog input signal into a digital word of a fixed bit length. For the sake of this presentation, we shall assume that it is also

N-bits wide. An analog-to-digital converter, also called an A/D converter or ADC, performs an almost-inverse operation to the D/A converter. Here the term "almost" is used to highlight that algorithms used to perform D/A conversions cannot be solved in reverse order to provide equations for A/D conversion. A functional block diagram illustrating the input-output signals to the A/D is shown in Figure 3. Fundamentally, D/A conversion is a one-to-one mapping whereas A/D conversion is a many-to-one mapping.

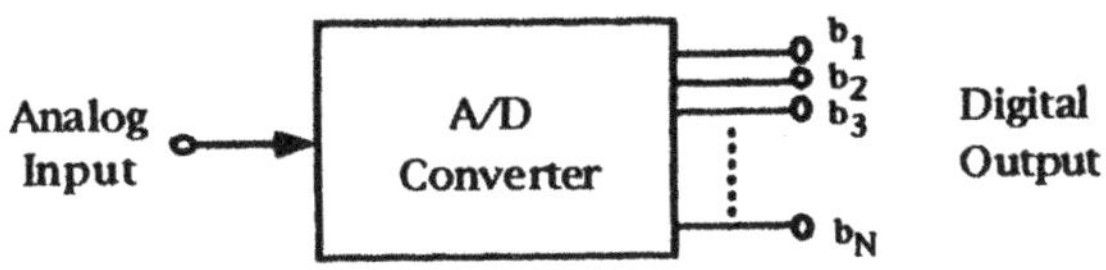

Figure 3 Basic A/D converter.

The ideal transfer characteristic for a 3-bit A/D is shown in Figure 4. Along the horizontal axis we have the analog or continuous input level and along the vertical axis we have the set of all possible output digital code words.

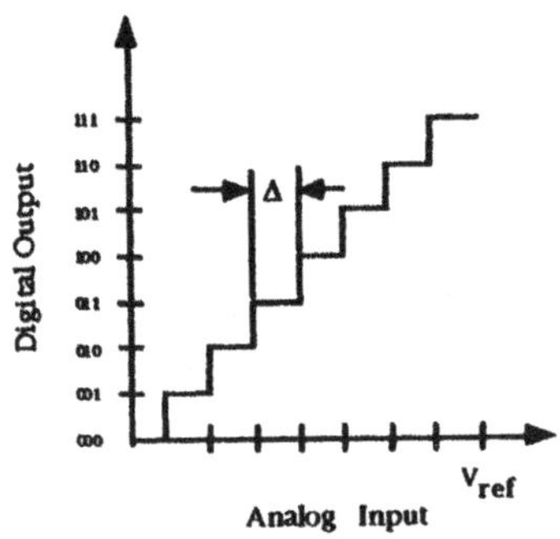

Figure 4 Ideal transfer characteristic of 3-bit A/D converter.

Like the ideal D/A, the ideal A/D has a discrete output set, but for each output code there is a range of input levels that could have caused it. An A/D converter has one-way uncertainty; if a specified analog level is applied to the input of a perfect A/D converter we would know precisely what its output code word is. But, in the reverse situation, that is, if we are told what the output code word is, we cannot tell exactly what the input signal is, only its range. This interval is usually referred to as the quantization step Δ. Its magnitude depends on the full-scale (FS) value and the number of bits associated with the converter in the following manner,

$$\Delta = \frac{FS}{2N} \tag{4}$$

It is also common practice to refer to Δ as one LSB as they have equivalent forms.

To better understand the effect of the uncertainty associated with an A/D converter, let us consider the following experiment: An analog signal is first converted to digital form using an ideal *N*-bit A/D converter and then converted back to analog form using an ideal *N*-bit D/A converter.

Let us further consider that the input signal is a ramp input, beginning at zero and rising to the A/D full-scale value. Figure 5 depicts such a situation.

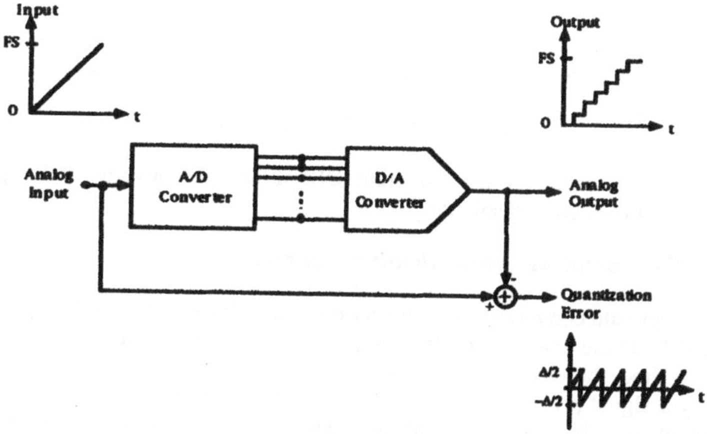

Figure 5 Illustrating the uncertainity present in A/D converters.

If the conversion process was distortionless then we should at the end of the two conversions have the same analog value that we started with, i.e. a ramp output. Instead, the output is a staircase signal with 2^N steps. It is the difference between the input and output signals which quantifies the uncertainty associated with the A/D conversion process. In this particular case, the difference between the input and output signals is a sawtooth waveform with an average value of zero and having maximum signal excursions of $\pm\Delta/2$. This waveform is commonly referred to as the quantization error or noise waveform of the A/D converter. For different natured input signals, the quantization noise waveform will take on different shapes but will be bounded by $\pm\Delta/2$. This suggests that the maximum absolute quantization error that can occur at any one time is $\Delta/2$. For quantization noise uniformly distributed between $\pm\Delta/2$ (such as the case for the sawtooth waveform), the RMS value of this error signal will be

$$e_{rms} = \frac{\Delta}{\sqrt{12}} \qquad (5)$$

or

$$e_{rms} = \frac{FS}{\sqrt{12}\,2^N} \qquad (6)$$

For an input ramp signal traversing the full range[1] of the A/D converter, its rms value would be $FS/\sqrt{12}$, thus the maximum signal-to-noise ratio of the A/D converter in this situation can be quantified as

$$SNR_{max} = 20\,Log\,\left(\frac{\frac{Fs}{\sqrt{12}}}{\frac{Fs}{\sqrt{12}\,2^N}}\right) \qquad (7)$$

or, when simplified, as

[1] Here the A/D full-range means FS-Δ.

$$SNR_{max} = 6.02\,N \quad dB \tag{8}$$

Similarly, for a full-scale sinusoidal input, the maximum SNR is found to be

$$SNR_{max} = 6.02\,N + 1.761 \quad dB \tag{9}$$

Here we see that for each case the maximum SNR of the A/D converter will improve by 6 dB for each additional bit of resolution.

10.3.2.2 Oversampling noise-shaping converters

Another class of data converters that are very popular today is oversampling noise-shaping converters [5]. These converters offer a means of exchanging resolution in amplitude for resolution in time, thereby circumventing the need for precision analog circuits. This is achieved by a signal processing operation that separates the quantization noise from the inband set of signals, e.g. if the inband signals have a lowpass spectrum then the quantization noise is made to have a high-pass spectrum with very little frequency overlap. By subsequent brick wall filtering, the quantization noise can be removed from the output signal thereby improving the overall SNR of the data conversion. The signal processing operation that is used to shape the quantization noise is known as delta-sigma or sigma-delta modulation.

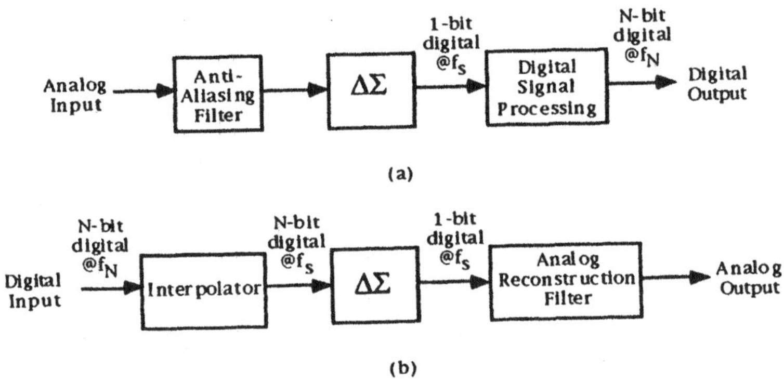

Figure 6 Oversampling noise-shaping data converters: (a) A/D (b) D/A.

Delta-sigma modulation is at the core of both D/A and A/D converters. Let us first describe the makeup of an oversampling noise-shaping A/D converter. Its basic structure is illustrated in Figure 6a where it is divided into three separate blocks: the input anti-aliasing analog filter, the delta-sigma modulator, and the digital signal post-processing. The input anti-aliasing filter is usually a low-order passive filter circuit whose 3 dB frequency is set at some frequency far above the Nyquist interval (i.e. twice the signal bandwidth) of the input signal. This allows for variations in the 3 dB frequency of the filter without causing any signal degradation. The delta-sigma modulator block performs two important functions: One function is to modulate the bandlimited analog input signal into a one-bit digital code at a frequency f_s much higher that the Nyquist rate f_N. The other function is to noise-shape the quantization noise such that the low-frequency input signal is separable from the quantization noise. Subsequently, the digital signal post-processing acts to suppress the quantization noise, resulting in an N-bit digital output occurring at the Nyquist rate f_N.

The basic structure of an oversampling noise-shaping D/A converter is illustrated in Figure 6b. It has been divided into three separate blocks: the input interpolator, the delta-sigma modulator and the analog reconstruction lowpass filter. The interpolator acts to up-sample an N-bit digital word at the Nyquist rate of f_N to an equivalent set of bits at the oversampling rate of f_S. Similar to the A/D case, although a multi-bit digital signal is the input, the delta-sigma modulator converts the N-bit input digital word to a digital one-bit stream and shapes the corresponding quantization noise. This one-bit digital stream is then converted to analog form where the quantization noise is suppressed by the analog filter.

The overall signal-to-noise behavior of the oversampling noise-shaping converter depends directly on the noise-shaping ability of the delta-sigma modulator, which in turn depends on the structure of the delta-sigma modulator. For a first order modulator (to be described in the next section) the maximum SNR ratio with a full-scale sinusoidal input is approximately

$$SNR_{max} = 9\ (\log_2 OSR\ ) - 5.2 \quad dB \tag{10}$$

where OSR denotes the oversampling ratio f_S/f_N. Similarly, it is easily shown that the maximum SNR for a second-order modulator is

$$SNR_{max} = 15\ (\log_2 OSR\ ) - 13 \quad dB \tag{11}$$

Here we see that by increasing the OSR by a factor of 2 will result in a 9 or 15 dB improvement in the maximum available SNR, depending on the order of the modulator. This is in contrast to a Nyquist-rate converter where doubling the resolution improves the SNR by only 6 dB. Higher-order delta-sigma modulators are also possible, but their stability behavior is not yet well understood.

10.3.3 Data converter algorithms

In the literature there are a vast number of D/A and A/D conversion techniques, some implemented as hybrids and others implemented in a single integrated form. Of the ones presented here, we limit our discussion to those that have and can be implemented using SI techniques. In particular, we shall describe below the algorithm associated with Nyquist-rate and oversampling data converters. These algorithms will be presented from a signal processing point-of-view rather than from a circuit's point-of-view. Subsequent sections will illustrate how these algorithms are converted to SI circuits. Our discussion of converter circuits will begin with a description of D/A converter algorithms. As the reader shall see, D/A algorithms are easier to understand than A/D converter algorithms, but more importantly, many A/D conversion algorithms involve D/A algorithms as part of their conversion schemes.

10.3.3.1 Nyquist-rate D/A converters

We begin our discussion of D/A conversion techniques [6] by first considering the mathematical equation that describes the input-output behavior of a D/A converter (equations (1) and (2)), i.e.

$$A_{out} = K_{ref}\left(\frac{b_1}{2^1} + \frac{b_2}{2^2} + \frac{b_3}{2^3} + \dots + \frac{b_N}{2^N}\right) \tag{12}$$

or when expanded

$$A_{out} = K_{ref}\frac{b_1}{2^1} + K_{ref}\frac{b_2}{2^2} + K_{ref}\frac{b_3}{2^3} + ... + K_{ref}\frac{b_N}{2^N} \qquad (13)$$

A block diagram illustrating the order of operations performed by the above expression is shown in Figure 7. This figure illustrates how the reference level K_{ref} is multiplied by each bit associated with the input digital word, $D = b_1, b_2 ... b_N$, and how each output is multiplied by different factors of power-of-two and then combined to form the analog output signal. Since each binary signal takes on values of either 0 or 1, the multiplier can be implemented very simply as a single-throw double-pole switch connecting the output between K_{ref} or ground, as shown in Figure 8.

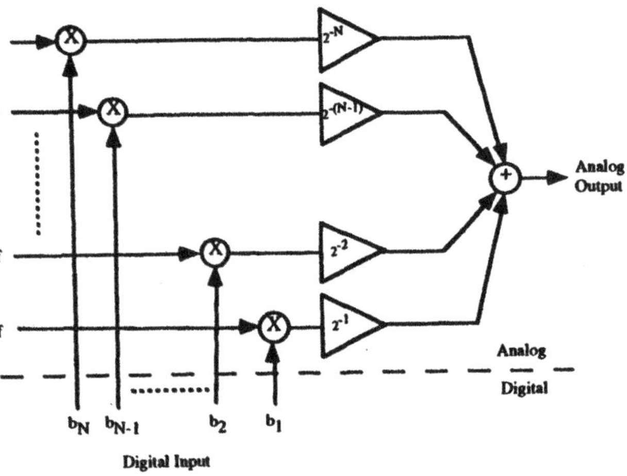

Figure 7 Block diagram of a basic D/A converter (parallel form).

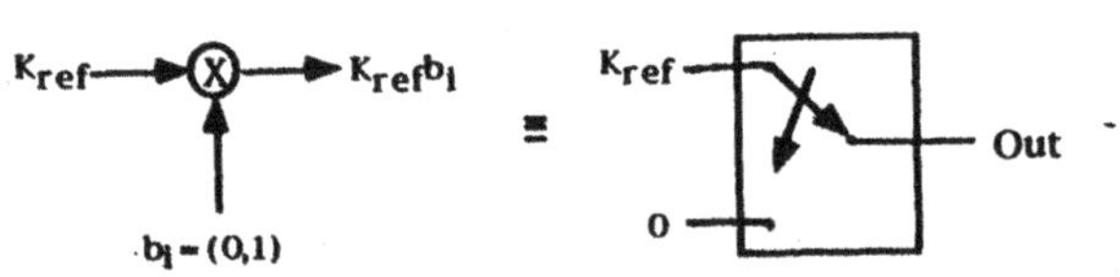

Figure 8 Analog implementation of multiplier circuit.

By simply re-arranging the order of multiplication between b_i and 2^{-i}, for any i, a different block diagram for D/A conversion can result. Figure 9 illustrates the block diagram when the order of multiplication is reversed. Although such an algebraic operation is trivial, the resulting circuit implementation can be drastically different.

To illustrate yet another algebraic manipulation that will result in a very different circuit implementation, consider re-writing equation (13) in nested form as follows:

$$A_{out} = ((\ ...((\ 0 + K_{ref}\,b_N\,)\,2^{-1} + K_{ref}\,b_{n-1}\,)\,2^{-1} + \ ....$$
$$+ K_{ref}\,b_2\,)\,2^{-1} + K_{ref}\,b_1)\,2^{-1} \qquad (14)$$

The reader can expand this expression and show that it is exactly the same as that given by equations (12) or (13). A block diagram illustrating the order of operations performed by equation (14) is shown in Figure 10. As can be seen from this figure, the algorithm is made up of *N*-identical stages consisting of a one-bit multiplier (or multiplixer), a summer and an amplifier with a gain of 1/2 (divide-by-two stage). The result of the leftmost stage is propagated to the right and combined with the next stage. This continues until the final analog result appears at the output of the rightmost stage. A single digital word conversion will therefore depend on the time it takes a signal to propagate from left to right through each stage. For an N-bit converter, the total time for a digital word conversion will be NT_{delay}, where T_{delay} is the delay time associated with each stage.

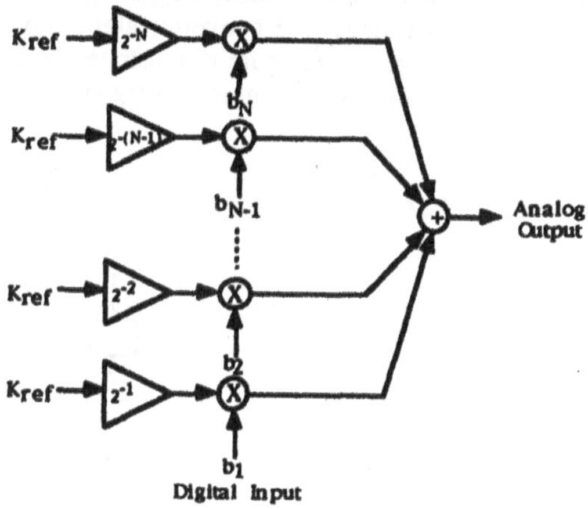

Figure 9 Alternative form to Figure 7.

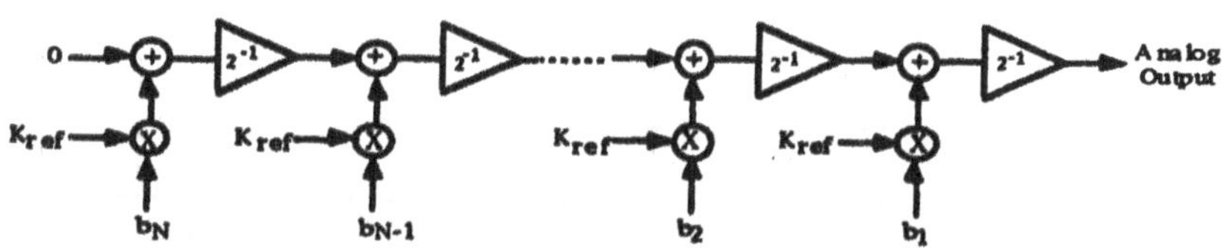

Figure 10 D/A conversion algorithm in nested form.

Clock Cycle	LSB	Stage .... MSB				Analog Ouput
1	b_N	X	...	X	X	Not Valid
2	X	b_{N-1}	...	X	X	Not Valid
.	.	.	.	.	.	.
.	.	.	.	.	.	.
.	.	.	.	.	.	.
N-1	X	X	...	b_2	X	Not Valid
N	X	X	...	X	b_1	Valid

Figure 11 Timing diagram for nested D/A conversion algorithm.

Figure 11 illustrates the timing and signal propagation through this serial D/A converter as a function of the number of clock iterations. The *X* mark is used to indicate that a stage is idle. As is evident from this figure only one stage is active at any one time, as well, once a stage has completed its bit conversion, it is no longer involved in the bit conversion process. This observation has lead to two different types of implementations; pipelined and algorithmic. Let us first describe the pipelined D/A algorithm.

The pipelined D/A conversion algorithm utilizes those stages that are idle during the conversion process to start the conversion of the next digital word to analog form. In this way, high D/A throughput is possible; in fact, one D/A conversion per clock cycle resulting in an *N*-times factor in speed improvement. By synchronizing the output of each stage by adding a controlled analog delay and by appropriately staggering the time that a single-bit of a digital word is applied to each stage, the converter algorithm of Figure 9 can be transformed into a pipelined D/A converter. Figure 12 illustrates the block diagram of such a converter and Figure 13 demonstrates the timing and signal propagation through this converter.

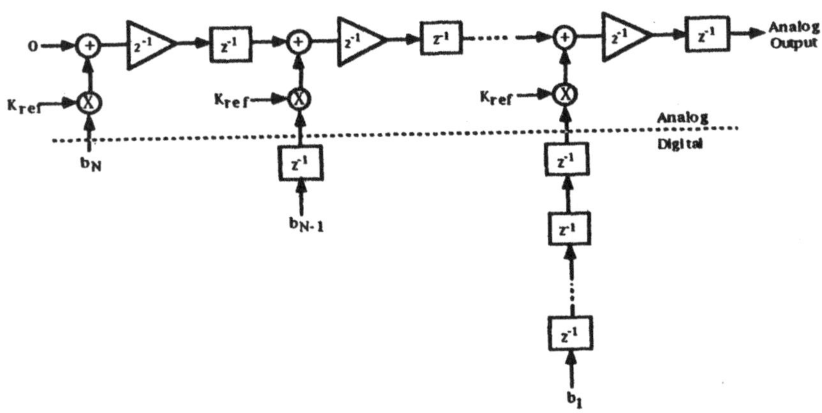

Figure 12 Pipelined version of a D/A converter.

Clock Cycle	Stage LSB			...	MSB	Analog Output
K+1	b_N^K	b_{N-1}^{K-1}	b_{N-2}^{K-2}	...	b_1^{K-N+1}	K-N+1
K+2	b_N^{K+1}	b_{N-1}^{K}	b_{N-2}^{K-1}	...	b_1^{K-N+2}	K-N+2
K+3	b_N^{K+2}	b_{N-1}^{K+1}	b_{N-2}^{K}	...	b_1^{K-N+3}	K-N+3
.	.	.	.	.	.	.
.	.	.	.	.	.	.
.	.	.	.	.	.	.
K+N	b_N^{K+N-1}	b_{N-1}^{K+N-2}	b_{N-2}^{K+N-3}	...	b_1^{K}	K

Figure 13 Timing diagram for pipelined D/A conversion algorithm.

Rather than distribute the computational load of the D/A conversion algorithm over time and hardware by pipelining, one can instead time-share the hardware of a single stage between different phases of the D/A algorithm since it is exactly the same. A block diagram of such an algorithm is shown in Figure 14. Implementing D/A conversion this way has the advantage of requiring less hardware, but has a slower throughput than a pipelined converter. D/A converters of this type are known as Algorithmic or Cyclic D/A converters

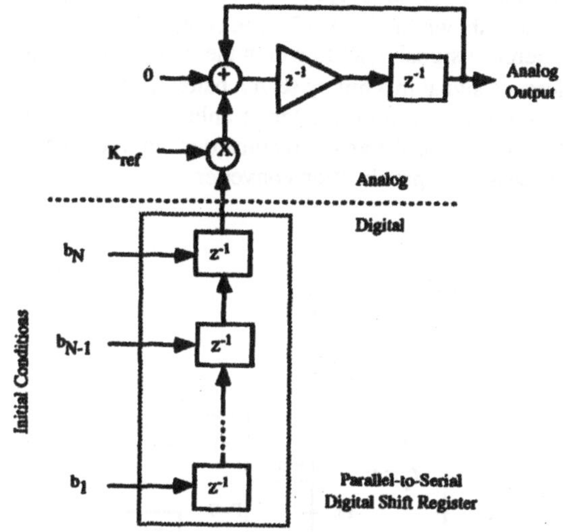

Figure 14 Algorithmic D/A conversion procedure.

10.3.3.2 Nyquist-rate A/D converters

As is the case for D/A converters, there exists a wide variety of A/D algorithms [6]. However, we limit our discussion to those A/D converters that are realizable using SI techniques [7-15].

Mathematically, the problem of A/D conversion can be stated as the following: For a given analog input A_{in}, what are the coefficients $b_1, ..., b_N$ that best satisfy the following expression

$$A_{in} = K_{ref}\left(\frac{b_1}{2^1} + \frac{b_2}{2^2} + \frac{b_3}{2^3} + ... + \frac{b_N}{2^N}\right) \tag{15}$$

It is self-evident that equation (15) cannot simply be rearranged in a form that would allow for each binary coefficient to be isolated and solved. Instead, methods have been devised in which to identify the values of these coefficients in a systematic way.

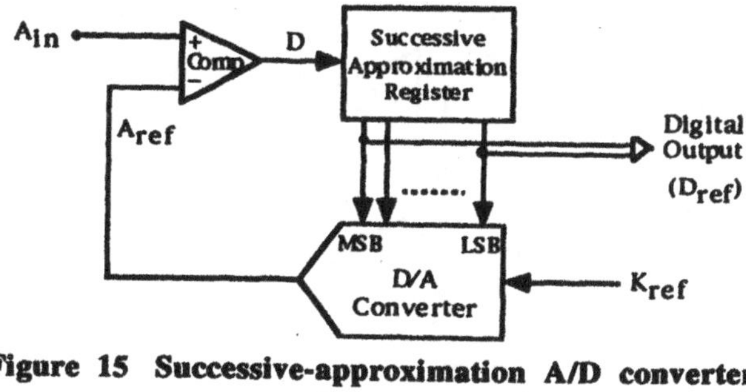

Figure 15 Successive-approximation A/D converter.

One of the most popular algorithms is the successive approximation A/D conversion method. This method simply performs a binary search identifying the coefficients that best approximate A_{in}. This technique employs a D/A converter, a comparator and some control circuitry according to that shown in Figure 15. The control circuit and the D/A converter are used to generate a reference signal A_{ref} to which the input analog signal A_{in} is compared. By adjusting the reference signal in a binary search pattern, beginning midway between full-scale, a successive approximation converter is able to converge on the correct set of coefficients using N comparisons for an N-bit converter. Figure 16 demonstrates the search process for a 3-bit successive approximation converter.

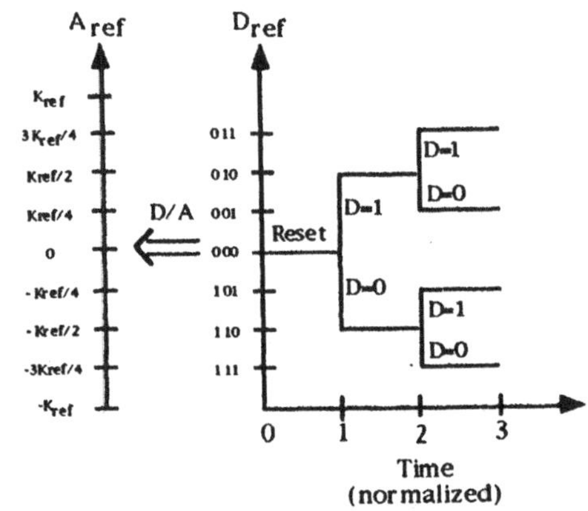

Figure 16 An example of the successive-approximation process (3-bit signed-magnitude conversion).

An alternative technique, similar to the successive approximation approach, is the algorithmic conversion technique. It consists of much the same circuitry as the successive approximation converter, except that the high-resolution D/A circuit has been replaced by a subtraction circuit, multiply-by-two circuit and an analog sample-and-hold stage. The block diagram of such a converter is shown in Figure 17.

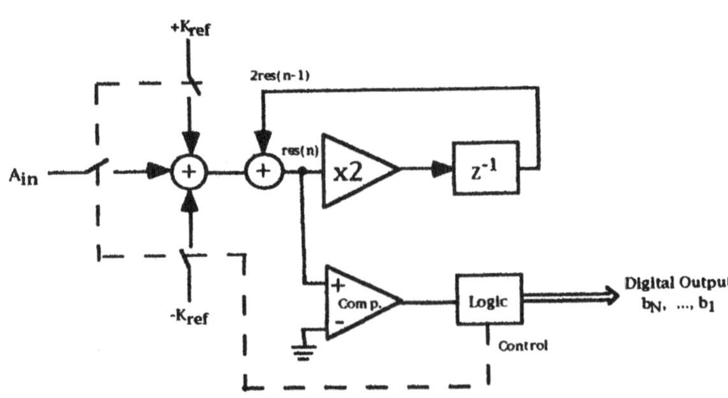

Figure 17 Algorithmic A/D converter.

The operation of the algorithmic converter is very similar to the successive approximation converter in that a comparison is made between an input signal and a reference in order to decide the direction of the binary search. The difference lies with the manner in which the comparison is made. In the case of the successive approximation converter the reference level is made to approximate the input signal level through *N* steps. In contrast, in the algorithmic converter, the reference level is held constant and the input signal level is made to approximate the reference level through the use of the subtraction and multiplication circuits in *N* steps. A flow chart describing the decision process is provided in Figure 18.

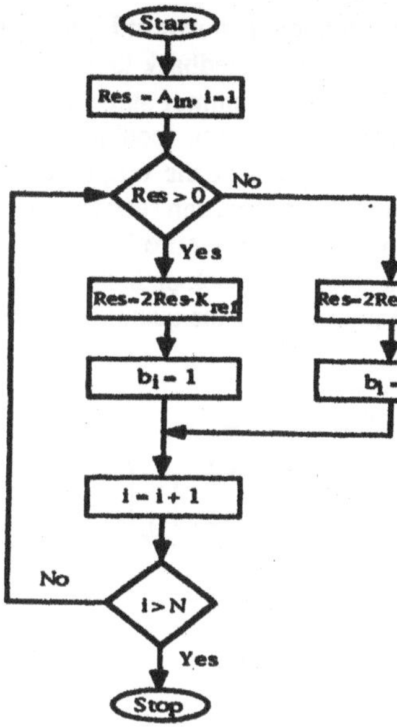

Figure 18 Algorithmic A/D conversion process.

If faster conversion speed is required then it is possible to distribute the computational burden over *N* stages by implementing a pipelined structure. Figure 19 illustrates the block diagram of a pipelined implementation of the algorithmic A/D converter. Like the D/A pipelined implementation, an *N*-times factor improvement in throughput is possible with this design.

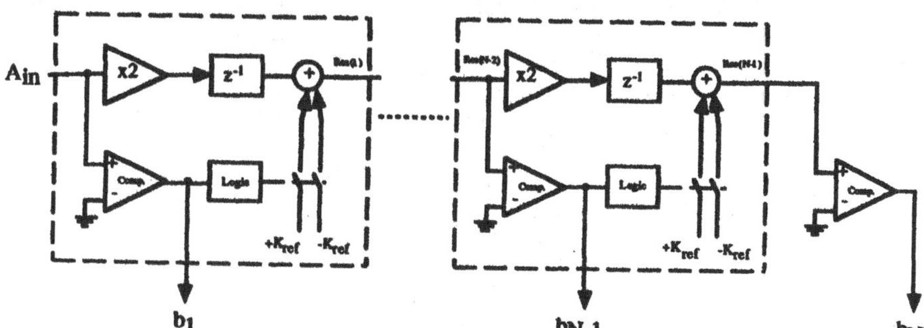

Figure 19 Pipelined-version of the algorithmic A/D converter.

10.3.3.3 Oversampling noise-shaping data converters

A simple view of noise-shaping data converters is that they consist of several filtering operations and a delta-sigma modulation process. As digital and analog filtering are well-understood operations, we shall focus our attention exclusively on the delta-sigma modulation operation. It is this operation which largely determines the overall SNR of the data converter. Finally, as both A/D and D/A converters involve the same conceptual model of a delta-sigma modulator, we shall limit our discussion here to delta-sigma modulators for A/D converters.

A first-order delta-sigma modulator is shown in Figure 20a. It consists of an analog delay z^{-1} and a summer stage in an internal feedback loop, which together forms an analog integrator circuit. Also present is another summer stage, a comparator and a one-bit D/A circuit. The one-bit D/A circuit is necessary to feedback a current to the input summer instead of the voltage signal that appears at the output of the comparator. To better understand the operation of this modulator, we can rearrange the structure of the modulator and obtain the new block diagram shown in Figure 20b. Here the block diagram for the delta-sigma modulator contains a single feedback path clearly indicating that the only signal being fed back to the input of the modulator is the quantization error made by the comparator circuit.

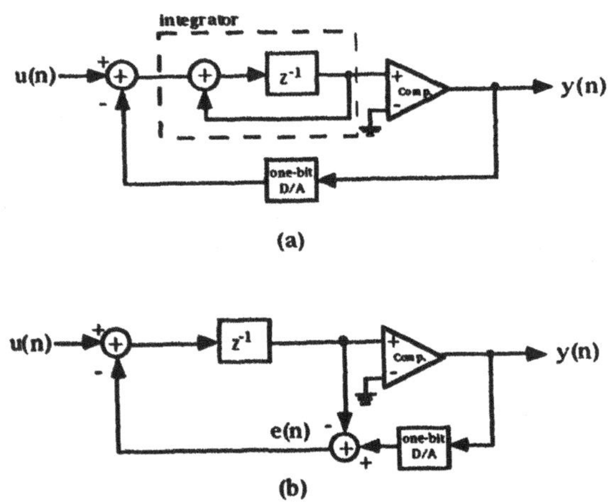

Figure 20 (a) First-order delta-sigma modulator (b) alternative form.

The operation of the first-order delta-sigma modulator can be described by following the input u through to the output y. To begin, let us assume that the quantization error e is zero at this instant. The input then appears at the input of the comparator and is compared to a zero reference. The output y is then forced into two possible states. Here we shall assume ± 1. Clearly then, if y is supposed to represent u, it is in error by the quantization term e, i.e.,

$$y(n) = u(n-1) + e(n) \tag{16}$$

By feeding back this error and subtracting it off the input, the output of the comparator on the next clock cycle will be

$$y(n+1) = [\, u(n) - e(n) \,] + e(n+1) \tag{17}$$

or

$$y(n+1) = u(n) + [\, e(n+1) - e(n) \,] \tag{18}$$

Likewise, future output samples will take on a similar form. Equation (18) highlights the operation of the delta-sigma modulator. On account of the feedback, the delta-sigma modulator is cancelling the present quantization error $e(n+1)$ by subtracting off the quantization error made on the previous cycle, i.e. $e(n)$. Since adjacent samples in a highly oversampled signal are highly correlated, the difference, $e(n+1) - e(n)$ is nearly zero and we see that $y(n+1) \sim u(n)$.

An alternative perspective of noise-shaping delta-sigma modulation, and one than lends it name to the term noise-shaping, can be had from a frequency domain perspective. Consider taking the z-transform of equation (18) above. On doing so we can write,

$$Y(z) = z^{-1}\, U(z) + (1 - z^{-1})\, E(z) \tag{19}$$

where $Y(z)$, $U(z)$ and $E(z)$ are the z-transforms of discrete signals $y(n)$, $u(n)$ and $e(n)$, respectively. According to equation (19) we see that the input is multiplied by z^{-1} while the quantization noise is weighted by $(1-z^{-1})$. For a bandlimited input signal $U(z)$, we see that its spectra will be unaltered by the modulation process. In the case of the quantization noise $E(z)$, it can have a wide variety of spectra, depending on the statistics of the input signal. To provide some frequency domain insight into the operation of the delta-sigma modulator, it is customary to assume that $e(n)$ is white noise. Thus, the quantization noise that appears at the output of the modulator will depend directly on the frequency characteristics of the term $(1-z^{-1})$. For physical frequencies, $z=e^{j\omega T}$, and ωT much smaller than unity to correspond to the highly oversampled situation, the magnitude response of $(1-z^{-1})$ can be very well approximated by ω. Thus, at low frequencies, the quantization noise is made insignificant, whereas at high frequencies it is greatly increased. We can therefore conclude that the transfer function from the quantization noise input $E(z)$ to the output $Y(z)$, commonly referred to as the Noise Transfer Function (NTF), is a highpass function and acts to "shape" the bulk of the noise power in the frequency region where the input signal is not located. Subsequent filtering can then separate the input signal from the quantization noise.

Higher-order noise-shaping delta-sigma modulators operate on exactly the same principle with the exception that more past samples of the quantization error are used to reduce the present quantization error. In effect, a higher-order NTF is used to significantly reduce the inband quantization noise power. Formulae that quantify the conversion quality (SNR) have been given in equations (8) and (9) for a first and second-order modulator, respectively.

10.3.4 SI building blocks

The presentation of the above data converter algorithms were all derived from a signal processing point-of-view. Common to most of these algorithms are a set of processing elements, such as a delay or sample-and-hold cell, a multiply-by-two or divide-by-two circuit. It is therefore a simple matter to map these algorithms into a chosen technology by providing realizations for these basic operations. In the following we shall provide a set of SI signal processing elements and describe their basic operation. Subsequently, the next section will demonstrate how to realize several of the converter algorithms described previously using SI circuit techniques.

10.3.4.1 The current memory cell

The most fundamental building block of SI circuits is the current copier or current memory cell [1-3,16]. Two different forms of this circuit have emerged since the introduction of the SI technique; the first and second-generation current memory cell. Circuit schematics for these two cells are shown in Figure 21. The first-generation memory cell is shown in Figure 21a and the second-generation cell in Figure 21b. Although they both operate on similar physical principles, they each have different advantages.

The operation of the first-generation memory cell is such that on clock phase 1, the switch controlled by this clock will be closed allowing for the output current i_{out} to track the input current i_{in}. However, when the switch is opened, a constant current is maintained at the output equal to the input current at the moment the switch was opened. This is due to the charge resident on the gate-source capacitance of transistor M_B being held constant and not allowed to discharge. Thus, with v_{GS} maintained constant, i_D is held constant. By scaling the aspect ratio of transistor M_B with respect to M_A, different current gains are possible with this circuit. By simply moving the switch interconnecting transistors M_A and M_B to the position where it diode-connects M_A alone, and by adding a switch in series with the drain terminal of M_A , the first-generation cell can be transformed into a second-generation cell. By doing so, it is now possible to access the memorized current from two different ports; the drain terminal of M_A or M_B. Although the change from the first-generation cell to the second-generation circuit is quite subtle, having the ability to access the memorized current directly from the same transistor doing the memorization eliminates any matching error that can occur between M_A and M_B. The drawback to the simple current memory cell is that it is not very accurate. Currents memorized by the memory cell may differ from those that are retrieved from the cell at a later time owing to the low output resistance. As in most circuit applications, to minimize the effect of the output resistance, cascoding techniques are used [17].

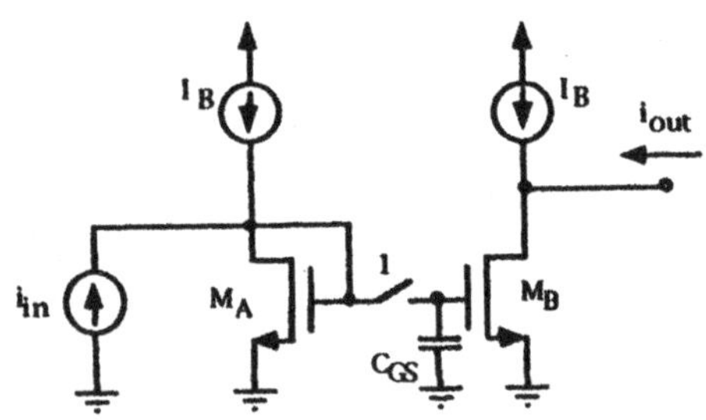

(a)

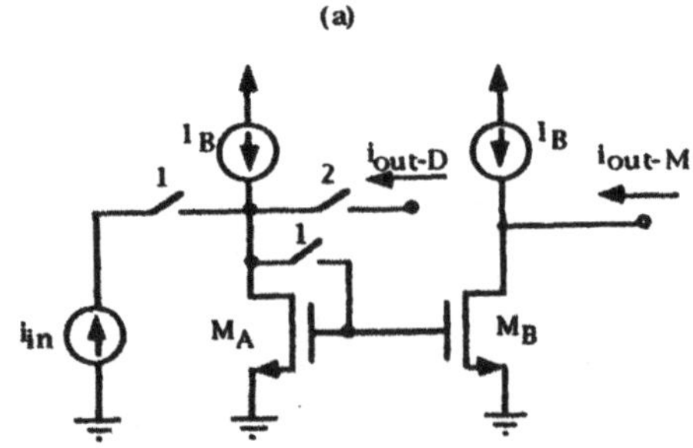

(b)

Figure 21 SI memory cells: (a) first-generation (b) second-generation.

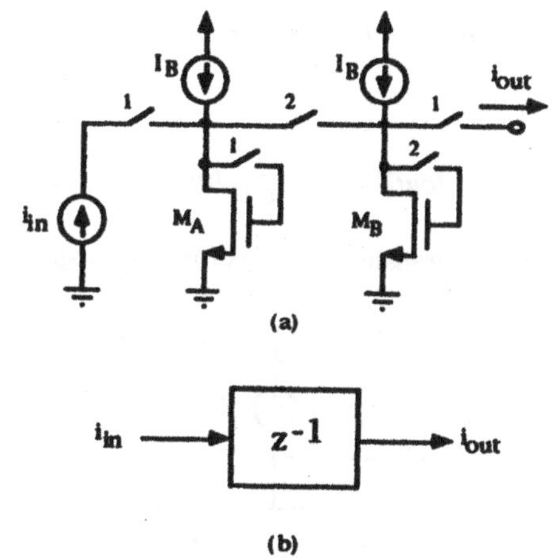

Figure 22 (a) Delay cell (b) equivalent block diagram.

Interested readers can refer to chapter 10.1 for further details about cascoding techniques and how they are incorporated with the SI technique.

10.3.4.2 SI delay cell

The delay or sampled-and-hold cell [3] is shown in Figure 22 with its corresponding equivalent block diagram. It consists of two second-generation memory cells connected in cascade with the first memory cell clocked on phase 1 and the second clocked on phase 2. Clock phases 1 and 2 are assumed non-overlapping. A similar delay cell can be realized using first-generation cells but is not shown here. If the output current is sampled on phase 1 then the output current will be equal to the input current sampled one clock period previously, i.e.

$$i_{out}(n) = i_{in}(n-1) \tag{20}$$

or, in the z-domain,

$$I_{out}(z) = z^{-1}\, I_{in}(z) \tag{21}$$

10.3.4.3 SI integrator circuit

A sampled-and-hold cell can be very easily turned into an discrete integrator circuit by providing a feedback path that interconnects the input with its output. One such circuit is shown in Figure 23a made from second-generation memory cells [16]. If the output current is mirrored from transistor M_A with scale factor K, we can write the output current i_{out1} as

$$i_{out1}(n) = -K\, i_{in}(n) + i_{out1}(n-1) \tag{22}$$

or, in the z-domain,

$$I_{out1}(z) = -K \frac{z}{z-1}\, I_{in}(z) \tag{23}$$

Similarly, when the output is mirrored from transistor M_B, we can write

$$i_{out2}(n) = K\, i_{in}(n-1) + i_{out2}(n-1) \tag{24}$$

or, equivalently

$$I_{out2}(z) = K \frac{1}{z-1} I_{in}(z) \tag{25}$$

A block diagram illustrating the transfer functions for this integrator is provided in Figure 23b.

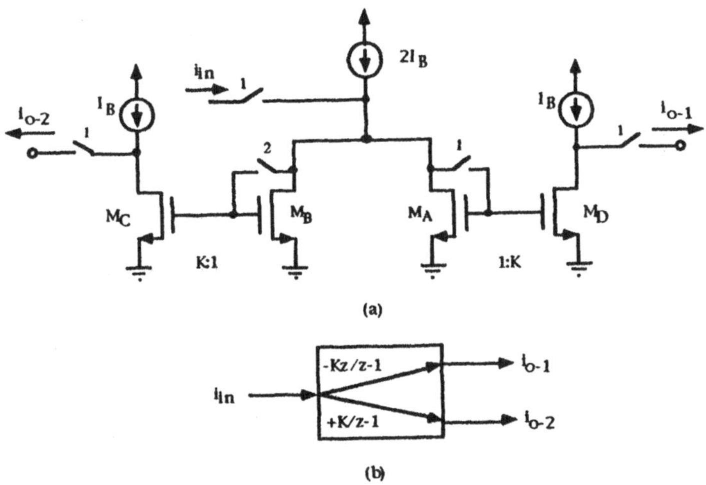

Figure 23 (a) Discrete-time integrator circuit (b) equivalent block diagram.

10.3.4.4 Scaling circuits

Scaling circuits that either increase or decrease the signal level by a factor of 2 are used extensively in data converter circuits. Below we describe two SI circuits that perform these operations that do not rely on transistor matching.

(i) Multiply-by-two circuit

A multiply-by-two circuit [8,9] and its equivalent block diagram is shown in Figure 24. The SI circuit of part (a) consists of three current memory cells operated over three non-overlapping clock phases. During clock phase 1, transistor M_A is made to memorize a sample of the input current. Likewise, during the next clock phase 2, transistor M_B is made to memorize the same sample of input current. During clock phase 3, the memorized current of M_A and M_B are combined and stored in transistor M_C. This current is then made available during clock phase 1, allowing for a full-period hold between the time the input current is first sampled and the time the output is made available. Thus, we can write

$$i_{out}(n) = 2\, i_{in}(n-1) \tag{26}$$

or, equivalently, in the z-domain

$$I_{out}(z) = 2\, z^{-1} I_{in}(z) \tag{27}$$

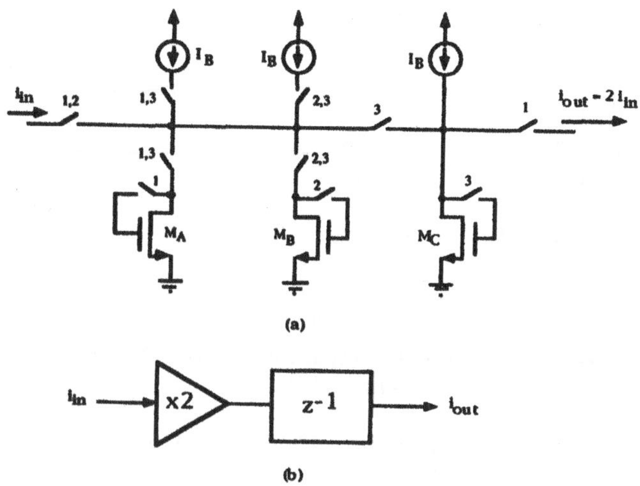

Figure 24 (a) Multiply-by-two circuit (b) equivalent block diagram.

(ii) Divide-by-two circuit

A divide-by-two circuit [18] and its equivalent block diagram is shown in Figure 25. The circuit of Figure 25a provides an output current that is exactly one-half that of the input current. A calibration circuit and an iterative algorithm is used to accomplish this task.

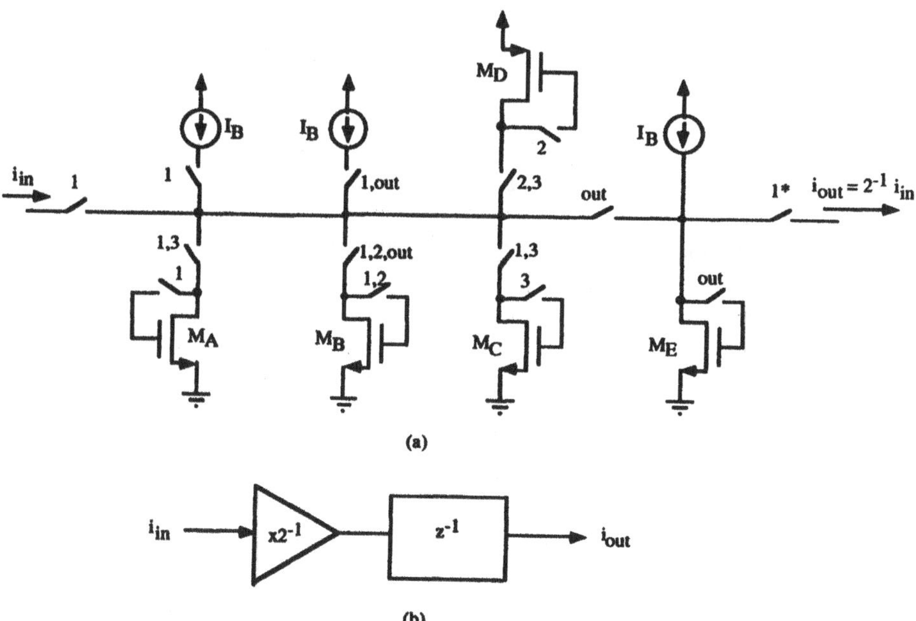

Figure 25 (a) Divide-by-two circuit with calibration (b) equivalent block diagram.

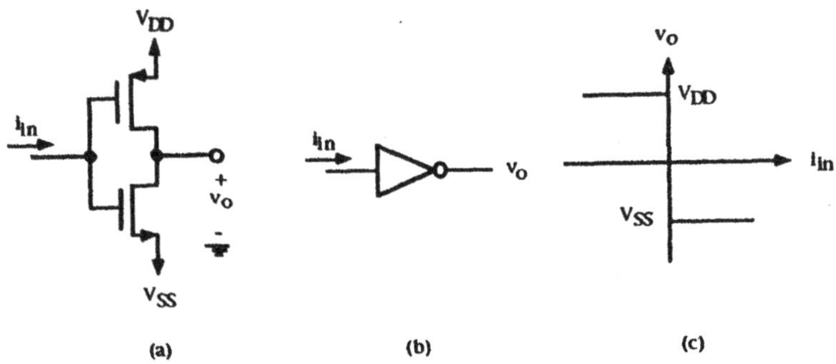

Figure 26 (a) CMOS current comparator circuit (b) symbol (c) transfer characteristic.

The algorithm begins with the input current being sampled by two memory cells (M_A and M_B) connected in parallel. Once the sampling is complete, the two memory cells are separated, thereby splitting the total gate-source charge between the two memory cells. Due to unavoidable transistor mismatches, this splitting will not be exact. The calibration circuit (M_C and M_D) corrects for mismatch errors by first computing the difference between the two values stored in memory cells (M_A and M_B), and feeding this difference back into the first phase where the input current is memorized by the two memory cells connected in parallel. This process is then repeated until the desired level of accuracy is reached. This usually takes about five iterations. The divided input signal is then transferred and held by the final memory cell (M_E) for later access. The difference equation describing this circuit is

$$i_{out}(n) = \frac{1}{2}\, i_{in}(n-1) \tag{28}$$

or, equivalently, in the z-domain

$$I_{out}(z) = \frac{1}{2} z^{-1}\, I_{in}(z) \tag{29}$$

The block diagram for this circuit is provided in Figure 25b.

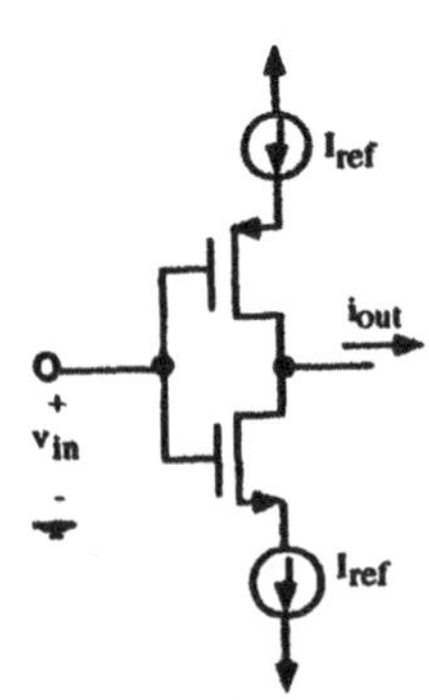

Figure 27 A one-bit D/A circuit (voltage input, current output).

10.3.4.5 Current comparator circuit

The CMOS inverter circuit shown in Figure 26 can act as a simple current comparator. For positive input currents, i.e. current is forced into the input node of the inverter, the voltage at the input of the inverter will ramp up until it hits the upper voltage rail V_{DD} of the circuit causing the output of the inverter to go low. On the other hand, for negative input currents, the input is pulled towards V_{SS} and the output is forced high.

10.3.4.6 One-bit D/A circuit

Figure 27 illustrates a one-bit D/A circuit. Depending on the state of the switch, a current of magnitude I_{ref} is either sourced or sunk at the output terminal of the circuit.

10.3.5 SI circuit examples

In the following we shall demonstrate how one can implement several data converter algorithms described in section 10.3.3 using SI circuit techniques. Other algorithms are straightforward extensions of the ones shown here.

10.3.5.1 Algorithmic D/A converter

The first design that we will consider here will be based on the algorithmic D/A converter scheme of Figure 14. Using the divide-by-two circuit of Figure 25, together with a current source to represent the reference level, we can replace each block in the signal flow diagram of Figure 14 and arrive at its circuit implementation in Figure 28. As is clearly evident, the resulting implementation is quite simple. This is typical of most SI circuits; as current summing is a rather trivial operation requiring only that two nodes be physically connected.

In the circuit of Figure 28 some simplifications were made to keep the schematic simple. The calibration subcircuit of the divide-by-two circuit has been intentionally left off the diagram. In addition, a single block is used to represent the logic for controlling the application of the reference level to the input of the divide-by-two circuit. For the most part, it consists of a series of shift registers and some controlling logic.

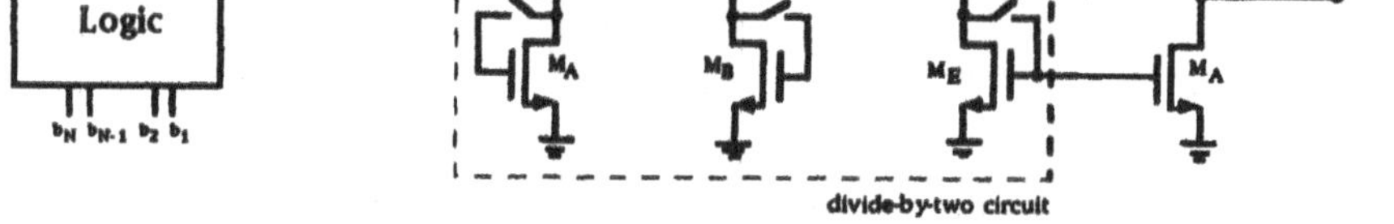

Figure 28 SI algorithmic D/A converter circuit.

The operation of the D/A of Figure 28 begins with the MSB controlling the application of the reference current to the input of the divide-by-two circuit. This current is then combined with the current that is fed back from the output of the divide-by-two circuit. Initially, this current is set to be zero. The net current is then applied to the divide-by-two circuit and reduced by a factor of two. This process is then repeated N-1 times, corresponding with the remaining number of bits in the input digital word. The order in which the switches are operated are indicated alongside each switch in Figure 28. On completion of the conversion process, an output current is available at the current mirror output.

10.3.5.2 Algorithmic A/D converter

The next scheme that we shall translate from block diagram to SI circuit is the algorithmic A/D converter shown in Figure 17.

Replacing the multiply-by-two and full-period-hold blocks with the SI circuit of Figure 24, together with a current comparator and a set of reference currents, we obtain the SI circuit shown in Figure 29. Once again, a relatively simple circuit results.

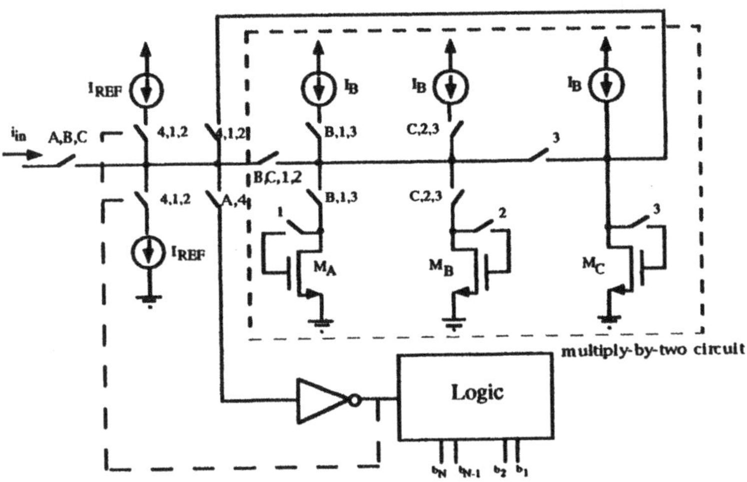

Figure 29 SI algorithmic A/D converter circuit.

The operation of this circuit begins with the polarity of the input current being first determined by the current comparator circuit. This is accomplished on clock phase A and stored in the series of shift registers denoted as logic in Figure 29. This first bit represents the MSB of the converted digital code. Next, the input current is routed into the multiply-by-two circuit via the switches controlled on clock phases B and C. Here the multiply-by-two circuit is operated over clock phases B, C, and 3. Subsequently, the output of the multiple-by-two circuit is fed back on clock phase 4 to the input of the A/D converter where it combines with either $\pm I_{ref}$, depending on the previous decision of the comparator. The polarity of this residual current is then determined by the current comparator circuit and digitally stored. Once this is complete, the residual current is re-applied to the multiply-by-two circuit on clock phase 1 and 2, and doubled on phase 3. The output of the multiply-by-two circuit is then fed back to the input on clock phase 4 to combine with the appropriate reference current to form the next residual current. The polarity of this signal is then determined and digitally stored. This cycle is then repeated over clock phases 1, 2, 3, and 4 to complete the N-bit conversion.

10.3.5.3 Noise-shaping A/D circuit

The final conversion scheme that we shall consider here is the first-order delta-sigma modulation scheme presented in Figure 20a. The approach taken here is identical to that seen previously. Utilizing an SI integrator with transfer function *1/(z-1)*, a current comparator with an additional inverter circuit to ensure negative feedback and a one-bit current-mode D/A circuit, we can translate the block diagram of Figure 20a into the SI delta-sigma modulator shown in Figure 30. The operation of this circuit is rather straightforward, operating on clock phases 1 and 2. An additional switch has been added at the input of the integrator circuit in order to provide a shunt path for the input and feedback current on clock phase 2. Implementing higher-order delta-sigma modulators follow the exact same train of thought, interested readers can consult references [11-15] for more information.

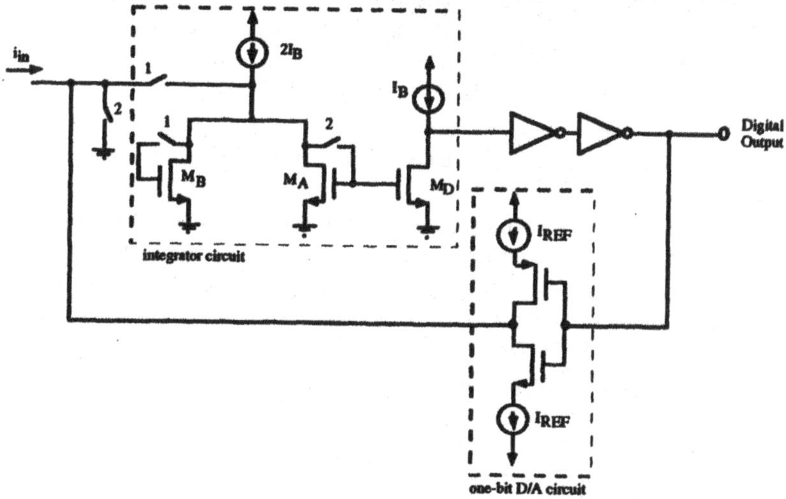

Figure 30 SI first-order noise-shaping delta-sigma modulator circuit.

10.3.6 Conclusion

Various data conversion algorithms and their SI implementations have been described at an introductory level. Although specific implementions have not been discussed, SI circuit techniques enable the realization of high performance data converters using a standard digital CMOS process with very low power and area requirements. As a result, the SI circuits are ideally suited for mixed-signal applications.

References

[1] E. A. Vittoz, "Dynamic analog techniques", in *Design of MOS VLSI Circuits for Telecommunications*, Y. Tsividis and P. Antognetti (Editors), Prentice-Hall Inc. Englewoods Cliffs, New Jersey, 1985.

[2] J. Daubert, D. Vallancourt and Y. P. Tsividis, "Current copier cells", *Electronics Letters*, Vol. 24, No. 25, pp.1560-1562, Dec. 1988.

[3] B. Hughes, N. C. Bird, and I. C. Macbeth, "A new technique for analog sample data signal processing", *Proc. IEEE International Symposium on Circuits and Systems*, Portland, Oregon, pp. 1584-1587, May 1989.

[4] C. Toumazou, J. B. Hughes and N. Battersby (Editors), *Switched-Current Techniques for Analogue VLSI*, Peter Peregrinus Limited, London, England,1993.

[5] C. Candy and G. C. Temes, "Oversampling Methods for A/D and D/A conversion", in *Oversampling Delta-Sigma Data Converters: Theory, Design, and Simulation*, J. C. Candy and G. C. Temes (Editors), IEEE Press, New York, 1991.

[6] M. Gordon, "Linear electronic analog/digital conversion architectures, their origins, parameters, limitations, and applications", *IEEE Trans. Circuits and Systems*, Vol. CAS-25, pp. 391-418, July 1978.

[7] W. Groeneveld, H. Schouwenaars, C. Bastiaansen and H. Termeer, Continuous Calibration D/A Conversion, in *Switched-Current Techniques For Analogue VLSI*, Toumazou, J. B. Hughes and N. Battersby (Editors), Peter Peregrinus Limited, London, England, pp. 350-380, 1993.

[8] D. G. Nairn and C. A. T. Salama, "A ratio-independent algorithmic analog-to-digital converter combining current-mode and dynamic techniques", *IEEE Trans. Cicuits and Systems*, Vol. CAS-37, pp. 319-325, March 1990.

[9] P. Deval, J. Robert and M. J. Declercq "A 14 bit CMOS A/D based on dynamic current memories", in *Proc. IEEE Custom Integrated Circuits Conference*, pp. 24.2.1-24.2.4, 1991.

[10] S. W. Kim and S. W. Kim, "Current-mode cyclic ADC for low power and high speed applications", *Electronics Letters*, Vol. 27, pp. 818-819, 1991.

[11] H. Schouwenaars, W. Groeneveld, C. Bastiaansen and H. Termeer, "An oversampled multi-bit CMOS D/A Converter for digital audio with 115 dB dynamic range', in *Proc. ISSCC Digest of Technical Papers*, pp. 72-73, Feb. 1991.

[12] J. Daubert and D. Vallencourt, "A transistor-only current-mode sigma-delta modulator", *Proc. IEEE Custom Integrated Circuits Conference*, pp. 24.3.1-24.3.4, May 1991.

[13] P. J. Crawley and G.W. Roberts, "Switched-Current Sigma-Delta Modulation for A/D Conversion", *Proc. IEEE International Symposium on Circuits and Systems*, San Diego, California, pp. 1320-1323, May 1992.

[14] P. J. Crawley and G.W. Roberts, "A Component Invariant Second-Order Switched-Current Sigma-Delta Modulator", *Proc. IEEE International Symposium on Circuits and Systems*, San Diego, California, pp. 1324-1327, May 1992.

[15] G. W. Roberts and P. J. Crawley, "Building Blocks for Switched-Current Sigma Delta Modulators", in *Switched-Current Techniques For Analogue VLSI*, C. Toumazou, J. B. Hughes and N. Battersby (Editors), Peter Peregrinus Limited, London, England, pp. 350-380, 1993.

[16] B. Hughes, I. C. Macbeth, and D. M. Pattullo, "Second Generation Switch-Current Signal Processing", *Proc. IEEE International Symposium on Circuits and Systems*, New Orleans, Louisiana, pp. 2805-2808, August 1990.

[17] C. Toumazou, J. B. Hughes, and D. M. Pattullo, "Regulated Cascode Switched-Current Memory Cell", *Electronics Letters*, Vol. 26, No. 5, pp. 303-305, March 1990.

[18] J. Robert, P. Deval and G. Wegmann, "Very accurate current divider", *Electronics Letters*, Vol. 25, pp. 912-913, July 1989.

Chapter 10.4

Switched-Current Cellular Nonlinear Networks

Angel Rodríguez-Vázquez

10.4.1 Summary

There are many signal processing and computational tasks which can be formulated as a multidimensional nonlinear state equation

$$x_{n+1} = f(x_n, P, u) \ , \quad n = 0, 1, 2, \ldots$$

where x represents the state vector, u is the input vector, and P is a design parameter vector. Illustrative examples include nonlinear filtering [1], ΣΔ modulators [2], periodic signal generation [3, 4], noise and chaos generation [5, 7] optimization [8] and problem solving [9], parallel processing of sensory signals and pattern recognition [10-12], etc. Since many applications related to these tasks involve signals and require on-line processing, a need for dedicated hardware arises. This is especially pertinent for array sensory signals, due to the large amount of data to handle.

The tasks mentioned above share a set of operators (delay, addition, multiplication, rectification, sorting, signal normalization, etc.) whose associated hardware can be realized by either digital or analog circuits. Analog solutions are mandatory for some tasks, for instance, ΣΔ modulation. In other cases, like nonlinear filtering, optimization, or problem solving, the question of wether to use digital or analog circuits will long remain an issue of controversy. This tutorial advocates analog computation based on:

a) its larger functional efficiency per silicon area than digital counterparts - a consequence of the full exploitation made by analog circuits of the many functional capabilities of MOS and BJT transistors [13-17];

b) the intrinsically larger speed/power ratio of analog circuits [14], which is of primary importance for neural networks VLSI systems;

c) the possibility to interface sensory signals directly, and to include sensory circuitry together with processing circuitry, on the same silicon substrate [18, 19].

Obviously, these premiums of analog computation are not free; the penalty is limited accuracy. However, for applications which tolerate accuracy levels below 7-8 bits, many of which are related to neural networks, analog computation may be the more convenient choice.

The tutorial which is summarized herein presents current-mode circuits to realize the above nonlinear state equation following a basic building block approach. The only circuit primitives of the proposed technique are CMOS active devices (MOSTs and CMOS-BJTs) what render it better suited for standard digital technologies than the switched-capacitor approach. Also, since currents, rather than voltages, are the physical support of information, larger dynamic range and operation speed can be expected by using the new technique.

Special emphasis is placed upon operators required for cellular systems, i.e., systems containing many identical nonlinear units which basically interact linearly [20, 21]. These cellular systems have a pletora of applications already demonstrated for motion detection, control, character recognition, or the simulation of physical systems described by lattice equations [22-32].

As an extension of the nonlinear circuits used in these cellular system, this tutorial presents circuits for function generation in current-mode domain [5], with application in such diverse topics as signal generation, sensor linearization, or fuzzy inference [33]. We present also collective computation current-mode circuits to maximum, minimum, normalization [33], etc., and a discussion of circuit strategies for programmability - an issue of primary significance for the tasks covered in the tutorial [34-39]. We cover not only ideal circuits, but a discussion of nonidealities and a presentation of circuit and layout strategies to cope with them [5, 18, 40-48].

Measurements from several CMOS prototypes chips are given to demonstrate validity of the approach. These chips include sensory/processing system with cellular architecture for frame images, periodic and aperiodic signal generators, and collective computation circuits for fuzzy inference.

This summary does not provide details on the topics covered, but just mention them to give the reader a general background. More details can be found in the presentation slides and the references associated to each topic. Copies of the slides, as well as copies of the quoted articles can be found from the author upon request.

Acknowledgments

Most of the material in this tutorial is based on research works carried out by my colleagues Manuel Delgado-Restituto, Servando Espejo, Rafael Domínguez-Castro, Ricardo Carmona and Fernando Vidal.

References

[1] I. Pitas and A. N. Venetsanopoulos, *Nonlinear Digital Filters: Principles and Applications*, Boston: Kluwer Academics, 1992.

[2] J. C. Candy and G. C. Temes, *Oversampling Delta-Sigma Converters*, New York: IEEE Press, 1992.

[3] B. J. Hosticka *et al.*, "Design of Nonlinear Analog Switched-Capacitor Circuits Using Building Blocks", *IEEE Trans. Circuits and Systems*, Vol. 31, pp. 354-367, April 1984.

[4] P. E. Fleisher, *et al.*, "A Switched-Capacitor Oscillator with Precision Amplitude Control and Guaranteed Start-Up", *IEEE J. Solid-State Circuits*, Vol. 20, pp. 641-647, April 1985.

[5] A. Rodríguez-Vázquez and M. Delgado-Restituto, "Generation of Chaotic Signals using Current-Mode Techniques", *Journal of Intelligent and Fuzzy Systems*, Vol. 2, pp. 15-37, January 1994.

[6] A. Rodríguez-Vázquez *et al.*, "Chaos from Switched-Capacitor Circuits Discrete Maps", *Proceedings of the IEEE*, Vol. 75, pp. 1090-1106, August 1987.

[7] M. Delgado-Restituto and A. Rodríguez-Vázquez, "Switched-Current Chaotic Neurons", *Electronics Letters*, Vol. 30, pp. 429-430, March 1994.

[8] A. Rodríguez-Vázquez *et al.*, "Nonlinear Switched-Capacitor Neural Networks for Optimization Problems", *IEEE Trans. Circuits and Systems*, Vol. 37, pp. 384-398, March 1990.

[9] A. Cichocki and R. Unbehauen, *Neural Networks for Optimization and Signal Processing*, Chichester: John Wiley & Sons, 1993.

[10] J. M. Zurada, *Introduction to Artificial Neural Networks*, St. Paul:West Publishing, 1992.

[11] J. C. Russ, *The Image Processing Handbook*, Boca Raton: CRC Press, 1992.

[12] L. O. Chua *et al.*, "Signal Processing Using Cellular Neural Networks", *J. of VLSI Signal Processing*, Vol. 3, pp. 25-51, January 1991.

[13] E. A. Vittoz, "The Design of High Performance Analog Circuits on Digital CMOS Chips", *IEEE J. Solid-State Circuits*, Vol. 20, pp. 657-665, June 1985.

[14] E.A. Vittoz, "Future of Analog in the VLSI Environment", *Proc. 1990 IEEE Int. Symp. Circuits and Systems*, pp. 1372-1375, May 1990.

[15] Z. Wang, *Current-Mode Analog Integrated Circuits and Linearization Techniques in CMOS Technology*, Konstanz: Hartung-Gorre Verlag, 1990.

[16] R. Wiegerink, "Analysis and Synthesis of MOS Translinear Circuits", PhD Dissertation, University of Twente, 1992.

[17] E. Seevinck, "Analysis and Synthesis of Translinear Integrated Circuits", PhD Dissertation, University of Pretoria, 1981.

[18] A. Rodríguez-Vázquez and S. Espejo, "Current-Mode Techniques for the Implementation of Continuous-Time and Discrete-Time Cellular Neural Networks", *IEEE Trans. Circuits and Systems II*, Vol. 40, pp. 132-146, March 1993.

[19] S. Espejo and A. Rodríguez-Vázquez, "Smart-Pixel Cellular Neural Networks in Analog Current-Mode CMOS Technology", *IEEE J. Solid-State Circuits*, Vol. 29, July 1994 (to appear).

[20] L. O. Chua and T. Roska, "The CNN Paradigm", *IEEE Trans. Circuits and Systems I*, Vol. 40, pp. 147-156, March 1993.

[21] H. Harrer, "Discrete-Time Cellular Neural Networks", PhD Dissertation, Technical University of Munich, Aachen:Verlag Shaker, 1992.

[22] L. O. Chua and L. Yang, "Cellular Neural Networks: Applications", *IEEE Trans. Circuits and Systems*, Vol. 35, pp. 1273-1290, Oct. 1988.

[23] T. Matsumoto *et al.*, "CNN Cloning Template: Connected Component Detector", *IEEE Trans. Circuits and Systems*, Vol. 37, pp. 633-635, May 1990.

[24] T. Matsumoto *et al.*, "CNN Cloning Template: Hole Filler", *IEEE Trans. Circuits and Systems*, Vol. 37, pp. 635-638, May 1990.

[25] T. Matsumoto *et al.*, "CNN Cloning Template: Shadow Detector", *IEEE Trans. Circuits and Systems*, Vol. 37, pp. 1070-1073, August 1990.

[26] T. Roska (ed.), *Proc. of the IEEE 1990 Int. Workshop on Cellular Neural Networks and Their Applications*, Budapest, December 1990.

[27] J.A. Nossek (ed.), *Proc. of the IEEE 1992 Int. Workshop on Cellular Neural Networks and Their Applications*, Munich, October 1992.

[28] K. Slot, "Cellular Neural Network Design for Solving Specific Image-Processing Problems", *Int. J. Circuit Theory and Applications*, Vol. 20, pp. 629-637, September 1992.

[29] T. Roska *et al.*, "Detecting Moving and Standing Objects using Cellular Neural Networks", *Int. J. Circuit Theory and Applications*, Vol. 20, pp. 613-628, September 1992.

[30] S. Paul *et al.*, "Mapping Nonlinear Lattice Equations onto Cellular Neural Networks", UCB/ERL M92/42, University of California at Berkeley, May 1992.

[31] L. O. Chua and B. Shi, "Multiple Layer Cellular Neural Network: A tutorial", in *Algorithms and Parallel VLSI Architectures*, North Holland, 1991.

[32] T. Roska and L. O. Chua, "Cellular Neural Networks with Non-linear and Delay-type Template Elements and Non-uniform Grids", *Int. J. Circuit Theory and Applications*, Vol. 20, pp. 469-482, September 1992.

[33] F. Vidal and A. Rodríguez-Vázquez, "Analog CMOS Design of Singleton Fuzzy Controllers", *IEEE Transactions on Fuzzy Systems* (submitted)

[34] R. Domínguez-Castro and A. Rodríguez-Vázquez, "Analog Neural Programmable Optimizers in CMOS VLSI Technologies", *IEEE J. Solid-State Circuits*, Vol. 27, pp. 1110-1115, July 1992.

[35] S. Espejo, "Analog VLSI Design of Cellular Neural Networks", PhD Dissertation, University of Seville, 1994.

[36] R. Domínguez-Castro *et al.*, "Programmability Issues in Current-Mode Cellular Neural Networks", *Circuit Theory and Design 93*, pp. 39-44, Elsevier 1993.

[37] K. Bult and H. Wallinga, "A Class of Analog CMOS Circuits Based on the Square-Law Characteristics of an MOS Transistor in Saturation", *IEEE J. Solid-State Circuits*, Vol. 22, pp. 357-365, June 1987.

[38] E. A. Klumperink and E. Seevinck, "MOS Current Gain Cells with Electronically Variable Gain and Constant Bandwidth", *IEEE J. Solid-State Circuits*, Vol. 24, pp. 1465-1467, October 1989.

[39] X. Arreguit, "Compatible Lateral Bipolar Transistors in CMOS Technology: Model and Applications", PhD Dissertation, EPFL, Lausanne, 1989.

[40] Z. Wang, "Novel Pseudo RMS Current Converter for Sinusoidal Signals using a CMOS Precision Current Rectifier", *IEEE Trans. on Instrumentation and Measurement*, Vol. 39, pp. 670-671, August 1990.

[41] A. Rodríguez-Vázquez *et al.*, "High Resolution CMOS Current Comparators: Design and Applications to Current-Mode Function Generation", *Analog Integrated Circuits and Signal Processing*, 1994 (to appear).

[42] C. Toumazou *et al.*, *Analogue IC Design: The Current-Mode Approach*, London: Peter Peregrinus, 1990.

[43] C. Toumazou *et al.*, *Switched-Currents: An Analog Technique for Digital Technology*, London: Peter Peregrinus, 1993.

[44] E. Sackinger and W. Guggenbuhl, "A High-Swing High-Impedance MOS Cascode Circuit", *IEEE J. Solid-State Circuits*, Vol. 25, pp. 289-298, February 1990.

[45] C. Eichenberger, *Charge Injection in MOS-Integrated Sample-and-Hold and Switched-Capacitor Circuits*, Zurich:Hartung-Gorre Verlag, 1989.

[46] S. Espejo *et al.*, "A Modified Dummy-Switch Technique for Tunable Feedthrough Cancellation in Switched-Current Circuits", *Proc. of the 1993 European Solid-State Circuits Conference*, pp. 270-272, September 1993.

[47] C. Michael and M. Ismail, "Statistical Modeling of Device Mismatch for Analog MOS Integrated Circuits", *IEEE J. Solid-State Circuits*, Vol. 27, pp. 154-166, February 1992.

[48] M. J. M. Pelgrom *et al.*, "Matching Properties of MOS Transistors", *IEEE J. Solid-State Circuits*, Vol. 24, pp. 1433-1440, October 1989.

Chapter 10.5

Analysis of Switched-Current Filters

Antônio Carlos Moreirão de Queiroz

10.5.1 Introduction

Switched-current filters operate, ideally, as periodically switched linear networks, where the circuit reaches a static steady state between the switching instants. This mode of operation is essentially the same of the switched-capacitor filters, and results in an analog circuit that processes signals in sampled-data mode, as a digital filter. In switched-capacitor filters, the signal is represented by capacitor voltages, and the computations are done by charge balancing at the switching instants. In switched-current filters, the signal is represented by currents, and the computations are done by current balancing between the switching instants.

Direct frequency-domain simulations of switched-current circuits, ignoring non-linear effects, can be obtained with the techniques used to analyze general periodically switched linear circuits [8,10], or non-ideal switched-capacitor circuits [9]. This approach is useful in the evaluation of the switching frequency limitations of a filter. It is, however, unnecessarily complex when effects as incomplete current balancing between switching instants are not significant, or when the simulation is used as a tool for design verification.

This tutorial describes a procedure for the frequency-domain analysis of switched-current filters, similar to the ones used for ideal switched-capacitor circuits [2, 3]. The method is described as it was implemented in a computer program named ASIZ, that analyzes multi-phase switched-current filters, obtaining transfer functions in z-transform, poles and zeros, frequency responses, transient responses, and sensitivities [1]. The method is also applicable to switched-capacitor circuits, as a particular case.

10.5.2 Signals in switched-current filters

In the operation of a switched-current circuit, a switching period is divided in a number f of phases (usually two). Any signal X_i in the filter can be considered as composed by f components $X_{i,m}$, each one for the $m=1,\ldots,f$ phases. Each of these components is composed by another f components $X_{i,mk}$, each for one of the $k=1,\ldots,f$ phases of the input signal, that add by superposition to form $X_{i,m}$. These $f \times f$ components add together to form X_i as shown in Figure 1, for the case of two phases.

As the components add in the time domain, they can also be added in the frequency domain. Considering the length of all the phases identical, and neglecting the time needed for the circuit stabilization after each switching operation, the frequency-domain representation of $X_i(t)$, $X_i(j\omega)$ can be obtained as (1), where T is the switching period. The term multiplying the summation is a "sampling function" corresponding to a hold time equal to T/f seconds in a period of T seconds.

The frequency-domain expressions for the signals can be obtained by analysis in the z-transform domain, followed by the substitution $z=e^{j\omega T}$. The expression (1) can be interpreted as a transfer function, if a unitary input signal is used in the analysis. It gives the ratio between output and input phasors, considering only the frequency component of the output that has the same frequency of the input.

$$X_i(j\omega)=\frac{1-e^{j\omega T/f}}{j\omega T}\sum_{m=1}^{f}\sum_{k=1}^{f}X_{i,mk}(j\omega) \tag{1}$$

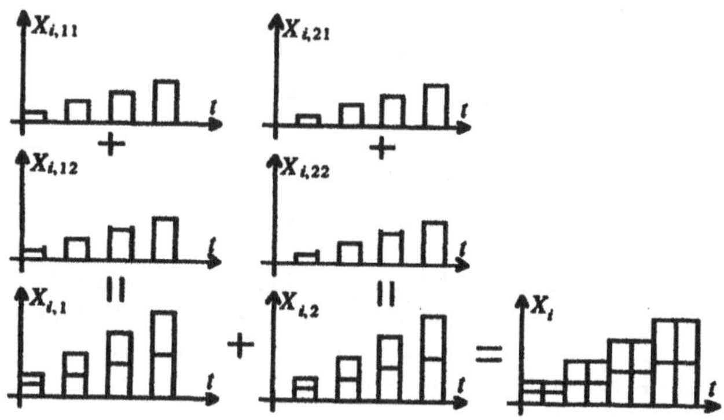

Figure 1 Signal composition in a two-phases switched-current filter.

10.5.3 Z-domain analysis of ideal switched-current filters

Simple ideal SI circuits can be analyzed by writing linear equations in z-transform involving only AC transistor currents. For example, Figure 2 depicts a basic SI "second-generation" integrator cell [4]. The two possible outputs are the currents in the transistors, that are usually copied to other circuit points by current mirrors. Ignoring the biasing current source, the following equations can be written

$$\begin{aligned}&\text{Phase 1:}\quad I_{1,11}+I_{2,11}=I_{in,1}\qquad I_{2,11}-I_{2,21}z^{-1/2}=0\\&\text{Phase 2:}\quad I_{1,21}-I_{1,11}z^{-1/2}=0\qquad I_{1,21}+I_{2,21}=0\end{aligned} \tag{2}$$

where the terms in $z^{-1/2}$ indicate a delay of one phase ($T/2$). The third subscript is always the same, because there is no input in phase 2 (and so all $I_{i,m2}$=0).

These equations can be solved, being obtained four partial values for each output current

$$\begin{aligned}&I_{1,11}=\frac{I_{in,1}}{1-z^{-1}}\qquad I_{1,12}=0\quad I_{1,21}=\frac{z^{-1/2}I_{in,1}}{1-z^{-1}}\qquad I_{1,22}=0\\&I_{2,11}=-\frac{z^{-1}I_{in,1}}{1-z^{-1}}\quad I_{2,12}=0\quad I_{2,21}=-\frac{z^{-1/2}I_{in,1}}{1-z^{-1}}\quad I_{2,22}=0\end{aligned} \tag{3}$$

The backward ($I_{1,1}/I_{in}$) and forward ($I_{2,1}/I_{in}$) Euler integration can be obtained with this circuit. The frequency responses are obtained from (1)

$$\frac{I_1}{I_{in}}(j\omega)=\frac{1-e^{j\omega T/2}}{j\omega T}\frac{1+e^{-j\omega T/2}}{1-e^{-j\omega T}}=\frac{1}{j\omega T}$$
$$\frac{I_2}{I_{in}}(j\omega)=-\frac{1-e^{j\omega T/2}}{j\omega T}\frac{e^{-j\omega T}+e^{-j\omega T/2}}{1-e^{-j\omega T}}=-\frac{e^{-j\omega T/2}}{j\omega T} \qquad (4)$$

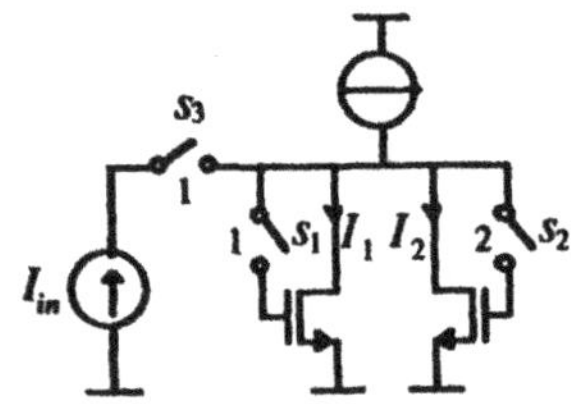

Figure 2 Basic second generation integrator circuit used as example.

10.5.4 Linearized modelling of switched-current circuits

The equations (2) are in the form of nodal equations, but for ideal circuits there is no need to express the variables in terms of nodal voltages. When imperfections as parasitic conductances and capacitances are considered, however, the fundamental assumption that the circuit stabilizes between the switching instants can be retained, but voltages must be considered.

In switched-current filters, nodal voltages are generally a non-linear function of the input, with only AC currents being linear functions of the input, if proper design techniques are used. However, if the circuit is considered as operating with small signals, the non-linearity of the transconductances and other elements can be neglected, and AC voltages and currents can be considered as linear functions of the input. The transistor currents become linear functions of their V_{gs} voltages, that represent the signals inside the filter as well as the currents. Regular nodal analysis can then be used to analyze the circuit at transistor level, using the basic elements that constitute a switched-current filter, shown in Figure 3:

- Ideal switches: their function is to short-circuit two nodes when closed, leaving them unconnected when open.
- Constant current sources: their function is just to bias the MOS transistors. The value of their DC currents has no effect in the small-signal AC behaviour of the filter, and so they are ignored.
- Signal current sources: they feed AC signals to the filter, and so their currents are significant.
- Resistors: model resistive losses, as output resistances of current sources and MOS transistors.
- Transconductors: used in the modelling of several elements.
- MOS enhancement mode transistors: for small-signal AC analysis, the two kinds of MOS transistors can be represented by the same model. The output circuit is modelled by a transconductance controlled by V_{gs}, and a possible non-zero output conductance. The input circuit consists of a C_{gs} capacitance. A parasitic C_{gd} capacitance can also be included.

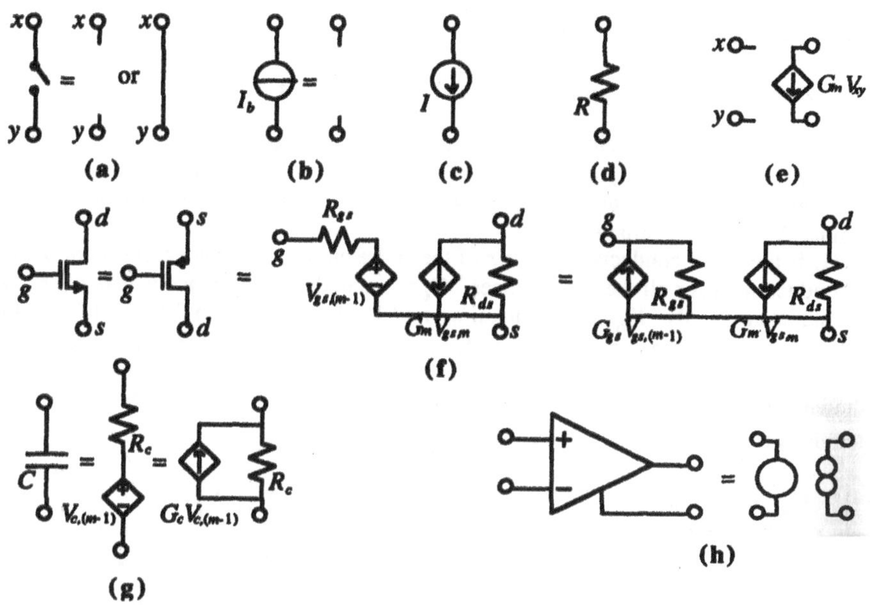

Figure 3 Elements of switched-current filters, and their AC small-signal models used in the analysis method (a) Switch (b) Biasing current source (c) Signal current source (d) Resistor (e) Transconductor (f) MOS transistor, including the C_{gs} capacitance, where G_{gs}=1/R_{gs} (g) Capacitor, where G_c=1/R_c (h) Ideal operational amplifier, a nullator-norator pair.

- Capacitors: model the inputs of the MOS transistors, parasitic capacitances, and are basic elements in switched-capacitor filters.
- Ideal operational amplifiers: used in switched-capacitor filters, and can be used in precision switched-current filters. Also useful in modelling voltage sources.

Capacitances must be modelled by an equivalent circuit that presents no load (infinite resistance) when fed by low impedance circuits, and acts as a voltage memory when disconnected from the remaining circuit. Additionally, the effects of charge redistribution, essential for switched-capacitor filters and circuits with parasitic capacitances, must also be correctly modelled. These functions can be realized by representing a capacitance C by a "large" resistor $R_C = C_0/C$ in series with a voltage source that remembers the value of V_C just before the last change in the configuration of the switches ($V_{C,(m-1)}$ in Figure 3). In the z-transform domain, the value of the memory voltage source can be expressed by $V_{C,m}(z) = V_{C,(m-1)}(z)z^{-1/f}$, where m is the number of the present phase. Note that m–1 is defined as f if m=1.

The restriction that the capacitors cannot load resistive circuits is satisfied without causing numerical problems (that occur if simply a large value is used for C_0) by ignoring terms due to the capacitances in any nodal equation where resistive terms appear ("low-impedance equations"), and considering them only in equations without resistive terms ("high-impedance equations"). Different normalizing factors can be used for conductances and capacitances, allowing the generation of a well-conditioned system of equations. This technique works correctly, as long as there are no resistive subcircuits connected to the remaining circuit by capacitors only. This would result in a singular system, but never occurs in practical circuits.

10.5.5 Z-domain nodal analysis of switched-current filters

The basic unknowns in the circuit are the partial nodal voltages $E_{i,mk}(z)$, for the $i=1,...,n_t$ nodes, for the $m=1,...,f$ output phases, and for the $k=1,...,f$ input phases. Considering at first all the switches open, by modelling the elements as in Figure 3 the system of linear equations (5) can be constructed.

$\mathbf{G}_n$ is the nodal conductance matrix of the entire circuit, with the exception of the G_C transconductances for all the capacitances. $\mathbf{G}_c$ is the nodal conductance matrix of the G_C transconductances only. $\mathbf{E}_{mk}$ are the unknown nodal voltage vectors in the z domain for the $m=1,..., f$ output phases and for the $k=1,..., f$ input phases. The signal inputs at the $k=1,...,f$ input phases are represented by $f\ \mathbf{I}_k$ input current vectors, initially equal, arranged as shown. Unitary values are used in the frequency response calculation. The order of the system is $n_t \times f$, and all the submatrices and subvectors are of order n_t.

$$\begin{bmatrix} \mathbf{G}_n & \mathbf{0} & \cdots & \mathbf{0} & -\mathbf{G}_c z^{-1/f} \\ -\mathbf{G}_c z^{-1/f} & \mathbf{G}_n & \cdots & \mathbf{0} & \mathbf{0} \\ \vdots & \vdots & \ddots & \vdots & \vdots \\ \mathbf{0} & \mathbf{0} & \cdots & \mathbf{G}_n & \mathbf{0} \\ \mathbf{0} & \mathbf{0} & \cdots & -\mathbf{G}_c z^{-1/f} & \mathbf{G}_n \end{bmatrix} \begin{bmatrix} \mathbf{E}_{1k} \\ \mathbf{E}_{2k} \\ \vdots \\ \mathbf{E}_{(f-1)k} \\ \mathbf{E}_{fk} \end{bmatrix} = \begin{bmatrix} \mathbf{I}_1 \\ \mathbf{0} \\ \vdots \\ \mathbf{0} \\ \mathbf{0} \end{bmatrix}, \begin{bmatrix} \mathbf{0} \\ \mathbf{I}_2 \\ \vdots \\ \mathbf{0} \\ \mathbf{0} \end{bmatrix} \cdots \begin{bmatrix} \mathbf{0} \\ \mathbf{0} \\ \vdots \\ \mathbf{I}_{f-1} \\ \mathbf{0} \end{bmatrix}, \begin{bmatrix} \mathbf{0} \\ \mathbf{0} \\ \vdots \\ \mathbf{0} \\ \mathbf{I}_f \end{bmatrix} \quad (5)$$

The effect of a closed switch is to short-circuit two nodes in the phases where it is closed. In the system (5), the effect of a closed switch between nodes x and y, closed at phase m, is to add the rows and the columns x and y in the submatrices and subvectors corresponding to the m-th phase (if x or y is zero, representing a connection to the signal ground, the other column and row are eliminated). The unknowns $E_{x,mk}$ and $E_{y,mk}$, $k=1,..., f$, are reduced to single ones (or eliminated if $x=0$ or $y=0$). The order of the system is reduced by the number of switches.

Ideal operational amplifiers can be processed in a similar way: for each phase, the columns corresponding to the input terminals are added, eliminating one unknown, and the rows corresponding to the output terminals are also added, eliminating one equation (for single-input or single-output amplifiers, the input column or the output equation is removed from the system). As this is done for all the phases, each operational amplifier eliminates f equations and $f \times f$ unknowns. The order of the system is reduced by $f \times$(number of op-amps).

The reduced system so obtained can be solved, for a particular numerical value of z, by the LU factorization of the system matrix and f substitutions, one for each of the $k=1,...,$ f input phase excitation vectors, being obtained the $f \times f$ partial nodal voltage vectors $\mathbf{E}_{mk}(z)$.

This analysis method corresponds to the analysis of a periodically switched *RC*-active network, where the circuit is allowed to stabilize between the switching instants. It can be used with switched-current circuits, switched-capacitor circuits, or any combination of them.

As an example, the same circuit is used (Figure 4), but with two imperfections added: a parasitic resistance $R=1/G$, and a parasitic capacitance C_{gd} (only one to keep the expressions not too complex). The switch s_4 must be added to provide a path for the input current in phase 2.

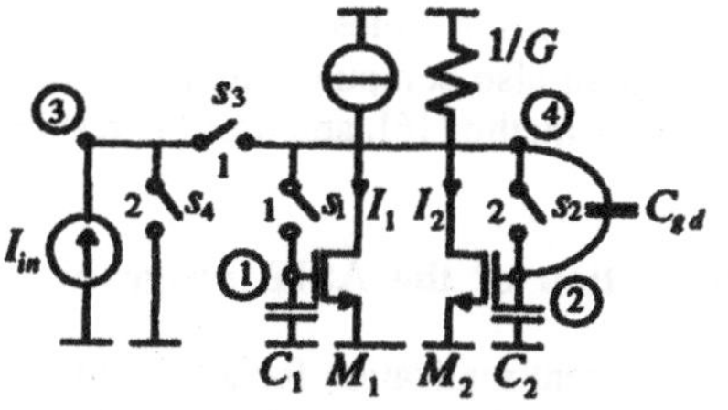

Figure 4 Example circuit for the nodal analysis.

The system of equations (5) takes the form shown in (6), with all the stamps of the elements in place. The switches cause the addition of several rows and columns, causing the columns and equations to be mapped as indicated by the column and row pointers into the final system (the seventh equation and column are eliminated by s_4). The final system is shown in (7), where the capacitive terms are maintained only in the "high-impedance equations" (marked). The input at phase 2 is null, and so all the $E_{i,m2}$=0. Four unknowns remain to be computed. The other ones are null or equal to one of the four remaining. The solution of (7) results in (8).

$$\left[\begin{array}{c|cccc:cccc} & 1 & 2 & 1 & 1 & 3 & 4 & - & 4 \\ \hline 1 & C_1 & & & & -C_1 z^{-\frac{1}{2}} & & & \\ 2 & & C_2+C_{gd} & & -C_{gd} & & -(C_2+C_{gd})z^{-\frac{1}{2}} & & C_{gd}z^{-\frac{1}{2}} \\ 1 & & & & & & & & \\ 1 & Gm_1 & Gm_2-C_{gd} & & G+C_{gd} & & C_{gd}z^{-\frac{1}{2}} & & -C_{gd}z^{-\frac{1}{2}} \\ \hdashline 3 & -C_1 z^{-\frac{1}{2}} & & & & C_1 & & & \\ 4 & & -(C_2+C_{gd})z^{-\frac{1}{2}} & & C_{gd}z^{-\frac{1}{2}} & & C_2+C_{gd} & & -C_{gd} \\ - & & & & & & & & \\ 4 & & C_{gd}z^{-\frac{1}{2}} & & -C_{gd}z^{-\frac{1}{2}} & Gm_1 & Gm_2-C_{gd} & & G+C_{gd} \end{array}\right] \begin{bmatrix} E_{1,11} \\ E_{2,11} \\ E_{3,11} \\ E_{4,11} \\ \hdashline E_{1,21} \\ E_{2,21} \\ E_{3,21} \\ E_{4,21} \end{bmatrix}, \begin{bmatrix} E_{1,12} \\ E_{2,12} \\ E_{3,12} \\ E_{4,12} \\ \hdashline E_{1,22} \\ E_{2,22} \\ E_{3,22} \\ E_{4,22} \end{bmatrix} = \begin{bmatrix} \\ \\ I_{in,1} \\ \\ \cdots\cdots \\ \\ \\ \\ \end{bmatrix}, \begin{bmatrix} \\ \\ \\ \\ \cdots\cdots \\ \\ I_{in,2} \\ \\ \end{bmatrix} \tag{6}$$

$$\left[\begin{array}{c|cc:cc} & Gm_1+G & Gm_2 & & \\ * & -C_{gd} & C_2+C_{gd} & & -C_2 z^{-\frac{1}{2}} \\ \hdashline * & -C_1 z^{-\frac{1}{2}} & & C_1 & \\ & & & Gm_1 & Gm_2+G \end{array}\right] \begin{bmatrix} E_{1,11} \\ E_{2,11} \\ \hdashline E_{1,21} \\ E_{2,21} \end{bmatrix}, \begin{bmatrix} E_{1,12} \\ E_{2,12} \\ \hdashline E_{1,22} \\ E_{2,22} \end{bmatrix} = \begin{bmatrix} I_{in,1} \\ \\ \\ \\ \end{bmatrix}, \begin{bmatrix} \\ \\ \\ \\ \end{bmatrix} \tag{7}$$

$$\frac{E_{1,11}}{I_{in,1}} = \frac{\dfrac{(C_2+C_{gd})(Gm_2+G)}{(Gm_2+G)\left((C_2+C_{gd})(Gm_1+G)+Gm_2C_{gd}\right)}}{1-\dfrac{Gm_2Gm_1C_2}{(Gm_2+G)\left((C_2+C_{gd})(Gm_1+G)+Gm_2C_{gd}\right)}z^{-1}} \quad \frac{E_{1,21}}{I_{in,1}} = \frac{\dfrac{(C_2+C_{gd})(Gm_2+G)}{(Gm_2+G)\left((C_2+C_{gd})(Gm_1+G)+Gm_2C_{gd}\right)}z^{-1/2}}{1-\dfrac{Gm_2Gm_1C_2}{(Gm_2+G)\left((C_2+C_{gd})(Gm_1+G)+Gm_2C_{gd}\right)}z^{-1}}$$

$$\frac{E_{2,11}}{I_{in,1}} = \frac{\dfrac{C_{gd}(Gm_2+G)-Gm_1C_2z^{-1}}{(Gm_2+G)\left((C_2+C_{gd})(Gm_1+G)+Gm_2C_{gd}\right)}}{1-\dfrac{Gm_2Gm_1C_2}{(Gm_2+G)\left((C_2+C_{gd})(Gm_1+G)+Gm_2C_{gd}\right)}z^{-1}} \quad \frac{E_{2,21}}{I_{in,1}} = \frac{\dfrac{-Gm_1(C_2+C_{gd})}{(Gm_2+G)\left((C_2+C_{gd})(Gm_1+G)+Gm_2C_{gd}\right)}z^{-1/2}}{1-\dfrac{Gm_2Gm_1C_2}{(Gm_2+G)\left((C_2+C_{gd})(Gm_1+G)+Gm_2C_{gd}\right)}z^{-1}} \tag{8}$$

The transistor currents can be recovered by multiplying E_1 and E_2 by Gm_1 and Gm_2 respectively. Note that the presence of C_{gd} causes an error in the current that M_2 stores in phase 2. The frequency response is still computed by (1).

The ASIZ program implements this algorithm [1], using FFT interpolation [5] to compute efficiently the partial and global nodal voltages as ratios of polynomials in z. From the polynomials, frequency responses, poles and zeros, and transient responses can be quickly computed. The program also computes frequency-domain sensitivities, using an extension of the adjoint network method [6], and uses the results to predict error margins in frequency responses.

10.5.6 Example of utilization of the ASIZ program

Figure 5 shows a switched-current realization for a 6th-order elliptic band-pass filter [7]. The schematic representation is in the form for input to the program, drawn in the EdFil schematic capture program. Note that the bias sources are omitted. The output is taken as a nodal voltage over a load resistor. The filter was designed for a passband with 1 dB maximum ripple from $0.16F_s$ to $0.24F_s$, where F_s is the switching frequency, and a minimum stopband attenuation of 40 dB.

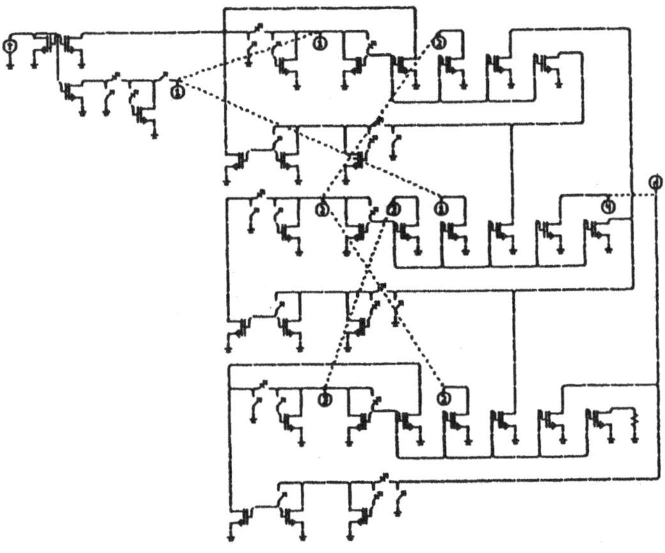

Figure 5 Schematic representation of an SI realization of a band-pass filter based on Euler integrators. Screen of the EdFil schematic capture program.

Figure 6 shows the ASIZ screen showing the computed frequency response, for a normalized switching frequency of 5 Hz (Note that the program does not take into account the time needed for stabilization of an actual circuit. 5 MHz would result in the same curves). The frequency response is plotted in the background window. The dotted margins are statistical deviations resulting from uncorrelated 5% errors in the transistor ratios, computed by sensitivity analysis. The peaks at the transmission zeros are artifacts of the use of gain sensitivities to compute the error (the transmission zeros are poles of the sensitivity function for the elements involved in their formation). The other curve is what results if a G_{ds} conductance equal to $0.01Gm$ and a C_{gd} capacitance equal to $0.01C_{gs}$ is considered for each transistor. This is an example of the high sensitivity of basic SI circuits to these parasitic elements. The other windows show the poles and zeros of the ideal transfer function in $z^{1/2}$ (this is necessary, because the period for the repetition of the transmission zeros, in the general case, is f times the switching frequency), and the transient response to a unity step.

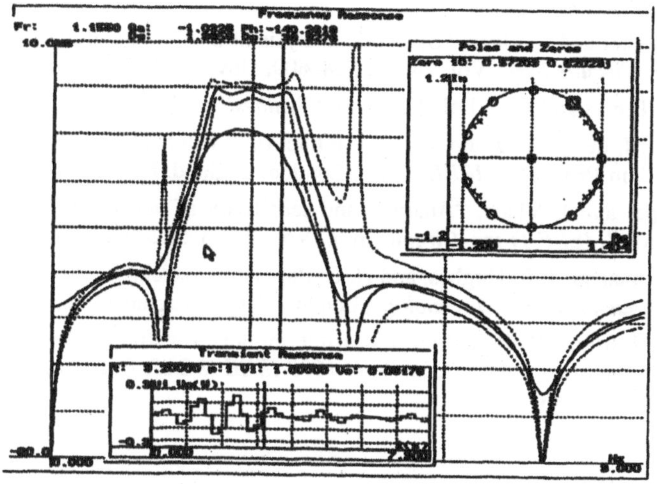

Figure 6 Screen of the ASIZ program, showing frequency response with error margins and the effect of imperfect transistors, poles and zeros in $z^{1/2}$, and a step response.

10.5.7 Conclusions

The small-signal AC analysis of switched-current filters, at transistor level, can be done in a way similar to the ones used for ideal switched-capacitor filters, that are a particular case. In switched-current filters, the described algorithm can model precisely the effects of drain conductances of MOS transistors and other resistive losses, and also the effects of parasitic capacitors, as C_{gd} capacitances. The described algorithm is ideal for design verification and investigation of structures for SI filters. For the study of other imperfections, more elaborated algorithms are also available. The effects of incomplete circuit stabilization in the switching intervals, and also of noise, can be investigated by the use of programs that analyze general periodically switched linear circuits. Non-linear effects, especially in the time domain, can be investigated with classical programs as SPICE, but some care must be taken with the models used, that must model correctly the non-linear capacitances involved, for example, in clock feedthrough effects.

References

[1] A. C. M. de Queiroz, P. R. M. Pinheiro, and L. P. Calôba, "Nodal analysis of switched-current filters", *IEEE Trans. Circuits Syst. II*, Vol. 40, pp. 10-18, Jan. 1993.

[2] J. Vlach, K. Singhal, and M. Vlach, "Computer oriented formulation of equations and analysis of switched-capacitor networks", *IEEE Trans. Circuits Syst.*, Vol. 31, pp. 753–765, Sept. 1984.

[3] A. C. M. de Queiroz and L. P. Calôba, "CAPZ – A PC program for switched-capacitor filter analysis in the z-transform domain", in *Proc. 34th MWSCAS*, Monterey, USA, pp. 134-137, May 1991.

[4] J. B. Hughes, I. C. Macbeth and D. M. Pattullo, "Second generation switched-current signal processing", in *Proc. IEEE ISCAS*, New Orleans, USA, pp. 2805–2808, May 1990.

[5] K. Singhal and J. Vlach, "Symbolic analysis of analog and digital circuits", *IEEE Trans. Circuits Syst.*, Vol. 24, pp. 598–609, Nov. 1977.

[6] G. C. Temes and J. W. LaPatra, *Circuit Synthesis and Design*, McGraw-Hill, 1977.

[7] A. C. M. de Queiroz and P. R. M. Pinheiro, "Switched-current ladder band-pass filters", in *Proc. 1994 IEEE ISCAS*, London, England, May 1994.

[8] A. Opal and J. Vlach, "Analysis and sensitivity of periodically switched linear networks", *IEEE Trans. Circuits Syst.*, Vol. 36, pp. 522–532, April 1989.

[9] C. Toumazou, J. B. Hughes, and N. C. Battersby, *Switched-Currents, an Analogue Technique for Digital Technology*, London: Peter Peregrinus Ltd., 1993.

[10] J. A. Barby, "Switched-current filter models for frequency analysis in the continuous-time domain", in *Proc. 1993 IEEE ISCAS*, Chicago, USA, pp. 1427-1430, May 1993.

Chapter 10.6

Current Monitoring for Test

Kevin R. Eckersall, Gaynor E. Taylor, Ian M. Bell
and Chris Toumazou

10.6.1 Introduction

Advances in the fabrication process of CMOS circuits have enabled the complexity of digital devices to increase enormously, with millions of transistors possible on a single chip. These digital devices have evolved to system level complexity using computer aided design (CAD) tools that not only allow relatively cheap routes to silicon, but considerably reduce the time for the design cycle. With the widespread availability of these CAD tools, non-specialist design engineers can design digital devices with little knowledge of the silicon layout rules.

The range of CMOS digital devices which is currently available is diverse, including programmable circuits which are immune to both noise and process variations. The advantage of the latter enables the design engineer to source a design to many foundries, allowing the decision to be influenced by economic as well as technical specifications.

The impact of MOS technology in digital ICs was not matched in the analogue functions used to interface the digital processing devices to the real (analogue) world. The early analogue MOS devices could not compete with bipolar devices which offered better speed, improved noise characteristics and more precise device characteristics. Recent developments in the CMOS process have made analogue MOS ICs more usable, with high value capacitors realised from double polysilicon layers in a much reduced area compared with the original metal-polysilicon capacitors. It is now possible to fabricate both the analogue interface circuits and the digital processing circuits on the same chip; these complex devices are known as mixed signal or mixed analogue-digital integrated circuits. The common analogue functions used in these circuits include input conditioning filters, analogue to digital and digital to analogue converters, signal reconstruction filters and precision gain mixers. Two technologies are currently used, one based on the switched-capacitor technique, the other being the voltage domain technique using operational amplifiers, comparators and buffers. More recently interest has been emerging in the use of the switched-current technique [1] which uses analogue designs that can be directly implemented in a sub-micron digital CMOS process.

Whilst the complexity of ICs has greatly increased, the number of available IC pins i.e. the access to the circuit has, in comparison, increased only marginally. Nevertheless the testing phase must ensure both the quality and reliability of the product with the test specification often providing a vital marketing tool. In order to minimise these test costs the design engineer must consider the testability requirements at all stages of the design cycle. The cost to correct any design error not only substantially increases as the design nears the production phase, but also delays the marketing of the final product and forces a higher retail price.

This tutorial is concerned with switched-current design and, in particular, this chapter will look at the problems posed by the need for post-fabrication test of such designs. Recent work at the Universities of Hull and Imperial College [2,3] has suggested a possible approach based on more general work at Hull on testing mixed signal devices. The general approach will be described first and it will then be shown how it may be refined and used in a particularly simple way in the switched-current case.

10.6.2 Testing mixed signal circuits

Digital devices are commonly tested through a series of functional and/or structural assessments in which primary output voltages are monitored during the application of a suitable test sequence. Testing can occur at either the wafer or packaged stage with the device commonly monitored under different environmental and electrical conditions. Whilst functional testing seeks to discover whether the circuit is working the alternative, structural testing, attempts to ascertain if the circuit is faulty. Structural test pattern generation directs attention to physical defects within circuits that cause functional failure. The most common model used for the generation and assessment of structural tests is the stuck-at fault model [4]. In this it is assumed that a physical defect manifests as a single net in the circuit permanently at logic "1" or logic "0".

The testing of analogue devices is considerably less well developed. Analogue functions pose problems concerning how to actually define or measure circuit performance. Different analogue sections may require transient or frequency analysis, with a range of environmental conditions (e.g. temperature) necessary. Other factors such as stability and leakage currents are also of concern. The whole area of analogue test has been termed context specific [5].

Mixed signal circuits are usually designed as application specific integrated circuits (ASICs), produced in relatively low volumes and often incurring high unit costs. They pose all the test problems offered by purely digital or purely analogue circuits plus additional difficulties such as propagating test stimuli and test outcomes from analogue to digital blocks and vice versa. At present there are few techniques available that can adequately test such designs, although the area has attracted considerable interest in recent years. The future growth of the mixed signal market will require acceptable test coverage which can be achieved within reasonable time and economic constraints. The identification of acceptable stimuli to test analogue and mixed signal circuits is also an area of great research interest.

In an attempt to reduce test costs in the digital case two approaches have been investigated; firstly, the development of more sophisticated test generation algorithms, and secondly, the introduction of Design For Test (DFT) structures on the IC itself. DFT structures may, however, introduce difficulties with the addition of on-chip test structures which could affect functionality. The requirement for extra silicon, also increases production costs and may decrease yield. A third approach, which is, at present, attracting a great deal of interest, is to monitor supply current rather than functional outputs in order to observe faults.

10.6.3 Test solutions

Early proposals for methods by which to test mixed signal ICs were derived from approaches used for digital ICs. The methods suggested relied on modelling of analogue blocks as digital equivalents and then applying conventional digital test techniques. The use of circuit partitioning in which the analogue and digital portions are reconfigurable for test has also been suggested. A number of approaches have been proposed including signal multiplexing [6], analogue test tables [7], analogue scan path [8] and hybrid built-in self test (HBIST) [9]. Whilst such methods can be used to test the individual partitions, the correct operation of the combined units can not be guaranteed. A comprehensive review of early approaches can be found in [10].

Work at Hull has concentrated, in particular, on a different and unified approach which explores the possibilities of supply current monitoring for fault detection in both the analogue and digital parts of the device. This will be described next.

10.6.4 Supply current monitoring

10.6.4.1 Background

The stuck-at fault model referred to above has been widely used to represent the faults occurring in digital devices. The value of this model has been questioned however for certain technologies [11, 12]. It was known to be adequate for TTL and nMOS devices, but, with the advent of CMOS technology using ever shrinking geometries, it does not adequately represent the defects which occur in practice. In the early technologies (TTL, nMOS) the presence of a bridging fault around a transistor would cause the component to latch to one of the power supply rails. The resulting stuck-at fault could then be propagated to an output and be logically testable. In the small micron CMOS devices widely used today the types of defects occurring are changing, with bridging faults an increasing problem (due to the shrinking geometry) along with others such as gate oxide shorts and floating gate failures [13]. Bridging faults are believed to represent up to 40% of all faults with this figure likely to increase as the geometries continue to shrink. However, such faults in digital CMOS devices do not cause the latch up that was known to occur in nMOS and TTL, but may cause the output to become logically indeterminate. The detection of these faults by traditional logic testing is not possible.

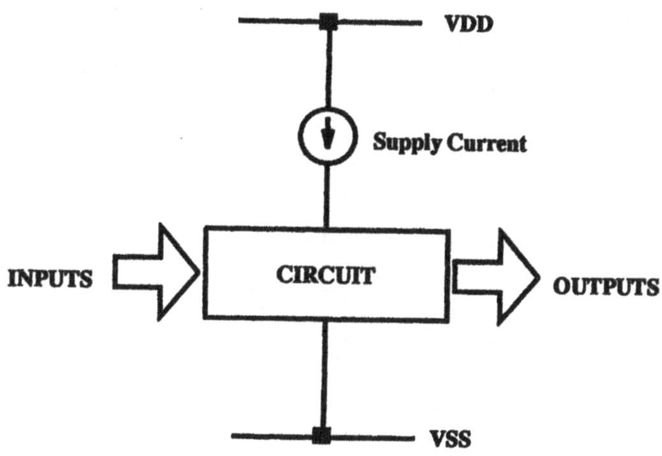

Figure 1 The IDDQ test method.

As indicated above, supply current monitoring as an approach to the test of purely digital CMOS circuits had recently attracted considerable interest [14-17]. In this approach the current passing through the V_{DD} or V_{SS} terminal is monitored during the application of test vectors. In the quiescent state such circuits draw a very low current, typically nano-amperes. This is raised by several orders of magnitude in the presence of certain defects thus offering a technique for detecting these. The technique offers the advantage of direct detection of the defect obviating the necessity to propagate a defective voltage to an output port. It is also able to detect certain defects missed by traditional voltage test approaches based on the stuck-at fault model.

10.6.4.2 Application to mixed signal devices

Given the availability of existing results concerning digital devices, initial research at Hull concentrated on establishing the validity of current monitoring as a means of detecting faults in common analogue macros. Faults in analogue devices can be classified as hard (causing circuit malfunction) or soft (altering circuit performance). Detection of hard faults through current monitoring was initially attempted by applying simple DC test signals. Deviation from the nominal current value was not, in general, as marked as that experienced in the digital case. The differences between the nominal (fault-free) case and particular fault cases were not order of magnitude changes (as was observed in I_{DDQ} monitoring), but they were deemed sufficient to differentiate between good and faulty devices with, typically, micro-amperes values recorded. Currents were as likely to decrease as to increase fróm the nominal values.

A second, more sophisticated approach examined dynamic testing (monitoring current transients rather than the steady state) and experimented with a number of classes of possible test vectors including pseudo random binary sequences (PRBS) and complementary sequences in addition to ramps and other simple signals. Good coverage of both hard and soft faults was achieved indicating that the approach offered a tool in the context of analogue as well as digital test. The fault models investigated included both the hard faults mentioned above and gate oxide shorts [18], for which it was shown that faults undetectable by functional testing were detectable by supply current test. This parallels results from the digital domain where it has been shown that devices exhibiting the type of failure under current test are highly likely to fail functionally during early lifetime. These results re recorded in detail in [19] and are supported by measurements made on real devices by an industrial collaborator. Finally, coverage in simple mixed devices such as A-D convertors was assessed.

10.6.4.3 Simulation results

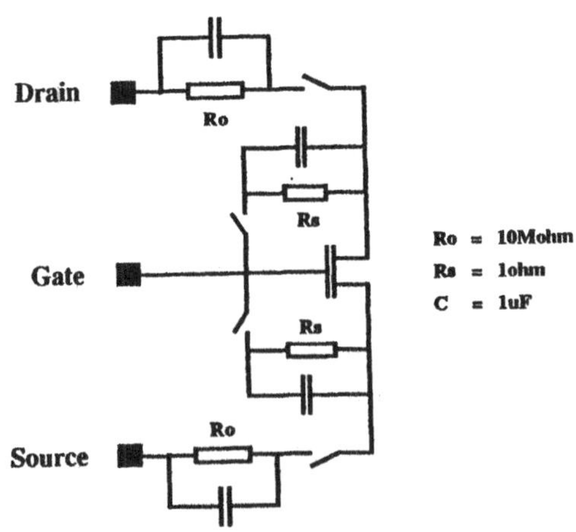

Figure 2 Hard fault transistor model.

To simulate the faulty behaviour in the devices, each transistor is replaced by the more complex model shown in Figure 2. Choice of switch operation allows simulation of the four catastrophic transistor faults namely drain open, source open, gate drain short and gate source short.

The fault model (Figure 3) used to describe the occurrence of a single unidimensional gate oxide short [20] includes a rectification barrier to provide the non-ohmic current injection described in [13], characterised by a threshold voltage and a breakdown voltage. The shunt resistance, R_S models the contact resistance of the short with high values representing minor or weak defects, whilst low values represent the more extensive hard defects.

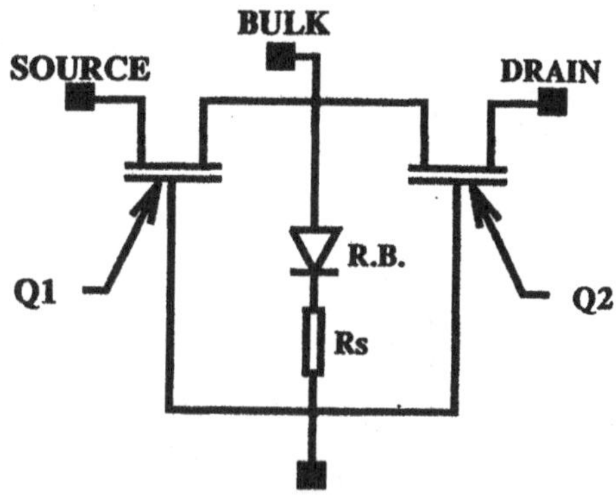

Figure 3 The gate oxide short fault model.

Figure 4 shows the basic method of fault detection employed. Each of the possible faults is individually injected into the circuit under test by selecting appropriate switch positions. The resulting faulty circuit is then simulated and the data extracted. This is then subtracted from the fault free data leaving the sampled *error signal*. Fault free and error data are then analysed to give detectability measures. A number of circuits were tested with a variety of test signals, including pseudo random sequences, sine waves and ramp signals [21]. For the sake of brevity only the three stage analogue band pass filter (Figure 5) will be reported here. For purposes of analysis a fault was said to be detected if the monitored current value in the presence of that fault differed from the fault free value by a sufficient amount. A difference of 5 μA was generally used since this value is achievable using on chip built in current sensors [22].

Figure 6a shows that all four test signals achieved detection of all faults. The test signals used included

- CSS Complementary Signal Set. An aperiodic pseudo random binary signal known to have good auto-correlation properties and used in the field of communications to test systems by a method of linear system identification.
- PRBS Pseudo Random Binary Sequence. A periodic maximal length signal was used.
- RAMP Ramp Signal, with the voltage linearly ramped between V_{SS} and V_{DD} power rails.

SINE The frequency of the sine wave signal was set to a value within the pass-band of the filter.

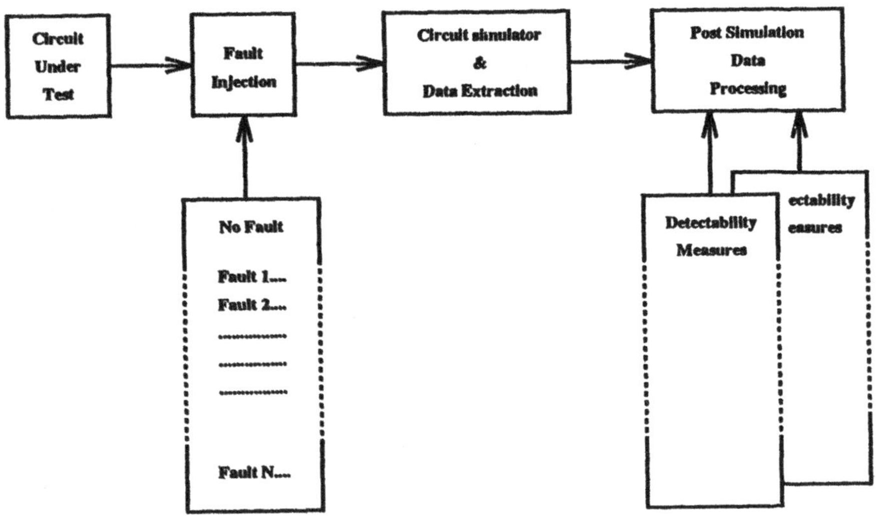

Figure 4 General fault detection method.

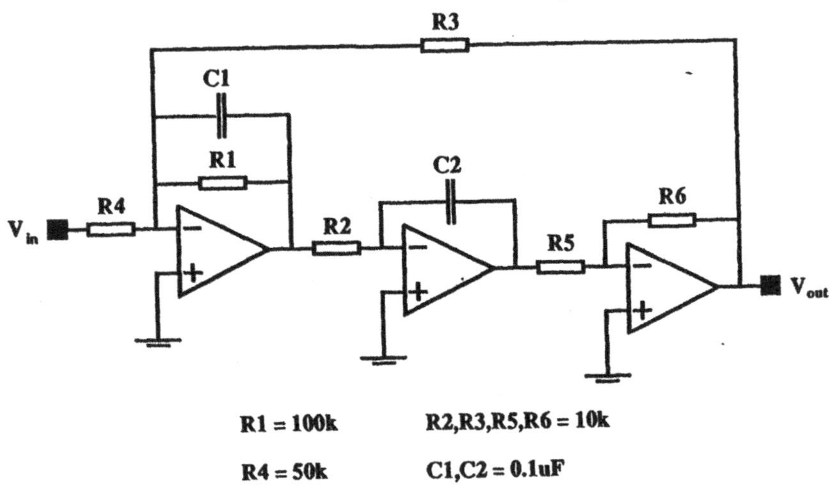

Figure 5 Analogue band pass filter.

Figure 6b shows the normalised sample counts (NSC) of all faults. The NSC is a confidence measure of fault detection, describing the quantity of fault samples found to sufficiently differ from the fault free circuit. The graph shows that gate-drain defects exhibited higher NSC values and can be said to be more detectable than drain open defects and gate oxide defects. Little difference was seen between different test signal types.

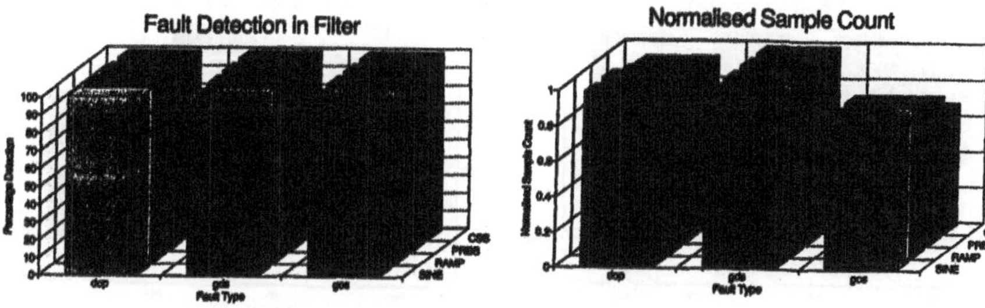

Figure 6 Fault detection of faults in BPF.

For commercial reasons it is common to design a circuit that can be manufactured at different silicon houses. For a CMOS digital device the differences imposed on device performance from one foundry to another are slight with the designs tolerant to process variations. For the case of analogue and hence mixed signal devices, there can be a degraded performance with process spread. The majority of foundries will achieve required performance, but there will be foundries at the extremities. For each foundry three modes of test are generally performed using different supply voltages and operating conditions. For example

Typical case	V_{DD} = 4.5 Volts	Temperature = 27 °C
Best case	V_{DD} = 3.0 Volts	Temperature = 125 °C
Worst case	V_{DD} = 6.0 Volts	Temperature = 55 °C

Thus a spread of performance versus mode can be used to quantify the effect of process variation as described in Figure 7.

Simulations were performed using process models representing the ***best-best*** and ***worst-worst*** foundries from the multi-sourceable library of an industrial collaborator. Figure 8 shows that faults introduced into the best type circuits were easier to detect with NSC values approximately 10% greater than similar faults in circuits simulated with the worst type process information.

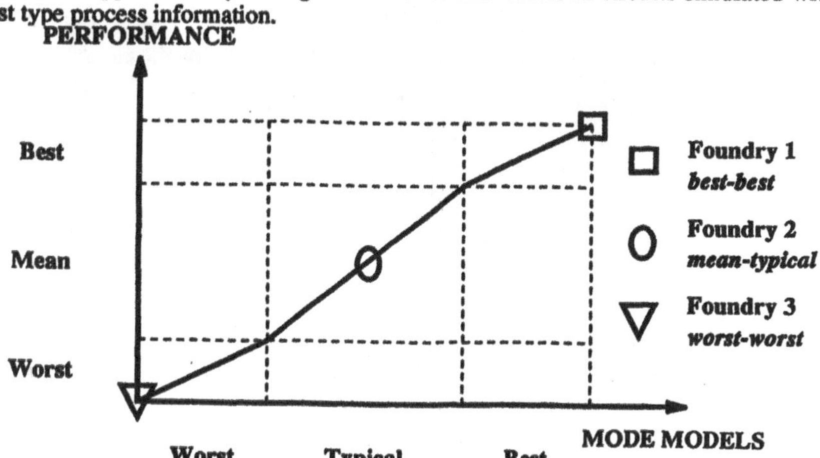

Figure 7 Performance versus mode evaluation.

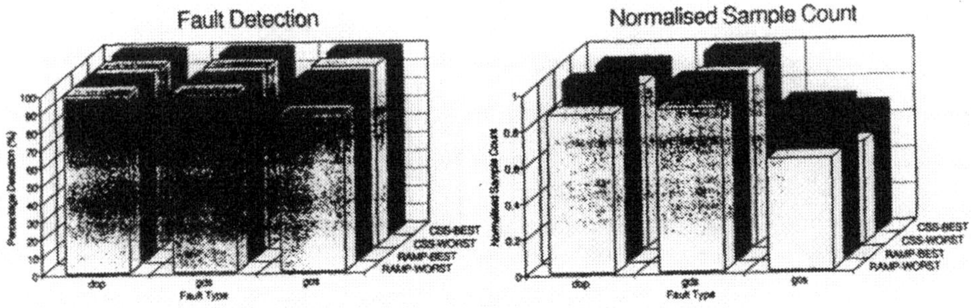

Figure 8 Mode detection of faults in BPF.

Post processing schemes were devised to analyse data from any circuit under test. Instead of relying on fault free signatures, the use of window comparison formed from the best-best and worst-worst case signatures were explored. Four methods were generally used

(1) Envelope Method	The envelope formed by the best and worst case spread was called the safety region. Responses wholly occurring within this region were undetectable. Responses either partly or preferably wholly outside the region were detectable.
(2) Difference Method	This again uses the envelope technique but the envelope is determined from a single process.
(3) Mean Value (Envelope)	The mean value of the fault signature samples was calculated and compared to the values of the best and worst simulations.
(4) Mean Value (Difference)	Mean value compared against same process fault free mean value.

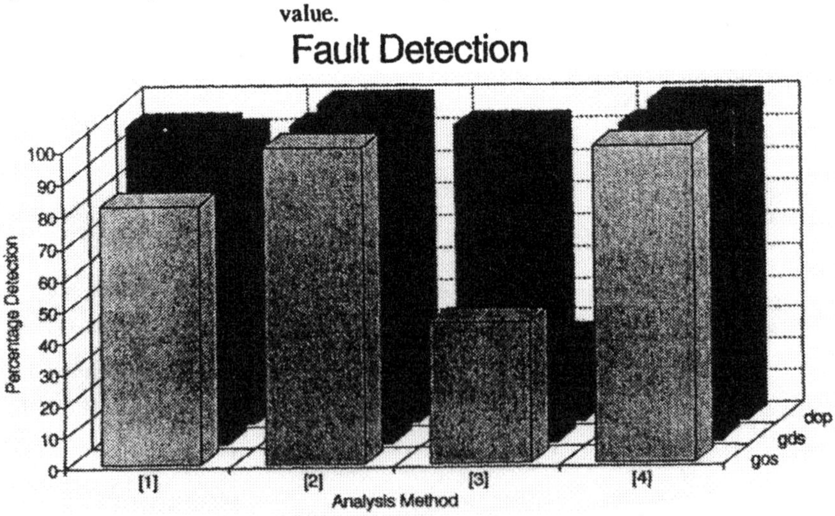

Figure 9 Comparison of analysis methods.

Figure 9 shows detection rates using these different analysis methods. All faults were detectable by some method, although in the context of practical testing it is uneconomical to encompass several techniques. As one would expect the methods based on difference techniques delivered higher results. The envelope methods produced variable results with drain open defects harder to distinguish. However such defects are considered less likely to occur in practice [11].

The work showed that supply current monitoring is an effective method by which faults can be detected in mixed circuits. The technique can be improved through the use of different test signals and analysis methods. PRBS signals were shown to be useful. They are known to be easily generated and could be incorporated on-chip as part of a fully BIST structure.

10.6.5 Built-in current sensing

The I_{DDQ} currents observed in faulty digital devices are known to be micro-ampere magnitudes. Similar currents in analogue or mixed signal circuits will depend on the nature and application of the analogue portion. Micro-ampere current levels, although large enough for discrimination, are difficult to measure using off-chip sensors. The larger currents in the integrated circuit pad drivers will swamp these smaller supply current signatures. The use of on-chip current sensing devices is required.

The application of on-chip built in current (BIC) sensing devices requires the user to have knowledge of the optimum positioning of the device, thus preventing the monitor circuit imposing a change in functionality on the circuit under test. The use of compact structures is desirable, minimising the silicon area overhead. The simplest form of sensors are current-mirrors which can be produced in CMOS or BiCMOS processes. For the case of a MOS designed mirror, the accuracy of the device is dependent on the matching accuracy between transistors and channel length modulation.

Work carried out at Hull [23] has used on-chip current monitoring techniques to detect faulty circuits. Current-mirrors were inserted between the power connections of some analogue components of a mixed device. The circuit was simulated with faults to study the effectiveness of the mirrors. The results are described in Figure 10.

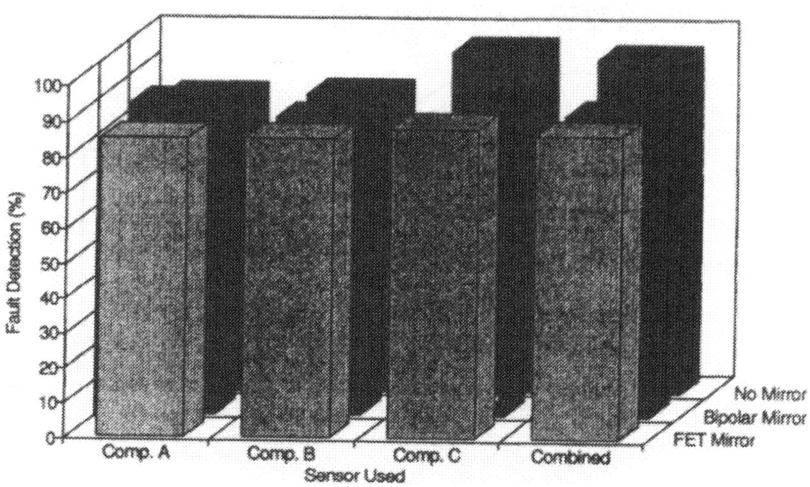

Figure 10 Fault detection.

10.6.6 Testing switched-current cells.

There is currently considerable interest in the application of switched-current techniques to the design of analogue sampled-data circuits and systems [3]. Switched-current techniques, unlike their predecessor switched-capacitors, offer full compatibility with CMOS digital processing and the potential to function with reduced supply voltages at high speed. Since the capacitors in cells do not transfer charge, they only have to hold charge, double polysilicon linear capacitors are not required with the technique hence its explotation as Digital Technology. Since the switched-current technique [24-27] is a sampled-data current domain process, stored information can be summed very easily rendering the technique well suited to current mode self-test.

As shown above circuit test based upon current monitoring essentially relies on comparison of current signatures from a known good source and the device under test. This comparison is basically summation of the two currents, an operation which is fundamental to current-mode signal processing and the switched-current technique. Thus there is clearly opportunity for test processing to be carried out on chip rather than externally. To allow investigation of these ideas a number of simple switched-current memory cells have been simulated with appropriate faults and test patterns.

The basic switched-current cell and the more practical regulated cascode cell are described in full elsewhere in this volume. The design of switched-current circuits uses repeated structures of the memory cell interconnected via timing control and feedback switches, implemented with minimal size MOS devices. For the purposes of testing these feedback paths are broken to form identical chains of memory cells. That is, the circuit is *reconfigured* for test purposes.

10.6.6.1 Test patterns

The initial test vector was a 2 MHz sinusoid of 30 μA amplitude and no DC offset. The transient current was sampled every 10 nanoseconds over the period 0 to 2 microseconds, giving 200 points per test vector.

10.6.6.2 Circuit examples

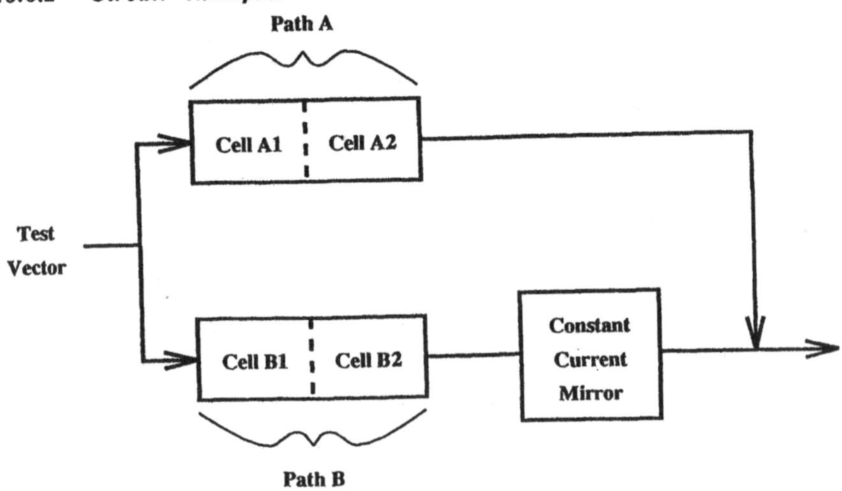

Figure 11 Block diagram of test circuit.

Initial experiments concentrated on fault detection in a single memory cell. Figure 11 shows the block diagram of another early example, which comprises four SI cascode memory cells and a current mirror which serves as a current invertor. The circuit thus provides two identical paths *A* and *B*, such that the two currents subtract at the output. Thus

$$I_{out}(n) = I_A(n-1) - I_B(n-1)$$

where *n* is a single clock cycle of Φ_1, Φ_2.

The circuit is simple in design but serves to demonstrate the principle behind a possible method of self test, as it is conceivable for a complex circuit to be designed to be reconfigurable in this way for test purposes.

If I_A, I_B are fed from the same signal, then for an ideal fault free circuit under test $I_{out} = 0$. In the practical case, as discussed above, process variation leads to differences in the two signal paths and hence as previously, a tolerance band is required.

10.6.6.3 Simulation results

The circuit was simulated using HSPICE under the level 2 parameters of a typical 5 Volt 2 micron digital CMOS process. Device dimensions were typical of those in [27]. A tolerance band of 5 μA was used for discrimination. The output current from the cells was used for the purpose of monitoring faults under test.

The graph shown in Figure 12 shows the detection rate of drain open (dop) and gate drain short (gds) faults in a single memory cell and two cascaded cells. For the case of drain open defects, approximately 80-85% manifest as detectable for both circuits. However, the method was able to detect all gate drain short defects. The normalised sample count results show that less than half the samples portrayed the defects. The circuits function on two phases with the input current restricted to only half the test duration. Therefore the sample count figures are misleadingly low, with a maximum value of 0.50 achievable.

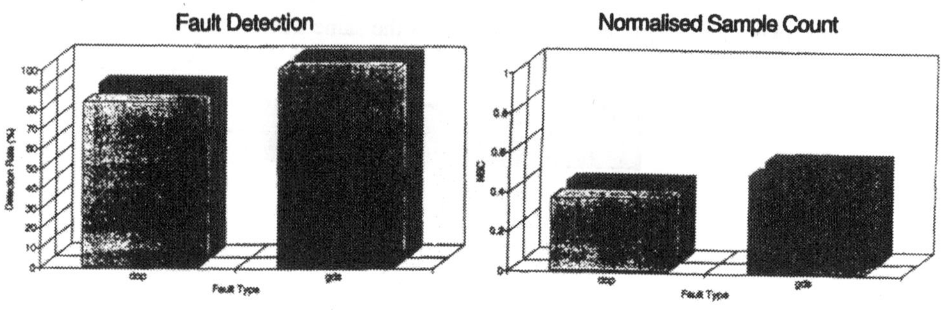

Figure 12 Fault detection in switched-current cells.

Figure 13 describes the mean difference of samples found to differ from the fault free response. The first two columns describe the current differences in the single cell circuit, whilst the second two columns describe the cascaded results. For both circuits gate drain defects were found to exhibit the larger differences. The scale on the figure allows a display of maximum current difference (per sample detected) of 200 μA. In fact the maximum current differences of gate drain defects was above 300 μA per sample. This is reflected in the mean difference, appearing larger for gate drain defects.

Comparison of results between the two circuits is interesting. For gate drain defects there is a definite decrease in current difference when two cells are cascaded. However for drain open defects the opposite occurs, with an increase recorded. In the latter case, it can be assumed that the defects are causing a loading effect, thus forcing larger current flow through the circuit. From these early results it could be concluded that current monitoring was a convenient technique, achieving high detection, with the capability for built in self test.

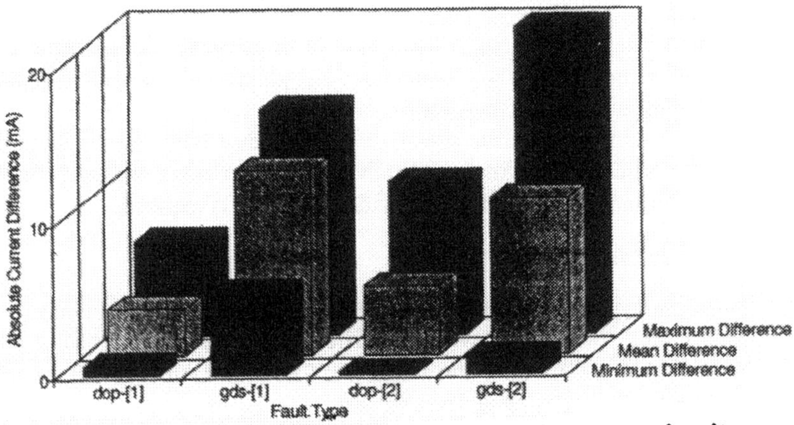

Figure 13 Current differences in switched-current circuits.

10.6.7 Built in self test for SI cells

Switched-current circuits are comprised entirely from memory cells with feedback loops and some additional (biasing) transistors providing specific functionality. It was demonstrated above that cascades of such cells are highly testable with standard patterns, and thus there is a clear immediate gain in testability if global feedback loops can be disconnected for test. Rather than comparing the signature of the device under test with simulated values of a fault free circuit, two equal length cascades of cells on the same device can be compared one against the other as shown schematically in Figure 14.

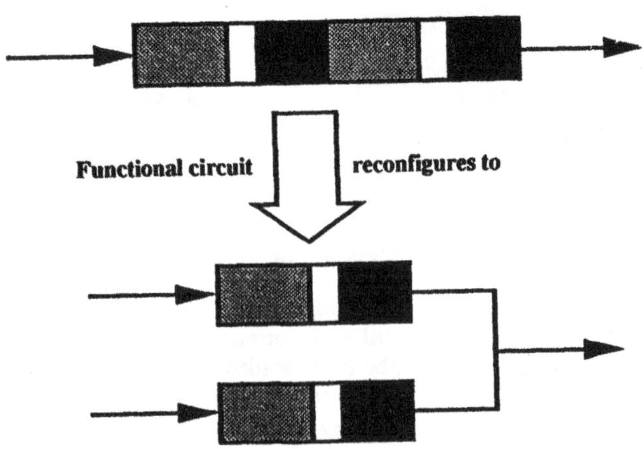

Figure 14 Block reconfiguration.

10.6.7.1 Functional inversion test circuit

Modifying the circuit given in [27] for a low pass second order filter, to give self testing capabilities requires five extra in-circuit test switches, as shown in Figure 15. *A* severs the unswitched feedback path, *B* and *C* connect the output of the first integrator's current copier to the output of the second with the same transistor ratios forming a summing node and giving that summing node access to the outside world. The *D* and *E* switches toggle between the scaled input to the second integrator and an input equal to that of the primary input. Other modifications to the circuit are the logical combination of the test line and the clock signals, and extra routing for the combined clock/test lines. The test switches are minimal size transistors and therefore introduce a negligible overhead. A block diagram of the circuit in normal and test modes is shown in Figure 16.

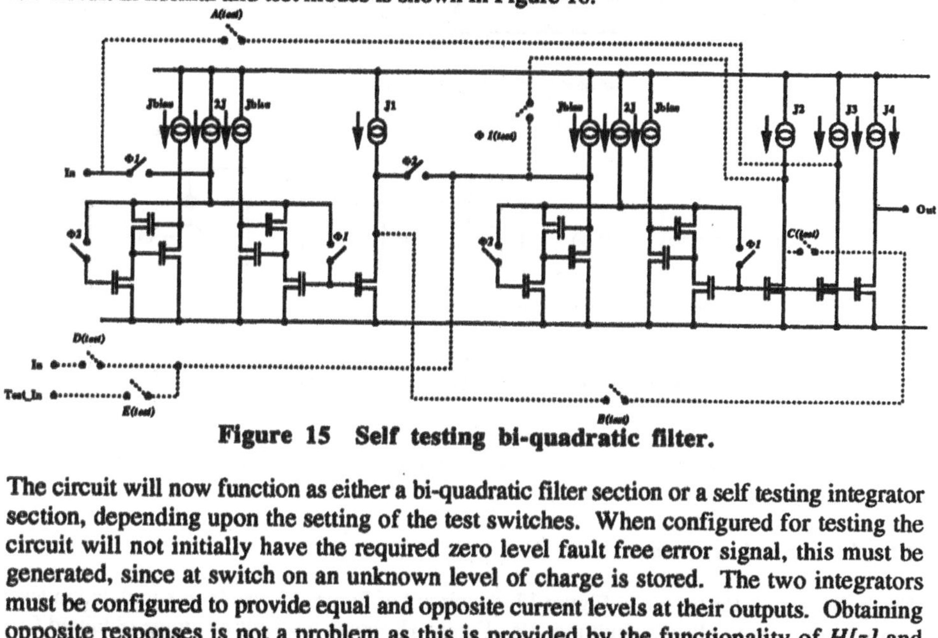

Figure 15 Self testing bi-quadratic filter.

The circuit will now function as either a bi-quadratic filter section or a self testing integrator section, depending upon the setting of the test switches. When configured for testing the circuit will not initially have the required zero level fault free error signal, this must be generated, since at switch on an unknown level of charge is stored. The two integrators must be configured to provide equal and opposite current levels at their outputs. Obtaining opposite responses is not a problem as this is provided by the functionality of *H[z]* and *-H[z]*, but obtaining equal levels is not so easy, there are only two definite levels for an SI circuit element, these being the positive and negative saturation levels, but with the implemented class A cells the absolute positive and negative saturation levels are unequal. So although it is feasible to saturate the two devices by the application of a suitably large amplitude current, the error signal remains non zero, but is now of a known difference

$$Error[n] = \left(h[n]_{+SAT}\right) + \left(-h[n]_{-SAT}\right)$$

Removing this difference is achieved simply by changing the input stimulus half way through a clock cycle, since *H[z]* reads on Φ_2 and *-H[z]* on the opposite phase, a change in the input signal between *H[z]* and *-H[z]* reading causes an effective subtraction step. Hence providing a step equal in magnitude to the error difference for half a clock cycle negates the error leaving the required zero level fault free signal. At this point the device is ready to accept the test stimuli, providing two restrictions on maintaining the zero error signal are observed. Firstly at no time must either integrator be allowed to saturate.

Secondly input the must only change on the beginning of a clock cycle with a step, hence linear ramps and sinusoids are not permitted unless quantized. Figure 17 gives an example test stimuli, and shows the fault free envelope for *Error[Z]*. *Error[Z]* must be defined in an envelope because of integrator drift δ.

$$Error[Z] = H[z] + (-H[Z]) = \delta \neq 0$$

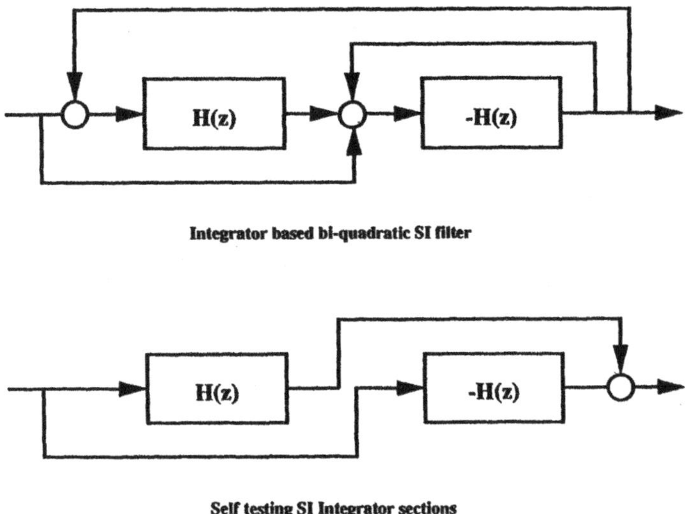

Figure 16 Circuit functionality, normal and test operation.

Thus the zero level fault free error signal degrades with time, this can be recovered to a degree by applying a large change in input stimulus forcing one integrator up to (but not into) saturation.

10.6.7.2 Fault coverage

Using the test vector shown in Figure 17, the circuit of Figure 15 was fault simulated in test configuration, with all four fault models. Figure 18 shows the fault coverage obtained for all of the MOSFET's directly in current flow, i. e. the current copiers isolated with the severance of feedback were not fault injected. The detection envelope about *Error[Z]* was defined as ±10 μA, just bracketing the fault free error signal, thus any point outside this envelope can be regarded as erroneous, and hence a fault has been detected, partial detection of a fault is denoted as less than 5% of the total sample points being in error.

	Detectable (%)	Partially Detectable (%)	Non Detectable (%)
GDS	69	22	9
GSS	50	37	13
SOP	50	11	39
DOP	60	0	40
Total	57	18	25

Figure 18 Reconfigurable bi-quadratic section fault coverage.

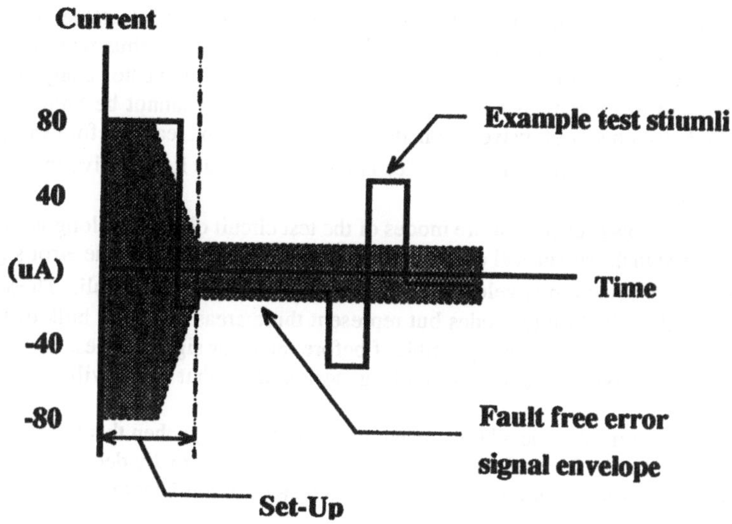

Figure 17 Example test stimuli.

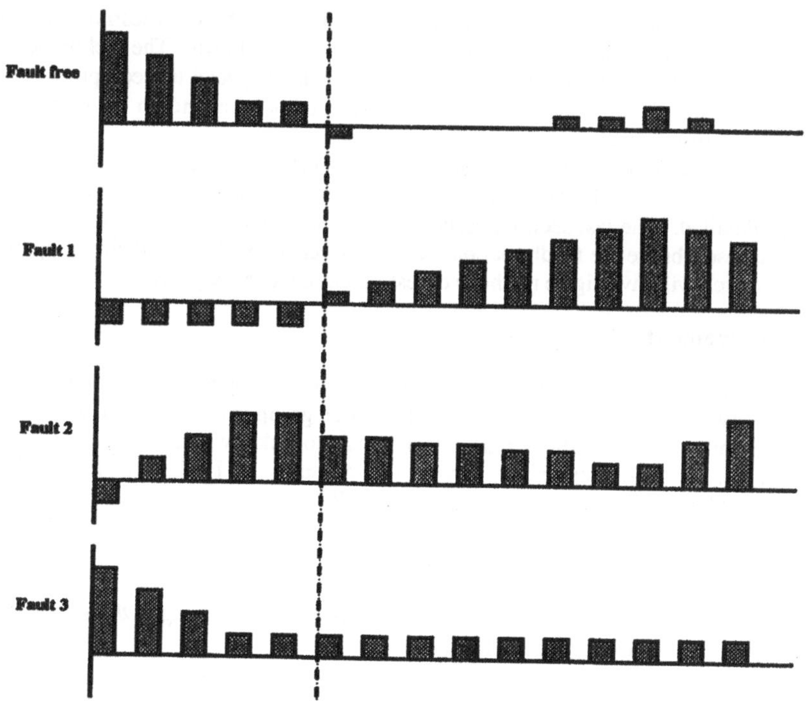

Figure 19 Failure modes in reconfigurable bi-quadratic sections.

From the table in Figure 18 it can bee seen that overall coverage including partially detectable faults is 75%, with the bulk of these faults being readily detectable. It is the case that a different profile test stimulus will highlight different faults, thus with several such stimuli the coverage will be increased. The earlier work on test vector analysis although applicable with this circuit is restricted, as saturation levels cannot be breached, thus a compromise between stimuli wavelength and amplitude must be reached, for example a ±50 μA stimulus of wavelength 16 clock cycles is inappropriate as it will drive the integrators into saturation.

Figure 19 shows several failure modes of the test circuit observed, along with the fault free zero error signal, the vertical dashed line represents the point when the error signal falls within the 10 μA detection envelope, after 2.7 μs of the 10 μs test stimuli. These failures do not cover all of the failure modes but represent three areas wherethe bulk of faults are observed. Fault detection is only possible if before the error signal settles, hence if faulty behaviour is displayed only in the settling period then that fault will be classed as undetectable.

Should the current copiers be included in the fault coverage then this falls as only 40% of faults are detectable, 26% of which must be declared as marginally detectable. This is an area that requires further work as noted in the conclusions of this chapter.

10.6.8 Conclusions

The work described in this chapter has shown that the use of supply current monitoring is a useful technique by which to test voltage domain mixed signal devices. The work has been extended to the testing of switched-current circuits. The possibility of testing most of the transistor circuitry through simple reconfiguration has been shown. The technique offers the advantage of in-circuit test signature generation with no special test equipment needed for go/no-go testing. One cascade of cells is tested against a second on the same device, thus reducing errors associated with process variation.

Further work is required to establish what the effects on detection may be due to factors such as process variation. Rules are also required to characterise the reconfiguration processes detailed. The last section, describing the reconfiguration of a bi-quadratic filter for test purposes, broke the feedback connections associated with normal circuit operation. Further research will investigate methods of testing these feedback paths.

Acknowledgement

The work described in overview here, covers results obtained by a number of researchers in the University of Hull, VLSI Design and Test research group in collaboration with Imperial College, London. The authors particularly wish to acknowledge the work of P. L. Wrighton, B. R. Bannister and J. Povanez (University of Hull), and E. G. Saether (Imperial College). The financial support of SERC grant number GR/J 14530 is also acknowledged.

References

[1] C. Toumazou, F. J. Lidgey, D. G. Haigh, *Analogue IC Design: the current-mode approach*. Peter Peregrinus Ltd., 1990.

[2] G. E. Taylor, C. Toumazou, P. L. Wrighton, N. Battersby, "Mixed Signal Test Considerations for Switched Current Signal Processing", *Proceedings Midwest Symposium on Circuits and Systems*, August 1992.

[3] C. Toumazou, J. B. Hughes, N. Battersby, *Switched Currents: an analogue technique for digital technology*, IEE Press, Chapter 18, 1993.

[4] B. R. Wilkins, *Testing Digital Circuits*.

[5] M. Rumsey, "Behavioural Modelling of Mixed Analogue/Digital Circuits", *Proceedings of 7th International Conference on Custom & Semi-custom ICs*, pp 49.

[6] K. D. Wagner, T. W. Williams, "Design For Testability of Mixed Signal Integrated Circuits", *Proceedings International Test Conference*, 1988, Paper 39.1.

[7] P. P. Fasang, D. Mullins, "The Design for Testability of Mixed Analog/Digital ASICs", *IEEE Custom IC Conference*, Paper 16.5, 1988.

[8] C. L. Wey, "Built in Self Test (BIST) Structure for Analog Circuit Fault Diagnosis", *IEEE Transactions on Instrumentation & Measurement*, Vol. 39, No. 3, 1990.

[9] M. J. Ohletz, "Hybrid Built in Self Test (HBIST) For Mixed Analog/Digital IC's", *Proceedings of European Test Conference*, pp. 307-316, 1991.

[10] J. Galiay, Y. Crouxet, M. Vergniault, "Physical versus fault models in MOS LSI circuits: impact on their testability", *IEEE Transactions on Computers*, Volume C-29, pp. 527-531, 1980.

[11] K. Baker, Technical Discussion, Phillips Research Laboratories, Netherlands.

[12] M. Ohletz *et al.*, "Mixed Signal and Analogue Testing", *ESPRIT III Archimedes, First Open Workshop*, Hannover, June 8-9 1993.

[13] L. Balado, R. Rodriguez-Montanes, J.A. Segura, V.H. Chapmac, J. Figueras, J.A. Rubio, "On-line Current Testing (BIST) for Bridges, GOS and Floating Gate Faults in Basic Combinational Modules", *CEC Deliverable*, June 1991.

[14] K. R. Eckersall, P. L. Wrighton, G. E. Taylor, I. M. Bell, B. R. Bannister, "Testing Mixed Signal ASICs through the Use of Supply Current Monitoring", *Proceedings of European Test Conference*, 1993.

[15] D. A. Camplin, I. M. Bell, G. E. Taylor, B. R. Bannister., "Investigations into Current Sensing Strategies", *IEE Colloquium* No. 1992/118.

[16] D. A. Camplin, I. M. Bell, G. E. Taylor, B. R. Bannister, "Supply Current Testing of Mixed Analogue and Digital ICs", *Electronics Letters*, Vol. 27, August 1991.

[17] D. A. Camplin, I. M. Bell, G. E. Taylor, B. R. Bannister, "Can Supply Current Monitoring be Applied to the Testing of Analogue and Digital Portions of Mixed ASICs ?', *Proceedings IEEE Test Conference*, pp. 538-542, 1991.

[18] C. F. Hawkins, J. M. Soden, "Reliability and Electrical Properties of Gate Oxide Shorts in CMOS ICs", *Int. Test Conference 1986*, Paper 13.4, pp. 443-451.

[19] D. Camplin, K. R. Eckersall, P. L. Wrighton, G. E. Taylor, B. R. Bannister, I. M. Bell, "MIXIT Project : Second Deliverable", Internal document, 1992.

[20] R. Rodriguez-Montanes, J. A. Segura, V. H. Chapmac, J. Figueras, J. A. Rubio, "Current vs Logic Testing of Gate Oxide Short, Floating Gate and Bridging Failures in CMOS", *International Test Conference 1991*, Paper 19.3, pp 510-519

[21] D. Camplin, K. R. Eckersall, P. L. Wrighton, G. E. Taylor, B. R. Bannister, I. M. Bell, "MIXIT Project : Final Deliverable", Internal document, 1993.

[22] A. H. Bratt, R. J. Harvey, A. P. Dorey, A. M. Richardson, "Aspects of Current Reference Generation and Distribution for IDDX pass/fail Current Threshold Determination", *IEE Colloquium*, Digest No. 1993/240.

[23] P. L. Wrighton, "Current Mirrors in Mixed Devices for the Purpose of On Chip Monitoring", Internal report.

[24] S. J. Daubert, D. Vallancourt, Y. P. Tsividis, "Current Copier Cells", *Electronics Letters*, Vol. 24, pp. 1560-1562, 1988.

[25] W. Groeneveld, H. Schouwenaars, H. Termeer, "A Self Calibration technique for Monolithic High Resolution D-A Converters", *ISSCC Digest of Technical Papers*, pp. 22-23, Feb. 1989.

[26] J. B. Hughes, N. C. Bird, I. C. Macbeth, "A New Technique for Analog Sampled Data Signal Processing", *Proceedings IEEE ISCAS*, pp. 1584-1587, 1989.

[27] G. Wegmann, E. A. Vittoz, "Very Accurate Dynamic Current Mirrors", *Electronics Letters*, Vol. 25, pp. 644-646, 1989.

[28] N. C. Battersby, C. Toumazou, J. B. Hughes, *Analogue-digital ASICs: Current Techniques and Design Tools*, Chapter 5.

Tutorial 10

Current-mode analog signal processing

Part 2

Current-Conveyors

Organised by: John Lidgey, *Oxford Brookes University*

Chapter 11.1

Current-Conveyor Basics and Applications

John Lidgey and Chris Toumazou

11.1.1 Introduction

The current-conveyor is a versatile analog device which is intended to be used with other circuit components to implement many analog signal processing functions. It is an analog circuit building-block in much the same way as a voltage op-amp, but it presents an alternative method of implementing analog systems which traditionally have been based on voltage op-amps. This alternative approach leads to new methods of implementing analog transfer functions, and in many cases the conveyor-based implementation offers improved performance to the voltage op-amp-based implementation in terms of accuracy, bandwidth and convenience. Circuits based on voltage op-amps are generally easy to design since the behaviour of a voltage op-amp can be approximated by a few simple design rules. This is also true for current-conveyors, and once the appropriate design rules are understood, the application engineer is able to design conveyor based circuits just as easily.

Although the current-conveyor concept has been around for a very long time, it is only in the past five years general purpose monolithic devices have been available. The emergence of these has resulted in a renewed awareness and an improved understanding of the benefits of designing analog systems using curent-conveyors and it is likely that current-conveyors will become very much more common-place in the near future.

11.1.2 Background

The first generation current-conveyor or CCI was proposed by Smith and Sedra in 1968 [1] and the more versatile second generation current-conveyor, or CCII, was introduced by the same two authors in 1970 [2], as an extension of their first generation conveyor. The CCII is without doubt a much more valuable and adaptable building block and we will concentrate mostly on this device throughout this chapter. Figure 1a shows the voltage~current describing matrix for the CCII and Figure 1b the schematic normally used for the CCII with the power supply connections omitted.

$$\begin{bmatrix} I_Y \\ V_X \\ I_Z \end{bmatrix} = \begin{bmatrix} 0 & 0 & 0 \\ 1 & 0 & 0 \\ 0 & \pm 1 & 0 \end{bmatrix} \begin{bmatrix} V_Y \\ I_X \\ V_Z \end{bmatrix}$$

(a) Describing I~V matrix

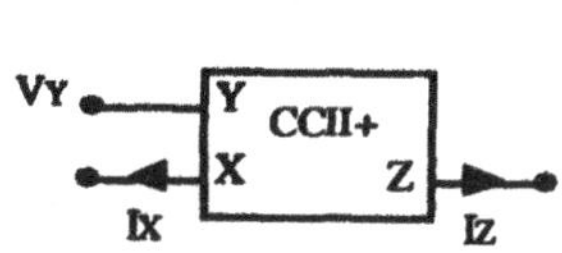

(b) Schematic

Figure 1 The CCII current-conveyor.

The voltage at the low impedance input node *X* follows that at the high impedance input node *Y*, while the input current at node *X* is mirrored or 'conveyed' to the high impedance output node Z. The ± sign indicates the polarity of the output current with respect to the input current; by convention, a positive sign indicates that both the input and output currents simultaneously flow into or out of the device, thus Figure 1b illustrates a CCII+. For the first generation conveyor, or CCI, the input current at node X was reflected to input Y, that is, the two inputs had equal currents. In the case of the second generation conveyor input Y draws no current, and this second generation, or CCII formulation has proved to be much more adaptable and versatile than its first generation predecessor. Because of the combined voltage and current following properties, CCIIs may be used to synthesise a number of analog circuit functions which are not so easily or accurately realisable using voltage op-amps.

At the time of the conception of the current-conveyor in 1968 the semiconductor industry was absorbed with the development of the monolithic voltage op-amp and without any clearly stated advantages there was no motivation for the development of a monolithic current-conveyor. However, since then several hundred papers have been published on the theory and applications of current-conveyors. Some of these application areas are shown in Figure 2 and circuit realisations are given throughout this chapter. As current-conveyors become more readily available and circuit designers become more familiar with the versatile of current-conveyors it is certain that further ingenious uses will be devised.

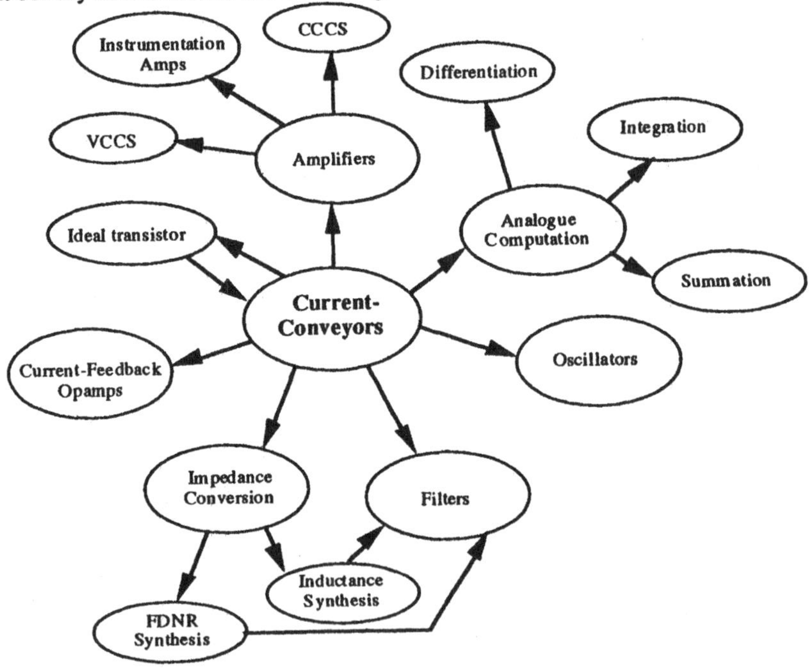

Figure 2 Current-conveyor applications.

11.1.3 Amplifiers built with current-conveyors

The current-conveyor has precise unity current gain between Z and X, rather than the high ill-defined open-loop gain of the voltage op-amp, therefore in amplifier applications the current-conveyor is generally used without any overall negative feedback. The advantage of this approach is that the traditional closed-loop gain-bandwidth conflict of negative feedback voltage op-amp circuits is avoided. Of course, the benefits of global negative feedback, for example noise reduction, improvement in input and output impedance levels, can no longer be exploited, but the absence of overall negative feedback generally results in wider bandwidths at higher levels of gain. However, to maintain a high level of accuracy without the use of negative feedback, a high quality current-conveyor integrated circuit realisation is required.

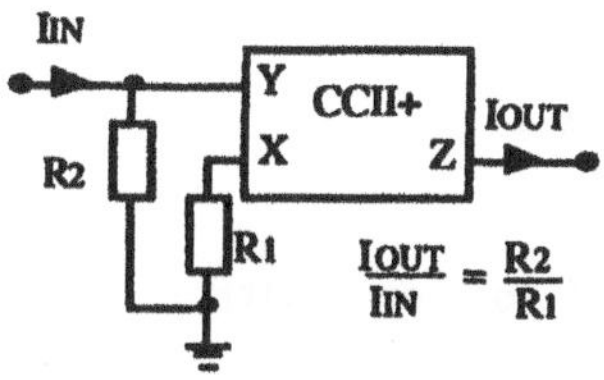

(a) Current amplifier (CCCS)

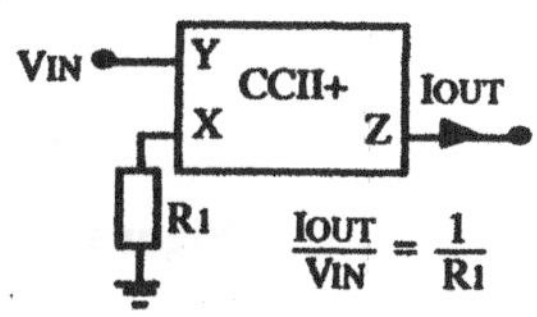

(b) *V* to *I* converter (VCCS)

Figure 3 Current output amplifiers.

The CCII can easily be used to configure the two current output amplifiers, as shown in Figure 3. In Figure 3a, R_2 must be much less than the input impedance at node *Y* for current transfer accuracy, that is, $R_2 << R_{INY}$, and this factor will limit the maximum possible gain of the circuit. In both circuits R_1 forms a pole with parasitic capacitance to ground at node *X*, therefore the value of this resistor should be kept low to ensure that this parasitic pole does not dominate the frequency response.

The current-conveyor can be used to advantage in current-mode circuits and probably the most useful utility building-block being the unity current gain current amplifier, shown in Figure 4. The *X* node has a low input impedance and conveys the input current through to the high output impedance *Z* node. Hence the current gain is unity, negative for a CCII+ and positive for a CCII-. This circuit is the antithesis of a voltage-follower and is referred to as a current buffer or current-follower [3] and can be used to advantage as an interface between voltage and current-mode circuits. For example, adding a current-follower before the amplifier of Figure 3a provides the circuit with a greatly reduced input impedance and improves the integrity of the current drive into the *Y* node of the second current-conveyor.

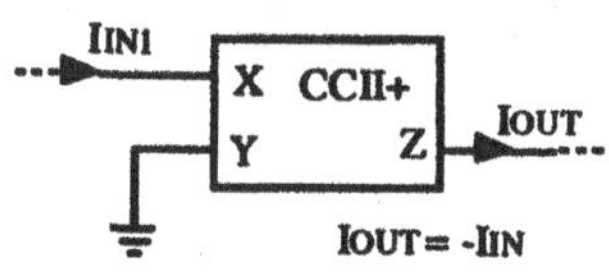

Figure 4 Current-buffer or current-follower.

Voltage amplification can be obtained by adding a voltage buffer to the VCCS of Figure 3b. The circuit of Figure 5 is such an amplifier, with a second CCII+ being used as a voltage-follower.

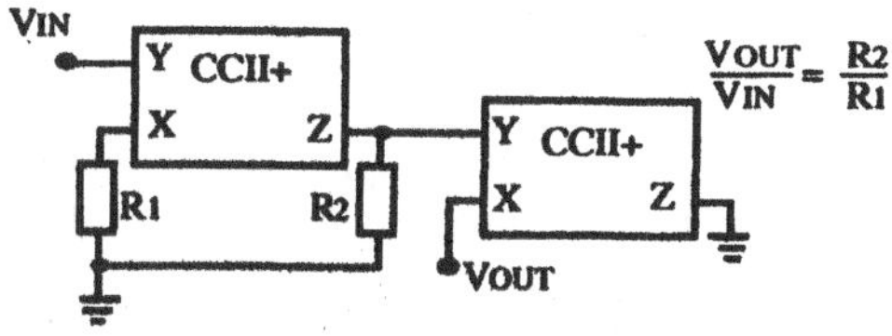

Figure 5 Voltage amplifier.

Using two CCIIs, Toumazou and Lidgey [4] described a differential voltage-to-current converter as shown in Figure 6.

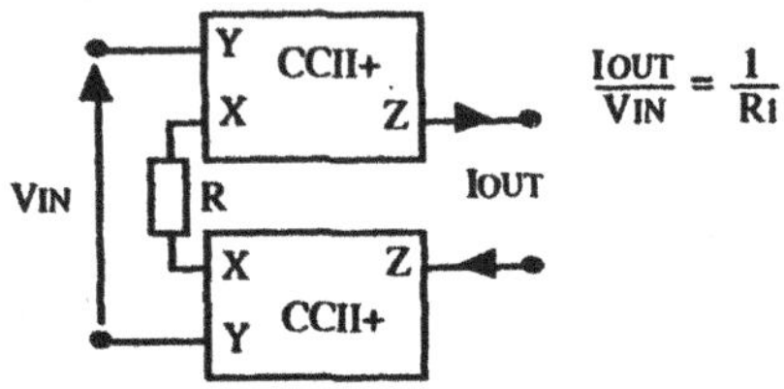

Figure 6 Differential *V* to *I* converter.

Any differential input voltage appears across resistor R_1 to generate a current which is then conveyed to the two outputs. The common-mode gain of the circuit will be zero provided that both devices are well matched and have a precise voltage-following action between their respective *Y* and *X* inputs. The circuit does not rely on any external resistor matching for high CMRR, however the CMRR will roll-off at higher frequencies due to parasitic capacitance to ground at the *X* nodes, which will then give rise to common-mode currents. The CMRR can be improved by increasing the differential gain, that is, by reducing the value of *R*. The ultimate limit on CMRR will be determined by the mismatch in open-loop bandwidth of the two conveyors.

By converting the output current back to a single-ended voltage, this differential transadmittance cell can be extended to produce a high performance instrumentation amplifier [5, 6], as shown in Figure 7. This instrumentation amplifier has a high CMRR, requires no external component matching and exhibits a bandwidth independent of gain. Experimental results of this architecture instrumentation amplifier are included later in section 11.1.10.

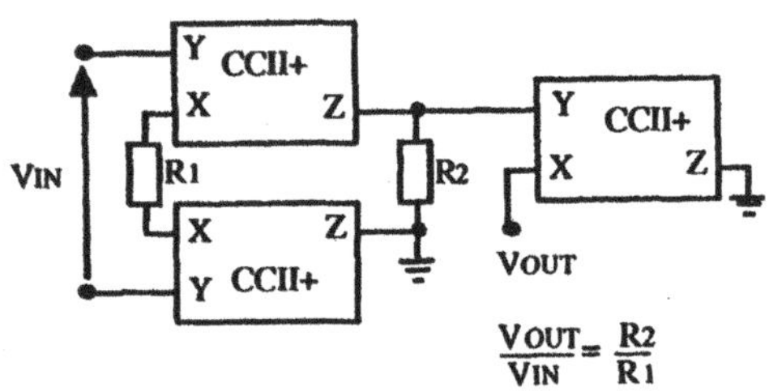

Figure 7 Instrumentation amplifier.

This instrumentation amplifier has proved to be particularly suitable for the implementation of EMG amplifiers with shutdown control [7]. Shutdown control is necessary to isolate the EMG amplifier during the period that an electrical stimulation is applied to the muscles, since this large stimulus will otherwise saturate the amplifier. The amplifier would then require a recovery period before it is able to measure the much weaker EMG muscle response. This shutdown control is most easily implemented by providing a switch control at the amplifier inputs. Conventional EMG amplifiers based on voltage op-amps require switching at a high impedance node, and this tends to cause transient spikes which themselves can saturate the amplifier. Using the instrumentation amplifier of Figure 6, the switching control can be moved to a low impedance node, so that resistor R_1 is either disconnected or connected between the two low impedance *X* inputs. This conveyor-based

EMG amplifier demonstrates no saturation or recovery problems, and is able to measure the EMG signal accurately, immediately after the stimulation pulse.

11.1.4 Analog computation

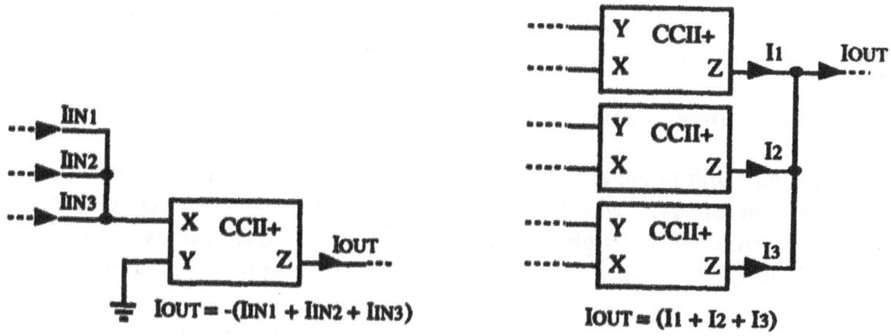

(a) Input summing. **(b) Output summing.**

Figure 8 Current summing with current-conveyors.

Current summation is readily achieved with a current-conveyor, either at the low input impedance *X* node or at the high output impedance *Z* node, as shown in Figure 8.

In their original papers presenting the current-conveyor, Sedra & Smith [1,2] outlined a number of current-mode analog computational elements, ranging from integrators and differentiators to more complex function generators. Current differentiators and integrators are easily implemented, as shown in Figure 9. In Figure 9a, resistor *R* should be fairly small to minimise the effect of stray capacitance at node *X*, meanwhile in Figure 9b, *R* should be limited in size to minimise current transfer errors, that is, $R << R_{INY}$.

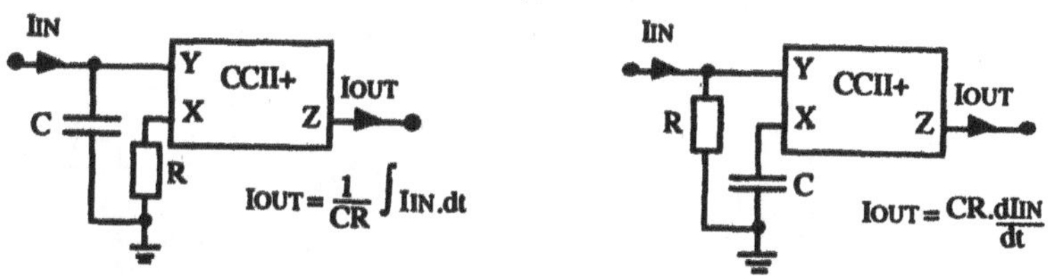

(a) Current integrator. **(b) Current differentiator.**

Figure 9

For voltage differentiation and integration, an additional output voltage buffer is required to supply load current, as shown in Figure 10.

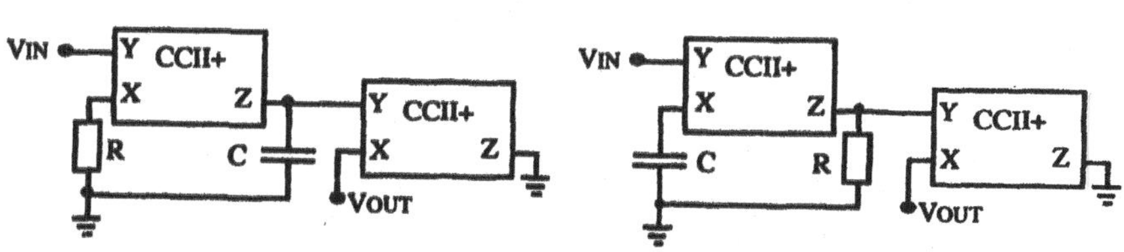

(a) Voltage integrator. **(b) Voltage differentiator.**

Figure 10

In many cases circuits containing voltage op-amps with capacitive feedback components must be designed very carefully since a capacitor contributes phase lag, which can result in circuit instability. However, because of the absence of global negative feedback, these conveyor-based integrator and differentiator circuits do not suffer the same potential instability problems. The circuits of Figures 9 and 10 also utilise grounded rather than floating capacitors, which is an advantage for integrated circuit realisations.

11.1.5 Impedance conversion

Much of the research in the area of current-conveyors has focused on the implementation of impedance converters. Many voltage op-amp-based impedance converters require complicated circuitry and tight component matching requirements. With its combined voltage and current following properties, the CCII is a more flexible alternative to implement these functions.

Impedance converters are used in many areas of active circuit design, for example in active filters and inductance simulation. The frequency dependent negative resistor, or FDNR, is a circuit element used in higher order filter design which is synthesised using impedance converters. Components with negative impedance are used to cancel out positive lossy components in certain circuits by generating an identical negative component and these components also are simulated using a negative impedance converter, [4, 5].

11.1.5.1 Grounded negative impedance converters (NIC)

Figure 11a shows a negative impedance converter presented by Sedra & Smith [2]. By terminating input *Y* with an impedance Z_T, as shown in Figure 11b, a current-controlled NIC, or negative admittance converter is obtained. Similarly, by terminating input *X* with an impedance of Z_T, as shown in Figure 11c and driving input *Y* with a voltage, a voltage controlled NIC is obtained.

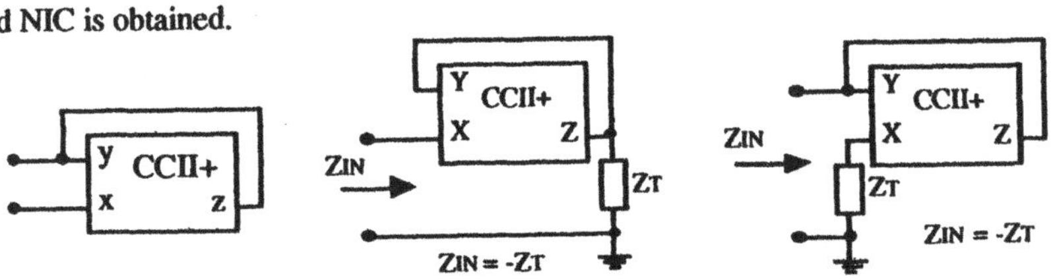

(a) NIC (b) Current-controlled (c) Voltage-controlled

Figure 11 Grounded negative impedance converters.

11.1.5.2 Grounded generalised impedance converter (GIC)

Using two CCIIs as shown in Figure 12, more general impedance conversion can be implemented, depending on the nature of the circuit components.

If both conveyors are of equal polarity, that is both CCII+ or CCII-, then a negative GIC is obtained. By terminating V_2 with a grounded impedance Z_3, the resulting input impedance at V_1, $Z_{IN} = -(Z_1Z_2/Z_3)$. Conversely, if the two conveyors are of opposite polarity then a positive GIC is obtained, and the addition of a grounded impedance Z_3 at V_2 will result in an impedance $Z_{IN} = (Z_1Z_2/Z_3)$ at V_1. By choosing Z_1 and Z_2 as resistors and Z_3 as a capacitor as shown in Figure 13, a grounded inductor where $L = R_1R_2C$ can be implemented. Similarly a grounded FDNR can be obtained by choosing Z_1 and Z_2 as capacitive and Z_3 as resistive.

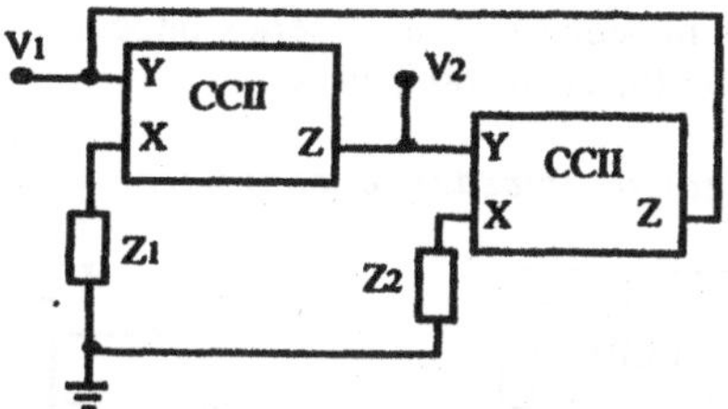

Figure 12 Generalised impedance converter (GIC).

More recently, systematic methods have been presented for the implementation of any required immitance function using current-conveyors. By extending the circuit of Figure 12, several researchers [8-11] present methods for synthesising series and parallel immitance combinations, and also implement immitance functions of any order by 'nesting' stages of the basic circuit. These complex immitance functions can then be used to implement filter circuits which are practically insensitive to active parameter variations.

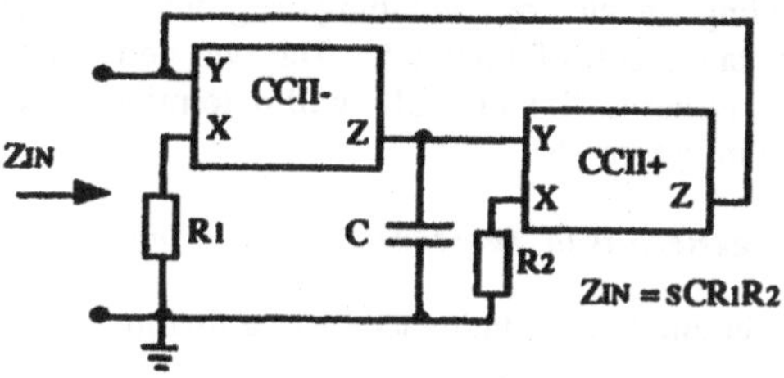

Figure 13 Simulated grounded inductance.

Himura et al [13, 14] have presented series and parallel impedance simulators which require only a single CCII and Figure 14 shows such a series impedance simulator.

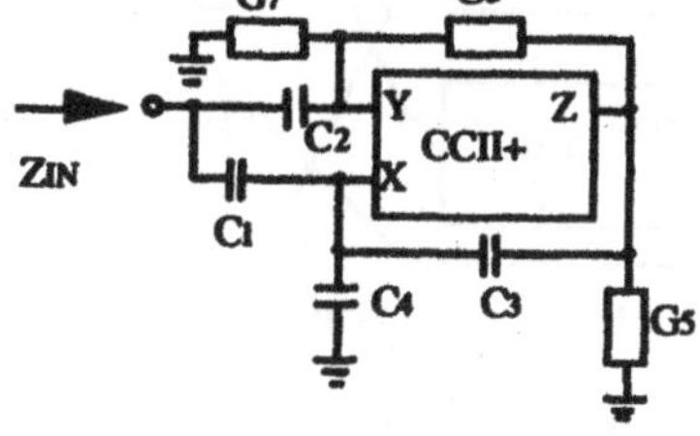

Figure 14 Series impedance simulator.

When the matching condition $C_4G_6 = 2C_3G_7$ is met, then the input impedance of the circuit becomes a series connection of inductance L_S, resistance R_S and capacitance C_S, where

$$L_S = (2C_2C_3) / [(C_1 + C_2)(G_7G_5 + G_6G_7 + G_5G_6)]$$

$$R_S = (G_5C_2 + G_6C_2 - G_6C_1) / [(C_1 + C_2)(G_7G_5 + G_7G_6 + G_5G_6)]$$

$$C_S = (C_1 + C_2).$$

By altering these component values, the value of R_S can be made positive, negative or zero, and thus the circuit of Figure 14 can implement either an *L-C-R* series impedance, an *L-C* series impedance ***(note that this configuration does requires tight component matching)***, or an *L-C-negative R* series impedance. These series impedances are useful in designing circuits such as active ladder filters and oscillators.

11.1.5.3 Floating negative impedance converter

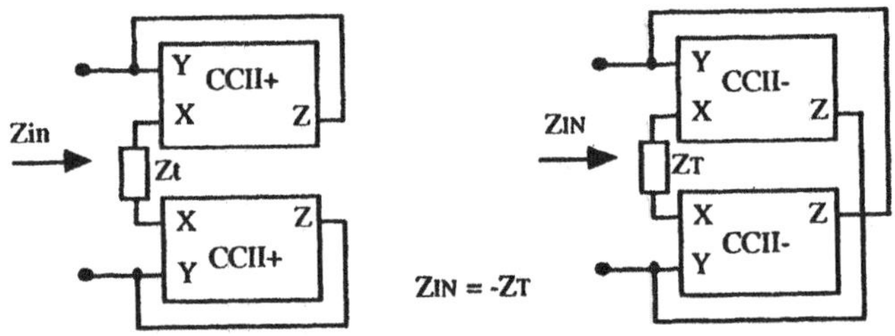

(a) Floating NIC using CCII+ (b) Floating NIC using CCII-

Figure 15 Floating negative impedance converters.

A floating NIC can be implemented by combining the grounded NIC of Figure 11c with the differential transadmittance cell of Figure 6. The two conveyors must have the same polarity; the choice of either CCII+ or CCII- will determine the nature of the feedback connections, as shown in Figure 15.

11.1.5.4 Floating generalised impedance converter

By combining two differential transadmittance cells, a floating GIC can be implemented as shown in Figure 16.

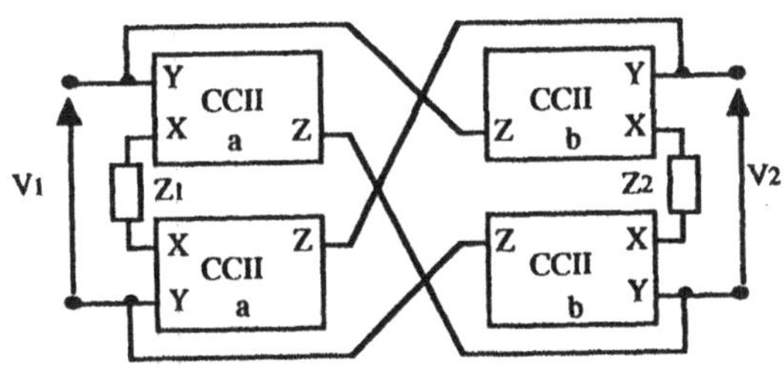

Figure 16 Floating GIC.

If all the conveyors are of equal polarity, that is, all CCII+ or all CCII-, then a positive GIC is obtained. By connecting an impedance Z_3 across V_2, the resulting impedance at V_1, is given by $Z_{IN} = (Z_1Z_2/Z_3)$. If the two conveyors marked *'a'* are of equal polarity while those marked *'b'* are of the opposite polarity, then a floating negative GIC is obtained. Alternatively, a negative GIC can be implemented by reversing one set of feedback connections in the circuit of Figure 16, as shown in Figure 17. This negative GIC implementation has the advantage that all the conveyors are of the same polarity. The circuit would therefore be especially suitable for monolithic fabrication, since four devices of the same type on the same silicon wafer would have a high degree of matching.

Floating inductors and FDNR implementations based on only two CCIIs have also been reported [8-10]. These two-conveyor implementations generally require an increased number of external components, many of which have to be tightly matched, and in some cases they also suffer from high component sensitivities. One circuit with low component sensitivities is the two-conveyor implementation reported by Senani [15], and implemented practically by Wilson [16], which is shown in Figure 18.

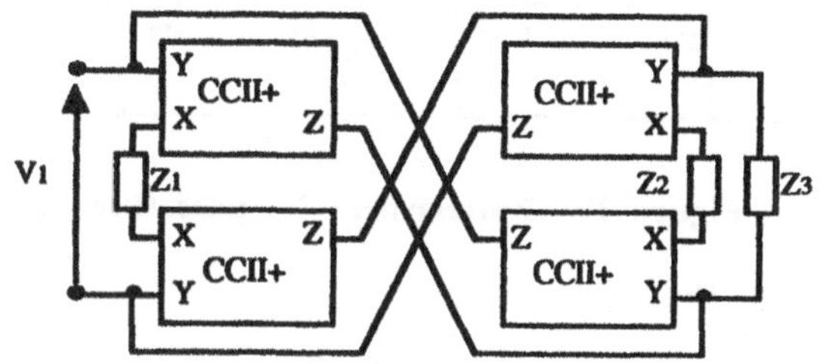

Figure 17 Floating negative GIC.

With either Z_2 or Z_4 capacitive and all other elements resistive, a floating inductor can be synthesised, where $L = R_1C_2R_3R_5/R_4$.

With two capacitive components (Z_1 and Z_3 or Z_1 and Z_5 or Z_3 and Z_5), and all other elements resistive, a floating FDNR is obtained, that is

$$Z_{IN} = -\omega^2 C_1 C_3 R_5 / R_2 R_4.$$

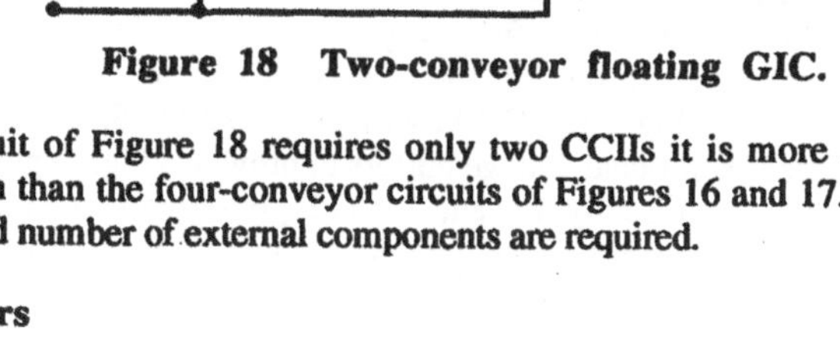

Figure 18 Two-conveyor floating GIC.

Since the circuit of Figure 18 requires only two CCIIs it is more suitable for discrete implementation than the four-conveyor circuits of Figures 16 and 17. However in Figure 18, an increased number of external components are required.

11.1.6 Filters

Current-conveyors have received considerable attention from designers concerned with the implementation of active filters, since voltage op-amp-based filters are bandwidth limited by the voltage op-amps themselves. Many of the earlier CCII-based filter designs required floating capacitors, had poor sensitivity values and were only able to realise one particular filter type [8-10]. The emergence of GICs described in the previous section allowed for a wider range of filters to be implemented based on simulated inductors, for example, elliptic ladder filters as well as various types of Butterworth filters.

A 3rd-order low-pass Butterworth filter reported by Nandi [17] which is particularly attractive, due to its use of grounded capacitors and equal-valued components, is shown in Figure 19.

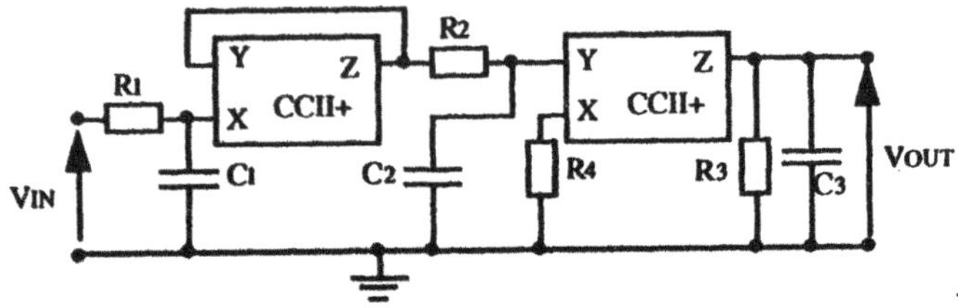

Figure 19 Butterworth lowpass filter.

For the circuit of Figure 19

$$\frac{V_{OUT}}{V_{IN}} = \frac{R_3 / R_4}{a_3 s^3 + a_2 s^2 + a_1 s + 1}$$

where $a_3 = R_1C_1R_2C_2R_3C_3$, $a_2 = R_1C_1R_2C_2 + R_2C_2R_3C_3 + R_1R_3C_3(C_1\text{-}C_2)$,

and $a_1 = R_1(C_1\text{-}C_2) + R_2C_2 + R_3C_3$.

With equal-valued components, that is $R_1=R_2=R_3=R$ and $C_1=C_2=C_3=C$, a 3rd-order transfer function is obtained

$$\frac{V_{OUT}}{V_{IN}} = \frac{1}{R^3C^3s^3 + 2R^2C^2s^2 + 2RCs + 1}$$

The resulting filter is practically insensitive to active parameters, that is insensitive to the current tracking errors of the current-conveyors.

Toumazou and Lidgey [18] proposed and demonstrated the first universal CCII biquad filter, based on an earlier circuit reported by Nawrocki and Klein [19]. The implementation used only grounded capacitors and resistors, but required seven CCIIs. Depending on how the CCIIs are realised in practice, this relatively large number of current-conveyors used to achieve a 2nd-order filter response can be reduced significantly using hardware reduction techniques in the full filter implementation. A similar performance universal active filter requiring three CCIIs was proposed by Singh and Senani [20], however this circuit needs additional voltage buffers to provide a high input impedance and low output impedance. More recently, Lui and Tsao [21] have presented a family of biquad filters with high input impedance which require only a single CCII and four external passive components. Figure 20 shows the general circuit implementation.

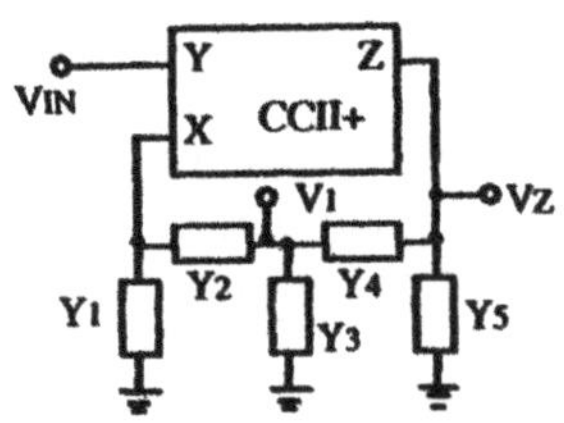

$$\frac{V_Z}{V_{IN}} = \frac{-Y_1(Y_2+Y_3+Y_4) + Y_2Y_3}{Y_5(Y_2+Y_3+Y_4) + Y_3Y_4}$$

$$\frac{V_1}{V_{IN}} = \frac{Y_1Y_4 - Y_2Y_5}{Y_5(Y_2+Y_3+Y_4) + Y_3Y_4}$$

Figure 20 General biquad filter architecture.

The nature of the admittances Y_1 to Y_5 determines the filter response. For example, with $Y_1=0$, $Y_2=1/R_2$, $Y_3=SC_3$, $Y_4=sC_4$ and $Y_5=1/R_5$, a low-pass characteristic is obtained at V_1 and a band-pass characteristic is obtained at V_Z, with

$$\omega_o = (C_3C_4R_2R_5)^{-1/2} \quad \text{and} \quad Q = \frac{1}{C_3+C_4}(R_5C_3C_4/R_2)^{1/2}.$$

High-pass, band-pass, all-pass and notch filters can also be implemented with selection of the appropriate admittances Y_1 to Y_5, and all filters exhibit a low sensitivity to the CCII current tracking error, and practically zero sensitivity to the voltage tracking error.

The filter designs discussed so far have been voltage-processing circuits; that is, they are an alternative method of implementing voltage op-amp-based filter designs. However, because of its current-following capability, the CCII is also able to implement current-processing filter circuits [22-26]. Similar to their voltage-mode biquad filters, Lui et al [23] have presented a family of current-mode biquads, based on a single CCII with five external passive components. The general filter configuration is shown in Figure 21.

Again selection of the appropriate admittances Y_1 to Y_5 gives the required filter response. For example, with $Y_1=1/R_1$, $Y_2=\infty$, $Y_3=1/R_3$, $Y_4=sC_4$ and $Y_5=sC_5$ results in a low-pass filter with $\omega_o = (R_1R_3C_4C_5)^{-1/2}$ and $\omega_o/Q = (C_4+C_5)/R_1C_4C_5$. Depending on the value of the admittances, an input current buffer may be necessary to prevent excessive loading of the current source input.

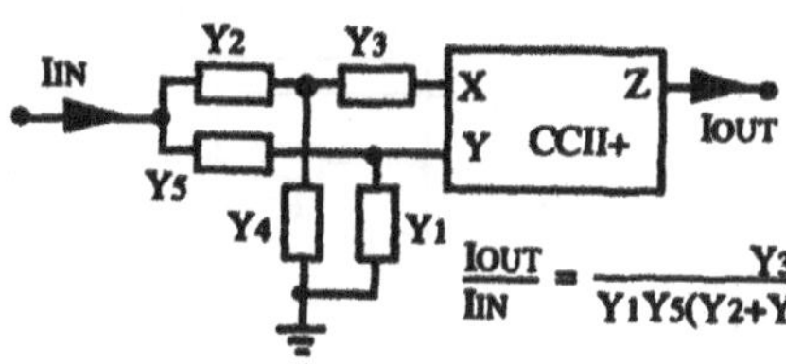

$$\frac{I_{OUT}}{I_{IN}} = \frac{Y_3(Y_1Y_2-Y_4Y_5)}{Y_1Y_5(Y_2+Y_3+Y_4)+Y_1Y_2(Y_3+Y_4)+Y_2Y_4Y_5}$$

Figure 21 General current-mode biquad filter.

A universal active current filter based on a single CCII has been presented by Chang [25], as shown in Figure 22.

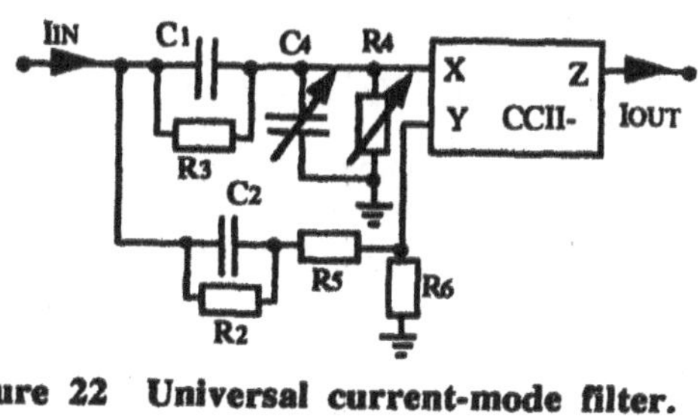

Figure 22 Universal current-mode filter.

With R_2 open circuit, this architecture can implement notch, all-pass and low-pass filter characteristics by altering the values of C_4 and R_4. With resistor R_2 present, band-pass and high-pass characteristics can be obtained, again by tuning the values of C_4 and R_4. The resulting filters have low passive sensitivities, and are almost insensitive to the current tracking error of the conveyor. An alternative method of deriving current-filtering circuits from already existing voltage filters has been presented by Roberts and Sedra [26].

Many voltage-mode *active-RC* filters use op-amps configured as unity gain buffers to maximise the useable bandwidth of the circuit, and current-mode filters can be derived from these voltage buffer-based filters using the adjoint circuit method. To find the adjoint circuit of any network, a replica of the original circuit must be constructed first. All elements in this replica circuit are then replaced by their adjoint elements; passive components such as resistors and capacitors remain the same, while unity gain voltage buffers are replaced by unity gain current buffers, or current-conveyors. Voltage input and output signals are replaced by current output and input signals, so the two adjoint circuits are exactly equivalent in the voltage-mode and current-mode. The new current-mode circuit will have the same component sensitivities as the voltage-mode original, so low sensitivity current-mode circuits can be obtained from low sensitivity voltage-mode circuits which have already been developed. The Sallen-Key active biquad (SAB) of Figure 23a would be therefore transformed to the CCII-based circuit of Figure 23b by the adjoint circuit method.

It should be noted that in practice an input DC current path must be provided for all active devices employed as for all analog circuits and in the circuits described in the Impedance Converter and Filter sections above any such additional DC bias components have not been included in the circuit diagrams.

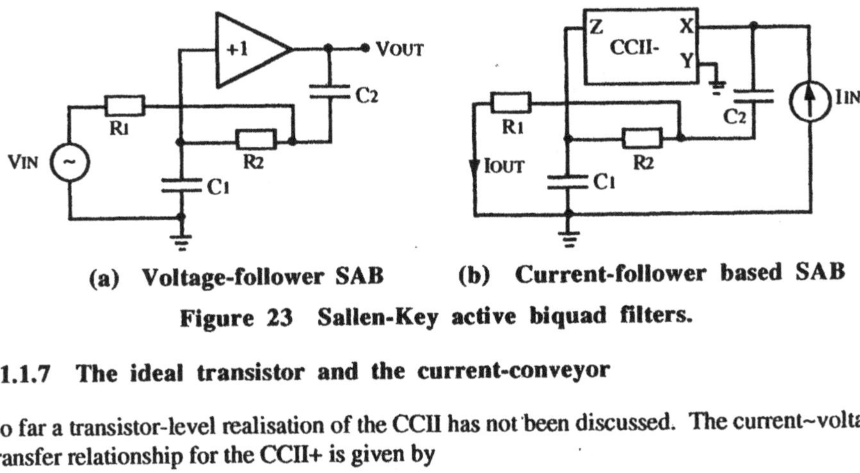

(a) Voltage-follower SAB **(b) Current-follower based SAB**

Figure 23 Sallen-Key active biquad filters.

11.1.7 The ideal transistor and the current-conveyor

So far a transistor-level realisation of the CCII has not been discussed. The current~voltage transfer relationship for the CCII+ is given by

$$V_X = V_Y, \; I_Y = 0 \; \text{ and } \; I_Z = I_X.$$

These equations show that there is a simple voltage-following action between input node Y and output node X and that there is a simple current-following action between input node X and output node Z. Also these characterisic equations tell us that the impedance relationship for the current-conveyor are

$$Z_{INY} = \infty, \; Z_X = 0 \; \text{ and } \; Z_{OUTZ} = \infty.$$

Figure 24 shows a schematic representation of a CCII- built with a single BJT and on reflection it is clear that the current-conveyor is effectively an ideal transistor, with infinite β and infinite g_m.

Driving into the base of a BJT gives almost unity voltage gain from input base to output emitter, with high input impedance and low output impedance, and driving into the emitter of a BJT gives almost unity current-gain from emitter input to collector output, with low input impedance and high output impedance. Drawing the comparision further the high input impedance *Y* node corresponds to the base (or gate) of a transistor, the low input impedance *X* node corresponds to the emitter (or source) of a transistór, and the high output impedance *Z* node corresponds to the collector (or drain) of a transistor. Clearly one transistor cannot function alone as a complete current-conveyor as an unbiassed single transistor can only handle unipolar signals and the high accuracy unity voltage and unity current gain required for a high performance current-conveyor cannot be obtained. However the generic relationship between the current-conveyor is valid and it does provide valuable insight into the development and operation of monolithic current-conveyors described in the next section.

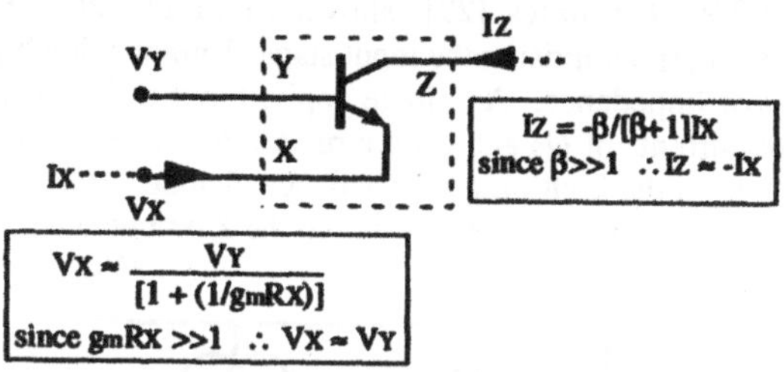

Figure 24 Single BJT CCII-.

11.1.8 Supply-current sensing, current-feedback op-amps and the CCII

Over the last decade current-converter research and current-conveyor research has proceeded in parallel with little cross reference, despite the fact that the two areas are very closely related. The difference between the two research areas is really only one of semantics and not substance, as the current-conveyor, or part current-conveyor, can be identified easily in nearly all the current-converter work. Toumazou and Lidgey worked extensively on current-converter realisations based on the technique of power supply-current sensing on voltage-mode op-amps, described in [3, 5], to achieve current output performance. The circuit performances obtained were extremely encouraging, with useful practical circuits reported on current-followers and follower-based amplifiers, high speed precision rectifiers, instrumentation amplifiers, and current amplifiers [5, 27].

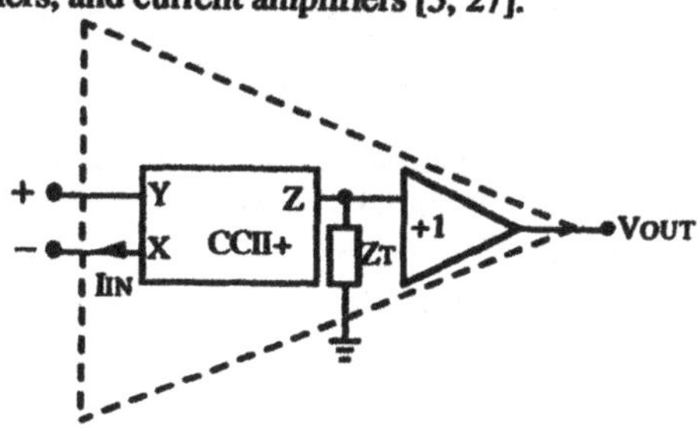

Figure 25 Current-feedback op-amp structure.

Monolithic current-feedback op-amps have relatively recently become commercially available as high-speed alternatives to voltage op-amps [28]. Examination of the internal structure of the current-feedback op-amp is revealing as the device is essentially a CCII+ with the internal high impedance *Z* node connected to an output voltage-follower, as shown in Figure 25. Any current flowing at the low impedance inverting input is conveyed to the gain node (Z_T), and the resulting voltage is buffered to the output. Z_T is thus the open-loop transimpedance gain of the current-feedback op-amp, which in practice is equal to the parallel combination of the CCII+ output impedance, the voltage buffer input impedance and any compensation capacitance at the gain or Z_T node.

Generally the gain node is not connected to an external pin and so the Z node of the CCII+ cannot be accessed. However, simply removing the output buffer from a current-feedback op-amp is all that is required to obtain a high performance CCII+.

11.1.9 CCII01 current-conveyor

The first commercially available current-conveyor based on the current-feedback op-amp is the CCII01 from LTP Electronics [29], shown in Figure 26. The circuit is fully symmetrical, with Q_1 to Q_4 comprising the input stage, Y input being high impedance and X input/output being low impedance. A voltage applied to the Y input is followed by the X input. The collector currents of Q_2 and Q_3 are reflected and recombined through current-mirrors CM_1 and CM_2 to the high output impedance Z node. Hence any input into the X node is conveyed with unity current gain through to the Z output node.

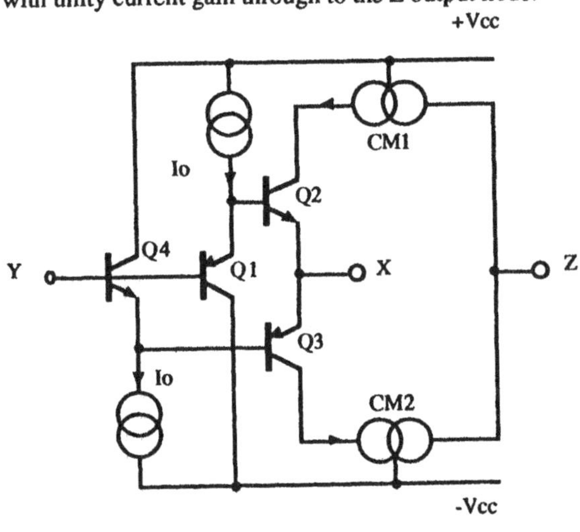

Figure 26 Internal schematic of the CCII01 current-conveyor.

The CCII01 is built on a high speed dielectric isolation fully complementary bipolar process and supplied as dual device in an 8-pin DIL package. The device features an equivalent slew-rate of 2000 V/μs and a 100 MHz bandwidth. The equivalent open-loop gain is 80 dB and the CMRR performance is better than 53 dB at 1 MHz. The maximum output current from the device is ±10 mA and it operates from ±5 V to ±15 V supplies.

11.1.10 IA and PFWR application examples of the CCII01

11.1.10.1 CCII01 instrumentation amplifier

A particularly useful and elegant application of the current-conveyor is the current-mode instrumentation amplifier (IA) employing the differential V to I converter of Figure 6. The circuit schematic and measured CMRR performance is shown below in Figure 27. The key advantage of this architecture IA is the high CMRR achievable at high frequencies without any requirement for an accurately matched resistor network.

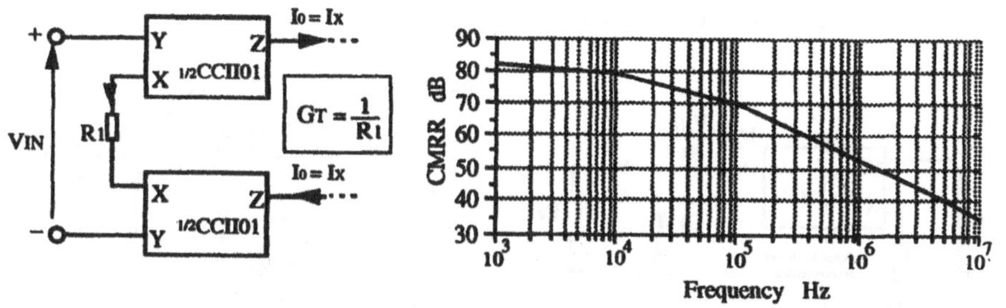

Figure 27 CCII01 IA and measured CMRR performance.

The output signal is taken from either Z node. Conversion of the output Z node current into a voltage is readily achieved using the conventional inverting transresistance stage shown in Figure 28. Alternatively the ouput signal current can be taken from the two conveyor Z nodes and fed into a differential tranresistance amplifier to achieve double the gain.

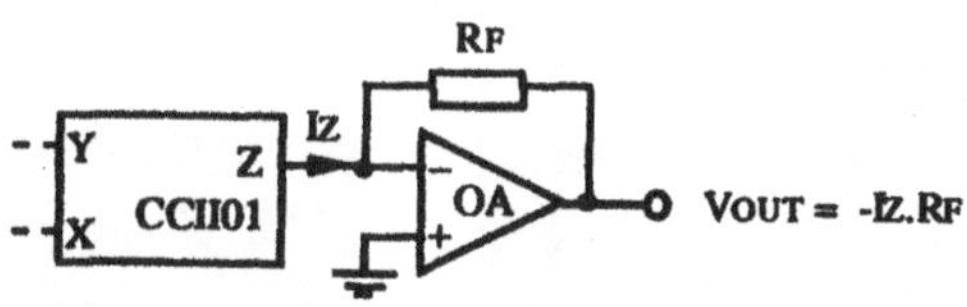

Figure 28 Transresistance conversion of I_Z.

11.1.10.2 CCII01 precision full-wave rectifier

Another equally elegant use of the current-conveyor is for high speed precision rectification. The classical problem with conventional precision rectifiers based on diodes and op-amps is that during the non-conduction/conduction transition of the diodes the op-amps have to recover with a finite small-signal *dV/dt* resulting in significant distortion during the zero-crossing of the input signal. Using high slew-rate op-amps does not solve this fundamental drawback since it is a small-signal transient problem [30]. Conventional rectifiers are thus limited to a frequency performance well below the gain-bandwidth product of the amplifier.

Improvements [30] have been made to rectifier high frequency performance by the use of current-mode techniques primarily based upon employing the power supply rails of the op-amp as a current rectification path. However, a problem encountered with such schemes is that signal levels need to be significantly higher than the supply bias to guarantee precision rectification at high frequency and so again loss of precision occurs at signal zero-crossings. Even using high speed current-feedback amplifiers the performance is still limited to some tens of kiloHertz [30], which is significantly below the f_T of the current-feedback amplifiers used.

A full-wave precision rectifier can be configured easily using two CCIIs, as shown in Figure 29. Both the CCIIs form a differential voltage to current converter such that during the positive input cycle, the output currents of value V_{IN}/R flow out of the Z-node of CCII(a) and into the Z-node of CCII(b), thus making only D_4 and D_2 active. Because D_2 is active, the current from the Z-node of CCII(a) flows into the output resistor R making $V_{OUT} = V_{IN}$. During the negative input cycle, only D_3 and D_1 are active thus the output current of

CCII(b) is driven into R making $V_{OUT} = V_{IN}$. Clearly the magnitude of gain is R_2/R_1 and this can be increased from unity by making $R_2 > R_1$.

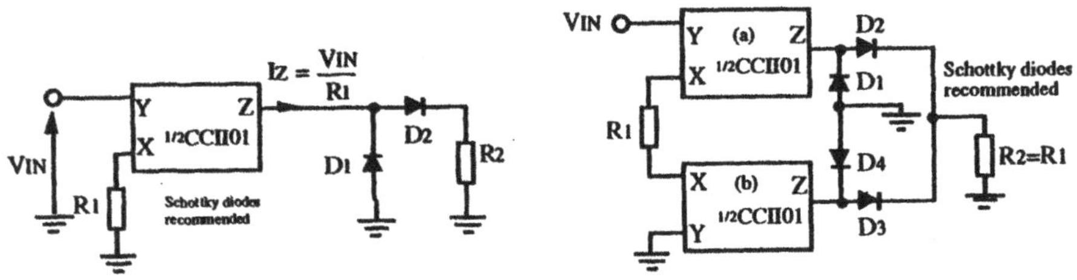

(a) Half-wave precision rectifier **(b) Full-wave precision rectifier**

Figure 29 Current-conveyor precision rectifiers.

The circuit was built with 100 Ω resistors and Schottky diodes and Figure 30 shows typical performance for the half-wave precision rectifier shown in Figure 29a at two operating frequencies, 100 kHz and 1 MHz.

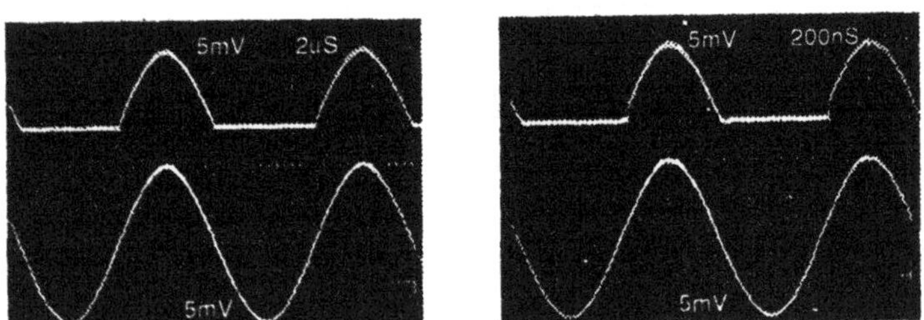

● V_{IN}=100mVpk at 100kHz ● V_{IN}=100mVpk at 1MHz

All measurements taken with a x10 probe

Figure 30 Precision half-wave rectifier.

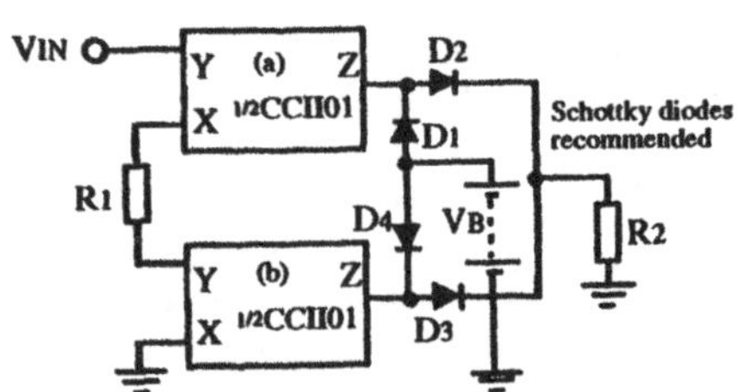

Figure 31 Improved current-conveyor PFWR

The performance is good but with the CCII01 exhibiting an f_T of approximately 100 MHz, one would naturally expect the circuit to work close to the f_T of the device. However, this is not the case since at very low signal levels all the diodes are off, and as a result the differential voltage to current converter is transformed into a high gain differential voltage amplifier. Although the CCII01 exhibits a very high slew-rate, in the region of 2000 V/μs, it is the small signal *dV/dt* that limits performance at zero-crossings. The solution to this problem is to modify the circuit by offsetting the output of the conveyors by biassing all the diodes appropriately. The new scheme [31] is shown in Figure 31.

In the circuit, the voltage at the anode of diodes, D_1 and D_4, is biassed by a low impedance voltage source, V_B, at approximately 0.6 V, allowing 0.3 V forward bias for each Schottky diode. Thus when D_1 and D_4 conduct the voltage at the subsequent Z-terminal is approximately +0.3 V and the output voltage is zero. This condition ensures that the load impedance presented to the Z–terminal is kept low at all times, especially as the diode pairs D_1/D_3 and D_2/D_4 swap conduction roles. The net effect is that the current-conveyor outputs do not go into slew at each zero-crossing. Effectively the improved scheme provides a class AB voltage bias so that all the diodes are on the edge of conduction during the zero input condition. Note that the battery, V_B, may be replaced by an appropriately biassed diode voltage source. The improved rectifier operates well up to a frequency of 30 MHz, as shown in [31]. Quite an acceptable rectifier performance is achieved and clearly is substantially less distorted than the non-diode biassed circuit which begins to show significantly poor performance at about 4 MHz. Beyond 30 MHz, the rectifier integrity of the final circuit does start to deteriorate. This is due to the roll-off in V_X/V_Y with frequency. However the measured results are encouraging and represent a significant improvement.

11.1.11 The future of the current-conveyor

Although 20 years old, the current-conveyor is a circuit concept whose time has now come. There is no doubt that the current-conveyor is both versatile and convenient for many applications. The applications described demonstrate the ease with which complex analog functions can be realised using current-conveyors. Often fewer parts are needed than the op-amp counter part, due to the very attractive combined voltage/current capability of the current-conveyor. There are many applications where op-amp realisations are significantly less practical than the equivalent current-conveyor realisation. Further more, the current-conveyor can provide high performance operation, particularly in terms of speed and bandwidth, due to the inherent local feedback of the follower-based structure of the device.

Commercially available current-conveyors are a welcomed and valuable addition for the application engineer, complementing the ubiquitious op-amp. The semiconductor industry has now recognised the potential of current-mode techniques, attested by the successful current-feedback op-amp which is a close relative to the curent-conveyor. All the ingredients for a high performance current-conveyor realisation are already accessable to the semiconductor industry, so it is only a matter of time before the current-conveyor takes its place as a standard part in the application designers tool-kit.

Acknowledgements

The authors would like to acknowledge the design staff of Elantec Inc. USA for technical support relating to this work.

References

[1] K. C. Smith and A. Sedra, 'The Current-Conveyor - A New Circuit Building Block', *IEEE Proc.*, 56, pp. 1368-1369, 1968.

[2] A. Sedra and K. C. Smith, 'A Second Generation Current-Conveyor and its Applications', *IEEE Trans.*, CT-17, pp. 132-134, 1970.

[3] F. J. Lidgey and C. Toumazou, 'Current-followers and a universal operational amplifier', *Electronics World and Wireless World*, 90 (1577), pp. 40-43, 1984.

[4] C. Toumazou and F. J. Lidgey, 'Floating Impedance Convertors Using Current-Conveyors', *Electron. Lett.*, 21 (15), pp. 640-642, 1985.

[5] C. Toumazou, F. J. Lidgey and C. A. Makris, 'Extending Voltage-Mode Op-amps to Current-Mode Performance', *IEE Proc.* Pt. G, 137 (2), pp. 116-130, 1990.

[6] C. Toumazou and F. J. Lidgey, 'Novel Current-Mode Instrumentation Amplifier', *Electron. Lett.*, 25 (3), pp. 228-230, 1989.

[7] E. Bruun and E. U. Haxthausen, 'Current-conveyor Based EMG Amplifier with Shutdown Control', *Electron. Lett.*, 27 (23), pp. 2172-2174, 1991.

[8] U. Kumar, 'Current-Conveyors : A Review of the State of the Art', *IEEE Circuits and Syst. Mag.*, 3, pp. 10-14, 1981.

[9] U. Kumar and S. K. Shukla, 'Recent Developments in Current-Conveyors and Their Applications', *Microelectron. J.*, 16, pp. 47-52, 1985.

[10] B. Wilson, 'Recent Developments in Current-Conveyors and Current-Mode Circuits', *IEE Proc.*, Pt. G, 137 (2), pp. 63-77, 1990.

[11] M. Ishida, M. Higashimura, Y. Fukui and K. Ebisutani, 'Synthesis of Immittance Function Using Current-Conveyors', *Proc. ISCAS*, pp. 2681-2684, 1988.

[12] M. Higashimura and Y. Fukui, 'Novel Method for Realising Higher-Order Immittance Function Using Current-Conveyors', *Proc. ISCAS*, pp. 2677-2680, 1988.

[13] A. Himura, Y. Fukui and M. Ishida, 'Series Impedance Simulators Using One CCII', *Electron. Lett.*, 26 (4), pp. 269-270, 1990.

[14] A. Himura, Y. Fukui, M. Ishida and M. Higashimura, 'Immittance Function Simulator Using a Single Current-Conveyor', *Trans. IEICE*, Japan, J72-E (12), 1989.

[15] R. Senani, 'Floating Ideal FDNR Using Only Two Current-Conveyors', *Electron. Lett.*, 20 (5), pp. 205-206, 1984.

[16] B. Wilson, 'Floating FDNR Employing a New CCII- Conveyor Implementation', *Electron. Lett.*, 21 (21), pp 996-997, 1985.

[17] R. Nandi, 'Equal-Valued Earthed-Capacitor Realisation of a Third-Order Lowpass Butterworth Characteristic Using Current-Conveyors', *Electron. Lett.*, 14 (22), pp. 215-216, 1978.

[18] C. Toumazou and F. J. Lidgey, 'Universal Active Filter Using Current-Conveyors', *Electron. Lett.*, 22 (12) pp. 662-664, 1986.

[19] R. Nawrocki and U. Klein, 'New OTA-Capacitor Realisation of a Universal Biquad', *Electron. Lett.*, 22 (1), pp. 50-51, 1986.

[20] V. K. Singh and R. Senani, 'New Multifunction Active Filter Configuration Employing Current-Conveyors', *Electron. Lett.*, 26 (21), pp. 1814-1816, 1990.

[21] S. I. Liu and H. W. Tsao, 'The Single CCII Biquads with High Input Impedance', *IEEE Trans.*, CAS-38, pp. 456-461, 1991.

[22] A. Fabre, F. Martin and M. Hánafi, 'Current-Mode Allpass/Notch and Bandpass Filters with Reduced Sensitivities', *Electron. Lett.*, 26 (18), pp. 1495-1496, 1990..

[23] S. I. Liu, H.-W. Tsao and J. Wu, "Cascadable Current-Mode Single CCII Biquads', *Electron. Lett.*, 26 (24), pp. 2005-2006, 1990.

[24] M. Alami and A. Fabre, 'Insensitive Current-Mode Bandpass Filter Implemented from Two Current-Conveyors', *Electron. Lett.*, 27 (11), pp. 897-899, 1991.

[25] C.-M. Chang, 'Universal Active Current Filters Using Single Second-Generation Current-Conveyor', *Electron. Lett.*, 27 (18), pp. 1614-1617, 1991.

[26] G. W. Roberts and A. S. Sedra, 'All Current-Mode Frequency Selective Circuits', *Electron. Lett.*, 25 (12), pp. 759-761, 1989.

[27] C. Toumazou and F. J. Lidgey, 'Universal Current-Mode Analogue Amplifiers', Chapter 4 in *Analogue IC design: the current-mode approach*, edited by C. Toumazou, F. J. Lidgey and D. G. Haigh, IEE Circuits and Systems Series 2, Peter Peregrinus, April 1990.

[28] D.F. Bowers, 'A Precision Dual Current-Feedback Operational Amplifier', *IEEE Bipolar Circuits and Technology Meeting*, pp. 68-70, 1988.

[29] Data sheet, 'CCII01 Current-Conveyor Amplifier ', LTP Electronics Limited, 27 Park End Street, Oxford, OX1 1HU, UK., tel/fax +44-865-200767.

[30] F. J. Lidgey, K. Hayatleh and C. Toumazou, 'New Current-Mode Precision Rectifiers', *Proc IEEE Int Symp on Circuits and Systems*, Chicago, pp. 1322-1325, May 1993.

[31] C. Toumazou, F. J. Lidgey and S Chattong, 'High Frequency Current-Conveyor Precision Full-Wave Rectifier', *Electron. Lett.*, 13 (10), 1994.

Chapter 11.2

Practical Integrated Current-Conveyors

Alison Payne and Chris Toumazou

11.2.1 Introduction

The linear analog semiconductor industry is currently witnessing dramatic changes. Systems which were once analog are now almost totally digital, bringing new levels of precision, reliability and flexibility. System designers are now looking to the analog sections of the circuit to provide the same level of performance, and applications engineers are becoming increasingly demanding in their choice of parts.

Let us imagine that you, an academic, have spent some time researching an idea which was first proposed over twenty years ago. This idea is the current-conveyor, a device which boasts hundreds of applications in the literature. Your theory shows that circuits based on the current-conveyor have the potential to outperform more conventional circuit implementations based on commercial parts (such as the voltage op-amp). In many situations, using a current-conveyor can reduce the complexity of the system implementation. So now you want to try and sell this idea to industry. What's the input offset? What's the noise floor? What's the gain accuracy? These are the questions that you are likely to face from potential marketing directors, whose primary concern is 'will it sell?'. After twenty years of research, only two current-conveyors are presently commercially available, produced by small entrepreneurial companies who believe they have seen the light.

It is our view that the current-conveyor is a device which shows great potential for high performance and high precision applications. Industry however is less concerned with theory and potential, and more interested in performance and sales. And at present, as a piece of silicon, the performance of the current-conveyor seems to fall apart in the more interesting and specific applications which should really benefit from its use.

While most of the other chapters are concerned with generic applications of the current-conveyor, this chapter concentrates on some of the architectural design issues. Practical performance limitations are discussed, and in some cases compensation and enhancement techniques are suggested. Although this chapter is not intended to be a 'how-to' manual for current-conveyor design, it is hoped that it will stimulate discussion and more importantly, the development of high-performance current-conveyor integrated circuits. This will not happen overnight - after all, the voltage op-amp is still evolving after more than twenty years. But with just a little effort, the availability of practical current-conveyors could at last allow the full potential of these devices to be realised.

11.2.2 The current-mode/voltage-mode dilemma

Many of the current-conveyor theories and applications discussed in other chapters have been tested out in practice using 'breadboard' conveyor circuits, due to the lack of availability of a commercial device. Some researchers have built current-conveyors from matched transistor arrays, but the most common way of implementing a fairly high performance current-conveyor has been based on the use of supply-current-sensing on a voltage op-amp [1], as shown in Figure 1. The high resistance op-amp input provides the current-conveyor *Y* node, while the action of negative feedback provides the low resistance *X* node. Current-mirrors in the op-amp supply leads copy the current at node *X* to node *Z*.

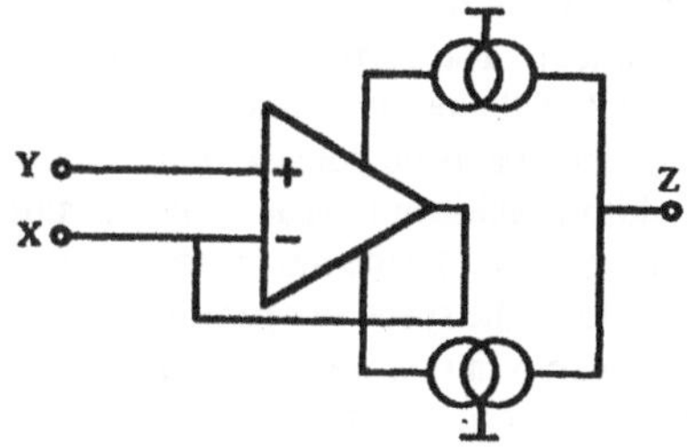

Figure 1 Supply-current-sensing on a voltage op-amp.

For researchers using this type of architecture, certain interesting facts soon became apparent. In some applications, the voltage op-amp seemed to operate well above its expected frequency range! [2] This is illustrated by the examples shown in Figures 2a and 2b.

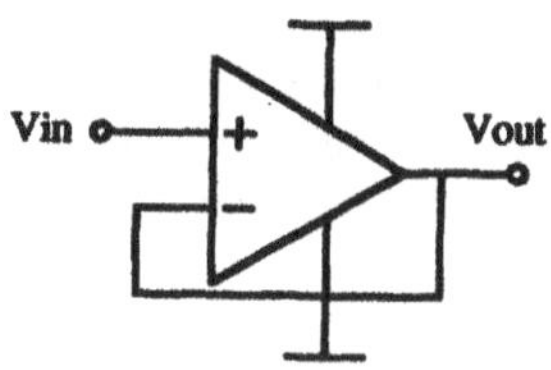

Figure 2a Voltage-follower.

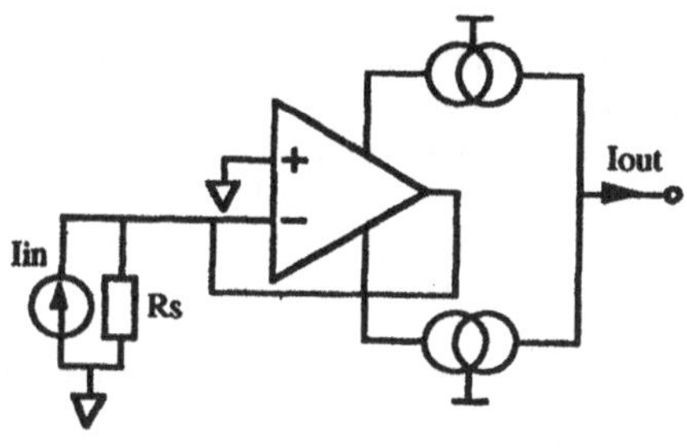

Figure 2b Current-follower.

In Figure 2b, R_S represents the output resistance of the current source. The open-loop gain of an op-amp can generally be written

$$\frac{V_{out}}{V_{in}} = \frac{A_o}{1+j(f/f_o)} \tag{1}$$

where A_o is the open-loop DC gain magnitude and f_o is the open-loop 3 dB bandwidth. Since A_o>>1, the transfer function of the voltage-follower of Figure 2a can be written as

$$\frac{V_{out}}{V_{in}} \approx \frac{1}{1+j(f/GB)} \tag{2}$$

where $GB=A_of_o$. From equation (2), the 3 dB bandwidth of the closed-loop voltage-follower is equal to the open-loop gain-bandwidth product (*GB*) of the op-amp. If the op-amp is configured instead to give a closed-loop voltage gain *A*, it is well known that the closed-loop bandwidth correspondingly reduces by the factor *A*.

The transfer function for the current-follower circuit of Figure 2b can be written as

$$\frac{I_{out}}{I_{in}} \approx \lambda \frac{1+j(f/GB)}{1+j(f/kGB)} \tag{3}$$

λ is the current transfer ratio of the current-mirrors and $k = \frac{R_s + r_O / A_O}{R_s + r_O}$, where r_O represents the output resistance of the op-amp Since $A_O \gg R_S \gg r_O$ then $k \approx 1$, and the pole and zero in equation (3) cancel. The current-follower circuit thus operates well above the gain-bandwidth product GB of the op-amp, and the 3 dB frequency of this circuit will be determined by higher frequency parasitic poles within the current-mirrors.

This 'extra' bandwidth is achieved because the op-amp is being used with input and output nodes held at virtual ground. Inside the op-amp there will be no large voltage swings at high impedance nodes to limit the speed of operation. In normal 'voltage-mode' operation you would never use the op-amp in this way, but in this 'current-mode' application the op-amp operates with a much higher maximum frequency. Of course, there are disadvantages to the current-mode circuit of Figure 2b; for example the λ term in equation (3) will limit the transfer accuracy. But these circuits illustrate the point that often it is the application, rather than the embedded device, which determines the overall performance. This statement is particularly true in the case of the current-conveyor, since the device can be used in both current-mode and voltage-mode applications, due to the high and low resistance Y and X inputs. It is thus important to consider the complete system when optimising the design of an integrated current-conveyor. For example, the fastest current-conveyors for high frequency applications are going to be based on simple open-loop voltage and current-followers, and examples of this type of architecture are discussed in the following chapter. On the other hand, the use of negative feedback often leads to other performance improvements, and so closed-loop current-conveyor architectures are also covered. Therefore before choosing the current-conveyor architecture, the 'mode' of operation of the complete circuit should be considered carefully.

11.2.3 First generation current-conveyors

Smith and Sedra's original paper [3] presenting the first generation current-conveyor (CCI) showed a transistor-level implementation based on discrete devices, shown in Figure 3. Assuming that transistors Q_3 - Q_5 and resistors R_1 - R_3 are matched, then (to a first order) the currents through these matched components will be equal. Transistors Q_1 and Q_2 are thus forced to have equal currents, and equal V_{BE}s. Input nodes X and Y therefore track each other in both voltage and current.

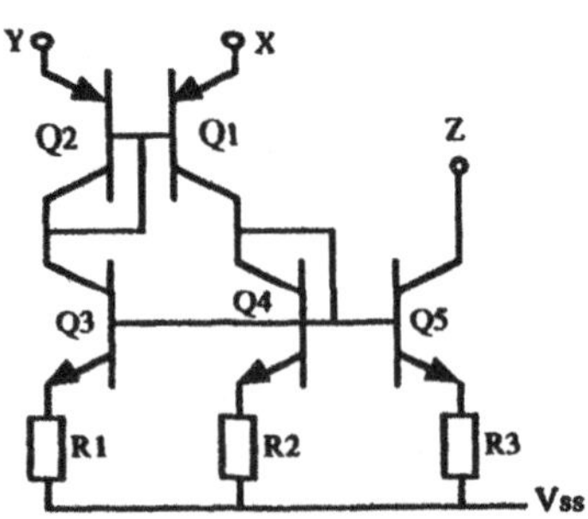

Figure 3 First generation current-conveyor (CCI).

In practice there will be slight differences in the collector currents of different transistors, due to the finite beta of the devices. These differences can be reduced, for example, by using more elaborate current-mirrors. The polarity of the output current at node Z can easily be inverted by using an additional mirror stage, and also the entire circuit can be inverted by replacing NPN transistors with PNPs, and vice versa. Connecting two complementary current-conveyors as shown in Figure 4 results in a class AB circuit capable of bipolar operation. Note that in practice this circuit may require additional components to guarantee start-up.

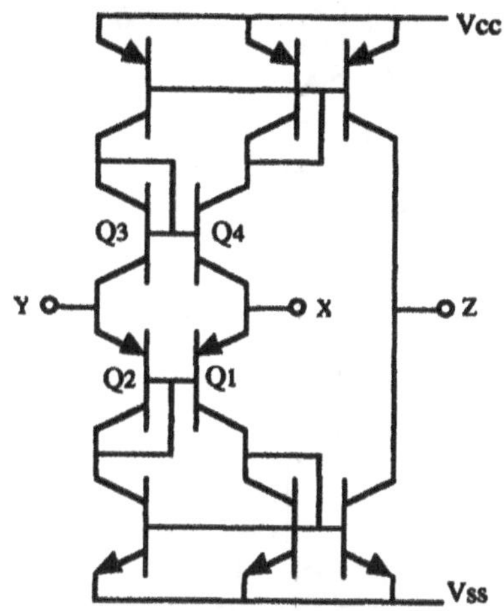

Figure 4 Class AB current-conveyor (CCI).

An integrated current-conveyor based on the architecture shown in Figure 3 is commercially available as the PA630 [4] (also in chapter 11.3), and the basic topology of this device is shown in Figure 5. An NPN Wilson mirror (Q_1 - Q_3) and a PNP Wilson mirror (Q_4 - Q_6) are used to provide the current and voltage following properties between inputs X and Y, similar to the circuit of Figure 3. Taking a second output from the PNP current-mirror to provide the Z_{output} would destroy the base-current compensation scheme of the Wilson mirror. Therefore a second NPN Wilson mirror (Q_7 - Q_9) is used to perform a current-splitting action: the combined emitter current of Q_7 and Q_8 is divided in two, with one half being shunted via Q_9 to the supply rail, and the other half driving an output PNP Wilson mirror (Q_{10} - Q_{12}). This results in an output current at node Z which is (to a first order) exactly equal to that at the X and Y inputs. Q_{13} is included to ensure that the device always starts up when turned on. The complete architecture of the PA630 CCI also includes frequency compensation to ensure stability, and modified output current-mirrors which use the 'wasted' collector current of Q_9 to effectively double the output resistance at node Z. A full description of the architecture and operation of this device can be found in [5].

11.2.4 Second generation current-conveyors

The current-conveyor architecture shown in Figure 5 includes both NPN and PNP transistors in the signal path, and thus the bandwidth and current-handling capability of this device will be poor if only lateral PNPs are available. The development of complementary bipolar processes, with vertical PNP as well as NPN transistors, has made possible the implementation of high performance integrated circuit current-conveyors such as this.

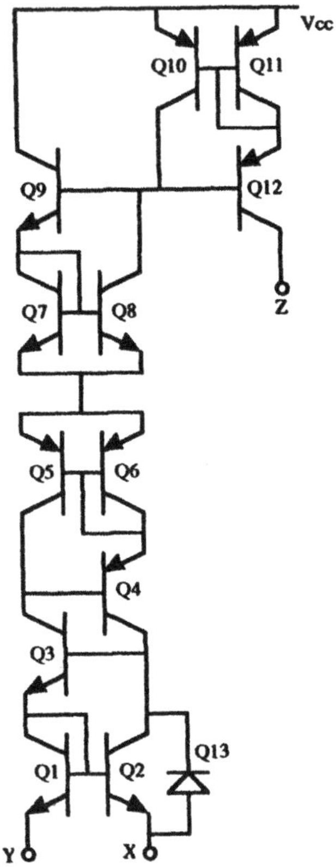

Figure 5 Simplified PA630 current-conveyor (CCI).

A second generation current-conveyor, CCII, can also be implemented simply on a complementary bipolar process by replacing the diode at the *Y* input with a transistor, and taking the input from the high resistance base terminal, as shown in Figure 6. A class AB version of this cell is shown in Figure 7.

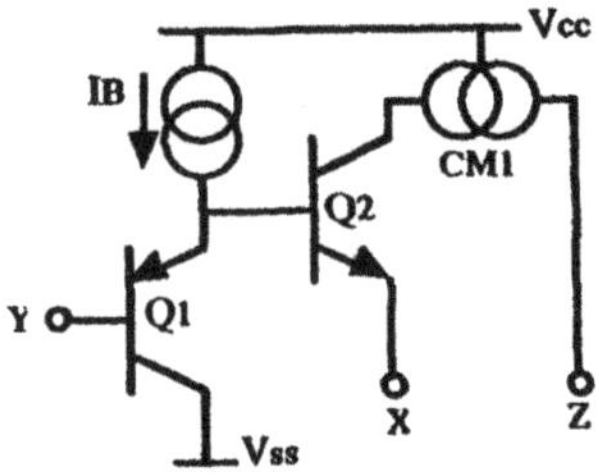

Figure 6 Class A CCII.

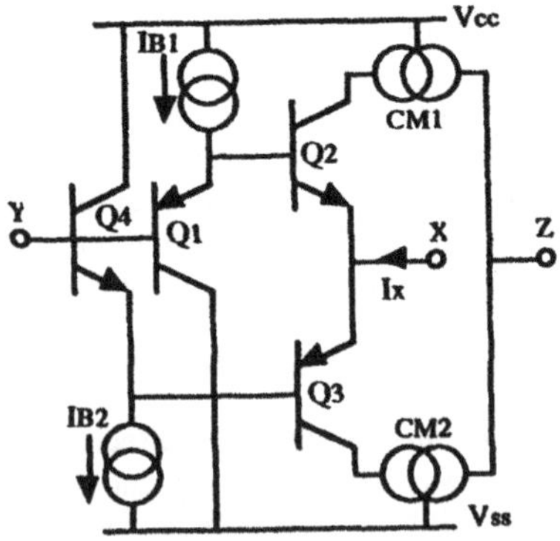

Figure 7 Class AB CCII.

Referring to Figure 7, transistors Q_1 - Q_4 act as a voltage buffer which transfers the voltage at node Y to node X. The current source and sink ($I_{B1} = I_{B2} = I_B$) provide the quiescent bias current for these input transistors. Any input current (I_X) at node X is split between Q_2 and Q_3, and is copied by current-mirrors CM_1 and CM_2 to the output node Z. This CCII architecture forms the basis of the commercially available CCII01 current-conveyor [6] introduced in chapter 11.1. It is also used as the basic input stage of the current-feedback op-amp, which has emerged as a high-speed alternative to the more conventional voltage op-amp [7].

The simple CCII architecture of Figure 7 will clearly exhibit a quiescent voltage offset between nodes X and Y due to the mismatch between the V_{BE}s of the NPN and PNP transistors Q_1/Q_2 and Q_3/Q_4, since

$$V_Y - V_X = V_{BE}(p) - V_{BE}(n)$$

$$= V_T \ln (I_{sp}/I_{sn}) \qquad (4)$$

where I_{sp} and I_{sn} are the reverse saturation currents of the PNP and NPN transistors respectively, and V_T is the thermal voltage. This process-dependent voltage offset can be reduced by including additional matching diodes in the input stage, as shown in Figure 8.

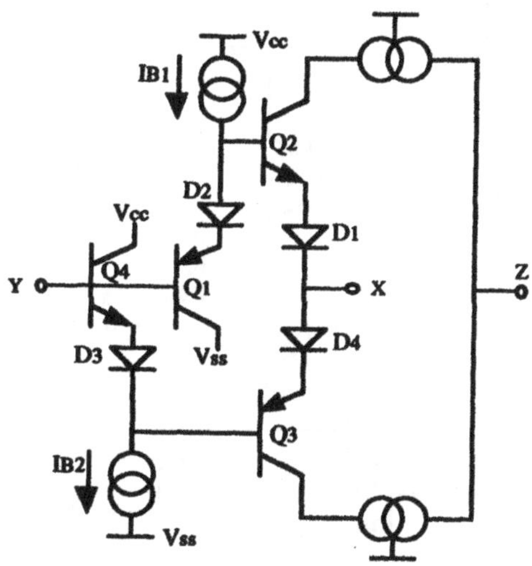

Figure 8 CCII with input matching diodes.

Referring to this diagram

$$V_Y - V_X = V_{BE}(Q_1) + V_{D2} - V_{BE}(Q_2) - V_{D1}$$

$$V_Y - V_X = [V_{BE}(Q_1) - V_{D1}] - [V_{BE}(Q_2) - V_{D2}] \qquad (5)$$

The inclusion of these diodes will clearly reduce the quiescent input voltage offset, provided that D_1 is matched to Q_1, D_2 is matched to Q_2, and so on. However, the addition of diodes D_1 and D_2 has several disadvantages. Firstly, the input voltage dynamic range of the circuit will be reduced by a diode voltage drop. Secondly, the small-signal input resistance seen looking into node X will be double that for the basic architecture given in Figure 7. This non-zero input resistance at node X (R_X) will compromise the performance of the current-conveyor, especially in applications where a non-zero input voltage is applied at node Y. The effect of the small-signal input resistance R_X is to produce a signal-dependent voltage offset V_d between nodes X and Y, where

$$V_d = R_X \cdot I_X \tag{6}$$

Since the value of R_X is determined by the small-signal resistance ($r_{e2} + r_{d2}$) in parallel with ($r_{e3} + r_{d3}$), its value could be reduced by increasing the value of the quiescent bias current I_B. However, an increase in bias current will lead to an increase in the total power consumption, as well as a possible increase in offsets, and so is certainly not an ideal solution.

11.2.5 Input bias-current scaling

The increase in R_X due to the inclusion of the matching diodes can be overcome using a method proposed by Koullias [8], shown in Figure 9. In this architecture, the matching diodes are eliminated to reduce R_X, and the quiescent input voltage offset is reduced by scaling the source and sink bias currents (I_{B1} and I_{B2}), forcing the NPN and PNP V_{BE}s to be equal.

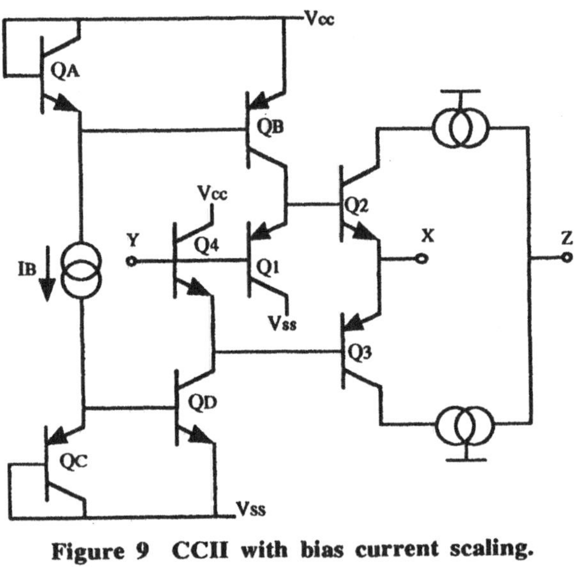

Figure 9 CCII with bias current scaling.

Referring to Figure 9, transistors Q_A and Q_B ensure that Q_1 develops the same V_{BE} as an NPN transistor with collector current I_B. Q_1 is therefore forced to match to the NPN transistor Q_2. Similarly Q_4 is made to 'look like' a PNP transistor with collector current I_B, and so this device matches to Q_3. The quiescent offset voltage between nodes X and Y clearly will be reduced without having increased the value of R_X. However, a signal-dependent input voltage offset will still be present, since when the input current I_X is non-zero, the collector currents of Q_2 and Q_3 are no longer equal to I_B. Although this bias-current scaling method reduces the DC voltage offset, the signal-dependent offset is still present, due to the finite value of R_X. This current-conveyor architecture is still an open-loop structure, and so it is almost impossible to reduce the value of R_X towards zero.

11.2.6 Local negative feedback

One method of reducing the input resistance at node *X* is to apply local negative feedback to the input stage [9], as shown in Figure 10, where CM_1 and CM_2 are dual output current-mirrors.

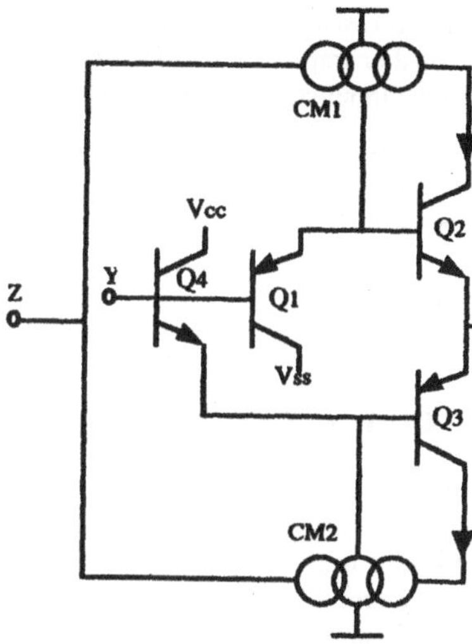

Figure 10 CCII with local negative feedback.

In this circuit, any input current I_X is mirrored to transistors Q_1 and Q_4, and so the V_{BE} change in Q_2 and Q_3 is followed by Q_1 and Q_4. The resulting voltage offset between nodes *X* and *Y* is therefore close to zero for all input signal levels, hence

$$R_X = \left(\frac{V_Y - V_X}{I_X}\right) \approx 0 \tag{7}$$

If a PNP/NPN scaled current-mirror, similar to that shown in Figure 9, is used to mirror the input current back to Q_1 and Q_4, then the resulting voltage offset between *X* and *Y* will be very low. However in this case the current transfer accuracy between nodes *X* and *Z* may be degraded. An alternative implementation is to use a more conventional current-mirror to provide the current feedback, and to include matching diodes to eliminate the process-dependent quiescent voltage offset between nodes *X* and *Y*. The inclusion of the matching diodes will not significantly increase the small-signal input resistance at node *X*, due to the use of local negative feedback.

This local feedback will unfortunately have the effect of reducing the input resistance (R_Y) at node *Y*, since the input signal current at this node will be equal to I_X/β, and thus

$$R_Y = \frac{V_Y}{I_Y} = \beta\frac{V_X}{I_X} = \beta Z_X \tag{8}$$

where Z_X is the value of the external impedance connected to node *X*. Of course in many applications this value of R_Y may be perfectly adequate, but in other situations the reduced input resistance at node *Y* may cause loading of the signal source. In this case, the value of R_Y can be increased, for example, by using a compensation circuit to provide the base current required by transistors Q_1 and Q_4.

11.2.7 A closed-loop current-conveyor

The use of local negative feedback within the current-conveyor input stage demonstrates how it is possible to reduce the input resistance R_X to very low values. The performance of an IC current-conveyor can be further improved by extending this principle and applying global negative feedback around the input stage of the conveyor, as shown in Figure 11.

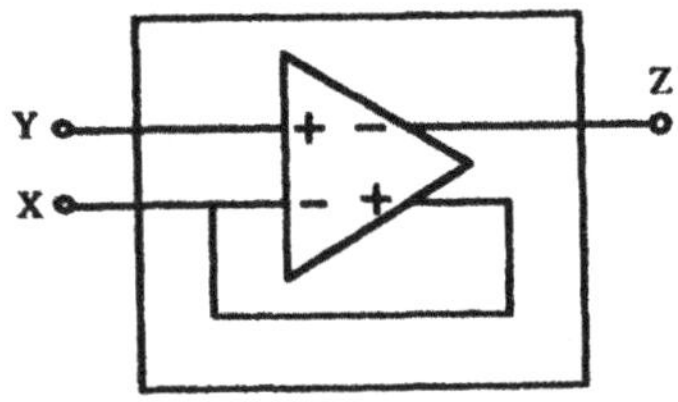

Figure 11 CCII with closed-loop input stage.

In this architecture the input stage is implemented as a closed-loop buffer, rather than an open-loop buffer as in the previous examples shown. The input resistance at node *X* will again be very small due to the use of negative feedback, and will be equal to R_X/A, where R_X is the open-loop input resistance and *A* is the open-loop gain of the operational amplifier. The action of negative feedback will similarly increase the input resistance at node *Y* by the open-loop gain *A*.

There are various possible implementations for the high gain op-amp imbedded within the closed-loop conveyor. The requirements for this device are

- differential current outputs to allow the provision of a current output at node *Z, and*
- high open-loop gain, to ensure zero input voltage offset between nodes *Y* and *X*, and zero input current at node *Y*.

A particularly ideal yet very simple architecture is based on the use of two basic CCII devices, as shown in Figure 12. The high open-loop gain is provided at node Z_1/Y_2, where the output current from the first CCII feeds into the high resistance input of the second CCII, giving a high open-loop transresistance gain. Frequency compensation to ensure closed-loop stability is provided by capacitor C_C.

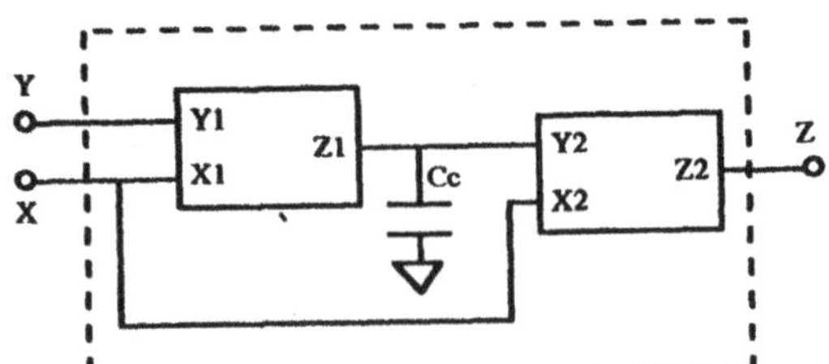

Figure 12 Dual current-conveyor.

This architecture results in a very high performance current-conveyor, since the low input resistance at X_1 is further reduced by the action of negative feedback, and similarly the high input resistance at node Y_1 is further increased. This leads to very low input offsets, and consequently a very high accuracy in the resulting current-conveyor application. The reduced input resistance at node X can be appreciated by analysing the equivalent macro-model of the closed-loop conveyor circuit of Figure 12, which is shown in Figure 13.

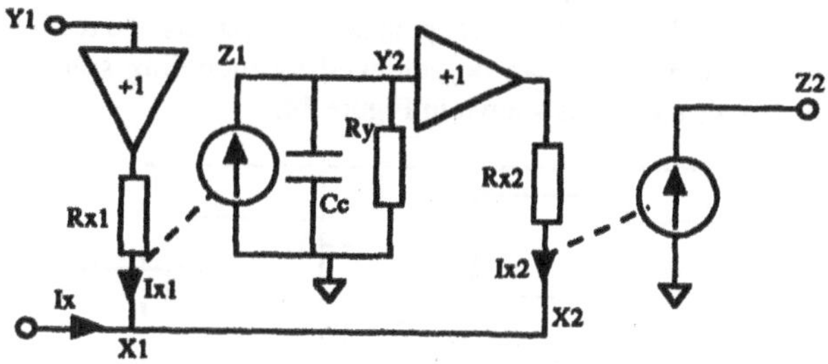

Figure 13 Closed-loop current-conveyor macro-model.

In the macro-model of Figure 13, ideal voltage buffers are used to represent the voltage-following action between nodes Y_1 & X_1, and nodes Y_2 & X_2. Ideal current-controlled current sources are used to represent the current-following action between X_1 & Z_1, and X_2 & Z_2. R_{X1} and R_{X2} represent the (low) open-loop input resistance at node X of the first and second current-conveyors respectively. R_Y represents the high open-loop input resistance at node Y of the second CCII ; this has been shown as an explicit component since it determines the magnitude of the open-loop gain. Referring to this macro-model

$$V_{Y1} - V_{X1} = I_{X1} \cdot R_{X1}$$

$$V_{Y2} - V_{X2} = I_{X2} \cdot R_{X2}$$

$$V_{Y2} = V_{Z1} = I_{X1} \cdot Z_t \qquad \text{(where } Z_t = R_Y || C_C\text{)}$$

$$I_{X1} + I_{X2} + I_X = 0$$

Rearranging these equations allows us to express V_{X1} as a function of V_{Y1} and I_X

$$V_{X1}\,(R_{X1} + R_{X2} + Z_t) = V_{Y1}\,(R_{X2} + Z_t) + I_X(R_{X1}R_{X2}) \tag{9}$$

To simplify this equation, assume that the closed-loop current-conveyor is being used as a current-input device with V_{Y1} grounded, then

$$V_{X1} = I_X\left(\frac{R_{X1}R_{X2}}{R_{X1} + R_{X2} + Z_t}\right)$$

or

$$R_{inX} = \frac{V_{X1}}{I_X} = \left(\frac{R_{X1}R_{X2}}{R_{X1} + R_{X2} + R_Y}\right)\left(\frac{1}{1 + sC_c[R_Y || (R_{X1} + R_{X2})]}\right) \tag{10}$$

Assuming $R_{X1} \approx R_{X2} \approx R_X$, then the small-signal input resistance will be approximately equal to $(R_X{}^2/R_Y)$. Since $R_Y >> R_X$, the net input resistance (R_{inX}) will be very low.

From equation (10), the closed-loop dominant pole is given by $s = 1/(C_C[R_{X1}+R_{X2}])$ (since $R_Y >> R_{X1}, R_{X2}$), and this dominant pole must be lower in frequency than any parasitic poles to ensure an adequate gain and phase margin. The value of the compensation capacitor C_C should thus be increased until closed-loop stability is guaranteed. For an integrated circuit implementation, it may be undesirable to have a fairly large on-chip capacitor. From equation (10), the dominant pole position can also be reduced by increasing the effective value of R_{X2}. This can be achieved by including some series feedback resistance R_F between X_1 and X_2 as shown in Figure 14.

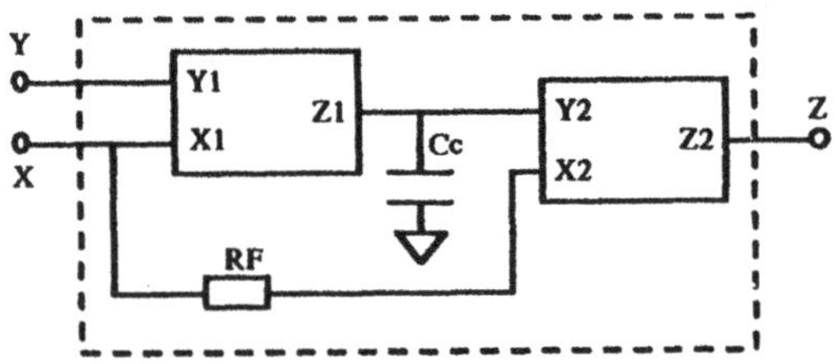

Figure 14 Closed-loop CCII with internal feedback resistor.

In this circuit, the dominant pole position will now be determined by $1/C_C R_F$ (assuming that $R_F >> R_{X1}, R_{X2}$). The compensation capacitance C_C can now be kept to a minimum, with the value of R_F being used to determine stability. However, it should be noted from equation (10) that this method of compensation will tend to increase the input resistance at node X. If the conveyor is to operate with a fairly high input current then it is better to use node X_2 as the X input of the dual conveyor, because the voltage drop across the feedback resistor R_F will be minimised, although the voltage offset between Y and X will now be increased by $R_F I_{X1}$.

The improved performance of the closed-loop CCII architecture can be demonstrated by interconnecting the two simple CCII structures available in the dual CCII01 [6]. Figure 15 shows that the input resistance at node X for the dual closed-loop current-conveyor is two orders of magnitude less than that of the single open-loop conveyor.

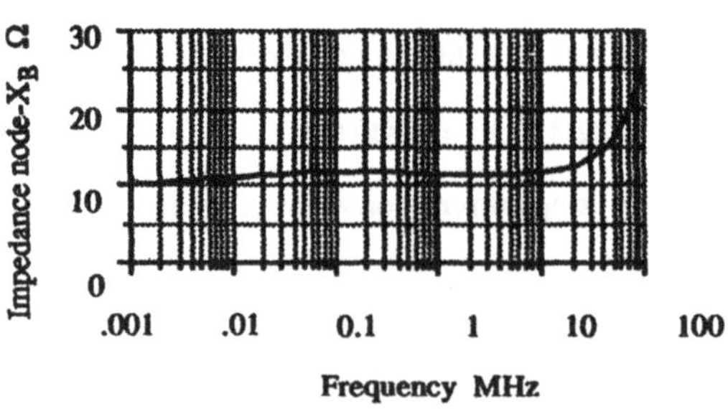

Figure 15a Single CCII.

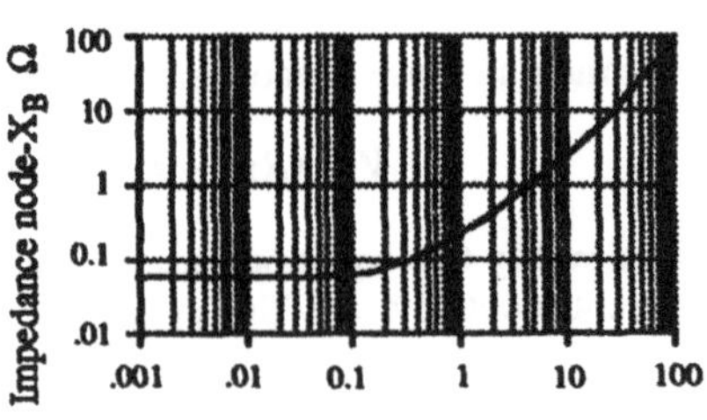

Figure 15b Dual CCII.

The main disadvantage of the closed-loop current-conveyor architecture is the reduced device bandwidth when compared to the more simple open-loop structure, due to the need for additional frequency compensation to guarantee stability. However, the bandwidth of the closed-loop device can be maximised if the architecture is optimised as a single integrated circuit. When all interconnections are made on chip, the parasitic pin capacitances are eliminated, allowing a smaller value of compensation capacitance to be used to guarantee closed-loop stability.

11.2.8 A general purpose 'dual' current-conveyor

A very versatile device is obtained if both X_1 and X_2 in Figure 12 are made accessible, as shown in Figure 16. By terminating node X_2 with a resistor R_1, a current-conveyor with a controllable current gain $(1 + R_F/R_1)$ between X_1 and Z_2 is obtained. Alternatively, by terminating node X_1 with a resistor R_1, a current-conveyor with a controllable voltage gain $(1 + R_F/R_1)$ between Y_1 and X_1 is obtained.

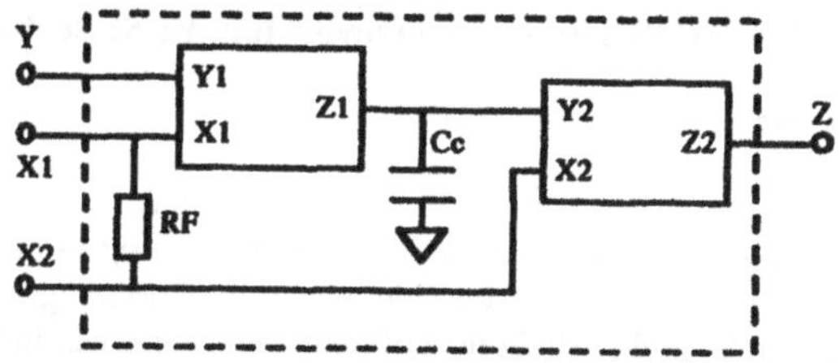

Figure 16 Four terminal 'dual' current-conveyor.

In fact this gain-controllable dual current-conveyor can be used to implement all of the four basic amplifier types, as shown in Figure 17. This figure serves to underline the versatility and ease of use of the current-conveyor.

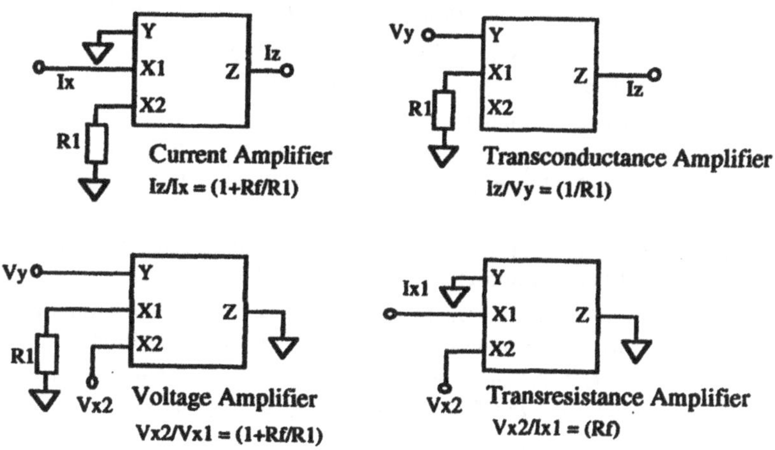

Figure 17 'Dual' conveyor closed-loop amplifiers.

Although this discussion has concentrated on bipolar implementations of the current-conveyor, interest is also growing in the realisation of MOS current-conveyors and this subject will discussed more fully in Chapter 11.5. Such a device would prove to be a versatile addition to VLSI analog cell libraries for future design automation. One such MOS implementation [10] is shown in Figure 17, and provides a very low offset voltage between nodes X and Y by using an offset-cancellation circuit similar to that of Figure 9.

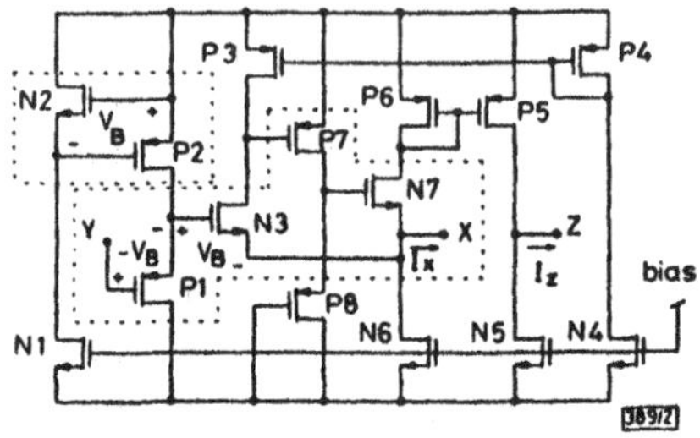

Figure 17 MOS current conveyor for VLSI technology.

11.2.9 Conclusions

Until high performance current-conveyors are widely available, these devices will continue to develop in theory rather than in practice. Process technologies and design techniques have now advanced to the stage where the implementation of an integrated current-conveyor is both desirable and viable, and a whole host of applications are waiting for its arrival.

References

[1] B. Wilson, "High Performance Current Conveyor Implementation", *Electron. Lett.*, Vol.20(24), pp 990-991, 1984.

[2] C. Toumazou, F. J. Lidgey and C. Makris, "Extending Voltage-Mode Opamps to Current-Mode Performance", *Proc. IEE*, Vol.137(2), Part G, pp116-130, 1990.

[3] K. C. Smith and A. S. Sedra, "The Current Conveyor - A New Circuit Building Block", *Proc. IEEE*, Vol. 56, pp 1368-1369, Aug. 1968.

[4] PA630 Data Sheet, Photronics Co.

[5] C. Toumazou, F. J. Lidgey and D. Haigh (Ed), *Analogue IC Design - The Current-Mode Approach*, Peter Peregrinus, 1990, Exeter UK, Chapter 15 (D.C. Wadsworth).

[6] CCII01 Data Sheet, LTP Electronics Limited, 27 Park End Street, Oxford, OX1 1HU, UK., tel/fax +44-865-200767.

[7] D. F. Bowers, "A Precision Dual Current-Feedback Operational Amplifier", *IEEE Bipolar Circuits and Technology Meeting (BCTM)*, pp 68-70, 1988.

[8] I. A. Koullias, "A Wideband Low-Offset Current-Feedback Op Amp Design", *IEEE Proc. Bipolar Circuits and Technology Meeting (BCTM)*, pp 120-123, 1989.

[9] A. Payne, "Current-Feedback Techniques for Analogue Circuits and Systems", PhD Thesis, 1992.

[10] M. C. H. Cheng and C. Toumazou, "3V MOS Current Conveyor Cell for VLSI Technology", *Electronics Letters*, Vol. 29, No. 3, pp 317-319, Feb. 1993.

Chapter 11.3

Commercial Applications of the Current-Conveyor

Doug C. Wadsworth

11.3.1 Introduction

Over twenty-five years ago [1], the current-conveyor was introduced as a new circuit building block. As the other contributors to this tutorial have indicated, this circuit configuration has subsequently become an essential element in the field of Current-Mode Analog Signal Processing. In fact, based on the extensive academic literature on the subject of current-conveyor theory and applications, its importance to current-mode techniques appears to rival the prominence that the ubiquitous operational amplifier enjoys in conventional design approaches.

However, the above assumption is severely undermined by the following observations: Unlike the op-amp, the current-conveyor does not seem to be that familiar to the "working engineer" (i.e. outside of academic circles). There are hundreds of standard product op-amp integrated circuits available, but to the author's knowledge, there are only two standard ICs [2, 3] on the market that are specifically designated as current-conveyors. The use of op-amps in virtually all types of consumer, industrial, and military electronic equipment is well known, but despite the many academic papers on current-conveyors, one is hard pressed to find references to their use in products that are actually in production.

Of course, all of the previous comments are interdependent. The design engineer drives both the demand for standard ICs, and the utilization of specific circuit techniques in end products. After 25 years of academic interest in current-conveyors and current-mode topology, should there not be engineers who require such ICs and use such techniques in products? The author believes that many designers use current-conveyors without knowing (or remembering) that this is the case. Even those familiar with the nomenclature may choose not to specifically identify such blocks for commercial reasons, or simply for expediency. This low profile approach is exemplified by the standard IC known as the current-feedback op-amp. This device is actually a current-conveyor with a unity gain buffer on its output, but it is seldom described as such in commercial data sheets. Semiconductor company marketing departments seem to prefer the familiarity of the op-amp designation over a more accurate description of the part.

The demand for standard current-conveyor ICs may be suppressed due to the fact that discrete components or transistor array ICs can be used in simpler applications. But more significantly, where product volumes warrant, the designer is likely to incorporate his current-mode ideas directly into an Application Specific Integrated Circuit (ASIC). Due to the high complexity of many of these analog ASICs, specific blocks such as current-conveyors may not be identified in the overall functional description. Therefore, there are probably many high volume products with unknown examples of current-mode design buried in custom silicon.

In light of the preceding, and in the context of this chapter, it would be useful to examine a large number of examples of the commercial application of current-conveyors. However, as has been previously indicated, identifying and obtaining information on such products is difficult. Therefore, by way of example, the author will focus on two products that are well known to him. Specifically, a telephone set that incorporates current-conveyors on two custom ASICs, and a digital audio hi-fi product with a current-conveyor stage will be discussed.

11.3.2 VISTA family of telephone sets

The VISTA family is a product line of electronic telephone sets that is being produced in high volume by Northern Telecom in Canada. These telephones [4] offer a number of advanced features including an LCD display, calling party number display and storage in a call log, non-volatile frequently called number memory with alpha-numeric storage, automatic correlation with storage and display of the alpha-numeric memory data corresponding to incoming calls, automatic loop equalization (gain), automatic network tuning (sidetone optimization), and detection and indication of parallel sets and disconnected line cords. This product line has received an enthusiastic response in the marketplace [5], and since its introduction in 1992, over a million have been shipped. A photograph of one of the more fully featured telephones in this family, the VISTA 200, is shown in Figure 1.

Figure 1 VISTA 200 telephone set.

The architecture of the VISTA telephones is based on a silicon chip set that consists of an analog ASIC, a digital ASIC, an electronic switch ASIC, a microprocessor, and memory chips. The three ASICs were designed by Northern Telecom and Bell Northern Research to specifically provide the sophisticated features outlined above. Both the analog ASIC and the electronic switch ASIC incorporate current-conveyors and their application will be discussed in the following sections.

11.3.2.1 VISTA analog ASIC

The analog ASIC used in VISTA telephones is fabricated on a 2 μm BiCMOS process and it is approximately 185 mils per side. There are well over 1000 active devices on this chip and a photograph of the IC is shown in Figure 2. The die is assembled in a 44 pin PLCC package. Current-conveyors are employed in three sections of this ASIC: the Shunt Regulator, the Series Regulator, and the Transmit Output Stage.

Figure 2 VISTA analog ASIC.

(i) Shunt regulator

The analog ASIC uses a shunt regulator configuration to define the impedance characteristics of the telephone set. An equivalent schematic of part of this block is shown in Figure 3, where transistor Q_1 is the shunt element driven by an impedance control circuit. Other sections of the chip require a scaled replica of the current flowing in Q_1, and this function is provided by a first generation [1] current-conveyor (Q_2-Q_7) and resistors R_1 and R_2. As has been discussed in other sections of this tutorial, port X of the current-conveyor will be driven to the same potential as port Y. Therefore, equal voltages will appear across R_1 and R_2. Since the shunt element current flowing in R_1 will define the voltage across it, and the voltage across R_2 will set the current in the conveyor, currents proportional to the shunt current and scaled by the ratio of R_1 to R_2 will appear at output ports Z_1 and Z_2.

The main advantages of the current-conveyor in this application are its simplicity and low voltage operation. On long subscriber loops, the voltage on the line can fall to almost 1 Volt. This configuration will operate down to this level and it provides convenient scaling set by the resistor ratios.

One characteristic that must be considered when using current-conveyors is that they can have a stable state at zero current. This means that they may not start up when power is first applied. This effect is not usually apparent with computer simulation since parameters which exacerbate the problem such as transistor offset and leakage are often not modelled. Fortunately, the solution is generally simple in that a small current can be injected into the conveyor with a resistor or current source to skew the balance towards the "on" state.

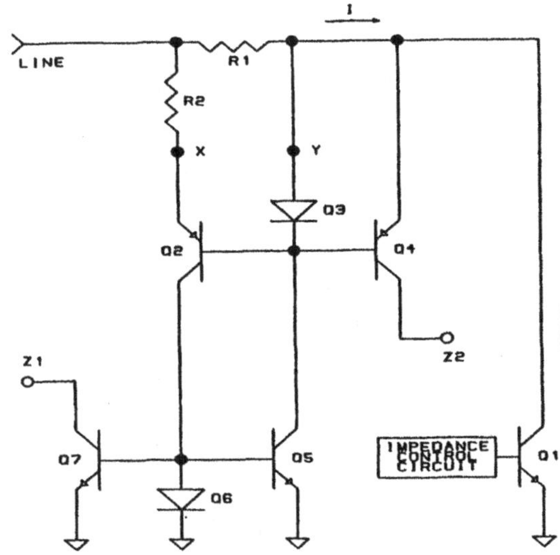

Figure 3 Analog ASIC shunt regulator.

(ii) Series regulator

A stable voltage rail, V_{dc}, is required in the ASIC for powering various blocks. This is generated from the line with a series pass transistor controlled by a regulator circuit. It is advantageous to monitor the voltage across this series pass element (V_{CE}) to determine if the device is close to saturation.

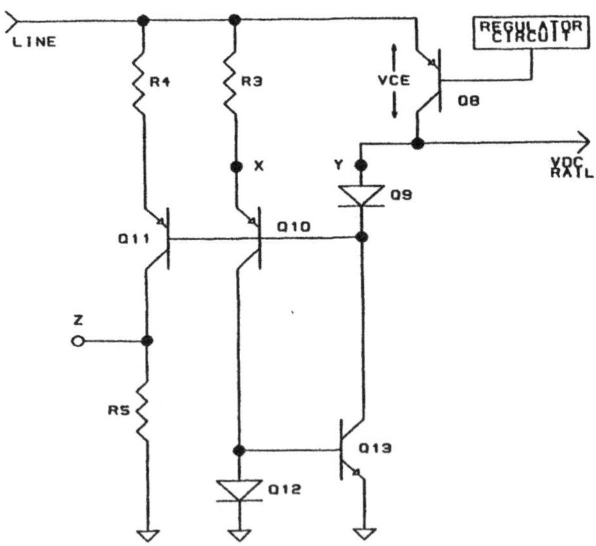

Figure 4 Analog ASIC regulator.

As shown in Figure 4, a current-conveyor (Q_9-Q_{13}) approach results in an efficient realization. As before, port *X* will be driven to the same potential as port *Y*. Therefore, the V_{CE} voltage of series pass transistor Q_8 will appear across R_3, and this will define the current in the conveyor. Since R_3 and R_4 are of equal value, and Q_{10} and Q_{11} are matched, the current into output port *Z* is also constrained to this value. This output current flows in resistor R_5 to produce a voltage that is referenced to ground. The ratio of R_3 and R_4 to R_5 is selected to provide voltage gain which simplifies interfacing due to the larger resulting signal level.

As with the shunt regulator, the current-conveyor offers the advantages of simplicity and low voltage operation. Since the V_{CE} voltage can be relatively small (200 mV) and it is referenced to the most positive potential available, it is difficult to use it as the direct input to a conventional comparator. The conveyor allows us to easily increase its amplitude and re-reference (level shift) it to ground for more convenient and efficient processing.

(iii) Transmit output stage

The current-conveyor in the transmit output stage (*TX*) is employed in essentially the same manner as previously described with the shunt regulator. A scaled version of the current flowing in the *TX* stage is required by two other blocks on the ASIC, and operation must be guaranteed down to almost 1 Volt. As shown in Figure 5, the *TX* current generates a voltage across resistor R_6, which is reflected across R_5. As a result, appropriate currents appear at output ports Z_1 and Z_2. Since currents from both Q_{15} and Q_{16} flow in resistor R_5, the ratio of the currents includes an additional factor of two, relative to the shunt regulator case.

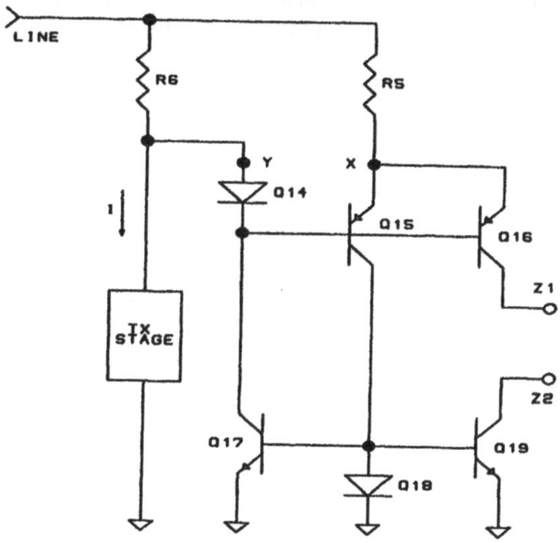

Figure 5 Analog ASIC transmit output stage.

Like the preceding examples, the current-conveyor offers here a straightforward solution to the requirements of the *TX* output stage, with a minimum expenditure of chip area.

11.3.2.2 Electronic switch ASIC

The electronic switch ASIC in the VISTA products is fabricated on a special CMOS process that is optimized for this specific application. The chip is about 150 mils by 100 mils and a photograph of the device is shown in Figure 6. The die is assembled in a 16 pin SOIC package. About 80% of the silicon area consists of a large PMOS switch transistor, with the remainder being control circuitry.

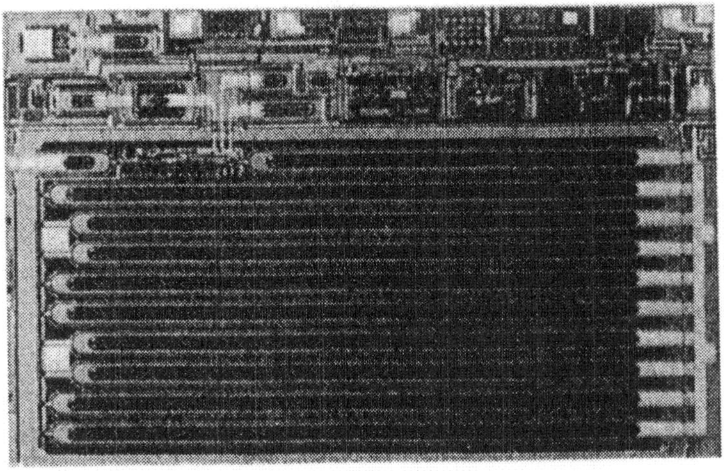

Figure 6 VISTA electronic switch ASIC.

The current through the PMOS switch must be monitored, and a proportional current made available to the control circuit. However, since considerable silicon area and process enhancements have been used to minimize the on-resistance of the switch, it is not possible to use a series resistance to sense the current. Low voltage operation is also required.

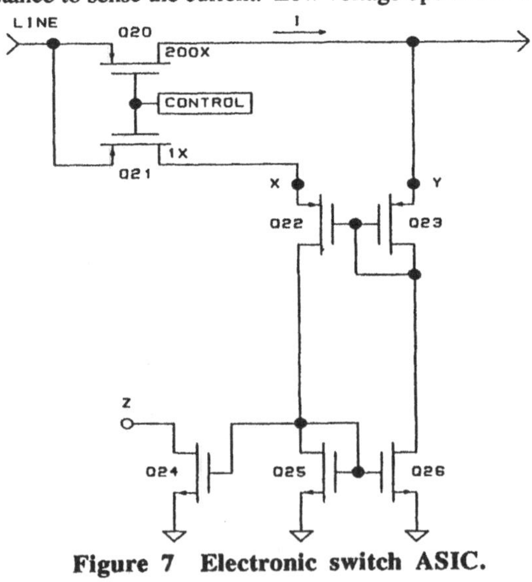

Figure 7 Electronic switch ASIC.

As shown in Figure 7, a CMOS first generation current-conveyor ($Q22$-$Q26$) is used to meet these requirements. The current sensing scheme exploits the matching between the large PMOS switch transistor ($Q20$), and a device ($Q21$) which is about 200 times smaller.

Since their sources and gates are connected together, and assuming adequate drain to source voltage (V_{DS}), the current in the smaller device would mirror that in the larger transistor on a 1:200 ratio. However, this switch is specifically designed to have a low on-resistance, and V_{DS} will be necessarily driven to a low value. In this region, the current in the MOS transistors depends not only on the gate to source voltage, but also on the value of V_{DS}. As a result, the drain potentials of $Q20$ and $Q21$ must be equal, otherwise gross errors in the current mirroring will result.

Since port X of the current-conveyor ($Q22$-$Q26$) will be driven to the same potential as port Y, equal V_{DS} voltages are imposed on $Q20$ and $Q21$, and the current into port X is conveyed to port Z. As the source, gate, and drain of $Q20$ and $Q21$ are now at correspondingly equal potentials, the current matching will hold in all regions of operation. A current of about 1/200 of that in large PMOS switch transistor $Q20$ is therefore available at port Z.

In essence, this application requires that a "virtual ground" be created at a low voltage which is also the most positive potential in the circuit. This makes it difficult, if not impossible, to employ conventional approaches such as an operational amplifier. Once again, however, the current-conveyor provides a simple and cost effective solution.

11.3.3 Digital audio hi-fi applications

At first glance, the use of current-conveyors in hi-fi audio applications would not appear to be particularly advantageous. Generally, these systems involve the processing of low frequency voltage signals within adequate supply rails, and a plethora of integrated circuits such as operational amplifiers are available to efficiently and economically address these requirements. From both a commercial and technical standpoint, these conventional ICs are being employed successfully and extensively in mass-produced hi-fi products.

However, within the hi-fi audio community, there is a school of thought (the "high-end") that believes that other design approaches such as discrete transistors and vacuum tubes (valves) provide superior sonic performance. These opinions are based on subjective listening comparisons, and efforts to scientifically verify these claims with double-blind tests and technical measurements have been inconclusive. Nevertheless, there are many "high-end" hi-fi manufacturers, dealers, magazines, and audiophiles who are actively marketing and purchasing very expensive products which incorporate these alternative architectures.

By way of rationalization, "high-end" enthusiasts point out that the use of op-amps in place of transistor and tube (valve) stages usually means that the short local feedback loops characteristic of these older designs are replaced with a global negative feedback loop that controls a single high gain block. It is claimed that such global feedback techniques can introduce dynamic distortion [6] such as transient intermodulation (TIM) and slewing induced distortion (SID). Skeptics counter that audio signals are generally band-limited and therefore may not contain the fast edges necessary for the generation of these types of non-linearities, but this is not necessarily true in all circumstances. For example, the analog stages of digital audio equipment may be subjected to transient artifacts (glitches) generated by the digital to analog converters (DACs) from which they are driven. As a result, the use of op-amps in the analog stages of CD players and DAT recorders has been severely scrutinized and criticized, and many such products now employ expensive ultra high speed op-amps, discrete transistors, or even tube (valve) topologies.

Given the above concerns, the current-conveyor can offer a performance advantage in these situations, as it can be used to implement amplifiers, inverters, I to V converters, and virtual ground inputs that do not employ any global negative feedback. This approach is therefore inherently free of dynamically induced distortion mechanisms. This application of the current-conveyor was proposed by the author in 1989 [7,8], and it has been subsequently investigated in the UK [9] and Scandinavia with promising results. A product based on the UK work is not presently commercially available, but the resulting Swedish digital audio product [10] is enjoying considerable success in the international marketplace. This unit will be discussed in the following sections.

11.3.3.1 Sentec DiAna audio DAC

DiAna is an audio digital to analog converter (DAC) that is manufactured by Sentec/Antemi in Sweden. A photograph of the unit is shown in Figure 8.

Figure 8 DiAna audio DAC.

An audio DAC is a hi-fi component which contains the digital filtering, D/A conversion, analog filtering, and output stages of a CD player (or DAT recorder). Since most CD players have a digital output which is situated prior to their own internal version of this processing, an audio DAC can be used to bypass this functionality with newer more advanced circuitry. As well, CD transports are available that do not include these blocks, and they are designed to be used exclusively with the electronics in these products.

DiAna employs a multi-bit architecture which incorporates a Crystal CS8412 receiver, a 20 bit DF1700/ SM5813 digital filter, two 20 bit Burr-Brown PCM63P-K Colinear dual DACs, and two Phototronics PA630 current-conveyor ICs. The unit has completely separate analog and digital power supplies that include four diode bridges and nine regulators. Figure 9 shows the printed circuit board with the two current-conveyor parts (16 pin DIPs) visible in the center-right section of the photograph.

Figure 9 DiAna printed circuit board.

This audio DAC has been evaluated and reviewed by a number of European and North American magazines and audio organizations [11-16], and it has been consistently rated as one of the best. A common observation in these reports is that DiAna's sonic reproduction is extremely clean and very comfortable to listen to. Although its digital architecture is similar to that used in a number of other units, its current-mode analog topology is unique. Therefore, it is not unreasonable to speculate that these reported improvements are due, in large part, to the desirable dynamic distortion characteristics of the PA630 current-conveyors.

11.3.3.2 PA630 current-conveyor IC

A functional schematic of the PA630 current-conveyor IC [7, 8, 17-20], which is used in the DiAna audio DAC, is shown in Figure 10. This patented [21] device consists of two unity gain buffers, a current-mirror, and a current-conveyor, and it is packaged in a 16 lead DIL plastic package. The chip is fabricated on a complementary bipolar process that produces high performance NPN and PNP transistors.

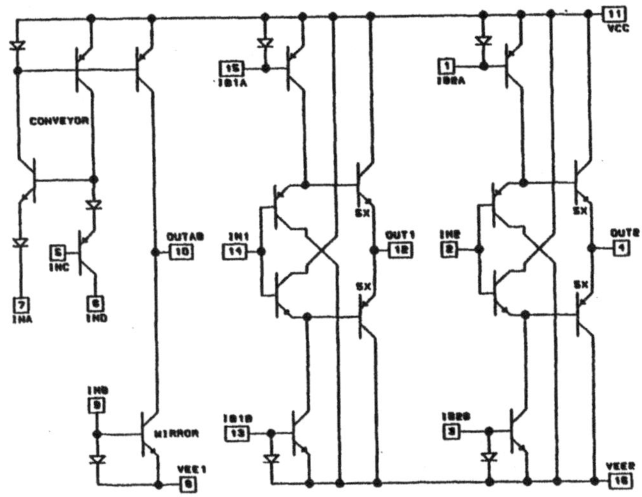

Figure 10 PA630 current-conveyor IC.

Each unity gain buffer on the PA630 consists of four emitter followers in complementary symmetry and two current sources for biasing. The quiescent operating point of each can be set independently with an external resistor.

The current-conveyor section shown in Figure 10 will be recognized as a simple first generation conveyor with a diode and a transistor added in series with the *X* and *Y* ports respectively. These additional devices allow the block to be arranged as a second generation current-conveyor [22] if required. As used in DiAna, the base and collector of this transistor are connected together, and the configuration reverts to a basic conveyor with a diode in series with each port.

The current-mirror can be interfaced with the conveyor to create a bi-directional current-mode block that is extremely versatile. This arrangement is employed in DiAna to implement a virtual ground node without the global negative feedback inherent with op-amp realizations. Although functionally equivalent to the schematic of Figure 10, the actual on-chip circuitry of the mirror and the current-conveyor incorporates significant enhancements which will be described in a following section.

11.3.3.3 DiAna analog architecture

Digital audio components such as CD players, DAT recorders, and audio DACs all require analog stages to interface with the digital electronics, and to provide analog filtering and output drive capability. The digital to analog interface usually takes the form of a virtual ground, which is commonly required by the actual DAC IC to sum its output currents. This operation, along with the filtering and output buffering, can be conveniently and economically implemented with op-amps, but not without the dynamic distortion concerns previously discussed. The PA630 Current Conveyor IC provides an alternative.

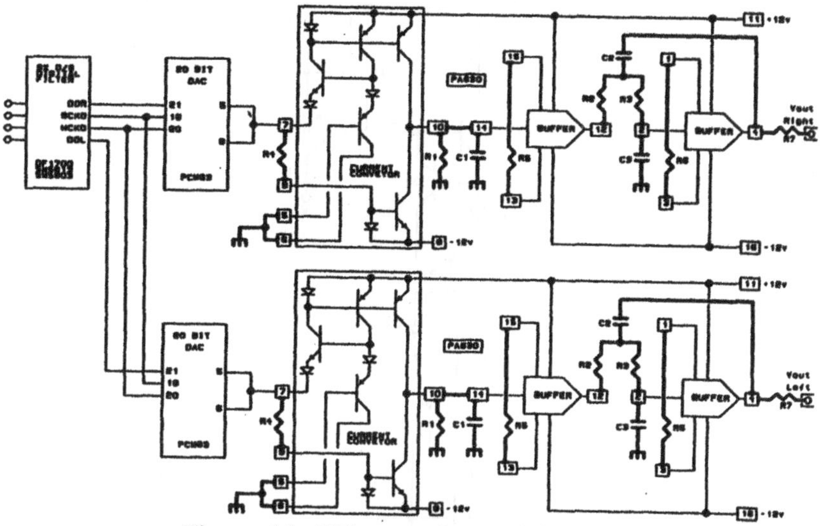

Figure 11 DiAna analog architecture.

The novel analog architecture of DiAna is shown in Figure 11. Two PA630s are used with Burr-Brown PCM63 DAC ICs and a DF1700 digital filter chip to produce a high performance stereo product. The PA630 current-conveyor section is configured to create a virtual ground at pin 7. R_4 sets the quiescent current to about 2.1 mA, which due to class A operation, must be greater than the maximum full scale output current of the PCM63 (2.04 mA). Similarly, R_5 and R_6 set the quiescent current in the buffers. R_7 defines the output impedance and protects against output short-circuits. DC coupling is maintained to the output, and offset is controlled with a DC servo (not shown), which nulls the voltage at pin 4 by feeding a correcting current into pin 14. Following the current-conveyor *I* to *V* stage, a 3rd order Sallen and Key Bessel analog filter is implemented with the buffers, resistors R_1-R_3, and capacitors C_1-C_3.

While the lack of global feedback eliminates the possibility of dynamic perturbations, the steady state distortion performance of DiAna at full scale will depend on the characteristics of the current-conveyor. Unlike op-amp realizations, this "open loop" topology cannot rely on feedback to minimize any non-linearities. Therefore, the current-conveyor must be inherently linear if this audio DAC is to be competitive in this area. This is not an insignificant challenge, as "closed loop" approaches employing 100 dB of gain can guarantee total harmonic distortion (THD) levels of better than 0.0015%. In the next section, the accurate current-conveyor block in the PA630 will be discussed in detail.

11.3.3.4 Accurate current conveyor topology

While the simple current-conveyor shown in Figures 10 and 11 offers the basic functionality required, base current errors and Early voltage (output conductance) effects result in inaccuracies which make its distortion characteristics unacceptable for hi-fi audio applications. Therefore, the conveyor circuitry on the PA630 chip actually consists of an innovative connection of two NPN and two PNP Wilson [23] current-mirrors. The basic structure is shown in Figure 12. This configuration takes advantage of the base current error cancellation scheme and enhanced output impedance of the Wilson circuit, which in turn greatly improves the accuracy and distortion performance.

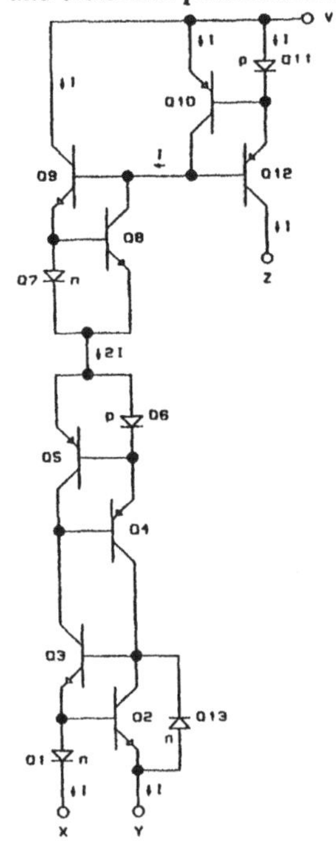

Figure 12 Wilson conveyor topology.

There is a possibility that the two cross-coupled Wilson mirrors (Q_1, Q_2, Q_3 and Q_4, Q_5, Q_6) might not "start" when the circuit is first powered up. This situation is a result of the fact that Q_3 generates its own base drive through the upper PNP mirror (Q_4, Q_5, Q_6). If Q_3 is off, it needs base drive to turn on, but it cannot produce base drive until it has turned on. Leakage currents would probably overcome this contradiction, but the addition of device Q_{13} ensures that, if necessary, a turn-on path would be established when port X fell three diode drops below port Y, thereby forcing current to flow through Q_{13} into the base of Q_3. Once the mirrors start up, Q_{13} is reverse biased and has no effect on the normal operation of the current-conveyor.

A characteristic of the Wilson mirror that must be considered is that it tends to ring or overshoot on fast pulse waveform edges. This is undesirable by itself, but it becomes a serious problem with the two cross-coupled mirrors (Q_1, Q_2, Q_3, Q_4, Q_5, Q_6) in Figure 12. The overshoot is actually a result of the fact that the current gain of the Wilson mirror becomes greater than unity at high frequencies. Such a situation, along with parasitic capacitance on port X, can cause the loop gain around these two mirrors to exceed unity, with oscillation as a potential result. The solution is to employ matching emitter degeneration resistors in each of the Wilson mirrors and then use capacitors to roll off the high frequency gain. The RC time constant is chosen to optimize the transient response to a critically damped characteristic, consistent with the parameters of the transistors utilized. These resistors have the added advantage of improving the accuracy of the mirrors. As well, an additional diode connected in series with the collector of one of the transistors in the Wilson mirror will also improve its accuracy. This device can therefore be added in the collector leads of *Q*2, *Q*5, *Q*8, and *Q*10 in Figure 12. The current-mirror block on the PA630 also uses this optimized Wilson architecture.

The basic structure of Figure 12, along with the above improvements and the following two enhancements, make up the complete accurate current-conveyor topology shown in Figure 13. The Wilson current-mirror has an output resistance that is much improved over a simple mirror. However, this is still only half that which can be obtained with a pure cascode circuit. As it is desirable to make the output resistance of port *Z* as high as possible in order to minimize distortion, the actual PA630 circuit incorporates a novel scheme whereby the performance of a pure cascode is obtained without introducing any significant errors into the Wilson base current cancellation arrangement. Referring to Figure 13, PNP transistors *Q*15 and *Q*20 have been added to effect this improvement. *Q*20 is the cascode transistor which raises the output resistance of the modified PNP Wilson mirror consisting of *Q*16, *Q*17, *Q*18 and *Q*19. The base of *Q*20 is biased from the NPN current splitting mirror (*Q*11, *Q*12, *Q*13, *Q*14) in such a way that its base current is shunted up to the rail with the unused half of the split current. As a result, the accuracy of the NPN splitter mirror is not compromised. There is, however, a potential for a major inaccuracy in the output current, as base current is subtracted from it as it flows from the emitter to the collector of *Q*20. This is reduced to a negligible error by introducing a balancing base current with *Q*15. Since this transistor and its emitter resistor match *Q*17 and *Q*18, two base currents are injected into the mirror at *Q*18 to cancel the first loss in *Q*19 and the second loss in *Q*20. The collector of *Q*15, through which a "wasted" current flows, can be connected to the collector of *Q*13 which carries the unused half of the split current. These two extraneous currents are very nearly equal, and so virtually cancel. This allows the collectors of *Q*13 and *Q*15 to be tied to any convenient bias point about two diode drops below rail, even if it is relatively high impedance, as only a small difference current needs to be supported.

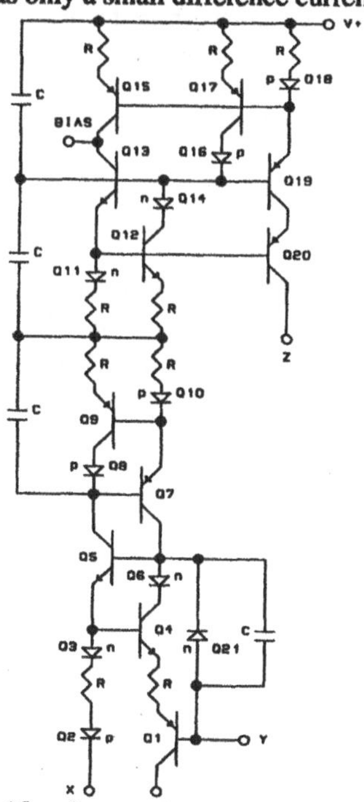

Figure 13 Accurate conveyor topology.

As mentioned earlier, additional devices Q_1 and Q_2 in Figure 13 provide flexibility in the application of the conveyor. If the collector and base of Q_1 are shorted together, then a first generation current-conveyor is produced with port X at the emitter of Q_2 and port Y at the connection of the base and collector of Q_1. The operation of the block is normal, except that an additional diode drop has been added in series with ports X and Y respectively. If however, Q_1 is utilized as a transistor, a second generation current-conveyor is realized. In this configuration, the collector of Q_1 is returned to a separate more negative rail and the base becomes port Y. Relative to the simpler arrangement, the current at port Y is reduced by the current gain of Q_1, and this can be advantageous in some applications.

In summary, this topology provides high accuracy, optimized transient response, and enhanced output impedance. As a result, the PA630 is able to deliver full scale THD performance in the order of 0.01% without employing any global negative feedback. Therefore, this current-conveyor architecture makes it possible to offer a digital audio product which eliminates the possibility of dynamic distortion at the expense of a moderate (and almost certainly inaudible) degradation in steady state THD. Ultimately, the validity of this trade-off will be determined by discerning listeners, and DiAna appears to be receiving their endorsement on a global basis.

11.3.4 Conclusions

While the preceding descriptions relate to two particular products and configurations, they should nevertheless aid in developing a more general understanding and appreciation of the utility and advantages of this versatile building block. As many researchers and authors can attest, the current-conveyor has already received critical acclaim on a theoretical level. Its commercial success is certainly more difficult to evaluate, and in a sense, the review is in the process of being written. If we extrapolate from the previous examples, it is probable that many successes based on custom current-mode ASICs are unsung, and that these techniques are incorporated into a significant number of lucrative products. Astute designers are likely to nurture and expand this trend in the future.

VISTA is a trademark of Northern Telecom

References

[1] K. C. Smith and A. Sedra, "The Current Conveyor: A New Circuit Building Block", *Proc. IEEE*, Vol. 56, pp. 1368-1369, August 1968.

[2] CCII01 Data Sheet, LTP Electronics Ltd., 27 Park End Street, Oxford, OX1 1HU, UK., tel/fax +44-865-200767.

[3] PA630/PA630A Data Sheet, Phototronics Co. Regd., Manotick, Ontario, Canada.

[4] VISTA Product Brochure, Northern Telecom Ltd., 2920 Matheson Blvd., Mississauga, Ontario, Canada.

[5] A. Lavia and P. Hollink, "Call Management Service", *Proc. Int. Switching Symp.*, Yokohama, Japan, October 1992.

[6] M. Otala, "Transient Intermodulation Distortion in Commercial Audio Amplifiers", *J. Audio Engr. Soc.*, Vol. 22, pp. 244-246, May 1974.

[7] D. C. Wadsworth, "Accurate Current Conveyor Integrated Circuit", *Electronics Letters*, Vol. 25, No. 18, pp. 1251-1253, August 1989.

[8] D. C. Wadsworth, "A Professional Audio Integrated Circuit", presented at the *Audio Engineering Society 87th Convention* (AES Preprint available), New York, October 1989.

[9] Dacula Audio DAC Data Sheet, Select Systems Ltd., Feltham, Middlesex, UK.

[10] DiAna Audio DAC Data Sheet, Sentec AB, Kopparberg, Sweden.

[11] *Musik & Ljudteknik Journal*, Vol. 34, No. 1, May 1993, LTS Society, Stockholm, Sweden.

[12] *Hifi & Musik Magazine*, No. 6/7, Summer 1993, Stockholm, Sweden.

[13] *Audio Review Magazine,* Vol. 13, No. 10, October 1993, Mensile, Italy.

[14] *Audio Ideas Guide Magazine*, Vol. 13, No. 2, Winter 1994, Toronto, Canada.

[15] *High Fidelity Magazine*, Vol. 7, No. 2, February/March 1994, Uppsala, Sweden.

[16] *The Audio Critic Magazine*, Vol. 21, Spring 1994, Quakertown, PA, USA.

[17] D. C. Wadsworth, "Accurate Current Conveyor Topology and Monolithic Implementation", *IEE Proceedings*, Part G, Vol. 137, No. 2, pp. 88-94, April 1990.

[18] D. C. Wadsworth, "The Performance and Applications of a New Current Conveyor Integrated Circuit", *Proc. IEEE Int. Symp. on Circuits and Systems*, Vol. 1, pp. 287-290, May 1990.

[19] C. Toumazou *et al* (ed.), *Analogue IC Design: The Current-Mode Approach*, Chapter 15, pp. 535-568, Peter Peregrinus Ltd., London, 1990.

[20] Applications, "Audio Current Conveyor", *Electronics and Wireless World*, Vol. 96, No. 1649, pp. 246-250, March 1990.

[21] US patent 4,983,930; UK patent 2,227,139; Canadian patent 1,281,386; PCT Application CA9000049.

[22] K. C. Smith and A. Sedra, "A Second Generation Current Conveyor and its Applications", *IEEE Trans. Circuit Theory*, Vol. CT-17, pp. 132-134, February 1970.

[23] G. R. Wilson, "A Monolithic Junction FET-NPN Operational Amplifier", *IEEE J. Solid-State Circuits*, Vol. SC-3, pp. 380-387, December 1968.

Chapter 11.4

Current-Mode Oscillators

Sittichai Pookaiyaudom

11.4.1 Introduction

Recently there is a strong interest in applying the current-mode technique [1] to electronic circuit designs. One important class of circuit that one can apply the current-mode technique advantageously to is the oscillator. The following are the main reasons for the present interest in current-mode oscillators or CMOs

1. It is relatively simple to generate high level high frequency oscillations in CMOs. Slew-rate limitations associated with more traditional voltage-mode oscillators or VMOs, especially in the operational amplifier based class of circuits are not significant in most current-mode circuits.
2. Since only fairly small voltage swings are needed across current-mode circuits; it is relatively easy to design CMOs that will operate with low power supply voltages of only a few V_{BE}s. Realization of micro-power oscillators is also quite straight-forward using the current-mode technique.
3. CMOs usually require less complicated circuits when compared to the more traditional VMOs. Thus, they are more suitable for use as parts of a larger system in an integrated circuit.
4. Lots of new ideas are still possible. The CMOs offer a very fertile field of research. In the author's opinion, the "best" CMO has not yet been invented!

11.4.2 Current-mode building-blocks

In addition to some passive circuit components, CMOs usually employ one or more of the following active building-blocks, namely, current-mirrors (CMs), current-conveyors (CCIIs), current followers (CFs), differential pairs, and some form of translinear cells [1]. However, the two most important building-blocks are the current-mirrors and the current-conveyors of both the positive and the negative types.

In high frequency oscillators, the simplest CMs consisting of only a matched transistor pair is preferred, even if at low frequencies, the output impedances of such CMs are quite low of the order of the Early-effect resistance, r_o, of the output transistors. However, the output impedance level changes with frequency more slowly than other more complicated CMs, which may possess much higher low frequency output impedances, of the order of $\beta_o r_o$ of the output transistor, where β_o is the low frequency current gain of the output transistor. However, the magnitudes of the output impedances of these more complicated CMs drop drastically when the frequencies are higher than only a few tens of kHz. Figure 1 shows the plots of the magnitudes of the output impedances of several types of CMs that are widely used in CMOs. This problem of rapid decrease of the magnitudes of the output impedances with frequencies is also inherent in the CCII's, and has to be taken into account in high frequency applications of CCII based circuits.

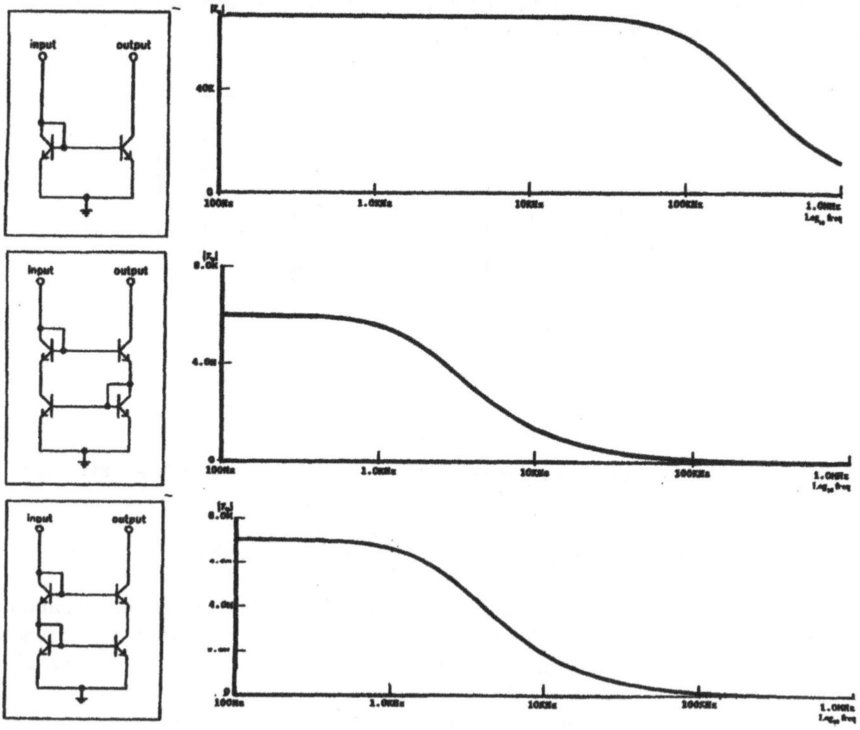

Figure 1 Magnitudes of the output impedances versus frequencies of several types of CMs.

CMOs generally employ differential pairs and/or translinear cells and it is these building-blocks that are usually used to adjust the loop gain of the oscillator. Another method of gain adjustment is to shunt signal currents away from the inputs of CMs. Figure 2 suggests some ideas of getting the required gain increase or decrease using only unity-gain CMs. Electronically adjustable gain blocks for automatic loop-gain adjustments will always be needed in practical oscillators, since any slight disturbances to the oscillation, for example, when the frequency is being changed, will cause amplitude variations and even ceasation of oscillation. The requirement for some form of automatic gain adjustment in oscillators is an inconvenient fact of life, which must always be taken into account in realizing real oscillators as opposed to the usually simple theoretical proposals of novel oscillator configurations reported in many publications.

Two techniques reported recently which appear to be promising for applications in CMOs are the log-domain circuit technique [2] and the switched-current technique [3]. In the first approach distortionless responses from networks consisting of forward biased PN-junctions and capacitors are used to realize so called log-domain filters. Such an approach also appears to be ideal for CMO applications, since the oscillations will be virtually distortionless and electronic frequency tuning can be easily incorporated. The usefulness of the switched-current approach is based upon the analogy to existing switching capacitor techniques, which have proven to be effective in the realizations of stable VMOs. Thus various frequency selective blocks realized by switched-current techniques should also be useful in the realizations of CMOs, but potentially at higher frequencies than are obtainable from switched-capacitor based counterparts.

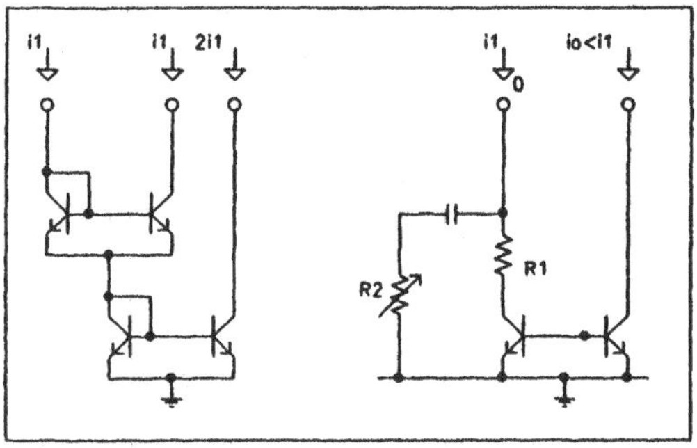

Figure 2 Some ideas on providing gain adjustments using only unity-current-ratio CM's.

11.4.3 Fundamental operations in current-mode circuits

The main difference in the operation between voltage-mode and current-mode circuits can best be illustrated with an example to obtain the order *RC* low-pass and high-pass functions. Figures 3a and 3b show first order low-pass and high-pass voltage-mode circuits respectively, where it is seen that the *RC* components in both circuits basically function as voltage dividers. However, in the current-mode version as is shown in the circuit of Figure 3c, the parallel connection of the *RC* components function as a current divider, where the current flowing through the resistance is the low-pass current, while the current flowing through the capacitance is the high-pass current. Therefore, the free ends of either the resistor or the capacitor must be connected to a low impedance point, for example, the input of a CM, and the extracted low-pass current or high-pass current can then be put to use via the output of the CM. The ability to visualize basic operations with currents will be important in the design of proper current-mode circuits, including CMO's.

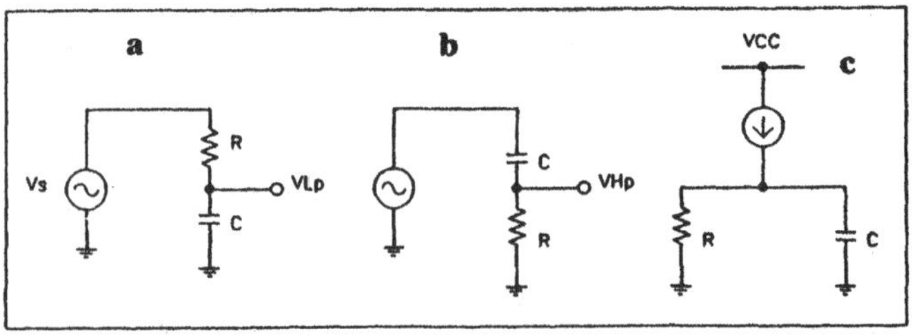

Figure 3 First order low-pass and high-pass transfer function using voltage-mode and current-mode circuits.

Furthermore, in designing CMOs, we should consider two ranges of frequencies; namely, the high frequency range, where the oscillation frequencies are higher than a few hundreds of kHz to a few tens of MHz, and the low frequency range, where the oscillation frequency is usually less than 100 kHz. In the first case, the high frequency output impedance of the building-blocks of the CMO is quite low, and careful attention must be paid to ensure that all passive impedances are quite low if accurate operations of the CMOs are to be expected. However, in the latter case where the magnitudes of the output impedances of the active building-blocks are still very high, there are lots of opportunities in using the high output impedance magnitudes of the active building-blocks in novel and clever circuit techniques.

11.4.4 Basic theory of CMOs

The complete analysis of any oscillator is quite difficult, since it is inherently a non-linear circuit. However, traditionally one usually assumes quasi-linear operating conditions, that is the loop-gain is assumed to be almost equal to unity; therefore, the resulting non-linear level-saturation effects are very small. Assuming such conditions, then the analysis is fairly simple to derive the required gain conditions in the active building-blocks and the resulting oscillation frequency.

In circuits where the forward-gain part and the feedback part can be identified easily, one can write the loop-gain equation, and let the equation be equal to unity, which by definition is the necessary condition for steady-state sinusoidal oscillation. If we let $A(\omega)$ be the forward gain and $B(\omega)$ be the feedback factor and assuming that the input impedances are infinite and the output impedances are zero for both circuit parts, then we can write the loop-gain equation as:

$$A(\omega)B(\omega)=1 \tag{1}$$

where both $A(\omega)$ and $B(\omega)$ may be complex numbers.

Equation (1) implies that the real part and the imaginary part must be simultaneously equal to unity and zero respectively. Usually the solution of both conditions will yield the oscillation frequency and the conditions that must be satisfied for the loop-gain to be exactly unity to maintain a steady oscillation amplitude. Although such an approach is quite straight forward, it does not give much insight into the operation of the circuit, especially the circuit mechanism for the determination of the oscillation frequency and the required gain adjustments. In addition if the forward gain part and the feedback part cannot be separated, then such an approach breakdowns.

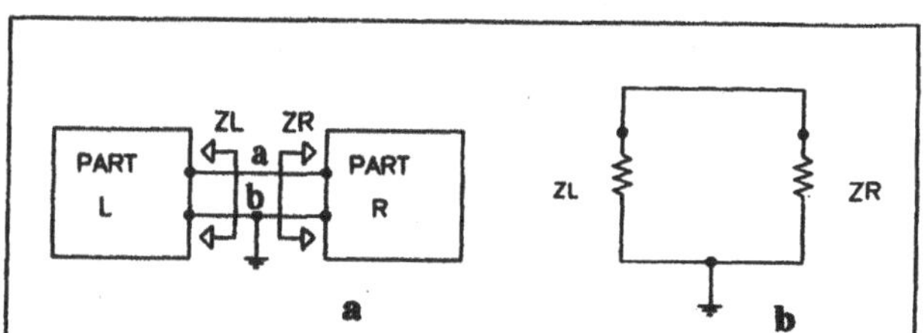

Figure 4 Arbitrary separation of an oscillator into two parts.

A more general and insightful method preferred by the author is to use the matched impedance approach, where the oscillator is separated at an arbitrary point into two parts, namely, parts *L* and *R*. These two parts are still connected at points *a* and *b*, where *b* is also the common-point of the oscillator circuit, as is shown in Figure 4a. Let Z_L and Z_R be the impedances looking into parts *L* and *R* respectively. These two impedances can be derived most conveniently by inserting a test voltage source across the points *a* and *b*, and then determine the resulting currents flowing into the LHS and the RHS parts respectively. In most oscillators, the first order approximated impedances can be obtained with ease. However, when more detailed expressions are required, a symbolic network analysis program may be used to obtain the results without excessive derivation time and involved mathematics.

The resulting equivalent circuit of the oscillator is shown in Figure 4b. Both Z_L and Z_R may be functions of the generalized frequency *s*. Furthermore, if we assume a steady state sinusoidal oscillation mode, that is, both the frequency and the amplitude of oscillations are constant, then using Kirchoff's Voltage Law, with a common current flowing through Z_L and Z_R, we can write

$$Z_L(j\omega_o)+Z_R(j\omega_o)=0 \qquad (2)$$

where for steady-state sinusoidal oscillation condition we have $s=j\omega_o$, and ω_o is the angular frequency of oscillation.

From equation (2), we can equate both the real and the imaginary parts to zero simultaneously to obtain the equation of the angular frequency of oscillation ω_o and also the conditions for unity loop-gain to occur. Usually, the imaginary part of equation (2) consists of the reactive parts of Z_L and Z_R which are frequency dependent, whereas the real part of equation (2) consists of the resistive parts of both positive and negative types which may or may not be dependent upon the oscillation frequencies. Using such an approach, one can usually draw a conceptual equivalent circuit showing how the frequency of oscillation is selected by the equivalent components of the circuit and how the unity loop-gain condition can be maintained. Examples 1 and 2 show the use of the impedance matching method in deriving the oscillation frequency and the required gain of a traditional Wien-Bridge VMO and a simplified and idealized delay-line controlled oscillator respectively. Although the two examples do not deal directly with CMOs, they are chosen to illustrate convincingly the usefulness of the impedance matching approach.

11.4.4.1 Example 1

Figure 5 shows the conventional operational amplifier based Wein-bridge VMO, showing arbitrarily chosen circuit separation points *a* and *b*. Using routine circuit analysis we can show that Z_L and Z_R are given by

$$Z_L=\frac{R}{(1-A)}+\frac{1}{s(1-A)C'} \qquad (3)$$

$$Z_R=\frac{1}{1+sRC'} \qquad (4)$$

where $A=1+R_1/R_2$ is the closed-loop voltage gain of the amplifier which is greater than unity. Therefore, we can see immediately that Z_L consists of a series connection of a negative resistance and a negative capacitance, whereas Z_R is simply a parallel connection of a resistance and a capacitance.

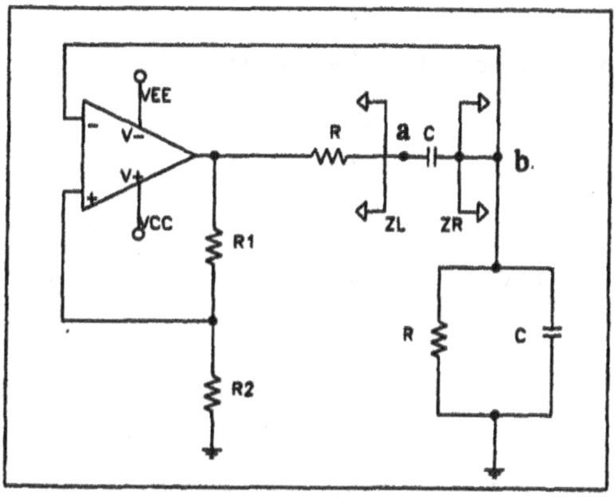

Figure 5 A conventional Wien-bridge VMO.

Upon equating the imaginary and the real parts of $Z_L + Z_R$ to zero, we can derive the well known expression for the angular oscillation frequency as $\omega_0 = 1/RC$, and the required voltage gain of 3 for A.

Since the point a is chosen arbitrarily, it can be verified readily that one can obtain the same end results by chosen other convenient points to separate the circuit.

11.4.4.2 Example 2

Figure 6 shows the simplified equivalent circuit of an ideal delay-line controlled oscillator, where the portion enclosed by the dotted line represent a delay-line having a delay-time of τ, with R_I and R_O the input and the output resistances respectively. The delay-line bandwidth has been assumed to be infinite.

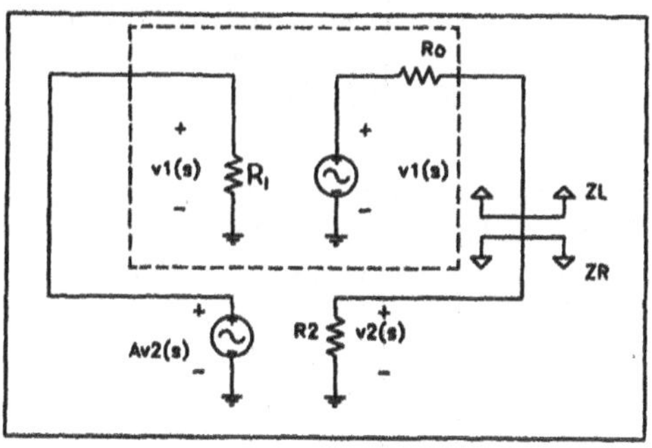

Figure 6 Equivalent circuit of an ideal delay-line controlled oscillator.

Again using routine circuit analysis, we can show readily that Z_L and Z_R are given by

$$Z_L = \frac{R_O}{[1 + A_v exp(-s\,\tau)]} \tag{5}$$

$$Z_R = R_2 \tag{6}$$

where R_2 is the input resistance of the ideal voltage amplifier. Upon substituting $j\omega_o$ in place of s, and equating both the real and the imaginary parts of $Z_L + Z_R$ to zero, then we can derive the oscillation frequency as:

$$\omega_o = \frac{n\pi}{\tau} \tag{7}$$

where n is an integral number denoting the mode number of oscillations. The required voltage gain is given by

$$A_v = \pm\left(1 + \frac{R_o}{R_2}\right) \tag{8}$$

where A_v is positive for even n and negative for odd n.

11.4.5 Approaches to the invention of new CMOs

Before discussing some of the useful CMO structures, one should mention some of the approaches that are being employed by researchers in deriving new oscillator structures, in particular new CMOs.

1. The first approach will be called the exhaustive search method, where for a given structure with a pre-defined number of active building-blocks and passive components, exhaustive permutation is made to every possible structure, and some useful configurations can then be selected. Unfortunately, such an approach is tedious mathematically and counter-intuitive to circuit designers. Nevertheless, one does come across such derivations from time to time in the literature! Fortunately, with the recent availability of several symbolic network analysis programs, such exhaustive search methods become more insightful and thus assist in generating new oscillator structures. A symbolic network analysis program [4] effectively removes the mathematical drudgeries involved in solving a large number of circuits. In addition, one can be more confident on the solutions, as opposed to the results from hand derivations.

2. The second approach is an intuitive one, using one's understanding of how the oscillation frequency is to be determined and how the loop-gain magnitude can be adjusted, which often results in the invention of elegant new structures. This approach is preferred by the author.

3. The third approach is to convert an existing VMO structure with existing desirable properties into an CMO, with the hope that additional performance features associated with a current-mode approach can be transferred to the new oscillator design, while preserving all the original oscillator properties. However, such an approach sometimes does not work for example, when a very good VMO in the form of a cascade of two first order phase-shifters is converted into its corresponding current-mode version, the original advantageous property inherent in using a very high loop-gain the operational amplifier is lost. This is due to the fact that extremely large current gain is usually not available from the basic building-blocks of current-mode circuits. However, closed loop system gain accuracy in voltage-mode circuits one could argue is analogeous to current transfer accuracy in CMs associated with current-mode circuits.

With the availability of symbolic network analysis programs, the combinations of all three approaches seem to be logical in the future invention of new CMOs. For example, often the first two approaches can be used to create some new CMOs, and using the exhaustive search approach to ensure that every possible related circuit in the family is identified and examined.

11.4.6 Some desirable oscillator properties

With the rapid advance in CMO research, new CMO structures are being invented all the time. The following are some of the significant properties that should be aimed for in any new family of CMOs

1. Frequency tuning should be done with a single non-reactive component, usually a resistor. However, for high frequency operations, and in some crystal controlled CMOs the use of varactor diodes may have to be tolerated. If more than one component are to be tuned, then accurate tracking of component values should not be necessary.
2. One should be able to control the loop-gain magnitude independently of the oscillation frequency. There are many structures that offer such a property, at least within the first order approximations of the circuits. However, most structures still exhibit some dependency between the loop-gain control and the frequency of oscillation control via some second order circuit properties.
3. Wide tuning range ability should be provided. This property will most likely demand more complex circuit designs than the basic core circuits.
4. All capacitors should be grounded. This is to enable fabrication as an integrated circuit. However, many attractive CMO structures do exist that do not possess such property. Also there are the high frequency properties of using grounded capacitors.
5. Some easy means of automatic oscillation amplitude control should be provided. This is an extremely important requirement for any oscillator, since without automatic oscillation amplitude control, the slightest changes in the loop-gain magnitude can cause the oscillation amplitude either to saturate or to terminate!
6. For *RC* oscillators, it is relatively meaningless to talk about short term frequency stabilities, since none of them are very stable. However, one should ensure that that the frequency of oscillation does not depend upon some active element parameter, since active element parameters will usually have strong dependences upon the ambient temperature, and as a rule, they do not age consistently.

11.4.7 Automatic amplitude control (AAC)

As has been mentioned in the previous section, AAC must always be incorporated into sinusoidal oscillators. This remains one of the most difficult aspects of oscillator design. Current-mode techniques offer some real advantages in the accurate detection of the high frequency oscillation amplitude due to the invention of new rectifier circuits [1]. However, the irony of the story is that in almost every oscillator, the AAC portion of the circuit is usually more complicated than the oscillating core itself. Two significant properties the oscillator designer should look for in good AAC circuits are

1. Fast amplitude detection. This usually means that the detection of the peak amplitude should be able to be completed within a single or less oscillation cycle, which is a

very difficult problem indeed especially at high oscillation frequencies. When the oscillation frequencies are low and if the amplitude detection time of the AAC is slow, then the oscillator may exhibit amplitude hunting phenomenon during the changing of the oscillation frequencies.

2. Large gain control range. Once the oscillation amplitude has been detected, the resulting signal is used to adjust the loop-gain magnitude, usually by changing the dynamic resistances of a FET or a diode, or by controlling the division of currents between the two outputs of a differential pair. Large gain changing range in the AAC circuit is desirable since it can help in facilitating a wider frequency tuning range.

Surprisingly, after many decades of development, there still exists many potential new ideas in the AAC circuit design. Applying current-mode techniques should give a new impetus to the future design of high-performance AAC circuits.

11.4.8 Examples of CMOs

In this section we will briefly discuss several practical examples of CMOs. All of the circuits discussed have been experimentally tested.

11.4.8.1 Current-mirror phase-shifter CMO

Figure 7 shows the circuit diagram of a high performance current-mirror phase-shifter oscillator [5] operating in current-mode, since only unity gain CMs have been employed.

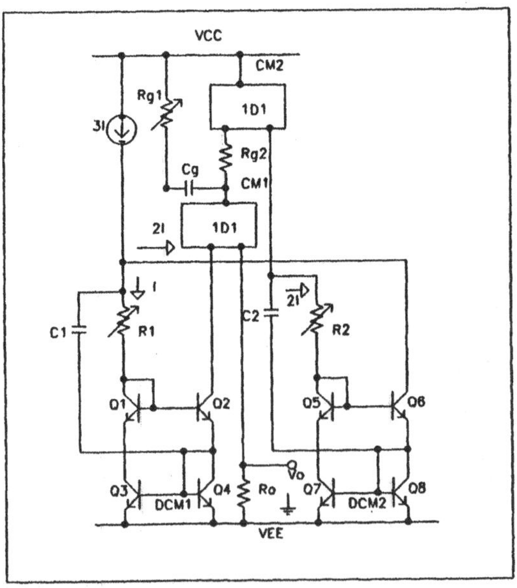

Figure 7 A current-mirror phase-shifter oscillator.

The two lower CMs (Q_1 to Q_4) and (Q_5 to Q_6) together with (R_1, C_1) and (R_2, C_2) respectively, function as first order phase-shifters, having the following current transfer function

$$\frac{i_o}{i_i} = \frac{(1 - s\tau)}{(1 + s\tau)} \qquad (9)$$

where τ is approximately equal to either $(R_1 C_1)$ or $(R_2 C_2)$. At around the oscillation frequencies, the two cascaded phase shifters provide a phase shift of π, and the two upper CMs provide the additional π phase shift, as well as providing some additional gain due to circuit imperfections. It can be shown that the angular oscillation frequency is given by

$$\omega_o = \frac{1}{\sqrt{R_1 R_2 C_1 C_2}} \qquad (10)$$

The present CMO can easily operate up to a frequency range of several MHz. One salient feature of such CMO is that very little AAC action is required, since the amplitude of the two phase-shifters are quite constant and close to unity. Frequency tuning can be done by changing either R_1 or R_2 or both at the same time, with no necessity for value tracking between R_1 or R_2. Alternatively, an electronically tunable current-mode phase-shifter [13] can also be implemented which will enable the resulting CMO to be electronically tunable.

11.4.8.2 RC current-mirror CMO

Figure 8a shows an example of practical *RC*-CMO [6], using two unity gain CMs, (Q_1 to Q_4) and (Q_5 to Q_7), where it can be shown that the current gain from the two CM's in conjunction with R_1 and R_2 can be given approximately by

$$A_i = \frac{i_1}{i_i} \cong 2\left(1 + \frac{R_1}{R_2}\right) \qquad (11)$$

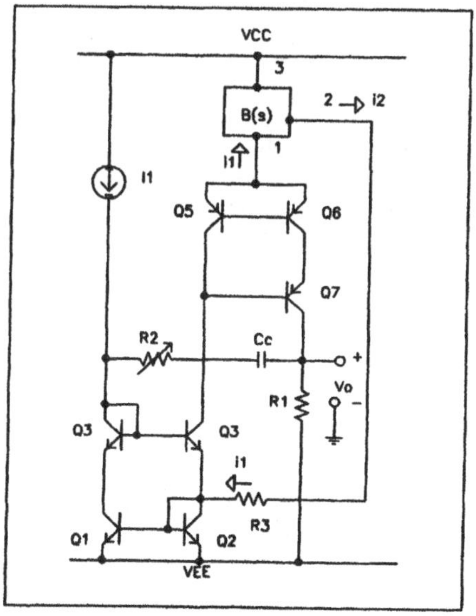

Figure 8a A current-mirror based CMO.

Figure 8b shows three possible feedback networks all possessing the same Wien-like band-pass characteristic of the form

$$B(s)=\frac{s\tau}{s^2\tau^2+3s\tau+1} \tag{12}$$

where $\tau = RC$.

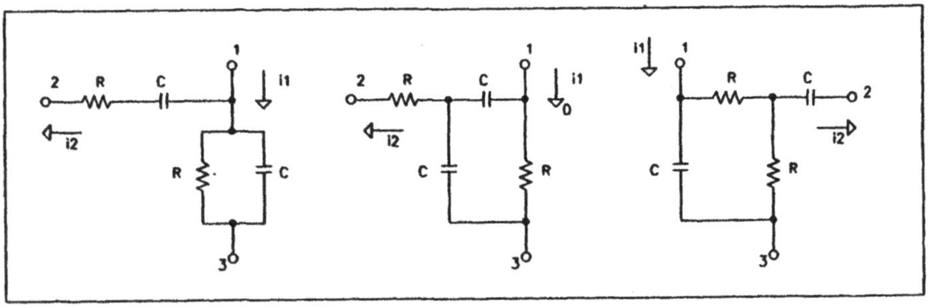

Figure 8b Three possible feedback networks all possessing the same Wien-like band-pass

The angular oscillation frequency is given by $\omega_o = 1/RC$, and unity loop-gain occurs when the ratio $R_2/R_1 = 1$. Readers should refer to reference [6] for a more detailed operation of the circuit, including a simpler conceptual circuit.

One major disadvantage of the CMO in Figure 8 is that the two resistors in the feedback network must track each other. Therefore, the CMO cannot be used without an AAC circuit.

11.4.8.3 Dynamic resistance controlled CMO

Figure 9 shows an electronically tunable CMO [7], using the small signal dynamic resistance of forward biased pn-junctions, where the differential pair Q_1 and Q_2 in conjunction with the capacitance C_1 creates the high-pass current, and the differential pair Q_3 and Q_4 in conjunction with the capacitance C_2 forces differential low-pass currents to flow into the emitters of Q_3 and Q_4 respectively. Thus the overall response possesses band-pass characteristics. Transistors Q_5 and Q_6 provide the required gain for steady state sinusoidal oscillations.

Using first order circuit approximations, we can readily show that the angular oscillation frequency is given by

$$\omega_o = \frac{I_1}{2V_T C} \tag{13}$$

where V_T is the thermal voltage associated with a forward biased PN-junction, I_1 is the bias current, and we have assumed matched capacitances, $C = C_1 = C_2$.

The frequency of oscillation can be tuned electronically over a range of four orders of magnitude. An upper frequency of oscillation of around 50 MHz was obtainedwith this design using transistor arrays. Also the bias currents can be designed to possess positive temperature coefficient to compensate for the temperature dependence of the dynamic resistance of the forward biased PN-junctions. The distortion level has been demonstrated experimentally to be lower than 0.1 percent [7]. The main disadvantage of the CMO of Figure 9 is its low amplitude of oscillation, which requires high voltage gain in the output stage to boost the signal up to a useful level, resulting in a relatively high output noise level and hence low signal to noise ratio.

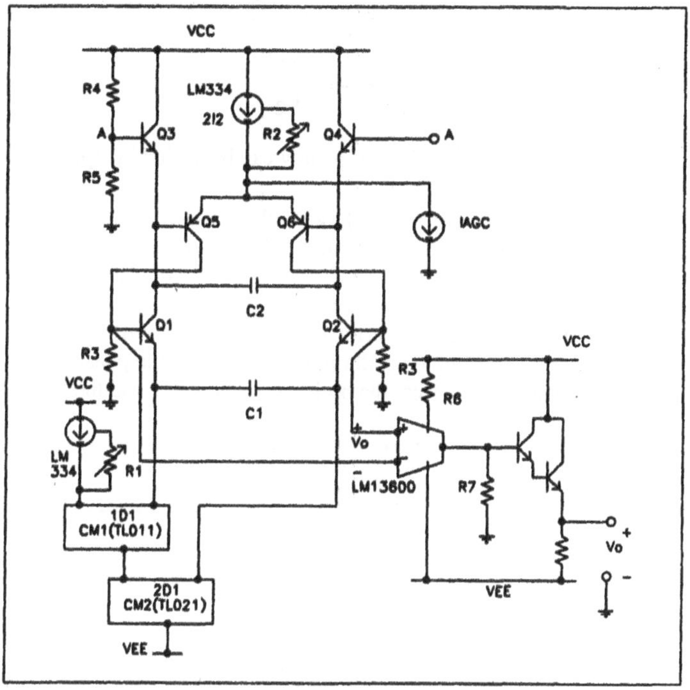

Figure 9 An electronically tunable CMO.

11.4.8.4 Current-conveyor CMO

Figure 10 shows a high performance single CCII+ [12] based CMO. Presently, CCII based CMO realizations are receiving much interest from several research groups, for both the single and double CCII based structures.

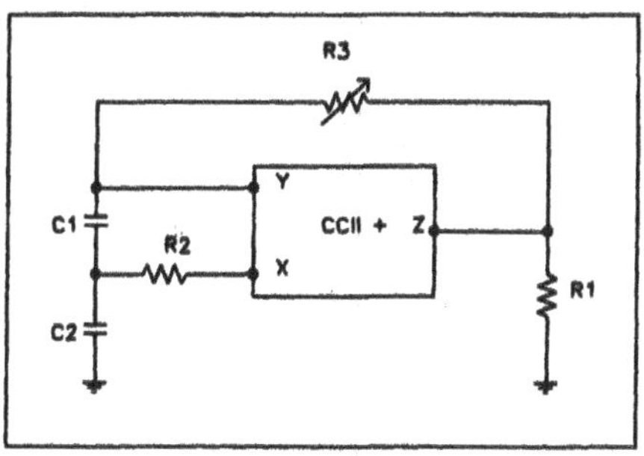

Figure 10 A current-conveyor based CMO.

Using the matching impedance analysis concept, and assuming ideal CCII+ properties, we can readily show that the angular oscillation frequency is given by

$$\omega_o = \frac{1}{\sqrt{C_1 C_2 R_2 (R_1 + R_3)}} \tag{14}$$

and the condition for unity loop-gain requirement is given by

$$\frac{R_1}{R_2} = 1 + \frac{C_1}{C_2} \tag{15}$$

where it can be seen that the oscillation frequency can be varied independently of the unity loop-gain condition by changing only R_3.

Experimentally, using a commercially available CCII+ [8], an oscillation frequency of around 30 MHz can be obtained readily. It should be noted here that Figure 10 shows only the core-CMO. Additional output signal buffering is required; also an AAC circuit will be required in the practical oscillator. Therefore, equations (14) and (15) represent only the idealized circuit. In real circuits various imperfections will force increasing gain requirements for constant amplitude oscillation. In addition, the phase shift contributed by the CCII at high frequencies will affect the oscillation frequency and the gain requirement.

11.4.8.5 Unity-gain current-mirror CMO

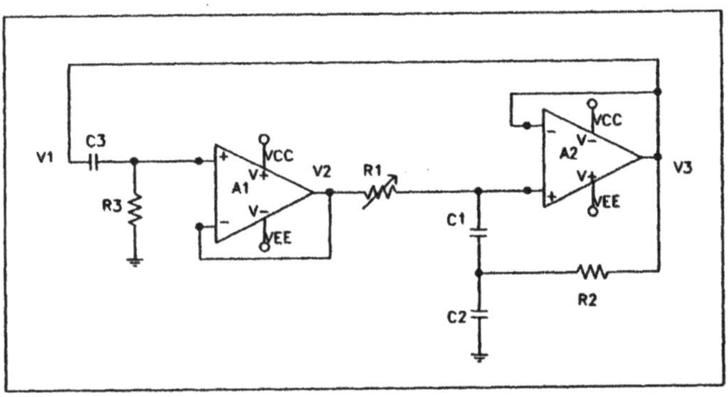

Figure 11a A unity-gain only VMO.

Figures 11a and 11b show an example of the conversion of a VMO [9] in Figure 11a, to a CMO [10] in Figure 11b. Assuming ideal device behaviors, and using matching impedance method, one can show that the angular frequency of oscillation is given by

$$\omega_o = \frac{1}{\sqrt{R_1 R_2 C_1 C_2}} \tag{16}$$

and the unity loop-gain condition requirement is given by

$$R_3 C_3 = R_2 (C_1 + C_2) \tag{17}$$

where, again, it is seen that the frequency of oscillation will be independent of the unity loop-gain condition when R_1, in this case, is used for frequency tuning.

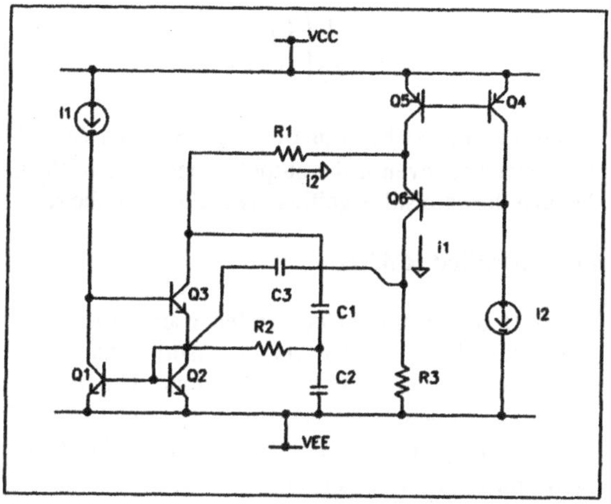

Figure 11b The correponding current-mode version.

The oscillator of Figure 11b also possesses another interesting property, within the frame of a first order approximation, that only unity gain active devices are necessary. In the practical version at high frequencies, however, additional gain stages may be required.

11.4.8.6 LC-CMO

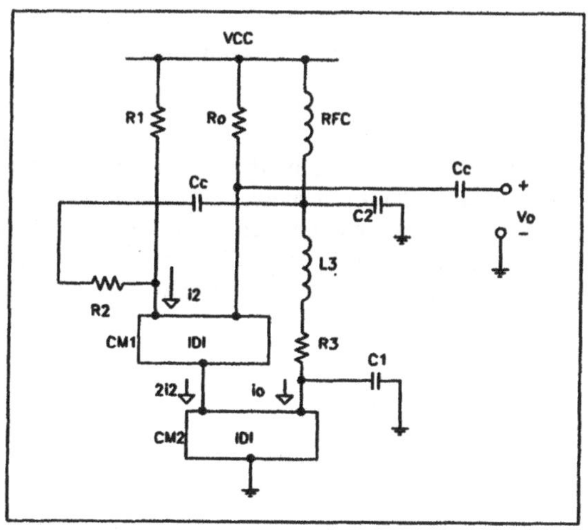

Figure 12 A current-mode *LC* oscillator.

Figure 12 shows a Colpitts-type *LC*-CMO, where it seen that CMs offer ideal properties for *LC* oscillators, their input impedance is low and their output impedances are high. Again using impedance matching analysis, we can show that the square of the angular frequency of oscillation is given by

$$\omega_o^2 = \frac{1}{L_3}\left(\frac{C_1 + C_2}{C_1 C_2}\right) \tag{18}$$

and the oscillation gain can be set by adjusting R_2. Such an LC-CMO can be used to generate quite high frequencies, even if the output impedances of the CMs are low at high frequencies, provided that they are still significantly higher than the reactance of C_1.

11.4.8.7 Crystal controlled CMOs

Figure 13 shows a CCII based crystal controlled oscillator, where using impedance matching analysis one can show that the crystal assumes an inductive property when resonates with the series connection of the two capacitors. The frequency dependent series resistance of the crystal is overcome by the negative resistance from an FDNR generated by the CCII in conjunction with C_1, C_2 and R_2. With the availability of commercially based CCII chips, such realizations are very practical.

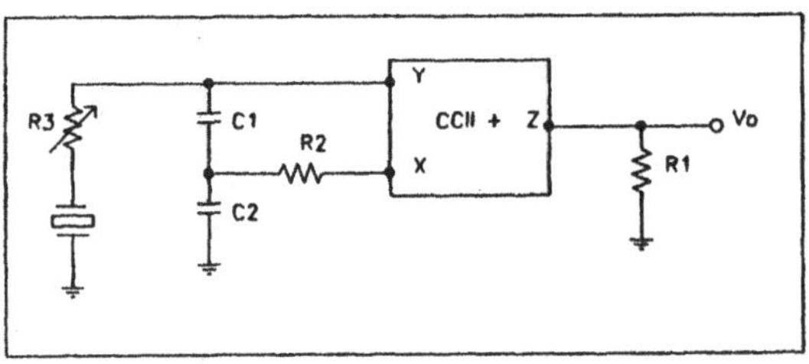

Figure 13 A crystal controlled CMO.

11.4.9 Conclusions

In this chapter we have described some of the advantages of using the current-mode approach in the realization of oscillators, namely, the potential for high frequency operation, and the simplicity in circuit design compared with the more classical voltage-mode counterparts. Research in the area of CMOs is still very fruitful. However, we are still at the tip of the iceberg, with much potential for further development and innovation of practical CMOs.

References

[1] C. Toumazou, F.J. Lidgey and D.G. Haigh, *Analogue IC Design: the Current-Mode Approach*, IEE Circuits and Systems Series 2, Peter Peregrinus Ltd., 1990.

[2] D. Frey, "Current Mode Class AB Second Order Filter", *Electronics Letters*, Vol. 30, No. 3, pp. 205-206, 1994.

[3] C. Toumazou, J. B. Hughes and N. C. Battersby, *Switched-Currents: An Analogue Technique for Digital Technology*, IEE Circuits and Systems Series 5, Peter Peregrinus Ltd., 1993.

[4] S. Jantarang and S. Pookaiyaudom, "SNAP - Symbolic Network Analysis Program, Users' Handbook", Mahanakorn University of Technology, No. 51 Cheum-Sampan Road, Nong-Chok District, Bangkok 10530, Thailand, 1994.

[5] S. Pookaiyaudom and K. Samootrut, "Current-Mirror Phase-Shifter Oscillator", *Electronics Letters*, Vol. 23, No. 1, pp. 21-23, 1987.

[6] S. Pookaiyaudom and K. Samootrut, "Current-Mirrors RC-Oscillators", *Int. J. Electronics*, Vol. 62, No. 6, pp. 803-808, 1987.

[7] S. Pookaiyaudom, B. Srisuchinwong and W. Kurutach, "A Current-Tunable Sinusoidal Oscillator", *IEEE Transactions on Instrumentation and Measurement*, Vol. IM-36, No. 3, pp. 725-729, 1987.

[8] CCII01 - Current Conveyor Amplifier Data Sheet, LTP Electronics Ltd., 27 Park End Street, Oxford OX1 1HU, UK, tel/fax +44-865-200767.

[9] R. Senani, "New RC-Active Oscillator Configuration Employing Unity-Gain Amplifiers", *Electronics Letters*, Vol. 21, No. 20, pp. 889-891, 1985.

[10] S. Pookaiyaudom and J. Ngarmnil, "Transistor -Level Consideration of an RC Current-Mode Oscillator Using Unity-Gain Current-Mirrors",to be presented at the *1994 International Symposium on Circuits and Systems*, 30 May-2 June, London.

[11] S. Pookaiyaudom and K. Saivichit, "RC Phase-Shifter Sinusoidal Oscillators Using Initial Condition-Restoration Amplitude Control", *IEEE Transactions on Instrumentation and Measurement*, Vol. IM-39, No. 6, pp. 1038-1044, 1990.

[12] S. Pookaiyaudom and K. Samootrut, "Efficient Circuit Implementation of Current Conveyors, Negative Impedance Converters and Nonlinear Impedance Converters Using Operational Transconductance Amplifiers", *Int. J. Electronics*, Vol. 64, No. 6, pp. 941-945, 1988.

[13] S. Pookaiyaudom and K. Samootrut, "A Differential-Current Electronically-Variable Current-Mirror Phase-Shifter", *Int. J. Electronics*, Vol. 65, No. 1, pp. 59-65, 1988.

Chapter 11.5

CMOS Current-Conveyors

Erik Bruun

11.5.1 Introduction

Most implementations of current-conveyors presented in the literature or as commercial products have been realized in bipolar technology, [1-4]. However, as CMOS is the preferred technology in mixed analog/digital systems the question arises whether CMOS is a viable technology for the implementation of current-conveyors. Certainly, CMOS has the important feature of providing complementary devices but there are also some problems caused by fundamental CMOS characteristics such as the body effect, the threshold voltage mismatch between N-channel and P-channel transistors, and the low value of gate transconductance. In this chapter some of these problems are reviewed and design directions are given which may provide useful solutions to the problems. We shall consider the implementation of second generation current-conveyors only, i.e. CCII+ and CCII-, and emphasis will be put on the presentation of different circuit configurations rather than on a detailed analysis of design equations or experimental results.

11.5.2 Basic CMOS implementations

Fundamentally, the current-conveyor is a combination of two different building blocks, a voltage follower and a current follower (CCII-) or a current inverter (CCII+). The basic MOS transistor configurations for these functions are the source follower, the common gate stage, and the current mirror stage, respectively. Combining the functions together, the configurations shown in Figure 1 result. The CCII- conveyor is simply a MOS transistor with the source as the *X*-input, the gate as the *Y*-input, and the drain as the *Z*-output. With respect to small signals - and assuming no bulk effect - it performs the voltage following action from *Y* to *X* and the current following action from *X* to *Z*. With ideal bias current sources we find the following characteristics: the voltage gain from *Y* to *X* is

$$A_{YX} = \frac{g_m}{g_m + g_{mb} + g_{ds}} \tag{1}$$

where g_m, g_{mb}, and g_{ds} are the transistor gate transconductance, bulk transconductance, and output conductance, respectively. The term g_{mb} in the denominator of (1) is avoided if bulk and source is connected. The current gain from *X* to *Z* is (with the direction of current flow taken positive into the conveyor)

$$A_{XZ} = -1 \tag{2}$$

The *Y*-input resistance is infinite. The *X*-input resistance is

$$R_X = \frac{(1 + g_{ds} R_{LZ})}{g_m + g_{mb} + g_{ds}} \tag{3}$$

where R_{LZ} is the resistance connected to the Z-output. Note, that an infinite value of R_{LZ} will cause the X-input resistance to be infinite.

Assuming ideal bias current sources, the Z-output resistance is

$$R_Z = \frac{1}{g_{ds}} + R_{LX}\left(1 + \frac{g_m + g_{mb}}{g_{ds}}\right) \tag{4}$$

where R_{LX} is the load (source) impedance connected to the X-input.

The high frequency properties of this basic CCII- conveyor are limited by the transistor cut-off frequency $2\pi f_T = g_m / C_{gs}$, the drain and source parasitic capacitances and the externally connected components.

For the CCII+ a current mirror must be added to perform the current inversion from X to Z. For this design we find the following characteristics:

$$A_{YX} = \frac{g_m}{g_m + g_{mb} + g_{ds}} \tag{5}$$

$$A_{XZ} = \lambda_m = 1 + \varepsilon_m \tag{6}$$

where λ_m is the current mirror gain (ideally 1) and ε_m is the relative gain error of the current mirror.The Y-input resistance is infinite. The X-input resistance is

$$R_X = \frac{1}{g_m + g_{mb} + g_{ds}} \tag{7}$$

The Z-output resistance is equal to the parallel combination of the current mirror output resistance and the bias currrent source output resistance. Note that the current mirror provides an impedance isolation between the Z-output and the X-input.

The high frequency properties of this design are determined in part by the transistor cut-off frequency $2\pi f_T = g_m / C_{gs}$, the drain and source parasitic capacitances and in part by the current mirror input pole. Note that the current mirror input pole frequency is typically at least a factor of two lower than the transistor cut-off frequency due to the parallel connection of the two transistor gates in the mirror.

Concerning DC input errors the most obvious and important error with the designs shown in Figure 1 is a level shift between Y and X. To a first order approximation this level shift is given by

$$V_{XY} = V_T + \sqrt{\frac{2 I_{ss} L}{K' W}} \tag{8}$$

where V_T is the transistor threshold voltage, K' is the transconductance parameter, and W and L are transistor channel width and length, respectively. Other DC input errors are error currents to the Y and X inputs. The Y-input error current is negligible as the Y-input is only connected to a transistor gate. The X-input error current (the input current required to produce a Z-input current of zero) is caused by mismatch between the bias currents and for the CCII+ also by current mirror mismatch.

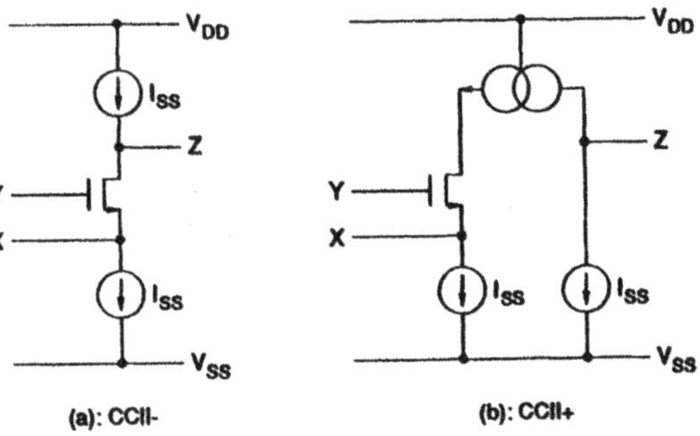

Figure 1 Basic MOS conveyor configurations.

Obviously, the configurations shown in Figure 1 may not be very useful as general purpose current-conveyors in high performance current-mode signal processing systems. Immediately apparent shortcomings are the DC voltage difference between *Y* and *X*, the *X*-input resistance (especially for the CCII-), and the problems with bulk effect. Some possible improvements will be considered in the next section.

11.5.3 Improved configurations

The first improvement to be introduced is a level shift compensation between *X* and *Y* in order to compensate for the DC voltage difference.

This can be accomplished either by an input level shift or by an output level shift to the source follower. As P-channel and N-channel transistors do not match (different threshold voltages) the same transistor types must be used for the source follower and the level shifter. In passing, it is noted that in bipolar technology NPN and PNP transistors can be made to match which makes it feasible to use a compensation based on a combination of NPN and PNP transistors, eg. an emitter follower cascade [5]. This method is not applicable in CMOS [6]. Figure 2 shows basic configurations of Figure 1 augmented with level shifting transistors. It is noted that the level shift transistor also compensates for level shift and gain error caused by the bulk effect provided the source follower transistor and level shift transistor are subject to the same bulk effect.

Comparing the *Y*-input and *X*-input level shifting, we note the following:

- The *Y*-input level shifting introduces a *Y*-input error current equal to the mismatch between the two bias sources for the level shift transistor. Also, the *Y*-input resistance is no longer infinite. Rather, it is equal to the parallel combination of the bias source output resistances. In terms of small signal high frequency performance, the level shift transistor introduces an *RC* stage in series with the *Y*-input with *RC* being of the order of $1/(2\pi f_T)$.
- The *X*-input level shifter does not alter the *Y*-input characteristics but adds a series resistance of $1/g_m$ to the *X*-input resistance. The *X*-input error current is also modified but still relies on the matching between the bias current sources and the current mirror matching. In terms of small signal high frequency performance, the

level shift transistor introduces an *RC* stage in series with the *X*-input with *RC* being of the order of $1/(2\pi f_T)$.

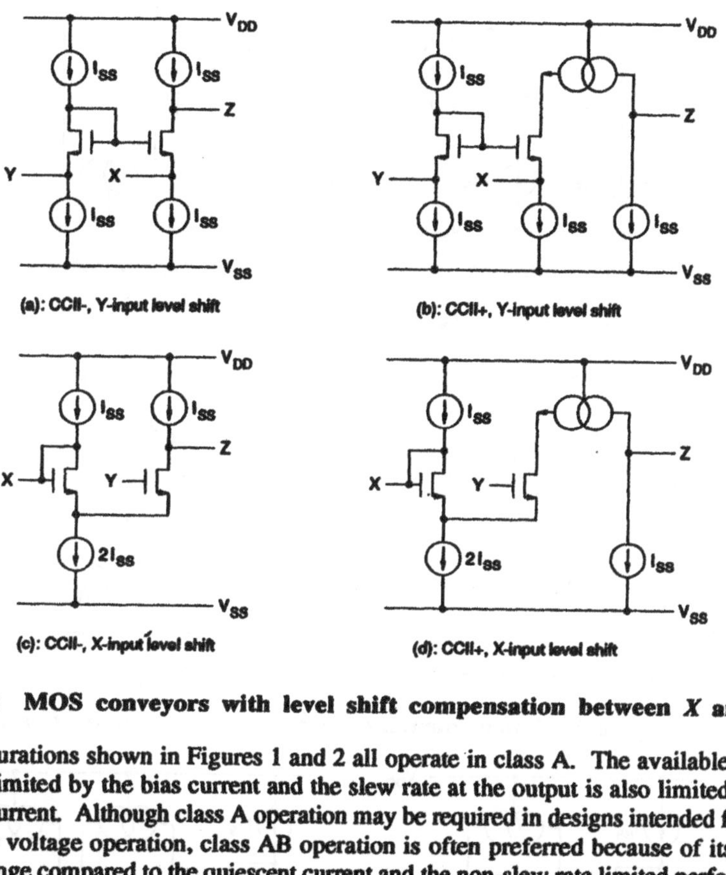

Figure 2 MOS conveyors with level shift compensation between *X* and *Y*.

The configurations shown in Figures 1 and 2 all operate in class A. The available output current is limited by the bias current and the slew rate at the output is also limited by the quiescent current. Although class A operation may be required in designs intended for very low supply voltage operation, class AB operation is often preferred because of its larger dynamic range compared to the quiescent current and the non-slew rate limited performance which can be obtained. For class AB operation, complementary versions of the designs of Figure 2 are derived. These are shown in Figure 3. The basic properties with respect to input errors, gain and impedance relations, and high frequency properties of these designs are similar to those of Figure 2. While the CCII+ conveyor with *Y*-input level shifting can accomodate *X*-input and *Z*-output currents in excess of the quiescent current, the maximum current in the other conveyor configurations is still limited by the bias current level. A true class AB operation requires a low impedance current path from *X* to each of the supply voltages. For the CCII- this can be obtained through cross-coupled current mirrors as shown in Figure 4. No easy way has been identified to achieve class AB operation of the conveyors with *X*-input level shifters.

The choice between the different conveyors shown in Figures 2 to 4 depends on the importance of different characteristics such as DC input errors, gain precision, small signal high frequency performance, slew rate performance, and input impedance characteristics. A qualitative comparison of the relative merits of these conveyor configurations is given in Table 1.

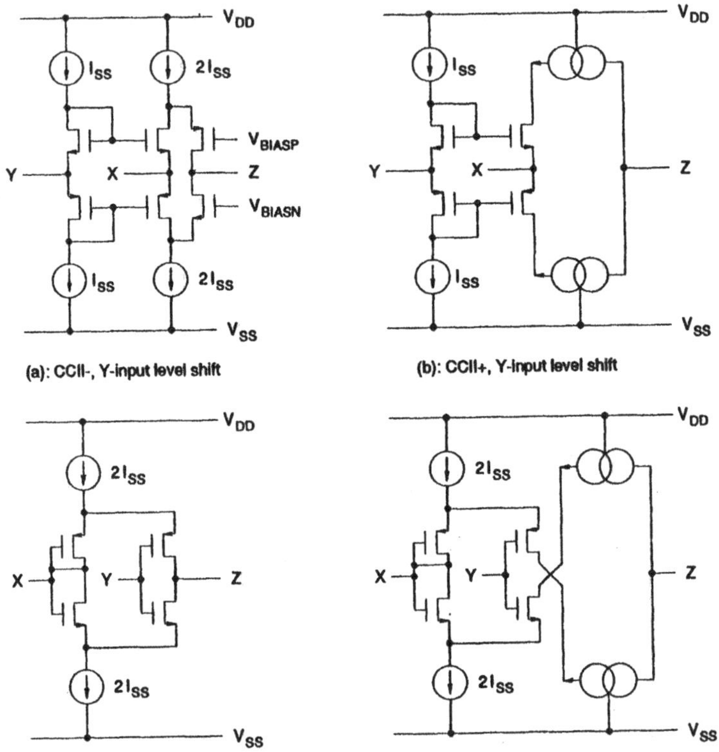

(a): CCII-, Y-input level shift

(b): CCII+, Y-input level shift

(c): CCII-, X-input level shift

(d): CCII+, X-input level shift

Figure 3 Complementary MOS conveyor configurations.

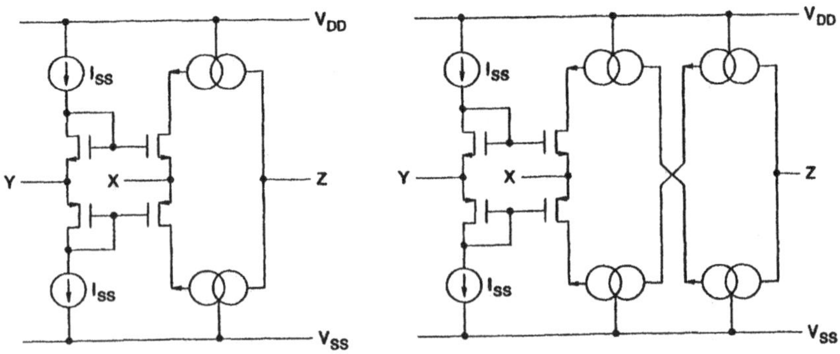

(a): CCII+ Class AB conveyor

(b): CCII- Class AB conveyor

Figure 4 Class AB MOS conveyor configurations.

Not shown in any of the Figures 1 to 4 is the implementation of current sources and current mirrors. Evidently, in order to obtain high output impedance from these, they must be implemented using some form of cascoding such as regulated cascodes [7] or low voltage cascodes [8] whenever high output resistances are required.

Implementation feature	*Figure reference*	*Advantage*	*Disadvantage*
Basic structure	1	Simple configuration, high speed	Large *X-Y* voltage offset
X-input level shift	2c 2d	No *Y*-input error current	Increase *X*-input resistance
Y-input level shift	2a 2b	Low *X*-input resistance	Non-zero *Y*-input error current
Complementary versions	3	Low *X*-input resistance	More complex design, high supply voltage
Non-complementary versions	2	Low supply voltage,simple design	Only class A operation possible
Class AB configurations	4	High slew rate, high current drive capability	Decreased linearity, decreased gain precision

Table 1 Comparison of conveyor configurations.

11.5.4 Configurations with feedback

It is a well known fact that feedback can be employed in order to improve the performance of unity gain amplifiers with respect to impedance levels and gain precision. Many options exist for utilizing feedback to improve the conveyor performance. These include the use of local feedback to obtain a low *X*-input resistance [9] or global feedback to improve the overall performance of the voltage and/or current buffers. Only a few options will be described here. For the voltage follower from *Y* to *X* it is natural to consider a unity gain amplifier obtained through the use of a voltage mode op-amp with feedback from the output to the inverting input. The current following or current inverting action required to form the *Z*-output is then obtained by a sensing of the supply current to the output stage of the op-amp. The resulting configuration is shown in Figure 5. In fact, some of the first CMOS current-conveyors reported were based on this approach [10].

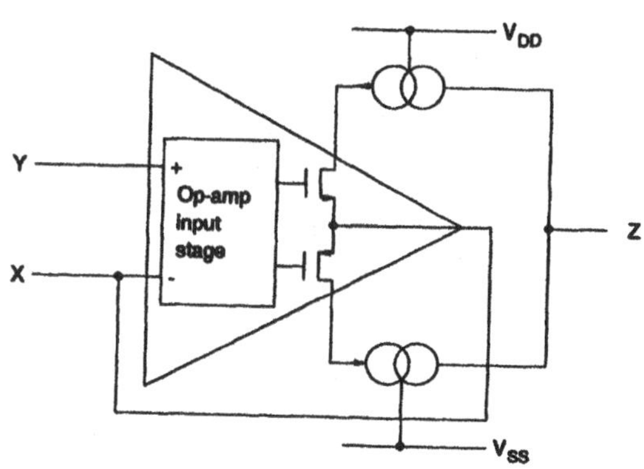

Figure 5 CCII+ with op-amp based voltage follower from *Y* to *X*.

Assuming an idealized op-amp model with an open loop output resistance of R_{ol} and an open loop gain of $A_{V,ol} = 2\pi GBW_V/s$ where GBW_V is the unity-gain bandwidth of the op-amp we find an inverting input impedance of

$$Z_X = s\frac{R_{ol}}{2\pi GBW_V}\frac{1}{\left(1+\frac{s}{2\pi GBW}\right)} \tag{9}$$

At high frequencies the input impedance approaches R_{ol} but at low frequencies a reduction of the input impedance is, in fact, obtained. However, it is noted that the input impedance is inductive at moderate frequencies with an equivalent inductance of $L_X = \frac{R_{ol}}{2\pi GBW}$. This may cause stability problems in systems with feedback to the conveyor *X*-input because of the additional phase shift caused by the inductance [6]. Another problem with the use of a feedback *Y* to *X* buffer is the slew rate. With a slew rate limited op-amp for the unity gain buffer the slew rate at the conveyor *Z*-output is also limited. Hence, in order to maintain the non-slew rate limited property of the class AB conveyors shown in Figure 4 a non-slew rate limited op-amp must be used for the unity gain buffer. This leaves only a few implementation principles for an input buffer with feedback. One is to use a current feedback op-amp [11]. Another is to utilize a voltage mode op-amp with a class AB input stage [12-14]. The latter approach has some advantages over the current feedback op-amp in that the voltage mode op-amp can be designed to have zero input error currents and better performance with respect to distortion and linearity [6,14]. Figure 6 shows a transistor level schematic of a CCII+ current-conveyor using this approach.

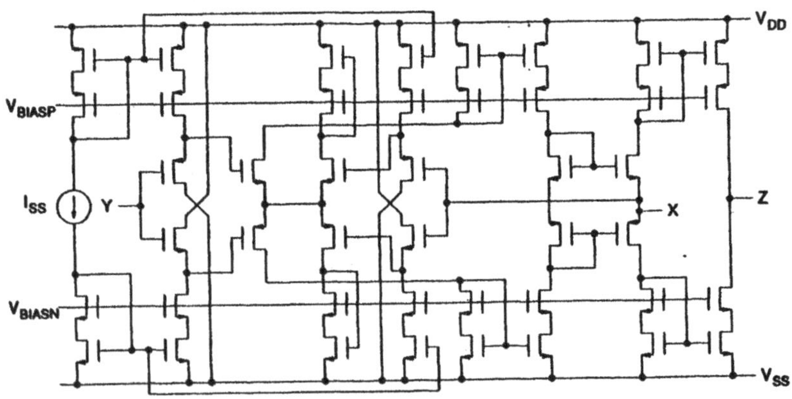

Figure 6 Transistor diagram of CCII+ conveyor with op-amp based voltage follower from *Y* to *X*. The circuit for generating the bias voltages V_{BIASN} and V_{BIASP} is not shown.

For the *X* to *Z* current follower (or inverter) a feedback configuration may be employed through the use of a current op-amp with feedback as shown in Figure 7.

The current op-amp is assumed to be a single input, multiple output op-amp [15,16]. The direction of current flow is defined as positive when the current flows into the op-amp. The current op-amp has in itself an input voltage reference which can be used directly as the conveyor *Y*-input. With complementary outputs from the current op-amp a CCII- results. With multiple in-phase outputs (I_{OUT} flowing into the op-amp when I_{IN} flows into the op-amp) a CCII+ results. With a current op-amp open loop gain of $A_{C,ol} = 2\pi GBW_C / s$ where GBW_C is the gain-bandwidth product of the current op-amp we find

$$Z_X = \frac{Z_{ol}}{1 + A_{C,ol}} = s\,\frac{Z_{ol}}{2\pi GBW_C}\,\frac{1}{\left(1 + \frac{s}{2\pi GBW_C}\right)} \tag{10}$$

where Z_{ol} is the input impedance of the current op-amp (without feedback). We note that with Z_{ol} resistive, the resulting *X*-input impedance is inductive, just as was the case when the input impedance was lowered by the use of a voltage buffer with feedback.

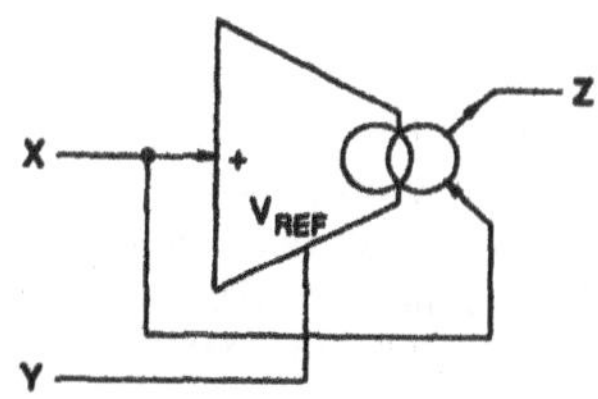

Figure 7 CCII- with op-amp based current follower from *X* to *Z*.

Figure 8 shows a transistor level schematic of a complementary output conveyor using this approach. In Figure 8 the current op-amp has been implemented using the conveyor type shown in Figure 4c followed by a multiple output complementary floating transconductance stage [17]. This current op-amp has an open loop input impedance which is mainly capacitive due to the capacitive loading of the conveyor *Z*-output, compare (3).

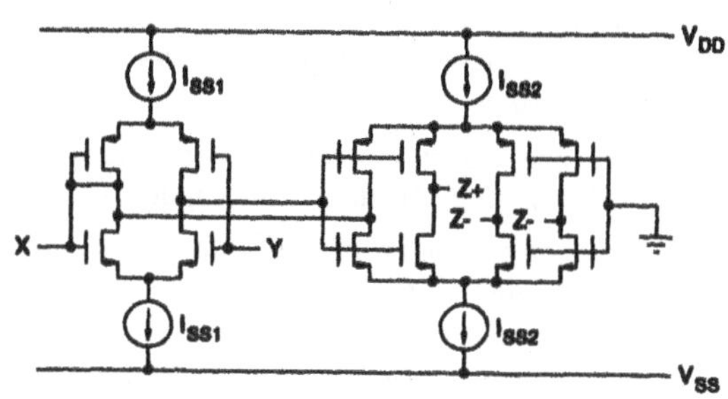

Figure 8 Complementary-output CCII with op-amp based current follower from *X* to *Z*.

Hence, from (10) we find that the resulting *X*-input impedance is resistive at low frequencies. It can be shown that the input resistance is on the order of $g_{ds1}/(g_{m1}g_{m2})$ where g_{m2} is the transconductance of output stage transistors and g_{m1} is the transconductance of the input stage transistors. A requirement for correct operation of the configuration of Figure 8 is that the conveyor *Z*-outputs are loaded by finite load impedances. If the load impedances are infinite the loop gain of the current op-amp feedback loop is zero and the conveyor *X*-input impedance is infinite. A situation with infinite load impedances at the outputs of a current op-amp is equivalent to a situation with a short circuited output of a voltage mode op-amp.

Topologically, the circuit of Figure 8 has the interesting feature of being composed merely of CMOS inverters supplied from constant current sources rather than from a constant supply voltage. This makes the design very compatible with standard digital CMOS circuits such as gate arrays or even standard CMOS inverters from the 4000-series or the 74HCU-series.

The previous examples showed how feedback could be employed either to the *Y* to *X* voltage follower or to the *X* to *Z* current follower/inverter. Of course, one could also envisage a design with feedback applied to both of these followers/inverter. This is an exercise left to the reader.

11.5.5 Conclusion

Several CMOS configurations of current-conveyors have been presented. Compared to bipolar technology CMOS suffers from a lack of matching between N-channel and P-channel devices and from considerably lower values of transconductance leading to high values of *X*-input resistance. Circuit configurations to overcome the first of these shortcomings have been shown, and it has been demonstrated how feedback can be employed to overcome the problems related to the input resistance. Nevertheless, it is obvious that the implementation issues related to CMOS for current-conveyors are far from being fully explored. Hopefully, the circuits presented here can serve as a source of inspiration in the process of designing new and improved circuits.

References

[1] A. S. Sedra and G. W. Roberts, "Current-conveyor theory and practice", Chapter 3 in: C. Toumazou, F. J. Lidgey, and D. G. Haigh (Ed.), *Analogue IC design: the current mode approach*, London, Peter Peregrinus Ltd. on behalf of IEE, 1990.

[2] Analog Devices, AD844 Data Sheet, 1990.

[3] Burr-Brown, OPA660 Data Sheet, 1990.

[4] LTP Electronics, CCII01 Data Sheet, 27 Park End Street, Oxford OX1 1HU, UK., tel/fax +44-865-200767, 1993.

[5] D. F. Bowers, "A precision dual current feedback operational amplifier", *Proc. 1988 Bipolar Circuits and Technology Meeting*, Minneapolis, U.S.A., pp. 68-70, 1988.

[6] E. Bruun, "CMOS Technology and Current-Feedback Op-Amps", *Proc. ISCAS'93*, pp. 1062-1065, Chicago, May 1993.

[7] E. Säckinger and W. Guggenbüühl, "A high-swing, high-impedance MOS cascode circuit", *IEEE J. Solid-State Circuits*, Vol. 25, pp. 289-298, February 1990.

[8] P. J. Crawley and G. W. Roberts, "High-Swing MOS Current Mirror with Arbitrarily High Output Resistance", *Electronics Letters*, Vol. 28, pp. 361-362, 1992.

[9] I. Mucha, "Low output impedance CMOS voltage buffer", *Electronics Letters*, Vol. 28, pp. 2071-2072, October 1992.

[10] C. Toumazou and F. J. Lidgey, "Wide-band precision rectification", *IEE Proc.*, Part G, Vol. 134, No. 13, pp. 7-14, 1988.

[11] A. Payne and C. Toumazou, "High frequency self-compensation of current-feedback devices", *ISCAS'92*, pp. 1376-1379, San Diego, May 1992.

[12] E. Bruun, "High speed, current-conveyor based voltage mode operational amplifier", *Electronics Letters*, Vol. 28, pp. 742-744, April 1992.

[13] I. Mucha, "Fully differential, current-conveyor based CMOS operational amplifier", *J. Electronics*, Vol. 74, pp. 697-703, 1993.

[14] E. Bruun, "CMOS high speed, high precision current-conveyor and current feedback amplifier structures", *Int. J. Electronics*, Vol. 74, pp. 93-100, 1993.

[15] E. Bruun, "A differential-input, differential-output current mode operational amplifier", *Int. J. Electronics*, Vol. 71, pp. 1047-1056, 1991.

[16] E. Bruun, "Design considerations for a high speed CMOS current opamp", *Proc. Eleventh NORCHIP Seminar*, pp. 24-31, Trondheim, Norway, 1993.

[17] A. F. Arbel and L. Goldminz, "Output Stage for Current-Mode Feedback Amplifiers, Theory and Applications", *Analog Integrated Circuits and Signal Processing*, Vol. 2, pp. 243-255, 1992.

Index

NOTE: Bold page numbers indicate figures and tables

D

G

H

Q

R

S

U

V

Printed and bound by CPI Group (UK) Ltd, Croydon, CR0 4YY
16/09/2026
14963445-0002